40th UIT International Heat Transfer Conference (UIT 2023)

Journal of Physics: Conference Series Volume 2685

Assisi, Italy
26-28 June 2023

ISBN: 978-1-7138-8963-2
ISSN: 1742-6588

Printed from e-media with permission by:

Curran Associates, Inc.
57 Morehouse Lane
Red Hook, NY 12571

CURRAN ASSOCIATES INC.
proceedings
.com

Some format issues inherent in the e-media version may also appear in this print version.

Institute of Physics
Dirac House, Temple Back
Bristol BS1 6BE UK

Phone: 44 1 17 929 7481
Fax: 44 1 17 920 0979

techtracking@iop.org

Additional copies of this publication are available from:

Curran Associates, Inc.
57 Morehouse Lane
Red Hook, NY 12571 USA
Phone: 845-758-0400
Fax: 845-758-2633
Email: curran@proceedings.com
Web: www.proceedings.com

TABLE OF CONTENTS

Preface

The 2023 UIT (Italian Union of Thermo-Fluid Dynamics) International Conference, hereafter referred as 40th UIT 2023, was organized by the Department of Engineering at the University of Perugia (Italy), in collaboration with the UIT, on June 26-28 2023, at Palazzo Bernabei, Assisi.

The annual UIT Conference was held in Assisi for the first time. The scope of the Conference covered a range of topics in computational fluid-dynamics and heat transfer; thermophysical properties; heat and mass transfer for sustainable energy systems; experimental techniques for heat and mass transfer; multiphase fluid-dynamics; natural, forced and mixed convection.

The 40th UIT Heat Transfer Conference 2023 program scheduled three keynote lectures by international recognised scientists: Ibrahim Dincer from Ontario Tech. University (Oshawa, Ontario) about the role of thermodynamics in integrated energy systems; Gary Neil Coleman from NASA Langley Research Center (Hampton, USA) about numerical studies of turbulent supersonic plane-channel flows; and Sauro Filippeschi from University of Pisa (Italy) about wickless two phase heat transfer devices. A total of 97 papers were submitted to the 40th UIT Heat Transfer Conference 2023, 75 of which presented by the Authors in oral sessions and the remaining 22 in one poster session.

About 150 researchers participated to the 40th UIT 2023. The Conference was a useful occasion to stimulate discussion, further understanding about heat transfer and related phenomena, present the state-of-the-art of some topics, discuss emerging trends, and promote collaborations.

The Organizing Committee hopes that the event results constituted a significant contribution to the knowledge in the fields of thermos-fluid dynamics and heat transfer.

A special thanks to UIT, all the people who contributed to the success of the event, the International Advisory Committee, the Local Organizing Committee, and to all the Conference attendees.

With Kind Regards
Franco Cotana, Università di Perugia, Italy
Federico Rossi, Università di Perugia, Italy
Cinzia Buratti, Università di Perugia, Italy

Journal of Physics: Conference Series **2685** (2024) 011001 doi:10.1088/1742-6596/2685/1/011001

Committees

Co-chairs

Franco Cotana, Università di Perugia, Italy
Federico Rossi, Università di Perugia, Italy

International Advisory Committee

Cinzia Buratti, Università degli Studi di Perugia, Italy
Giorgio Baldinelli, Università degli Studi di Perugia, Italy
Sara Rainieri, Università degli Studi di Parma, Italy
Alfonso Niro, Politecnico di Milano, Italy
Dario Ambrosini, Università degli Studi dell'Aquila, Italy
Fausto Arpino, Università degli Studi di Cassino e del Lazio Meridionale, Italy
Pietro Asinari, Politecnico di Torino, Italy
Giovanni Sebastiano Barozzi, Università degli Studi di Modena e Reggio Emilia, Italy
Gian Piero Celata, ENEA
Michele Ciofalo, Università degli Studi di Palermo, Italy
Davide Del Col, Università degli Studi di Padova, Italy
Marco Dell'Isola, Università degli Studi di Cassino e del Lazio Meridionale, Italy
Paolo Di Marco, Università degli Studi di Pisa, Italy
Andrea Gasparella, Libera Università di Bolzano, Italy
Praveen Linga, NUS, Singapore
Nicola Massarotti, Università degli Studi di Napoli Parthenope, Italy
Gian Luca Morini, Università degli Studi di Bologna, Italy
Vincenzo Naso, Università degli Studi di Napoli Federico II, Italy
Enrico Nobile, Università degli Studi di Trieste, Italy
Giorgio Pagliarini, Università degli Studi di Parma, Italy
Fabio Polonara, Università Politecnica delle Marche, Italy
Stefania Proietti, Università Telematica Guglielmo Marconi, Sindaco di Assisi
Renato Ricci, Università Politecnica delle Marche, Italy
Gianluca Rossi, Università degli Studi di Perugia, Italy
Eugenia Rossi Di Schio, Università di Bologna, Italy
Massimo Santarelli, Politecnico di Torino, Italy
Adriano Sciacovelli, University of Birmingham, England
Giovanni Tanda, Università degli Studi di Genova, Italy
Ding Yulong, University of Birmingham, England

Local Organising Committee

Cinzia Buratti, Università degli Studi di Perugia, Italy
Giorgio Baldinelli, Università degli Studi di Perugia, Italy
Beatrice Castellani, Università degli Studi di Perugia, Italy
Franco Cotana, Università degli Studi di Perugia, Italy
Alessia di Giuseppe, Università degli Studi di Perugia, Italy
Mirko Filipponi, Università degli Studi di Camerino, Italy
Francesca Merli, Università degli Studi di Perugia, Italy
Elisa Moretti, Università degli Studi di Perugia, Italy
Andrea Nicolini, Università degli Studi di Perugia, Italy
Anna Laura Pisello, Università degli Studi di Perugia, Italy
Federico Rossi, Università degli Studi di Perugia, Italy

Contact details for the declaration

- ✓ Franco Cotana, Università degli Studi di Perugia, Italy, franco.cotana@unipg.it
- ✓ Federico Rossi, Università degli Studi di Perugia, Italy, federico.rossi@unipg.it
- ✓ Cinzia Buratti, Università degli Studi di Perugia, Italy, cinzia.buratti@unipg.it

Peer Review Statement

All papers published in this volume have been reviewed through processes administered by the Editors. Reviews were conducted by expert referees to the professional and scientific standards expected of a proceedings journal published by IOP Publishing.

- **Type of peer review**: Single Anonymous
- **Conference submission management system**: Morressier
- **Number of submissions received**: 74
- **Number of submissions sent for review**: 74
- **Number of submissions accepted**: 74
- **Acceptance Rate (Submissions Accepted / Submissions Received × 100)**: 100
- **Average number of reviews per paper**: 1
- **Total number of reviewers involved**: 38
- **Contact person for queries**:
  Name: Francesca Merli
  Email: francesca.merli@unipg.it
  Affiliation: Department of Engineering – University of Perugia

Accuracy in evaluating convective heat transfer coefficient by RANS CFD simulations in a rectangular channel with high aspect ratio and 60° tilted staggered ribs

M Corti[1], P Gramazio[1], D Fustinoni[1] and A Niro[1,*]
[1] Politecnico di Milano, Department of Energy via Lambruschini 4, 20156 Milano, Italy
E-mail: *alfonso.niro@polimi.it

Abstract. A numerical analysis of heat transfer characteristics of an air flow through a narrow rectangular channel of 1:10 aspect ratio with ribbed surfaces has been carried out, by means of a CFD commercial code, exploiting Reynolds Averaged Navier-Stokes (RANS) equations. The channel is 120 mm wide, 840 mm long, with 60° tilted staggered ribs. Ribs have a square cross section with two different side heights (2 mm and 4 mm) and three different values of dimensionless pitch (10, 20 and 40). The numerical results have been compared with experimental data obtained by the authors inside a ribbed channel both with same geometry and operating conditions as one numerically modelled. Agreement between numerical and experimental data on convective global heat transfer coefficient, i.e., averaged over the whole channel, is discussed for the six different configurations considered. Moreover, performances are presented by considering at the same time both heat transfer enhancement and pressure-drop penalization, highlighting strengths and weaknesses per each configuration. This work is aimed at finding a suitable configuration of a CFD model with RANS that will allow authors to apply it to the range of dimensionless pitches and side heights during early design-phases of parametric studies.

Keywords: forced convection, heat exchanger, rectangular channel, ribbed surfaces, experimental tests, CFD validation, RANS

1. Introduction

Considering the current and future scenario, aiming to reduce energy consumption and improve efficiency in the context of the green economy, heat transfer plays a significant role. In many engineering and electronic applications, innovative solutions are applied and studied to enhance heat transfer and increase efficiency. Regarding forced convection, this process has been largely studied in traditional fields like for the cooling of turbine blades [1] or for promoting heat transfer in nuclear reactors. In addition to these well-established and extensively-studied areas, newer engineering applications have been lately investigated, like solar collectors [2] or more recently for automotive batteries cooling [3] and high-performance CPU cooling [4,5]. These novel industrial applications that require heat exchangers with large contact areas and very small heights can benefit from the use of ribs to enhance heat transfer associated to high aspect ratio rectangular channels with ribbed surfaces [6]. The result is more efficient and effective heat transfer, which can translate into significant energy savings and improved performance.

In literature, studies regarding former cases are quite extensive, although the documentation is somewhat fragmented or related to some specific application. Therefore, a systematic review has been conducted at Politecnico di Milano, over years, trying to establish general guidelines. Authors have conducted experimental analysis on fluid-dynamics and heat transfer characteristics of forced convection inside a ribbed channel with an aspect ratio of 1:10 [7], studying how varying ribs pitch-to-rib-side ratio, cross-section, or blockage ratio affects heat transfer. To reduce the experimental testing time, since CFD is a useful tool for studying fluid flow in ducts with rib roughened walls, authors have also developed CFD models using the RANS $k-\omega-$ SST model, which has been widely used for similar cases compared to other RANS models given its effectiveness [8], and the LES model to achieve greater accuracy.

In this paper, results on convective heat transfer coefficient computed from RANS simulations are presented. The aim of this study is to continue with previous works and identify a suitable CFD model configuration, which can be applied to a range of ribs configurations.

2. Problem configuration and setup

2.1. Experimental setup

The experimental setup consists in a rectangular channel of 1:10 aspect ratio with 60° staggered ribs on the upper and the lower walls. Lower and upper walls of the experimental channel are kept at a fixed temperature, while the side walls and the ribs are adiabatic. The air flows at Reynolds number ranging from 700 to 8000, in this specific study Reynolds 7550 has been chosen. The experimental apparatus is described in details in [7] and the main geometric parameters are summarized in Table 1. Ribs analysed in this paper have a squared cross section, and a section side s of 2 mm and 4 mm. The ribs are 60° tilted with respect to the stream flow and staggered each other in each configuration proposed in this study. The pitch-to-side ratio p/s considered have been 10, 20 ,40. For simplicity configurations considered will be named respectively P10S4, P20S4, P40S4 for ribs with width of 4 mm and P10S2, P20S2, and P40S2 for ribs with s=2mm. An overview of the channel can be seen in Figure 1a.

2.2. Numerical procedure

Numerical simulations have been carried out by means of a finite volume solver using the software Ansys Fluent 19.1. The solution domain consists in a 3D rectangular channel as described previously, to which two additional domains are merged: one dedicated to the entrance and one dedicated to the outlet. The latter is recommended to avoid reverse flows problems [9], frequent effects while dealing with geometries with obstacles such as ribs inside channels. The working fluid is air, incompressible, with negligible differences in specific heat at the temperatures considered.
The internal fluid flow has been analysed at Re 7550, which is in a turbulent fluid flow regime. For this reason, the mesh adopted respects the parameters requested by the solver and by the problem, which will be introduced later. A sensitivity analysis has been carried out to obtain a grid independent solution analysing the three directions. It has been observed that the sensitivity to the mesh is greater in the y-direction or that in the direction of the channel height. Summarizing, several grids have been tested maintaining the same topology between them with a number of elements ranging from $60 \cdot 10^6$ to $100 \cdot 10^6$ and the grid finally chosen for the discretization of the examined geometries has about $80 \cdot 10^6$ elements. Clearly the exact number varies based on the channel configuration. Using a medium quality mesh instead of a fine mesh leads in average to an error of 1% on the analysed parameters but in any case is lower than 4%. A sketch of the computational grid is shown in Figure 1b.

2.3. Boundary conditions

At the inlet of the computational domain a fully developed velocity profile, previously generated and based on the selected Reynolds number, is applied. The working fluid is air and it enters the channel at a uniform temperature, per example in most of the

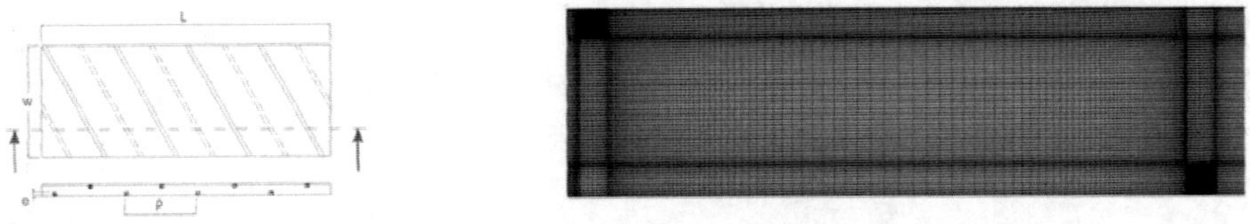

(a) P20S4 channel - geometry. **(b)** P20S2 channel - mesh.

Figure 1. Geometry and mesh overview.

Table 1. Main channel parameters.

Main geometric parameters	Value	Main geometric parameters	Value
Channel height, 2h=H	12 mm	Ribs height, e	2 mm and 4 mm
Channel width, 2w=W	120 mm	Ribs width, s	2 mm and 4 mm
Channel length, L	840 or 880 mm	Adimensional pitch p/s	p/s=10, 20, 40
Hydraulic diameter, Dh	21.28 mm	Reynolds number	7550

configuration is $T_{in} = 296K$. The temperature profile will develop inside the considered channel since the fluid flow is not thermally developed yet. At the outlet a pressure outlet with $P_{out} = 0$ is applied. At walls a no-slip velocity condition is applied and all the walls (upper, lower and laterals) are maintained at a fixed constant temperature $T_{wall} = 313.15K$. Also walls of ribs have a no-slip velocity condition applied, but all of them are considered adiabatic, imposing $q = 0$. The choice to build adiabatic ribs is needed since authors want to highlight the beneficial effect of the ribs at a fluid-dynamic level, thus dissociating ribs from their possible fin behavior. The applied boundary conditions follow exactly the setup of the experimental apparatus with the exception of the boundary condition of the lateral walls. In the experimental setup they should be considered adiabatic while in the numerical model lateral walls are heated at a constant temperature. It has been seen through previous studies [10] that this boundary condition is the most suitable to ensure that the numerical problem best represents the experimental configuration.

2.4. Solver settings and models

For pressure-velocity coupling the Semi-Implicit Method for Pressure Linked Equation-Consistent (SIMPLEC) segregated algorithm is used. This choice follows a comparison of other algorithms in previous trials showing a better and faster convergence, since in SIMPLEC algorithm the pressure-correction under-relaxation factor is generally set to 1.0. The gradients are discretized using a Least Squares Cell-Based scheme. The spatial discretization of pressure, momentum and energy equations is performed with a second order (upwind) scheme as for turbulent kinetic energy and specific dissipation rate. The flow has been solved with RANS, choosing the $k-\omega-$SST model. This choice has been motivated by evidence in literature [8] and by previous analysis [10].
The $k-\omega-$SST model uses as default the Enhanced Wall Treatment, requiring a careful sizing of the computational grid at walls that has to ensure a $y^+ = 1$ at walls. Given the grid dimensions and the cores availability, in order to not exceed 96 hours of calculation per each single simulation, RANS simulations stop when the continuity residual is lower than $1 \cdot 10^{-5}$.

3. Results

In this study the air flow within the channels considered occurs at a Reynolds number equal to 7550. The choice of this Reynolds number is mainly due to two reasons. Firstly, it is currently the highest Reynolds number that has been tested experimentally with the current experimental setup, secondly, in agreement with previous analyses [10], it has been observed that using the RANS model $k - \omega - SST$, it improves its accuracy as the Reynolds number increases, considering $Re > 5000$.

Per each configuration considered, the heat transfer characteristics have been compared with the available experimental data that can be found in [7]. Values of the friction factor and the average Nusselt number vs Reynolds number for staggered 60° tilted-ribs are presented in the reference [7] in Figure 6 and Figure 7, respectively. In the cited paper it's also possible to read about the error analysis and uncertainty that affects experimental data used as comparison in this work.

3.1. Bulk temperature and heat exchanged

Given the will to calibrate the numerical model on the experimental one, from experimental data is possible to directly compare the temperature measured at the outlet of the channel, T_{out}. The agreement is good because $\Delta T_{experimental}/(T_{out} - T_{in})_{CFD} = 1.05$ although the percentage error remains lower than 1%. For this reason, the error on the thermal power could be representative. The heat exchanged Q by the air along the considered channel is calculated as $Q = \dot{m}c_p(T_{out} - T_{in})$, introducing some errors due to $\dot{m}$. Error remains lower than 5% for channel with ribs with s=4mm and ranges from 12% to 15% for channel with ribs with s=2mm. Figure 2 shows a summary of the data considered. Temperatures have been measured with a precision of $0.05K$, while the overall experimental uncertainty on Q is 2.3% calculated according to Moffat [11]. In the following figures the experimental uncertainties are represented by error bars.

3.2. Convective heat transfer coefficient

As usual approach, a conventional mean convective heat transfer coefficient is calculated from the energy balance as $h_{m,ln} = \dot{m}c_p(T_{out} - T_{in})/(A\Delta T_{m,ln})$ when considering a channel with imposed wall temperature and a developed temperature profile. Comparing the mean convective heat transfer coefficients experimentally founded and numerically evaluated, the latter underestimate the experimental values, as can be seen in Figure 3a. Causes can be found in both the use of RANS model and the assumptions made since in the considered channels it is not certain that the temperature reaches the fully developed condition. Discrepancies increase while the channel complexity increases and

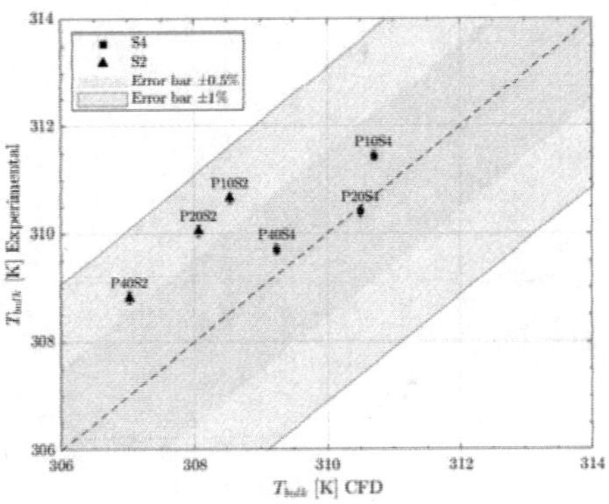

(a) Comparison of T_{bulk}.

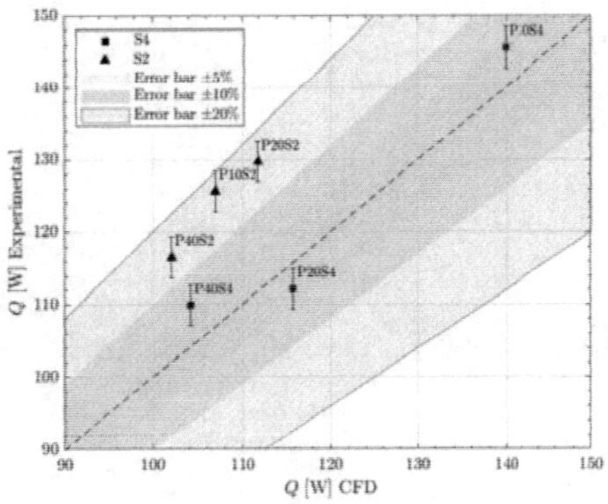

(b) Comparison of Q.

Figure 2. Comparison of experimental data versus CFD data, T_{bulk} and Q .

especially when rib height decreases, showing that the RANS model probably can not model correctly flow and heat transfer associated. The uncertainty associated to the experimental convective heat transfer coefficient is 3%.

Another quantity that can be analyzed, and that is interesting to propose, is the actual convective heat transfer coefficient, h_m (1), evaluated on the basis of numerical results.

$$h_m = \frac{1}{\sum_{z=1}^{n_{sections}} \Delta\ A_z} \left(\sum_{z=1}^{n_{sections}} h_z A_z \right) \quad (1)$$

Each contribution of the local convective heat transfer coefficient h_z has been weighed on the affected heat transfer surface ΔA_z and sum up. This actual convective heat transfer coefficient provides a more comprehensive evaluation of the heat transfer coefficient with respect to $h_{m,ln}$ since h_m considers also the effect of the channel entrance and the changes of h_z along the channel. To calculate the local heat transfer coefficient h_z, some sections, equidistant from each other, have been built along the channel length; an example of them can be seen in Figure 4a. The number of sections selected ensures that the actual convective heat transfer coefficient is no more influenced by the number of them. For each section the bulk temperature T_b and the heat flux q have been computed as mean section variables and mean line variables, respectively. Having this quantities, the local convective heat transfer coefficient can be computed at any sections as $h_z = q_z/(T_w - T_{b,z})$. Figure 4b shows the trend of the local convective heat transfer coefficient h_z plotted as a function of the length z of the channel. In the graph the trend of the h_z of a plane channel, without ribs, is also plotted as comparison. The presence of ribs shows a significant change in the trend for all configurations with s=4mm, whereas for 2 mm height ribs, when p/s is 10 and 20 a little change in the trend can be seen, but for the configuration P40S2 the trend is almost the same as the plane channel.

Lastly, comparing the values of $h_{m,ln}$ and h_m in an area (downstream z=0.274m) that could be considered far enough from entrance region, errors remain lower than 10% for channels with 4 mm height ribs, and lower than 5% for channels with rib height of 2 mm as can be seen in Table 2. This can be an indication on how much could be the error using $h_{m,ln}$ instead of $h_{m,z}$ for characterizing convective heat transfer coefficient in these channels.

Table 2. Comparison of convective heat transfer coefficient.

Channel Configuration	CFD $h_{m,ln}$ $[W/m^2/K]$	CFD $h_{m,z}$ $[W/m^2/K]$ downstream z=0.274m	Error [–]
P10S4	101.69	92.68	9%
P20S4	85.05	77.81	9%
P40S4	64.20	61.07	5%
P10S2	64.55	61.29	5%
P20S2	59.10	57.56	3%
P40S2	48.63	46.54	4%

3.3. Pressure Drops

Friction factors f obtained from pressure drops ΔP, deriving from numerical results through equation 2, have been compared with experimental data, measured with 7% uncertainty. Comparisons are shown in Figure 3b; notice that the experimental data of the configuration P10S4 is no available. Errors remain below the 5% for P10S2, below the 20% for P20S2 and P20S4 while for p/s=40 errors are higher than 25%.

$$f = \frac{2\Delta P}{\rho u^2}\left(\frac{D_h}{L}\right) \tag{2}$$

4. Discussion and Conclusions

In this study a CFD model has been developed for a rectangular channel with high aspect ratio and 60° tilted staggered ribs on upper and lower walls. The $k-\omega-$SST RANS model has been chosen and it has shown good agreement with experimental data at both fluid-dynamic and thermal levels. However, the increase in errors at lower ribs height could mean that RANS is not capable to capture with high accuracy the

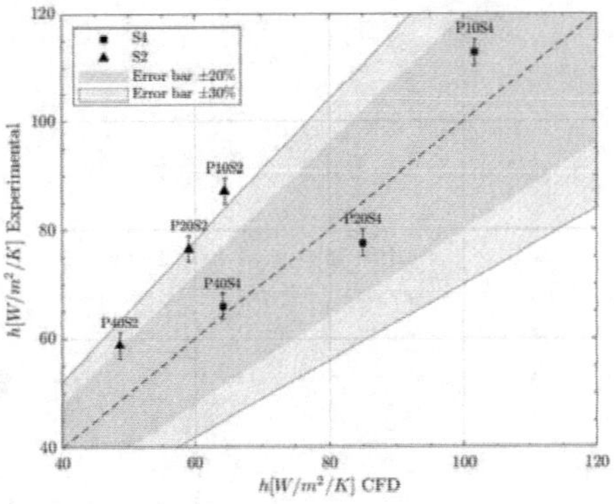

(a) Comparison of heat transfer coefficient.

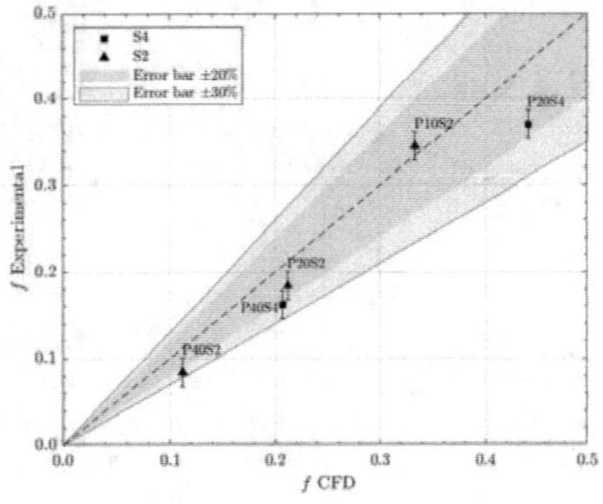

(b) Comparison of friction factor coefficient.

Figure 3. Comparison of experimental data versus CFD data, $h_{m,ln}$ and f.

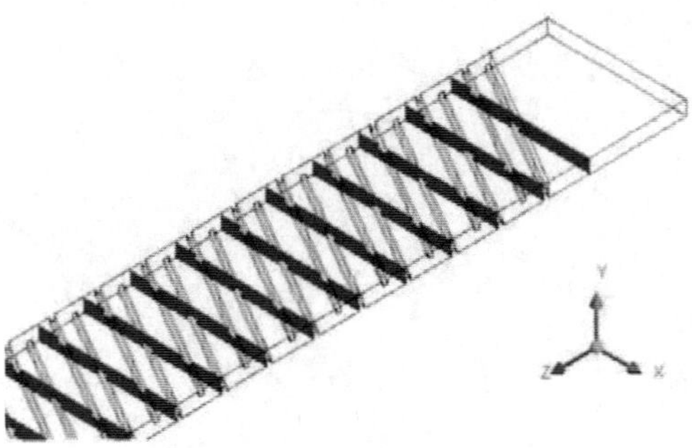

(a) Exemplary section-planes.

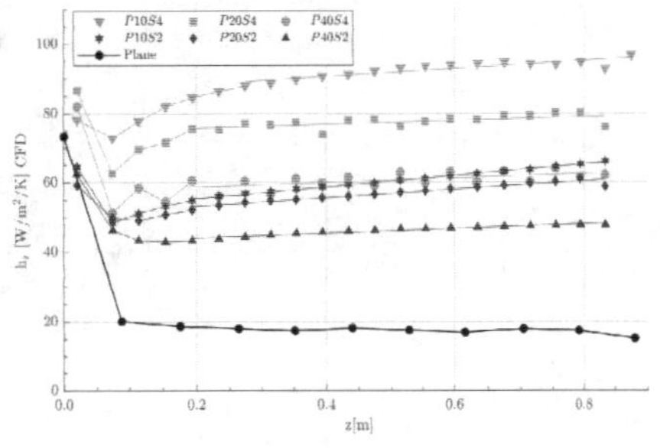

(b) Trend of the local heat transfer coefficient.

Figure 4. Local convective heat transfer coefficient, construction planes and trend.

swirling flows caused by little ribs. However, the reliability of $k-\omega-SST$ model with respect to other RANS models, to Reynolds stress model, and to v^2-f model has been proved in literature also for ribs with very small height, such as microribs with $e/H \simeq 0.1$ [12]. The presented results are slightly less accurate than results presented in [13], where the ribs were 90° staggered, meaning that increasing the complexity of the channel RANS accuracy decreases. Nevertheless errors are acceptable, so this CFD model can be exploited as a starting point to expand the study of thermo-fluid dynamics of high aspect ratio channel with ribbed surfaces, limiting the need of experimental measurements. The intention is to build a database for some parametric studies or for the creation of correlations useful for the design-phase and the optimization of rib-roughened channels.

References

[1] Nourin F N and Amano R S Review of Gas Turbine Internal Cooling Improvement Technology 2021 *J. Energy Resour. Technol.* **143** 080801

[2] Jain P K, Lanjewar A, Rana K B and Meena M L Effect of fabricated V-rib roughness experimentally investigated in a rectangular channel of solar air heater: a comprehensive review 2021 *Environ Sci Pollut Res* **28** 4019–4055

[3] Huynh D and Mahjoob S 2020 Numerical Analysis of Structured Ribbed Channels for Electronic Cooling Applications *2020 19th IEEE* pp 766–771

[4] Zhao N, Guo L, Qi C, Chen T and Cui X Experimental study on thermo-hydraulic performance of nanofluids in CPU heat sink with rectangular grooves and cylindrical bugles based on exergy efficiency 2019 *Energy Convers. Manag.* **181** 235–246

[5] Deng Z, Zhang S, Ma K, Jia C, Sun Y, Chen X, Luo Y, Li B and Li T Numerical and experimental study on cooling high power chips of data centers using double-side cooling module based on mini-channel heat sink 2023 *Appl. Therm. Eng.* **227** 120282

[6] Bhattacharyya S, Vishwakarma D K, Srinivasan A, Soni M K, Goel V, Sharifpur M, Ahmadi M H, Issakhov A and Meyer J Thermal performance enhancement in heat exchangers using active and passive techniques: a detailed review 2022 *J Therm Anal Calorim* **147** 9229–9281

[7] Fustinoni D, Gramazio P, Colombo L P M and Niro A First average and local heat transfer measurements on a forced air-flow at low Re-numbers through a rectangular channel with ribbed surfaces 2012 *J. Phys.: Conf. Ser.* **395** 012042

[8] Sharma S K and Kalamkar V R Computational Fluid Dynamics approach in thermo-hydraulic analysis of flow in ducts with rib roughened walls – A review 2016 *Renew. Sust. Energ. Rev.* **55**

[9] Pope S B 2000 *Turbulent Flows* 1st ed (Cambridge University Press)

[10] Corti M, Gramazio P, Fustinoni D, Vitali L and Niro A Accuracy in evaluating heat transfer coefficient by RANS CFD simulations in a rectangular channel with high aspect ratio - Part 1: benchmark on a channel with plane walls 2023 *J. Phys.: Conf. Ser.* **2509** 012009

[11] Moffat R J Describing the uncertainties in experimental results 1988 *Exp. Therm. Fluid Sci.* **1** 3–17

[12] Li X, Zhang S, Zuo J, Wei J, Zhou X and Bao W Flow and heat transfer characteristics of supercritical hydrogen in unilateral heated channels with micro-ribs 2023 *Appl. Therm. Eng* **221** 119900

[13] Corti M, Gramazio P, Fustinoni D, Vitali L and Niro A Accuracy in evaluating heat transfer coefficient by RANS CFD simulations in a rectangular channel with high aspect ratio - Part 2: channel with ribbed walls 2023 *J. Phys.: Conf. Ser.* **2509** 012010

Journal of Physics: Conference Series **2685** (2024) 012002 doi:10.1088/1742-6596/2685/1/012002

Heat transport in a non-homothermal turbulent particle-laden flow in the collisional regime

Hamid Reza Zandi Pour* **and Michele Iovieno**

Dipartimento di Ingegneria Meccanica e Aerospaziale, Politecnico di Torino, Corso Duca degli Abruzzi 24, 10129 Torino, Italy

* Corresponding author: hamid.zandipour@polito.it

Abstract. In this investigation, we explore the impact of particle collisions on turbulent heat flux within a temporally developing thermal mixing layer arising from the interaction of two homothermal regions driven by a statistically homogeneous and isotropic velocity field. Employing two-way thermally coupled Eulerian-Lagrangian Direct Numerical Simulations (DNSs) at a Taylor microscale Reynolds number up to 124, we examine the influence of particle collisions for a Stokes number from 0.2 to 3, maintaining a fixed thermal to kinetic relaxation times ratio of 4.43. Our study quantifies the reduction in average heat transport induced by particle-to-particle collisions, which disrupt the temperature-velocity correlation created by the initial temperature difference. Notably, while collisions diminish this correlation, our analysis reveals their overall impact remains minor, even at the highest simulated Stokes number. Additionally, we present statistics of the temperature difference among colliding particles across various flow conditions.

1. Introduction

The ubiquity of particle-laden flows in almost every application makes them a still very active field of research for many disciplines. Furthermore, the mechanical and thermal impacts of inter-particle collisions in several industrial and natural turbulent flows carrying inertial heavy particles play a significant role even in dilute regimes. For instance, understanding the droplet collisions in clouds and particle-particle collisions in sandstorms are essential in order to study these multi-physics phenomena [1]. Given the difficulties of Lagrangian measurements, in particular when the thermal interplay between the two phases is considered, a lot of insight has been obtained through numerical simulations. Fluid-particle thermal interplay has been studied mostly under the point particle assumption, which is valid for particles much smaller than Kolmogorov size, both in wall-bounded turbulent flows [2, 3, 4] and in homogeneous turbulent regimes [5, 6, 7].

Saffman and Turner [8] developed the theory for small droplet collision in clouds. They found that collision rate of identical sub-Kolmogorov droplets depends only on the turbulent dissipation rate, fluid viscosity and droplet radius, and they also showed that the collision frequency is independent of droplet thermophysical properties. The effect of particle parameters such as particle response time, number density and size on collision rate was investigated by Sundaram and Collins. They found that collision rate depends highly on Stokes number [9]. Moreover, some other works have studied the inter-particle collision effect in particulate turbulent flow by developing kinetic-based methods [10, 11, 12, 13]. In the last decade some works have extensively studied the impact of particle-particle collision along with particle-wall collision in the wall-bounded flows [14, 15].

Collisions between particles influence drastically the transport of heat in solid-fuel combustors, because particles much denser than the fluid can escape the near-wall zone, thus transferring heat between the boundaries and zones away from the wall. Even if the volume fraction of these applications usually is far above the validity range of the two-way coupling regime, and often beyond the validity of point-particles approach, [16], they suggest that fluid-particle thermal interactions can lead to a significant modification of the ability of a flow to transport heat also in dilute suspensions. The impact of collisions is normally not considered, when heat transfer is under investigation, even if it is evident that at such high volume fractions the collisions likelihood increases leading to collision-induced heat transfer modifications. There exist only a small number of theoretical studies involving collisions, like the work of [17] who assesses in an analytical way the heat transfer during a non-instantaneous elastic collision event with a solid wall. Carbone et al. [6] have assessed the effect of inter-particle collisions on the fluid temperature structure functions in isotropic turbulence when particles are one-way thermally coupled, showing that temperature statistics at small-scale barely are influenced by collisions. However, collisions reduce the acceleration of small particles while large particles are accelerated more, resulting in intensified velocity scattering. When a high fluid temperature gradient exists, such scattering may enable them not only to transfer heat near their position, but also to transfer heat to regions quite distant from their neighbourhood. This mechanism is responsible for a heat transfer enhancement in two-phase particulate flow by amplifying the interphase exchange of heat between two phases along particle Lagrangian trajectories [18], even in presence of an isotropic velocity field. Recently we have studied the thermal interplay of fluid and particle in a non-isothermal flow. This basic configuration is also suitable to be employed to improve the existing turbulence models of two-phase flows. One- and two-way thermally coupled particle-laden flows have been numerically investigated [18] in a dilute particulate flow without considering inter-particle interaction.

To explore the contribution to the heat transport of colliding particles in dilute suspension, we carry out Eulerian-Lagrangian DNSs to analyze the effect in the heat transport of non-cohesive collisions between particles in the two-way thermally coupled regime in the simplest inhomogeneous flow configuration, the time-evolving mixing layer between two regions at a different temperature subject to a homogeneous and isotropic turbulent velocity field. We complement our previous work [18] by examining the role of collisions for Stokes numbers from 0.2 to 3, for particles with a ratio between momentum and thermal relaxation times is fixed as 4.43, in a flow with a Reynolds number based on Taylor microscale up to 124. We quantify the extent to which particle collisions attenuate the average heat transfer relative to the flow regime with non-colliding particles and demonstrate that the overall impact is insubstantial even for large inertial particles. We compare our findings with the same flow configuration in absence of collisions, so that the statistics presented in [18] are used a baseline, providing valuable insights into the effects of the collisions among particles. We show that collisions bring a reduction of particle temperature variance with respect to fluid temperature variance, while fluid-to-fluid and particle-to-particle variance ratios slightly increase by collisions especially when particle inertia increases. Due to particle thermal back reaction, collisions increase the variance of the temperature of the carrier flow higher than the one of the particles, especially at the highest simulated Stokes numbers, even if the difference is small and it could become considerable only at a much higher volume fractions [19]. The ability of inertial particles to carry heat to zones distant from their position is indicated also by the mean temperature difference between colliding particles conditioned over the normalized position across the mixing layer.

2. Physical model

2.1. Governing equations

Within the Eulerian-Lagrngian method, the motion of a non-homothermal divergence-free flow seeded by small heavy inertial particles is governed by the Navier-Stokes equations coupled with the Lagrangian equations for the dynamics of discrete particles. Temperature variations are assumed small enough not to be producing relevant density changes, so that temperature behaves as an advected passive scalar. Hence, the continuous fluid phase field equations are

$$\partial_i u_i = 0 \tag{1}$$

$$\partial_t u_i + u_j \partial_j u_i = -(1/\rho_0)\partial_i p + \nu \partial_j \partial_j u_i + f_{u,i}, \tag{2}$$

$$\partial_t T + u_j \partial_j T = \kappa \partial_j \partial_j T + (1/\rho_0 c_{p0}) C_T. \tag{3}$$

Here $\boldsymbol{u}(t,\boldsymbol{x})$, $p(t,\boldsymbol{x})$, and $T(t,\boldsymbol{x})$ are fluid velocity, pressure, and temperature, while the (constant) fluid density, kinematic viscosity, thermal diffusivity and isobaric specific heat capacity are denoted by ρ_0, ν, κ, and c_{p0}. Additionally, $\boldsymbol{f}_u(t,\boldsymbol{x})$ is an external force on the fluid imposed to keep the velocity of the carrier fluid in a statistically steady state, while C_T is the fluid-particle interphase heat transfer per unit mass per unit time. No particle momentum feedback has been taken into account, as in [18, 19, 20], because, as shown in [6], in a dilute suspension it has a insignificant impact on temperature statistics. Particles are modelled as non-deformable spheres whose radius R is much smaller than the turbulence dissipation lengthscale η (i.e. Kolmogorov microscale), so that it's possible to consider them as material points. When their density is much higher than the fluid density, their motion is dominated by the Stokes drag, allowing to neglect all other forces in the Maxey-Riley equation [21]. By using analogous arguments, an equation for the evolution of the temperature of small particles, with a Biot number much smaller than unity, can be derived, thus leading to the following system of equations,

$$\frac{\mathrm{d}^2\boldsymbol{x}_p}{\mathrm{d}t^2} = \frac{\mathrm{d}\boldsymbol{v}_p}{\mathrm{d}t} = \frac{\boldsymbol{u}(t,\boldsymbol{x}_p) - \boldsymbol{v}_p}{\tau_\nu}, \qquad \frac{\mathrm{d}\vartheta_p}{\mathrm{d}t} = \frac{T(\boldsymbol{x}_p,t) - \vartheta_p}{\tau_\vartheta}, \tag{4}$$

where $\boldsymbol{x}_p(t)$, $\boldsymbol{v}_p(t)$ and $\vartheta_p(t)$ are the the position, velocity and temperature of the p-th particle. Coefficients τ_ν and τ_ϑ in equation (4) are the momentum and temperature relaxation times of particles, defined by

$$\tau_\nu = \frac{2}{9}\frac{\rho_p}{\rho_0}\frac{R^2}{\nu}, \qquad \tau_\vartheta = \frac{1}{3}\frac{\rho_p c_{pp}}{\rho_0 c_{p0}}\frac{R^2}{\kappa}, \tag{5}$$

where the density of a single particle is ρ_p and c_{pp} denotes its specific heat at constant pressure. Collisions are the only direct particle-particle interaction that is included in this study. Thus, the particle back reaction per unit volume per unit time on the fluid temperature field reads

$$C_T(t,\boldsymbol{x}) = \frac{4}{3}\pi R^3 \rho_p c_{pp} \sum_{p=1}^{N_p} \frac{\mathrm{d}\vartheta_p(t)}{\mathrm{d}t}\delta(\boldsymbol{x} - \boldsymbol{x}_p). \tag{6}$$

2.2. Numerical method and setup

The main objective of this work is the turbulent transport of heat occurring at the interface between two flow regions which are subject to homogeneous and isotropic turbulence with uniform temperatures, viz. T_1 and $T_2 < T_1$. To simulate it, the governing equations presented in Section 2.1 are numerically solved within a parallelepiped domain with dimensions $L_1 = L_2$ and $L_3 = 2L_1$ along the x_1, x_2, and x_3 directions. We set the initial temperature be be equal to T_1 in the $x_3 < L_3/2$ half-domain and equal to T_2 in the other half domain, $x_3 > L_3/2$. Temperature of both fluid and particles is decomposed into the sum of a linear part $T_1 + (T_2 - T_1)(x_3/L_3)$ and a residual part, following [18], so that periodicity can be imposed on the residual part, while particles which exit the domain reenter on the opposite side with the same residual temperature.

Equations (1-6) are rendered non-dimensional by using as reference the length $L_1/(2\pi)$, a velocity scale derived from ε, the kinetic energy dissipation rate imposed by the forcing, and the temperature difference $T_1 - T_2$ between the two zones, as described in detail in [18]. To make results more physically insightful, the Taylor microscale λ is used as reference length in the definition of dimensionless parameters, so that, in the adimensional form, the flow is governed by the Reynolds number $\mathrm{Re} = u'\lambda/\nu$, the Prandtl number $\mathrm{Pr} = \nu/\kappa$, and the thermal mass fraction $\varphi_\vartheta = \varphi(\rho_p c_{pp})/(\rho_0 c_{p0})$, where φ denotes the volume fraction of particles. Analogously, it is more insightful to relate the particle relaxation times, the only dimensional parameters in particle equations, to the timescale of small-scale fluid fluctuations, the Kolmogorov timescale $\tau_\eta = (\nu/\varepsilon)^{1/2}$, more than to the adimensionalization timescale, so that their dynamics is governed by the small scale Stokes number $\mathrm{St} = \tau_v/\tau_\eta$ and thermal stokes number $\mathrm{St}_\vartheta = \tau_\vartheta/\tau_\eta$. All simulations have been performed at the same volume fraction $\varphi = 4 \times 10^{-4}$, with $\rho_p/\rho_0 = 1000$, $c_{pp}/c_{p0} = 4.16$, $\mathrm{Pr} = 0.71$, so that $\mathrm{St}_\vartheta/\mathrm{St} = 4.43$, and, in dimensionless form, $\tau_\eta = 0.098$, $\eta = 0.0153$, $\varepsilon = 0.25$. We carried out three different sets of simulations with Taylor Reynolds number $\mathrm{Re}_\lambda = 56$, 86, 124 by forcing wavenumber $\kappa_f = 5, \sqrt{6}, \sqrt{3}$ respectively. Their corresponding dimensionless integral lengthscale, Taylor lengthscale and velocity fluctuation root mean square are respectively $\ell = 0.4$, 0.74, 0.94, $\lambda = 0.226$, 0.290, 0.350, and $u' = 0.59$, 0.71, 0.85. In addition, we have carried out simulations over a large range of Stokes numbers, from 0.2 to 3. A pseudospectral method, fully dealiased with the 3/2 rule, has been used to discretize the carrier flow equations (1–3). A linear deterministic forcing f_u, given in the Fourier space by

$$\hat{f}_{u,i}(t,\boldsymbol{\kappa}) = \varepsilon \frac{\hat{u}_i(t,\boldsymbol{\kappa})}{\sum_{||\boldsymbol{\kappa}||=\kappa_f} ||\hat{u}_i(t,\boldsymbol{\kappa})||^2} \delta(||\boldsymbol{\kappa}|| - \kappa_f), \tag{7}$$

where κ_f is the forced wavenumber has been used in (2). The injected energy per unit mass and time ε is equal to the mean energy dissipation rate when statistically steady conditions are achieved. A recently proposed numerical method [22, 23], which uses forward and inverse non-uniform fast Fourier transforms with a basis of fourth order B-splines, is implemented to interpolate fluid variables at particle positions and to compute the particle thermal feedback (6). An explicit exponential integrator of second order is used for the time integration of both the continuous phase, equations (1–3), and the discrete phase, equations (4). Please, refer to ref. [22, 23] for more details.

In this work, the only direct interaction between the particles occurs through elastic collisions. A binary collision takes place when the distance between the centres of a couple of particles matches their diameter. To incorporate collisions into the numerical simulation, we employ a first-order reconstruction of the particle trajectories, so that the p-th and q-th particles collide within a time step, $t \in [t_n, t_{n+1})$, if equation

$$||(1-\tilde{t})(\boldsymbol{x}_p(t_n) - (\boldsymbol{x}_q(t_n)) + \tilde{t}(\boldsymbol{x}_p(t_{n+1}) - (\boldsymbol{x}_q(t_{n+1}))|| = 2r_p, \tag{8}$$

where $\tilde{t} = (t - t_n)/\Delta t$, has a real solution $\tilde{t}$ in the interval $[0, 1)$. When this happens, particles continue their motion for the remaining part of the time-step with the velocity obtained after the collision, determined by momentum and energy conservation. It is supposed that particles do not exchange heat during the impact. Computationally, a direct detection of collisions can be highly expensive and impractical when the number of particles N_p is large, because the number of operations is of order $O({N_p}^2)$. Therefore, we use an alternative approach, which involves sorting particles into smaller boxes and searching for collisions within each box [24], a more practical and computationally feasible solution. Any potential missing arising from the box boundaries is avoided by repeating the algorithm after having shifted the boxes. Non binary collisions are not taken into consideration by this procedure.

3. Results and discussion

In this section, we compare the turbulent heat transport in the presence of colliding and non-colliding particles in the parameter range described in section 2.2 in the two-way coupling regime. A snapshot of a thin layer normal to the initial interface, showing particles coloured according to their temperature, is illustrated in figure 1. Particles move across the layer separating the two regions at different temperatures, advected by the flow and gaining and losing heat in the process. After about one timescale $\tau = \ell/u'$, this

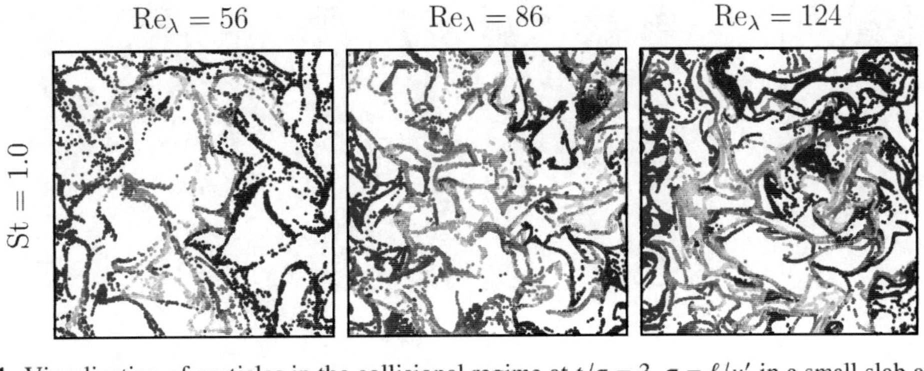

Figure 1. Visualization of particles in the collisional regime at $t/\tau = 3$, $\tau = \ell/u'$ in a small slab around a plane of (x_1, x_3). Particles are drawn not to scale and coloured in accordance with their temperature, from blue (lowest temperature) to red (maximum temperature).

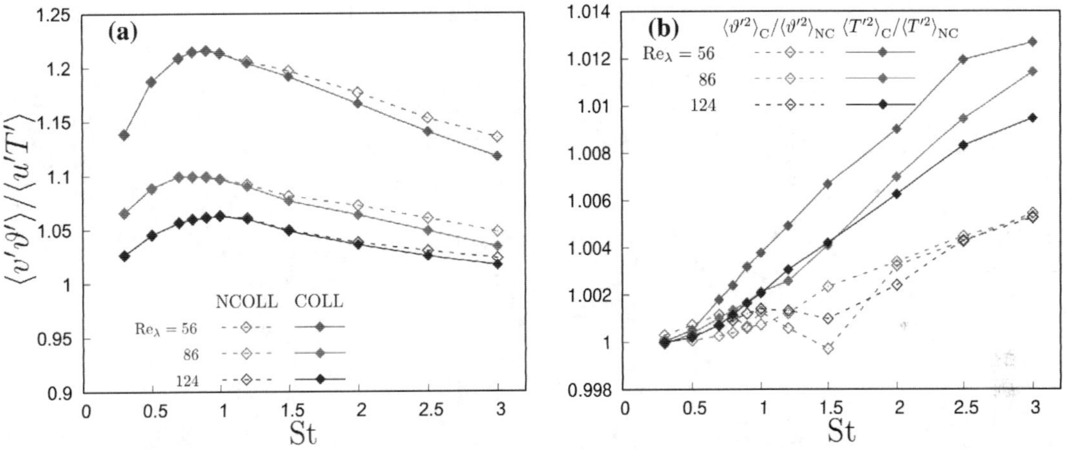

Figure 2. (a) Particle to fluid velocity-temperature correlation ratio as a function of St at different Re_λ (solid lines: collisional regime, dashed lines: collisionless). (b) Effect of collisions on fluid (continuous line) and particle (dashed lines) temperature variance as a function of St.

flow becomes self-similar: all the fluid and particle temperature statistics collapse, up to the numerical uncertainty, when they are rescaled with the thickness of mixing layer δ extracted from the mean temperature distribution, $\delta(t) = (T_1 - T_2)/\max\{|\partial T/\partial x_3|\}$, which shows an almost diffusive growth [18, 19]. The high fluid and particle temperature variance generated in the initial timescale decays as $1/\delta$ in this stage of evolution, while their ratio remains constant. Non-binary collisions are not taken into consideration.

Collisions produce a slightly reduction of the variance of particle temperature when compared to fluid temperature variance, in particular for particles with higher inertia [19]. The particle scattering induced by collisions changes the temperature variance. As illustrated in figure 2(b), in presence of collisions, the particle thermal feedback produces a more pronounced increase in fluid temperature variance compared to particle variance. This effect is particularly evident at large Stokes numbers. Although the difference is relatively small in present simulations, it becomes significant in terms of its implications at higher volume fractions, where collisions are more frequent and thermal feedback more intense. From equation

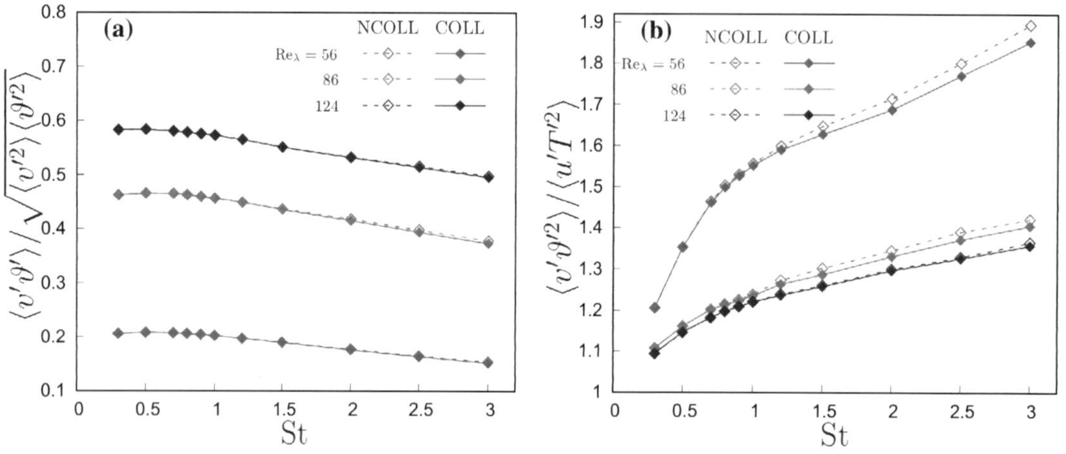

Figure 3. (a) Particle velocity-temperature correlation function as a function of the St. (b) Normalized particle temperature variance flow in the inhomogeneous direction x_3.

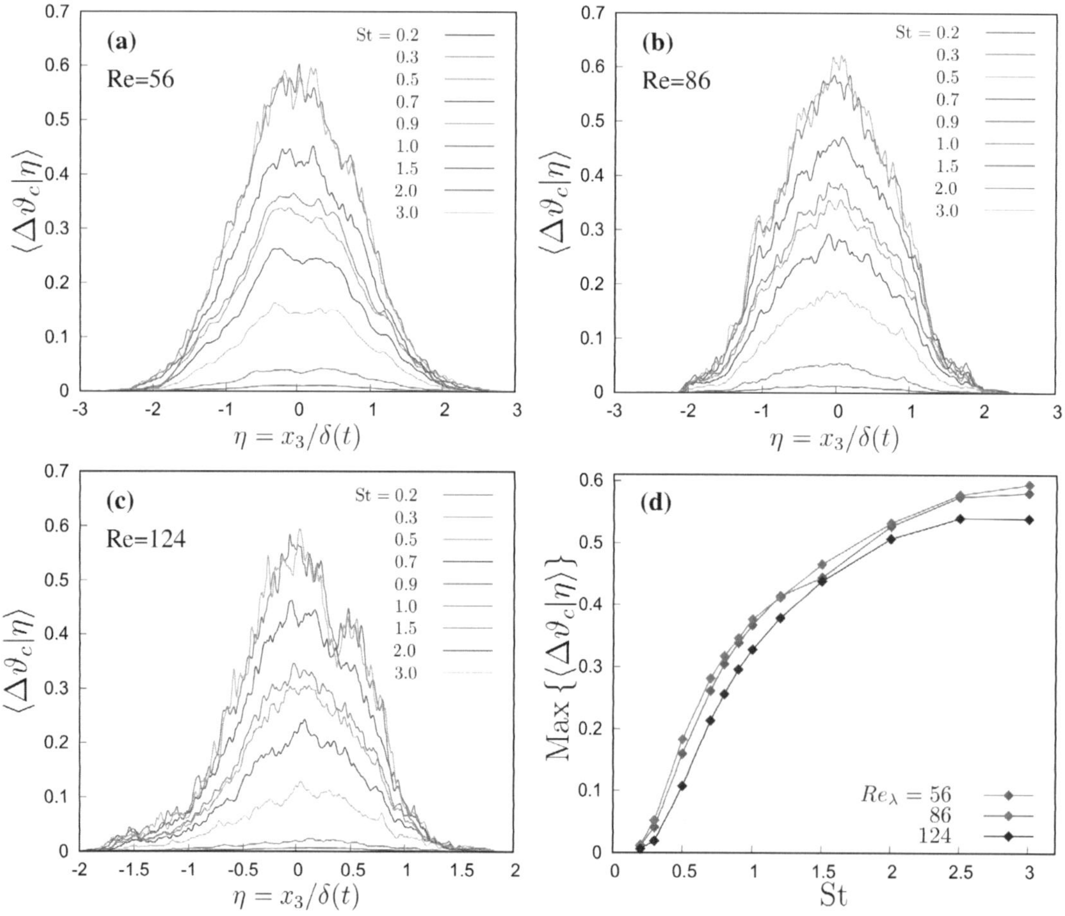

Figure 4. (a),(b),(c) average of $\Delta\vartheta_c = |\vartheta_a - \vartheta_b|$, where ϑ_a and ϑ_b denote the temperatures of the colliding particles, conditioned at the normalized position η of the collision, (d) maximum value of $\langle\Delta\vartheta_c|\eta\rangle$ as a function of St at different Re$_\lambda$.

(3), the overall mean heat flux in the inhomogeneous direction x_3 is

$$\dot{q} = -\lambda \frac{\partial \langle T \rangle}{\partial x_3} + \rho_0 c_{p0} \langle u_3' T' \rangle + \varphi \rho_p c_{pp} \langle v_3' \vartheta' \rangle \tag{9}$$

that is, the sum of the diffusive heat transfer, the fluid convective part, and the particle contribution [18]. The ratio between particle and fluid temperature-velocity correlation gives the particles relative contribution to the mean heat transfer. Figure 2(a) shows this ratio, computed in the centre of the mixing layer where these correlations have their maximum, as a function of St for three simulated Re_λ. The maximum occurs for a Stokes number around one at all Reynolds numbers both in collisional and collisionless regimes, in correspondence of the more intense clustering. Since the ratio between thermal and inertial relaxation times is fixed, in the tracer limit particles behave both dynamically and thermally as passive tracers and we have $\langle v_3' \vartheta' \rangle / \langle u_3' T' \rangle \to 1$ as $\mathrm{St} \to 0^+$. On the other side, in the infinite inertia limit, $\mathrm{St} \to \infty$, particles velocity and temperature tend to become independent of local flow structures, leading to a gradual reduction of $\langle v_3' \vartheta' \rangle / \langle u_3' T' \rangle$. Particle collisions have a negligible influence on fluid convective heat transfer. This implies that, within the examined volume fraction range, the collision-induced deviation from particles Lagrangian path has a mild effect on their average thermal feedback on the fluid. This is in agreement with Carbone et. al [6], who suggest two-way coupling and collision have a very weak effect on heat exchange between fluid and particle at high St. However, collisions do reduce the correlation between particle velocity and temperature, thereby limiting their capacity to transport heat over distances of the order of the mixing thickness. Given the increase in the collision frequency of larger particles which has been observed in [20], the impact on the turbulent heat flux in direction x_3 becomes more apparent as St increases [19, 20].

We can relate the observed attenuation of the correlation of particle temperature-velocity to the sling effect that produces interpenetration of particle trajectories from far zones in the inhomogeneous direction, therefore with very different temperatures, enhancing collisions which produces a scatter of the v_3 velocity component, thus reducing the $\langle v_3' \vartheta' \rangle$. Moreover, for large particles, this is associated, as explained by Carbone et. al [6], with the formation of thermal caustics, where particles with a very different temperature are gathered. Since we fixed the thermal Stokes to Stokes number ratio, these two simultaneous mechanisms act in the same way resulting in a dynamical and thermal decorrelation from local flow. Consequently, even if the ability of particles heat exchange with each other increases due to the sling effect and thermal caustic, but their ability to transfer heat with fluid decreases and we can see the reduction in overall heat transfer in direction x_3 in collisional regime at large St. On one hand, due to high thermal inertia, particles avoid transferring heat with surrounding fluid, collision is also boosting this manner, on the other hand, particle momentum exchange with other particles reduces their ability to carry heat to the other distant regions, then causing a heat flux attenuation in thermal mixing layer. Information about the particle ability to transport heat can also be inferred from the velocity-temperature correlation function $\langle v_3' \vartheta' \rangle / \sqrt{\langle v_3'^2 \rangle \langle \vartheta'^2 \rangle}$, which reduces with St, as shown in figure 3(a). The reason is that larger particles with higher inertia tend not to be correlated with local fluid velocity field and resulting in lower correlation function. Conversely, the correlation function has the highest value for lowest Stokes number implying that particles at this Stokes number are highly correlated with fluid field. Particle temperature variance flux becomes much larger than fluid temperature variance, figure 3(b) as the Stokes number increase, as particles can move over longer distance keeping their temperature. Even in this case, the collisional scattering reduces this flux. Indeed, an increase of the colliding particles temperature difference with the St can be observed, see figures 4(a),(b),(c) an indication that a higher particle inertia makes possible collisions between particles coming from very far zones, letting them pass through the fluid temperature fronts [6], while they can keep their temperature due to their thermal inertia. The temperature difference at collision has a minor dependence on the Reynolds number, but increases with the Stokes number, approaching a constant value for St larger than 2 (figure 4(d)). As observed in [6], inertial particles relative temperature difference depends non locally on the fluid small-scale temperature structure function experienced along their respective trajectories so that even at small scale separation their temperature difference can significantly exceed the small-scale local fluid temperature difference. Since it has been found that collisions event between particles with larger inertia are, as expected, more likely, the impact on the temperature-velocity correlation increases as St increases [20].

4. Conclusion

This study delves into some aspects of the modification of turbulent heat transport induced by particle collisions. Collision-induced scattering significantly amplifies fluid temperature variance compared to particle variance, notably at larger Stokes numbers, through particle feedback, and decrease the correlation between particle velocity and temperature, limiting their long-distance heat transport efficiency, but the effect is globally a minor one. With increasing Stokes numbers, collision frequency rises, influencing turbulent heat flux. This decline in particle temperature-velocity correlation is linked to the sling effect, boosting collisions and diminishing heat transfer capacity. Overall, our findings highlight the intricate interplay of particle collisions, inertia, and their interaction with fluid, revealing nuances in heat transport mechanisms. They underscore the role of collision dynamics and particle characteristics in shaping turbulent heat flux within particle-laden flows.

5. References

[1] Pumir A and Wilkinson M 2016 *Annu. Rev. Condens. Matter Phys.* **7** 141–170
[2] Zonta F, Marchioli C and Soldati A 2008 *Acta Mech.* **195** 305–326
[3] Kuerten J G M, van der Geld C W M and Geurts B J 2011 *Phys. Fluids* **23** 123301
[4] Rousta F and Lessani B 2020 *Int. Commun. Heat Mass Transf.* **112** 104475
[5] Béc J, Homann H and Krstulovic G 2014 *Phys. Rev. Lett.* **112**(23) 234503
[6] Carbone M, Bragg A D and Iovieno M 2019 *J. Fluid Mech.* **881** 679–721
[7] Saito I, Watanabe T and Gotoh T 2022 *J. Fluid Mech.* **931** A6
[8] Saffman P G and Turner J S 1956 *J. Fluid Mech.* **1** 16–30
[9] Sundaram S and Collins L r 1997 *J. Fluid Mech.* **335** 75–109
[10] Abrahamson J 1975 *Chem. Eng. Sci.* **30** 1371–1379
[11] Zaichik I I and Pershukov V A 1995 *Fluid Dyn.* **30** 49–63
[12] Wang L P, Wexler A and Zhou Y 2000 *J. Fluid Mech.* **415** 117 – 153
[13] Sommerfeld M 2001 *Int. J. Multiph. Flow* **27** 1829–1858
[14] Johnson P 2019 *Int. J. Multiph. Flow* **124** 103182
[15] Kuerten J and Vreman A 2016 *Int. J. Multiphase Flow* **87** 66–79
[16] Jiang F, Wang H, Liu Y, Qi G, Al-Rawni A E, Nkomazana P and Li X 2021 *Powder Technol.* **381** 55–67
[17] Sun J and Chen M M 1988 *Int. J. Heat Mass Transfer* **31** 969–975
[18] Zandi Pour H R and Iovieno M 2022 *Fluids* **7** 345
[19] Zandi Pour H R and Iovieno M 2023 *8th World Congress on Momentum, Heat and Mass Transfer*
[20] Zandi Pour H R and Iovieno M 2023 *J. Fluid Flow Heat Mass Transf.* **10** 140–149
[21] Maxey M R and Riley J J 1983 *Phys. Fluids* **26** 883–889
[22] Carbone M and Iovieno M 2018 *WIT Trans. Eng. Sci.* **120** 237–248
[23] Carbone M and Iovieno M 2020 *Int. J. Saf. Sec. Eng.* **10** 191–200
[24] Onishi R, Takahashi K and Vassilicos J C 2013 *J. Comput Phys.* **242** 809–827

Acknowledgements

The authors would like to gratefully acknowledge PRACE for having awarded them access to high performance computing and special storage resources at HPC Vega-IZUM, Maribor, Slovenia under the PRACE EuroHPC Development Access project 2023D02-007.

Journal of Physics: Conference Series **2685** (2024) 012003 doi:10.1088/1742-6596/2685/1/012003

Mixed convection in turbulent particle-laden channel flow at $Re_\tau = 180$

D Zaza[1,*] and M Iovieno[1]

[1] Dipartimento di Ingegneria Meccanica e Aerospaziale, Politecnico di Torino, Corso Duca degli Abruzzi 24, 10129 Torino, Italy

* Corresponding author: domenico.zaza@polito.it

Abstract. A numerical investigation of mixed convection in a turbulent particle-laden channel flow is presented. Using Eulerian-Lagrangian Direct Numerical Simulations (DNSs) within the point-particle approach, the interaction between wall turbulence, particle inertia and thermal inertia, and buoyancy in the two-way coupling regime has been studied. The flow dynamics is controlled by the friction Reynolds number, Richardson number, particle Stokes and thermal Stokes numbers, and the Prandtl number. The effects of particle inertia and buoyancy on fluid and particle statistics are shown for a friction Reynolds number of 180, Stokes numbers ranging from 0.6 to 120, Richardson numbers ranging from 2.72×10^{-4} to 27.2 at a single Prandtl number, 0.71. The results indicate that the effect of particle inertia is significant on particle statistics, but not as pronounced on fluid statistics at the relatively low volume fraction considered. Particle inertia modulates the overall heat flux in a non-monotonic way.

1. Introduction

Particles suspended in turbulent flows are a common occurrence in many natural and industrial processes. Natural examples include the transport and deposition of aerosols, the formation of droplets in clouds, sand and dust storms, volcanic eruptions, soil erosion processes, and sediment transport in watercourses. On the other hand, engineering processes such as pharmaceutical sprays, liquid fuel combustion, and combustion in solid propellant endoreactors also involve particle-laden flows. Direct Numerical Simulation (DNS) has become the main tool for improving the physical understanding of these flows due to recent advances in high-performance computing. This approach allows for the study of the details of the flow, which are often difficult to observe experimentally, particularly when additional physical phenomena such as heat transfer, buoyancy, mass transfer due to condensation or evaporation, chemical reactions, and acoustic waves play a role. When the particle size is smaller than the Kolmogorov scale, the Lagrangian point-particle method is the most commonly used approach [1]. In this approach, particles are treated as point-like, and their dynamics is determined by solving a set of ordinary differential equations for each particle. This method reduces the computational burden as it does not require the detailed simulation of the flow around each particle. Over the past fifty years, since the pioneering work by [2], point-particle DNS has allowed for the simulation of very large numbers of particles in different flow configurations, leading to fundamental insights into the subject. In inhomogeneous turbulent flows, for example, the phenomenon of preferential concentration of particles, known as turbophoresis, is crucial [3]. Particles tend to migrate from regions of high turbulence intensity toward lower-intensity regions due to the property of inertial particles to disperse more quickly where the turbulent intensity is higher. In particular, in wall-bounded flows, large-scale clustering and preferential concentration of particles are observed in the near-wall region [4] and are strongly influenced by the local Stokes number [5]. However, this phenomenon is weakened in the two-way coupling regime, i.e. when the full momentum exchange between the two phases is taken into account, as a consequence of the damping performed by the particles on the wall-normal and spanwise velocity fluctuations. Indeed, for particle volume fractions greater than 10^{-6}, the modulating effects of particles on turbulence cannot be neglected [6]. In this regime, the momentum and heat exchange between the phases has to be taken into account. Inertial particles have been found to enhance heat transfer as their thermal inertia grows [7], while they tend to suppress fluid temperature increments as the particle thermal response time increases [8].

Mixed convection occurs when both forced convection (driven by a pressure gradient) and natural convection (driven by buoyancy forces) are present in a fluid flow, and can significantly impact on the fluid flow and, consequently, the heat transfer characteristics of a system. The presence of particles can further modulate the flow, due to the interplay between preferential concentration and small-scale features. While there has been research on both particle-laden flows and mixed convection flows separately, to the best of our knowledge, the combination of the two phenomena has not been widely explored for finite inertia particles. All studies focus on nanoparticles and use a single-phase Eulerian approach or a two-phase Eulerian-Eulerian method, which can provide a description of the flow only for very small Stokes numbers [9]. Investigating the combined effects of these two phenomena could lead to a better understanding of the complex behaviour of particle-laden flows in both nature and industrial processes. In this exploratory study, we used Lagrangian-Eulerian direct numerical simulations to evaluate the role of particles with finite thermal inertia on the mixed convection in an horizontal channel flow heated from below.

2. Physical model

In this section, we present the physical model for the non-isothermal particle-laden flow in an horizontal plane channel, when heat transfer by natural convection is also considered. A phase consisting of small (sub-Kolmogorov), heavy, spherical, and collisionless particles is dispersed in the fluid. The Eulerian-Lagrangian approach is adopted: the continuous phase is described in an Eulerian way, whereas the dispersed phase is followed in a Lagrangian way, solving a set of equations for each particle. The two-way coupling between particles and fluid is here considered for both the momentum and the internal energy.

The flow is driven by a uniform pressure gradient parallel to the walls, which defines the streamwise direction, and is subject to buoyancy effects due to a temperature difference between the upper and lower walls combined with a uniform wall-normal gravitational field. The temperature is considered uniform over each wall, with the bottom wall warmer and the upper wall colder to obtain an unstable stratification, as in [10]. Computationally, the flow is solved in a parallelepiped domain, with size $2\pi h$ in the streamwise direction x, $2h$ in the wall normal direction y, and $2\pi h$ in the spanwise direction z, where h is the channel half-height. Periodicity is imposed in the streamwise and spanwise directions, exploiting the statistical homogeneity in these directions, while at the walls fluid velocity satisfies no-slip conditions and the temperature is set equal to the wall temperature. Since the flow is driven by both the mean pressure gradient, $\partial\langle p\rangle/\partial x$, and the temperature difference, $T_1 - T_2$, between the lower and upper walls, the equations are made dimensionless using h as the reference length-scale, the friction velocity $u_\tau = [(h/\rho)\,|\partial\langle p\rangle/\partial x|]^{1/2}$ as the reference velocity scale, the carrier flow mean density as the density (and mass) reference, and the half temperature difference $(T_1 - T_2)/2$ as the reference temperature variation. The governing equations are detailed below in dimensionless form; all variables denote dimensionless quantities unless explicitly stated.

2.1. *Fluid phase*

The fluid phase is governed by the incompressible Navier-Stokes equations with an advection-diffusion equation for the fluid temperature. Momentum and temperature are coupled by buoyancy forces through the usual Boussinesq approximation, so that the fluid phase is governed by the following equations in tensorial notation ($x_1 = x, x_2 = y, x_3 = z$),

$$\frac{\partial u_j}{\partial x_j} = 0, \tag{1}$$

$$\frac{\partial u_i}{\partial t} + \frac{\partial (u_i u_j)}{\partial x_j} = -\frac{\partial p}{\partial x_i} + \frac{1}{\mathrm{Re}_\tau}\frac{\partial^2 u_i}{\partial x_j^2} + \mathrm{Ri}_\tau \vartheta \delta_{i,2} + \delta_{i,1} - C_{u,i}, \tag{2}$$

$$\frac{\partial \vartheta}{\partial t} + \frac{\partial (\vartheta u_j)}{\partial x_j} = \frac{1}{\mathrm{Re}_\tau \mathrm{Pr}}\frac{\partial^2 \vartheta}{\partial x_j^2} - C_\vartheta. \tag{3}$$

Here, $u_i(t, \boldsymbol{x})$ is the i-th component of the fluid velocity field, $p(t, \boldsymbol{x})$ stands for the fluctuating pressure field, and $\vartheta(t, \boldsymbol{x})$ is the temperature field. In this form, the wall boundary conditions for the dimensionless temperature are $\vartheta(t, x, \pm 1, z) = \mp 1$. The three non-dimensional governing parameters arising from the choice of reference scales are the friction Reynolds number, $\mathrm{Re}_\tau = u_\tau h/\nu$, the Prandtl number, $\mathrm{Pr} = \nu/\kappa$, and the friction Richardson number, $\mathrm{Ri}_\tau = h\alpha g \Delta T_{\mathrm{ref}}/u_\tau^2$, where g is the gravity acceleration and ν, κ and α are respectively the kinematic viscosity, the thermal diffusivity, and the thermal expansion coefficient of the fluid. Thus, the term $\mathrm{Ri}_\tau \vartheta \delta_{i,2}$ in equation (2) represents the net buoyancy force acting against the gravitational field. Finally, the terms $C_{u,i}$ and C_ϑ are the particle feedback on the momentum and the internal energy of the carrier phase, which will be soon discussed.

2.2. *Dispersed phase*

We consider the flow seeded with N_p identical, rigid, small, heavy, and spherical particles, with radius r much smaller than any flow length scale. In such a situation, particle dynamics is described using a Lagrangian approach, where the position, velocity, and temperature of each particle are determined by external forces and heat exchanged with the carrier fluid. The equations of motion for a spherical particle in a fluid flow were formulated by Maxey and Riley in [11], where it was shown that for particle densities much larger than that of the fluid, only the Stokes drag force needs to be considered, with the added mass, Basset history, and pressure gradient forces being negligible. To avoid sedimentation, no buoyancy is considered for particles. The motion of the j-th particle is described by the dimensionless equations:

$$\frac{\mathrm{d}\boldsymbol{x}_j}{\mathrm{d}t} = \boldsymbol{v}_j(t), \qquad \frac{\mathrm{d}\boldsymbol{v}_j}{\mathrm{d}t} = \frac{\boldsymbol{u}(t, \boldsymbol{x}_j) - \boldsymbol{v}_j(t)}{\tau_u}\left(1 + 0.15\,\mathrm{Re}_j^{0.687}\right), \tag{4}$$

where $\boldsymbol{x}_j(t)$ and $\boldsymbol{v}_j(t)$ denote respectively the position and velocity of the particle, $\boldsymbol{u}(t, \boldsymbol{x}_j)$ is the fluid velocity at the particle position and time, and τ_u is the particle relaxation time, given in dimensionless form by

$$\tau_u = 2/9\, \rho_p r^2 \mathrm{Re}_\tau. \tag{5}$$

The term within parenthesis in equation (4) represents a semiempirical correction to the Stokes drag coefficient devised in [12] for spherical particles. This correction is based on the Reynolds number of the fluid-particle relative motion, $\mathrm{Re}_j = 2r\left|\boldsymbol{u}(t, \boldsymbol{x}_j) - \boldsymbol{v}_j(t)\right|\mathrm{Re}_\tau$.

An equation for the temperature of the j-th particle, ϑ_j, can be obtained applying the first law of thermodynamics to the particle, and assuming quasi-steady conduction and convection over particle scale as the main mechanisms through which heat is transferred between the two phases:

$$\frac{\mathrm{d}\vartheta_j}{\mathrm{d}t} = \frac{\vartheta(t, \boldsymbol{x}_j) - \vartheta_j(t)}{\tau_\vartheta}\frac{\mathrm{Nu}_j}{2}. \tag{6}$$

Here, the particle Nusselt number, Nu_j, is given by the correlation introduced in [13] for a spherical particle:

$$\mathrm{Nu}_j = 2 + 0.6\,\mathrm{Re}_j^{1/2}\mathrm{Pr}^{1/3}. \tag{7}$$

Similarly to τ_u, the quantity τ_ϑ is the particle thermal relaxation time. In dimensionless form, it is given by

$$\tau_\vartheta = 1/3\ \rho_p c_p r^2 \mathrm{Re}_\tau \mathrm{Pr}, \tag{8}$$

where c_p is the particle specific heat, normalized with the fluid specific heat at constant pressure. The relaxation times τ_u and τ_ϑ describe particle inertia and thermal inertia, and their ratios with the fluid flow scales, called Stokes numbers, determine their dynamics [1]. In a turbulent flow, the most important flow timescale for particle motion is the smallest one, which in a turbulent wall bounded flow is the viscous timescale. In dimensionless form, it is $\tau_\eta = 1/\mathrm{Re}_\tau$, so that the particle Stokes number and thermal Stokes number are defined as

$$\mathrm{St} := \frac{\tau_u}{\tau_\eta} = \frac{2}{9}\rho_p r^2 \mathrm{Re}_\tau^2, \qquad \mathrm{St}_\vartheta := \frac{\tau_\vartheta}{\tau_\eta} = \frac{1}{3}\rho_p c_p r^2 \mathrm{Re}_\tau^2 \mathrm{Pr}. \tag{9}$$

Particle-particle interactions are disregarded, and elastic collisions are assumed when particles hit the walls. Moreover, if a particle leaves the domain through a periodic boundary, it is reintroduced at the opposite side with the same velocity and temperature. When the volume fraction is above 10^{-6}, particle feedback on the carrier flow, represented by $\boldsymbol{C}_u$ and C_ϑ in equations (2) and (3) cannot be neglected. These terms are equal to the momentum and heat exchanged per unit volume by all the N_p particles with the fluid, so that, for point particles they have the following form:

$$\boldsymbol{C}_u = \frac{4}{3}\pi\rho_p r^3 \sum_{j=1}^{N_p} \frac{\mathrm{d}\boldsymbol{v}_j}{\mathrm{d}t}\delta(\boldsymbol{x}-\boldsymbol{x}_j), \qquad C_\vartheta = \frac{4}{3}\pi\rho_p c_p r^3 \sum_{j=1}^{N_p} \frac{\mathrm{d}\vartheta_j}{\mathrm{d}t}\delta(\boldsymbol{x}-\boldsymbol{x}_j). \tag{10}$$

The time derivatives $\mathrm{d}\boldsymbol{v}_j/\mathrm{d}t$ and $\mathrm{d}\vartheta/\mathrm{d}t$ are given by equations (4) and (6), respectively.

3. Numerical method

The carrier flow equations (1-3) are solved using the formulation proposed by [14], which is convenient because the Poisson equation for the pressure field does not need to be solved. This formulation uses the wall-normal component of the vorticity vector, $\eta = \partial u/\partial z - \partial w/\partial x$, and the wall-normal component of the velocity field, v, as main variables which are advanced in time, and reformulates continuity and momentum equations as a second order equation for η and a fourth order equation in v. The equations for the fluid phase are discretized in space using a pseudo-spectral method: the Fourier-Galërkin approach is applied in the two homogeneous directions x and z, whereas the Chebyshev-tau method is used to treat the wall-normal direction y. The convective non-linear terms are computed pseudo-spectrally using a de-aliasing procedure based on the 2/3-rule. Time advancement of the fluid equations is performed using the explicit two-step Adams-Bashforth scheme for the convective, the particle feedback, and the buoyancy terms, and an implicit Crank-Nicolson scheme for the diffusive terms. The application of these schemes, along with the Influence Matrix method [15], reduces the problem of determining the velocity and the temperature fields to solving, at each time-step and for each couple of Fourier modes, three one-dimensional Helmholtz equations for η, $\varphi = \nabla^2 v$, and ϑ. The Chebyshev-tau scheme applied to these equations leads to quasi-tridiagonal algebraic systems in which the coefficients of the even and the odd Chebyshev polynomials are decoupled [16]. Thus, each algebraic system can be split in two smaller problems which are solved by means of the Thual algorithm [17]. Then, v is determined from $\varphi = \nabla^2 v$ exploiting the Influence Matrix method, and u and w are obtained from η and the incompressibility constraint. Finally, the equations for the mean flow in the streamwise and the spanwise directions are solved. Equations (4) and (6) describing particle dynamics are integrated in time using the same Adams-Bashforth scheme employed for the fluid phase. The evaluation of the fluid velocity and temperature at the particle location is carried out through a tri-linear interpolation, while the feedback terms are evaluated through the tri-linear projection onto the eight grid points of the computational cell the particle belongs to.

4. Results and discussion

In order to analyse the role of particles in mixed convection, we conducted eighteen simulations on a 192×201×192 grid in the streamwise, wall-normal and spanwise directions. These simulations were performed using a constant friction Reynolds number ($\mathrm{Re}_\tau = 180$) and a given Prandtl number ($\mathrm{Pr} = 0.71$). The resolution is therefore $\Delta x^+ = \Delta z^+ \simeq 5.9$, and Δy^+ between 0.02 and 2.9. The number of particles ($N_p = 11{,}782{,}400$) and the particle non-dimensional radius ($r = 4 \times 10^{-4}$) were fixed, resulting in a volume fraction equal to 4×10^{-5}. Different Stokes numbers were obtained by changing particle density. The simulations can be categorized into three groups. The first group involved simulating the unladen flow, considering both forced convection and mixed convection for five friction Richardson numbers ranging from 2.72×10^{-4} to 27.2. These simulations served as benchmarks to assess the influence of particles on the flow. The second group consisted of simulations of the two-way coupled particle-laden flow, considering three different Stokes numbers: $\mathrm{St} = 0.6$, 60, and 120. Since the ratio between the specific heats of the two phases ($c_p = 4.17$) remained constant, the ratio between the thermal Stokes number and the Stokes number also remained constant. These simulations exclusively accounted for forced convection ($\mathrm{Ri}_\tau = 0$). Finally, in the third group, we examined the combined effect of mixed convection and particle inertia by conducting nine particle-laden flow simulations with two-way coupling. These simulations considered three friction Richardson numbers (2.72×10^{-2}, 2.72×10^{-1}, 2.72) and three Stokes numbers ($\mathrm{St} = 0.6, 60, 120$).

Figure 1 illustrates the impact of mixed convection on the unladen channel flow. For friction Richardson numbers below one, the influence of natural convection on the mean flow and mean temperature field is negligible. The curves closely align with the reference curve, which corresponds to the case of forced convection ($\mathrm{Ri}_\tau = 0$). However, as the value of Ri_τ exceeds one, natural convection becomes more significant, leading to a substantial change in the flow structure. Out of the wall layer ($y^+ > 30$), the mean streamwise velocity increases for $\mathrm{Ri}_\tau = 2.72$ and then decreases for $\mathrm{Ri}_\tau = 2.72 \times 10^1$ compared to the reference case. Throughout the range of $30 < y^+ < 120$, a recognizable logarithmic law of the form $\langle u \rangle = 1/k \log y^+ + B$ is still observed for all cases, with the exception of $\mathrm{Ri}_\tau = 2.72$, where no clear logarithmic layer can be identified. When data are fitted with a logarithmic profile, for $\mathrm{Ri}_\tau < 1$ we obtained $k \simeq 0.393$ and $B \simeq 5.335$, the classical values of the coefficients in a boundary layer, while for $\mathrm{Ri}_\tau = 27.2$, we obtained $k \simeq 0.626$ and $B \simeq 8.346$. The presence of natural convection is also evident in the abrupt change of the mean temperature profile (figure 1(b)) and temperature variance profile (figure 1(d)). Natural convection flattens the mean temperature profile in the central region of the channel, confining temperature gradients to a narrower zone near the walls, whose thickness reduces with Ri_τ. Furthermore, for the two highest simulated Ri_τ values, an inflection point with negative temperature gradients is observed near the centerline of the channel, as reported in [18] for Ri_τ values of 10^2 and 10^3. The temperature variance profile (figure 1(d)) confirms this abrupt change in flow regime. In the $\mathrm{Ri}_\tau < 1$ range, the temperature variance exhibits a maximum at the center of the channel and a secondary peak at $y^+ \simeq 20$, in the upper region of the buffer layer. As the influence of natural convection increases, the temperature variance at the center decreases and the curve flattens, while the intensity of the peak near the wall grows, shifting closer to the wall. When natural convection dominates over forced convection, the temperature variance decreases throughout the channel, and the peak further migrates toward the wall. For higher Richardson numbers, the trend of the temperature variance is expected to approach that of pure Rayleigh-Bénard flow. Figure 1(c) presents the wall-normal temperature-velocity correlation, which is directly proportional to the convective heat flux by turbulent fluctuations in the y-direction.

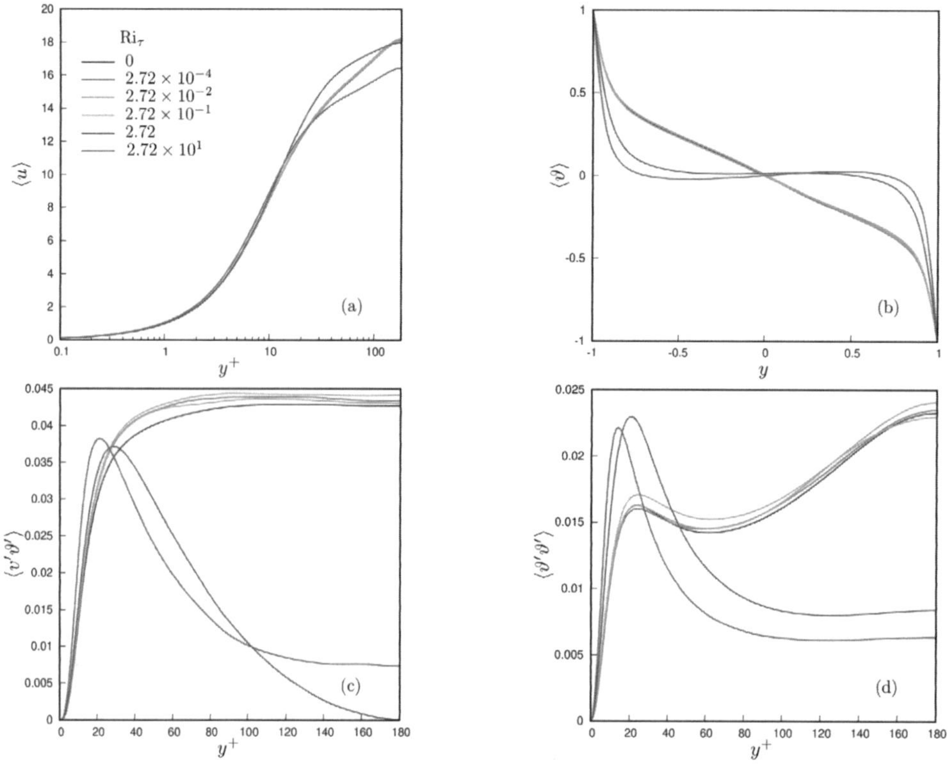

Figure 1. Unladen flow, effect of the Richardson number, Ri_τ, on flow statistics: (a) mean streamwise velocity, (b) mean temperature, (c) wall-normal turbulent transport, (d) temperature variance.

Figure 2 illustrates the particle and flow statistics obtained from a two-way coupled particle-laden flow, focusing specifically on forced convection conditions. Due to the relatively low particle volume fraction of 4×10^{-5}, the influence of particle inertia on flow statistics is not as significant as on particle statistics, in agreement with existing literature [1]. One notable effect observed is turbophoresis, as revealed by the concentration profiles (figure 2(b)). The concentration of particles normalized with the mean concentration exhibits a maximum in the viscous sublayer ($y^+ < 5$), which strengthens and moves towards the wall as the Stokes number increases from 0.6 to 60. This behavior is expected since particle inertia becomes more pronounced, leading to enhanced particle accumulation near the wall. However, it is important to note that the trend of the maximum concentration is non-monotonic: as particle inertia further increases and the Stokes number reaches 120, the maximum concentration decreases. This reduction can be attributed to the damping effect of particles on the wall-normal and spanwise velocity fluctuations, resulting in a net reduction of turbulent kinetic energy. Regarding the impact on fluid properties, the effect of particles on the mean streamwise velocity and the mean temperature is found to be negligible at the volume fraction simulated. However, some modulation is observed in the fluid temperature variance. When the Stokes number increases to 60, the maximum temperature variance at the center of the channel increases by approximately 12% compared to the reference unladen flow. Conversely, when the Stokes number reaches 120, the maximum temperature variance decreases by approximately 6%. These changes in temperature variance are in line with the interplay between particle inertia and fluid turbulence, where enhanced mixing and heat transfer initially increase the variance, but the damping effect of particles subsequently reduces the turbulent kinetic energy and, thus, the temperature variance. Furthermore, the particle temperature variance is consistently larger than the fluid variance and tends to reach its maximum value near the wall. This behavior can be attributed to particles being ejected from the wall region, where heat transfer is dominant, while retaining their temperature.

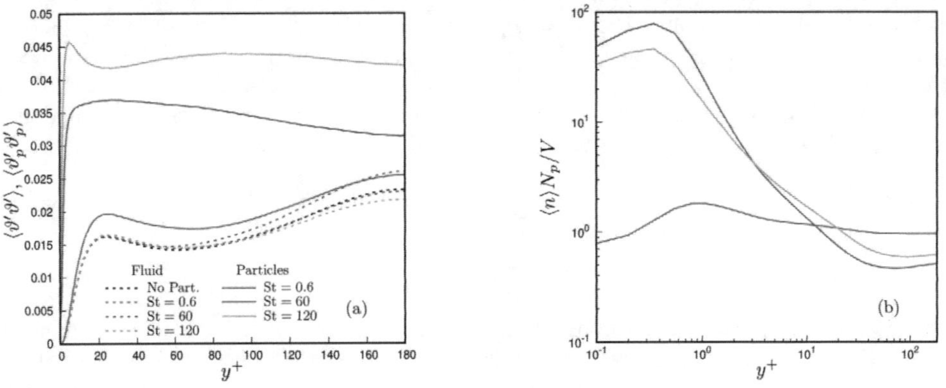

Figure 2. Forced convection in the two-way coupled particle-laden flow. Effect of the Stokes number, St: (a) fluid and particle temperature variance, (b) particle concentration.

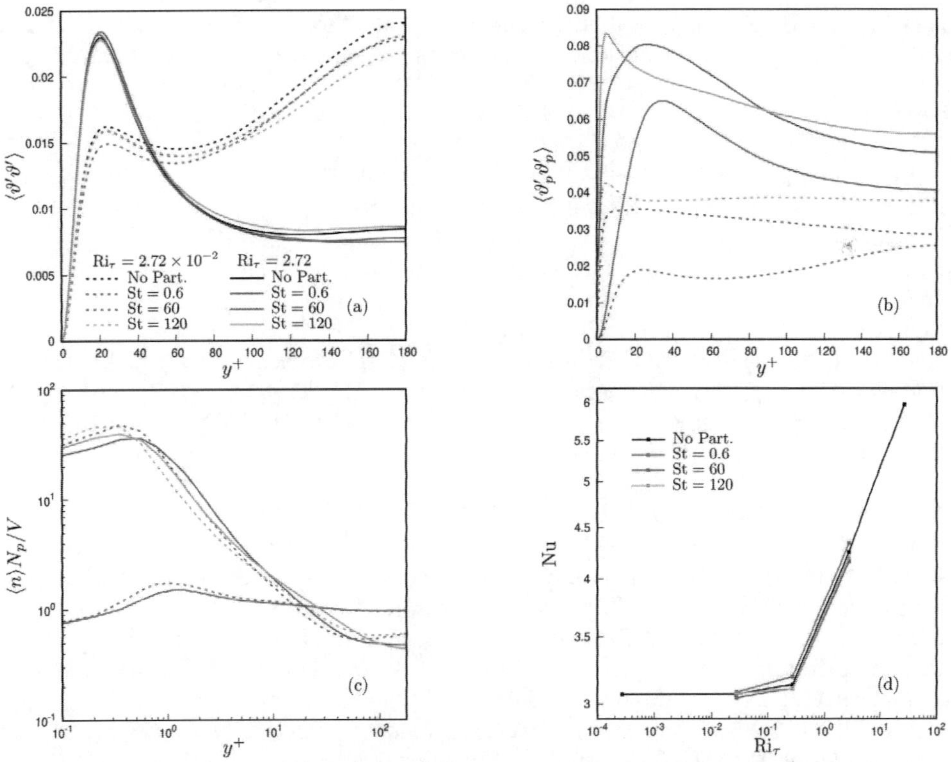

Figure 3. Mixed convection in the two-way coupled particle-laden flow. Effect of both the Stokes number, St, and the Richardson number, Ri_τ, on flow and particle statistics: (a) fluid temperature variance (b) particle temperature variance, (c) particle concentration, (d) average Nusselt number, $\mathrm{Nu} = 1/2\ (\partial\langle\vartheta\rangle/\partial y)_w$.

Figure 3 shows the influence of the Stokes number in the two-way coupled particle-laden channel flow when also natural convection is considered at $\mathrm{Ri}_\tau = 2.72 \times 10^{-2}$ and 2.72, representative of the two flow regimes. Unlike the case of pure forced convection, particle feedback tends to damp the variance of the carrier flow temperature (figure 3(a)) for low Richardson numbers. This effect is milder for $\mathrm{Ri}_\tau > 1$. The enhancement of the wall-normal motion due to buoyancy increases particle temperature variance (figure 3(b)) due to a more frequent particle ejection from the wall region which also reduces the concentration peak near the wall (figure 3(c)), in particular at intermediate Stokes numbers. When the Richardson number exceeds one, the influence of natural convection becomes significant, enhancing heat transfer and increasing the Nusselt number (figure 3(d)). A non-monotonic modulation of the Nusselt number operated by particle inertia is observed.

5. Conclusions

This study provides a comprehensive exploration of mixed convection in turbulent particle-laden channel flows. The intricate dynamics revealed through eighteen simulations underscore the substantial influence of particle inertia on particle statistics, and a milder effect on fluid statistics due to the relatively low particle volume fraction considered. Notably, natural convection leads to distinct changes in flow structure and thermal characteristics as the friction Richardson numbers exceeds one. Turbophoresis is mainly observed under forced convection, with concentration profiles exhibiting non-monotonic behaviour as the Stokes numbers increase. The interplay of particle inertia and buoyancy in the presence of natural convection further modulates the Nusselt number, revealing a non-monotonic trend influenced by particle characteristics. These findings contribute valuable insights into the complex interactions between particles and fluid dynamics, essential for refining models in engineering applications, and into systems in which mixed convection play a pivotal role in heat transfer.

6. References

[1] Brandt L and Coletti F 2022 *Annu. Rev. Fluid Mech.* **54** 159-189
[2] Riley J J and Patterson Jr G S 1974 *Physics of Fluids* **17** 292-297
[3] Reeks M W 1983 *Journal of Aerosol Science* **14** 729-739
[4] Marchioli C and Soldati A 2002 *Journal of Fluid Mechanics* **468** 283-315
[5] Perrone D, Kuerten J, Ridolfi L and Scarsoglio S 2023 *International Journal of Multiphase Flow* **165** 104489
[6] Elghobashi S 1994 *Applied Scientific Research* **52** 309-329
[7] Kuerten J G M, van der Geld C W M and Geurts B J 2011 *Physics of Fluids* **23** 123301
[8] Carbone M, Bragg A D and Iovieno M 2019 *Journal of Fluid Mechanics* **881** 679-721
[9] Harish R and Sivakumar R 2021 *Powder Technology* **378** 303-316 ISSN 0032-5910
[10] Pirozzoli S, Bernardini M, Verzicco R and Orlandi P 2017 *Journal of Fluid Mechanics* **821** 482-516
[11] Maxey M R and Riley J J 1983 *Physics of Fluids* **26** 883-889
[12] Morsi S A and Alexander A J 1972 *Journal of Fluid Mechanics* **55** 193-208
[13] Ranz W E and Marshall Jr W R 1952 *Chemical Engineering Progress* **48** 173-180
[14] Kim J, Moin P and Moser R 1987 *Journal of Fluid Mechanics* **177** 133-166
[15] Kleiser L and Schumann U 1979 *Notes on Numerical Fluid Mechanics* **2** 165-173
[16] Canuto C, Hussaini M Y, Quarteroni A and Zang T A 1988 *Spectral Methods in Fluid Dynamics* (Heidelberg: Springer) ISBN 978-3-540-52205-8
[17] Thual O 1986 *Transition vers la turbulence dans des systèmes dynamiques apparentés a la convection* (Ph.D Thesis, Universitè de Nice-Sophia Antipolis)
[18] Sid S, Dubief Y and Terrapon V E 2015 *8th International Conference on Computational Heat and Mass Transfer* (Istanbul, Turkey) p 190

Acknowledgments
The authors acknowledge the computer resources provided by CINECA under the ISCRA initiative, project HP10CIUF9S. Additional resources were provided by HPC@POLITO, a project of Academic Computing within DAUIN at the Politecnico di Torino (http://www.hpc.polito.it).

Topology optimization of a pseudo 3D microchannel heat sink: influence of the initial guess

N Bianco[1], A Fragnito[1,*], M Iasiello[1] and GM Mauro[3]

[1] Department of Industrial Engineering, Università degli Studi di Napoli Federico II, Piazzale Tecchio 80, 80125, IT

[2] Department of Engineering, Università degli Studi del Sannio, Piazza Roma 21, 82100 Benevento, IT

* Corresponding author: andrea.fragnito@unina.it

Abstract. Nowadays, the optimization of microchannel heat sinks involves using algorithms, *e.g.,* topology optimization (TO), to find the optimal layout of fluid channels to enhance certain objectives, *e.g.,* heat transfer, pressure drop. The outcome is an unintuitive material distribution in the design space, subject to constraints such as manufacturability, fluid flow requirements, and thermal performance. This approach has been shown to provide significant benefits compared to traditional design methods, leading to increased cooling performance. However, depending on the application, defining an initial (*i.e.*, starting) guess for the optimization problem is not trivial. Initial guess for TO, *i.e.*, starting design for the optimization process, can play a crucial role in determining the final optimized design. Its choice may affect the convergence rate, the performance of the final solution, and the computational effort, as it can be set in a variety of ways, *e.g.,* a random distribution, a preconceived design, or a combination of these two. In this study, we propose a heuristic approach for incorporating a genetic algorithm (GA) as outer algorithm into the TO routine in order to find the initial guess that ensures the lowest thermal resistance while limiting the pumping power. The aim is to correlate initial designs with final objectives and to define new useful insights on microchannel optimization, since TO has proved to be a promising area of research for the development of advanced thermal management systems.

1. Introduction

Topology optimization (TO) is a current mathematical technique used in several engineering branches to define the optimal material layout within a predefined space [1]. While TO has been widely used in structural design, its application in CFD problems is relatively new and presents intricate challenges, involving both purely numerical and conceptual aspects [2]. When coupling TO to CFD problems, the objective is to optimize shape and topology of the solid/fluid domain to achieve desired flow characteristics and fulfil thermal targets. In this frame, the choice of design variables, the selection of appropriate constraints and boundary conditions is a well-known challenge, as well as the complex physics of fluid flow, *e.g.,* turbulence, nonlinearity, and multiphase flow [3].

However, among these issues, one of the critical input parameters that can significantly impact the optimization process and the final design is the initial guess [4]. The initial guess is an initial estimate of the optimized design that serves as a starting point for the TO algorithm. The choice of the initial guess can affect the convergence rate, as well as the topology and performance of the optimized design. In this case the problem can be defined as initial guess dependent (IGD). This means that different initial designs result in diverse topologies, although having close objective values [5]. Since the initial guess can affect the convergence rate and the quality of the optimized design, a reasonable option should be a starting guess close to the optimal solution, with features of the expected design [2]. Thus, the initial guess should be chosen based on prior knowledge of the design problem, which is not always feasible, especially in the area of thermofluid-dynamics where designs tend to have aerodynamic or fluid-based shapes.

Thermofluid-dynamics topology optimization problems are quite non-convex and can easily converge to only locally optimal topologies depending on the starting guess or initial design [5], [6]. Thus, exploring different initial design options is mandatory to comprehensively explore the design domain and avoid trivial solutions. More specifically, regarding thermal management, microchannel heat sink optimization has proven to be IGD [4],[6]. An apriorist evaluation of various configurations with different fin placements can be time-consuming, which added to the computational cost of TO makes the process lengthy and tedious. To this end, the aim of this work is to apply a heuristic approach for integrating a genetic algorithm (GA) as an outer algorithm into the TO routine: GA generates multiple pin fins' configurations that are used as initial guesses by TO to find a final design able to minimize the thermal resistance between the channel layer and the base plate layer, and the pressure drop. By doing this, starting guesses can be correlated with final objectives, defining new helpful insights on microchannel optimization.

2. Materials and methods

The model is based on a pseudo-3D formulation [6]. A full 3D optimization is computationally expensive since topology optimization (TO) typically requires several hundred iterations before convergence to a final design. Accordingly, the pseudo 3D heat sink model approximates the original 3D one with two 2D thermally coupled problems, which are not independent themselves. Figure 1 shows a 3D sketch of the forced convection heat sink here considered. The base plate – the one concerned by the heat source – is thermally coupled to the channel layer thanks to an additional term. At the same time, this term is balanced on the channel layer side through a complementary one. Consequently, this approach – even being 2D – considers out-of-plane effects. The design domain, *i.e.*, channel layer, is also depicted in figure 1. It consists of a rectangular domain, with an internal rectangular subdomain (Ω_d) where topology takes place. The outer ones (Ω_f) serve as inflow and outflow sections. Therefore, Ω_d represents the domain where the initial guess has to be set.

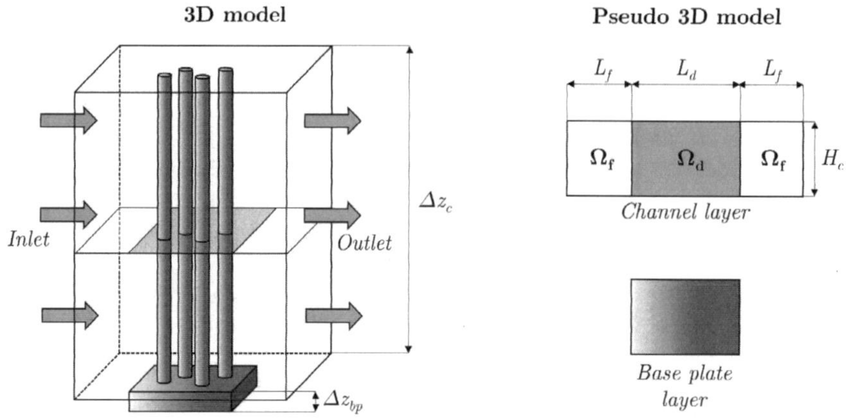

Figure 1. Sketch of 3D heat sink design and its reduction to a pseudo 3D model.

The starting design is chosen to be a microchannel heat sink with round pins. The arrangement and the number of round pins are not defined as constraints, whereas the initial volume constraint, *i.e.,* the ratio between the area of the solid material and the total of Ω_d, is fixed to a value of 0.2. Note that this constraint is only set on the initial design and not on the final one. Therefore, the aim of this work is to investigate the effect of varying pins position and size as initial guess for the TO problem.

2.1. Governing equations

To define the model equations, a continuous design field, $\gamma(x)$ is defined in Ω_d. Its values range from 0 to 1, and the upper value represents the fluid, whereas the lower one represents heat sink fin material. The reason why $\gamma(x)$ takes one value over the other is specified in Section *2.2*.

An incompressible, laminar, and steady-state flow is assumed throughout this case. The fluid and heat transfer problems in the thermofluid-dynamic design layer are modeled two-dimensionally, as described in the previous section. The 2D assumption is motivated by the fin height, significantly larger than the baseplate one. In addition, since the base plate's extension is much larger than its height, the original 3D thermal diffusion problem in the base plate is simplified to a 2D problem. Under these assumptions [6], the governing equations for the fluid are written as:

$$\rho_f(\nabla\cdot u)=0 \tag{1}$$

$$\rho_f(u\cdot\nabla)u=-\nabla p+\mu_f(\nabla^2 u)+F \tag{2}$$

$$F(\gamma)=-\bar{\alpha}uI_\alpha(\gamma)=-\frac{\mu_f}{Da\,L_c^2}uI_\alpha(\gamma) \tag{3}$$

where ρ_f is the fluid density, u is the fluid velocity vector, p is the pressure field, μ_f is the dynamic viscosity of the fluid, and F is the Brinkman friction term, which is used in TO flow problems to penalize flow through solid areas within the design domain. $\bar{\alpha}$ is the maximum inverse permeability of the porous medium and $I_\alpha(\gamma)$ a suitable function for the inverse permeability interpolation. Da is the Darcy number and L_c is a characteristic length which corresponds to the design domain width. It is worth noting that there is no Brinkman friction on the fluid outside of the design domain.

The 2D thermal convection-diffusion equations with/without a heat source are solved in the thermofluid design layer inside/outside the design domain:

$$\rho_f c_f u\nabla T_f-\nabla\cdot(k_f\nabla T_f)=0 \quad in\ \Omega_f \tag{4}$$

$$\gamma\rho_f c_f u\nabla T_f-\nabla\cdot(k_f I_k(\gamma)\nabla T_f)=\frac{\dot{q}_{inter}(\gamma)}{\Delta z_{channel}} \quad in\ \Omega_d \tag{5}$$

where T_f is the temperature field in the thermofluid design layer, c_f is the specific fluid heat capacity, k_f the thermal conductivity of the fluid, $I_k(\gamma)$ is a function that interpolates between thermal conductivities. $\Delta z_{channel}$ is the height of the air channel and fins, and $q_{inter}(\gamma)$ is the heat transferred from the solid base plate to the thermofluid design layer.

A 2D heat conduction problem is solved in the metal base plate:

$$\nabla\cdot(k_s\nabla T_s)=-\frac{\dot{Q}_{source}}{V_{bp}}+\frac{\dot{q}_{inter}(\gamma)}{\Delta z_{bp}} \quad in\ \Omega_d \tag{6}$$

where k_s is the thermal conductivity of the base plate, T_s is the temperature field in the base plate, Q_{source} is the prescribed heat production rate, V_{bp} is the base plate volume, and Δz_{bp} is the base plate height. The base plate is assumed to produce heat at a constant rate in this study. The conductive heat transfer from the base plate into the heat sink fins, as well as the convective heat transfer from the base plate to the fluid, must be represented flexibly in the out-of-plane heat transfer between base plate and thermo-fluid design layer, $q_{inter}(\gamma)$. This is accomplished by interpolating between a high conductive heat transfer into the fins and a lower convective heat transfer to the fluid using a heat transfer coefficient:

$$\dot{q}_{inter}(\gamma)=h_f I_h(\gamma)(T_s-T_f) \tag{7}$$

where h_f is the convective heat transfer coefficient to the fluid, and $I_h(\gamma)$ is a function that interpolates between h_f and the heat transfer coefficient from the base plate into the heat sink fins, h_s. The conductive heat transfer resistance in the z-direction in the fins is represented by the parameter h_s, which is empirically determined. All walls are adiabatic, except for the inlet where the temperature T_{in} is fixed. The thermophysical properties of air that are used in this study are: c_{air} = 1006 J/(kg K), k_{air} = 0.024 W/(m K), $\mu_{air} = 1.94 \cdot 10^{-5}$ Pa s, and ρ_{air} = 1.204 kg/m^3.

Pressure drop between inlet and outlet, Δp, is computed as difference of pressure at the inlet and outlet and a normal laminar inflow is set at the inlet boundary Γ_{in}. A symmetry boundary condition for the fluid flow is applied on bottom and upper boundaries.

2.2. *Topology optimization*

Density based TO is here applied [1]. This method supports regularization via Helmholtz filter. The filter size is defined as two times the mesh size, which is set as 0.02 mm. Interpolation functions are used to represent intermediate values in the given design problem. An interpolation function is used in this work for the Brinkman friction term (Eq. 3):

$$I_\alpha(\gamma)=\frac{1-\gamma}{1+b_\alpha\gamma} \tag{8}$$

where b_α is a parameter that determines the interpolation's convexity. A rational approximation of material properties (RAMP)-style interpolation is used to interpolate the thermal conductivity within the design layer and the heat transfer between the heat sink base plate and the thermo-fluid design layer:

$$j(\gamma)=\frac{\gamma(C_j(1+b_j)-1)+1}{C_j(1+b_j\gamma)} \tag{9}$$

where b_j is the interpolation convexity parameter and C_j is defined in the respective case as k_f/k_s and h_f/h_s.

2.2.1. *Problem formulation.* The following optimization problem has been solved:

$$\begin{cases} \underset{\gamma\in\mathbb{R}^n}{minimize} \quad \phi=w_{Rth}\,\phi_{Rth}+w_{\Delta p}\,\phi_{\Delta p} \\ \phi_{Rth}=\dfrac{T_{bp,avg}(\gamma)-T_{in}}{\dot{Q}_{source}} \\ \phi_{\Delta p}=p|_{\Gamma_{in}}-p|_{\Gamma_{out}} \end{cases} \tag{10}$$

As described in equation (10), ϕ is the composition of 2 terms, ϕ_{Rth} and $\phi_{\Delta p}$, which are weighted through the terms w_{Rth} and $w_{\Delta p}$. Minimizing both the objectives defines a multi-objective problem. To reduce the thermal resistance, the heat transfer area has to increase, while to minimize the pressure drop less flow resistance is required. In this work, the weighting factors have been considered equals. It is worth specifying that in order to consider pressure drop as part of the objective, only inlet velocity, *i.e.,* 0.1 m/s, has been specified as fluid boundary condition. Further, no volume constraint has been defined, apart from the one related to the initial guess.

2.3. Genetic algorithm

A genetic algorithm (GA) is run as outer algorithm into the TO routine to investigate the initial design condition that simultaneously fulfils thermal and fluid dynamic objectives. The design variables used in this work are the minimum circle radius (r_{min}) and the minimum distance between the circles (s_{min}).

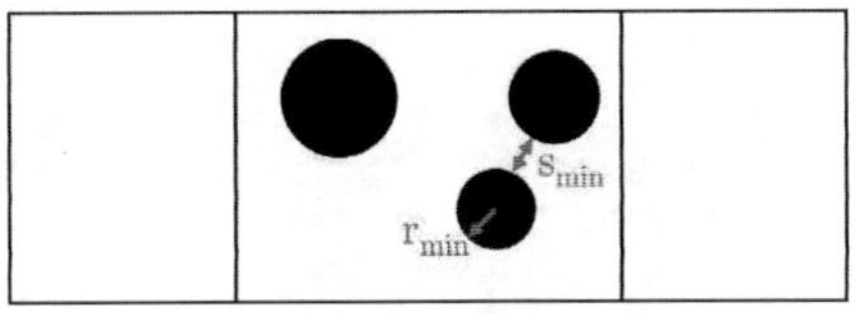

Figure 2. Parameters affecting the round pins size and displacement (r_{min}, s_{min}) used as design variables for the GA.

The parameter r_{min} can vary from 0.06 mm to 0.13 mm, each 0.005 mm, whereas s_{min} varies from 0.04 mm to 0.08 mm, each 0.01 mm. The reason for these ranges will be explained later. However, s_{min} is set to be higher than the maximum mesh size. By doing this, an excessive mesh refinement between two circles – which should compromise the TO convergence – is avoided. All settings for GA tuning are taken from [7].

2.3.1. Coupling GA with TO. The fitness function code generates a figure of randomly positioned circles within the rectangular domain. The code takes as input the dimensions of the rectangular domain, the minimum radius and distance between the circles, and a vector of target ratios between the total area of the circles and the area of the rectangle. For each target ratio, the code generates circles randomly until the total area of the circles reaches or exceeds the target ratio times the area of the rectangle. The code uses a while loop to generate circles one at a time. For each new circle, the code generates a random radius and position, and checks whether it intersects with any of the existing circles or exceeds the boundary limits of the rectangle. If the new circle is valid, it is added to the figure and the total area of the circles is updated. The loop continues until the total area of the circles reaches or exceeds the target ratio times the area of the rectangle, or the maximum number of iterations, *i.e.,* 100, is reached. The code also sets up a figure window with a fixed size and saves the final image as a PNG file with a name that includes the target ratio. Note that the code assumes that the aspect ratio of the rectangular domain is 1.7:1 – equal to the design domain one – and that the circles are black with a solid fill. By doing this, the image can be imported in the COMSOL file and processed as initial guess. The method of moving asymptotes (MMA) is used to perform TO.

3. Results and discussions

This section reports the validation of the presented model under a Δp equal to 3 Pa and minimizing the thermal resistance R_{th}, and the results on the investigation of the initial guess by optimization using GA + TO. Each GA simulation lasts around 40 min, using a continuation scheme for the TO algorithm similar to the reference one [6]. The number of overall iterations has been reduced from 1160 (reference) to 400, without losing accuracy (as proved in the following).

3.1. Model validation

Before running simulations about the optimization problem of equation (10), the TO reliability is tested replacing the optimization problem solved in [6]. Thus, the model validation is assessed comparing qualitatively (figure 3) the final design, velocity magnitude field, temperature field of

Journal of Physics: Conference Series **2685** (2024) 012004 doi:10.1088/1742-6596/2685/1/012004

channel layer and base plate, with the one used in the reference work, and quantitatively through the value of the thermal resistance, R_{th}. The one reported in the reference is 144.6 (K/W), while the one evaluated in this work is 144.9 (K/W).

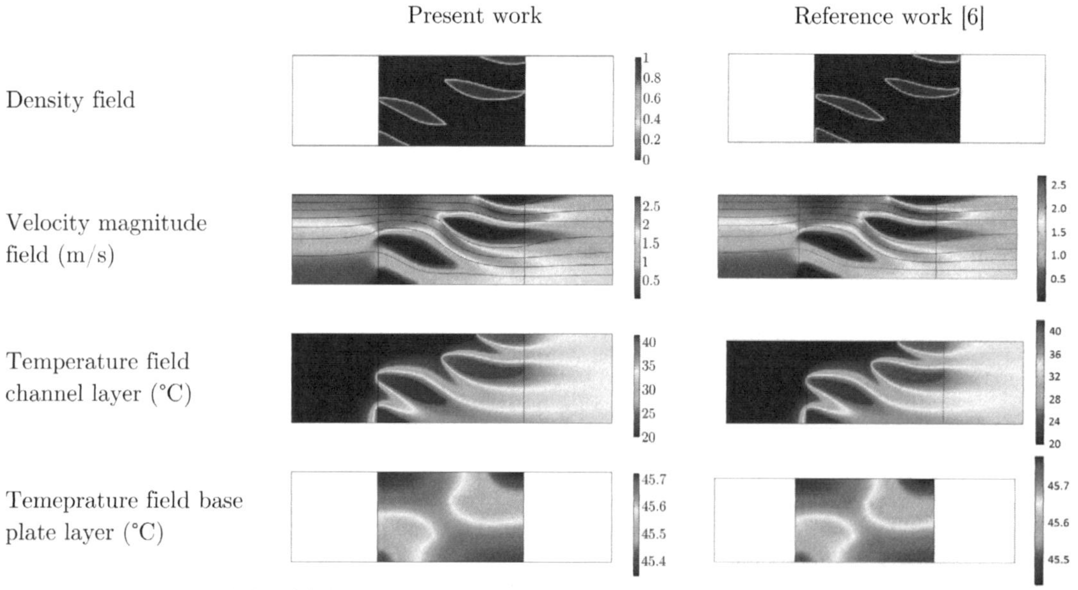

Figure 3. Model validation results: design domain; velocity field, design plate temperature; baseplate temperature field.

The main difference that can be spotted in the final design is the presence of material on the vertical boundaries, since in the present work there is no prescribed density condition on them.

3.2. Initial guess influence

170 Individuals computed through the GA + TO routine are depicted in figure 4. Among these, 5 relevant individuals have been highlighted: the optimal solution (ϕ = 61.95) and four worst solutions. In detail, in order to visualize a hypothetical dependency between the value of the objective function and the initial design, the plot of the initial design is shown for each of these individuals.

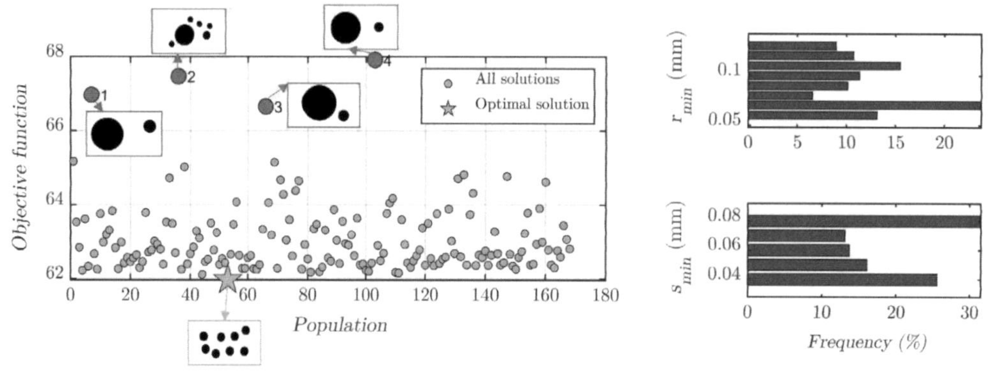

Figure 4. left: GA population's objective function (ϕ) history: detection of the optimal solution and four individuals with worst ϕ, and related initial designs. Right: Design variables distribution.

By doing so, it can immediately be seen that the presence of a pin with a larger radius leads to a reduction in performance, while the presence of several pins with a smaller radius results in an optimal solution. To confirm what is presented through the plot of the objective function, the values of the design variables are grouped according to population and the frequency with which they occurred (figure 4). It shows that the GA preferred smaller radii and longer distances. Nevertheless, with regards to r_{min}, a non-negligible number of individuals can be noted, *i.e.,* 26 designs with $r_{min} = 0.11$ mm, to which correspond to individuals with intermediate ϕ. At this point, it is useful to check the advantage of adopting a proper initial design (Optimal solution), as opposed to one of those "worst" highlighted above (Case 4). The comparison (figure 5) is made in terms of initial guess, final design, channel layer temperature field and velocity magnitude field.

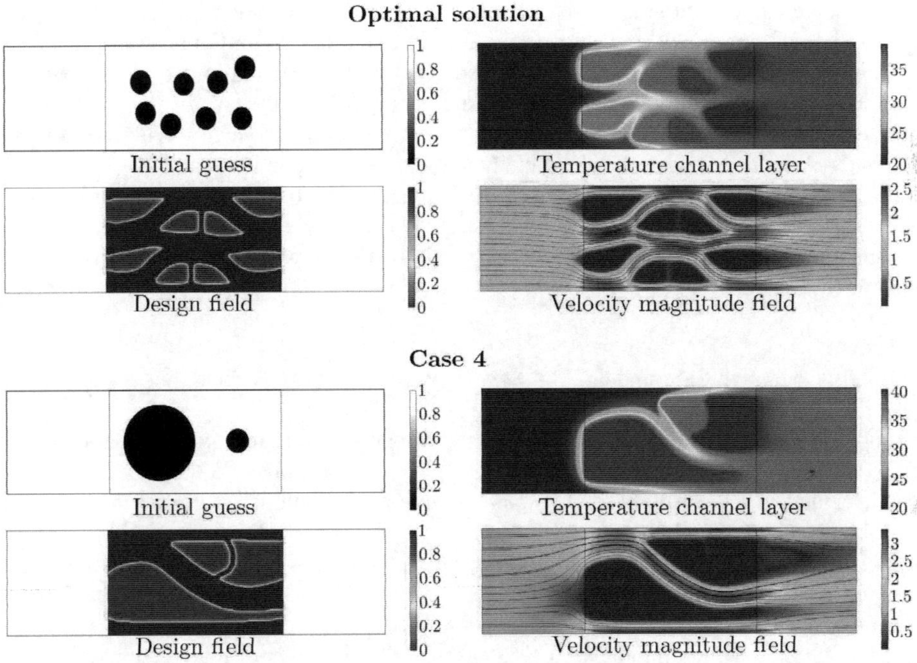

Figure 5. Optimal solution and "case 4": initial guess, final design, temperature channel layer and velocity modulus field.

Starting the optimization with more pins with lower radii allows to reach a better final solution. The increase in the heat transfer allows to reduce the thermal resistance (table 1), and more aerodynamically shaped fins ensure a lower pressure drop than in Case 4, *i.e.,* 12.46 Pa against 15.66 Pa. However, pressure drop values are consistently higher than the one used in the validation step, *i.e.,* 3 Pa, which explains why the thermal resistance is significantly lower.

Table 1. Model results comparison.

	Optimal solution	Case 4	Validation
R_{th} (K/W)	111.5	120.1	144.9
Δp (Pa)	12.46	15.66	3.00
$T_{max,c}$ (°C)	38.90	40.21	41.22
$T_{max,bp}$ (°C)	39.80	41.10	45.72

Moreover, as listed in table 1, the maximum temperature at both channel and base plate layers is lower, because of the reduced thermal resistance. It is worth noting that the resulting solid fraction – computed on the design domain Ω_d differs from the optimal solution (31%) to Case 4 (48%). This is an interesting outcome because even if the Case 4 design has more solid material available, its displacement does not imply improved thermal diffusion. Lastly, the difference between thermal resistance values is slight. This confirms the need for searching a proper initial guess before running the TO. Indeed, running TO without exploring the field of possible initial solutions would risk experiencing a local minimum, whose value is really close to the global one.

4. Conclusions

This work investigates the effects of using different initial designs in the use of topology optimization (TO) for a microchannel heat sink. The CFD model is based on a pseudo-3D approach, which allows to obtain unintuitive design without resorting to 3D simulations. The initial design is made of multiple pin fins, and the starting solid fraction in constrained to 0.2. To explore different pin dimensions (varying the radius) and arrangements (varying the distance), an outer algorithm, *i.e.,* a GA has been implemented. The aim of the TO is to find a final design able to minimize – with the same magnitude – the thermal resistance between the channel layer and the base plate layer, and the pressure drop.

Results confirm that both dimensions and displacements of pins have an influence on both objective functions and final design. In detail, starting with bigger pins leads to local minima, while using more pins with lower radii enables to reach a design with higher heat transfer and more aerodynamically shaped fins.

5. References

[1] Bendsøe M P and Sigmund O 2004 *Topology Optimization* (Berlin, Heidelberg: Springer Berlin Heidelberg)

[2] Sigmund O and Maute K 2013 Topology optimization approaches: A comparative review *Struct. Multidiscip. Optim.* **48** 1031–55

[3] Alexandersen J and Andreasen C S 2020 A review of topology optimisation for fluid-based problems *Fluids* **5** 1–32

[4] Li H, Ding X, Meng F, Jing D and Xiong M 2019 Optimal design and thermal modelling for liquid-cooled heat sink based on multi-objective topology optimization: An experimental and numerical study *Int. J. Heat Mass Transf.* **144** 118638

[5] Zhao X, Zhou M, Sigmund O and Andreasen C S 2018 A "poor man's approach" to topology optimization of cooling channels based on a Darcy flow model *Int. J. Heat Mass Transf.* **116** 1108–23

[6] Haertel J H K, Engelbrecht K, Lazarov B S and Sigmund O 2018 Topology optimization of a pseudo 3D thermofluid heat sink model *Int. J. Heat Mass Transf.* **121** 1073–88

[7] Bianco N, Fragnito A, Iasiello M and Mauro G.M 2021 A comprehensive approach for the multi-objective optimization of Heat Recovery Steam Generators to maximize cost-effectiveness and output power *Sustain. Energy Technol. Assessments* **45** 101162

Experimental data analysis and dimensionless exergy levels in a commercial stratified thermal storage

AV. Anacreonte[1,2,*], M. Musto[1,3], N. Bianco[1], R. Vitobello[2], R. Russo[3]

[1] Department of Industrial Engineering, University of Naples Federico II, P.le Tecchio, 80, Naples (Italy)
[2] TVP Solar SA, 10 rue du Pré-de-la-Fontaine ZIMEYSA, 1242 Satigny, GE, Switzerland
[3] Institute of Applied Sciences and Intelligent Systems, National Research Council of Italy, Via Pietro Castellino 111, Napoli, 80131, Italy

* Corresponding author: vitalianoalessandro.anacreonte@unina.it

Abstract. The continuously rising demand for renewable energy sources within the energy system has highlighted the need for technologies capable of mitigating the impact of renewable energy intermittency and addressing the generation-demand mismatch in the system. Thermal energy storage (TES) emerges as a versatile storage medium with a wide range of applications, spanning from solar energy utilization and power peak shaving to the storage of industrial waste heat. In this study, data collected from an operating commercial stratified tank are used to validate a 2-D axisymmetric CFD model. Temperature profiles are collected throughout one month with a one-minute refresh rate. The model replicating the tank is generated in COMSOL Multiphysics® and validated by emulating the registered charging phases of the real storage. The model is then employed to optimize the stratification capability of the tank, by varying the logics applied to pinpoint optimal values of both inlet water temperature and velocity. The study aims to maximize the exergetic efficiency of the system using dimensionless exergy, a parameter often utilized in literature to identify the ability of the storage to generate and preserve optimal temperature stratification. Finally, the experimental dimensionless exergy has been evaluated for the aforementioned temperature profiles.

1. Introduction

The escalating energy requirements resulting from industrialization and population expansion have spurred a global drive towards clean, dependable, cost-effective, and sustainable energy alternatives. It is widely acknowledged that energy holds the key to addressing numerous contemporary challenges and unlocking new prospects. The United Nations underscores the significance of ensuring universal availability of these energy resources. Among the array of renewable energy sources, solar power emerges as the most promising, owing to its extensive potential and versatility. However, the intermittent nature of solar energy can give rise to a discrepancy between energy supply and demand, necessitating the integration of solar energy storage systems. These storage systems play a pivotal role in various sectors, including power generation fields such as solar thermal, geothermal, and nuclear power plants.

Thermal energy storage (TES) systems, which rely on renewable thermal energy sources, represent a cost-effective and robust solution for large-scale applications. Recent advancements in materials have sparked a revolution in TES technologies [1]. TES typically involve the storage of thermal energy through the modification of a material's temperature (sensible heat) or the phase transition of a substance (latent heat), or a combination of both, resulting in hybrid TES technology that caters to a broader operating range [2]. Sensible thermal energy systems (STES) have gained popularity and commercial viability as TES methods due to their relative affordability and ease of operation and maintenance. STES utilizes heat transfer fluids that exhibit stability at high temperatures, and its effectiveness relies on several factors [3]. Stratified sensible TES (SSTES) has emerged as one of the most extensively studied sensible thermal energy storage methods over the past four decades, primarily due to its ability to reduce the overall setup cost and enable cost-effective power generation [4-7]. Moreover, the simplicity of this technology allows for efficient scalability to meet a wide range of power requirements, making it a preferred choice over alternative approaches such as the two-tank method [8].

SSTES overcomes existing technological limitations and holds substantial potential in revolutionizing the solar energy landscape.

Several numerical studies have been conducted on SSTES. Ievers and Lin [9] have investigated the flow dynamics using 3-D CFD models of storages with several geometrical aspect ratios. Kocijel et al. [10] numerically tested the influence of various geometrical and process parameters on the stratification capabilities of the storage. Nevertheless, several more or less accurate models have been developed using the finite differences method [11-13]. A great number of numerical studies have been validated on experimental data, but the vast majority of said studies have resorted to modelling and simulating an in-house experimental setup, thus under finely controlled testing conditions. However, the literature is lacking when it comes to models related to storages implemented into industrial systems and working under real load conditions.

In this study, dimensionless exergy ξ^* is evaluated for an experimental stratified thermal energy storage using pure water as a fluid working in a real industrial application. Subsequently, a 2-D axisymmetric CFD model has been generated in COMSOL Multiphysics® and validated by comparing numerical and experimental data of various charging phases of the system. The model has then been deployed to numerically test the tank under different realistic loading conditions to assess the capability to restrain its stratification under more severe boundary conditions.

2. Stratified thermal storage

2.1. Layout of the system

This study has been conducted using as a reference a Stratified Thermal Energy Storage (STES) utilized as a buffer tank in an industrial solar water-heating system located in Switzerland. The system is mainly formed of four parts: the production circuit, the thermal block, the storage circuit, and the consumption circuit. The production circuit consists of ten rows of High-Vacuum Flat Plate Collectors (HVFPC) connected in series, for a total of 96 solar units. The model of HVFPC is MT-Power v4 SK, manufactured by TVP Solar. In the production circuit flows a 30% mixture of water and glycol.

The storage circuit is a short loop, receiving thermal energy from the production circuit using a flat plate heat exchanger, as schematized in Figure 1. Thus, the fluid utilized in this loop differs from the one used in the production section of the system by being pure water. The pump that feeds the circuit (Calpeda TM-65E) works under inverter and has a maximal flow rate of $4.2\ m^3/h$. The flow rate is measured using an IFM SV8150 vortex flowmeter. Two thermocouples, model Endress+Hauser Easytemp® TMR31, are placed at the inlet and outlet section of the heat exchanger. Given the short length of the pipe leading to the tank, it is reasonable to assume the absence of heat losses in this trait, therefore assuming the latter as the inlet temperature of the storage.

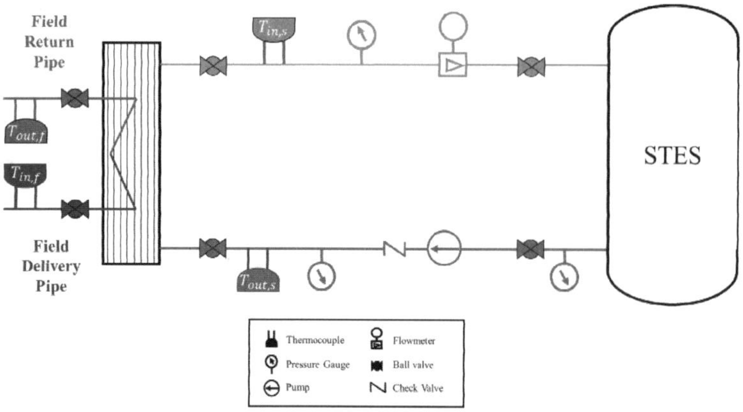

Figure 1. Layout of the system.

The STES utilized in the system is produced by Jenni Energietechnik AG and has a total capacity of $25030\ l$. It is a cylindrical vertical storage with a radius of $2.60\ m$ and a height of $5.10\ m$ from the base. The storage is realized in stainless steel, while the isolation is provided by using a

250 mm thick layer of rock wool. The temperature of the storing fluid is measured at five different heights using the thermocouples model RTD PT100 Endress+Hauser, as shown in Figure 2.

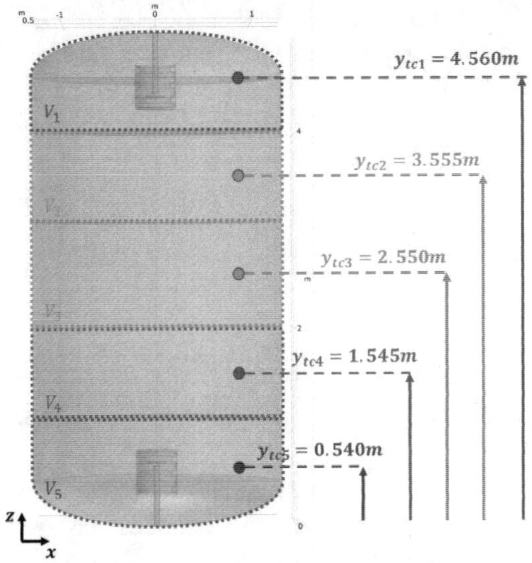

Figure 2. Thermocouples position in the analyzed STES.

Figure 3. Inlet and outlet diffusers' geometry. Image courtesy of Jenni Energietechnik AG.

The tank utilizes two identical diffusers for both the inlet and the outlet section, placed specularly. The diffuser is a large cylinder measuring 280 mm in diameter and 450 mm in height, and it is characterized by the presence of annular bulges on the sides, as displayed in Figure 3. The fluid enters the diffusers from a horizontal pipe with a diameter of 50 mm that splits into a T-shaped section with a diameter of 65 mm.

To not disclose information regarding the internal energy management of the company, the consumption circuit will not be displayed. Thus, the charging phases will be the only ones analysed in this study.

2.2. *Dimensionless Exergy*

To quantify the thermal stratification of the fluid in the thermal storage, a lot of approaches has been proposed in literature. Consul *et al.* [14] proposed a parameter called dimensionless exergy ξ^* and defined as:

$$\xi^* = \frac{\Xi_{str} - \Xi_{exp}}{\Xi_{str} - \Xi_{mix}} \tag{1}$$

where, for a given set of initial conditions, Ξ_{exp}, Ξ_{str}, and Ξ_{mix} are the instantaneous exergies of the experimental storage, a fully stratified equivalent, and a fully mixed equivalent, respectively. The instantaneous exergy can be defined as follows:

$$\Xi = \int_V \xi \varrho dV \tag{2}$$

where the specific exergy ξ is obtained from Equation (3), when neglecting kinetic and potential contributions.

$$\xi = (h - h_0) - T_0(s - s_0) \tag{3}$$

The reference temperature chosen for this analysis is 273.15 K to have a realistic ambient temperature reference while also being always lower than the temperatures reached by the storage fluid at any point. In this study, dimensionless exergy has been deployed to evaluate the resilience of the tank in maintaining its internal stratification under various inlet conditions related to real charging phases.

3. CFD Model

3.1. Mathematical model

The two-dimensional axisymmetric fluid dynamics of the tank is described by the Navier-Stokes equations, coupled with the Boussinesq assumption, and written in cylindrical coordinates along the radius r and the height z:

$$\frac{1}{r}\frac{\partial(r\rho u_r)}{\partial r}+\frac{\partial(\rho u_z)}{\partial z}=0 \tag{4}$$

$$\frac{\partial u_r}{\partial t}+u_r\frac{\partial u_r}{\partial r}+u_z\frac{\partial u_r}{\partial z}=-\frac{1}{\rho}\frac{\partial P}{\partial r}+\nu\left[\frac{1}{r}\frac{\partial}{\partial r}\left(r\frac{\partial u_r}{\partial r}\right)-\frac{u_r}{r^2}+\frac{\partial^2 u_r}{\partial z^2}\right] \tag{5}$$

$$\frac{\partial u_z}{\partial t}+u_r\frac{\partial u_z}{\partial r}+u_z\frac{\partial u_z}{\partial z}=-\frac{1}{\rho}\frac{\partial P}{\partial z}+\rho g+\nu\left[\frac{1}{r}\frac{\partial}{\partial r}\left(r\frac{\partial u_z}{\partial r}\right)-\frac{u_z}{r^2}+\frac{\partial^2 u_z}{\partial z^2}\right] \tag{6}$$

$$\frac{\partial T}{\partial t}+u_r\frac{\partial T}{\partial r}+u_z\frac{\partial T}{\partial z}=\alpha\left[\frac{1}{r}\frac{\partial}{\partial r}\left(r\frac{\partial T}{\partial r}\right)+\frac{\partial^2 T}{\partial z^2}\right] \tag{7}$$

3.2. CFD Model

Both the computational domain and the meshing were realized using the CAD Module of COMSOL MultiPhysics®, and the simulations were carried out using the same software. The model has been built as a 3-D representation, then a 2-D reduction was obtained using cut-plane properties. While the external shell of the storage is transposed into a computational domain with no modification, the diffuser had to be slightly modified for the axisymmetric to be verified. Indeed, the T-shaped tube at the end of the inlet and outlet sections could not be maintained in the 2-D transposition, thus needing to be modified into an axisymmetric circular section, while maintaining the overall area for the fluid to flow through.

The domain has been divided into five subdomains. Since the temperature values observable in the data are largely discretized and lack further details, each volume is initially ($t = 0s$) set to the temperature $T_{i,0}$ measured by the thermocouple at the start of the simulated charging phase. The turbulent flow is numerically approached using a RANS method, with a standard k-ε model. In fact, given the absence of flow separation and adverse pressure gradients, said model was chosen for its robustness and lower computational burden. The velocity scale parameter is set to 1 m/s and the length scale factor is 0.035. The time-dependent solver deployed was based on AMG (Algebraic Multigrid) method.

Table 1. Summary of the boundary conditions and initial conditions applied to the model.

Boundary	Boundary conditions	
Inlet section	$u = u_{in}$ $v = w = 0$	$T_{in} = T_{set}(t)$
Outlet section	$p_{out} = 0$ Pa	$Outflow$
Shell and Diffuser Walls	$u \cdot n = 0$	$\frac{\partial T}{\partial n} = 0$
	Initial conditions	
i-th Volume	$u = 0$ $v = 0$ $p = p_{atm}$	$T_{0,i} = T_i\|_{t=0}$

The boundary and initial conditions applied are summarized in Table 1. Considering that the charging phases simulated occupied short periods, averaging at around one hour per phase, the conduction along the walls was neglected to not burden the computational time needed. Therefore, both the external walls and the diffuser walls are set to adiabatic. The inlet temperature of the water is set as a time-dependent boundary condition, to reflect the actual temperature of input of the fluid. Since the volume flow at the inlet section is nearly stationary, a single value of inlet velocity was imposed. For

Journal of Physics: Conference Series **2685** (2024) 012005 doi:10.1088/1742-6596/2685/1/012005

the validation phase, given the average volume flow rate of $3\ m^3/h$ and an inlet section of $7.85 \cdot 10^{-3}\ m^2$, the inlet velocity of the fluid was set to $0.42\ m/s$.

In the validation phase, the time simulated per charge phase has been set to one hour. The applied time step was $30\ s$. Initially, the fluid in the storage is supposed to not have residual kinetical energy, thus being still. An initial temperature $T_{i,0}$ is set for each node, without the presence of gradients. Nonetheless, it must be noted that COMSOL automatically generates small gradients between the various bodies of fluid at different temperatures within the first minutes of simulated time to stabilize the simulation. Said gradients could be slightly overestimated, but the discretized experimental data do not allow for a more thorough investigation.

4. Results and discussion

4.1. Validation

The model was validated by simulating three real charging phases and comparing the results, as shown in Figure 4. The discriminating factor in the choice of the charging phases has been to select those which were followed by a steep decline of the ξ^*, thus selecting three case studies related to portions of charging phases occurred on December 6th, 2022, December 20th, 2022, and January 7th, 2023, which will be subsequently referred to as Charge Phase 1, Charge Phase 2, and Charge Phase 3, respectively.

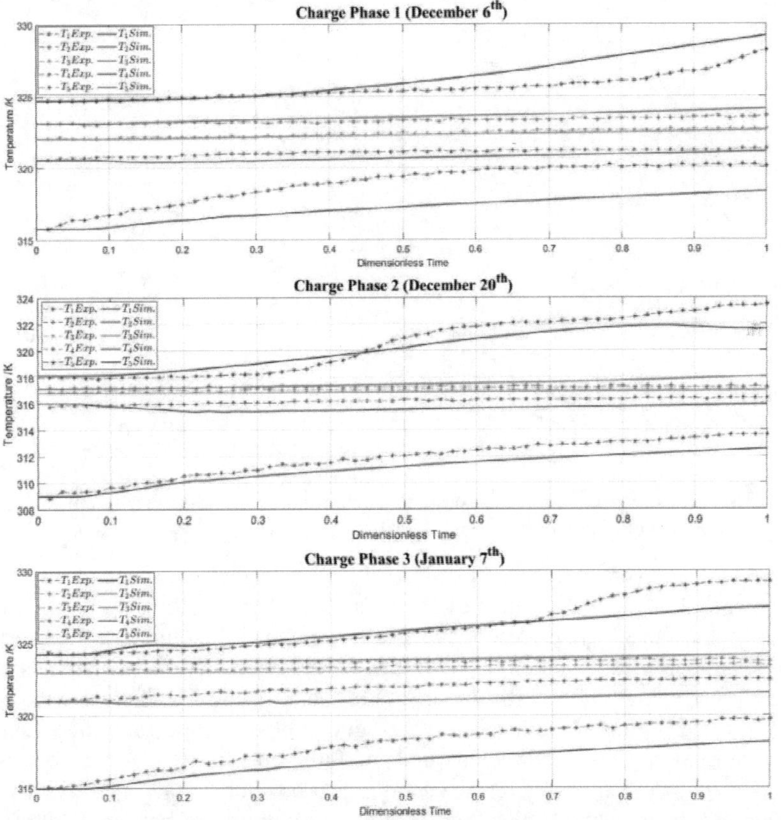

Figure 4. Comparison between field-acquired data and numerical results for the validation of the model in the three charging phases used as case studies.

The model proves to be capable of predicting the actual evolution of the temperature layers in the tank in all the scenarios analysed. The maximum deviation registered between field data and numerical results equals to 2.05°C and can be attributed to slight inaccuracies in the definition of the initial conditions in the model, inherently dependent on the coarse distribution of the thermocouples and, therefore, of the acquired data. Nonetheless, the fit is precise in most of the cases and the model can be considered validated.

4.2. Case studies

Charge Phase 1 has been adopted as a case study baseline. The temperature distribution inside the tank (i.e., the initial conditions) has been kept unchanged. Four case studies have been analysed to assess the behaviour of the storage when charged at different inlet temperatures (Figure 5a). The temperature at which the water enters the STES has been varied with a step of $15\ K$ between $318\ K$ and $363\ K$, the latter being the operating temperature limit of the physical storage. In all the case studies the length of the charging phase has been set to two hours and the volumetric flow entering the tank is $Q_{in} = 3\ m^3/h$.

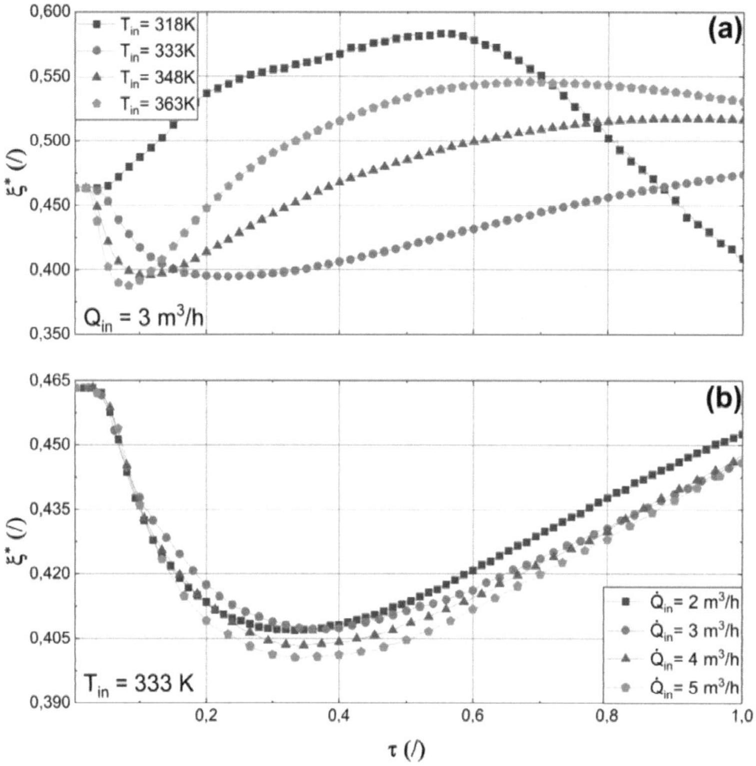

Figure 5. "Charge Phase 1" dimensionless exergy variation for different inlet water temperature boundary conditions with $Q_{in} = 3\ m^3/h$ **(a)** and for different inlet volumetric flow boundary conditions and $T_{in} = 333\ K$ **(b)**.

The case with an average inlet temperature of $318\ K$ emulates a scenario in which the storage circuit pump activates once the field temperatures are barely higher than the ones registered at the bottom of the tank. This case is the only one presenting a positive slope of the dimensionless exergy during the initial period of charge. It must be noted that, while the overall quantity of thermal energy stored in the tank is still barely increasing, the overall time length of the charging phase plays a major role. In fact, as the duration of the charging phase increases, the volume of cold fluid (at a lower temperature than the inlet temperature) in the tank will be exhausted and fully replaced by a volume of water at the inlet temperature conditions. Conversely, given the geometrical position of the inlet, this volume of colder fluid has to cross the other volumes of hotter water on its path to the bottom of the tank, inevitably triggering phenomena such as thermal diffusion and heat exchange (Figure 6a). Therefore, for longer charging phases, this process would inevitably prompt a decrease in the overall exergy related to the storage fluid. The cases with higher inlet temperatures ($T_{in} = 348\ K$ and $T_{in} = 363\ K$, respectively) show similar trends (Figure 6b). The dimensionless exergy value lowers at first, and then increases to higher values than the ones registered in the reference case ($T_{in} = 333\ K$). The time at the inflection point appears to be inversely proportional to the temperature, thus manifesting earlier when the inlet temperatures rise.

When varying the inlet velocity, a more regular pattern can be observed (Figure 5b). The volumetric flow of the fluid has been varied between 2 m^3/h and 5 m^3/h. Again, he initial temperature distribution in the STES is the one taken from the "Charge Phase 1" case. The inlet temperature is set to $T_{in} = 333\ K$. In each of the simulations, a total volume of 5 m^3 of water has been introduced in the storage and, as a result, the four charging phases have a different duration, ranging from one hour (case with $Q_{in} = 5\ m^3/h$) to two and a half hours (case with $Q_{in} = 2\ m^3/h$). A dimensionless time τ, as in Equation (8), has been employed to visually compare the results more clearly.

$$\tau = \frac{t}{t_{tot}} \tag{8}$$

The dimensionless exergy shows an inverse proportionality to the volumetric flow registered at the inlet. This is a direct effect of the laminarization of the flow at lower velocity regimes, which causes a lower amount of fluid mixing. Nonetheless, it must be noted that the inlet velocity and the inlet temperature are strictly related under the given temperatures of the fluid flowing into the production circuit. Also, considering the negligible differences among the values of ξ^* registered, all the inlet velocities imposed in the different cases appear to be in the operating range of the diffuser. In fact, with a slight exception for the 2 m^3/h case, all the curves collapse on very similar values when the objective volume of charge is reached ($\tau = 1$).

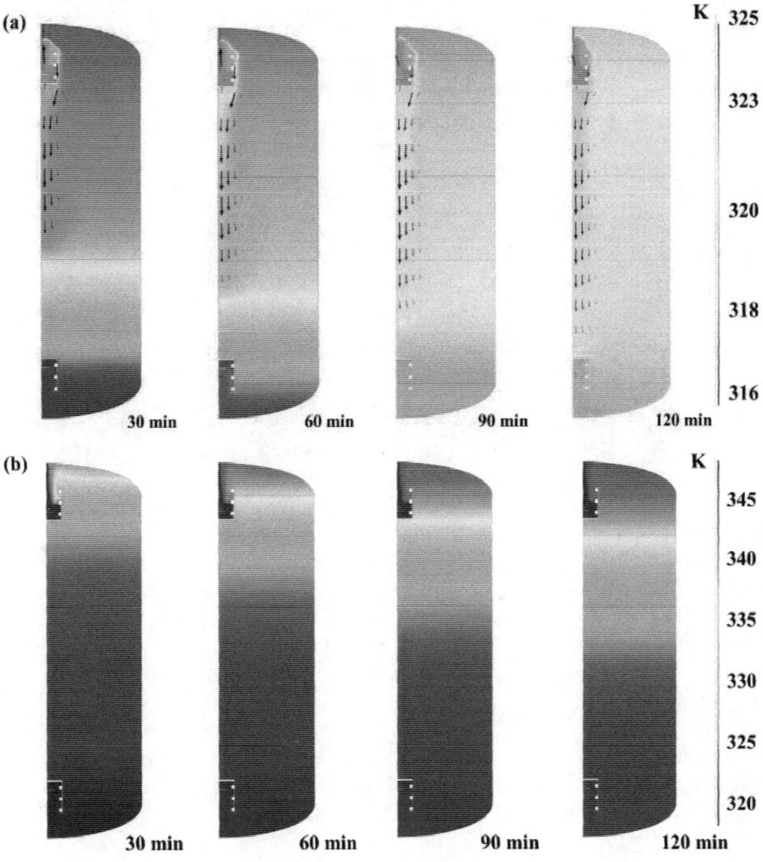

Figure 6. Temperature profiles at different time steps for $T_{in} = 318K$ **(a)** and $T_{in} = 348K$ **(b)**. Arrows indicate the direction of convective heat transfer phenomena.

Conclusions

In this study, information gathered from an industrial stratified tank has been used to validate a 2-D axisymmetric CFD model, generated in COMSOL Multiphysics® to mirror the design of the storage, by replicating the documented charging stages of the real storage. Once validated, the model is utilized to parametrically test various charging phase inlet conditions, varying both the inlet temperature and the inlet velocity of the fluid to trace behavioral characteristics and pinpoint optimal solutions.

The parametrical analysis conducted on the inlet temperature of the fluid did not evidence a very specific pattern. The temporal length of the charging phases appears to be a discriminating factor when evaluating optimal charging temperatures. The parametrical analysis of the inlet velocity of the fluid exhibits a more linear pattern, whereas lower velocities inevitably tend to result in a more refined stratification. Nonetheless, given the results obtained in the latter, plant logics that focus on the temperature as a control parameter rather than the inlet velocity can have a larger impact on obtaining optimal performances out of the thermal storage.

References

[1] Cabeza L 2019 Sensible Thermal Energy Storage at High Temperatures *Recent Advancement in Materials and Systems for Thermal Energy Storage* (*Green Energy and Technology*) ed A Frazzica and L Cabeza (Switzerland: Springer Nature) chapter 2 pp 3-7.

[2] Njoku H O, Ekechukwu O V and Onyegegbu S O 2014 *Heat Mass Transf.* **50** 1017.

[3] Gil A, Medrano M, Martorell I, Làzaro A, Dolado P, Zalba B and Cabeza L F 2010 *Renew. Sust. Energ. Rev.* **14** 31.

[4] Sliwinski B J, Mech A R and Shih T S 1978 *6th International Heat Transfer Conference*.

[5] Wood R J, Al-Muslahi S M, O'Callaghan P W and Probert S D 1981 *Appl. Energy* **9** 231.

[6] Van Berkel J, Rindt C C M and Van Steemhoven A A 1999 *Sol. Energy* **67** 65.

[7] Cabelli A 1977 *Sol. Energy* **19** 45.

[8] Han Y M, Wang R Z and Dai Y J 2009 *Renew. Sust. Energ. Rev.* **13** 1014.

[9] Ievers S and Lin W 2009 *Appl. Energy* **86** 2604.

[10] Kocijel L, Mrzljak V and Glazar V 2020 *Energy* **194** 1.

[11] Lago J, De Ridder F, Mazairac W and De Schutter B 2019 *Appl. Energy* **248** 640.

[12] Xu G, Hu L, Luo Y, Tian Z, Deng J, Yuan G and Fan J 2022 *Energy Build.* **275** 112459.

[13] Untrau A, Sochard S, Marias F, Reneaume J M, Le Roux G A C and Serra S 2023 *Appl. Energy* **333** 120614.

[14] Consul R, Rodrìguez I, Perez-Segarra C D, Soria M 2004 *Sol. Energy* **77** 179.

Nomenclature

A	area, m^2
c_p	specific heat, J/kgK
p	relative pressure, Pa
P	absolute pressure, Pa
Q	volumetric flow, m^3/h
r	tank radius, m
t	time, s
T	temperature, K
V	volume, m^3
u	velocity along z, m/s
v	velocity along x, m/s
y	vertical displacement, m
w	velocity along y, m/s
z	tank height, m
Greek Letters	
α	thermal diffusivity, m^2/s
ξ	exergy
ν	kinematic viscosity, m^2/s
ρ	density, kg/m^3
τ	dimensionless time
Subscripts	
avg	average
exp	experimental
in	inlet
mix	perfectly mixed
str	perfectly stratified
tot	total simulation time
tc	thermocouple
out	outlet
Superscripts	
*	dimensionless
Acronyms	
CFD	Computational Fluid Dynamics
HVFPC	High-Vacuum Flat Plate Collectors
SSTES	Sensible Stratified Thermal Energy Storage
TES	Thermal Energy Storage

Analysis of geometric uncertainties in 3D thermo-fluid problems solved by RBF-FD meshless method

R Zamolo, D Miotti[*] and E Nobile

Dipartimento di Ingegneria e Architettura, Università degli Studi di Trieste, via A. Valerio 10, 34127 Trieste, Italy

* Corresponding author: rzamolo@units.it

Abstract. The tolerances of production processes can lead to uncertainties in the behaviour and in the features of the manufactured products. From the point of view of the design of engineering components it is therefore of valuable practical interest to be able to quantify such uncertainties as well as the expected, i.e., averaged, performances. Such uncertainty quantification is carried out in this work by means of the Non-Intrusive Polynomial Chaos (PC) method in order to estimate the propagation of geometrical uncertainties of the boundaries, i.e., when the boundaries are described by stochastic variables. Existing deterministic solvers can be used with the PC method because of its non-intrusive formulation, allowing an accurate and practical prediction of the random response through a simple set of deterministic response simulations. The Radial Basis Function Finite Differences (RBF-FD) method is employed as a black box solver for the computation of the required set of responses defined over deterministic boundaries. The RBF-FD method belongs to the class of meshless methods which do not require a computational mesh/grid, therefore its main capability is to easily deal with practical problems defined over complex-shaped domains. The geometrical flexibility of the RBF-FD method is even more advantageous when coupled to the Non-Intrusive PC method for uncertainty quantification since different deterministic solutions over different geometries are required. The applicability of the proposed approach to practical problems is presented through the prediction of geometric uncertainty effects for a steady-state forced convection problem in a 3D complex-shaped domain.

1. Introduction

Manufacturing tolerances are pervasive in almost every area of engineering and their impact can often be crucial: indeed, tolerances in production processes lead to an uncertain behavior in the performances of the manufactured products. These uncertainties should be quantified to some extent early in the design phase, especially when dealing with robust design. With this in mind, it is essential to develop efficient and accurate numerical methods aimed at quantifying such uncertainties [1]. In this work we focus on the uncertainty quantification in the presence of geometric uncertainties [1-2] on the boundaries in 3D incompressible, laminar and steady-state thermo-fluid problems.

In this work the stochastic problem of uncertainty quantification is solved by means of the Non-Intrusive Polynomial Chaos method [3-6]. The non-intrusive formulation [7] allows the use of existing, deterministic solvers as black boxes, i.e., without the need for any adjustment or modification: the assessment of the sought stochastic response is reduced to the computation of a limited set of deterministic simulations. The Polynomial Chaos formulation, on the other hand, minimizes the number of required deterministic responses, allowing a fast and accurate estimate of the sought stochastic response [4-5]. The required simulations on deterministic geometries are computed in this work through the Radial Basis Function-generated Finite Differences (RBF-FD) meshless method [8]. This choice is due to the geometric flexibility and the ability of RBF-FD meshless method to easily deal with complex-shaped domains since a simple distribution of nodes is required only. This numerical approach, along with many other meshless/meshfree techniques [9], is recently gaining more and more popularity in practical applications since the traditional mesh/grid required by well-known mesh-based methods, e.g., Finite Volume or Finite Element methods, is no longer needed. Furthermore, the convergence rate of RBF-FD methods can be easily increased in order to provide high-fidelity numerical results with optimal computational effort [10]. The coupling of Non-Intrusive Polynomial Chaos and RBF-FD method has already been successfully employed for the prediction of geometric uncertainty in 2D thermo-fluid problems [11], highlighting the practical feasibility of such approach. In this work the aforementioned approach is employed for the quantification of the statistical moments, i.e., expected value μ and variance σ^2, of the local and global Nusselt number for the forced convection inside a 3D pipe with wavy surfaces whose amplitude is described by two stochastic variables with Gaussian distribution.

1.1. Governing equations

Let us consider an incompressible fluid with density ρ, kinematic viscosity ν, thermal diffusivity α and thermal conductivity κ. The resulting conservation equations of mass, momentum and energy are:

$$\nabla \cdot \boldsymbol{u} = 0 \tag{1}$$

$$\frac{\partial \boldsymbol{u}}{\partial t} + (\boldsymbol{u} \cdot \nabla)\boldsymbol{u} = -\nabla p + \frac{1}{\mathrm{Re}}\nabla^2 \boldsymbol{u} \tag{2}$$

$$\frac{\partial T}{\partial t} + \boldsymbol{u} \cdot \nabla T = \frac{1}{\mathrm{Re} \cdot \mathrm{Pr}}\nabla^2 T \tag{3}$$

In the above equations, length, velocity $\boldsymbol{u} = (u, v, w)$, time t, pressure p and temperature T are made nondimensional by taking L, U_0, L/U_0, ρU_0^2 and ΔT as reference quantities, respectively. $\mathrm{Re} = U_0 L/\nu = 500$ is the flow Reynolds number, $\mathrm{Pr} = \nu/\alpha = 0.71$ is the Prandtl number.

1.2. Domain

By considering 3D cylindrical coordinates $(r = \sqrt{x^2 + y^2}, \theta = \mathrm{arctg2}(y, x), z)$ with respect to the z axis, the domain is implicitly defined as follows:

$$\underbrace{R_1 + A_1 g(\theta, z)}_{R_{min}} \leq r \leq \underbrace{R_2 + A_2 g(\theta, z)}_{R_{max}} \tag{4}$$

$$0 \leq z \leq H \tag{5}$$

where $g(\theta, z) = (\cos(2\theta) - 1)(\cos(2\pi z/H) - 1)$, $L = R_2 - R_1$ is the chosen reference length, $H = 5L$, $A_1 = A_2 = L/4$ and $R_2/R_1 = 3/2$. From now on, nondimensional coordinates will be employed, i.e., $L = 1$.

1.3. Boundary conditions

The following boundary conditions are enforced:

- Inlet ($z = 0$): $u = v = 0, w = T = u_z(r)$ from equation (11) (see Appendix A);
- Outlet ($z = 5$): $p = 0, \partial T/\partial n = 0$;
- Wavy surfaces ($r = R_{min}, r = R_{max}$): $\boldsymbol{u} = \boldsymbol{0}$, $T = 0$.

The maximum velocity at the inlet (see Appendix A) is chosen as the reference velocity U_0 and the difference between the maximum temperature at the inlet and the temperature at the wavy surfaces is chosen as the reference temperature scale ΔT. Because of the double symmetry of both the domain and the previous boundary conditions with respect to the $x - z$ and the $y - z$ planes, only one fourth of the previous domain is considered for the following numerical computations, i.e., $x, y > 0$. The symmetry conditions are the following:

- $x - z$ plane ($y = 0$): $\partial u/\partial y = \partial w/\partial y = \partial T/\partial y = v = 0$;
- $y - z$ plane ($x = 0$): $\partial v/\partial x = \partial w/\partial x = \partial T/\partial x = u = 0$.

Graphical representations of the computational domain are shown in figure 2.

2. Numerical methods

2.1. Tensorial-expanded Polynomial Chaos

Given a vector of M independent random variables $\xi = (\xi_1,\ldots,\xi_M)$, under certain conditions [4] a random process $\lambda(\xi)$ with finite second-order moments can be expanded as follows:

$$\lambda(\xi) = \sum_{i=0}^{+\infty} \lambda_i \varphi_i(\xi) \tag{6}$$

where $\{\varphi_i\}$ is a suitable basis and λ_i are the corresponding expansion coefficients. In this work random variables with gaussian distribution $\xi_i \sim \mathcal{N}(0,1)$ will be considered only, for which the best choice for the basis $\{\varphi_i\}$ in terms of convergence rate for the sought statistical moments is given by Hermite polynomials (Hermite-Chaos). Indeed, 1D Hermite polynomials are orthogonal with respect to the gaussian probability density function (PDF), therefore the M-dimensional tensorial-expanded Hermite polynomials are orthogonal with respect to the M-dimensional gaussian PDF:

$$\omega(\xi) = \prod_{k=1}^{M} \frac{e^{-\xi_k^2/2}}{\sqrt{2\pi}} = \sqrt{\frac{e^{-\xi\cdot\xi}}{2\pi}} \tag{7}$$

The expansion (6), expressed in the Hermite-Chaos form and truncated to a finite number Q + 1 of terms, is:

$$\lambda(\xi) \approx \sum_{i=0}^{Q} \lambda_i \varphi_i(\xi) = \sum_{i_1=0}^{q_1} \cdots \sum_{i_M=0}^{q_M} \lambda_{i_1\cdots i_M} H_{i_1}(\xi_1) \cdots H_{i_M}(\xi_M) \tag{8}$$

where H_j is the Hermite polynomial of degree j. The statistical moments of λ, i.e., expected value/mean μ and variance σ^2, can be computed from the previous approximation:

$$\mu(\lambda) = \int_{R^M} \lambda(\xi)\omega(\xi)d\xi \approx \lambda_{0\cdots 0} \tag{9}$$

$$\sigma^2(\lambda) = \int_{R^M} \left(\lambda(\xi) - \lambda_{0\cdots 0}\right)^2 \omega(\xi)d\xi \approx -\lambda_{0\cdots 0}^2 + \sum_{i_1=0}^{q_1} \cdots \sum_{i_M=0}^{q_M} \lambda_{i_1\cdots i_M}^2 i_1! \cdots i_M! \tag{10}$$

In order to obtain the best approximation for μ and σ^2, we note that equations (9) and (10) are weighted integrals with a gaussian weight function $\omega(\xi)$. Therefore, the best choice for the $Q + 1 =$ $= (q_1 + 1) \cdots (q_M + 1)$ sample points, i.e., the points where $\lambda(\xi)$ has to be evaluated, are the roots of the Hermite polynomial of degree $q_k + 1$ along each of the $k = 1,\ldots, M$ dimensions, i.e., Gauss-Hermite quadrature. By using a collocation technique, the Hermite-Chaos expansion (8) is therefore made valid at each of these $Q + 1$ sample points in order to obtain the $Q + 1$ unknown coefficients $\lambda_{i_1\cdots i_M}$. In other words, the Hermite-Chaos expansion can be seen as an interpolant matching the random process $\lambda(\xi)$ at the sample points.

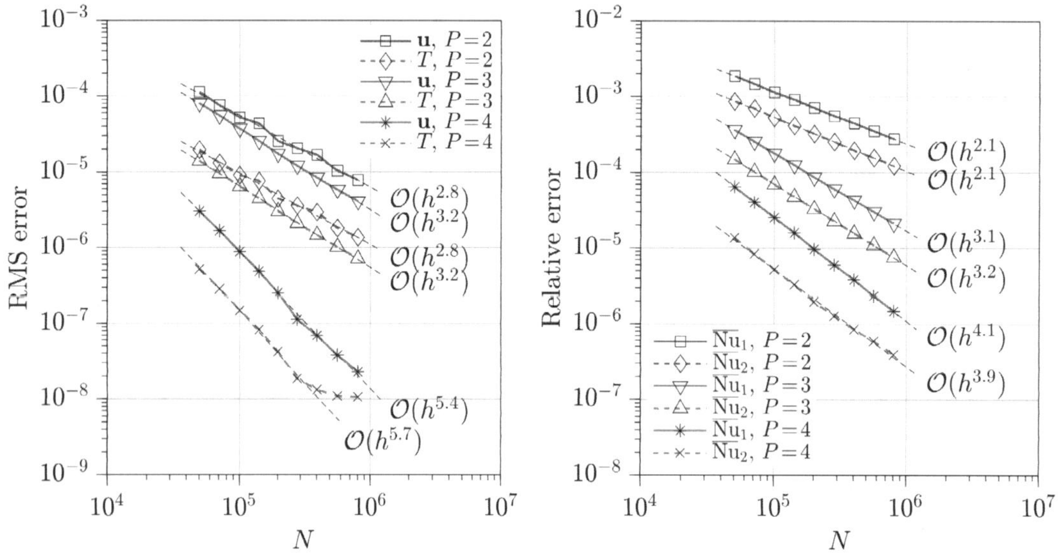

(a) (b)

Figure 1. Convergence curves, analytic solution given by equation (11): root mean square error over the domain (a), relative error for $\overline{\mathrm{Nu}}_1$ and $\overline{\mathrm{Nu}}_2$ (b).

2.2. *RBF-FD discretization and solution procedure*

The RBF-FD meshless method [7] is employed to discretize the governing equations on meshless distributions of N nodes scattered over the domain and on the boundary. Local interpolants with polynomial degrees $P = 2, 3, 4$, requiring $n = 20, 40, 70$ stencil nodes, respectively, have been used. Implicit ∇^4 hyperviscosity is employed to suppress advective and pressure instabilities [8,11]. More details can be found in [12].

At each time step, the computation of velocity, pressure and temperature through equations (1)-(3) is decoupled using a projection scheme with a three-level Gear scheme for the time discretization. Both the advective and diffusive terms are treated implicitly in equations (2)-(3), allowing the use of larger time steps to reach the steady-state solutions. BiCGSTAB iterative solver with an incomplete LU factorization as preconditioner are employed for the solution of the whole set of equations using a relative tolerance of 10^{-9}. The steady-state convergence is declared when $\left|\partial\overline{\mathrm{Nu}}_{i=1,2}/\partial t\right| < 5 \cdot 10^{-6}$, where $\overline{\mathrm{Nu}}_i$ are the average Nusselt numbers over the internal and external wavy surfaces.

The whole procedure is developed in Julia [13], a high-level, high-performance, dynamic programming language particularly suited for scientific computing.

2.3. *Verification of the RBF-FD deterministic solver*

The verification of the solver presented in subsection 3.2 is carried out by considering the analytical solution presented in Appendix A. The convergence curves are shown in figure 1 for node distributions with N ranging from 50,000 to 800,000 nodes. The root mean square (RMS) error over the whole domain for velocity and temperature is shown in figure 1a: the convergence rates of the RMS error for RBF-FD polynomial degrees $P = 2, 3, 4$ are $2.8, 3.2, 5.5$, respectively, showing a great advantage in the use of higher polynomial degrees, both in terms of convergence rate and absolute

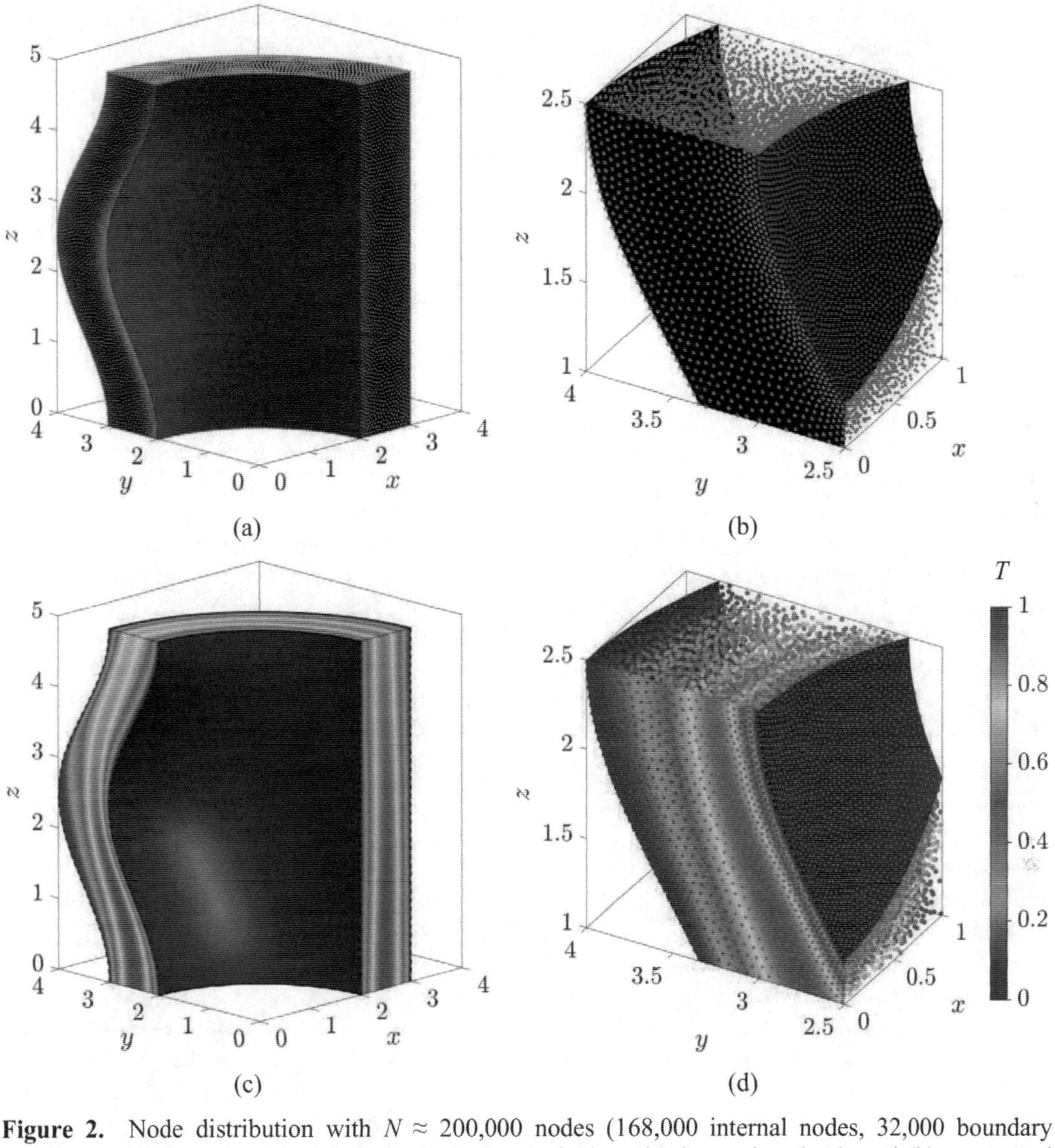

Figure 2. Node distribution with $N \approx$ 200,000 nodes (168,000 internal nodes, 32,000 boundary nodes): boundary nodes (a), enlarged view of a particular with internal nodes in red (b), computed temperature field on the boundary (c), enlarged view of a particular of the meshless temperature field with boundary nodes in red (d).

magnitude of the error. On the other hand, the increase from $P = 2$ to $P = 3$ highlights only a moderate improvement in the accuracy. The flattening of the RMS error of T around 10^{-8} for N > 300,000 nodes is due to the single precision accuracy in the generation of boundary nodes. The relative error for the average Nusselt numbers is shown in figure 1b, where the convergence rate and the absolute value of the error both improve by increasing the RBF-FD polynomial degree P, for which the convergence rate is approximately equal to P.

Table 1. Average Nusselt numbers for node distributions with *N* nodes and RBF-FD polynomial degree *P*, deterministic case.

	$\overline{\mathrm{Nu}}_1$			$\overline{\mathrm{Nu}}_2$			$\overline{\mathrm{Nu}}$		
N	*P*=2	*P*=3	*P*=4	*P*=2	*P*=3	*P*=4	*P*=2	*P*=3	*P*=4
80,000	3.884	3.801	3.761	2.531	2.464	2.430	3.087	3.014	2.977
125,000	3.851	3.785	3.769	2.511	2.459	2.444	3.062	3.005	2.989
200,000	3.824	3.783	3.780	2.495	2.461	2.458	3.041	3.005	3.001
320,000	3.807	3.781	3.784	2.487	2.465	2.468	3.030	3.006	3.009
500,000	3.801	3.783	-	2.482	2.467	-	3.024	3.008	-
800,000	3.795	3.784	-	2.479	2.470	-	3.020	3.010	-

3. Results

3.1. Deterministic simulations

In order to highlight the convergence properties of the deterministic RBF-FD solver in the case of the actual problem described in section 2, a convergence study is performed for node distributions with *N* ranging from 50,000 to 800,000 nodes and polynomial degrees $P = 2, 3, 4$. An example of node distribution with *N* = 200,000 nodes is depicted in figure 2a and 2b, while the computed temperature field is shown in figure 2c and 2d. From these figures it can also be noted how the node density is increased near the internal wavy surface highlighted in figure 2d, i.e., where the thermal boundary layer is thinner and more nodes are required to capture the gradients with accuracy.

The results of the convergence study are summarized in table 1 in terms of average Nusselt numbers $\overline{\mathrm{Nu}}_1$ and $\overline{\mathrm{Nu}}_2$, highlighting a somehow slow convergence in the case $P = 2$. Faster convergence is achieved with $P = 3$ and 4, for which satisfactory results can be obtained with much less nodes *N* than $P = 2$. Simulations for $P = 4$ with $N > 320{,}000$ nodes were computationally too expensive and they have not been carried out.

3.2. Uncertainty quantification

Gaussian uncertainty is introduced in the amplitudes A_1 and A_2 describing the internal and external wavy surfaces as defined in subsection 2.2. Two cases with 5% and 10% standard deviation on the amplitudes are considered, i.e., $A_{i=1,2} = L/4(1 + 0.05\xi_i)$ and $A_{i=1,2} = L/4(1 + 0.1\xi_i)$.

Equal polynomial degrees $q_1 = q_2 = q$ are employed for the Hermite-Chaos expansion over the $\mathrm{M} = 2$ uncertain variables, requiring $Q + 1 = (q + 1)^2$ deterministic RBF-FD solutions over the corresponding geometries. Based on the convergence study of subsection 4.1, $N \approx 200{,}000$ nodes and $P = 3$ are employed as a reasonable balance between accuracy and computational effort for the deterministic solutions over the $(q + 1)^2$ required geometries. We point out that each of these deterministic solutions is obtained by generating a completely new meshless node distribution, requiring a post-processing mapping to the reference geometry with $\xi_i = 0$.

The convergence of mean and standard deviation of the average Nusselt numbers $\overline{\mathrm{Nu}}_1$, $\overline{\mathrm{Nu}}_2$ and $\overline{\mathrm{Nu}}$ (averaged Nusselt number over both wavy surfaces) for four different Hermite-Chaos polynomial degrees $q = 1, 2, 3, 4$ is reported in table 2 and table 3 for 5% and 10% standard deviation of the amplitudes, respectively. The mean values μ are almost exactly the same to the fourth digit, while a

Table 2. Mean and standard deviation of the average Nusselt numbers, stochastic case with 5% standard deviation of A_1 and A_2.

	$\overline{\mathrm{Nu}}_1$		$\overline{\mathrm{Nu}}_2$		$\overline{\mathrm{Nu}}$	
q	μ	σ	μ	σ	μ	σ
1	3.794	0.0427	2.465	0.0692	3.011	0.0606
2	3.794	0.0427	2.465	0.0693	3.011	0.0607
3	3.794	0.0424	2.465	0.0694	3.011	0.0607
4	3.794	0.0430	2.465	0.0691	3.011	0.0606

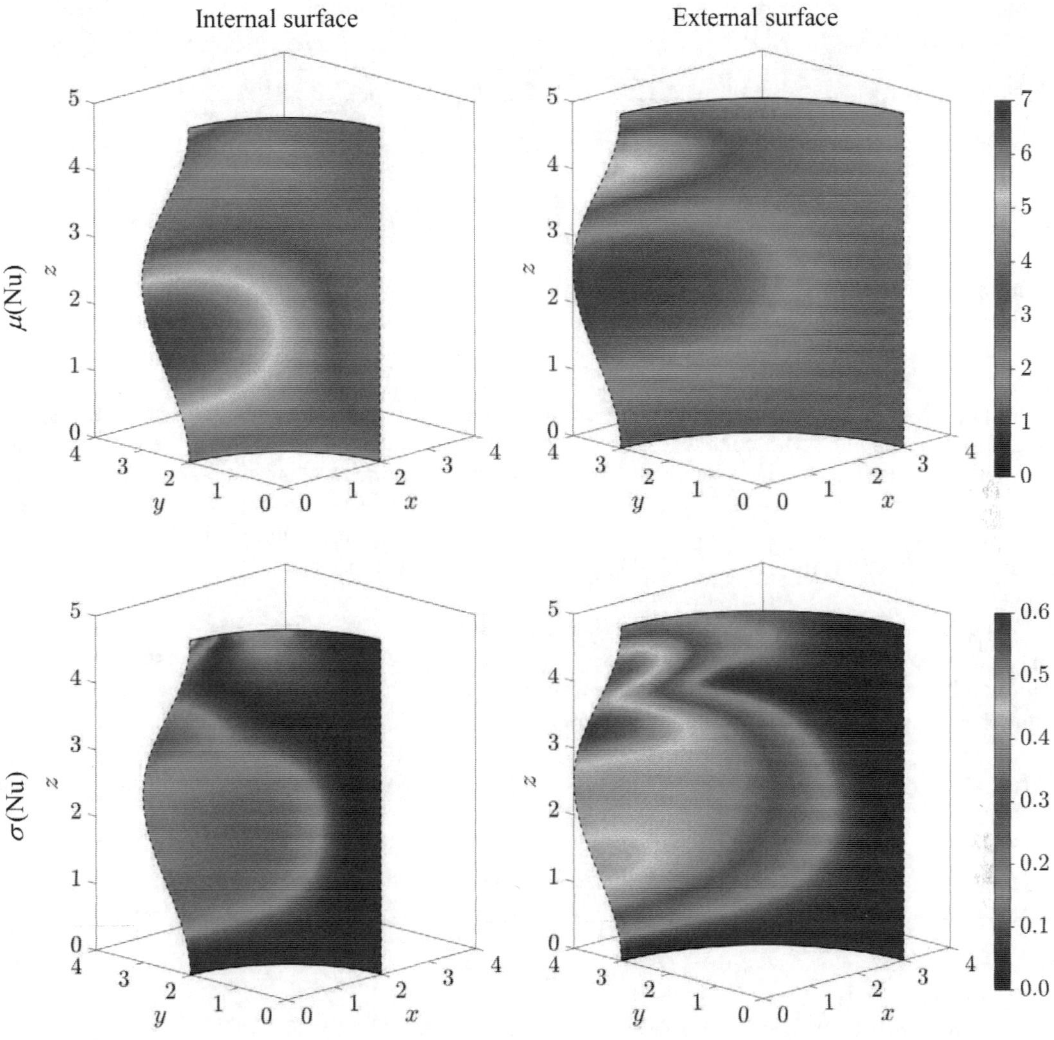

Figure 3. Mean and standard deviation of the local Nusselt number, stochastic case with 10% standard deviation of A_1 and A_2.

very fast convergence is achieved for the standard deviation σ in both cases and for each Nu value, confirming the accuracy of the tensorial-expanded Hermite Chaos method.

The computation of mean and standard deviation is then applied to the local Nusselt number over the wavy surfaces. The results for the case with 10% standard deviation of the amplitudes and $q = 4$ are shown in figure 3 for the internal and the external wavy surfaces. As expected, a moderate standard deviation of the local Nu number at the internal surface is found where the Nu value is high, but the highest values of the standard deviation are found at the external surface, especially in the upper part near $x = 0$ and $z = 3$.

4. Conclusions

In this work the Non-Intrusive Polynomial Chaos method is coupled to the RBF-FD meshless method for the accurate prediction of the uncertainties in laminar, incompressible and steady 3D thermo-fluid problems where the boundaries are described by stochastic variables. The main statistical properties

Table 3. Mean and standard deviation of the average Nusselt numbers, stochastic case with 10% standard deviation of A_1 and A_2.

	$\overline{\mathrm{Nu}}_1$		$\overline{\mathrm{Nu}}_2$		$\overline{\mathrm{Nu}}$	
q	μ	σ	μ	σ	μ	σ
1	3.799	0.0861	2.475	0.1412	3.019	0.1233
2	3.798	0.0880	2.475	0.1422	3.019	0.1245
3	3.799	0.0877	2.475	0.1424	3.019	0.1246
4	3.799	0.0879	2.475	0.1424	3.019	0.1246

of any primitive or derived flow quantity can be accurately computed by using the minimum number of deterministic fluid-flow simulations required by the tensorial-expanded Hermite-Chaos formulation. This approach is employed to compute the statistical moments of both local and average Nusselt numbers in a forced convection problem over a 3D pipe with wavy surfaces with uncertain amplitude. The main advantage of the presented meshless approach over classical mesh-based approaches is its capability to easily deal with complex 3D geometries, which is an additional advantage when coupling the RBF-FD method to the employed Non-Intrusive Polynomial Chaos method since multiple deterministic solutions over different geometries are required. Consistent results are found, suggesting that the presented approach can be used for the efficient quantification of the uncertainties in fluid flow problems on 3D geometries of practical interest.

5. Appendix A. Analytical solution in cylindrical coordinates

In the case $A_1 = A_2 = 0$, i.e., an annular section pipe where the internal and the external cylinders with nondimensional radii R_1 and R_2 are kept at temperatures T_1 and T_2, respectively, a steady-state solution of equations (1)-(3) in nondimensional cylindrical coordinates (r, θ, z) is given by:

$$u_z(r) = \frac{b + a \cdot \log r - r^2}{c}, \qquad u_r = u_\theta = 0, \qquad \frac{dp}{dz} = -\frac{4}{c \cdot \mathrm{Re}}, \qquad T(r) = d \cdot \log r + e \tag{11}$$

for which the maximum velocity is $u_z(\hat{r}) = 1$ for $\hat{r} = \sqrt{a/2}$. By choosing $\Delta T = T_2 - T_1 > 0$ as the reference temperature scale, the average Nusselt numbers over the internal and external surfaces are:

$$\overline{\mathrm{Nu}}_1 = \left(R_1 \log\frac{R_2}{R_1}\right)^{-1}, \overline{\mathrm{Nu}}_2 = \left(R_2 \log\frac{R_2}{R_1}\right)^{-1} \tag{12}$$

6. References

[1] Chinchalkar S and Taylor D L 1994 *Comput. Syst. Eng.*, **5**(2) 159-170 DOI
[2] Gumbert C, Newman P and Hou G 2002 Effect of Random Geometric Uncertainty on the Computational Design of a 3D Flexible Wing *20th Applied Aerodynamics Conference* DOI
[3] Hosder S, Walters R and Perez R 2006 A Non-Intrusive Polynomial Chaos Method For Uncertainty Propagation in CFD *44th AIAA Aerospace Sciences Meeting and Exhibit* DOI
[4] Xiu D and Karniadakis G 2002 *SIAM J. Sci. Comput.* **24**(2) 619-644 DOI
[5] Parussini L, Pediroda V and Poloni C 2010 *Comput. Fluids* **39**(1) 137-151 DOI
[6] Eldred M S 2009 Recent Advances in Non-Intrusive Polynomial Chaos and Stochastic Collocation Methods for Uncertainty Analysis and Design *50th AIAA/ASME/ASCE/AHS/ASC Structures, Structural Dynamics, and Materials Conference* DOI
[7] Witteveen J A S 2014 Non-intrusive uncertainty quantification *In T Magin & C Gorle (Eds.), Lecture notes for NATO-VKI Short Course on Uncertainty Quantification 2014* LINK
[8] Fornberg B and Flyer N 2015 *Acta Numerica* **24** 215-258 DOI
[9] Liu G R 2009 *Meshfree Methods: Moving Beyond the Finite Element Method*, CRC Press DOI
[10] Jančič M, Slak J and Kosec G 2021 *J. Sci. Comput.* **87**(9) DOI
[11] Zamolo R and Parussini L 2020 *J. Comput. Phys.* **421** 109730 DOI
[12] Miotti D, Zamolo R and Nobile E 2021 *Energies* **14**(5) 1351 DOI
[13] Bezanson J, Edelman A, Karpinski S and Shah V B 2017 *SIAM Rev.* **59** 65–98 DOI

Numerical analysis of the binder angle effect on convective heat transfer and pressure drop in drilled-hollow sphere architected foams

M Iasiello[1,#], G M Mauro[2], N Bianco[1], A Andreozzi[1], W K S Chiu[3] and V Naso[1]
[1] Dipartimento di Ingegneria Industriale - Università degli Studi di Napoli Federico II Piazzale Tecchio 80 - 80125 – Napoli – Italy
[2] Dipartimento di Ingegneria - Università degli Studi del Sannio – Piazza Roma 21 - 82100 – Benevento – Italy
[3] Department of Mechanical Engineering – University of Connecticut – 191 Auditorium Road – 06269 – Storrs (CT) - USA

[#] Email: marcello.iasiello@unina.it

Abstract. Due to their capability of generating customized microstructures, additive manufactured cellular materials are promising to being employed in heat transfer devices. To this aim, among printable cellular materials Drilled-Hollow Spheres Architected (DHSA) foams are investigated. However, at the present status, limited data on pressure drop and heat transfer in DHSAs are available. Starting from hollow spheres, a metal DHSA foam is generated with CAD software in this study. Forced air convective heat transfer in the foam is investigated numerically, under the assumptions of air incompressible laminar flow and uniform wall heat flux from the solid to the fluid phase. Mass, momentum and energy equations in the fluid region are written and solved numerically, for various values of the foam binder angle and the velocity of the inlet air. The convective heat transfer coefficients, the pressure drop and the friction factor are predicted. The effects of the binder angle and the air inlet velocity on heat transfer and pressure drop are highlighted.

Nomenclature

a	cubic cell side (m)
b	radius of curvature (m)
c_p	heat capacity (J/kg K)
h	minimum radius of the binder (m)
h_c	interfacial heat transfer coefficient (W/m^2 K)
h_v	volumetric heat transfer coefficient (W/m^3 K)
k	thermal conductivity (W/m K)
l	binder length (m)
p	pressure (Pa)
q	heat flux (W/m^2)
R	outer shell radius (m)
R_i	inner shell radius (m)
Re	Reynolds number
S	area of the solid surface (m^2)
T	temperature (K)
$\mathbf{u}$	velocity vector (m/s)
V	volume (m^3)
x, y, z	Cartesian coordinates (m)
Greek symbols	
ε	porosity (-)
θ	binder angle (°)
μ	dynamic viscosity (Pa s)
ρ	density (kg/m^3)
τ	shear stress (Pa)
Subscripts	
f	Fluid
in	inlet section
s	Solid

Journal of Physics: Conference Series **2685** (2024) 012007 doi:10.1088/1742-6596/2685/1/012007

1. Introduction

Cellular materials are a promising class of materials in applications where heat transfer plays a primary role, such as heat exchangers, thermal energy storage systems, and heat sinks for electronics [1]. Their large ratio of the heat transfer surface area to the volume, relatively high thermal conductivity, and tortuosity, that promotes flow mixing and heat transport, make them suitable to enhance heat transfer. Some examples of cellular materials are body-cubic centered structures, wire-meshes, honeycombs, open-cell foams [2]. Attention was focused on the prediction of convective heat transfer, which plays an important role in foams applications [3]. Younis and Viskanta [4] employed a single-blow transient technique, coupled with inverse heat transfer, to predict the volumetric heat transfer coefficients for different foams and volumetric flow rates. Numerical techniques have been used through the years to predict convective heat transfer. By using foam reconstruction from computed tomography, Bodla et al. [5] predicted heat transfer coefficients simulating the flow of air through a foam. The same technique was used by Iasiello et al. [6] on Kelvin's foam-based geometrical models, to evaluate the possibility of reconstructing the foam with the help of ideal foam models. Wu et al. [7] analyzed Kelvin's foams under a uniform surface temperature boundary condition and analyzed also the thermally-developing effects. Iasiello et al. [8] carried out a similar analysis by tuning the strut shape of Kelvin's foams as well.

Recent advances in additive manufacturing allowed for the design of cellular materials aimed to enhance heat transfer [9, 10]. 3D-printed structures based on well-established structures, such as body-cubic centered crystals [11], cubic lattices [12], Kelvin's foams [13], also triply-periodic minimal surfaces [14], were developed. 3D printing looks really promising in designing innovative heat transfer devices, even if there are still some limits because of printing costs and time [15] as well as surface roughness [16]; however, depending on the applications and on the operating conditions, surface roughness of 3D-printed parts can even be an advantage for heat transfer devices [17]. Drilled-hollow spheres architected foams were proposed for heat transfer applications [18] and looked promising compared to other porous materials.

The literature survey shows that the employment of additive-manufactured structures in heat transfer devices deserves further research. In this study, air forced convective heat transfer and pressure drop in drilled-hollow sphere architected foams are analyzed numerically. Different binder angles of the foams and inlet velocities of the entering air are taken into account. Their effects on heat transfer and pressure drop are pointed out.

2. Mathematical model

2.1. Foams design and governing equations

The structure and the generation of the herein investigated Drilled-Hollow Sphere Architected (DHSA) foams were described in [18]. The elementary cell, which, repeating itself periodically, generates the structure, and one eighth of it are reported in figure 1a, with a the side of the cubic cell,

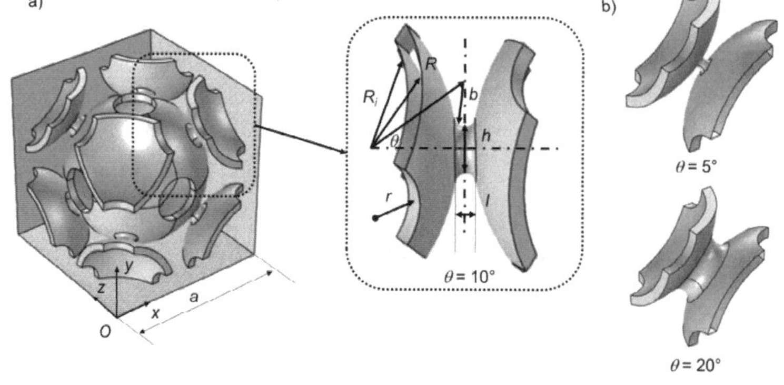

Figure 1. The typical elementary DHSA cell: a) the geometrical parameters; b) the binder shape for different binder angles.

R_i and R the inner and outer radii of the shell, respectively, r the perforation radius of the shell, h the minimum radius of the cylindrical binder, l its length, θ the binder angle, b the radius of curvature. The above geometrical parameters of the foam are correlated by the following equations:

$$R+l/2=(R+b)\cos\theta \tag{1}$$

$$b+h/2=(R+b)\sin\theta \tag{2}$$

Once R is chosen, the other parameters can be varied according to equations (1) and (2). The dependence of the binder shape on the binder angle is shown in figure 1b. Also the performance of a Kelvin's foam, whose generation was described in [6], will be reported in this study.

2.2 Numerical modeling

Air forced convection through the foam was simulated with the elementary cell surrounded by a bounding box. After carrying out Boolean operations, the air domain, which includes half-cell side long regions upstream and downstream of the cell, was simulated by the mathematical model. The investigated domain, for a $\theta = 10°$ binder angle, is sketched in figure 2.

The porosity, $\varepsilon = V_f/(V_f + V_s)$, and the ratio of the heat transfer area to the foam volume, $S/(V_f + V_s)$, computed via a numerical solver as a function of the binder angle, for all the investigated DHSA foams, are reported in figure 3. It shows that, in the investigated range, increasing the binder angle decreases both the porosity and the ratio of the heat transfer area to the volume because of the increasing binder volume; one can also remark that the porosity of the foams is fairly affected by the binder angle.

The governing equations for continuity, momentum and energy conservation, under the assumption of steady-state laminar incompressible flow are:

$$\nabla\cdot\mathbf{u}=0 \tag{3}$$

$$\rho\,\mathbf{u}\cdot\nabla\mathbf{u} = -\nabla p+\mu\,\nabla^2\mathbf{u} \tag{4}$$

$$\rho\,c_p\,\mathbf{u}\cdot\nabla T = k\,\nabla^2 T \tag{5}$$

2.3 Boundary conditions

Boundary conditions, presented in detail in [6, 8], are herein assumed. Air at uniform temperature and uniform velocity enters the foam; a zero gauge pressure is assumed at the domain exit section. The side walls of the domain are assumed to be adiabatic and symmetric, *i. e.* a slip condition that allows to analyze an elementary cell extracted from a larger set of cells. At the solid/fluid interface, a no-slip condition and a uniform solid-to-fluid heat flux are imposed; the heat relased to the air is such as to neglect temperature-dependent property effects. Equations were solved by means of a commercial finite-element code. For each case, about 800.000 of both tethraedral and boundary layer elements were employed. Deviations always lower than 3% on both interfacial heat transfer coefficient and pressure drop were found with 1,500,000 and 2,000,000 elements. The linear system was solved with the PARDISO algorithm, using crosswind and streamline stabilization tools.

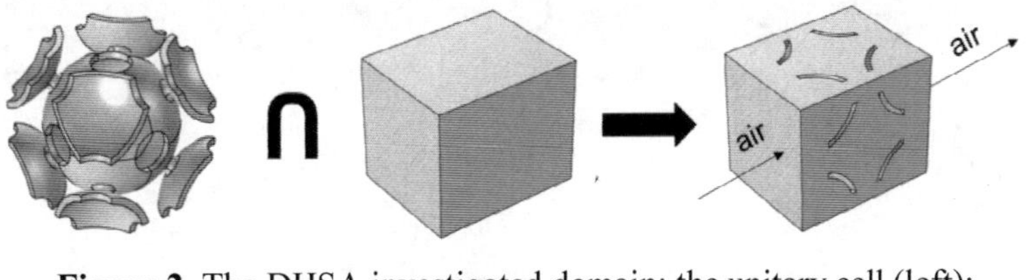

Figure 2. The DHSA investigated domain: the unitary cell (left); the bounding box (center), the computational domain (right).

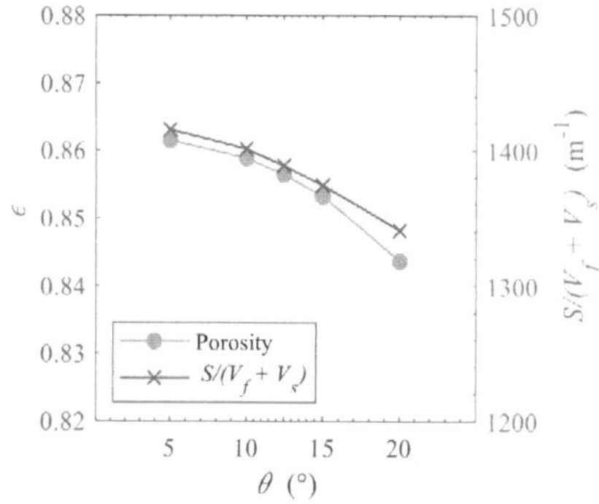

Figure 3. Porosity and surface-to-volume ratio *vs.* the binder angle.

3. Results and discussion

Velocity fields, convective heat transfer coefficients, friction factors were predicted for a cubic cell with a = 5.0 mm length of the cell side, θ = 5° - 20° binder angles, u_{in} = 0.2 m/s – 1.0 m/s inlet air velocities, T_{in} = 293.15 K inlet temperature of the air, q_{in} = 1,000 W/m^2 heat flux released to the air.

The air velocity fields in the cell at y = 2.5 mm, for u_{in} = 1.0 m/s, θ= 10° and θ= 20°, are reported in figures 4a and 4b, respectively. Both figures exhibit a significant increase in the velocity in the core of the cell because of the decreasing area of the cross-section available to the air; namely, the maximum velocity of the fluid is more than about 5 times higher than the air inlet velocity. The increase in the fluid velocity promotes both pressure drop and heat transfer. The comparison of figures 4a and 4b shows that the velocity increase in the foam with θ= 20° is larger than that occurring in the foam with θ= 10°, because the greater volume of the binder reduces the cross section available to the air.

The velocity fields in the cell, for u_{in} = 1.0 m/s, θ= 10° and θ= 20°, are presented in figures 5a and 5b, respectively. Both figures point out that in the regions close to the bounding box, where the area of the cross section available to the fluid is the smallest, the largest values of the air velocity, up to about 6 times higher than that in the inlet section, are attained.

In the following, reference will be made to the interfacial convection heat transfer coefficient, h_c:

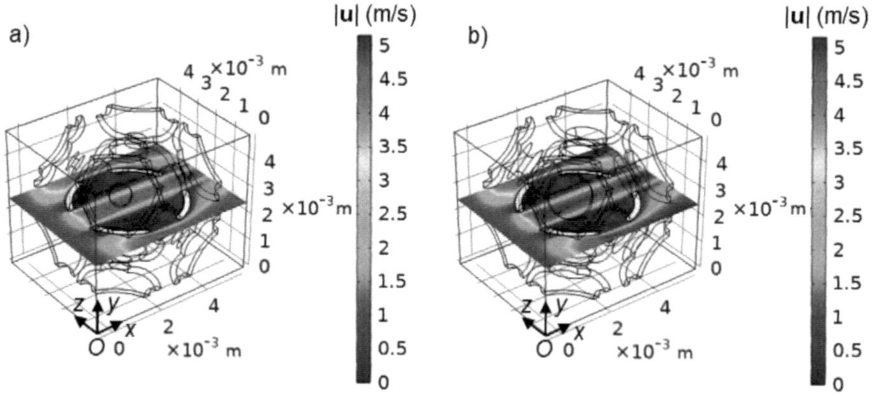

Figure 4. Velocity fields at y = 2.5 mm, for u_{in} = 1.0 m/s: a) θ= 10°; b) θ= 20°.

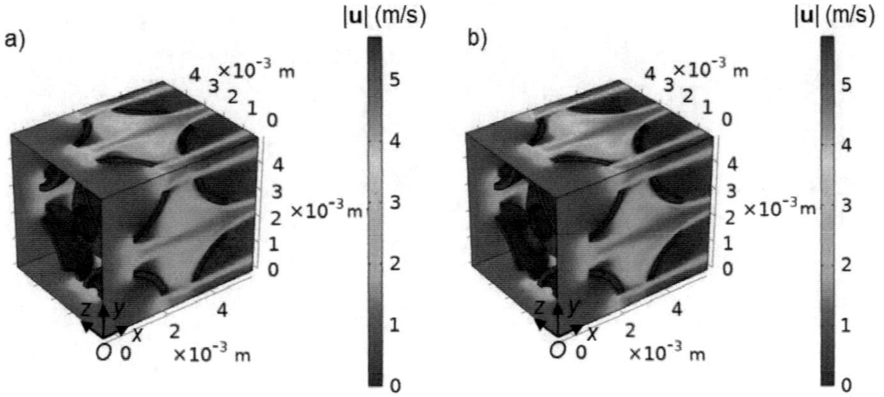

Figure 5. Velocity fields in the cell, for u_{in} = 1.0 m/s: a) θ = 10°; b) θ = 20°.

$$h_c = \frac{q_{in}}{T_{s,av} - T_{f,av}} \tag{6}$$

where $T_{s,av}$ and $T_{f,av}$ are the temperatures of the volumetric averaged solid/fluid interface and of the fluid phase, respectively. Reference will be made also to the volumetric interfacial convective heat transfer coefficient, h_{cv}, obtained by scaling h_c with $S/(V_f + V_s)$:

$$h_v = h_c \frac{S}{V_f + V_s} \tag{7}$$

that allows to account for the actual convection heat transfer at the fluid/solid boundaries, as well as to the Reynolds number:

$$\mathrm{Re} = \frac{\rho\, u_{in}\, a}{\mu} \tag{8}$$

the friction factor:

$$f = \frac{\Delta p}{a} \frac{2}{\rho\, u_{in}^2} \tag{9}$$

the shear stress at solid-fluid interface walls:

$$\tau = \mu \frac{\partial u}{\partial n} \tag{10}$$

where n is the outgoing direction normal to the surface.

The predicted interfacial heat transfer coefficient and friction factor in the investigated DHSAs are presented in the following. Simulations for a Kelvin's foam are also presented; consistently with the values of the porosity and the ratio of the heat transfer area to the volume in figure 3, reference was made to a 0.85 porosity Kelvin's foam. It is worth remarking that the $S/(V_f + V_s)$ = 843 m^{-1} of the chosen Kelvin's foam is far lower than that of the investigated DHSAs, implying that the DHSAs investigated should perform better; on the other hand, one might also expect higher pressure drop.

The interfacial and volumetric interfacial heat transfer coefficients as a function of the inlet air velocity with various binder angles, for the DHSA and the Kelvin's foam, are reported in figures 6 and 7. The figures exhibit both interfacial convection heat transfer coefficients decreasing at increasing binder angles and at any air inlet velocity, as it was also reported in [19] for Kelvin's foams. This occurs because a decrease in the binder volume will reduce the crossflow section of the solid phase

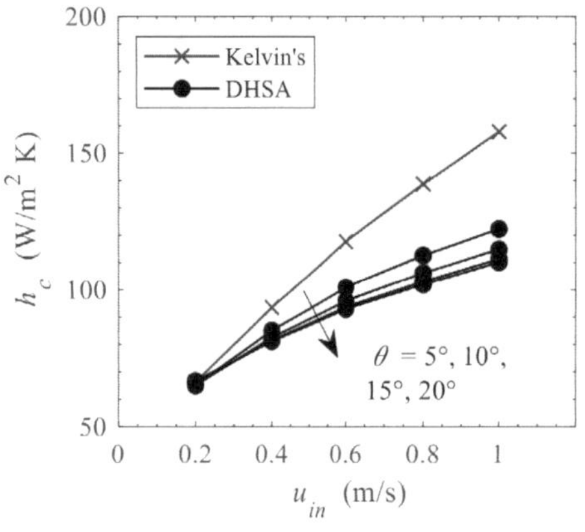

Figure 6. Interfacial heat transfer coefficient *vs.* inlet air velocity, for various binder angles.

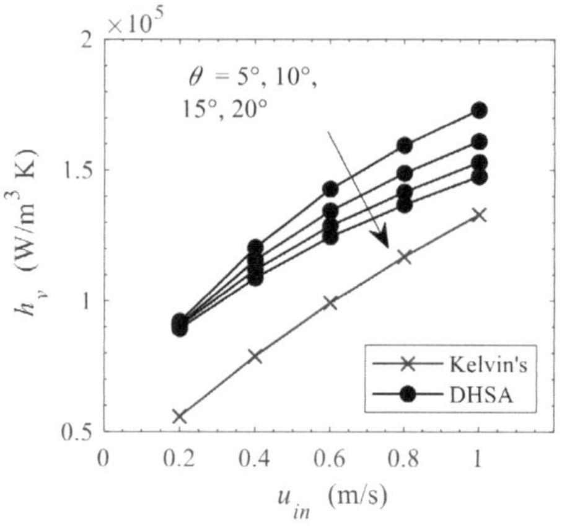

Figure 7. Volumetric interfacial heat transfer coefficient *vs.* inlet air velocity, for various binder angles.

and the wake region (see figure 1) and, therefore, less flow recirculation is available to enhance convective heat transfer. Moreover, decreasing the binder angle will reduce the free local cross-section to the air between two shells (see figure 1), thus increasing the velocity of the air and the heat transfer coefficient. It is also worth noting that, as it was shown in figures 4 and 5, increasing the angle and the volume of the binder reduces the free flow cross section, thus increasing the velocity and the heat transfer coefficient. One can notice that in the investigated DHSA foams, both heat transfer coefficients are fairly affected by the binder angle, with the volumetric heat transfer coefficient appearing to be more sensitive to binder angle.

Finally, the comparison between figures 6 and 7 shows that the interfacial convection heat transfer coefficients in the Kelvin's foam are higher than in DHSA foams, with larger differences observed at increasing binder angles. The opposite is observed to occur to the volumetric convection heat transfer coefficient. This is due to the ratio of the heat transfer area to the foam volume that is larger in the Kelvin's foam than in the DHSA foams, as it was remarked in sub-section 2.2.

Temperature contours on solid-fluid interface walls of the shell, for u_{in} = 1.0 m/s: and θ = 10° and 20°, are reported in figures 8a and 8b, respectively. The figures exhibit far closer contours for θ = 10° than for θ = 20°, suggesting that higher temperature gradients arise at smaller binder angles. Since temperature gradients are proportional to the convection heat transfer coefficients via the solid/fluid temperature difference, we can conclude that heat transfer is higher for θ = 10°, as depicted in figure 6.

The pressure drop per unit length as a function of the inlet air velocity and the friction factor as a function of the Reynolds number in DHSA and Kelvin's foams, for various binder angles, are reported in figures 9 and 10, respectively.

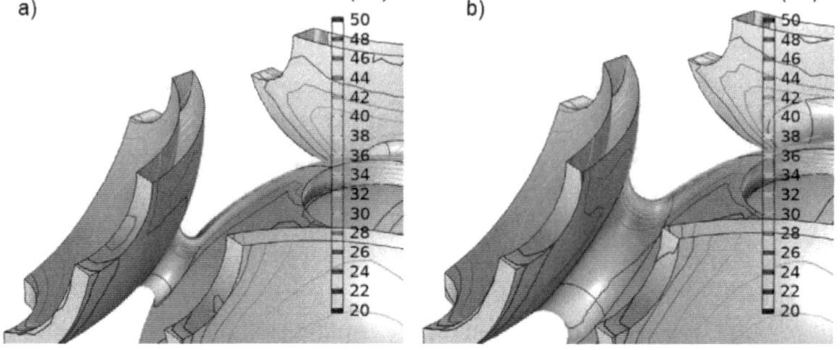

Figure 8. Temperature contours on solid-fluid interface walls, for u_{in} = 1.0 m/s: a) θ = 10°; b) θ = 20°.

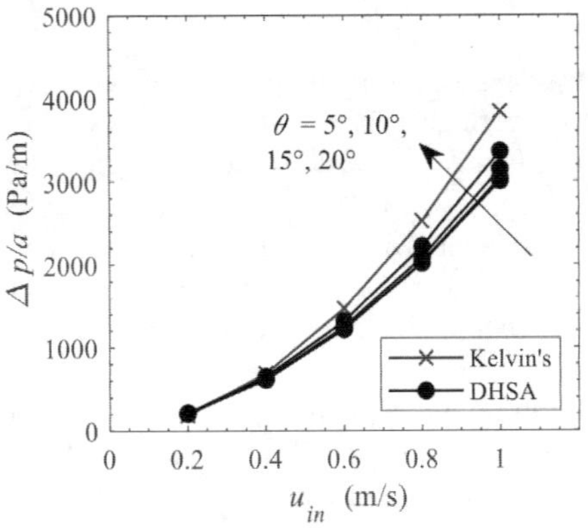

Figure 9. Pressure drop *vs*. inlet velocity for different binder angles.

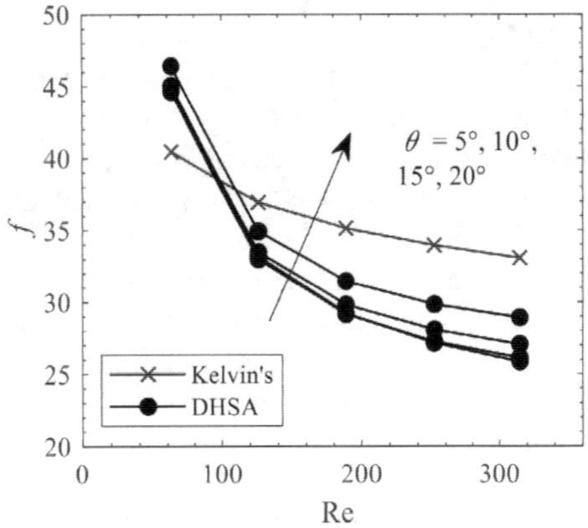

Figure 10. Friction factor *vs*. Reynolds number for different binder angles.

Figure 9 shows that at the range of inlet air velocities considered, the binder angle has a weak effect on the pressure drop in DHSA foams, with a larger angle resulting in higher pressure drop. This occurs because, even if an increase of binder angle results in a higher cross section area available for flow (see figure 1) and lower velocities, the wake region becomes smaller, thus increasing the dissipation and the momentum transfer between the adjacent shells. It is worth remarking that the rate of variation of the pressure drop in DHSA foams is practically the same as that in the Kelvin's foam as well as one can notice that the pressure drop in DHSA foams is slightly lower than in Kelvin's foam.

Figure 10 highlights that the friction factor in DHSA foams is essentially independent of the binder angle, apart from a fair increase at the largest angle. The figure points also out that the friction factor in the Kelvin's foam is less dependent on the Reynolds number than it is in DHSA, foams due to the Kelvin's foam's more uniform structure.

Shear stresses at solid-fluid interface walls, for u_{in} = 1.0 m/s and θ = 10° and 20°, are reported in figures 11a and 11b, respectively. Both figures show that the binders are more stressed than the shell, since the velocity of the air flow normal to the binders axis attains its largest values, thus causing highest dissipations. One can also notice that shear stresses are larger at the walls of the binder with the smaller angle (θ = 10°), their maximum values being attained at the mid-height external circumference of the binder. This occurs because a smaller angle implies a smaller length of the binder and a shorter distance between the shells, thus increasing the air velocity and the shear stresses. However, pressure drop are generally higher for higher binders because of the larger wake region downstream.

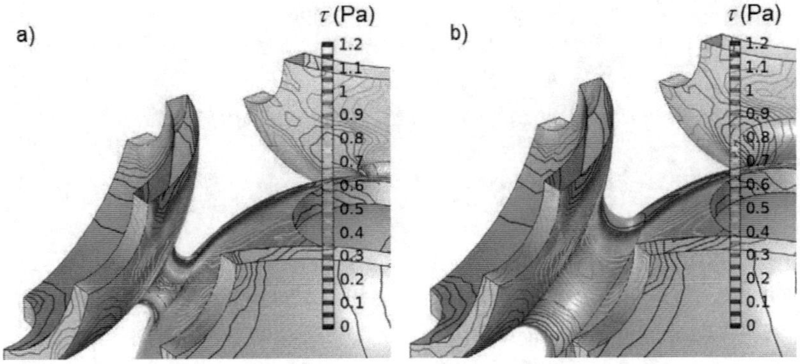

Figure 11. Shear stresses at solid-fluid interface walls, for u_{in} = 1.0 m/s: a) θ = 10°; b) θ = 20°.

4. Conclusions

The effect of the binder angle on forced air convective heat transfer and pressure drop in select drilled-hollow sphere architected (DHSA) foams has been analyzed numerically. The simulations showed that at any velocity of the air entering the DHSA foams the interfacial and volumetric convective heat transfer coefficients slightly decrease at increasing binder angles; they are fairly affected by the binder angle, the effects of the angle being larger on the volumetric heat transfer coefficient. Interfacial convection heat transfer coefficients in a Kelvin's foam are higher than in DHSA foams, with increasing differences at increasing binder angles; the opposite occurs to the volumetric heat transfer coefficient, due to the larger heat transfer area in DHSA foams.

In the range of investigated inlet air velocities, the binder angle has a weak effect on the pressure drop in DHSA foams; these pressure drop are higher with larger binder angles. The rate of variation of the pressure drop in DHSA foams is similar as that in the Kelvin's foam while pressure drop in DHSA is slightly lower than in Kelvin's foams. The friction factor in DHSA foams is independent of the binder angle, apart from a fair increase in the foam with the largest investigated angle. The friction factor in the Kelvin's foam is less dependent on the Reynolds number than it is in DHSA foams. Temperature gradients at the solid/fluid interface of DHSA foams are far higher at lower binders; the binders are more stressed than the shell, with shear stresses larger in the binder with a smaller angle. This can be attributed to the higher local velocities faced when the shells are closer.

5. References

[1] Dukhan N (Ed.) 2013 *Metal foams: fundamentals and applications* (DEStech Publications, Inc)
[2] Ashby M F and Gibson L J 1997 *Cellular solids: structure and properties* (Press Syndicate of the University of Cambridge, Cambridge, UK) p 175
[3] Bhattacharya A and Mahajan R L 2002 *J. Electron. Packag.* **124(3)** 155-163
[4] Younis L B and Viskanta R 1993 *Int. J. Heat Mass Tran.* **36(6)** 1425-1434
[5] Bodla K K, Mur J Y and Garimella S V 2010 *Numer. Heat Tr. A –Appl.* **58(7)** 527
[6] Iasiello M, Cunsolo S, Oliviero M, Harris W M, Bianco N, Chiu W K S and Naso V 2014 *J. Heat Trans.* **136(11)** 112601
[7] Wu Z, Caliot C, Flamant G and Wang Z 2011 *Int. J. Heat Mass Tran.* **54(7-8)** 1527
[8] Ambrosio G, Bianco N, Chiu W K S, Iasiello M, Naso V and Oliviero M 2016 *Appl. Therm. Eng.* **103** 333
[9] Kaur I and Singh P 2021 *Int. J. Heat Mass Tran.* **178** 121600
[10] McDonough J R 2020 *Therm. Sci. Eng. Prog.* **19** 100594
[11] Dixit T, Nithiarasu P and Kumar S 2021 *Int. J. Therm.l Sci.* **159** 106607
[12] Liang D, Bai W, Chen W and Chyu M K 2020 *Int. J. Heat Mass Tran.* **153** 119579
[13] Sun M, Yan G, Hu C, Zhao J Duan, F and Song Y 2023 *Appl. Therm. Eng.* **219** 119433
[14] Khalil M, Ali M I H, Khan K A and Al-Rub R A 2022 *Case Stud. Therm. Eng.* **38** 102313
[15] Dilberoglu U M, Gharehpapagh B, Yaman U and Dolen M 2017 *Procedia Manuf.* **11** 545
[16] Jamshidinia M and Kovacevic R 2015 *Surf. Topogr.* **3(1)** 014003
[17] Ventola L, Robotti F, Dialameh M, Calignano F, Manfredi D, Chiavazzo E and Asinari P 2014 *Int. J. Heat Mass Tran.* **75** 58
[18] Gurra E, Iasiello M, Naso V and Chiu W K S 2023 *J. Therm. Sci. Eng.. Appl.* **15(4)** 041002
[19] Iasiello M, Cunsolo S, Bianco N, Chiu W K S and Naso V 2018 *J. Enhanc. Heat Transf.* **25(4-5)** 333

Acknowledgments

This work was carried out within a Cooperation Agreement between the University of Connecticut (USA) and the Università degli Studi Federico II (Italy) and was supported by Italian Government MIUR Grant No. PRIN-2017F7KZWS.

Journal of Physics: Conference Series **2685** (2024) 012008 doi:10.1088/1742-6596/2685/1/012008

Assessment of a Flow-dependent Subgrid Characteristic Length for Large-Eddy Simulation on Anisotropic Grids

V D'Alessandro[1], Y Delorme[2], M Falone[1], M Wasserman[2] and R Ricci[1]

[1] Dipartimento di Ingegneria Industriale e Scienze Matematiche
Università Politecnica delle Marche
Via Brecce Bianche 12, 60131 Ancona (AN), Italy
[2] Toga Networks, a Huawei company, Tel–Aviv, Israel

E-mail: v.dalessandro@univpm.it

Abstract.
This paper presents the latest results of a long track development activity in the context of low-dissipative finite volume method for compressible flows. Specifically, here we focus our attention on the Large-Eddy Simulation (LES) approach which can be considered a good candidate for turbulent flow simulations over the next decades. One of the key ingredients of LES models is the subgrid length scale which is typically evaluated based on the local mesh size. This standard approach suffers from loss of accuracy on anisotropic grids that are commonly employed to obtain sufficient wall-normal resolution, whilst keeping the total cell count to a minimum. In order to avoid this issue, we assess the effectiveness of a velocity-gradient-based length scale, referred to as least square length (LSQ) [1]. In this paper, we present for the first time results obtained with the LSQ length scale in the context of compressible LES. The superiority of the LSQ approach over the standard cubic-root length scale is demonstrated in terms of accuracy and overall time to solution.

1. Introduction

Direct numerical simulations (DNS) of transitional/turbulent flows are still not achievable for practical applications due to the required computational effort. For this reason, the Large-Eddy Simulation (LES) approach is commonly adopted as the best compromise between solution accuracy and required resources. LES models require the evaluation of a sub–grid length scale which is commonly related to the local mesh size. Standard, geometry-based length scales can strongly affect the accuracy of the solution when anisotropic grids are employed [1]. This issue can be solved by increasing the overall number of cells, with the aim of reducing the cell anisotropy. However, this incurs a significant increase in the required computational resources. In order to propose a more affordable solution, Trias et al. [1] proposed a velocity-gradient dependent length scale, termed least square length (LSQ). The LSQ subgrid length scale, was demonstrated to minimize the effects of cell anisotropy on the accuracy of LES of incompressible flows. In this work, the effectiveness of the above described filter length was assessed in a compressible solver, for the well-established test case concerning the flow over a circular cylinder operating at a Reynolds number equal to 3900. This test case was selected thanks to the vast amount of available experimental and numerical results. The previously published results put in evidence

that there is a flow bifurcation in the very near wake region. So, it can be considered a reliable test for the sake of benchmarking the proposed numerical simulation technique.
This paper is organized as follows: in Sec. 2 the governing equations are presented, Sec. 3 provides a brief description of the discretization technique, Sec. 4 contains the numerical results and Sec. 5 summarizes the conclusions.

2. Governing equations

The flow model considered in this work is compressible LES in which continuity, momentum conservation and energy conservation equations are solved as follows:

$$\begin{aligned}
&\frac{\partial \overline{\rho}}{\partial t} + \boldsymbol{\nabla} \cdot (\overline{\rho}\tilde{\mathbf{u}}) = 0 \\
&\frac{\partial}{\partial t}(\overline{\rho}\tilde{\mathbf{u}}) + \boldsymbol{\nabla} \cdot (\overline{\rho}\tilde{\mathbf{u}} \otimes \tilde{\mathbf{u}}) = -\boldsymbol{\nabla}\overline{p} + \boldsymbol{\nabla} \cdot (\boldsymbol{\tau} - \boldsymbol{\tau}_{SGS}) \\
&\frac{\partial}{\partial t}\left(\overline{\rho}\tilde{E}\right) + \boldsymbol{\nabla} \cdot \left(\overline{\rho}\tilde{\mathbf{u}}\tilde{E}\right) = -\boldsymbol{\nabla} \cdot (\overline{p}\tilde{\mathbf{u}}) + \boldsymbol{\nabla} \cdot (\boldsymbol{\tau} \cdot \tilde{\mathbf{u}}) - \boldsymbol{\nabla} \cdot \overline{\mathbf{q}} + \dot{\tilde{E}}_{SGS}
\end{aligned} \tag{1}$$

where the overbar denotes the Reynolds averaging the tilde is the Favre averaging. In eq. 1 $\overline{\rho}$ denotes the density, $\tilde{\mathbf{u}}$ is the velocity vector, $\overline{p}$ is the pressure and $\tilde{E}$ denotes the total energy which is related to the other flow variables in the following fashion:

$$\overline{\rho}\tilde{E} = \frac{\overline{p}}{\gamma - 1} + \frac{1}{2}\tilde{\mathbf{u}} \cdot \tilde{\mathbf{u}} + \frac{1}{2}tr\left(\boldsymbol{\tau}_{SGS}\right). \tag{2}$$

In the implementation here presented, the last term in eq. 2 is not included in the solver since it produces a negligible impact on the obtained solution at low Mach numbers, [2, 3]. On the other hand, the closure of eq. 1 is achieved through the constitutive equation for Newtonian fluids, Fourier law and the equation of state for the perfect gas. Lastly, the $\dot{\tilde{E}}_{SGS}$ contribution in eq. 1 is handled in the following way, [2]:

$$\dot{\tilde{E}}_{SGS} = -\lambda\boldsymbol{\nabla}\tilde{T} \tag{3}$$

since sub–grid scales (SGS) turbulent diffusion, and the SGS viscous diffusion can be considered negligible, [2].
The anisotropic part of the SGS stresses, $\boldsymbol{\tau}_{SGS}$, is parametrized using a standard approach:

$$\boldsymbol{\tau}_{SGS} - \frac{1}{3}tr\left(\boldsymbol{\tau}_{SGS}\right)\mathbf{I} = -2\mu_t\left[\tilde{\mathbf{S}} - \frac{1}{3}\left(\boldsymbol{\nabla} \cdot \tilde{\mathbf{u}}\right)\mathbf{I}\right] \tag{4}$$

where $\tilde{\mathbf{S}}$ is the strain rate tensor and the SGS viscosity, μ_t, is obtained using the following equation:

$$\mu_t = C_S^2\Delta^2\overline{\rho}\left|\tilde{\mathbf{S}}\right|. \tag{5}$$

In eq. 5 C_s is the so–called Smagorinsky constant, Δ is the sub–grid length scale and $\left|\tilde{\mathbf{S}}\right|$ is the magnitude of the strain rate tensor. By contrast, the dynamic procedure proposed by Germano-Lilly, [4], was adopted to compute C_s. As regards to the sub–grid length scale, we have employed two different approaches. The first approach is based on the cube root of the cell volume:

$$\Delta_{cbrt} = \sqrt[3]{V_P} \tag{6}$$

While the second approach is the least square length (LSQ), proposed in [1]:

$$\Delta_{lsq} = \sqrt{\frac{\mathbf{\Delta_D G^T G} : \mathbf{\Delta_D G^T G}}{\mathbf{G^T G} : \mathbf{G^T G}}}. \tag{7}$$

In the above equations V_P is the cell volume. Differently, LSQ sub-grid scale is built starting from the velocity gradient tensor $\mathbf{G}$ and the length scales tensor $\mathbf{\Delta_D} = (\Delta x, \Delta y, \Delta z)\,\mathbf{I}$. It is also very important to put in evidence that the length scales $\Delta x, \Delta y, \Delta z$ appearing in $\mathbf{\Delta_D}$ should be the same as the discrete lengths used to obtain the discrete velocity gradient [5]. Using Green's theorem, the generic cell P velocity gradient components are calculated as follows:

$$\left(\frac{\partial u_i}{\partial x_j}\right)_P = \frac{1}{2V_P} \sum_{N \in N(P)} \left[(u_i)_P - (u_i)_N\right] S_{PN} \left(n_{x_j}\right)_{PN} \tag{8}$$

with $\left(n_{x_j}\right)_{PN} = [n_x, n_y, n_z]^T_{PN}$ denoting the outward–pointing unit normal vector to the interface PN which is shared by the cell P and its neighbour cell N, The term S_{PN} denotes the face area of the interface PN and $N(P)$ is the set of cell P's direct neighbors. Lastly, the corresponding cell length scales, $\Delta x_j = [\Delta x, \Delta y, \Delta z]$ are given by:

$$(\Delta x_j)_P = \frac{2V_p}{\sum_{N \in N(P)} S_{PN} \left|\left(n_{x_j}\right)_{PN}\right|}. \tag{9}$$

3. Numerical approximation

A standard, cell–centered finite volume method, available within the OpenFOAM framework, is adopted to discretize the governing equations described in Sec. 2. Convective terms are evaluated using Pirozzoli's second–order accurate, energy–conserving scheme [6], while a standard central-differencing scheme is adopted for the diffusive contributions.

For time integration, a compact-storage (2N) explicit Runge–Kutta (ERK) scheme was implemented thanks to its benefits for large scale computations. In particular we have considered a five-stage, fourth-order accurate, ERK scheme available in the open literature, [7]. Our computational experience showed that this scheme is extremely appealing since it allows to use a maximum Courant number equal to 1. Thus, it allows to exploit the good parallel performance of the explicit approach, without imposing a severe time–step restriction.

All the aforementioned numerical methods were implemented in a freely–available OpenFOAM-based solver, termed `caafoam`, which is extensively described in D'Alessandro et al. [8]

4. Results

The test case considered in this work concerns the flow past a circular cylinder operating at a Reynolds number, Re, based on the cylinder diameter D, equal to 3900 [9, 10], and free–stream Mach number equal to 0.2. This is a well-established benchmark for evaluating the simulation capabilities of separated and transitional flows. The flow is characterized by a separated shear layer that detaches from the cylinder surface in a laminar regime and becomes turbulent downstream of the bluff body, as depicted in Fig. 1.

Moreover, a wide array of experimental and numerical results put in evidence that there is a flow bifurcation in the very near wake region: one flow state is characterized by the presence of a V–shaped profile of the stream-wise velocity [11], while a second state reveals the presence of a U–shaped profile, [12]. Both states have been recognized in many LES and DNS computations. Recent literature studies, [13], confirmed that the U–shaped state is correct and that the V–shaped profile occurs both in experimental analyses and in numerical simulations because of an early laminar-to-turbulent transition of the separated shear layer. It is important to note that obtaining the correct solution of a U–shaped profile is not a trivial task. This specific flow problem has consequently become a popular case for testing LES solvers and associated numerical schemes.

In this work the computational domain extends for 38 D in the radial direction starting from the

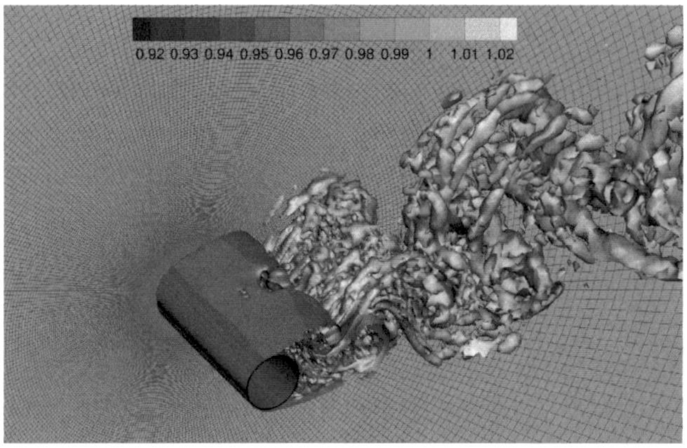

(a) Isosurface coulourd by pressure

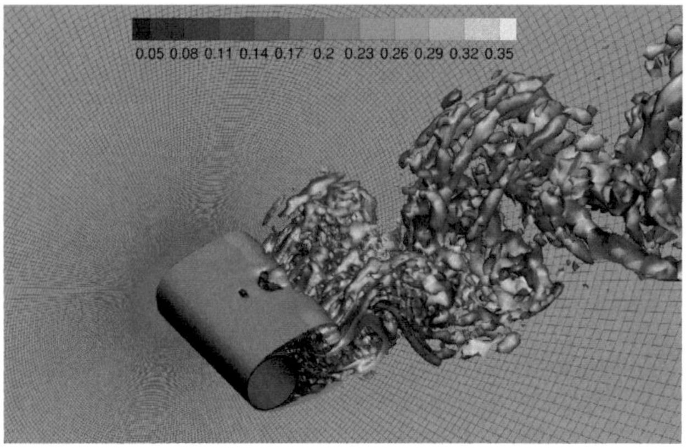

(b) Isosurface coloured by Mach number magnitude

Figure 1. Vortical structures, identified by the Q–criterion.

Table 1. Computational grids features.

Grid	N_θ	N_r	N_z	N_{tot}
$G1$	300	330	64	$6.336 \cdot 10^6$
$G2$	300	330	48	$4.725 \cdot 10^6$
$G3$	300	330	32	$3.168 \cdot 10^6$

cylinder centre and π D in the spanwise direction. A set of fully structured O–type grids were generated using $N_\theta = 300$ cells in the circumferential direction, $N_r = 330$ cells are employed in the radial one, while the span was discretized using a varying number of cells, N_z, ranging from 32 to 64, as summarized in Table 1. Computational cells were clustered near the cylinder surface and the dimensionless height of the first cell next to the wall was imposed to be 10^{-3}.

An artificial sponge layer was also adopted as non-reflecting boundary treatment to damp the solution to a reference value at the far-field, avoiding unphysical wave reflections. In the considered configuration, the sponge layer starts from the external boundary of the computational domain; its thickness, L_{sp}, is fixed to 13 D, leading to a dimensionless extension, $L_{sp} f / c_\infty \simeq 0.5$, where f is the vortices detachment frequency and c_∞ is the free-stream speed of sound. Moreover, as suggested in Mani et al. [14], a sponge strength η_{target} of 40 dB was chosen.

Finally, all the computations described below were run using 3 nodes of The TaiShan 200 cluster hosted by Huawei. Each node in the cluster contains 2x 64-core Huawei Kunpeng 920-6426 CPUs operating at 2.6 GHz (128 cores per node).

A first baseline computation was conducted using the G1 grid and adopting the dynamic Smagorinsky SGS model, in which the sub–scale filter length, Δ, is computed as the cubic root of the cell volume. The obtained results are compared with three different sets of experimental data from the literature, [11, 12, 15]. It is very clear that the above described approach is able to properly predict the U–shaped profile of the mean stream-wise velocity, Fig. 2, in the wake region at $x/D = 1.06$. Furthermore, the other velocity profiles are also in very good agreement with the reference literature data. Moreover, the effectiveness and the reliability of the proposed

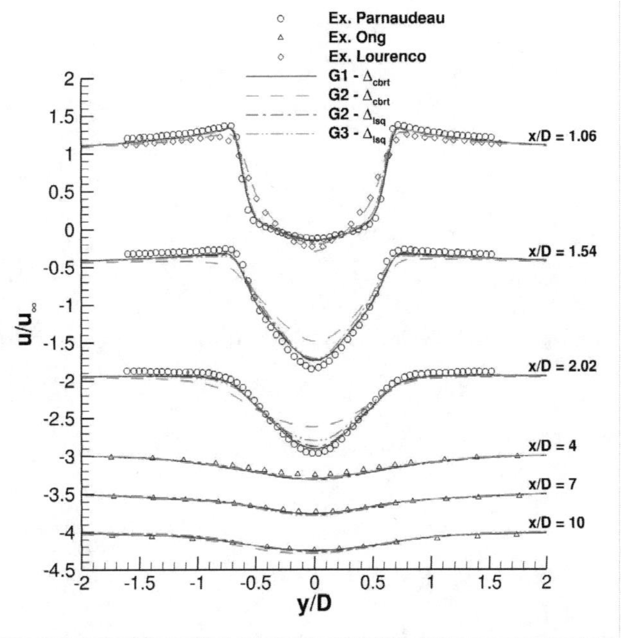

Figure 2. Mean streamwise velocity at different locations in the wake region.

solution technique is also confirmed by observing the mean transverse velocity (see Fig. 3), and the mean resolved streamwise Reynolds stresses in the wake region (see Fig. 4). It is worth noting that flow statistics were collected after the flow field was fully developed, by averaging a period of time corresponding to 500 non-dimensional units.

A second campaign of analysis was conducted adopting G2 and G3 grids, in order to investigate the possibility to reduce the computational load required to proper reconstruct the flow field developed around the 3D cylinder. The adopted SGS model was also the dynamic Smagorinsky in this case, while for what concerns the sub-scale filter length, it was evaluated both with the cubic root of the cell volume and the least square technique. Looking at the mean velocity profiles in Fig. 2, it is evident that the adoption of Δ_{cbrt} results in a V–shaped profile in the wake region at $x/D = 1.06$, and the produced solution does not match Parnaudeau's data for the mean transverse velocity. On the other hand, the least square filter length does not exhibit a similar sensitivity to spanwise resolution. The LSQ filter enables to increase the grid anisotropy without compromising the solution accuracy, thus resulting in a reduction of up to 50% in the total number of cells, with respect to those required with the cubic root filter length. The mean streamwise transverse resolved Reynolds stresses depicted in Fig. 4 confirm the effectiveness

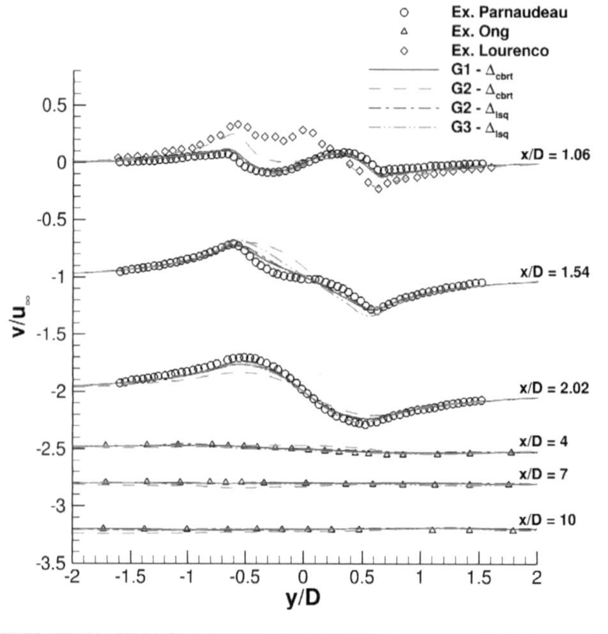

Figure 3. Mean transverse velocity at different locations in the wake region.

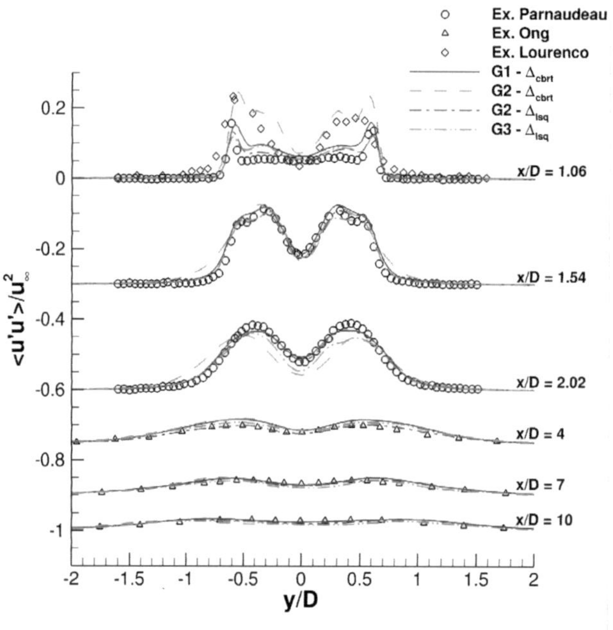

Figure 4. Mean transverse velocity at different locations in the wake region. Streamwise direction

Table 2. Drag and lift coefficients.

	$G1-\Delta_{cbrt}$	$G2-\Delta_{lsq}$	$G3-\Delta_{lsq}$	Lysenko et al.
$\langle C_D \rangle$	1.010	1.003	1.019	0.97
$C_{L_{rms}}$	0.186	0.161	0.171	0.090

of adopting Δ_{lsq} when the spanwise resolutiong is decreased. In fact, the coarse G3 grid also produces acceptable results, but it requires more than double of the computed dimensionless time to converge to a stable solution, compared with G1 and G2.
Finally, Tab. 2 summarizes the mean drag coefficient, $\langle C_D \rangle$, and root mean square of the lift coefficient, $C_{L_{rms}}$, deriving from the above–described computations. Both $\langle C_D \rangle$ and $C_{L_{rms}}$ are slightly overestimated if compared with Lysenko's results, in all the adopted approaches. Nevertheless, the overall agreement with the reference can be considered satisfactory.

5. Conclusions

In this paper we have presented several sets of LES computations of the flow field developing over a cylinder operating at Re = 3900 and a free–stream Mach number equal to 0.2. The main goal of the paper is to assess the impact of different sub-grid characteristic lengths on the compressible LES simulation accuracy. In this context, it was noted that the adoption of the flow-dependent, least square length introduced by Trias et. [1] offers a clear advantage over the standard, geometry-based (cube root) length scale, in terms of accuracy. Indeed, the evaluation of the LSQ length increases the number of floating-point operations (FLOPS) with respect to the simpler cube root length. However, it enables to significantly reduce the overall number of grid cells that are required to correctly capture the complex physics of separated flow. Overall, our measurements demonstrated that the total time-to-solution is in-fact reduced when using the LSQ length scale, considering the combination of the two aforementioned effects (i.e increase in FLOPS and reduction in cell count). This behavior may be extremely appealing especially in the context of large-scale, industrial LES.

References

[1] Trias, F X and Gorobets, A and Silvis, M H and Verstappen, R W C P and Oliva, A 2017 *Physics of Fluids* **29**

[2] Pino Martín M, Piomelli U and Candler G V 2000 *Theoretical and Computational Fluid Dynamics* **13** 361–376

[3] Garnier, E and Adams, N and Sagaut, P 2009 *LES Governing Equations* (Dordrecht: Springer Netherlands) pp 5–39

[4] Germano M, Piomelli U, Moin P and Cabot W H 1991 *Physics of Fluids A: Fluid Dynamics* **3** 1760–1765

[5] Mor-Yossef, Y and Kidron, Y and Wasserman, M and Levy, Y 2021 *Aerospace Science and Technology* **113** 106700

[6] Pirozzoli S 2011 *Annu. Rev. Fluid Mech.* **43** 163–94

[7] Kennedy C, Carpenter M and Lewis R 2000 *Appl. Numer. Math.* **35** 177–219

[8] D'Alessandro V, Falone M and Ricci R 2020 *Comput. Fluids* **203** 104517

[9] Lysenko, D A and Ertesvåg, I S and Rian, K E 2012 *Flow, Turbulence and Combustion* **89** 491–518

[10] D'Alessandro, V and Montelpare, S and Ricci, R 2016 *Computers & Fluids* **136** 152–169

[11] Lourenco L and Shih C 1993 Characteristics of the plane turbulent near wake of a circular cylinder, a particle image velocimetry study published in: Beaudan, P. and Moin, P., Report No. TF62, Thermosciences Division, Department of Mechanical Engineering, Stanford University

[12] Parnaudeau P, Carlier J, Heitz D and Lamballais E 2008 *Physics of Fluids* **20**

[13] Moin P 2014 *VORTICAL STRUCTURES AND WALL TURBULENCE* ed Verzicco R (Rome)

[14] Mani A 2012 *J. Comput. Phys.* **231** 704–16

[15] Ong L and Wallace J 1996 *Experiments in Fluids* **20** 441–453

Investigation of a thermomechanical improved tangential fan geometry

J Hauch[1,*], K Reich[1], D Büschgens[1] and H Pfeifer[1]

[1] RWTH Aachen University - Department for Industrial Furnaces and Heat Engineering, Kopernikusstr. 10, 52074 Aachen, GERMANY

* Corresponding author: hauch@iob.rwth-aachen.de

Abstract. In convection-dominated thermoprocessing plants, the current state of the art is to use centrifugal and axial fans in conjunction with flow straighteners to ensure a homogeneous inflow of the fluid and thus a homogeneous temperature distribution. However, flow straighteners lead to an additional pressure loss in the system that must be overcome. Tangential fans can deliver a homogeneous volume flow over the entire width of the fan even without flow straighteners. Therefore, they offer a possibility to increase the energy efficiency of these systems. Based on this motivation, the possibilities of increasing the thermomechanical long-term stability of tangential fans are being evaluated as part of a current research and development project. To this end, the effects of design changes on the generated flow and the mechanical loads are being investigated with the help of FEM and CFD simulations. The aim is to design a tangential fan that can withstand the mechanical and dynamic loads and deliver a consistently high and homogenous fluid flow at the same time. A geometry based on the results of the simulations was manufactured as a functional sample. This sample was investigated at the hot test stand of the IOB and is compared to a reference tangential fan.

1. Introduction

Precise temperature control to achieve optimum material properties is the main requirement of a modern thermoprocessing plant. To achieve this goal, high convection furnaces with forced circulation are used for the temperature range up to approx. t = 800 °C. These plants are used for heat treatment, especially in the aluminium and copper industries. Axial or centrifugal fans are used for forced circulation, which generate an inhomogeneous volume flow due to their mode of operation. For this reason, flow straighteners are used to ensure a homogeneous flow through the material and thus a homogeneous temperature distribution within the product. The flow straighteners reduce the cross-sectional area through which the flow passes. The resulting pressure losses lead to a reduced energy efficiency of the overall system [1-5].

Currently tangential fans are mainly used in room air-conditioning technology at ambient temperatures, in the food industry or in thermo processes with temperatures below t = 300 °C. They draw in the air radially and blow it out radially across the entire width. The resulting homogeneous volume flow appears optimal for the use in industrial furnaces without flow straighteners [6,7].

As part of a former research project, the possibility of using a tangential fan in thermoprocessing plants was investigated at the Department for Industrial Furnaces and Heat Engineering (IOB). For this purpose, a hot test stand was set up and the flow behaviour of a tangential fan was investigated up to a temperature of t = 500 °C. The results of the investigation showed that a tangential fan can be used without a flow straightener and the required homogeneous velocity distribution can be generated at a sufficiently high volume flow. Therefore, the investigation showed that a tangential fan can be used in a thermoprocessing plant overcoming the current limitations of t = 300 °C. After a short period of operation, the tangential fan failed due to dynamic loads such as vibrations and oscillations. This resulted in the need for design modifications to ensure thermomechanically stable long-term operation in thermoprocessing plants [8,9].

2. Simulation setup

All simulations are carried out with commercial simulation software. ANSYS Mechanical is used for the FEM simulations and ANSYS Fluent for the CFD simulations. With the help of FEM simulations, the effect of design changes on the static load situation of the tangential fan at n = 1500 rpm are investigated. An isolated 3D model of the cross-flow rotor is used for this purpose. The effects of the design changes are then investigated for their fluid mechanical influence using CFD. For this purpose, a 3D model is used that depicts the tangential fan as well as the guide housing and an inlet and outlet area. The boundary conditions for these simulations are a rotational velocity of n = 1500 rpm, a free pressure outlet and a pressure inlet with variable inlet pressure. The k-ω shear stress transport (SST) model is used for the simulation as it is less susceptible to turbulence at the pressure inlet than the standard k-ω model. Additionally, it is less prone to turbulence overestimation in the exit zone of the fan blades than the k-ε model [10].

3. Experimental setup

The hot test stand at the IOB, as shown in figure 1, was used for the investigations of two different types of tangential fans. The hot test stand is a thermally insulated closed flow channel. A motor controlled by a frequency inverter drives the fan. The resulting flow, shown here as a blue arrow, flows through a heating register that allows for heating the fluid flow to temperatures of up to t = 550 °C. There are five flow measuring points each in the lower as well as in the upper part of the flow channel. With the help of S-Pitot tubes, velocity fields can be recorded at these points. For this purpose, the S-Pitot tubes are moved in height by a linear traverse. By adjusting throttle valves by means of a linear actuator, different operating points of the fan can be investigated and fan characteristics can be recorded. To determine the pressure loss through the heating register and the pressure increase through the fan, the static pressure at the relevant positions is measured.

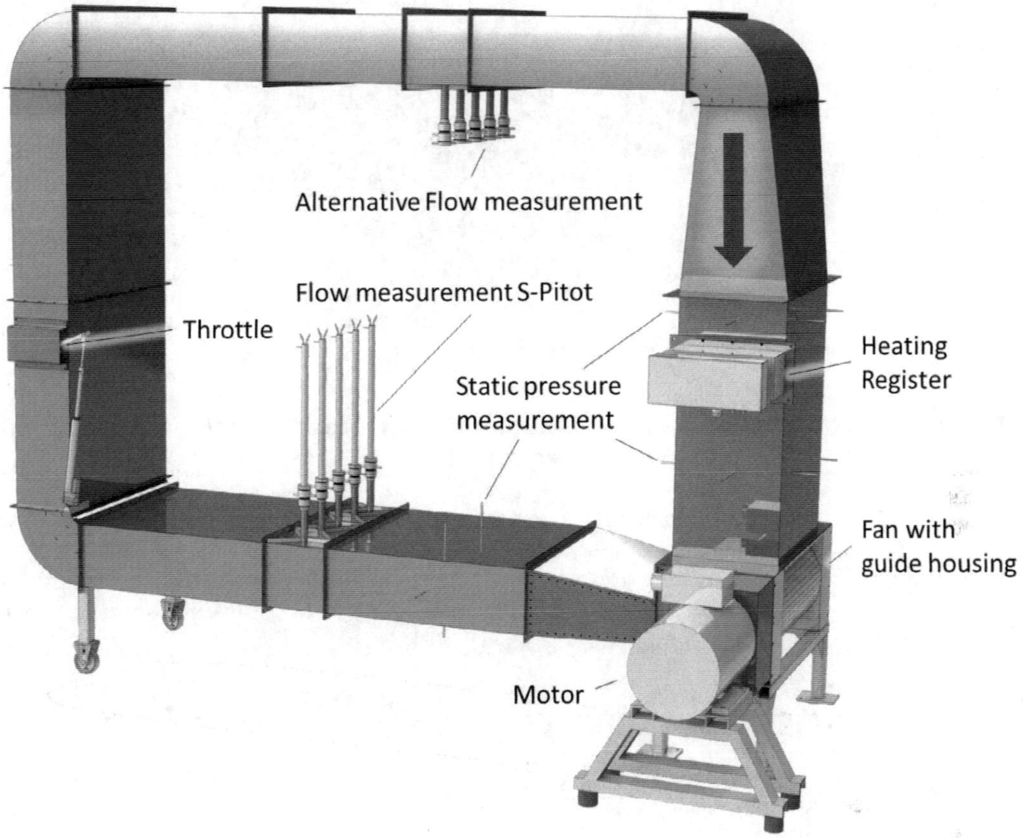

Figure 1. Hot test stand at the IOB

4. Results

This chapter presents selected results of the FEM and CFD simulations that led to the manufacture of a functional sample. Afterwards, the results obtained with this functional sample on the hot test stand are presented and compared with a reference tangential fan.

4.1. FEM Simulation

The reference tangential fan consists of a total of six support discs and 52 blades, with a thickness of 1.5 mm. This is driven on one side and mounted friction-free on both sides. Due to the free path length between blades and supporting disks, deformations of the blades occur between them during operation. This in turn leads to stress maxima at the clamping points in the support discs, as shown in figure 2. These stress maxima occur at all clamping points and amount to up to $\sigma_{V,max}$ = 119 MPa at the peak. Due to the equal distribution of the support discs, a maximum of deformation of the blades results between the outer support discs. To reduce this, the distance between the outer support discs can be reduced to 0.93 times the length of the inner support discs, resulting in a decrease in stress to $\sigma_{V,max}$ = 107 MPa.

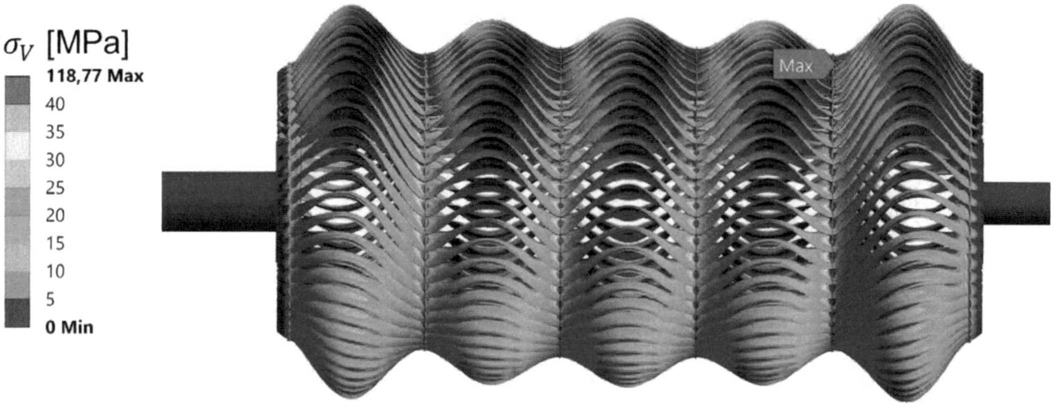

Figure 2. von Mises equivalent stress distribution with over scaled deformation at the reference tangential fan

The correlation between the number of support discs and the maximum equivalent stress is shown in figure 3 a). In this case the thickness of the support discs and blades was kept constant at 1.5 mm. It can be seen that the maximum equivalent stress decreases exponentially with increasing number of support discs. Increasing by four the six support discs of the reference fan to ten discs lowers the maximum equivalent stress by two-thirds to approx. $\sigma_{V,max}$ = 35 MPa.

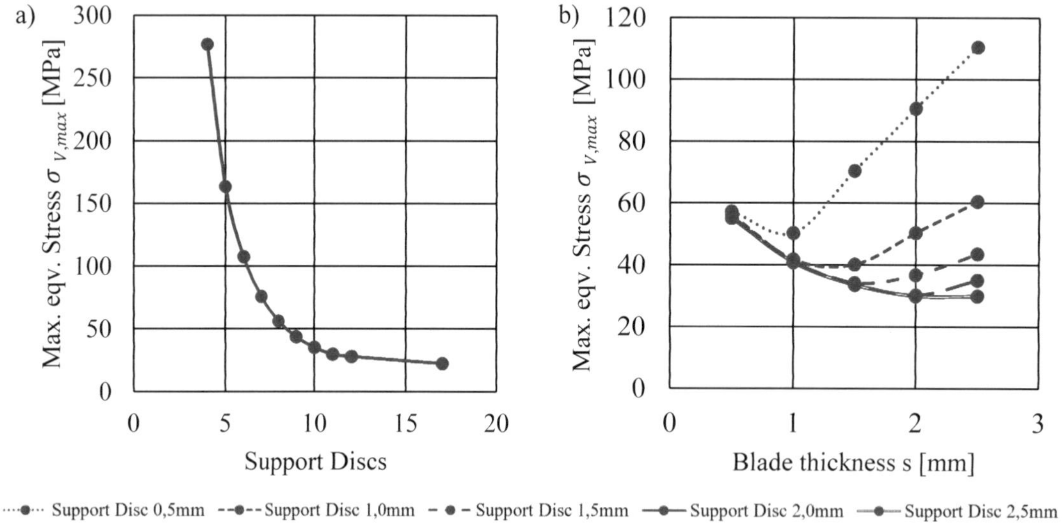

Figure 3 Maximum von Mises equivalent stress with a) different number of support discs and b) different steel sheet parameters

Based on a new model with ten support discs, the influence of the support disc and blade sheet thickness was investigated, shown in figure 3 b). With a support disc thickness of s = 1.5 mm or more, an optimum appears when both sheet thicknesses are identical. With lower support disc thicknesses, the optimum shifts to the next higher blade thickness. Overall, it can be seen that the combination of identical sheet thicknesses for supporting discs and blades is the best solution.

4.2. CFD Simulation

The averaged velocity distribution over the width of the reference tangential fan shown in figure 4 shows the typical S-shaped flow through the fan. The air is sucked in through the pressure inlet flows through the tangential fan and is blown out through the pressure outlet after a 90° deflection. Thereby two turbulence zones are formed. One in the upper area of the fan and one in the lower area which prevents a backflow into the inlet area due to the "nose" of the guide housing. The maximum flow velocity is reached in the lower area of the cross-flow fan at approx. v = 150 m/s and the flow is directed towards the flow channel ceiling. A volume flow rate of $\dot{V}$= 8.72 m³/s is achieved here.

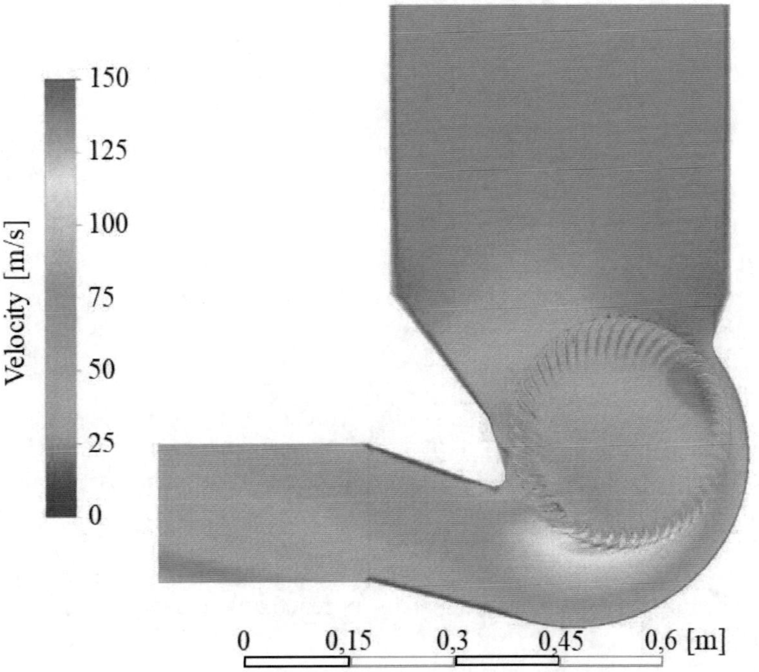

Figure 4. Velocity distribution of the reference tagential fan from the numerical model

As known from the literature, centre shafts in tangential fans reduce the flow rate to a not inconsiderable extent by blocking the S-shaped flow through the centre. The velocity distribution in a tangential fan with a centre shaft measuring d = 50 mm in diameter is shown in figure 5 a). It can be seen that the maximum velocity drops to approx. v = 110 m/s after the flow has passed the tangential fan.

As a new approach, a through-flow capability of such a centre shaft was examined. For this purpose, a centre shaft was designed as a tube and provided with holes. This should help to utilise the positive effects on the static and dynamic stability of a centre shaft on the tangential fan and at the same time reduce the flow-mechanical disadvantages. The result of this simulation is shown in figure 5b). The maximum flow velocity drops to approx. v = 130 m/s. Compared to that of the reference fan, the volume flow rate is approx. 18.5 % lower for a centre shaft and only 8 % lower for a perforated centre shaft.

 2685 (2024) 012009 doi:10.1088/1742-6596/2685/1/012009

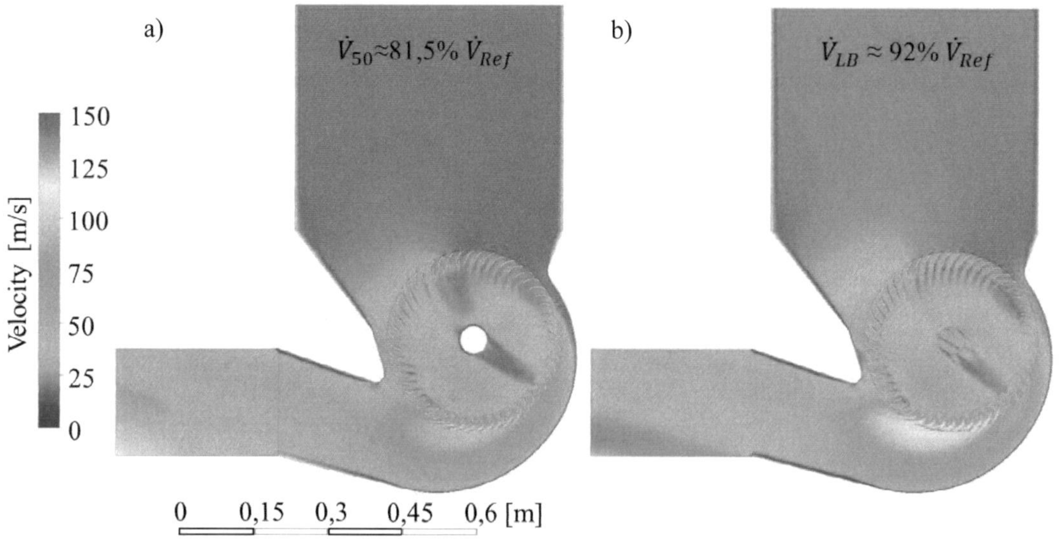

Figure 5. Velocity distribution of a tangential fan with a) a 50mm diameter centre shaft and b) a 50mm diameter centre shaft with punch holes

4.3. Experimental validation

Fan curves for the reference tangential fan were recorded at the IOB hot test stand for various temperatures and speeds; these are shown in figure 6. Also shown is the fan curve for the numerical model at n = 1500 rpm. It can be seen that the experimental data validates the numerical models, especially at high volume flow rates of approx. $\dot{V}$ = 5 m³/s and upwards. The deviation towards higher total pressure increases of the simulation model at lower volume flow rates can be explained by the deviating boundary conditions. While the numerical model uses a negative inlet pressure for the calculation, the operating range is set experimentally by generating counterpressure through the throttle valves at the discharge area.

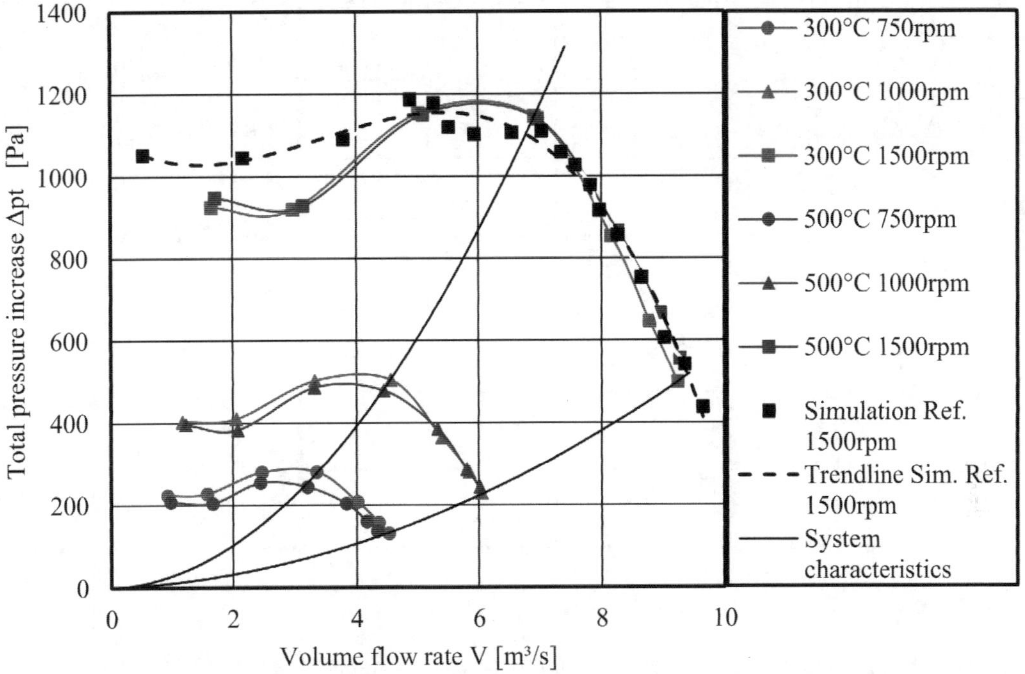

Figure 6. Experimental fan and system characteristics for the reference tangential fan in comparison to the numerical solution at n = 1500 rpm

Based on these results, a functional model was designed and built that, in addition to increasing the number of support discs and the sheet thickness of support discs and blades, includes a flow-through centre shaft. The fan characteristics of this functional model in comparison with the reference tangential fan and the numerical model of the functional model are shown in figure 7. All changes combined lead to a significant deterioration of the fan curve at all rotational velocities. This is mainly explained by the significantly higher sheet thickness of the blades, as these significantly increase the area blocked in the direction of rotation. Unfortunately, this had to be accepted for manufacturing reasons. The comparison between the numerical model and the experimental results at n = 1500rpm again shows sufficient congruence, but the numerical solution slightly overestimates the overall fan curve.

5. Conclusion

With the help of the results presented here, the influence of various reinforcing elements on the stability of a tangential fan could be shown. In addition, the influence of these reinforcing elements on the fluid flow was investigated and validated by means of experimental tests. The greatest innovation here was the use of a flow-through centre shaft. In order to be able to conclusively prove its effectiveness the centre shaft is to be blocked and investigated in further studies.

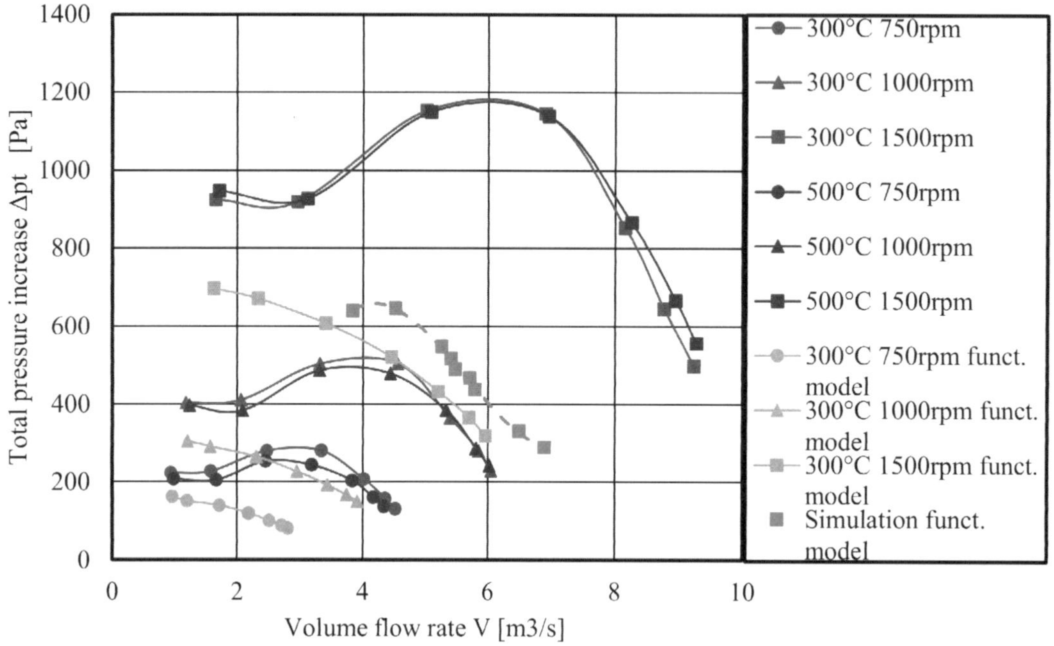

Figure 7. Experimental fan characteristics for the reference tangential fan and the functional model and the comparison to the numerical model of the functional model

6. References

[1] Bommes L *et al.* 2022 *Ventilatoren* vol 2, ed Bommes L, Fricke J, Grundmann R (Essen: Vulkan-Verlag)

[2] Dornstetter S 2002 *Numerische und experimentelle Untersuchungen an Querstromventilatoren* (Karlsruhe: University Karlsruhe Dissertation)

[3] Pfeifer H, Nacke B, Beneke F, 2018 *Handbook of Thermoprocessing Technologies, Volume 1: Fundamentals, Processes, Calculations* vol 3 (Essen: Vulkan-Verlag)

[4] Carolus T 2013 *Ventilatoren* vol 3, ed Carolus T (Wiesbaden: Springer Vieweg)

[5] Specht E and Becker F 2020 *Dubbel Taschenbuch für den Maschinenbau Volume 3 Maschinen und Systeme* vol 26, ed B Bender and D Göhlich (Wiesbaden Springer Vieweg) chapter 29-33 pp 603-662

[6] Klemm T 2005 *Numerische und experimentelle Untersuchungen an Ventilatoren mit hoher Leistungsdichte* (Karlsruhe: University Karlsruhe Dissertation)

[7] Sauter H and Engel F. 2018 *Querstromventilatoren im Industrieofenbau (Prozesswärme* vol 2*)* (Essen: Vulkan-Verlag) pp 82-88

[8] Strämke S and Pfeifer H 2018 *Numerische und experimentelle Untersuchungen von Querstromventilatoren (Prozesswärme* vol 4*)* (Essen: Vulkan-Verlag) pp 35-40

[9] Strämke S and Pfeifer H 2018 *Entwicklung von Querstromventilatoren für den Einsatz in Thermoprozessanlagen* (Aachen: RWTH Aachen University Dissertation)

[10] ANSYS Inc. 2022 *ANSYS Fluent Theory Guide, Release 2022 R1* ed ANSYS Inc. (Cannonsburg)

Acknowledgments

The project (project no. 20783 N) was submitted with the support of the Research Association of Industrial Furnace Manufacturers (FOGI) via the Forschungskuratorium Maschinenbau e.V. (FKM). It is funded via the German Federation of Industrial Research Associations „Otto von Guericke" (AiF) in the program for the Promotion of Industrial Research (IGF), initiated by the Federal Ministry for Economic Affairs and Climate Action due to a resolution of the German Bundestag.

Numerical and experimental study to examine methods to increase the convectional heat transfer of recuperator tubes

E Trampe*, J Hauch[1], D Büschgens[1], H Pfeifer[1]

[1] RWTH Aachen University - Department for Industrial Furnaces and Heat Engineering, Kopernikusstr. 10, 52074 Aachen, GERMANY

* Corresponding author: trampe@iob.rwth-aachen.de

Abstract. The use of structured tubes in tube bundle recuperators is intended to improve the energy efficiency of gas-fired industrial furnace systems with central air preheating. Currently, smooth tubes as well as smooth tubes with internal twisted tape are commonly utilised. The increased surface area of structured tubes leads to an increased heat transfer between the off-gas and the combustion air. However, structuring leads to an increased pressure loss. The aim is to find an optimal operating range in which the pressure losses are still acceptable and the greatest increase of heat transfer compared to the reference will be achieved. Smooth tubes and also smooth tubes with twisted tapes are considered as reference systems in comparison to two concave structured tubes. CFD simulations were used to investigate the influence of the operating parameters of combustion air velocity and off-gas velocity on heat transfer according to industrial scale. The simulations have shown that the heat flow increases with the honeycomb depth and strongly depends on the operating conditions. Experimental measurements in a test bench were carried out to validate the model and confirm the positive influence of structuring on heat transfer.

1. Introduction

One way to make industrial furnaces more energy efficient is to install an upstream recuperator. The principle of heat recovery in a recuperator is based on two separate flows, whereby the heat is transferred from the medium with higher temperature to the one with lower temperature. Regardless of industrial furnaces in the metal, ceramics, or chemical industry, the combustion air can be preheated [1-3]. One way to make air preheating more efficient is to increase the heat transfer area between the two fluid flows and to increase the degree of flow turbulence. In many cases, the supplied combustion air is led through the tubes which are surrounded by the off-gas from outside. To increase the heat transfer, turbulence generators, twisted tapes, can be installed in smooth tubes. Another possibility is to apply structures to conventional smooth tubes which enlarge the surface area and intensify the turbulence.

Several different techniques have been used in the past to structure tubes for heat exchangers. One way to improve the surface roughness is to punch dimples in a metal sheet, which is then formed into a tube shape by rolling. Maithani and Kumar [4] examined such a dimpled surface with regard to the development of the correlation of Nusselt number and friction factor. Nusselt number and friction factor increase with a growing ratio of dimple depth to pressure diameter up to the value 1, after where they decrease. Aroonrat and Wongwises [5] confirms this with their own investigations on dimpled pipes. The heat transfer coefficient and the friction pressure loss rise with increasing dimple depth.

Studies have also been conducted on the influence of different twisted tape geometries. In their experimental study, Bas and Ozceyhan [6] investigated the influence of the twist ratio and the clearance ratio on the heat transfer and flow friction. They found that the twist ratio has a greater influence on the above-mentioned variables. Zheng et al. [7] have carried out a numerical study of heat transfer which investigates the performance and flow characteristics of structured tapes. Heat transfer is greatly improved owing to the disturbance to flow structures by the tape dimples.

This study addresses the investigation of concave structured tubes and compares the performance of the structured tubes to the two common cases of smooth tubes with and without twisted tape under identical operating conditions of recuperators. The inward structuring was deliberately chosen in order to realise a simple manufacturing with a forming tool and to create a maximum heat transfer area.

2. Materials and Methods

The investigations are carried out on two structured tubes and two reference geometries. Smooth tubes with and without internal twisted tape are used as reference geometries as they are commonly installed in recuperators. The tubes to be examined have an outer diameter of d_{tube} = 42.4 mm and a length of l_{tube} = 1000 mm. The tube wall thickness is x_{tube} = 2.5 mm. The reference tubes are made of available stainless steel X5CrNi18-10 (1.4301), which are seamlessly welded. The twisted tape has a diameter of d_{tape} = 30 mm and a length of l_{tape} = 1000 mm. The pitch of the tape is p_{tape} = 125 mm.

The structured tubes, figure 1, have been additively manufactured from X5CrNiMo17-12-2 (1.4401) stainless steel. They have the same dimensions as the smooth tubes plus the structuring which corresponds to a concave honeycomb structure. The width of the honeycomb is x_w = 6 mm. The honeycomb depth x_d differs for the two structured tubes, one amounts $x_{d,1}$ = 3 mm and the other one $x_{d,2}$ = 6 mm.

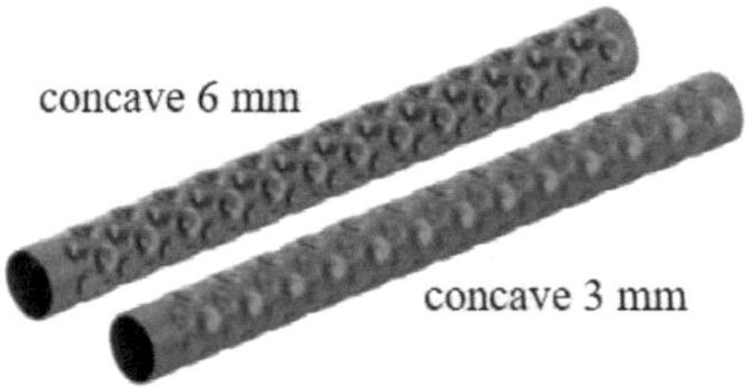

Figure 1. Sketch of the structured test tubes

The purpose of this investigation is to represent the heat exchange in a recuperator between the supplied combustion air and the off-gas. Since the experimental investigations cannot implement with a real off-gas composition, atmospheric air is used instead for both the experiments and the numerical model for comparability.

2.1. Numerical modelling

A simplified model of a tube bundle recuperator, figure 2, was used for the numerical investigations. It consists of a 3 x 3 tube bundle within a flow channel. The width of the channel corresponds to the tube length of l_2 = 500 mm. This means that the tubes are only half as long as in the experimental investigations. The reduced tube length is necessary due to computing time savings. The spacing between the tubes is 1.7 times the tube diameter d_t.

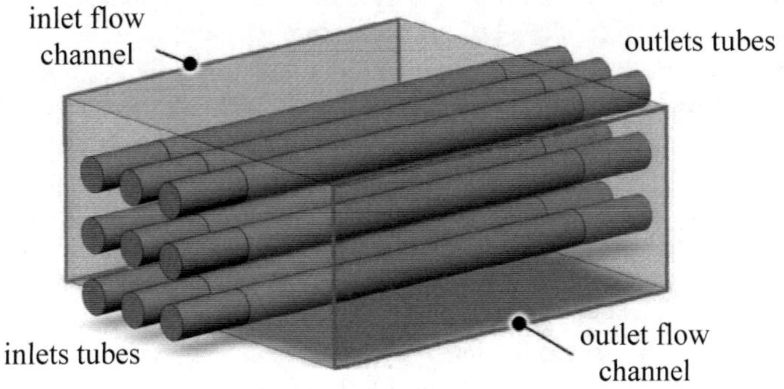

Figure 2. Geometry of the model; black/grey: wall, blue: velocity inlet, red: pressure outlet

The model describes two different fluid flows that are separated from each other. The off-gas enters the channel (blue rectangle) at a temperature $T_{off\text{-}gas}$ = 500 °C. The inlet velocity of the off-gas is varied between $v_{off\text{-}gas}$ = 1 m/s to $v_{off\text{-}gas}$ = 5 m/s. The red rectangle represents a pressure outlet of the off-gas domain. The combustion air is supplied at a temperature of T_{air} = 20 °C via the tube inlets (blue circles). This is followed by a 100 mm long intake section. Here, a uniform flow profile can be established before the actual test tube is flowed through. The combustion air velocity is varied in the range v_{air} = 1 m/s to v_{air} = 20 m/s. The air flows out of the tube via a pressure outlet. All pressure outlets have the boundary condition of a gauge pressure p_{gauge} = 0 Pa. The walls of the investigated tubes are modelled using shell conduction with one layer which have a virtual thickness of x_{layer} = 2.5 mm. All other walls are assumed to be adiabatic. For the numerical investigations, the realizable k-ε model of *Ansys Fluent 2022 R1* was used.

2.2. Experimental setup

For the investigation of the heat transfer between an off-gas and the combustion air, a flow channel is adapted with an exchangeable channel segment to examine one separate tube, see figure 3. The air flowing through the channel can be heated up to maximum of $T_{off\text{-}gas}$ = 500 °C by electrical heating elements inside the channel and is circulated via a cross-flow fan. More information about the hole test bench can be found in Strämke [8]. The existing measurement technology, consisting of thermocouples and pressure sensors in the channel, is used to ensure a homogeneous temperature throughout the flow of off-gas onto the test tube.

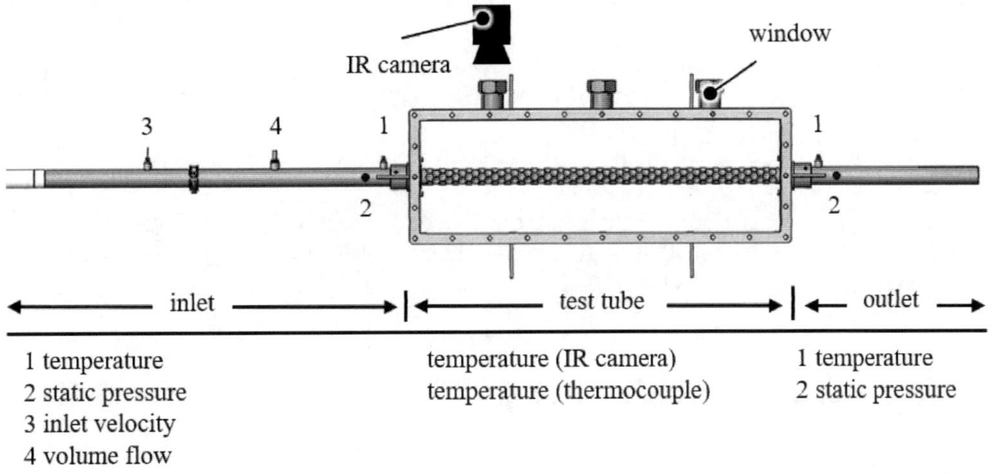

Figure 3. Cross-section of the channel segment with structured tube and measurement technology

At the beginning of the inlet section of a single tube, a side channel compressor generates an air flow into the tube. Within the inlet section, the volume flow $\dot{V}_{air}$, the flow velocity v_{air}, the inlet pressure $p_{air,in}$ and the inlet temperature $T_{air,in}$ into the structured tube are measured. In the test tube section, the tube wall temperature T_{tube} is measured by a welded-on thermocouple first. The built-in IR permeable windows at the top of the channel allow a non-contact temperature measurement of the tube surface with an IR camera at two more positions. This offers the advantage that the flow around the test tube is not affected and the total tube wall temperature can be observed. In the subsequent outlet section, the outlet temperature $T_{air,out}$ and the outlet pressure p_{out} are measured.

2.3. Calculation heat transfer coefficient

The local heat transfer coefficient at the inner tube wall is calculated via an energy balance. Based on the first law of thermodynamics, the change in the stored energy in a control volume, here the change of combustion air temperature, is set equal to the convectively exchanged heat flux. This assumption is admissible because the tube is considered to be thermally thin. The evaluation of whether a component is to be classified as thermally thin or thick is made via the Biot number, equation (1) [9]. If Bi << 0.1, then the component is considered to be thermally thin. In this case, thermally thin means that the wall temperature measured outside of the tube and the wall temperature on the inside may be assumed to be the same at the same time. The temperature gradients in the component are neglected.

$$\mathrm{Bi} = \frac{\alpha \cdot \frac{x_{tube}}{2}}{\lambda} \tag{1}$$

Due to the fact that the heat transfer coefficient is unknown, it must be replaced by the known heat flow according to the general equation of convective heat transfer, equation (2). Substituting equation (2) into equation (1) yields equation (3).

$$\dot{q} = \alpha \cdot \Delta T \tag{2}$$

$$\mathrm{Bi} = \frac{\dot{q} \cdot x_{tube}}{\Delta T \cdot 2 \cdot \lambda} \tag{3}$$

The minimum temperature difference ΔT between the combustion air and the off gas is $\Delta T = 80$ °C. The maximum heat flux density $\dot{q}$ for the heat transfer between the off gas and the combustion air via the tube wall is given as $\dot{q} = 13.040$ W/m^2. These experimentally determined values are deliberately set low for the present calculation in case of the temperature difference and high for the heat flow density in order to represent all experimental states. According to [10], the thermal conductivity λ of stainless steel is 16 to 20 W/(m·K) in a temperature range between 100 and 400 °C. The lowest thermal conductivity is assumed for the calculation. This results in a Biot number of Bi = 0.013, which is significantly smaller than 0.1 and the tube can be assumed to be thermally thin. The heat flow from the channel flow to the combustion air is calculated according to the simplified steady-flow thermal energy equation 4 [9].

$$\dot{Q}_{air} = \dot{m}_{air} \cdot c_{p,air} \cdot \left(T_{air,out} - T_{air,in}\right) \tag{4}$$

By rearranging the equation for calculating the convective heat transfer of flow-through tubes and the heat flow calculated by equation (4), the heat transfer coefficient on the inside of the tube can be determined [11].

$$\alpha = \frac{\dot{Q}_{air}}{A_{wall} \cdot \left(\frac{\left(T_{wall} - T_{air,in}\right) - \left(T_{wall} - T_{air,out}\right)}{\ln \frac{T_{wall} - T_{air,in}}{T_{wall} - T_{air,out}}} \right)} = \frac{\dot{m}_{air} \cdot c_{p,air} \cdot \left(T_{air,out} - T_{air,in}\right)}{A_{wall} \cdot \left(\frac{\left(T_{wall} - T_{air,in}\right) - \left(T_{wall} - T_{air,out}\right)}{\ln \frac{T_{wall} - T_{air,in}}{T_{wall} - T_{air,out}}} \right)} \tag{5}$$

3. Results

3.1. Numerical study

The numerical investigations show that the increase of the heat flux, the heat transfer coefficient and the pressure loss strongly depend on the operating parameters. Figure 4 shows the increase in heat flux of the smooth tube with tape and the concave structured tubes compared to the smooth tube. The smooth tube with tape shows an increase of approx. 20 % compared to the smooth tube overall operating points. In case of the concave structured tubes it is noticeable that the increase is dependent on the operating parameters. At an off-gas velocity of $v_{off\text{-}gas} < 1$ m/s, the advantage of structuring is only minimal. Here the increase is around 20 %. While at combustion air velocities $v_{air} < 5$ m/s the greatest increases are achieved compared to a smooth tube. With increasing structure depth, the increase in heat flux grows. Accordingly, the maximum increase is 45 % for the concave 3 mm structured tube and 80 % for the concave 6 mm structured tube.

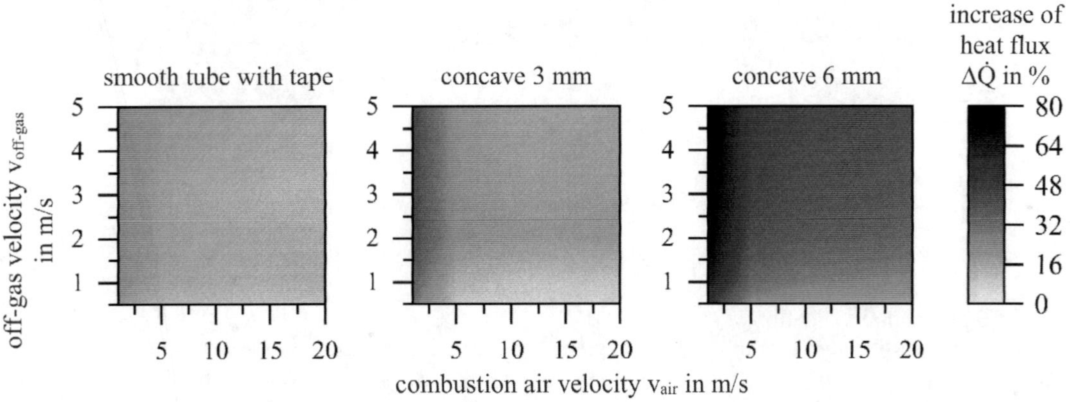

Figure 4. Increase of the heat flux $\Delta\dot{Q}$ across the full operating range

Figure 5 illustrates the increase of the heat transfer coefficient of the investigated tubes in the full operating range. The heat transfer coefficient is increased by an average of 30 % with the use of a tape compared to a smooth tube in the full operating range. For the concave structured tubes, the increase in heat transfer coefficient is non-uniform over the operating range. Here, a particularly large increase is seen when the combustion air velocity is $v_{air} <$ 5m/s. In this range, a doubling of the heat transfer coefficient is achieved with a concave 6 mm structuring.

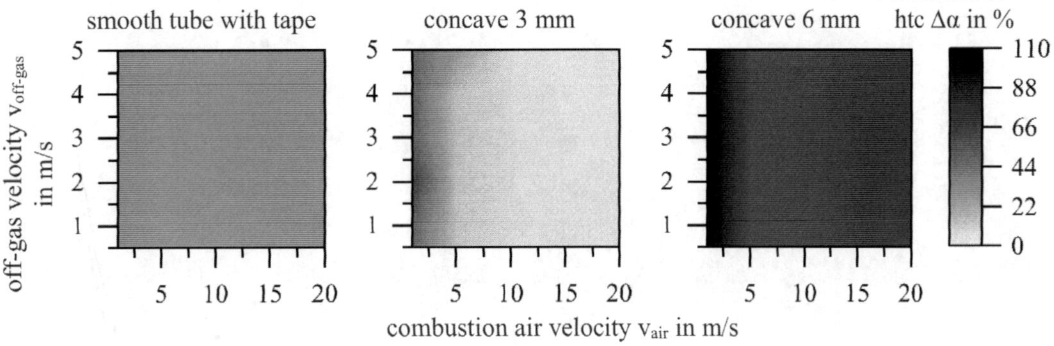

Figure 5. Increase of the heat transfer coefficient $\Delta\alpha$ across the full operating range

The changes in pressure loss of the investigated structured tubes compared to the smooth tube are shown Figure 6. The application of a structure or the use of a tape leads to an increase in pressure losses. Whereby the tape shows an increase of 210 % at combustion air velocities below 10 m/s and 260 % above. The pressure loss in the concave 3 mm structured tube increases significantly less than in the smooth tube with tape. Only in the range $v_{air} < 5$ m/s is a pressure increase of 130 % achieved. Beyond that, the increase is approx. 65 %. The deeper structuring in the concave 6 mm structured tube leads to a very significant increase in pressure loss compared to a smooth tube, which is 500 % on average. An exception is the range of $v_{air} = 5$ to $v_{air} = 10$ m/s, where the increase in pressure loss is between 380 % and 430 %.

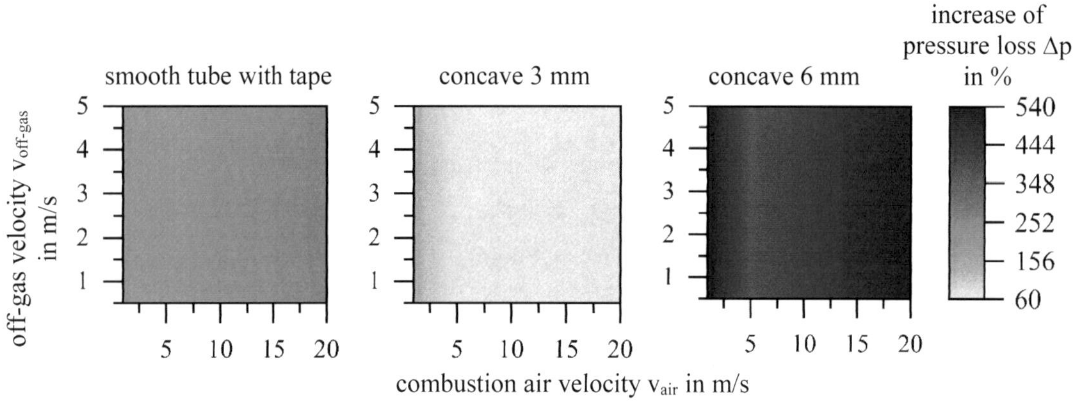

Figure 6. Increase of the pressure loss Δp across the full operating range

3.2. Experimental validation

With the test set-up presented in chapter 2.2, the tubes were examined with regard to the heat transfer and the pressure losses. Table 1 compares the results of the numerical simulations and the experimental measurements for one operating point. The operating conditions refer to an off-gas velocity $v_{off\text{-}gas} = 5$ m/s and a combustion air velocity $v_{air} = 10$ m/s. It should be noted that the simulations were carried out at an off-gas temperature $T_{off\text{-}gas,num} = 500$ °C and the experimental trails at $T_{off\text{-}gas,exp} = 400$ °C. The 500 °C off-gas temperature of the simulation corresponds to industrial specifications. However, it was not possible to maintain a uniform flow at this temperature in time during the experiment, which is why a lower off-gas temperature had to be selected in order to be able to guarantee identical conditions for all tests.

Table 1. Validation of the numerical results for the operating point $v_{off\text{-}gas} = 5$ m/s, $v_{air} = 10$ m/s

	increase of $\Delta\dot{Q}$		increase of $\Delta\alpha$		increase of Δp	
	num.	exp.	num.	exp.	num.	exp.
smooth tube with tape	19 %	45 %	33 %	116 %	240 %	155 %
	+ 26 %		*+ 83°%*		*- 85 %*	
concave 3 mm	18 %	55 %	18 %	124 %	65 %	405 %
	+ 37 %		*+ 106 %*		*+ 340 %*	
concave 6 mm	46 %	78 %	65 %	218 %	430 %	780 %
	+ 32 %		*+ 150 %*		*+ 350 %*	

[a] $T_{off\text{-}gas,num} = 500$ °C
[b] $T_{off\text{-}gas,exp} = 400$ °C

The smooth tube with the twisted tape experimentally shows an increase in heat flux and heat transfer coefficient compared to the smooth tube. The deviation from the numerical solution is 26 % for the heat flux and 83 % for the heat transfer coefficient. When determining the difference in pressure loss between the smooth tube and the tube with tape, it is shown that the pressure loss is 85 % lower in real terms than calculated numerically. This is less due to the numerical model than to the different installation of the tape. In the model, the tape was inserted into the centre of the tube in a floating position. In reality, the tape is welded at one point inside the tube. This means that the tape is placed at any angle in the tube and the area of flow between the tape and tube wall changes along the tube length. This was not numerically feasible, as different zones would have overlapped.

The advantages of the structured tubes compared to the smooth tubes turn out to be greater than the numerical results when studied experimentally. The increase in heat flux is experimentally measured to be 55 % better in the case of the concave 3 mm structured tube and 78 % for the concave 6 mm structured tube. This corresponds to an increase of about 35 % compared to the numerical results. The increase in the heat transfer coefficient is measured to be higher, by 106 % for the concave 3 mm structured tube and 150 % for the concave 6 mm structured tube, relative to the numerical solutions. Also, the increase in pressure loss is significantly higher when measured. There, the numerical model underestimates the pressure loss by 345 %.

A correction of the parameters due to the temperature difference of 100 °C and tube length difference of 500 mm in the numerical and experimental results was omitted, since the increases are always related to the smooth tube with the corresponding tube length and off-gas temperature. This allows the performance of the individual tubes to be compared with each other.

4. Discussion

The heat transfer is affected positively by a structuring or the use of a tape compared to a smooth tube. This is due to the increased turbulence in the fluid flow inside the tube. Within the structural patterns, backflows or recirculation are observed. The intensity of the backflows rises with increasing structural depth. Concave structures reduce the mean flow-through area of the tubes, creating an increase in the mean flow velocity inside the tube and also contributing to an increase in heat transfer. However, this also leads to an increase of pressure loss through the tube. The twisted tape increases turbulence through swirls, but no recirculation occurs here.

From the results it can be observed that the heat flux and the heat transfer coefficient do not rise equally, although the quantities are closely related. This behaviour is an indication that the increased heat transfer is not solely due to the increased tube surface. The heat flux is dependent on the tube surface, whereas the heat transfer coefficient is independent of it. The heat transfer coefficient is significantly influenced by the adjacent flow. If the degree of turbulence at the surface increases, the heat transfer coefficient rises. Thus, the heat transfer coefficient increases more than the heat flux with increasing structure depth.

The positive influence of structured tubes on the heat transfer and the increased pressure loss could be confirmed experimentally. It has been shown that the numerical model underestimates both the heat transfer and the pressure loss. Since this tendency can be seen repeatedly, it must be assumed that the turbulence of the airflow in the tube is underestimated by the simulation. The numerical investigations were all carried out with the identical model, corresponding to the same mesh properties, in order to compare the tubes with each other. Probably a general mesh for the different tube geometries is not sufficient to be able to obtain quantitative statements. Here, a specific mesh adjustment should be made in the structures depending on the structure depth. Nevertheless, the numerical results are lower, but reveal the same characteristics as observed experimentally.

Both the numerical and experimental investigations show that the tape with twisted tube results in a lower thermal advantage than the concave structured tubes. As the highest heat transfer is achieved with the concave 6 mm structured tube. Aroonrat and Wongwises [5] findings can be confirmed, with respect to the increased heat transfer at grater structure depth.

Based on the results, estimates can be made in which operating conditions smooth tubes should be substituted with structured tubes to improve the energy efficiency of gas-fired industrial furnace systems. Since the increase in heat transfer and pressure loss of smooth tubes with tape is independent of the operating conditions, they can be used flexibly in recuperators with a wide range of operating modes. For specialised applications, especially at low combustion air velocities, the concave structured tubes have advantages. Depending on the acceptable pressure losses, both concave $x_{d,1}$ = 3 mm and concave $x_{d,2}$ = 6 mm structured tubes offer a better heat exchange.

5. References

[1] M. Langer 2015 *Plants for reheating semi-finished products prior to rolling*, in: F. Beneke, B. Nacke, H. Pfeifer (ed.), *Handbook of Thermoprocessing Technologies*, *Volume 2: Plants, Compontens, Safety,* Vulkan-Verlag

[2] C. Bischoff, Central recuperators, in: F. Beneke, B. Nacke, H. Pfeifer (ed.) 2015 *Handbook of Thermoprocessing Technologies Volume 2: Plants, Compontens, Safety*, Vulkan-Verlag

[3] E. Specht, F. Becker 2018 Wiesbaden *Öfen für geformtes Gut, Taschenbuch für den Maschinenbau,* Springer-Verlag Berlin Heidelberg

[4] R. Maithani, A. Kumar 2019 *Correlations development for Nusselt number and friction factor in a dimpled surface heat exchanger tube*, Experimental Heat Transfer DOI 10.1080/08916152.2019.1573863

[5] K. Aroonrat, S. Wongwises 2019 *Experimental investigation of condensation heat transfer and pressure drop of R-134a flowing inside dimpled tubes with different dimpled depths,* International Journal of Heat and Mass Transfer, 128, 783-793 ISSN 0017-9310

[6] H. Bas, V. Ozceyhan 2012 *Heat transfer enhancement in a tube with twisted tape inserts placed separately from the tube wall,* Experimental Thermal and Fluid Science, 41, 51-58 ISSN 0894-1777

[7] L. Zheng, Y. Xie, D. Zhang 2017 *Numerical investigation on heat transfer performance and flowcharacteristics in circular tubes with dimpled twisted tapes using Al2O3-water nanofluid,* International Journal of Heat and Mass Transfer, 111, 962-981 ISSN 0017-9310

[8] S. Strämke, H. Pfeifer 2018 *Numerische und experimentelle Untersuchung von Querstromventilatoren,* Prozesswärme, 1, 35-42

[9] T. L. Bergman, A. S. Lavine 2017 Hoboken (NJ, USA) *Fundamentals of Heat and Mass Transfer,* 8th ed., Wiley

[10] M. Spittel, T. Spittel, H. Warlimont 2009 Wiesbaden *Thermal conductivity of steel, Metal Forming Data of Ferrous Alloys- deformation behaviour*, Springer-Verlag Berlin Heidelberg DOI 10.1007/978-3-540-44760-3_7

[11] P. Stephan, S. Kabelac, M. Kind, D. Mewes, K. Schaber, T. Wetzel 2019 Wiesbaden *VDI-Wärmeatlas,* Springer Berlin Heidelberg ISBN 978-3-662-52988-1

Acknowledgments

This research project is supported by the Federal Ministry of Economics and Climate Protection within the Central Innovation Program for small and medium enterprises on the basis of a decision by the German Bundestag under the AiF project no. KK5123801RH0.

CFD design of a novel device for temperature profile measurement in Waste-to-Energy plants

G. Grossi [1,*], F. Arpino [1], C. Canale [1], G. Cortellessa [1], G. Ficco [1], T. Lombardi [2]

[1] Department of Civil and Mechanical Engineering, University of Cassino and Southern Lazio, Via G. Di Biasio 43, Cassino, 03043, FR, Italy
[2] HERAmbiente S.p.a, Via dell'Energia, Pozzilli, 86077, IS, Italy

* Corresponding author: giorgio.grossi@unicas.it

Abstract. Monitoring the flue gas temperature is a crucial issue for Waste-to-Energy (WtE) plants companies, since it is essential to preserve the materials in the post-combustion chamber, the energy efficiency of the plants as well as to control the emissions of pollutants. As of today, the temperature of flue gases in such plants is commonly monitored by means of thermocouples, infrared pyrometers or aspirated thermocouples. However, all these instruments show limitations in terms of accuracy and reliability inside post-combustion chambers. In this paper, the authors present the thermo-fluid dynamics design of a novel device aimed at mitigating the issues of existing technologies, realised by using the modern CFD techniques. Numerical analyses, performed with the open-source OpenFOAM code, allowed to find a suitable shape for the device and to give a first estimate of the measuring errors, which are of the order of 2% (~26 K at the typical working temperature of WtE plants' post-combustion chambers).

Keywords: Waste-to-Energy plants, combustion chamber, temperature profile, thermocouple, CFD analysis

1. Introduction

The temperature of flue gases in the post-combustion chamber of Waste-to-Energy (WtE) plants must be maintained within a narrow range of values, for environmental (i.e., to control the emissions of pollutant agents), structural (i.e., to ensure the preservation of materials properties) and energy efficiency reasons [1]. The minimum value is set by the Legislative Decree No. 152/2006 and its subsequent amendments and supplementations, prescribing that after the last injection of combustion air, the temperature of the gaseous combustion products must be kept above 850 °C for at least 2 seconds (1100 °C if hazardous wastes with a content of more than 1% of halogenated organic substances, expressed as chlorine, are incinerated). Such temperature, referred to as T_{2S}, is not directly measurable and is typically estimated by employing advanced CFD techniques or empirical and semi-empirical correlations based on temperature, volumetric flow rate and oxygen content continuously measured in the post-combustion chamber. Monitoring the flue gas temperature is essential also to ensure that it is not overly high, as an excessive rise in the combustion temperature leads to an increased production of nitrogen oxides, results in a decrease of the plant efficiency and could also compromise the structural integrity and functionality of the materials and components employed [1].

The temperature of flue gases in WtE plants is commonly monitored by means of thermocouples, infrared pyrometers or aspirated thermocouples (also referred to as suction pyrometers or high velocity thermocouples) [1–4]. The first two techniques showed limitations in strong radiative environments such as the one inside the post-combustion chambers of a WtE plant, both in terms of accuracy and reliability [1,2,5]. The latter, albeit exhibiting greater accuracy and overcoming most of the issues affecting the performances of the other instruments, still presents criticalities in its management and reliability [4]. In addition, none of these measurement techniques allows to obtain a temperature profile with a single immersion by simultaneously measuring the temperature at different depths. In fact, to measure temperature in multiple points of the combustion chamber, essential for validating the CFD codes employed to verify the plant operating conditions, the same instrument must be moved back and forth resulting in an uncertainty of the position and measurements in correspondence of potentially different operating conditions of the WtE plant. In fact, the properties and the conditions of the flue gases may vary significantly over time according to the composition of the waste being burned.

In this paper, the authors present the design of a novel device to measure the temperature of flue gases evolving in the post-combustion chamber of WtE plants, designed for the HERAmbiente S.p.a. plant in Pozzilli (Italy), aiming to mitigate the issues of existing temperature measuring approaches. The novel device relies on thermocouples as sensing elements, shielded from radiative heat fluxes. The rising motion of the fumes is exploited to take measurements at three points of the chamber and near the wall simultaneously. This work presents the preliminary results of the thermo-fluid dynamics design of the device, realised by means of transient CFD analyses. The developed numerical tool considers the effects of thermal radiation, essential to correctly estimate the errors of measurements taken inside post-combustion chambers, and of conjugate heat transfer between fluid and solid media.

2. Materials and methods

The concept of the novel device is sketched in Figure 1. It is 2 m long, 0.035 m wide (limited by the maximum opening available in the membrane walls of the Pozzilli plant) and 0.38 m high.

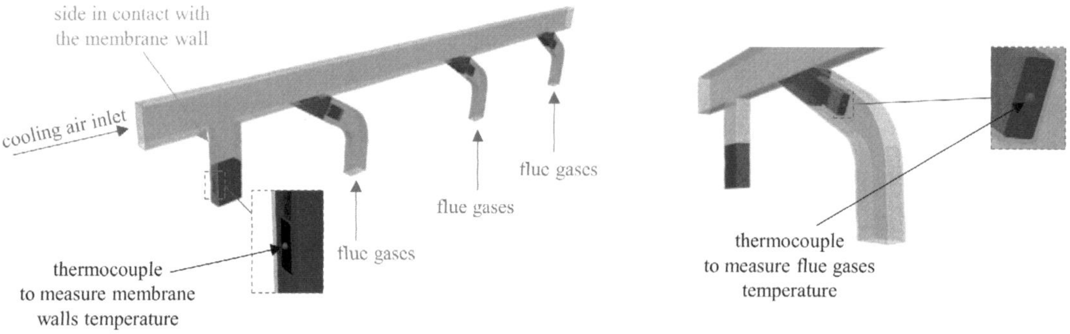

Figure 1. The novel measuring device.

It relies on the use of one thermocouple to measure the inner wall temperature and of three thermocouples to measure the temperature of flue gases flowing in the post-combustion chamber. Thanks to the designed shape, the framework itself shields the sensing elements from the radiative heat exchange between the sensors and the chamber walls. Cooling is provided by fresh air injected into the main duct and released in the flue gases stream; refractory cement (coloured blue in Figure 1) is employed to prevent contact between the hot gas stream and the thermocouples wirings. The piping is made of Inconel 625 alloy, whereas the refractory is made of a silicon carbide-based cement. More structural details of the device are shown in Figure 4. The solution here proposed has several advantages over the technologies currently existing on the market: (i) errors due to radiation are reduced thanks to the shape of the device; (ii) there is a clear reference for the position of the measurement points, since one side is in contact with the membrane wall and the relative distance between sensors is constant and known; (iii) temperature is measured at multiple points simultaneously, with a single immersion;

(iv) unlike suction pyrometers, cooling is provided by injecting air, not water, and flue gases are not drawn out of the post-combustion chamber; (v) the gas stream is not sped up (as it happens for suction pyrometers), conversely its natural vertical motion is exploited.

2.1. Steps in the thermo-fluid dynamics design of the novel device

The device shown in Figure 1 is the result of thermo-fluid dynamics analyses carried out on four different geometries, depicted in Figures 2-4. From first to fourth (i.e., the final) test, changes were made according to the results of CFD analyses, as described in Section 3. To save computational resources, at this stage of the design process a simplified geometry was analysed.

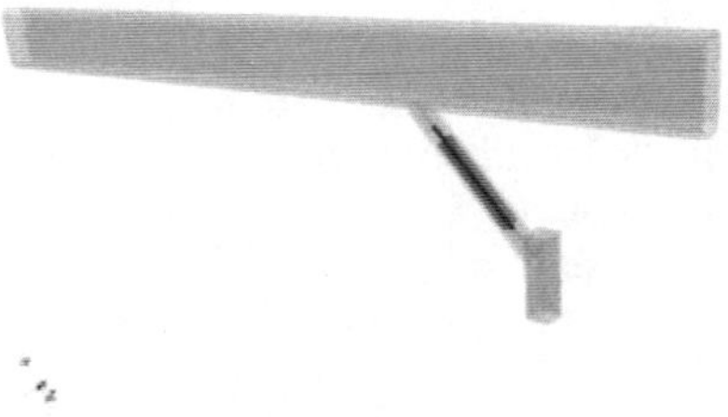

(a) first test

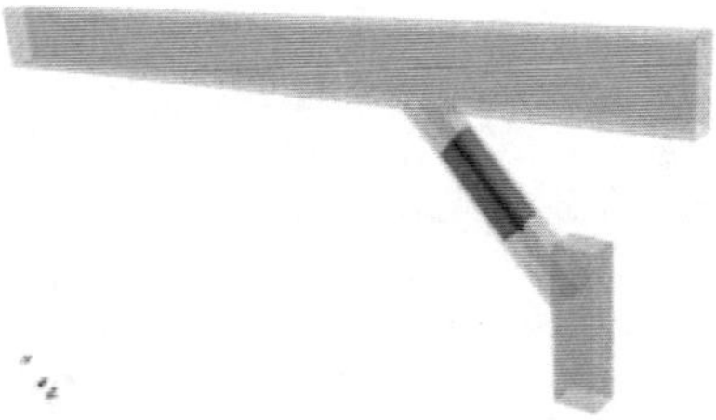

(b) second test

Figure 2. (a) first and (b) second tests for the design of the novel device.

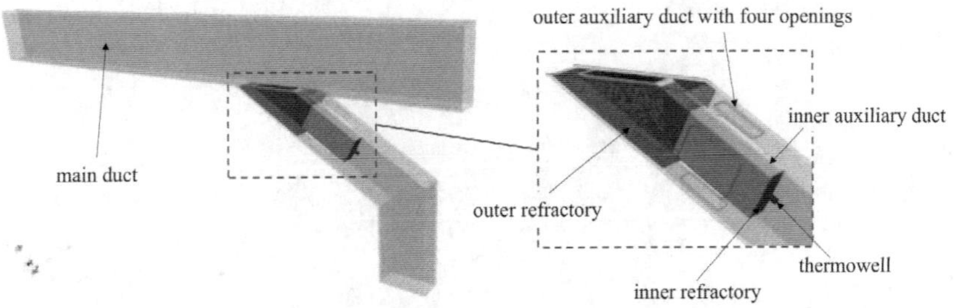

Figure 3. Third test for the design of the novel device.

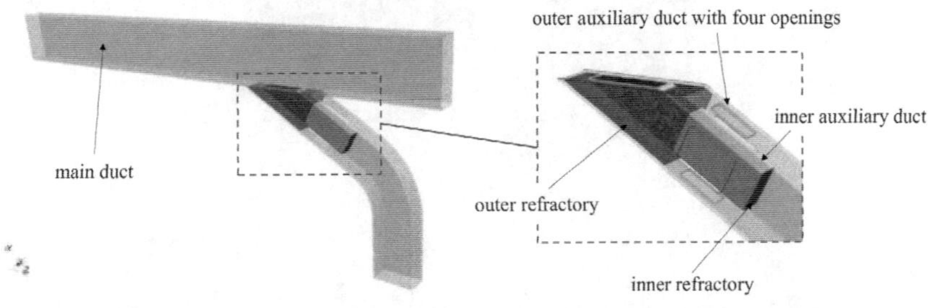

Figure 4. Fourth (and final) test for the design of the novel device.

The third and fourth geometries (Figure 3 and Figure 4) differ significantly from the initial concepts (Figure 2); an inner auxiliary duct brings the thermocouple and an outer auxiliary duct channels the flow, shields the thermocouple from radiation and allows the hot gases to flow out through four openings.

2.2. Numerical setup

Numerical investigations were performed employing the *chtMultiRegionFoam* solver available in the open-source OpenFOAM code, based on the finite volume formulation.

Velocity, pressure and temperature in the fluid region were predicted solving the mass, momentum and energy conservation equations under the assumption of three-dimensional, unsteady, turbulent and compressible flow with ideal gas behaviour. The unsteady Reynolds-averaged Navier-Stokes (URANS) approach was selected for turbulence modelling, solving the Shear Stress Transport (SST) k–ω model [6,7] as recommended by Costa et al. [8]. Pressure-velocity coupling was solved with the PIMPLE algorithm [9], processing the overall transient phenomenon as consecutive steady-state time steps; a time step of 0.005 s was adopted and as requirement for time step convergence the scaled residuals were set $<10^{-3}$ for pressure and $<10^{-5}$ for other variables. The effects of thermal radiation were taken into account using the Finite Volume Discrete Ordinates Method (fvDOM) model [7,10], solving the radiative transfer equation (RTE) for 60 discrete solid angles and considering a non-scattering medium. Once all the directional radiation intensities I (W m^{-2} sr^{-1}) are known, it is possible to calculate the radiative heat fluxes at surfaces, q_r (W m^{-2}), as well as the radiative source term in the fluid energy conservation equation. As regards the solid domains, the only variable is temperature, and it is determined by solving the energy conservation equation under the assumption of constant thermophysical properties within the regions. For a comprehensive description of the governing partial differential equations, the reader is referred to the existing literature [6,7,10].

Figure 5 shows the computational domain for the fourth (i.e., the final) test case. Its set-up is identical to the one used for the other scenarios and is composed of multiple sub-regions: (i) a fluid one, with two inlets (one for the flue gases, coloured green, and one for the cooling air, coloured blue), an outlet section (coloured red) and walls (coloured grey); (ii) separate solid regions for the piping and for refractories (other concepts also include a solid region for the thermowell, as visible in Figure 2 and Figure 3).

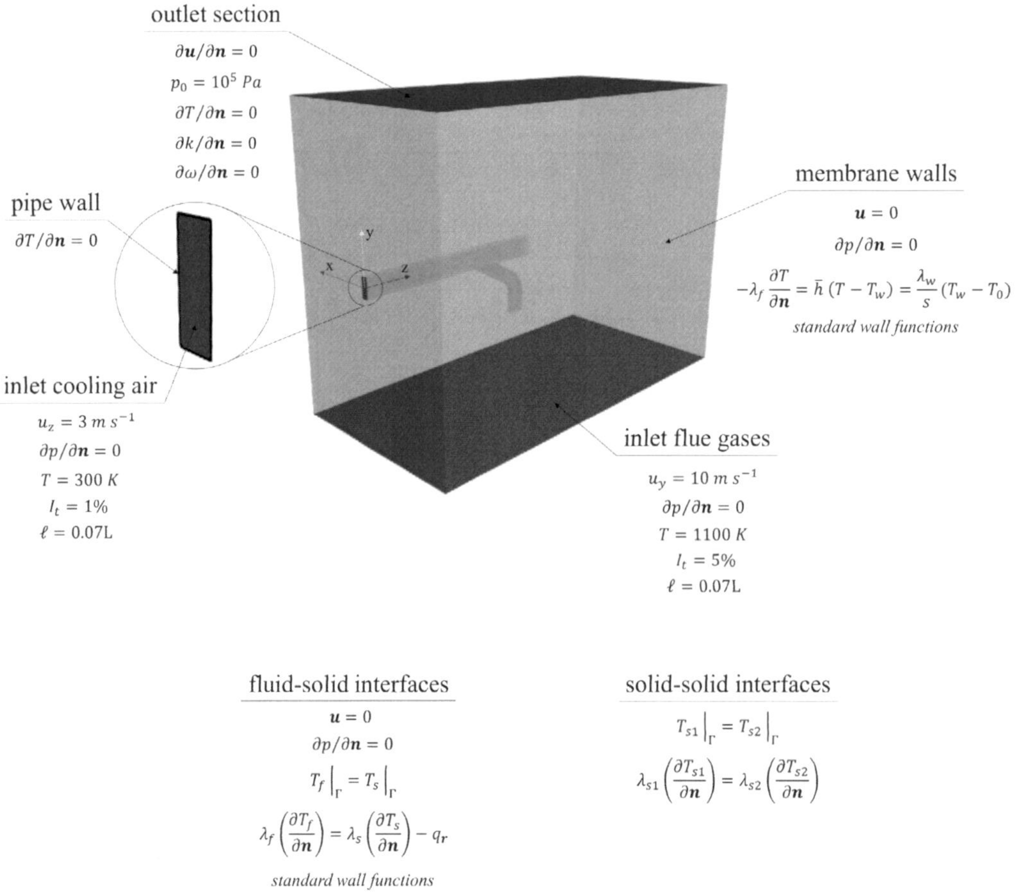

Figure 5. Computational domain and main boundary conditions employed for the simulation of the fourth geometry.

Figure 5 also presents the main boundary conditions set for numerical simulations. A velocity of 3 m s^{-1} was assumed for the cooling air whereas the velocity of flue gases (assumed to be air) was set equal to 10 m s^{-1} [1]. At the outlet section, the value of the total pressure p_0 is imposed. For turbulent quantities, the turbulence intensity I_t (%) and the turbulent mixing length ℓ (m) were specified. The latter was calculated as 7% of the characteristic length L (m), assumed equal to the width of the corresponding inlet section. For the resolution of temperature field, at the membrane walls heat transfer with the external environment is modelled: λ (W m^{-1} K^{-1}) is the thermal conductivity, $\bar{h} = 1.5\ \mathrm{W\,m^{-2}\,K^{-1}}$ the convection heat transfer coefficient, s (m) the wall thickness, T_W (K) the wall temperature and T_0 the temperature of saturated water at 40 bar (524.48 K). It should be remarked that the velocity and temperature values set at the inlet for the flue gases and the cooling air are preliminary values. These values are used to conduct simulations aimed at finding the best shape for the measuring device. Future investigations will replicate the actual conditions of the WtE plant and will involve a thorough analysis to determine the optimal velocity and temperature of the cooling water.

At the interface Γ between two regions (i.e., fluid-solid and solid-solid interfaces), the continuity of temperature and the balance of the heat fluxes are imposed. Finally, a grey-diffuse condition is provided for the radiation intensity [7,10], at all boundary patches.

The domain sketched in Figure 5, was meshed with an unstructured hexa-based grid, made up of 5,323,212 cells; the maximum non-orthogonality is equal to 56 and the maximum skewness is equal to 1.52. A detail of the grid, on the yz plane at x=0, is depicted in Figure 6. Grid sensitivity analysis, as well as the validation of the numerical model, will be performed in future studies.

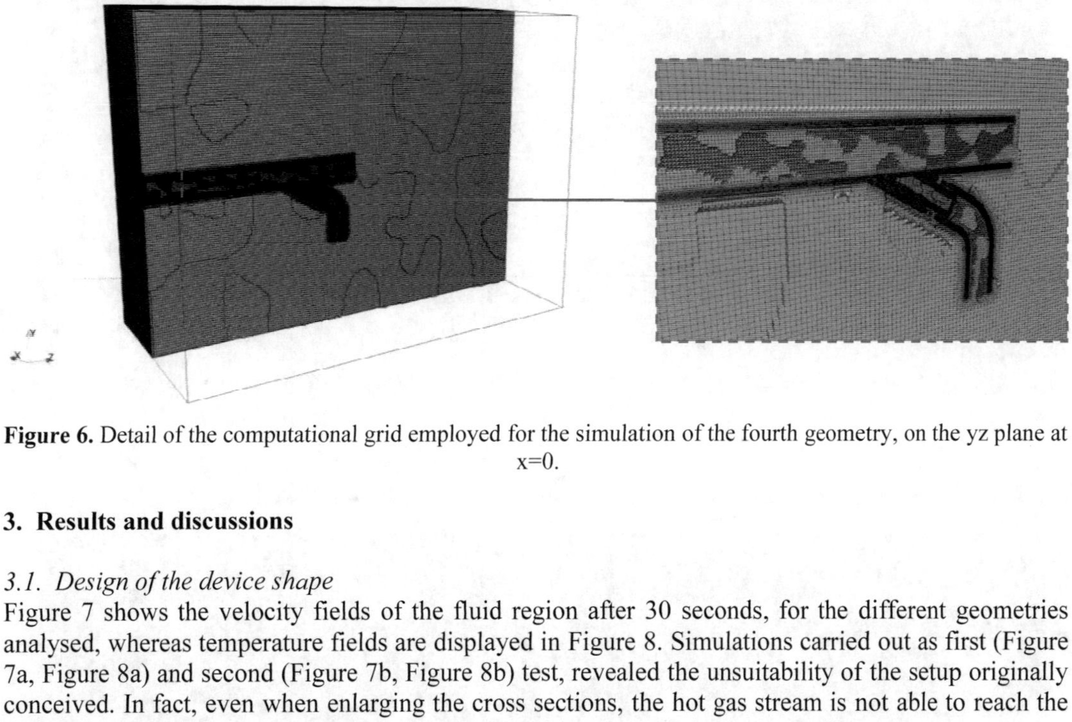

Figure 6. Detail of the computational grid employed for the simulation of the fourth geometry, on the yz plane at x=0.

3. Results and discussions

3.1. Design of the device shape

Figure 7 shows the velocity fields of the fluid region after 30 seconds, for the different geometries analysed, whereas temperature fields are displayed in Figure 8. Simulations carried out as first (Figure 7a, Figure 8a) and second (Figure 7b, Figure 8b) test, revealed the unsuitability of the setup originally conceived. In fact, even when enlarging the cross sections, the hot gas stream is not able to reach the thermowell and exchange efficiently, resulting in cold spots near the thermowells.

The third setup (Figure 7c and Figure 8c) is far better in terms of fluid dynamics and the flue gases are now able to reach the sensor. However, such solution still presents criticalities: (i) recirculation zones, due to the sharp edges of the outer auxiliary duct, are responsible for cold spots that must be avoided to improve the dynamic response of the device; (ii) building the instrument in this form would require the use of welding, not suitable for the extreme environments inside the combustion chambers. The above considerations led to the design of the fourth configuration (Figure 7d and Figure 8d), with a smoother shape. It presents no structural criticalities and avoids the formation of recirculation zones and hot spots; therefore, it is suitable for the construction of the novel pyrometer.

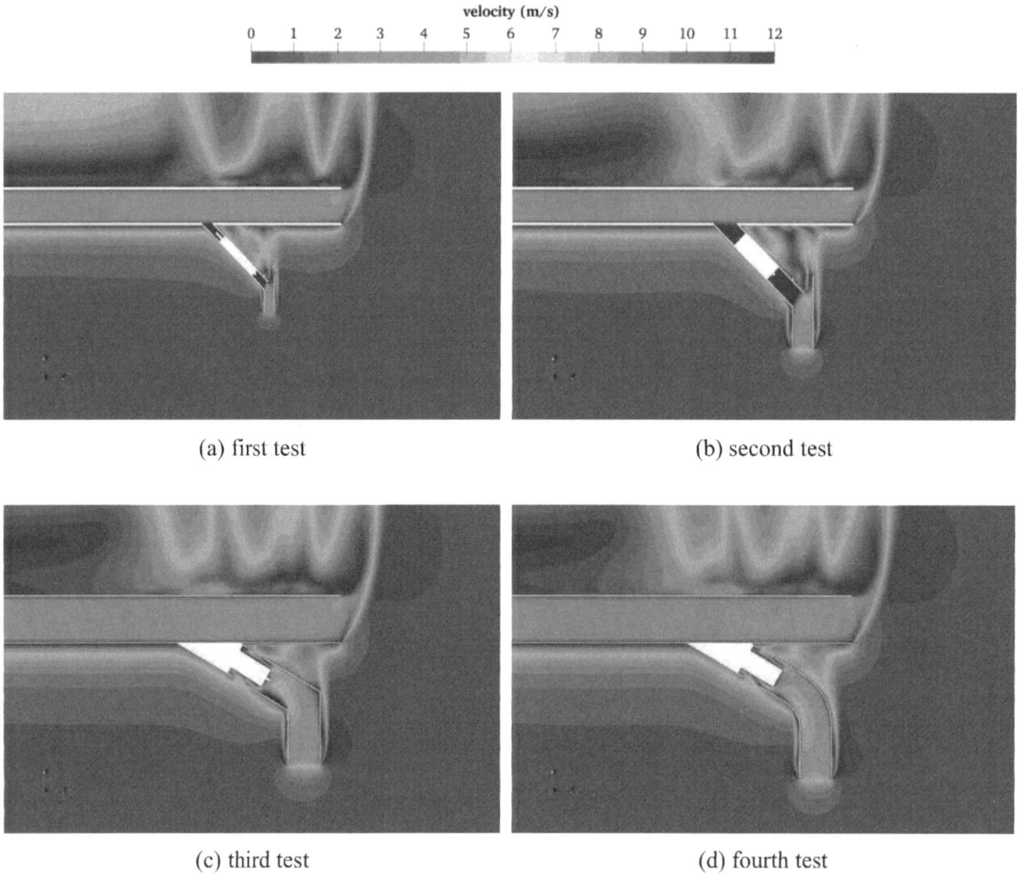

(a) first test (b) second test

(c) third test (d) fourth test

Figure 7. Velocity fields on the yz plane at x=0, after 30 seconds, for the different geometries considered.

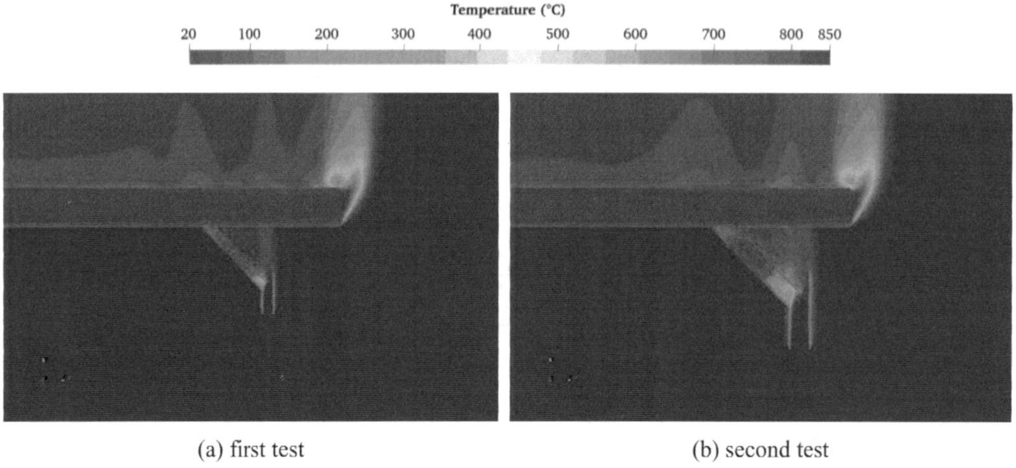

(a) first test (b) second test

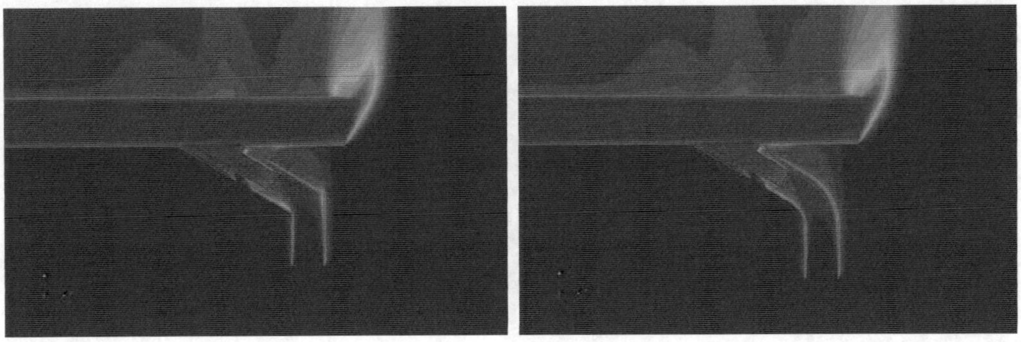

(c) third test (d) fourth test

Figure 8. Temperature fields on the yz plane at x=0, after 30 seconds, for the different geometries considered.

3.2. Estimation of the measuring errors

Once the best shape for the device was selected, further analyses were conducted to get a first estimate of the measuring errors. For this purpose, two virtual probes were put in the computational domain, one in the fluid region (where the flow enters the outer auxiliary duct) and one inside the refractory (where the thermocouple would be placed). Temperature was monitored over time and the relative error calculated as $\left(T_{fluid} - T_{refractory}\right)/T_{fluid}$; the results, shown in Figure 9, reveal that in quasi-steady state conditions, the error is of the order of 2% (~26 K).

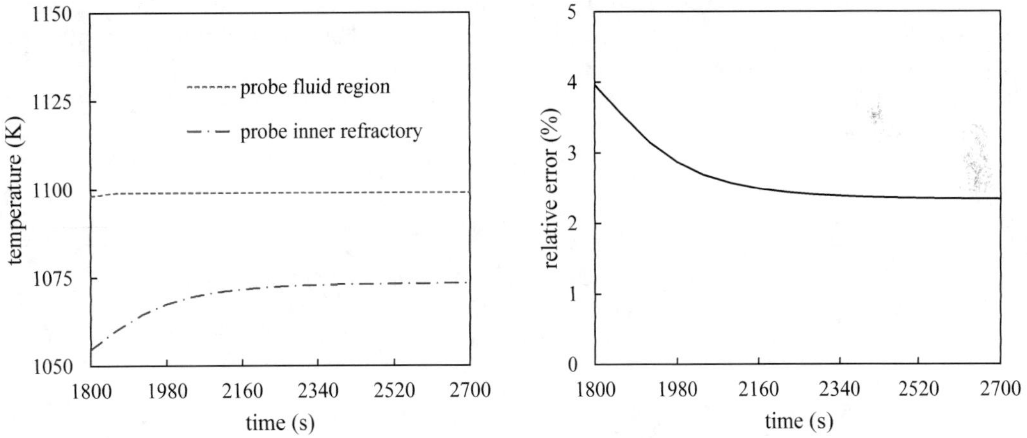

Figure 9. Temperature trends over time at the virtual probes and corresponding relative error.

Such a difference in temperature may be due to: (i) conductive fluxes within the solid media; (ii) ineffective convective heat transfer between the flue gases and the refractory.

The contribution of conduction and convection heat transfer mechanisms was assessed extracting a temperature profile straddling the fluid and the solid region, as depicted in Figure 10. As expected, a drop in temperature occurs at the interface; however, in the fluid and solid regions the temperature field is quite uniform, suggesting that the effect of conductive and convective heat transfer can be deemed negligible.

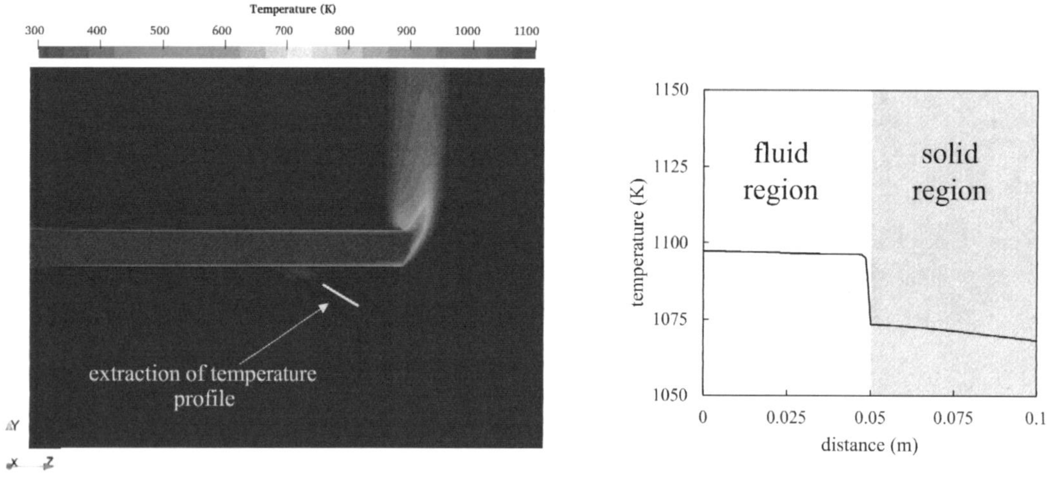

Figure 10. Temperature field and temperature profile extracted between the fluid and solid regions.

4. Conclusions

This work presents the preliminary results of the thermo-fluid dynamics design of a novel device aimed at measuring flue gases temperature inside WtE plants. The design is realised by means of transient CFD analyses, performed with the open-source OpenFOAM toolbox, taking into account the effects of thermal radiation and of conjugate heat transfer between fluid and solid media.

Numerical analyses allowed to find a suitable shape for the device, capable of shielding the sensing elements from incident radiation and of ensuring contact with the hot gas stream; they also allowed to give a first estimate of the measuring errors, which are of the order of 2% (i.e., 26 K).

Future simulations, reproducing the real operating conditions of the plant, will look at finding: (i) the time required to take measurements; (ii) the minimum flow rate of the cooling air, needed to preserve the structural integrity of the different components; (iii) the maximum temperature reached by the materials.

References

[1] F. Rinaldi, B. Najafi, Temperature Measurement in WTE Boilers Using Suction Pyrometers, Sensors. 13 (2013) 15633–15655. https://doi.org/10.3390/s131115633.
[2] A. Z'Graggen, H. Friess, A. Steinfeld, Gas temperature measurement in thermal radiating environments using a suction thermocouple apparatus, Meas. Sci. Technol. 18 (2007) 3329–3334. https://doi.org/10.1088/0957-0233/18/11/010.
[3] H. Sefidari, N. Razmjoo, M. Strand, An experimental study of combustion and emissions of two types of woody biomass in a 12-MW reciprocating-grate boiler, Fuel. 135 (2014) 120–129. https://doi.org/10.1016/j.fuel.2014.06.051.
[4] Jr. Baukal Charles E., ed., Industrial Combustion Testing, 0 ed., CRC Press, 2010. https://doi.org/10.1201/EBK1420085280.
[5] J.V. Nicholas, D.R. White, Traceable temperatures: an introduction to temperature measurement and calibration, 2nd ed, Wiley, Chichester, 2001.
[6] F.R. Menter, M. Kuntz, R. Langtry, Ten years of industrial experience with the SST turbulence model, Turbulence, Heat and Mass Transfer 4. (2003) 625–632.
[7] H.K. Versteeg, W. Malalasekera, An introduction to computational fluid dynamics: the finite volume method, 2nd ed, Pearson Education Ltd, Harlow, England ; New York, 2007.
[8] M. Costa, N. Massarotti, A. Mauro, F. Arpino, V. Rocco, CFD modelling of a RDF incineration plant, Applied Thermal Engineering. 101 (2016) 710–719. https://doi.org/10.1016/j.applthermaleng.2016.01.073.
[9] C.J. Greenshields, H.G. Weller, Notes on computational fluid dynamics: general principles, CFD Direct Limited, Reading, UK, 2022.
[10] Michael Modest, Radiative Heat Transfer, Elsevier, 2013. https://doi.org/10.1016/C2010-0-65874-3.

Direct Numerical Simulation of liquid metal forced and mixed convection in a square rod bundle

D Trane[1], M Grespan[1], E Stalio[2], and D Angeli[1,3]

[1] Department of Sciences and Methods for Engineering, University of Modena and Reggio Emilia, Via Amendola 2, 42122 Reggio Emilia, Italy
[2] Department of Engineering "Enzo Ferrari", University of Modena and Reggio Emilia, Via Vivarelli 10, 41125 Modena, Italy
[3] En&Tech Interdepartmental Center, Via Amendola 2, 42122 Reggio Emilia, Italy

E-mail: danila.trane@unimore.it

Abstract. This paper reports results of Direct Numerical Simulations of fully developed, forced and mixed convection of Liquid Lead Bismuth Eutectic (LBE) around four vertical cylindrical rods arranged in a square lattice at Pr = 0.031. The solutions provided here are placed in a context where very few data are available for low-Prandtl flows, especially in mixed convection conditions, and can hence be useful for the development and validation of advanced turbulent heat transfer models. The equations are discretized using a Finite Volume implementation of a second order projection method. The irregular cylindrical boundaries are handled with an original Immersed Boundary technique. A single friction Reynolds number value is set ($\mathrm{Re}_\tau = 180$) and buoyancy effects are accounted for by imposing the Rayleigh number, under the Boussinesq approximation. A Ra-value $\mathrm{Ra} = 2.5 \times 10^4$ is selected in order to investigate the main differences between forced and mixed convection at low Prandtl number values. Time-averaged velocity and temperature fields are shown, in order to discuss the main features of the flow and thermal fields. First order statistics are also presented, highlighting the effect of aiding buoyancy on turbulent heat and momentum transport.

1. Introduction and scope

Heat transfer around cylindrical rods is part of a highly multidisciplinary series of studies on liquid metal fast reactors, aimed at increasing their safety and improving the efficient use of resources. In this context, the application of Direct Numerical Simulations (DNS) fulfils the need for advanced tools and high-fidelity solutions. The advancements in computational methods and resources made it possible to overcome the relative complexity of the geometries involved and the challenges posed by the computational demands of DNS. Despite these efforts, several aspects concerning flow and heat transfer around bundles of heated rods still need clarification.

So far, cases of forced convection on vertical rod bundles have mainly been considered and not much has been documented about buoyancy effects. El-Genk carried out several experiments regarding convection of water in a uniformly heated square lattice, collecting data for different number of rods and variable pitch-to-diameter ratio [1, 2], from which he found that the rod arrangement negligibly affects the overall Nusselt number in both forced and natural convection regimes. In his review, Meyer [3] pointed out that a natural mixing phenomenon from one subchannel to another, through the gaps between the rods, enhances the heat transfer to the cooling fluid by reducing the temperature differences. More recently, Angeli et al. [4] performed

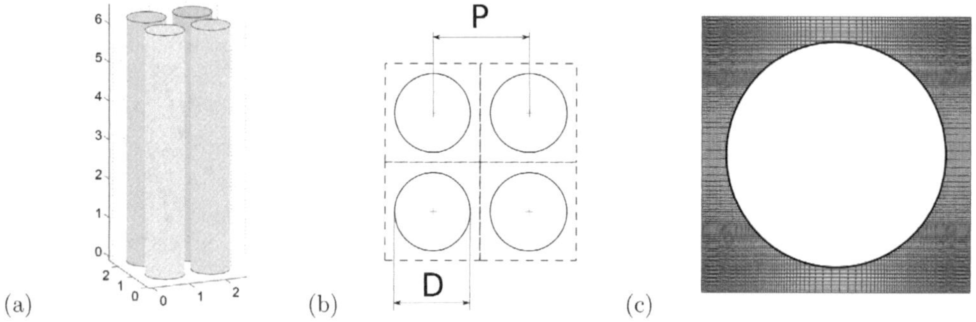

Figure 1. 3D view (a) and cross-sectional sketch (b) of the computational domain; (c) one-quarter of the 512^2 mesh in the $x-y$ plane.

DNSs for fully-developed forced and mixed convection around a bare triangular rod bundle, under a constant pressure gradient, pointing out that buoyancy effects tend to decrease turbulent fluctuations, both increasing the bulk Reynolds number and decreasing the heat transfer rate. However, the need for further understanding of the effects of buoyancy on the flow and heat transfer in this kind of domains is still significant, because of the impact that these effects can have on the operating conditions in actual reactor cores, especially in emergency conditions such as DHR (Decay Heat Removal). Furthermore, data for liquid metals or more generally for low Prandtl numbers, relevant for Gen-IV reactors, are even scarcer.

In the following, data of fully-developed flow and heat transfer around a bare vertical, square rod bundle are presented, obtained by means of Direct Numerical Simulations in both forced and mixed convection conditions. Simulations are performed applying a friction Reynolds number $\mathrm{Re}_\tau = 180$, considering Liquid Lead-Bismuth Eutectic (LBE) at $\mathrm{Pr} = 0.031$ as the coolant fluid. Buoyancy effects are introduced by imposing a Rayleigh number $\mathrm{Ra} = 2.5 \times 10^4$, corresponding to a Richardson number $\mathrm{Ri} \approx 0.24$.

2. Problem statement

A flow around 4 cylindrical rods is analysed. The rods are arranged in a square lattice, as shown in Figure 1(a). The rod diameter is denoted as D, while the centers are spaced by pitch P, see Figure 1(b). In the present case, the pitch-to-diameter-ratio is fixed at $P/D = 1.25$. Each bar is treated as uniformly heated by a constant heat flux q'' and the flow is considered fully developed. Periodic conditions enforced on the velocity field and the modified pressure field p_m well delineate the physical situation. Furthermore, the temperature field needs to be normalized so that periodic BCs can be also set on a modified temperature-like variable θ [5].

3. Mathematical formulation

The incompressible Navier-Stokes formulation is adopted. The Boussinesq approximation is adopted, all the fluid properties being consistently assumed as constant, with the exception of density in the buoyancy term.

Accordingly, the continuity equation and the momentum equation can be written as:

$$\frac{\partial u_i}{\partial x_i} = 0 \tag{1}$$

$$\rho\left(\frac{\partial u_i}{\partial t} + u_j\frac{\partial u_i}{\partial x_j}\right) = -\frac{\partial p}{\partial x_i} + \mu\frac{\partial^2 u_i}{\partial x_j \partial x_j} + \rho g_i \tag{2}$$

Journal of Physics: Conference Series **2685** (2024) 012012 doi:10.1088/1742-6596/2685/1/012012

Table 1. Flow and thermal parameters set for the two DNSs performed and relevant integral values calculated.

Quantity		**Forced case**	**Mixed case**
Friction Reynolds number	Re_τ	180	180
Rayleigh number	Ra	0	2.5×10^4
Prandtl number	Pr	0.031	0.031
Bulk Reynolds number	$\mathrm{Re_b}$	1785	1850
Bulk Peclet number	$\mathrm{Pe_b}$	55	57
Richardson number	Ri	0	0.24
Nusselt number	Nu	7.60	7.50

where u_i indicates the i-th component of the velocity field and the repeated index notation is implied.
The buoyancy force in Equation 2 is obtained by means of the Boussinesq approximation and the assumptions of the case as previously stated:

$$\rho g_i = \rho^o g_i - \rho^o g_i \beta^o \left[T - \overline{T}_\mathrm{b}(z)\right] - \rho^o g_i \beta^o a(z - z^o) \tag{3}$$

where $\overline{T}_\mathrm{b}(z)$ is the bulk temperature and the subscript "o" denotes reference condition.
In Equation 3, a denotes the temperature slope in z, which may be evaluated from an energy balance

$$a = \frac{q''\omega}{\dot{m}c_p} \tag{4}$$

with q'' the imposed wall heat flux over the wetted perimeter ω, $\dot{m}$ the mass flow rate and Ω the cross section.
With the gravity directed along z, ($g_1 = 0$, $g_2 = 0$, $g_3 = -g$), the first and the third terms in equation 3 may be incorporated into the pressure field giving place to a modified pressure:

$$p_m \equiv p - \rho^o g_3 z + \rho^o g_3 \beta^o a \frac{(z - z^o)^2}{2} \tag{5}$$

Due to fully developed conditions, the time averaged streamwise gradient of the modified pressure is constant [6], and, considering the pure natural convection requirement, such a constant equals zero. The heated rods investigated are characterised by a heat source of finite height and, as they are associated with a suitable cooling section, the flow induced by buoyancy is a non-zero mean flow [7]. As a consequence, the flow regime under consideration cannot be viewed as pure natural convection, because in that case the hypothesis of full development would entail a zero mean flow rate. To overcome this issue, a forcing term is added which realises a fully developed mixed convection flow, representative of the actual flow configuration.

The modified pressure term can be split into a linear part P_m, depending on the streamwise coordinate only, and a fluctuation p'_m, which is set as periodic in the present numerical procedure:

$$p_m(x, y, z) = P_m(z) + p'_m(x, y, z) \tag{6}$$

The pressure gradient term becomes:

$$-\nabla p_m = -\nabla p'_m + \sigma \tag{7}$$

where σ denotes the constant pressure gradient due to external forcing of the flow ($\sigma = -dP_m/dz$).

Table 2. Domain size and grid spacing for the present DNSs.

Quantity		$(\cdot)_x$	$(\cdot)_y$	$(\cdot)_z$
Domain size	$L_{(\cdot)}/D_h$	2.5	2.5	2π
no. of cells	$n_{(\cdot)}$	512	512	256
spacing	$\Delta(\cdot)^+$	2.6	2.6	4.4
resolution	$\Delta(\cdot)/\eta_{min}$	1.1	1.1	2.3

The fluid excess temperature is defined as

$$\theta(x, y, z) = \left[T(x, y, z) - \overline{T}_\mathrm{b}(z)\right] \tag{8}$$

where $\overline{T}_\mathrm{b}(z)$ is the time averaged bulk temperature. Given the fully developed conditions, $\overline{\theta}$ is independent of the streamwise coordinate [8].
Defining the temperature scale as $\Delta T_\mathrm{ref} = q''\ L_\mathrm{ref}/\lambda$ and the velocity scale as $u_\mathrm{ref} = \sqrt{\tau_w/\rho^o}$, the momentum equation can be written in its dimensionless form:

$$\frac{\partial u_i^*}{\partial t^*} + u_j^* \frac{\partial u_i^*}{\partial x_j^*} = -\frac{\partial p_m^*}{\partial x_i^*} + \frac{1}{\mathrm{Re}_\tau} \frac{\partial^2 u_i^*}{\partial x_j^* \partial x_j^*} + \left(\frac{\mathrm{Gr}}{\mathrm{Re}_\tau^2} \theta^* + \frac{\omega^*}{\Omega^*} \right) \delta_{i3} \tag{9}$$

The term ω^*/Ω^*, which represents the dimensionless constant pressure gradient due to external forcing of the flow, has to be modified to take into account the possible excess or defect of the buoyancy source. In statistically steady conditions, the volume average of the fluid-excess temperature $\langle\theta^*\rangle$ is time invariant and depends on the initial conditions only, in addition, $\langle\theta^*\rangle$ is not zero, even though $\langle u_z^* \theta^*\rangle = 0$, by definition of θ under isoflux conditions. If $\langle\theta^*\rangle \neq 0$, the term $\mathrm{Gr}/\mathrm{Re}_\tau^2 \langle\theta^*\rangle$ acts as an additional uniform momentum source. In order to ensure that the integral of the momentum source adds up to the same value both with or without the buoyancy source, the following constraint is imposed:

$$\frac{\omega^*}{\Omega^*} = \left(\frac{\omega^*}{\Omega^*} \right)_\mathrm{ref} - \frac{\mathrm{Gr}}{\mathrm{Re}_\tau^2} \langle\theta^*\rangle \tag{10}$$

where $(\omega^*/\Omega^*)_\mathrm{ref}$ is the reference value of the uniform momentum source when buoyancy is not active.

The energy equation, in its dimensionless form, can be written as

$$\frac{\partial \theta^*}{\partial t^*} + u_i^* \frac{\partial \theta^*}{\partial x_i^*} = \frac{1}{\mathrm{Re}_\tau \mathrm{Pr}} \frac{\partial^2 \theta^*}{\partial x_i^* \partial x_i^*} - \frac{1}{\mathrm{Re}_\tau \mathrm{Pr}} \frac{\omega^*}{\Omega^*} \frac{u_3^*}{u_\mathrm{b}^*} \delta_{i3} \tag{11}$$

where Pr is the Prandtl number and u_b^* denotes the mean bulk velocity.

Values of the dimensionless parameters Re_τ, Ra = Gr Pr and Pr imposed in the examined cases are reported in Table 1, together with the values of the bulk Reynolds number Re_b, Richardson number $\mathrm{Ri} = \mathrm{Gr}/\mathrm{Re}_\mathrm{b}^2$ and Nusselt number resulting from the DNS. The imposed value of $\mathrm{Re}_\tau = 180$ has been selected. Purely forced and mixed convection conditions are considered, in order to investigate the effect of buoyancy on the mean flow and turbulent quantities. To achieve $\mathrm{Ri} \approx 0.24$ (i.e. non negligible buoyancy effects), a Rayleigh number of 2.5×10^4 has been set. The Prandtl number value is $\mathrm{Pr} = 0.031$, corresponding to LBE at the reference temperature $T^o = 220°\mathrm{C}$, see [4].

No-slip conditions ($u_i|_v = 0$) and a constant heat flux ($-\lambda \partial\theta/\partial\eta|_v = q''$) are imposed at the rod wall, whilst periodic boundary conditions are enforced at all ends of the unit cell portrayed in Figure 1(b).

4. Numerical method

The Navier-Stokes equations are treated in their conservative form, making use of the Finite Volume method discretization on Cartesian grids. A second-order Projection Method [9] is implemented to decouple the pressure field from velocity. The time-advancement scheme is second-order accurate and it is the result of a three-level scheme, which is semi-implicit (explicit in the streamwise direction z) for the diffusive terms, and explicit (using Adams-Bashforth) for the advective terms. For spatial derivatives a second-order central difference scheme on staggered non-uniform Cartesian grids is implemented. The modeling of irregular cylindrical boundaries on Cartesian grids is accomplished thanks to an original scheme developed for convective cases, see [10] and [11]. The computational domain size and grid spacing are reported in Table 2. A constant grid spacing is set along the streamwise direction z, while the discretization is non-uniform along the two cross-flow directions (x and y). Considering the fully developed conditions outlined above, the bundle of rods can be assumed to have an infinite extension along the streamwise direction z. The conformity of the computational grid spacing to DNS standards is verified by a posteriori evaluation of the minimum Kolmogorov scale η_{min} (see again Table 2).

5. Results and discussion

Some relevant results obtained are presented in this section. For the sake of conciseness, asterisks denoting dimensionless variables will be dropped starting from now on.

Firstly, contours of the average streamwise velocity component $\overline{u_z}$ are displayed: in Figure 2 the different behaviour between the forced (Ri $= 0$) and mixed case (Ri $= 0.24$) is evident by observing the extent of the region of maximum velocity. Indeed, for Ri $= 0$ the streamwise velocity component $\overline{u_z}$ peaks at the center of the subchannel and decreases its value moving towards the narrow gap. For Ri $= 0.24$ the contour area for the maximum velocity value expands towards the narrow gap, as in Figure 2(b). The previous observation are reinforced by the profile plots in Figure 3, for the narrowest and largest gap between two adjacent rods. The mean velocity reaches its maximum value at the center of the subchannel for both cases and it assumes almost the same value (Figure 3(b)), whereas in the narrow gap (Figure 3(a)) the maximum velocity is notably higher for the mixed case.

Another parameter that reflects this behaviour is the bulk Reynolds number: the values obtained from the simulations are $\mathrm{Re}_b = 1785$ for the forced case and $\mathrm{Re}_b = 1850$ for the mixed one, see Table 1. The two Reynolds numbers do not differ much from each other, even with the effect of aiding buoyancy, which is mainly due to the friction Reynolds number of both simulations being set at 180 (a relatively low value).

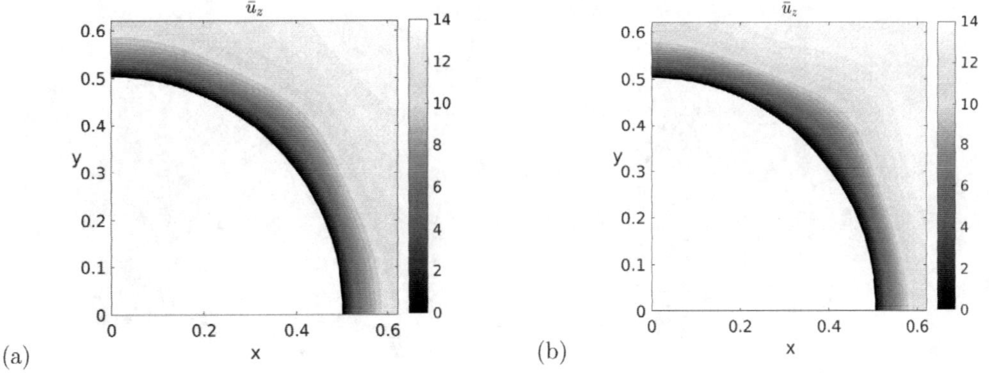

Figure 2. Contours of $\overline{u_z}$ for (a) Ri $= 0$ and (b) Ri $= 0.24$.

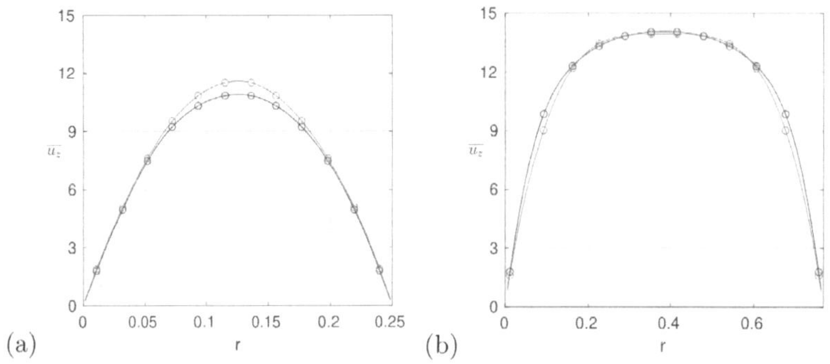

Figure 3. Profiles of $\overline{u_z}$ for narrow (a) and large gap (b): the black line stands for Ri $= 0$, while the red one for Ri $= 0.24$.

Contours of the time-averaged cross-flow velocity magnitude $\overline{u}_\perp = \sqrt{u_x^2 + u_y^2}$ are shown in Figure 4 buoyancy determines a decrease of the magnitude of $\overline{u}_\perp$. Moreover, in Figure 4 the streamlines of the cross-flow components are displayed: the circulation is driven towards the rods in the narrowest gap, while it is directed towards the center of the subchannel in the largest gap for Ri $= 0$. Otherwise, for the mixed case, a different trend is observed: the circulation is directed towards the center of the subchannel for both large and narrow gap.

To better understand the effect of aiding buoyancy, a brief analysis of the fluid excess temperature field is reported. For conciseness, in Figure 5(a) the contour of the average flow excess temperature is shown for the forced case only, since the mixed convection case for Ri $=$ 0.24 exhibited a similar qualitative trend. Considering the low Prandtl number of the working fluid, heat transfer is actually dominated by diffusion. The profiles for the narrow and large gap are presented as well in Figure 5(b)-(c): in general, for both cases, slightly higher values of $\overline{\theta}$ are found in the narrow gap; in the large gap (Figure 5(c)) temperatures for Ri $= 0.24$ are visibly lower with respect to Ri $= 0$. These rather small differences can be appreciated in the slight decrease of the Nusselt number, from Nu $= 7.60$ for Ri $= 0$ to Nu $= 7.50$ Ri $= 0.24$ (Table 1).

In Figure 6(a) and 6(b), contours of the time-averaged turbulent kinetic energy are displayed. At first glance it is easily understood that for Ri $= 0.24$ there is a significant decrease of fluctuation intensity (up to $\approx 80\%$). The result is in agreement with other studies carried out with different Richardson numbers [12], which highlight how the presence of aiding buoyancy leads to a decrease in kinetic energy over the entire domain. Taking into account the profiles in

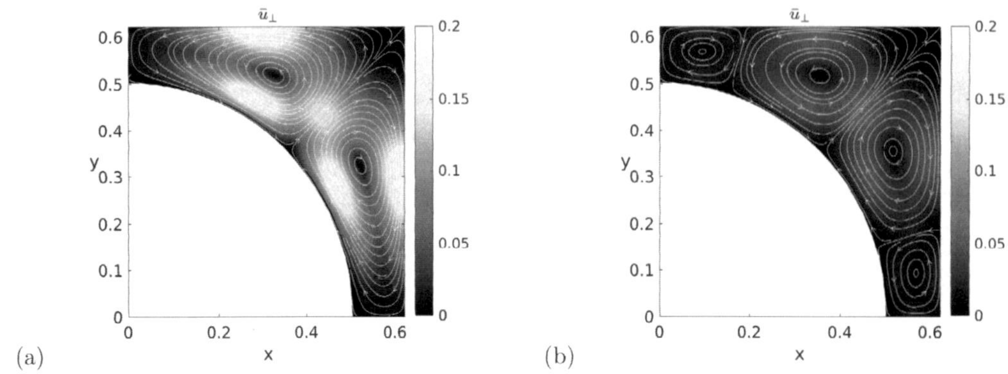

Figure 4. Contours of $\overline{u}_\perp$ for (a) Ri $= 0$ and (b) Ri $= 0.24$.

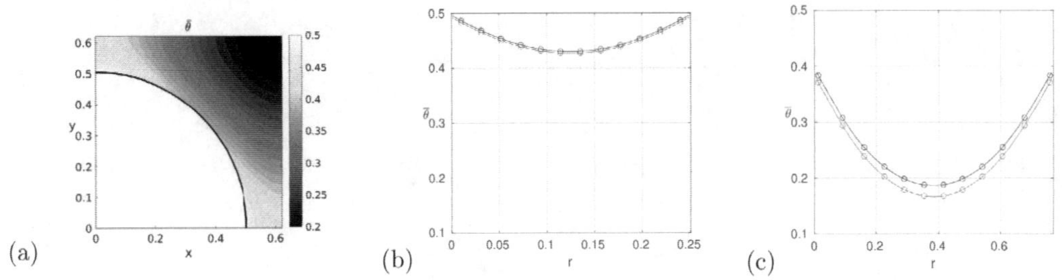

Figure 5. Contours of $\overline{\theta}$ for Ri = 0 (a). Profiles for narrow (b) and large gap (c): the black line stands for Ri = 0, while the red one for Ri = 0.24.

Figure 7, it can be pointed out that the decrease is more pronounced in the large gap region, see Figure 7(b). In addition, in Figure 7 the contributions of $\overline{u'_x u'_x}$, $\overline{u'_y u'_y}$ and $\overline{u'_z u'_z}$ are reported: the decrease of fluctuation intensity can be mainly ascribed to the z-component $\overline{u'_z u'_z}$.

The low Prandtl value considered in this study also entails that the intensity of temperature fluctuations is generally very low, as heat transfer is dominated by diffusion. The effect of aiding buoyancy further reduces temperature fluctuations, in agreement with what has been observed so far for the fluctuating velocity field. A proof comes from the maps of temperature variance in Figure 8: a significant decrease in the values of $\overline{k_\theta}$ can be appreciated over the entire domain, since the peak value is reduced by $\approx 93\%$ when aiding buoyancy is active.

6. Concluding remarks

Results are presented of Direct Numerical Simulations of turbulent flow around a square-arranged bundle of heated rods, cooled by Liquid Lead-Bismuth Eutectic (LBE) at Pr = 0.031. The numerical technique used is an original Cartesian immersed boundary method. The computational domain consists of four subchannels of a bundle with a pitch-to-diameter ratio of 1.25. Simulations for $\mathrm{Re}_\tau = 180$ are performed, in both forced (Ri = 0) and mixed convection conditions (Ri = 0.24); the bulk Reynolds number is $\mathrm{Re_b} = 1785$ for Ri = 0 and $\mathrm{Re_b} = 1850$ for Ri = 0.24. Statistics are reported in the form of 2D time-averaged distributions on one quarter of a single subchannel, together with with 1D profiles along the narrowest and largest gap between two adjacent rods. At these Reynolds number values the effect of buoyancy can be appreciated by looking at the average velocity profile (especially cross-flow components),

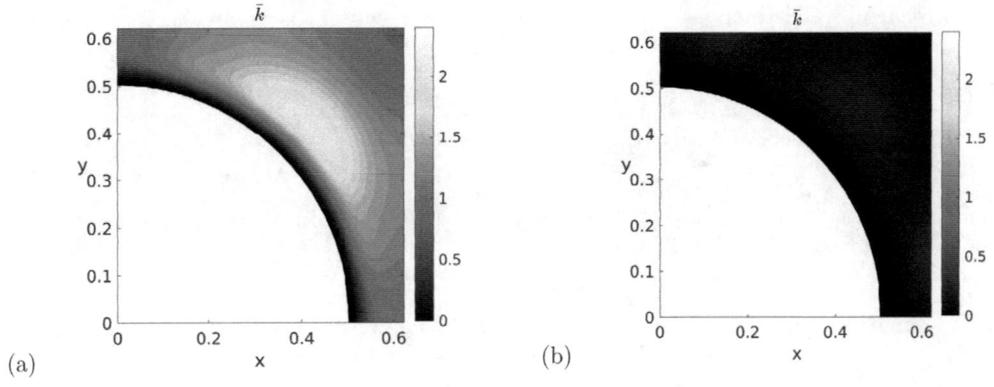

Figure 6. Contours of $\overline{k}$ for (a) Ri = 0 and (b) Ri = 0.24.

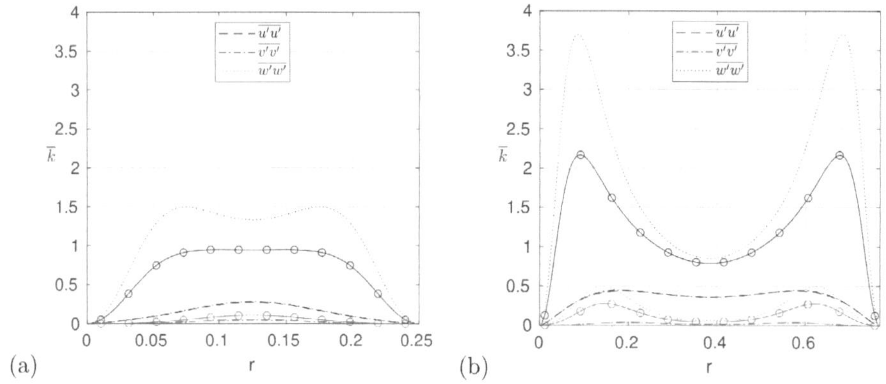

Figure 7. Profiles of $\overline{k}$ for narrow (a) and large gap (b): the black line with circles stands for Ri = 0, while the red one with circles for Ri = 0.24.

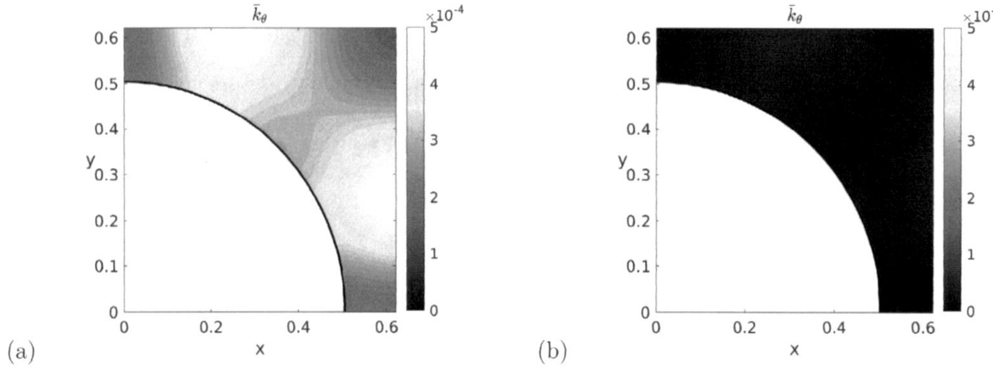

Figure 8. Contours of k_θ for (a) Ri = 0 and (b) Ri = 0.24.

but even more so by focusing on the fluctuations related to turbulent kinetic energy. Moreover, the effects on the mean temperature fields is almost negligible. The Nusselt number reveals a slight deterioration of the heat transfer rate for Ri = 0.24. This can be ascribed to the reduction of turbulent fluctuations of both velocity and temperature. The data presented herein are also intended to serve as a benchmark for the development and validation of turbulent heat transfer models, both current and to come.

References

[1] El-Genk M S, Bedrose S D and Rao D V 1990 *Heat Transfer Engineering* **11** 32–43
[2] El-Genk M, Su B and Guo Z 1993 *International Journal of Heat and Mass Transfer* **36** 2359–2374
[3] Meyer L 2010 *Nuclear Engineering and Design* **240** 1575–1588
[4] Angeli D, Fregni A and Stalio E 2019 *Nuclear Engineering and Design* **355**
[5] Piller M and Stalio E 2012 *International Journal of Heat and Mass Transfer* **55** 6506–6513
[6] Piller M, Polidoro S and Stalio E 2014 *International Journal of Heat and Mass Transfer* **79** 1014–1026
[7] Hallinan K and Viskanta R 1985 *International Journal of Heat and Fluid Flow* **6** 256–264
[8] Patankar S, Liu C and Sparrow E 1978 *International Journal of Heat and Mass Transfer* **21** 557–566
[9] Gresho P M 1990 *International Journal for Numerical Methods in Fluids* **11** 587–620
[10] Barozzi G S, Bussi C and Corticelli M A 2004 *Numerical Heat Transfer B* **46** 56–77
[11] Angeli D and Stalio E 2019 *Computers and Fluids* **183** 148–159
[12] Kasagi N and Nishimura M 1997 *International Journal of Heat and Fluid Flow* **18** 88–99

Effect of closed-loop nanochannels on the onset of explosive boiling: a molecular dynamics simulation study

R Fallahzadeh[1,*], F Bozzoli[1,2] and L Cattani[1]

[1] Department of Engineering and Architecture, University of Parma, Parma, Italy
[2] SITEIA.PARMA Interdepartmental Centre, University of Parma, Parma, Italy

* Corresponding author: rasoul.fallahzadeh@unipr.it

Abstract. As a primary boiling mode, explosive boiling has shown a promising future in many applications and received much research attention. The topology of the solid surface in contact with the liquid, particularly nanostructured surfaces, significantly affects the onset time of explosive boiling of a liquid nanofilm. Most studies investigated explosive boiling on non-closed-loop (parallel) nanochannel surfaces. Here, for the first time, explosive boiling in a closed-loop nanochannel was studied by the non-equilibrium molecular dynamics simulation method. Explosive boiling of liquid argon nanofilm on solid copper surfaces with different topologies, including an ideally smooth, a non-closed-loop, and a closed-loop nanochannel, was simulated. The results showed that, compared with the ideally smooth surface, the onset time of explosive boiling decreased for the non-closed-loop and closed-loop nanochannel surfaces. However, it turned out that compared to the non-closed-loop nanochannel, using the closed-loop nanochannel has an adverse effect on heat flux and the onset time of explosive boiling.

1. Introduction

There is significant scientific and practical value in investigating the phase transition processes of ultrathin liquid films on a solid surface. The surface temperature determines whether the phase transition occurs during normal or explosive boiling. Explosive boiling occurs when a liquid is heated above its saturation temperature but below its critical thermodynamic temperature; this causes a sharp pressure increase and the rapid ejection of liquid near the superheated surface, giving the phenomenon that looks like an explosion [1].

The utilization of explosive boiling can serve as an effective way to enhance the efficiency of heat transfer on heated surfaces. Therefore, enhancing the comprehension of this phenomenon is an imperative and crucial subject matter to improve thermal energy conversion efficiency.

Although nearly all solid surfaces display some degree of molecular-level roughness, creating multiple distinct nanopatterns on a flat surface is feasible by advancing nano-manipulation technology [2]. Although explosive boiling of ultrathin liquid films has been experimentally investigated in recent years [3-4], a profound comprehension of physical phenomena based on macroscopic experiments is not fully understood owing to the intricacy of physical phenomena at the nanoscale, the limits on the time scale, and extremely high heat fluxes and temperatures. Therefore, this phenomenon must still be better understood, especially in nanoscale. Recent developments in hardware and software have made it possible to use computational methods to probe explosive boiling at the nanoscale. Molecular dynamics simulation (MDS) method has proven to be an effective tool for investigating the dynamics of explosive boiling on nanostructured surfaces in recent years.

Table 1 summarizes and tabulates the results of several MDS research on using parallel (non-closed-loop) nanochannel surfaces in explosive boiling. Results have shown that using parallel nanochannels can decrease the onset time of explosive boiling [5-8] and increase heat flux [7-8].

Table 1. Literature overview of MDS studies of explosive boiling on parallel nanochannel surfaces.

Study	Mediums of Fluid / Solid	Nanostructure Topology	Substrate Temperature (K)
Zhang et al. [5]	Argon / Copper	Parallel cubical nanochannels	350
Liao and Duan [6]	Argon / Gold	Parallel cubical nanochannels	240
Bai et al. [7]	Water / Copper	Parallel cubical nanochannels	800
Bai et al. [8]	Water / Copper	Parallel cubical nanochannels	800

Therefore, the MDS studies proved that non-closed-loop nanochannel surfaces could drastically enhance the explosive boiling performance. To the best of the authors' knowledge, the effect of closed-loop nanochannel surfaces on explosive boiling has not been investigated within available literature and remains to be unveiled. However, using closed-loop nanochannels could alter the absorption of heat flux by the liquid nanofilm. Hence, they should also be of interest from a scientific viewpoint. The main novelty of this study is the investigation of the effect of closed-loop nanochannels, together with the parallel nanochannels and the ideally smooth surfaces, on explosive boiling behavior of liquid nanofilm using the MDS method.

2. Simulation method

Four simulation cases, including one liquid–vapor argon coexistence system (simulation Case I) and three explosive boiling simulation cases with distinct surface topologies (an ideally smooth, a non-closed-loop, and a closed-loop nanochannel), were studied to examine the model validation and impact of nanostructured surfaces on explosive boiling, respectively. The details of the simulation procedure, which are presented here, were similar for all but one. Simulation details for the exception, simulation Case I, are presented in section 3.1.

All simulations were carried out using the open-source MDS code LAMMPS (large-scale atomic/molecular massively parallel simulator) package [9]. Due to the lack of visualization, the Open Visualization Tool (OVITO) was used for the post-processing analysis [10].

The interactions between argon (Ar) and copper (Cu) atoms were modeled by the Lennard-Jones 12-6 function potential (U_{ij}) as follows [11]:

$$U_{ij}=4\varepsilon_{ij}\left[\left(\frac{\sigma_{ij}}{r_{ij}}\right)^{12}-\left(\frac{\sigma_{ij}}{r_{ij}}\right)^{6}\right] \tag{1}$$

where r_{ij} denotes the distance between interacting atoms (i and j), and ε_{ij} and σ_{ij} represent the energy parameter (the depth of the potential well) and the length parameter (the specified distance at which the interatomic potential is zero), respectively. Table 2 lists the energy and length parameters obtained from Refs. [12,13], the Lorentz combination rule [14], and a modified version of the Berthelot rule proposed by Din and Michaelides [15]. In order to guarantee the calculation's precision and accelerate the simulations, the cut-off radius was adjusted at 9×10^{-10} m. Moreover, to maintain the position of Cu atoms throughout the simulations, an individual spring force was applied to each atom; hence, the Cu atoms could vibrate around their initial lattice position. A force of magnitude -Kr was used at every single step, where K is the spring constant, and r is the distance between the atom's current location and its original location. The spring constant is closely related to Young's Modules; therefore, it is possible to estimate it as [16]:

$$K=Ed \tag{2}$$

where E is Young's Modules (128.20 GPa at 300 K [17]), and d is its corresponding lattice constant (3.61×10^{-10} m) of copper. The above data and correlation give the following value for the spring constant: K= 46.34 J/m^2.

Table 2. Force field parameters of different atom pairs.

Atom pairs	σ (m)	ε (J)
Cu-Cu	1.93×10^{-10}	32.80×10^{-21}
Ar-Ar	3.40×10^{-10}	1.67×10^{-21}
Ar-Cu	2.67×10^{-10}	1.04×10^{-21}

As presented in Figure 1, three explosive boiling simulation cases in this work employed a rectangular simulation cell consisting of three different regions: a solid copper surface (with different topologies), a liquid argon region, and a vapor argon region.

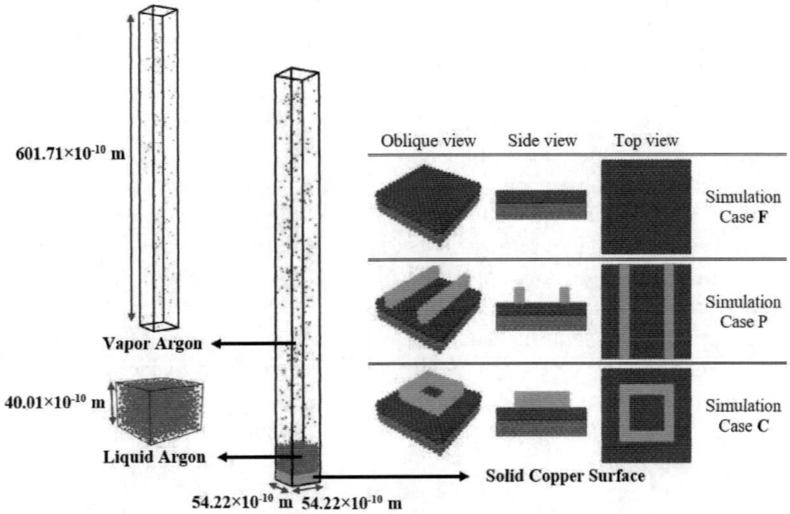

Figure 1. Snapshot of the initial configuration of explosive boiling simulation cases. Purple and yellow represent Ar and Cu atoms, respectively. Brown, red, and light/dark green represent the fixed, phantom, and conduction region of the solid copper surfaces, respectively.

The solid copper surface was created of copper (in face-centered cubic (FCC) with a lattice constant of 3.61×10^{-10} m which corresponds to the copper density of 8936.78 kg/m^3) and set at the bottom of the simulation boxes so that its 〈0 0 1〉 surface in touch with the liquid argon region. All the solid copper surfaces had the same dimensions in X– and Y– directions (shown in Figure 1). In order to better understand the influences of different topologies, a non-closed loop nanochannel (the surface P) and a closed-loop nanochannel (the surface C) with the same dimensions in nanowall width and height (3.61×10^{-10} m and 9.04×10^{-10} m, respectively) were constructed. The nanowalls were formed in the upper layer of the ideally smooth surface (the surface F). As can be seen in Figure 1, for all simulation cases, to create a more realistic substrate, the solid copper surfaces had three different regions from the bottom to the top, namely, the fixed region, the phantom region, and the conduction region. The first monolayer at the bottom of the surface (the fixed region) was kept stagnant to avoid any migration, penetration, and thermal deformation of the solid surface. The next four monolayers (the phantom region), whose temperature was regulated by a thermostat, acted as the heat source for generating heat flux. The remaining monolayers (the conduction region) were set as real atoms from which the liquid argon was heated, thus triggering explosive boiling process. The nanowall and ideally smooth surfaces are marked with different colours (light and dark green, respectively) to clearly show the topologies. The liquid argon region of the initial thickness of 40.01×10^{-10} m, containing 2476 atoms, was placed over the solid copper surfaces. The liquid argon density was 1396.20 kg/m^3 at 87.18 K [18]. The space above the liquid argon region was filled by 152 vapor argon atoms (with an initial thickness of 601.71×10^{-10} m) with a density of 5.70 kg/m^3 at 87.18 K [18] to simulate a real phase change system. Considering this, the size of the vapor space in the Z– direction was selected to be large (much greater than the liquid argon film thickness) so that the top boundary had almost no influence on explosive boiling process. In addition, the periodic boundary conditions were applied in the X– and Y– directions, so an infinite plane of explosive boiling is simulated, and the non-periodic fixed boundary condition was applied in the Z– direction. This configuration complies with actual experiment procedures.

The simulation process was divided into four steps: energy minimization, equilibrium preparation, equilibrium relaxation, and non-equilibrium explosive boiling. Step I (energy minimization): at the beginning of each simulation, after the initial configuration and the interaction force field were prepared, using the conjugate gradient (CG) algorithm, the energy of the whole simulation system was minimized. The stopping tolerances for energy and force were taken as 1.60×10^{-24} J and 1.60×10^{-9} N, respectively. Step II (equilibrium preparation): in this step, the uniform temperature of the simulation systems was set at 87.18 K, which is the boiling point of liquid argon at 1 bar [18] under the Langevin thermostat for 2×10^{-9} s. During this step, all simulation systems reached their equilibrium state. Step III (equilibrium relaxation): in this step, the simulation systems were run for another 2×10^{-9} s while the thermostat was changed into NVE (i.e., constant atom number, domain volume, and energy) ensemble for the fluid argon domain; however, the temperatures of the solid copper surfaces were still fixed at 87.18 K by the Langevin thermostat. Step IV (non-equilibrium explosive boiling): in this step, the phantom monolayer atoms on solid copper surfaces were abruptly elevated to a higher temperature (250.00 K), simulating ultrafast laser heating via Nose-Hoover thermostat, while the rest of atoms (Ar atoms as well as Cu atoms in the fixed and conduction monolayers) were free to interact with the NVE ensemble, and simulations were run for 1×10^{-9} s.

Usually, the onset of explosive boiling is defined as the first appearance of a complete layer of vapour film with a specific thickness (20.00×10^{-10} m in this work) over the solid surface. For the density studies, the simulation systems were divided into equal bins with a thickness of 2.00×10^{-10} m in the direction of the phase change (Z– direction). The heat flux between the solid copper surface to Ar atoms was calculated by [19]:

$$q=\frac{1}{A}\frac{\partial E_f}{\partial t} \tag{3}$$

where A represents the cross-sectional heat transfer area between solid copper surfaces and Ar atoms, projected onto the X-Y plane (the nominal surface area), and E_f stands for the total energy of whole Ar atoms.

3. Results and discussion

3.1. Model Validation: Simulation Case I

In order to determine the reliability of computational accuracies, simulation model validation for a liquid–vapor argon coexistence system (the simulation Case I) was carried out in this work. Since density is the common evaluation property for validation of MDS method, this property was calculated and compared with the NIST-data [18], a famous fluid properties database. The methodology, results, and discussion are presented here. A liquid–vapor argon coexistence system, as shown in Figure 2, was developed to validate the simulation model for two-phase equilibrium. The liquid film size was $40.00\times40.00\times40.00$ ($\times10^{-10}$ m)^3 (consisting of 1347 Ar atoms) with a $40.00\times40.00\times400.06$ ($\times10^{-10}$ m)^3 long vapor domain (consisting of 55 Ar atoms) on each side. Then it was set at 87.18 K (the boiling point of argon at 1 bar [18]) through NVT ensemble (constant atom number, constant domain volume, and temperature canonical ensemble) for 2.5×10^{-9} s.

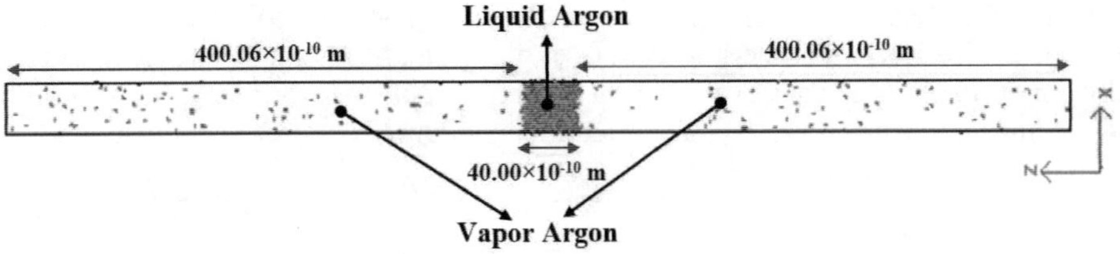

Figure 2. Snapshot of the initial configuration of the simulation Case I.

In order to extract the mean value of density distributions along the Z– direction, the system was divided into 210 bins with a height of 4.00×10^{-10} m. Figure 3 demonstrates a good agreement between the liquid and vapour densities from the simulation and the experimental data.

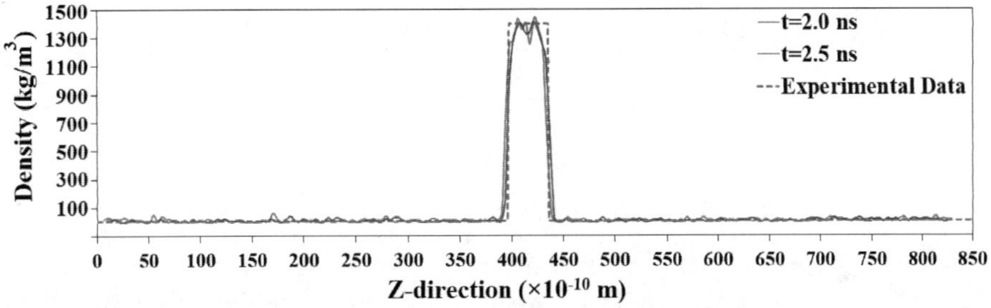

Figure 3. Comparison between density profile along the Z– direction (at 2 and 2.5×10^{-9} s) for the simulation Case I and the experimental data. The experimental data are from Ref. [18].

3.2. Effects of surface topology: Simulation Cases F, P, and C

Figure 4 illustrates the snapshots and onset time of explosive boiling for liquid argon nanofilms on solid copper surfaces with different topologies. As can be seen, adding nanochannels could expedite the explosive boiling occurrence. Moreover, the non-closed-loop nanochannel surface (the simulation Case P) shows a lower onset time than the closed-loop nanochannel surface (simulation Case C).

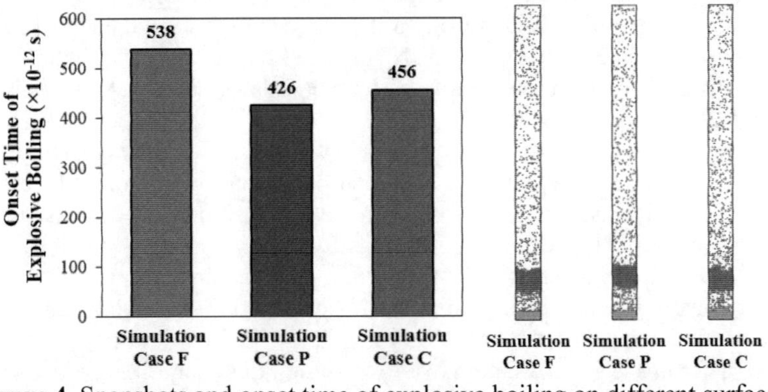

Figure 4. Snapshots and onset time of explosive boiling on different surfaces.

Figure 5 shows the unsmoothed and smoothed heat flux of the fluid argon domain over different solid copper surfaces. As can be seen, for all simulation cases, the heat fluxes reached the maximum as the solid temperature grew up continuously under the thermostat. Then they reduce to a relatively low value. The transition from the solid-liquid to the solid-vapor interface occurs during this time. When the vapor argon layer appears between the liquid argon cluster and the solid copper surface, it significantly increases the interfacial resistance to the heat flux (the Kapitza thermal resistance) because the Kapitza thermal resistance of solid-vapor is much higher than the solid-liquid one. Finally, heat fluxes reach almost zero after the departure of the liquid argon cluster and its vapor-liquid tail from the surface. Moreover, Figure 5 indicates that the non-closed-loop and closed-loop nanochannel surfaces show a higher value for the maximum heat flux for the maximum heat flux than the ideally smooth surface. This result is also compatible with the onset time of explosive boiling (Figure 4).

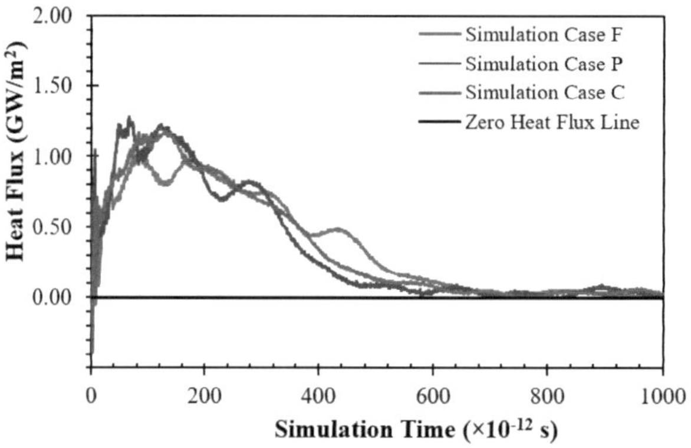

Figure 5. heat flux for different surface.

The presence of nanochannels increases the contact area between the liquid argon nanofilms and the solid copper surface. Therefore, more liquid Ar atoms will be at the solid-liquid interface, which in turn excels in heat transfer from the solid surface to the fluid argon domain, decreasing the onset time of explosive boiling. Furthermore, the potential energy interaction between liquid Ar atoms and copper atoms is crucial in heat transfer at the nanoscale. The potential energy between the atoms of argon and copper is most concentrated close to the surface of the copper, especially in the nanowalls. Consequently, using nanochannels enhances the interaction of potential energy between the atoms of liquid Ar and the atoms of the copper surface. Thus, it features a significant thermal vibrational coupling, which leads to greater heat flux and an earlier onset time of boiling.

The simulation Cases P and C have an equal wetted surface area; however, interestingly the simulation Case C has a lower heat flux and longer onset time. Therefore, the onset time of explosive boiling cannot be determined solely by the value of the interaction potential energy, and there must be other reasons. Herein, this unexpected behavior is attributed to the space for the motion of Ar atoms (movement space). Therefore, for further analysis, the motion of Ar atoms should be studied via their line trajectories plot.

Figure 6 indicates the snapshots of line trajectories of liquid Ar atoms between the nanowalls for Cases P and C after 4×10^{-10} s of beginning the simulation and before the onset of explosive boiling. Inspection of Figure 6 manifests that using closed-loop lead to a significant decrease in the mobility of Ar atoms. In other words, the movement of Ar atoms for simulation Case P is much better than the simulation Cases C. Therefore, other than the interaction potential energy, the effect of the movement space on the onset of explosive boiling is significant. When the solid–liquid interactions are very large, and on the other hand, the movement space is very small, the liquid Ar atoms cannot break their high potential energy barrier in a short time and move toward the liquid nanofilm bulk; therefore, accounts for the relatively long time required for the onset of explosive boiling. Therefore, the onset time of explosive boiling is delayed. Moreover, the lower movement of liquid atoms, the less they collide, so heat transfer occurs at a lower flux.

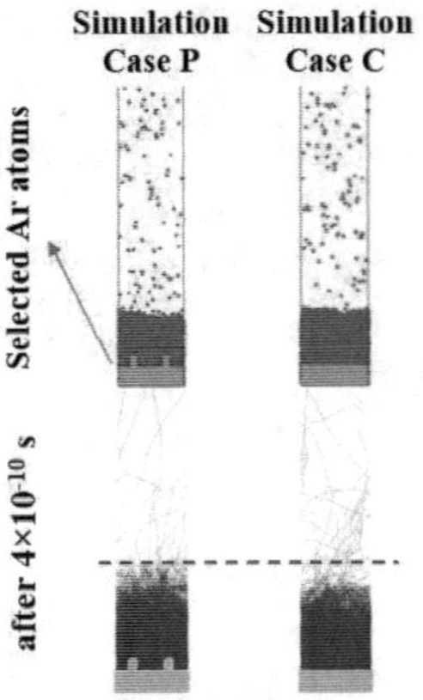

Figure 6. Projections of selected Ar atoms and their trajectory motion lines for simulation Cases P and C.

4. Conclusion

This study investigates the effect of using a closed-loop nanochannel on the nanoscale explosive bubbling process. By comparing the explosive boiling process on substrates with different topologies, including an ideally smooth surface, a non-closed-loop nanochannel, and a closed-loop nanochannel, it was found that the onset time of explosive boiling decreased for the non-closed-loop and closed-loop nanochannel surfaces when compared to the ideally smooth surface. In contrast to the non-closed-loop nanochannel, however, the use of the closed-loop nanochannel has a negative effect on heat flux and the onset time of explosive boiling. Based on the abovementioned results and analysis, closed-loop nanochannel surfaces should be avoided in the design of nanostructured surfaces.

5. References

[1] Bai P, Zhou L and Du X 2020 *International Journal of Heat and Mass Transfer* **162** 120375.
[2] Seyf HR and Zhang Y 2013 *ASME Journal of Heat and Mass Transfer* **135** 121503.
[3] Kudryashov SI and Allen SD 2001 *Journal of Applied Physics* **95** 5820–5827.
[4] Kudryashov SI and Allen SD 2006 *Journal of Applied Physics* **100** 104908.
[5] Zhang S, Hao F, Chen H, Yuan W, Tang Y and Chen X 2017 *Applied Thermal Engineering* **113** 208–214.
[6] Liao MJ and Duan LQ 2020 *Numerical Heat Transfer, Part A: Applications* **78** 94–105.
[7] Bai P, Zhou L and Du X 2021 *Journal of Molecular Liquids* **335** 116169.
[8] Bai P, Zhou L and Du X 2021 *International Communications in Heat and Mass Transfer* **120** 105036.
[9] Leite PR and Koning MD 2019 *Computational Materials Science* **159** 316–326.
[10] Stukowski A 2010 *Modelling and Simulation in Materials Science and Engineering* **18** 015012.
[11] Zarringhalam M, Ahmadi-Danesh-Ashtiani H, Toghraie D and Fazaeli R 2019 *International Journal of Heat and Mass Transfer* **141** 1–8.
[12] Liu H, Qin X, Ahmad S, Tong Q and Zhao J 2019 *International Journal of Heat and Mass Transfer* **145** 118799.
[13] Liu H, Deng W, Ding P and Zhao J 2021 *Nuclear Engineering and Design* **382** 111400.
[14] Delhommelle J and Millié P 2001 *Molecular physics* **99** 619–625.
[15] Din XD and Michaelides EE 1997 *Physics of Fluids* **9** 3915–3925.

[16] Chen Y, Zou Y, Sun D, Wang Y and Yu B 2018 *International Journal of Heat and Mass Transfer* **118** 1143–1151.
[17] Reed RP and Clark AF 1983 *Materials at low temperatures* (Michigan: American Society for Metals) p 8.
[18] NIST Chemistry WebBook, NIST Standard Reference Database, NIST Chemistry WebBook. 10.18434/T4D303.
[19] Ilic M, Stevanovic VD, Milivojevic S and Petrovic MM 2023 *Applied Thermal Engineering* **220** 119749.

Experimental validation of a heat exchanger model for thermoacoustic applications

A Di Meglio[1,*], N Massarotti[1] and A Piccolo[2]

[1] Department of Engineering, University of Naples "Parthenope", Centro Direzionale, Isola C4 – 80133, Italy.
[2] Department of Engineering, University of Messina, Contrada di Dio, 98166 S. Agata-Messina, Italy.

* Corresponding author: armando.dimeglio001@studenti.uniparthenope.it

Abstract. Thermoacoustics is a promising technology for energy conversion purposes. Among the bottlenecks limiting a large diffusion of thermoacoustic devices, there are heat exchangers, whose behaviour in oscillatory flows is rather different than those working in stationary flows. Furthermore, the classical linear acoustic theory in the frequency domain cannot predict with high-fidelity the thermo-fluid dynamics of such heat exchangers. In this article, a CFD model based on a standing wave device including a parallel plate heat exchanger is proposed. The setup is inspired by a similar prototype of a thermoacoustic engine in which the performance of the ambient heat exchanger was tested. The results of the CFD model are therefore compared, in terms of the temperature difference between fluid and solid wall in the heat exchanger, average heat flux and Nusselt number, with experimental data showing a satisfying agreement.

Keywords: *oscillatory flows, CFD, heat exchanger, thermoacoustic engine, porous media*

1. Introduction

Thermoacoustic engines are energy devices converting heat into electricity through acoustic waves [1]. They are based on thermoacoustic effect: if a temperature difference across a particular porous media called stack overcomes a threshold value (the onset temperature), pressure and velocity instabilities arise in the system [2]. This stack, the core where the energy conversion between heat and acoustic power takes place, is generally sandwiched between the hot Heat eXchanger (HX), the source at high temperature, and the cold one representing the thermal sink. The working principle and therefore the optimal design of heat exchangers in oscillatory flows is distinct from those one functioning in steady-state conditions. In oscillatory flows, the geometrical characteristics (the length and the pore dimension) of the heat exchangers are respectively bounded by the peak-to-peak-particle displacement, and the thermal penetration depth, strictly depending on the working frequency [3]. For this reason, both standard heat transfer correlations, which rely on frequency, and simple one-dimensional models, hindered by sudden changes in section area that impede fluid flow and the development of thermal fields, fail to accurately represent the observed physical phenomena [4]–[6].

In this article, a thermoacoustic heat engine model based on Computational Fluid Dynamics is presented and validated against experimental data of a finned tube heat exchanger (ambient one). A comprehensive review of CFD modelling in thermoacoustics is provided in ref. [7]. While the ambient heat exchanger is modelled microscopically taking into account the fluid flow across the fins, the hot heat exchanger and the stack are modelled by means of an already validated porous media macroscopic approach for oscillatory flows. This has been done in order to reduce the computational costs of the simulations [8].

2. Physical system and Experimental setup

In this section the experimental setup, described in a more detail in ref. [9], is briefly described. The physical system is made of a standing wave thermoacoustic engine. The working fluid is air at atmospheric. Its resonance frequency stands at about $f_0 = 145\,Hz$ and by an onset temperature difference of approximately 300 °C. A sketch of the standing wave engine is shown in Figure 1. The resonator is made of stainless-steel tube. The main components are the hot duct (between the right wall the hot HX) also known as thermal buffer tube (diameter of 10.5 cm and 10 cm long), the stack (diameter of 10.5 cm and length of 8 cm). The thermoacoustic core is a ceramic (cordierite) honeycomb stack 7.9 cm long in the oscillating flow direction. The pores have a squared shape with R_h= 0.28 mm while the cell density is near 400 Cell Per Square Inch. Even though it is previously shown that suck kind of stacks are not among the most efficient ones [10], they are commercial being used as filter in the automotive sector. An electric heater is adopted as input energy source into the system in place of a classical heat exchanger. The ambient HX is a finned-tube type 2 cm long in the oscillatory flow direction. The fins are made of strips of laminated copper 0.45 mm thick and 2 cm wide. These strips have been welded to the external walls of the pipe with a fin spacing of 1.1 mm (resulting in R_h= 0.55 mm), as shown in Figure 2. The porosity of this HX (ϕ) is calculated to be around 54%. A photograph of the presented finned-tube HX is pictured in Figure 2 .

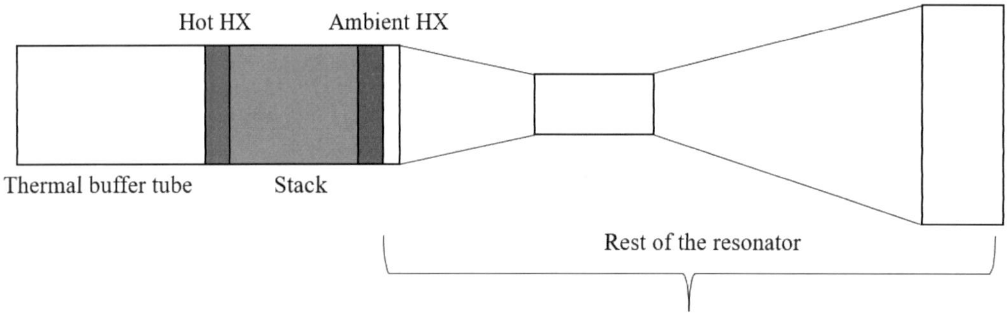

Figure 1 A sketch of the thermoacoustic engine

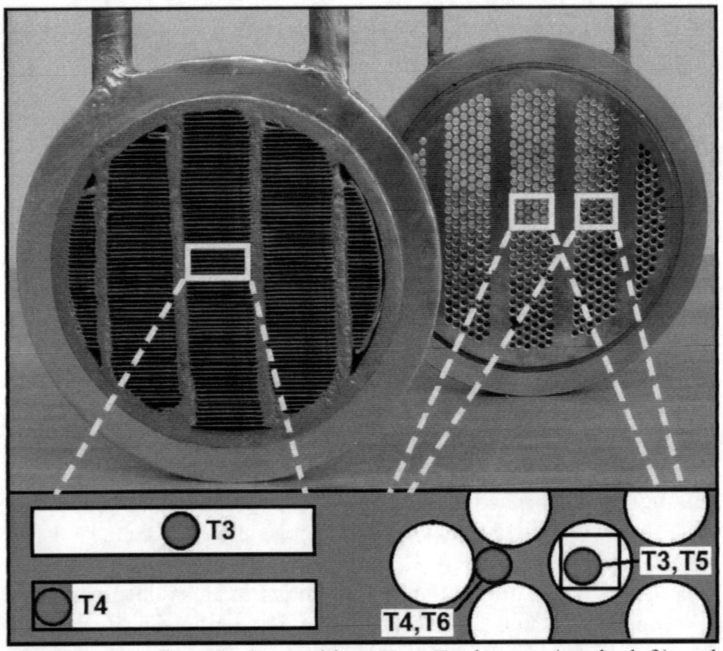

Figure 2 The finned tube Ambient Heat Exchanger (on the left) and position of the thermocouples

3. Governing equations and numerical model

Thermoacoustic phenomena are generally described the Navier-Stokes equations system. At the macroscopic level, in the porous media such as the stack and the HXs, specific source term and virtual mass/ thermal inertia corrections are respectively considered (C_{VM}, C_{TI}) as shown in the following equations (1-3) [8].

$$\frac{\partial \rho}{\partial t} + \nabla \cdot \boldsymbol{v_D} = 0 \tag{1}$$

$$\frac{1}{\phi}\rho \frac{\partial \boldsymbol{v_D}}{\partial t}(1 + C_{VM}\,\xi) + \frac{1}{\phi^2}(\boldsymbol{v_D}\cdot\nabla)\boldsymbol{v_D} = \nabla\cdot\left[-\langle p\rangle \boldsymbol{I} + \frac{K}{\phi}\right] - \frac{\mu}{r_h^2}\Re\left[\frac{1}{Da}\right]u_D\xi \tag{2}$$

$$K = \mu\left(\nabla\boldsymbol{v_D} + \nabla\boldsymbol{v_D}^{\mathrm{T}}\right) - \frac{2}{3}\mu\,(\nabla\cdot\boldsymbol{v_D})\mathbf{I}$$

$$\rho_f c_p(1 + C_{TI}\,\xi)\frac{\partial T}{\partial t} + \rho_f c_p \boldsymbol{v}\cdot\nabla T - \nabla\cdot(k\nabla T) = \frac{\partial p}{\partial t} + \boldsymbol{v}\cdot\nabla p + q_{sf}\xi \tag{3}$$

In order to generalize the model for porous media and free-fluid zones, a specific application solver has been developed in OpenFOAM environment. The variables density, Darcian velocity, pressure, and temperature are indicated as ρ, u_D, p, T respectively. The parameter ξ is equal to 1 in the macroscopic porous zone and zero otherwise. The virtual mass coefficient and thermal inertia correction factor are:

$$C_{VM} = \frac{1}{2}\left(\frac{\delta_v}{r_h}\right)^2 \Im\left[\frac{1}{Da}\right], C_{TI} = \frac{1}{2}\left(\frac{\delta_k}{r_h}\right)^2 \Im[Nu] \tag{4}$$

The viscous and thermal penetration depths depend on frequency and fluid properties, while the complex Darcy (Da) and Nusselt number (Nu) depend on also on the microstructure of the porous media through the thermos-viscous functions (f_v, f_k). For a square pore stack, such functions have an analytical expression [11].

$$f_{v,k} = 1 - \frac{64}{\pi^4}\sum_m \frac{1}{m^4 C_m}, \quad C_m = 1 - i\frac{\pi^2 \delta_{v,k}^2}{4r_h^4} m^2 r_h^2 \tag{5}$$

In order to calculate the thermal power exchanged and the temperature difference between the average fluid temperature and solid wall one, it is necessary to model the ambient Heat Exchanger at the microscopic level. In theory, a 3D unitary cell at least should be taken into consideration due to the presence of the water tubes. However, to reduce the computational cost of the simulations, three 2D unitary cells representing the fins of the HX are simulated, neglecting the three-dimensional effects of the water tubes. It was shown that three fluid channels represent a good compromise between accuracy and simulation cost [12]. On the other hand, the hot HX and the stack are modelled with the porous media macroscopic approach based on the thermos-viscous functions developed in a previous article (f_v,f_k) [8] and briefly summarized above. It must be noted that non-linear effects in the stack region are not considered as made in ref. [13] because Reynolds number values are below the threshold value for considering also Forchheimer effects The heat generation term in the equation (3) is proportional to the real part of Nusselt number and the temperature difference between fluid and solid phase:

$$q_{sf} = \Re[Nu]\frac{k_f}{r_h^2}(T - T_s) \tag{6}$$

To effectively reproduce the experimental conditions, the boundary conditions of the numerical model have been obtained from the experimental measurements. More specifically, for each simulation, the pressure amplitude of the microphone "P1" has been used on the right boundary to set a sinusoidal input boundary condition. The other side of the resonator has been set a solid wall. In terms of energy equation, a linear assumption has been assumed between the measured hot temperature T_h (hot temperature) in the heater and the wall temperature T_c (cold temperature) for the stack.

$$T_s(x) = T_h - \frac{T_h - T_c}{L_{stack}}(x - x_0) \tag{7}$$

where $L_{stack} = 0.09\ m, x_0 = 0.11\ m$ and the temperature data corresponding to each pressure are shown in Table 1. Similarly, the wall temperature of the ambient HX is prescribed to the experimental value T_c. Such experimental data, used as boundary conditions in the numerical model, are summarized in Table 1.

Table 1. Experimental data used in the numerical model as boundary conditions, data from ref. [9] and private conversation

p_1 (Pa)	T_c (°C)	T_h (°C)
2244.58	29.36	329.86
2917.42	30.97	332.36
3519.06	33.43	341.16
4110.02	36.32	351.24
4467.8	37.45	360.08
4823.8	39.47	365.34
5188.7	41.51	375.72
5518	44.39	383.92

4. Results and discussion

In the experimental setup, the heat exchanged between the thermoacoustic working fluid and cold water has been measured directly from inlet and outlet of water temperatures and its mass flow rate, as explained in a more detail in ref. [9]. On the other hand, from a numerical perspective, it is possible to calculate the average heat flux by integrating the normal temperature gradient along the solid wall of the heat exchanger and dividing over the total exchange area considered in the numerical model. While on the water side, after a transient phase, the steady-state condition is reached, on the air side, the thermoacoustic working fluid is oscillating. For this reason, the flux data have to be also averaged over an acoustic period of time $(1/f_0)$.

In formulas:

$$\langle\overline{q}\rangle = \int\int q(x,t)\, dAdt \tag{8}$$

This quantity can be compared to the experimental values once the thermal power available from the experimental data provided is divided over the total heat exchange area of the physical prototype. It is estimated to be around 0.39 m^2. The comparison between experimental and numerical heat flux is illustrated in Figure 3. The uncertainty bar included the experimental data ranges from 14% to 7%, related to the lowest and highest Reynolds number investigated. In most cases, the numerical data fall into the bar uncertainty of the experimental data. Both the curves show a quadratic dependence of the heat flux on the velocity inside the ambient HX, expressed in terms of Reynolds number.

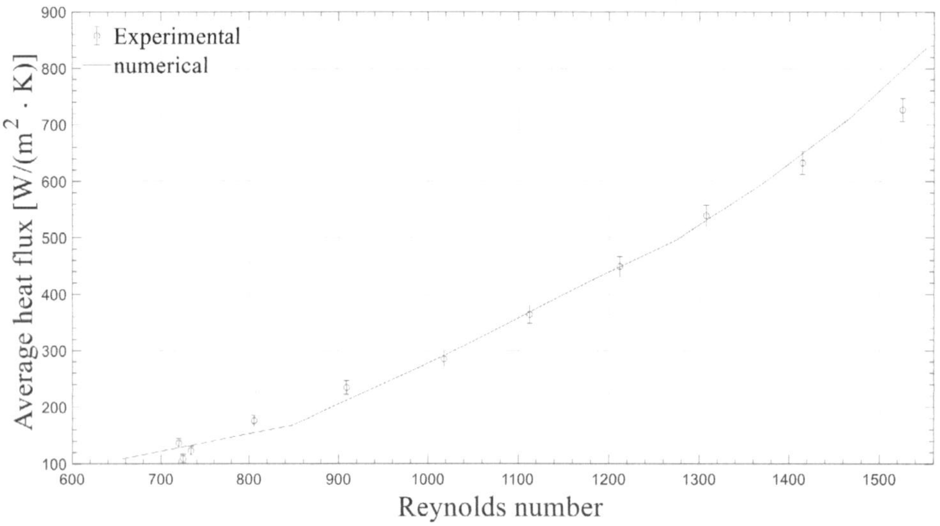

Figure 3 Comparison between numerical and experimental heat flux versus Reynolds number

Similarly, in Figure 4 the comparison between the numerical and experimental solid-fluid temperature differences is presented. The fluid temperature is calculated performing a space average along the section area in the middle of the stack and a time average over an acoustic period. In this case, the experimental uncertainty is constant with respect to Reynolds number and equal to $\pm 0.2°C$.

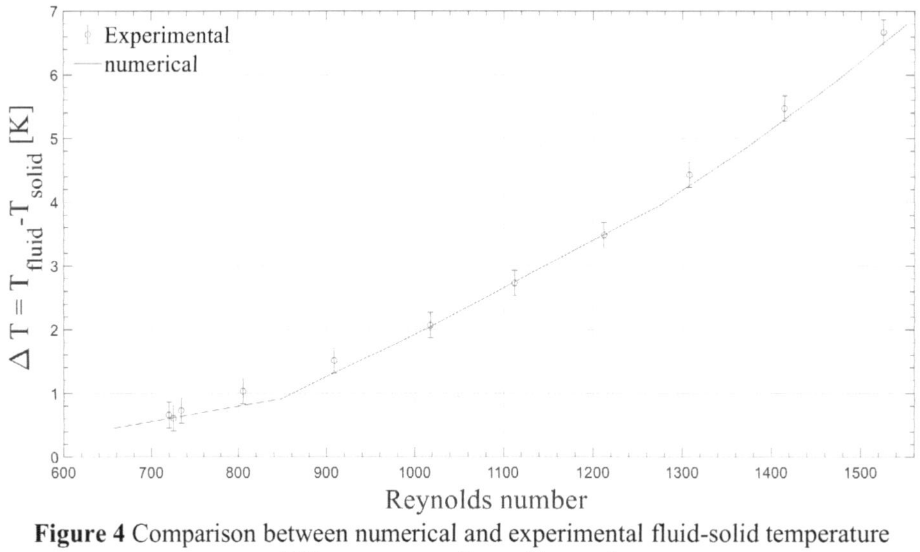

Figure 4 Comparison between numerical and experimental fluid-solid temperature difference versus Reynolds number

As a consequence of the previous results, the Nusselt number can be calculated from the ratio between the above heat fluxes and temperature difference using the hydraulic diameter D_h and the fluid thermal conductivity k_f as scaling factor. It has to be specified that this Nusselt number should be not confused with the one shown in ref. [4], in which a logarithmic temperature difference involving also the water temperature was considered. For this reason, a different absolute uncertainty model of the Nusselt number has to be considered in the present Nusselt calculation.

$$u_{Nu} = Nu \cdot \sqrt{\left(\frac{u_q}{q}\right)^2 + \left(\frac{u_{\Delta T}}{\Delta T}\right)^2} \tag{9}$$

where $u_{Nu}, u_q, u_{\Delta T}$ are the absolute uncertainties of Nusselt, heat flux and temperature difference respectively. From the trend pictured in Figure 5 it is possible to understand that, for Reynolds number larger than about 1000, the temperature difference and heat flux data fit almost the same quadratic dependence. Even if the trend for the Nusselt number (in which the logarithmic mean temperature difference method was applied) is opposite, it is visible that, for the same threshold Reynolds number, a saturation value is reached. According to the authors, this behaviour is justified by the fact that, for that Reynolds number, the particle displacement is approximately equal to the HX length [9].

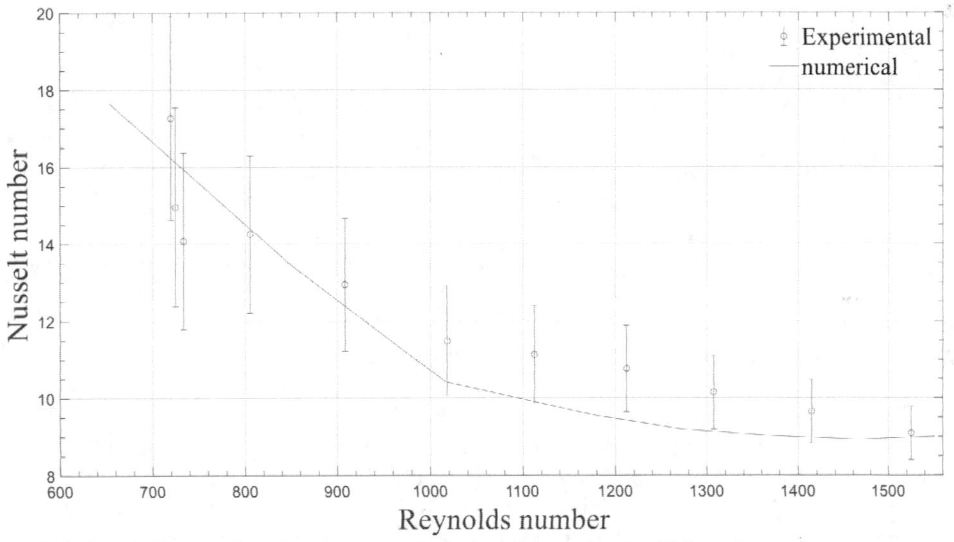

Figure 5 Comparison between numerical and experimental Nusselt number versus Reynolds number

5. Conclusions

In this article, a numerical model of a standing wave thermoacoustic engine is proposed. The main conclusions can be summarized by the following bullet points.

- The generalized model for macroscopic porous media, adapted for oscillatory flows and already validated in a previous article against experimental literature data, has been extended for a complex computational domain involving not only porous media but also free fluid regions.
- The experimental setup for the standing wave engines has permitted to perform a systematic comparison between numerical and experimental data, using pressure and temperature data as boundary conditions in the numerical model.
- The results show that numerical data almost always fall in the experimental uncertainty range both in terms of temperature and heat flux, as well as Nusselt number.

- The macroscopic model for porous media in oscillatory flows can be a reliable tool in assisting the design of thermoacoustic engines in order to the into account also nonlinear phenomena.

6. References

[1] G. W. Swift, "Thermoacoustic engines," *J. Acoust. Soc. Am.*, vol. 84, no. 4, pp. 1145–1180, 1988, doi: 10.1121/1.396617.

[2] G. Chen, L. Tang, B. Mace, and Z. Yu, "Multi-physics coupling in thermoacoustic devices: A review," *Renew. Sustain. Energy Rev.*, vol. 146, no. May, 2021, doi: 10.1016/j.rser.2021.111170.

[3] O. M. Ilori, A. J. Jaworski, and X. Mao, "Experimental and numerical investigations of thermal characteristics of heat exchangers in oscillatory flow," *Appl. Therm. Eng.*, vol. 144, no. January, pp. 910–925, 2018, doi: 10.1016/j.applthermaleng.2018.07.073.

[4] N. A. A. Qasem and S. M. Zubair, "Compact and microchannel heat exchangers: A comprehensive review of air-side friction factor and heat transfer correlations," *Energy Convers. Manag.*, vol. 173, pp. 555–601, Oct. 2018, doi: 10.1016/j.enconman.2018.06.104.

[5] A. Piccolo, R. Siclari, F. Rando, and M. Cannistraro, "Comparative performance of thermoacoustic heat exchangers with different pore geometries in oscillatory flow. Implementation of experimental techniques," *Appl. Sci.*, vol. 7, no. 8, 2017, doi: 10.3390/app7080784.

[6] A. Piccolo and G. Pistone, "Estimation of heat transfer coefficients in oscillating flows: The thermoacoustic case," *Int. J. Heat Mass Transf.*, vol. 49, no. 9–10, pp. 1631–1642, 2006, doi: 10.1016/j.ijheatmasstransfer.2005.11.009.

[7] A. Di Meglio and N. Massarotti, "CFD Modelling of Thermoacoustic Energy Conversion: A Review," *Energies*, 2022.

[8] A. Di Meglio, E. Di Giulio, R. Dragonetti, and N. Massarotti, "A novel model for macroscopic simulation of oscillating heat and fluid flow in porous media," *Int. J. Therm. Sci.*, vol. 181, no. August 2021, p. 107758, 2022, doi: 10.1016/j.ijthermalsci.2022.107758.

[9] A. Piccolo and A. J. Jaworski, "Experimental study of heat transfer characteristics of finned-tube and circular-pore heat exchangers in oscillatory flow," *Appl. Therm. Eng.*, vol. 181, p. 116022, 2020, doi: 10.1016/j.applthermaleng.2020.116022.

[10] E. Di Giulio, A. Di Meglio, N. Massarotti, and R. Dragonetti, "Effective Thermal Conductivity of Tetragonal Pin Array Stack," pp. 1–7, 2022, doi: 10.11159/enfht22.239.

[11] J. P. Clark, W. C. Ward, and G. W. Swift, "Design environment for low-amplitude thermoacoustic energy conversion (DeltaEC)," *J. Acoust. Soc. Am.*, vol. 122, no. 5, p. 3014, 2007, doi: 10.1121/1.2942768.

[12] A. J. Jaworski, X. Mao, X. Mao, and Z. Yu, "Entrance effects in the channels of the parallel plate stack in oscillatory flow conditions," *Exp. Therm. Fluid Sci.*, vol. 33, no. 3, pp. 495–502, Mar. 2009, doi: 10.1016/j.expthermflusci.2008.11.003.

[13] A. Di Meglio, N. Massarotti, S. Rolland and P. Nithiarasu, P (2023), "Analysis of non-linear losses in a parallel plate thermoacoustic stack", *International Journal of Numerical Methods for Heat & Fluid Flow*, Vol. ahead-of-print No. ahead-of-print. https://doi.org/10.1108/HFF-02-2023-0071.

Acknowledgments

This research has been partly funded by Italian Ministry of University and Research (MUR): Project number PRIN 2017JP8PHK.

A 3D numerical model for the performance analysis of a differential pressure flow meter in transient conditions for liquid fuels

C. Canale[1*], F. Arpino[1], G. Cortellessa[1], G. Ficco[1], G. Grossi[1], M. Huovinen[2] and A. Karvinen[2]

[1] Department of Civil and Mechanical Engineering, University of Cassino and Southern Lazio, Via Di Biasio 43, 03043 Cassino (FR), Italy

[2] Teknologian Tutkimuskeskus VTT Oy, Sokajrventie 9D3, Puristamo 9D3, Kajaani, 87100, Finland

[*] e-mail: christian.canale@unicas.it

Abstract. In the present paper, the metrological performance of a single-hole, sharped edge, and the orifice flow meter is numerically investigated employing different liquid fuels. Numerical investigations have been performed for a three-dimensional transient flow. Turbulence has been modeled employing the Realizable K-ε turbulence model, based on the Unsteady Reynolds-averaged Navier-Stokes (URANS). The present work is conducted in the context of the European SAFEST 20IND13 project, aimed at investigating the performance of the orifice flow meter numerical model in a wide range of temperatures, density, viscosity, and different liquid fuels. The numerical model, validated according to the ISO standard 5167-2 is employed to analyze the metrological performance of a test rig available at project partners' laboratories and was aimed at reproducing the fuel consumption curve of a light and heavy transport vehicle.

1. Introduction

Greenhouse gas emissions produced in the transport sector, especially for road and maritime facilities, consistently contribute to climate change, therefore these emissions need to be reduced. Green hydrogen and renewable fuels are expected to play a crucial role in achieving net zero greenhouse gas emissions by 2050, particularly in the hard-to-abate and transport sectors. Nevertheless, the extensive use of green fuels opens new challenges from the metrological point of view. In the past few years, impressive improvements to engines and transition to new fuel types have been made. The metrological sector also needs improvements, for the reason that the new engine efficiency and impact of new fuels depend directly on the quality of measurements. Flow meters, nowadays are employed typically to measure fuel flows in the transport sector, but could show inadequate metrological performance when measuring green fuels. Besides, particular attention must be paid to the meter's ability to measure extremely variable flow rates, as green fuels are expected to be largely employed to measure consumption in the transport sector. Looking at the current metrological scenario, it can be seen that different types of flow meters are employed, depending on the specific application and the size. Among other meters, differential pressure-based measuring instruments are widely used all over the sectors, due to simplicity, reliability, and low maintenance costs. Such meters have been widely investigated in the scientific literature, especially the optimization of the geometrical parameters and the choice of the best orifice among the different choices.

Hutagalung et al. explored the impact of the orifice plate's release coefficient on the water system. Their study delved into the correlation among the Reynolds number, β ratio, and discharge coefficient. The authors deduced that the values of the discharge coefficient are influenced by both the β ratio and the Reynolds number. They further emphasized that all other factors contribute to the overall error in measuring the release coefficient Cd [1]. A comparison of the performance between a single-hole (SHO) and multi-hole (MHO) orifice plate flow meter was conducted by Bikić et al., showing the limits of the SHO approach due to high-pressure difference, slower pressure recovery and lower discharge coefficient. The numerical investigation shows the potentiality of the CFD approach applied to the metering sector and the possibility of avoiding energy losses and test rig optimization [2]. A comparison of the performance of the MHO was conducted by Tomaszewski et al. [3], showing the impact of the holes' position on the pressure and energy losses in the test rig.

Nevertheless, significant investigations are still required to analyze the dependence of metrological performance on new green fuels' composition and physical properties.

The present work fits into a European project 20IND13 SAFEST, which aims to develop a metrological infrastructure to perform advanced measurements in the transport sector, and aims to characterize the interaction between the fluid and the meter. A numerical model has been realized in the OpenFOAM environment, analyzing the performance of the orifice flow meter numerical model for different operating conditions and performing parametric analysis, thus avoiding expensive experimental campaigns.

The collaboration with the SAFEST research groups has made possible the CFD modeling of a four-hole orifice plate flowmeter, whose dimension has been given by the Technical Research Centre of Finland (VTT). This work is a preliminary study in metrological modeling and it allows us to understand the behavior of these meters in a numerical environment. The validation process has been realized taking into account the ISO Standard 5167 as a reference for the research, considering that an experimental validation will be performed in collaboration with the VTT research group to analyze the metrological performance of a test rig available at project partners' laboratories and to reproduce the fuel consumption curve of a light and heavy transport vehicle. Numerical investigations have been performed for a three-dimensional transient flow. Turbulence has been modeled employing the standard k-ε turbulence model, based on the Unsteady Reynolds-averaged Navier-Stokes (URANS).

2. Material and methods

Measurements of the flow rate of liquids, gases, and vapors with orifice flowmeters have found wide use in industrial, scientific, and research measurements. A restriction fulfilling the function of a primary converter of the potential energy of the flow into kinetic energy by changing the cross-section of the pipe in which they are installed. The differential pressure flow meters are based on Bernoulli's working principle, which states that the pressure drop and subsequently the measured signal are functions of the square of flow speed under a specific hypothesis. The flow rate is linked directly to the pressure difference across the section and fluid characteristics, such as temperature, density, and viscosity. Due to high flexibility, great reliability of the measure, and low maintenance costs, the orifice plate flow meters have found great application in engineering and it has been chosen as the subject of the following numerical study.

2.1 The ISO Standard 5167 and geometrical characteristics

Modeling the orifice plate involves a systematic approach to ensure accurate flow measurement and establish uncertainty values. The ISO Standard 5167 part two is the metrological guideline for geometrical specifications and method of use, such as installation and operating conditions, of the orifice plates. As reported in the previous introduction, the measurement's working principle is based on installing a restriction into a pipeline, which causes a static pressure difference between the upstream and downstream sides of the plate. The mass flowrate, indicated as q_m, is reported in Equation (1):

$$q_m = \frac{C}{\sqrt{1-\beta^4}}\varepsilon\frac{\pi}{4}d^2\sqrt{2\Delta p\rho_1} \quad (1)$$

Where C represents the coefficient of discharge, dependant on the Reynolds number Re, and it is given by the Reader-Harris/Gallagher (1998) equation [4] and ε represents the expansibility factor and it is given by Reader-Harris equation [5] . The iterative process, given by the ISO standard 5167 part 1 and part 2, consents to the determination of the reference value of the mass flow rate (or the volumetric flow rate) in specified conditions, thus comparing the numerical simulation results to the ISO Standard reference.
As reported in the introduction, the studied geometry has been provided by the VTT institute and it is a multi-hole orifice plate flow meter, composed of four holes of 15.75 mm diameter, oriented at 45° from the horizontal axis. A representation of the analyzed case is reported in Figure 1, which reports the upstream and downstream faces of the meter:

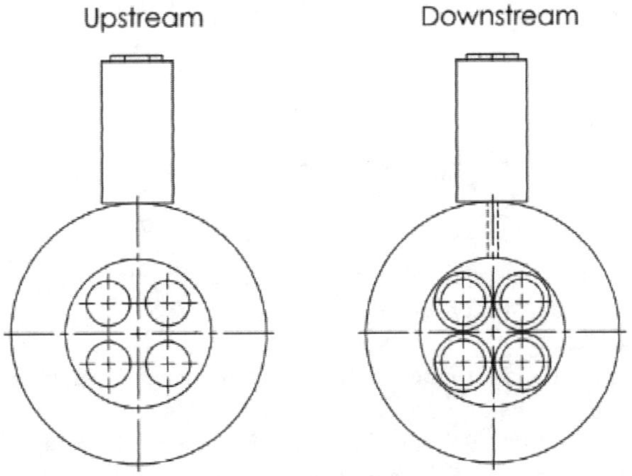

Figure 1 The orifice plate flowmeter chosen for the numerical campaign

As reported in ISO Standard 5167, the calculation of the β ratio is given by Equation 2.

$$\beta = \frac{d}{D} \tag{2}$$

Where the inner diameter d in the multi-hole case is calculated by doubling the distance from the center of the orifice, to the center of one of the four holes. The main geometrical characteristics of the flow meter are summarized in Table 1:

Table 1. Orifice plate dimensions and characteristics

Geometrical parameter	Dimensions [mm]
Pipe diameter	92.00
Orifice diameter	52.40
Four-holes diameter	15.75
45° edge length (downstream)	2.32
***β* ratio**	0.54

2.2 Numerical model

The numerical tool developed in the present study is a three-dimensional, time-variant, and Eulerian numerical model, in which the continuity and momentum equations are solved to simulate a flow meter's performance. The present study aims to define the behavior of these particular flow meters under different operating conditions, far from the laboratory testing conditions. A non-conventional fuel has been chosen to realize this research, to estimate the performance of the flowmeter with new green fuels. The fluid properties and characteristics have been provided by the National Metrology Institute of the Federal Republic of Germany, the so-called Physikalisch-Technische Bundesanstalt (PTB), and they are summarized in Table 2:

Table 2. Fluid properties due to the temperature variation

Fluid properties			
Temperature	25	40	$[°C]$
Density	780.401	750.08	$[kg/m^3]$
Dynamic Viscosity	1.24	1.02	$[mPa \cdot s]$
Kinematic Viscosity	1.58741E-06	1.36132E-06	$[m^2/s]$

The governing equations used in this study are the continuity equation and momentum equation, which are respectively reported in equations 3 and 4:

$$\frac{\partial \rho}{\partial t} + \nabla \cdot (\rho U) = 0 \tag{3}$$

$$\frac{\partial (\rho U)}{\partial t} + \nabla \cdot (\rho U \times U)) = \nabla \cdot \left(-\rho\delta + \mu(\nabla U + (\nabla U)^T)\right) + S_M \tag{4}$$

Where the capex T represents the transposition symbol of the matrix and the S_M represents the source term including the contributions due to body forces only.

As reported by Erdal and Andersson in [6], the authors showed a good agreement between the calculation of the $k - \varepsilon$ turbulence model and measurements. More recently, Shah et al. have proved the applicability of the standard k$-\varepsilon$ model and the better performance above the others, in the fully developed flow with the orifice plate. Hence, the standard $k - \varepsilon$ model has been chosen for the computations and the governing equations of the model are reported in equations 5 and 6:

$$\frac{\partial(\rho K)}{\partial t} + \nabla \cdot (\rho U k) = \nabla \cdot \left[\left(\mu + \frac{\mu_t}{\sigma_k}\right)\nabla k\right] + P_k - \rho\varepsilon \tag{5}$$

$$\frac{\partial(\rho e)}{\partial t} + \nabla \cdot (\rho U \varepsilon) = \nabla \cdot \left[\left(\mu + \frac{\mu_t}{\sigma_\varepsilon}\right)\nabla \varepsilon\right] + \frac{\varepsilon}{k}\left(C_{\varepsilon_1} P_k - C_{\varepsilon_2}\rho\varepsilon\right) \tag{6}$$

Where C_{ε_1}, C_{ε_2} , σ_k and $\sigma_\varepsilon$are the standard $k - \varepsilon$ turbulence model constants and P_k represents the turbulence production caused by viscous and buoyancy force and it is defined as:

$$P_k = \mu_t \nabla U \cdot (\nabla U + \nabla U^T) - \frac{2}{3}\nabla \cdot U(3\mu_t \nabla \cdot U + \rho k) + P_{kb} \tag{7}$$

Furthermore, the turbulence viscosity is modeled as a function of turbulent kinetic energy, turbulent kinetic energy dissipation rate, and turbulent viscosity constant, in the form of:

$$\mu_t = C_\mu \rho \frac{k^2}{\varepsilon} \tag{8}$$

The PIMPLE (Pressure Implicit Method for Pressure-Linked Equations) algorithm and an adjustable time step have been chosen to conduct the CFD analysis. This involves that the time steps are controlled by the Courant number, which is given by Equation 9:

$$\mathrm{Co} = \frac{|\mathrm{U}|\Delta\theta}{\Delta x} \tag{9}$$

Where |u| is the magnitude of the flow velocity, $\Delta\theta$ is the time step and Δx is the characteristic mesh size. A maximum Courant number of 2 was chosen, while Arpino et al. showed in [7] an average deviation of 1% concerning a maximum Courant of 2 for incompressible and turbulent flow. An average of 10^{-7} was fixed for the numerical residuals in order to simulate the orifice performance and to get comparable characteristics of a flow meter measurement.

3. Results

In this section, the results obtained from the CFD analysis concerning the performance of the orifice plate flowmeter are illustrated and discussed in detail. The results are expressed in terms of:

- Mesh sensitivity analysis, comparing three different grid sizes;

- The CFD analysis is conducted by variating the temperature and consequently the fluid properties. The objective of the variation is to highlight the critical conditions in which the orifice plate flowmeter drastically decreases the performance;
- A validation of the results has been conducted through ISO Standard 5167, which gives a guideline to calculate the reference value comparable to CFD results.

3.1 Mesh sensitivity analysis

The mesh has been realized with a hybrid method, that is based on the geometry construction through MECA-Salome® and the mesh production with the OpenFOAM's tool "snappyHexMesh". This process consents to better control of the cells near the wall and in the zone of interest. Three different grids have been realized to get the proper sensitivity analysis, thus reaching a proper average error from the reference below the 3.2% or pressure and velocity field. A representation of the chosen grid is reported in Figure 2 (a) and Figure 2 (b), which respectively highlight the orifice and pressure taps (in blue) zone:

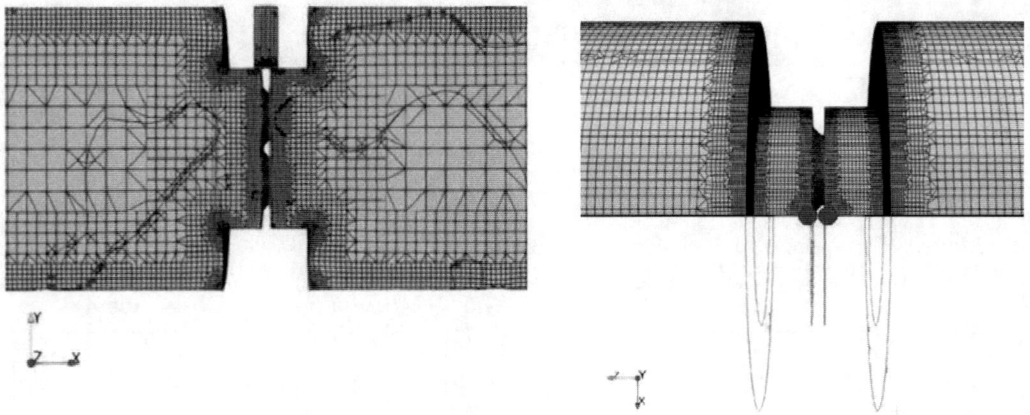

(a) (b)

Figure 2.(a) A x-y representation of the grid focusing on the holes. **(b)** A z-x representation focusing on the pressure taps.

The grid characteristics are reported in Table 3, described in terms of number of cells, max skewness, and non-orthogonality, which are the main parameters to get under control for a mesh:

Table 3. Max skewness and non-orthogonality for the three different grids used for the sensitivity analysis

	Number of cells	**Max Skewness**	**Non-Orthogonality**
Mesh 1	2 211 715	3.41	60.11
Mesh 2	3 984 202	3.12	60.06
Mesh 3	8 931 030	3.31	60.17

The sensitivity analysis is conducted by evaluating the velocity profiles one diameter upstream and a half diameter downstream of the orifice plate. The comparison between the velocity field is reported in Table 4 which highlights the good behavior of Mesh 2, composed of four million elements, against the reference grid of 8 million elements.

Table 4. The average error of the mesh from the reference grid.

	Upstream	**Downstream**
Avg. error (8M-4M)	1.87%	5.67%
Avg. error (8M-2M)	4.42%	82.38%

A representation of the average velocity reported in Figure 3. The primary device forces the fluid to pass into the four-hole section and an increasing velocity is clear downstream of the orifice. Consequently, the pressure drop is visible in Figure 3 following the principles described in the previous sections.

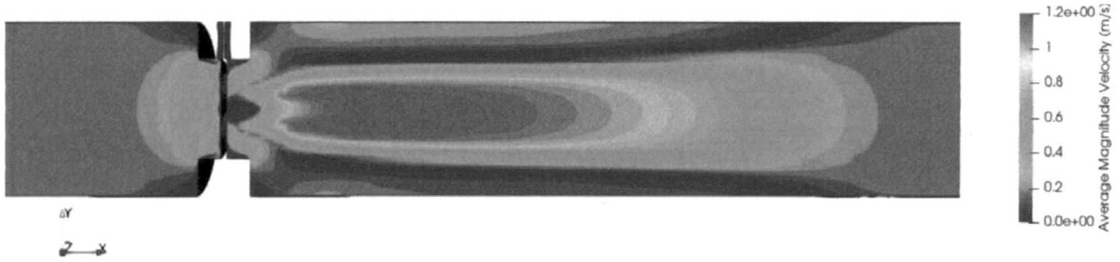

Figure 3. Average velocity field of the four million mesh in the orifice plate zone

A representation of the velocity profiles upstream and downstream of the orifice plate are reported in Figure 4a and Figure 4b, the profiles are extracted one diameter before (upstream) and an-half diameter after (downstream) the orifice plate.

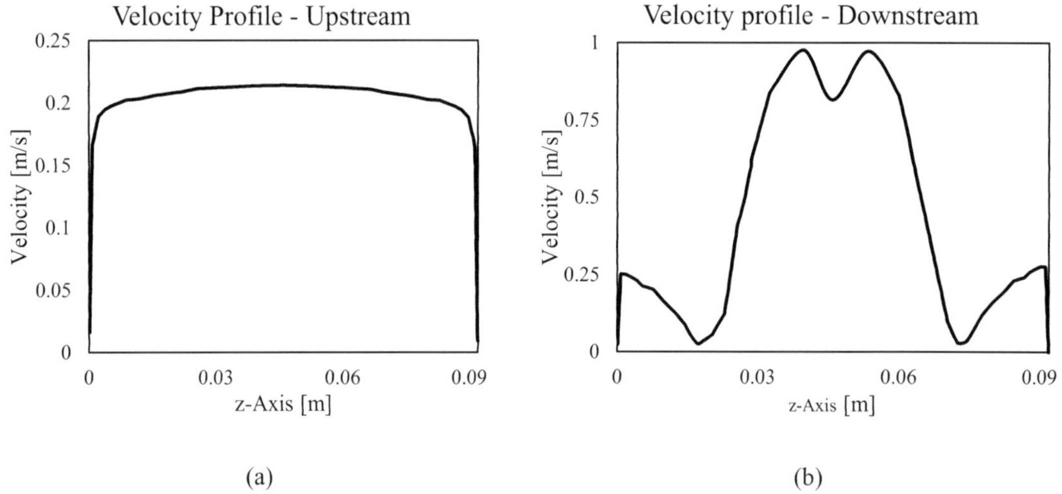

Figure 4 (a) velocity profile extracted one diameter before the orifice; **(b)** Velocity profile an-half diametr after the orifice

3.2 Fluid properties variation

Energy accounting is becoming more relevant in the energy transition process and energy-saving economy. According to 2022 IEA data [8], the transport sector weights 23% ca. on the CO_2 global emission of CO_2 and the transition process starts from the measurement chain and energy accounting. To become more aware of the problem of energy consumption, the measurement process is becoming increasingly relevant.

This section aims to evaluate the performance under different operating conditions, for the reason that the laboratory conditions in which the meters are calibrated are different from the ordinary working conditions of the meters. Furthermore, the test liquid used to calibrate an orifice plate is typically water. Water is a common choice because it is readily available, and its properties are well-known and standardized. The green fuels may have different properties, such as viscosity and density, compared to traditional fuels. The accuracy of the flow meter measurement is dependent on the geometrical characteristics of the orifice, which are usually fixed, and on the fluid properties used in the application. The accuracy of the measurement, due to the employment of the innovative fuels, is attributable to:

- Variability in fluid properties: the viscosity and the density of green fuels are the main parameters that affect the measurements;
- Composition changes: some alternative fuels may have a variable composition and this is another uncertainty factor in the behavior of the fluid, compared to the traditional fuels used in the transportation sector;
- Calibration challenges: Orifice plates are typically calibrated for specific fluids under specific conditions. Calibrating for innovative fuels may pose challenges due to the lack of standardized calibration data for these fuels.

In the present study, a well-known biodiesel has been provided by the Physikalisch-Technische Bundesanstalt (PTB), including the viscosity and density characteristics due to the temperature variation and therefore the interaction between the orifice measure and the fluid properties have been tested. A non-conventional bio-diesel has been considered for the numerical simulations, whose characteristics are well-known and summarized in Table 2 in section 2.2.

Two different temperatures of 25 and 40 °C have been considered, to understand in which condition the meters are not able to maintain the accuracy and precision levels. The temperature variation induces a density and viscosity variation in the fluid properties and it affects directly the measurement performance. These two temperatures were chosen because they are the minimum and the maximum measured values for the non-conventional fluid used in the numerical campaign. The results of the two numerical simulations are compared to the ISO Standard reference in the same hypothesis:

Table 5. Comparison of the pressure drop and volumetric flow rate at 25°C and 40°C

		$\Delta p\ [Pa]$	$q_m\ [kg/s]$	Avg. Error [%]
$T = 25°C$	**ISO Standard 5167 reference**	6893.13	1.044	-
$T = 40°C$		7171.69	1.044	-
$T = 25°C$	**Numerical Model**	6430.23	1.008	3.42
$T = 40°C$		6338.21	0.981	5.99

As presented in Table 5 the numerical model is able to predict in standard conditions the behavior of the fluid, but increasing the temperature and lowering the density of the fluid, presents some difficulties in predicting the correct pressure drop.

3.3 Validation of results through ISO Standard 5167

The validation of the numerical CFD results has been conducted against the ISO Standard 5167 as a reference for the comparison. The comparison has been realized taking into account the pressure drop obtained by the numerical simulations, presented in Table 5, and the theoretical value calculated through the ISO Standard Equation 1 in section 2.1. The results provided in Table 5 show a good agreement of the numerical model for the temperature of 25°C, with an average error from the ISO Standard reference of 3.42%, while the error slightly up to 5.99% increase when the temperature is 40°C. Although the numerical model can predict the fluid behavior when the standard conditions are provided, the numerical model suffers more when the temperature increases the difference from the reference also increases due to the lower viscosity of the fluid.

4. Conclusion and future development

In this paper, we evaluated the performance of the numerical model of a four-hole orifice plate flow meter in different operating conditions. The 3D Computational Fluid Dynamics (CFD) numerical simulations were

validated against the ISO Standard 5167 reference. The geometry and dimensions of the orifice plate were provided by the Technical Research Centre of Finland (VTT) in the context of the European 20IND13 SAFEST project. The performance of the flowmeter was evaluated by calculating the volumetric flow rate from the pressure drop across the section observed in the numerical simulations.
The results of the CFD analysis demonstrated that the CFD model can predict with good agreement with the reference value the pressure drop across the section, with an average error of about 4%. Future developments of the study will be focused on the accuracy of the model and an in-field experimental evaluation of the orifice plate, to properly validate the results of the CFD simulations against the experimental data set.

References:

[1] S. S. Hutagalung, «Effect of Release Coefficientof Orifice Plate on Water Fluid Flow Systems», *J. Phys.: Conf. Ser.*, vol. 1230, fasc. 1, p. 012086, lug. 2019, doi: 10.1088/1742-6596/1230/1/012086.
[2] S. Bikić, M. Đurđević, M. Bukurov, e S. Tašin, «Comparison of single-hole and multi-hole orifice energy consumption», *Advances in Mechanical Engineering*, vol. 14, fasc. 1, p. 168781402210754, gen. 2022, doi: 10.1177/16878140221075461.
[3] A. Tomaszewski, T. Przybylinski, e M. Lackowski, «Experimental and Numerical Analysis of Multi-Hole Orifice Flow Meter: Investigation of the Relationship between Pressure Drop and Mass Flow Rate», *Sensors*, vol. 20, fasc. 24, p. 7281, dic. 2020, doi: 10.3390/s20247281.
[4] M. J. Reader-Harris e J. A. Sattary, «The orifice plate discharge coefficient equation», *Flow Measurement and Instrumentation*, vol. 1, fasc. 2, pp. 67–76, gen. 1990, doi: 10.1016/0955-5986(90)90031-2.
[5] M. Reader-Harris, «Orifice Expansibility Factor», in *Orifice Plates and Venturi Tubes*, in Experimental Fluid Mechanics. Cham: Springer International Publishing, 2015, pp. 187–201. doi: 10.1007/978-3-319-16880-7_6.
[6] A. Erdal e H. I. Andersson, «Numerical aspects of flow computation through orifices», *Flow Measurement and Instrumentation*, vol. 8, fasc. 1, pp. 27–37, mar. 1997, doi: 10.1016/S0955-5986(97)00017-4.
[7] F. Arpino, G. Cortellessa, G. Grossi, e H. Nagano, «A Eulerian-Lagrangian approach for the non-isothermal and transient CFD analysis of the aerosol airborne dispersion in a car cabin», *Building and Environment*, vol. 209, p. 108648, feb. 2022, doi: 10.1016/j.buildenv.2021.108648.
[8] IEA (2023), «CO2 Emissions in 2022». IEA. [Online]. Disponibile su: https://www.iea.org/reports/co2-emissions-in-2022

Acknowledgment

This project (EMPIR 20IND06 PROMETH2O) has received funding from the EMPIR programme co-financed by the Participating States and from the European Union's Horizon 2020 research and innovation programme

Numerical modelling of an electric motor cooling jacket

M Grespan[1], L Campanelli[1], D Angeli[2] and R Freddi[3]

[1] DISMI - Dipartimento di Scienze e Metodi dell'Ingegneria, Università di Modena e Reggio Emilia, Via Amendola 2, Pad. Buccola, 42122 Reggio Emilia (Italy)

E-mail: mattia.grespan@unimore.it

[2] Centro Interdipartimentale EN&TECH, Piazzale Europa 1, 42124 Reggio Emilia (Italy)

[3] Fira Spa, Strada Statale 255, 374/B, 44047 Dosso Terre del Reno, Italy

Abstract. The automotive industry is amid a sweeping change of propulsion technology to meet the increasingly stringent limits on emissions and fuel consumption. Traditional combustion engines are gradually being replaced by electric motors or hybrid power trains. The efficiency and power density levels achievable by electric motors are entirely dependent on the effectiveness of the employed cooling solution. Therefore, extensive analyses are needed to determine and optimise the thermal performance of these systems. In this work a numerical model is developed to determine the friction losses and the heat transfer properties of an electric motor cooling jacket. The cooling channels, which coil along the circumferential direction within the motor casing, are studied by means of CFD analysis of a basic periodic module. Flow and temperature fields are determined by applying a 3D Finite Volume approach. Numerical solutions are obtained by means of a validated conjugate heat transfer solver. Integral flow field results are employed to derive the equivalent Darcy friction factor and side-wall specific Nusselt numbers for several flow regimes. A lumped parameter thermal model, based on the graph theory and aforementioned CFD results is also developed to determine the overall system performance. The equivalent thermal resistances are computed from geometric parameters and CFD results. Finally preliminary numerical results on friction losses and heat transfer are compared with available experimental data.

Keywords: electric motor, cooling jacket, forced convection, CFD, lumped parameter model, RANS

1. Introduction

Electric motors are increasingly being employed for automotive traction, due to more and more stringent standards on fuel consumption and pollutant emissions. The design and optimization of the cooling solution for electric motors is of utmost importance, as it strongly affects the achievable efficiency and power levels. Traditionally, thermal analysis of electric machines is carried out by means of thermal resistance models, whereas the use of more advanced numerical techniques such as computational fluid dynamics and finite element analysis is still limited.

One of the most relevant works on the development of lumped parameter models is that of Mellor *et al.* [1], in which the authors developed a thermal resistance network based on a general hollow cylindrical component, considering radial and axial conduction. Kral *et al.* [2] and Aglén *et al.* [3], developed similar models for water cooled electric machines, using empirical correlations to derive the convective heat transfer coefficients. Nategh *et al.* [4] performed simple

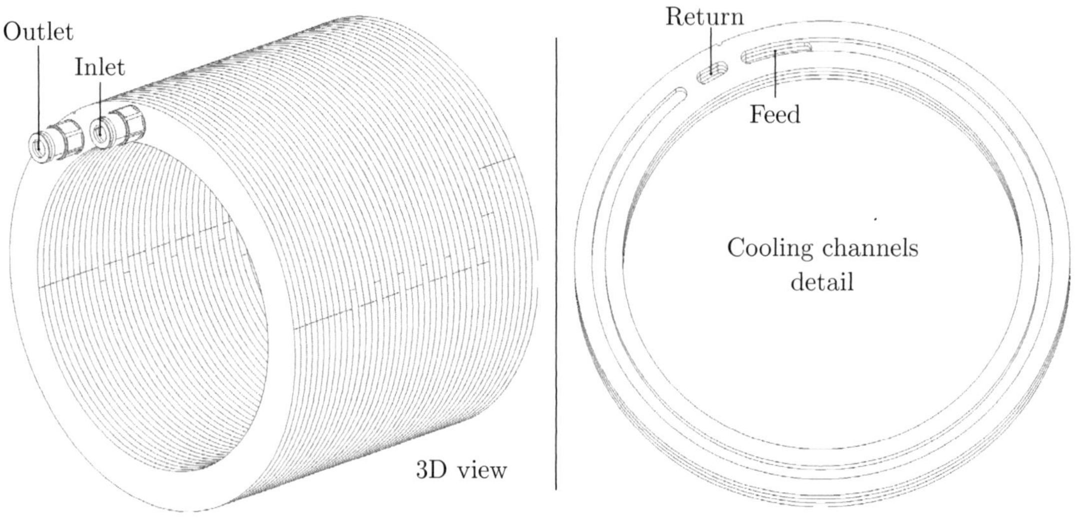

Figure 1. External geometry of the motor casing under examination (left), and a set of plates that clearly show the geometry of the feed and return channels (right).

CFD analyses to determine coolant flow distribution through the housing channels, and on the end winding surfaces. The resulting velocity values were then used to calculate the heat transfer coefficients, by employing empirical relationships.

In this work, a multiscale approach for the fluid dynamic and thermal analysis of an electric motor cooling jacket is presented. The cooling channels, which evolve in steps along the circumferential direction, are analysed by developing a CFD model for a periodic channel module under fully developed flow conditions. Numerical results on head losses and heat transfer, presented in terms of equivalent Darcy friction factor and sidewall-specific Nusselt numbers, are used to derive suitable correlations. A lumped parameter thermal model is also presented, with aim of studying the temperature distribution in the entire cooling jacket. The thermal resistances are determined from geometric parameters, physical principles, and the aforementioned correlations.

Preliminary global results on head losses and heat transfer, based on a simplified version of the model are compared with available experimental results and prior numerical studies.

2. Case study

This study is focused on an aluminum electric motor casing, constructed by stacking and brazing aluminium plates, which feature machined slots. This production process yields motor casings with great mechanical strength and cooling channels with intricate geometries, achieved by staggering and stacking several slotted plates. In this case, coolant flows along the casing circumference, reversing the direction when passing from one channel to the next. Then, coolant flows back towards the outlet port, positioned on the same side as the inlet, through a straight axial return channel. The motor casing features internal and external diameters of 147 and 183 mm, respectively. The slots constituting the cooling channels have a width of 5 mm and are positioned on the mean circumference. The axial length is 152 mm. Figure 1 shows a drawing of the cooling jacket, with a detail of the internal cooling channels, which are obtained by alternating plates with slots covering different angular spans.

3. Numerical model of the feed channel

Fluid flow and heat transfer analysis about the cooling jacket feed channel is conducted by means of CFD techniques. The geometry is divided into periodic modules which are repeated along the cooling jacket axis. Each module consists of the fluid and solid regions associated with a full turn of the cooling channels, so that the inlet and outlet sections are vertically aligned. The steady-state, incompressible, and fully developed flow and thermal fields about the unitary module are determined from the numerical solution of RANS and energy equations.

$$\begin{cases} \nabla \cdot (\, \vec{v} \otimes \vec{v} \,) = -\dfrac{1}{\rho} \nabla p + \nabla \cdot [\, (\nu + \nu_t) \nabla \vec{v} \,] + \sigma \\ \nabla \cdot \vec{v} = 0 \\ \nabla \cdot (\rho \vec{v} h) + \nabla \cdot \left(\rho \vec{v} \dfrac{|\vec{v}|^2}{2} \right) = -\nabla \cdot \vec{q} + \nabla \cdot (\vec{\tau} \cdot \vec{v}) \end{cases} \tag{1}$$

Temperature in solid regions is determined from the Laplace equation:

$$\nabla^2 T = 0 \, . \tag{2}$$

Eddy viscosity is determined by means of the k-ω SST turbulence model [5]. The momentum source term σ in Eqs. (1) stands for the external forcing needed to maintain the desired flow rate at the channel vents. The channel height y is chosen as the problem reference length, and it is used to scale down the computational domain. The flow is defined on the channel height based Reynolds and Prandtl numbers. Mean velocity over inlet and outlet sections, fluid density, and thermal conductivity are set to unitary reference values. Under the selected dimensionless framework the fluid dimensionless kinematic viscosity and specific heat are:

$$\mu^* = \nu^* = \frac{1}{\mathrm{Re}} \qquad c_p^* = \mathrm{Re}\,\mathrm{Pr} \, . \tag{3}$$

Dimensionless properties of solid regions are derived by scaling the thermophysical properties of aluminum by the ones of the adjacent fluid.

$$\lambda_s^* = \frac{\lambda_{\mathrm{alu}}}{\lambda_{\mathrm{fluid}}} \tag{4}$$

A fully developed flow solution is attained by enforcing periodic conditions between the inlet and outlet patches. At all fluid-solid interfaces no-slip conditions are applied to velocity and a zero-gradient condition is assigned to pressure; finally, continuity of temperature and heat fluxes is enforced.

Heat transfer effectiveness of the geometry under examination is evaluated by deriving an equivalent heat transfer coefficient for each sidewall of the channel. To this end, six sidewall specific sets of boundary conditions for wall temperature are defined: a uniform entering heat flux is applied to the sidewall under investigation, while all other walls are assumed adiabatic.

Numerical solution of the governing equations is obtained by means of a Finite Volume conjugate heat transfer solver based on the SIMPLE algorithm, and implemented in the OpenFOAM computational suite [6]. Advective terms are discretized by means of second-order upwind schemes. Instead, central schemes featuring explicit non-orthogonality correction are used for diffusive terms.

The computational domain consists of a conformal hybrid mesh, whose selected views are shown in Fig. 2, in which fluid and solid regions are clearly highlighted. In addition, a numeric ID is assigned to each sidewall for future reference.

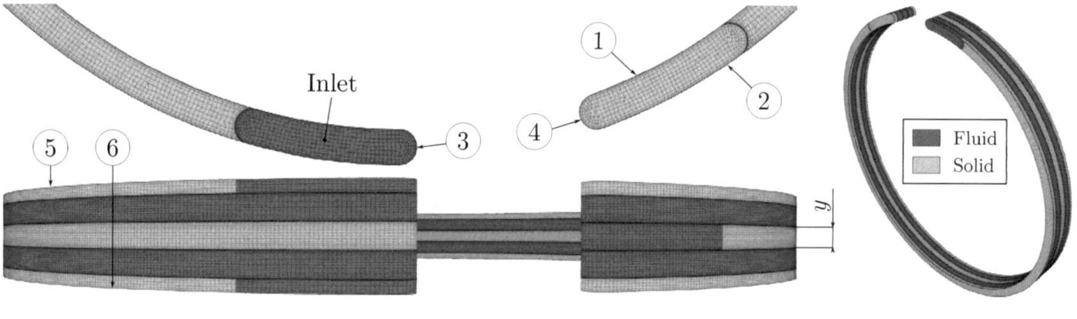

Figure 2. Selected views of the feed channel computational grid, along with identification of each sidewall via a numeric ID.

4. Lumped parameter thermal model

In the following a lumped parameter thermal model is presented, to determine the temperature distribution within aluminium walls, and the heat flux between feed and return channels. The motor casing is again divided in axial periodic modules, each of which includes the periodic feed channel section analysed in §3, a portion of the return channel, and the associated aluminium walls. Each periodic module is then divided into a set of hollow cylindrical sectors. The schematic of figure 3a shows the subdivision into sectors of a periodic casing module: the blue region represents the feed channel, thus includes the fluid and solid regions constituting the CFD model discussed in §3; the red region stands for the portion of the axial return channel associated with the periodic module, while aluminium walls are depicted in white.

The generic cylindrical sector is modelled by a network formed by seven nodes and six thermal resistances, under the assumption that the centre temperature defines heat flow along the radial, circumferential, and axial directions, and that the heat fluxes are independent of each other. In a solid region, the centre point temperature 0 represents the average temperature of the aluminium section, while in fluid regions T_0 stands for the average bulk temperature between inlet and outlet sections.

4.1. Solid element

The six thermal resistances that make up a generic solid cylindrical element are calculated on the basis of the conduction shape factor S and the average thermal conductivity λ_a [7]:

$$R_{ij} = \frac{1}{S_{ij}\,\lambda_\mathrm{a}} \,. \tag{5}$$

The cylindrical conduction shape factors are computed from geometric parameters following the general approach proposed by Yovanovich [8].

$$S_{01} = \frac{l\,\vartheta}{\ln\left(r_2/r_1\right)} \tag{6}$$

$$S_{02} = \frac{l\,\vartheta}{\ln\left(r_3/r_2\right)} \tag{7}$$

$$S_{03} = S_{04} = \frac{2\,l\,\ln\left(r_3/r_1\right)}{\vartheta} \tag{8}$$

$$S_{05} = S_{06} = \frac{\vartheta\left(r_3^2 - r_1^2\right)}{l} \tag{9}$$

Journal of Physics: Conference Series **2685** (2024) 012016 doi:10.1088/1742-6596/2685/1/012016

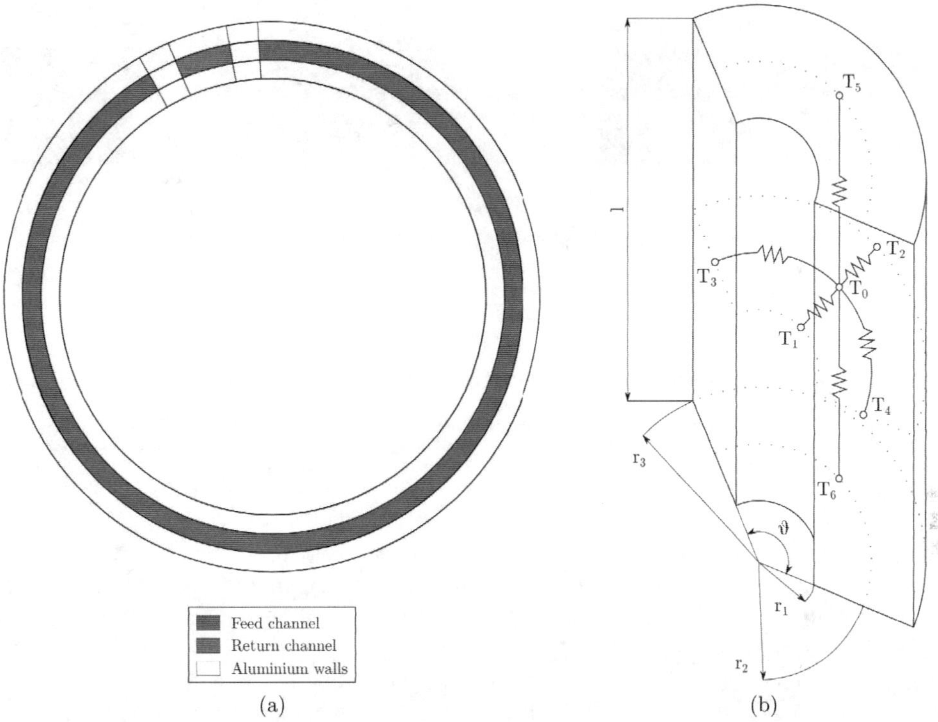

Figure 3. Schematics showing the set of cylindrical sectors constituting a casing periodic module (a), and the arrangement of nodes and thermal resistances modelling heat flux within a single cylindrical sector (b).

4.2. Feed channel

The feed channel is modeled by a thermal resistance network similar to the one depicted in Fig. 3b, with the addition of two nodes standing for the inlet and outlet bulk temperatures. The convective thermal resistance between the centre node, and each sidewall is derived from the classical relationship:

$$R_{ij} = \frac{1}{h_{ij}\, A_{ij}}\,, \tag{10}$$

where h_{ij} is determined from the CFD developed Nusselt number correlations presented in §5. The equivalent resistance between the centre node and the nodes associated with inlet and outlet bulk temperatures is given by:

$$R_{\mathrm{blk}} = \frac{1}{\dot{m}\, c_{\mathrm{p}}}\,. \tag{11}$$

4.3. Return channel

The return channel is represented by a seven-node thermal network in which the bottom and top nodes stand for the inlet and outlet bulk temperatures, respectively. The convective heat transfer coefficient is assumed uniform and determined from the Dittus–Boelter correlation

$$\mathrm{Nu}_{\mathrm{D}} = 0.023\, \mathrm{Re}_{\mathrm{D}}{}^{4/5}\, \mathrm{Pr}^{n}\,. \tag{12}$$

The resistances associated with the bulk temperature nodes are obtained from Eq. 11.

Journal of Physics: Conference Series **2685** (2024) 012016 doi:10.1088/1742-6596/2685/1/012016

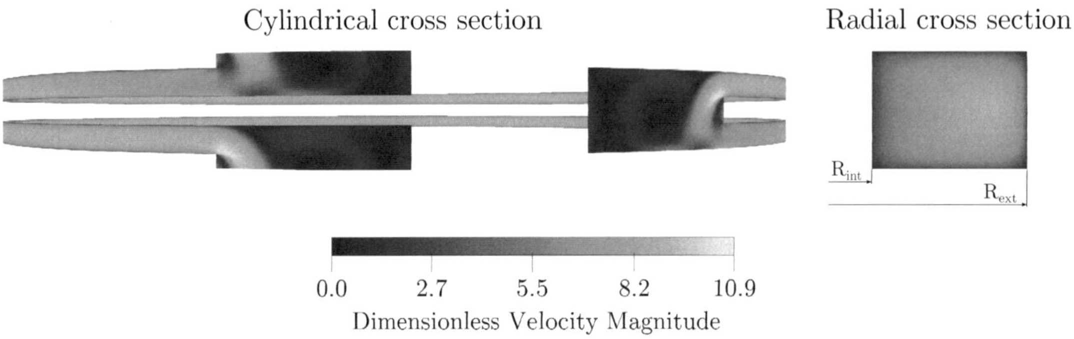

Figure 4. Colour maps of dimensionless velocity magnitude on cylindrical and radial sections, at Re=1800 and Pr=7.

5. Results

5.1. Feed channel

Flow field results are presented by means of dimensionless velocity colour maps about a radius-centred cylindrical section, and a radial cross section, shown in Fig. 4. At the intersection between two circumferential channels the flow turns sharply forming a small separation bubble on the inside of the turn. Subsequent impingement and flow redistribution on the channel base lead to a local high velocity region which improves heat transfer effectiveness. In the connecting region between the channels overall low fluid velocity leads to poor local heat transfer performance, as will be shown later. In the radial velocity section (Fig. 4) the effects of centrifugal forces are clearly spotted, as fluid velocity is greater towards the external side of the channel. The observed flow pattern leads to major head losses and decreased heat transfer effectiveness in the connecting region between channels, this is also confirmed by the integral results presented in the following.

Integral results on head losses and heat transfer effectiveness consist of an equivalent Darcy friction factor and wall-specific Nusselt numbers. The equivalent friction factor is derived from the Darcy-Weisbach formula, employing the external forcing term σ:

$$f = \frac{2\,\sigma\,y}{\rho\,\bar{u}_{\mathrm{I}}^{2}}\,, \tag{13}$$

where $\bar{u}_{\mathrm{I}}^{2}$ is integral mean velocity over the inlet section. The Nusselt number related to the i-th channel sidewall is derived from the classic convection thermal resistance definition:

$$\mathrm{Nu}_{i} = \frac{q''\,y}{\lambda\,(\bar{T}_{\mathrm{w},\,i} - \bar{T}_{\mathrm{b}})}\,. \tag{14}$$

In Eq. (14) $\bar{T}_{\mathrm{w},\,i}$ stands for the surface mean temperature over the i-th sidewall, while $\bar{T}_{\mathrm{b}}$ is the average bulk temperature between the inlet and outlet sections. Finally, q'' is the heat flux entering at the wall. Numerically derived f and Nu values are employed to derive Reynolds number based correlation functions:

$$f = -1.373 \times 10^{5}\,\mathrm{Re}^{5.038\times10^{-4}} + 1.380 \times 10^{5}\,, \tag{15}$$

$$\mathrm{Nu} = c_{1}\,\mathrm{Re}^{c_{2}} + c_{3}\,. \tag{16}$$

The c_i coefficients in Eq. (16), related to a single Prandtl number (Pr=7), are reported in Tab. 1. Figure 5 shows the numerical f and Nu values along with the related correlation functions. The

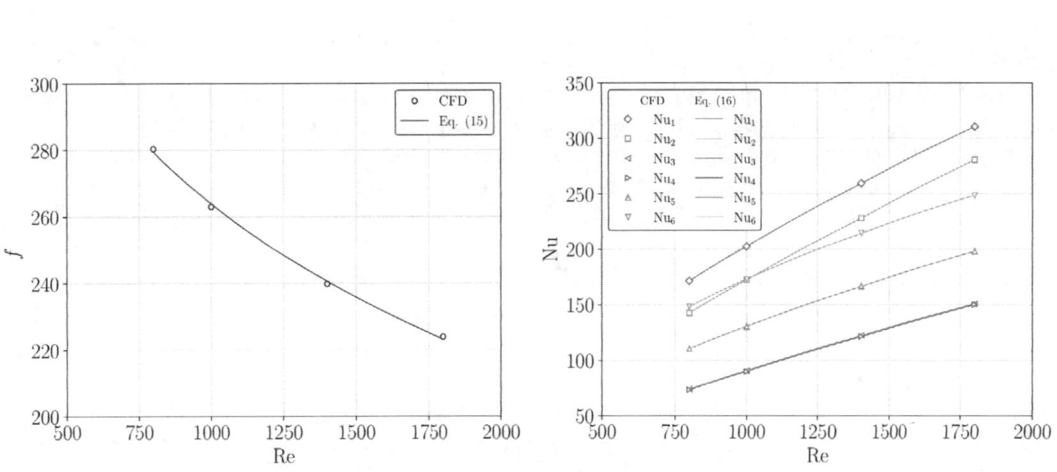

(a) (b)

Figure 5. Numerical results and the associated correlations for the equivalent Darcy friction factor (a), and Nusselt numbers related to each sidewall of the feed channel (b).

internal and external walls feature the highest Nusselt number values out of the six side walls, as at these walls most of the heat flux is applied directly to the fluid. The difference between the upper and lower Nusselt number values suggests the existence of an axial heat flux within the cooling jacket; this deviation is due to the previously discussed flow pattern in the connecting region between channels.

Finally, the lowest Nu-values are found for the two walls facing the return channel, due to the low velocity separated flow. In this case, a small heat transfer coefficient is beneficial, as it prevents heat from the return channel to enter. Finally, these two sidewalls feature an identical thermal behaviour, meaning that one less CFD simulation would be needed for each Reynolds number.

5.2. Preliminary results on overall head losses and heat transfer

In the following preliminary results on head losses and heat transfer are compared with available experimental data. Overall head losses are evaluated using the Darcy-Weisbach formula and deriving the friction factor value from Eq. (15). At a flow rate of 3 l/min of a water-glycol solution the numerically obtained head losses are 16% higher than the experimental ones, which

Table 1. Values of c_i coefficients in Eq. (16).

Sidewall ID	c_1	c_2	c_3
1	2.769	0.643	-31.75
2	0.862	0.778	-14.0
3	1.103	0.681	-30.96
4	0.858	0.710	-25.44
5	3.040	0.579	-35.49
6	23.67	0.372	-136.9

is considered acceptable.

A simplified version of the lumped parameter model presented in §4, which considers only the convective heat transfer at the feed channel internal wall, is used to obtain the inner wall temperature. The calculated temperature exceeds the reference value by 3.3 K; a slight overestimation is expected, as heat transfer with the external ambient air is not taken into account.

6. Conclusions

A multiscale modelling methodology for an electric motor cooling jacket was presented in this work. Cooling channels featuring a stepped circumferential geometry were studied by developing a CFD model. Numerical results were presented and discussed. Integral results on head losses and heat transfer were employed to derive correlation functions for the equivalent Darcy friction factor and sidewall-specific Nusselt number. A thermal resistance model was also presented to study the temperature distribution within the entire cooling jacket. Preliminary results based on a simplified version of the aforementioned model were compared with available experimental data showing a promising accuracy.

The developed approach can of course be applied to motor casings having cooling channels of any geometry. Furthermore, if circumferential periodicity of the cooling channels can be exploited the number of nodes of the lumped parameter model can be significantly reduced, allowing for a faster and more convenient implementation.

Future research efforts will be focused towards completing the development of the lumped parameter model and on the investigation of different fluids and flow regimes.

Acknowledgments

The research has been partially funded under the National Recovery and Resilience Plan (NRRP), Project "Sustainable Mobility Center (Centro Nazionale per la Mobilità Sostenibile - CMNS)", Project code CN00000023 - CUP E93C22001070001.

References

[1] PH Mellor D Roberts D T Lumped parameter thermal model for electrical machines of tefc design 1991 *IEE proc., Electr. power appl.* **138**(5) 205–218(13)
[2] Kral C, Haumer A and Bauml T Thermal model and behavior of a totally-enclosed-water-cooled squirrel-cage induction machine for traction applications 2008 *IEEE Trans. Ind. Electron.* **55** 3555–3565
[3] Aglen O and Andersson A 2003 Thermal analysis of a high-speed generator *38th Conf. Rec. Ind. Appl. Soc. IEEE-IAS Annu. Meet.* vol 1 pp 547–554 vol.1
[4] Nategh S, Huang Z, Krings A, Wallmark O and Leksell M Thermal modeling of directly cooled electric machines using lumped parameter and limited cfd analysis 2013 *IEEE Trans. Energy Convers.* **28** 979–990
[5] Menter F Two-equation eddy-viscosity turbulence models for engineering applications 1994 *AIAA J.* **32** 1598–1605
[6] Weller H G, Tabor G, Jasak H and Fureby C A tensorial approach to computational continuum mechanics using object-oriented techniques 1998 *Comput. phys.* **12** 620–631
[7] Incropera F P and DeWitt D P 2006 *Fundamentals of Heat Transfer* (Wiley)
[8] YOVANOVICH M A general expression for predicting conduction shape factors 1973 *11th Aerosp. Sci. Meet.*

Combined Experimental and Numerical Approach for the Thermal Heat Exchange Investigation of Li-Ion Cells for Automotive Applications

C. Karaca*[1], G. Baldinelli[1], L. Postrioti[1], F. Scrucca[2]

[1] 1 Department of Engineering - University of Perugia, Italy
[2] Department of Sustainability, ENEA - Italian National Agency for New Technologies, Energy and Sustainable Economic Development, Rome, Italy

* Corresponding author: cem.karaca@studenti.unipg.it

Abstract. Lithium-ion (Li-ion) battery is an advanced technology in the field of electrochemical energy storage, but its management constitutes one of the most intriguing challenges for electric vehicles. Many parameters need to be controlled and managed and many aspects need to be optimised. This work presents a methodology for laboratory characterization of Nickel Manganese Cobalt (NMC) Lithium-Ion batteries suited for automotive applications. The purpose consists of obtaining a detailed description of the electrical and thermal behaviour of a single battery cell to provide an accurate model (static, dynamic, and thermal) that could ensure optimized real-time battery management by a management system for several battery packs. A battery testing system was built using a bidirectional power supply and a software/hardware interface was implemented within the National Instruments LabVIEW environment that monitors current, voltage and temperature sensors. This dedicated laboratory equipment can be used to apply and report charging/discharging cycles according to the user-defined load profile. A bidimensional CFD dynamic condition/transient simulation in the Ansys FLUENT environment was performed to study the heat thermal fluxes generated by a determined current value in the battery cells, and the results have been compared to the experimental data for validation.

1. Introduction

Reduction of carbon dioxide emission is currently a main driving force to find sustainable and more efficient alternatives, rather than the traditional Internal Combustion Engines based on fossil fuels [1]. In this context, Electric Vehicles are quite promising and commercially feasible. The main difference between an Internal Combustion Engine Vehicle (ICEV) and an EV or HEV is the presence of a battery pack used to power the vehicle, where single cells are connected in series to increase the voltage, and in parallel to increase the capacity. Batteries are also the main limitation in electric vehicle spread due to their limited capacity and long charging time, lifetime, cost, and safety issues such as thermal runaway [2]. Furthermore, lithium-ion batteries have limited operating conditions restricted by temperature and voltage windows, due to performance and safety issues [3].

Therefore, it is crucial to monitor multiple parameters, such as the state of charge (SOC), temperature, load current, and to understand how battery cells would behave in certain applications. This is accomplished by implementing an electronic unit called Battery Management System (BMS).

A battery management system is a system control unit that reacts with external and internal events to safeguard the battery. It balances the cell voltage (i.e. keeping the cells at the same SOC level), overviews aging issues, and also takes corrective measures against any abnormal condition at the system infrastructure. To execute these tasks effectively, an accurate battery model, able to estimate and communicate its energy and power availability to the rest of the system from its terminal voltage and load-current history, must be implemented in the BMS online control strategy [3]. Thus, finding a fine method to characterize battery cell behaviors is useful in performance modeling and predicting degradation. For this purpose, BMS needs two essential parameters: the State of Charge (SoC) and the State of Health (SoH) [4,5].

Many studies have been released to explain the thermal conditions on batteries under different C rates and charge or discharge duration by comparing the results of CFD and experimental tests. Ö. Ekici et al. have been established their studies with the conditions as galvanostatic charge and discharge tests at 0.5C, 1.0C and 1.5C rates were conducted. Test durations were 30, 60 and 90 min, depending on the C rate values. Experiments were performed at three different operating temperatures, namely, 20°C, 35°C and 50°C [11]. Another study is a zero-dimensional numerical model based on the transient energy balances and the analogy between heat transfer and electrical transfer using resistances and capacitors. The simulations were carried at 3C discharging/0.5C charging rates and the obtained results were compared with both three-dimensional computational fluid dynamic (CFD) results and experimental results from the literature [12]. The Nickel Manganese Cobalt (NMC) lithium-ion battery cells characterization follows these steps: static characterization, dynamic characterization, and thermal characterization which are studied by a Capacity test, a Hybrid pulse power characterization test (HPPC), CFD simulations and their experimental validation.

In brief, the first test is performed since, even if a nominal capacity value is always provided by the manufacturer inside the battery cell datasheet (2.9 Ah in this case), this represents a worst-case estimate, therefore a test should be performed to estimate a more reasonable value to be employed inside the cell model. HPPC allows the proper characterization of the Electrical Equivalent Circuit (EEC) model by soliciting the battery with charging/discharging current pulses while observing its voltage response. The purpose is to define the impedance and its non-linear behavior depending on operating conditions [6] [7]. Finally, CFD simulations have been implemented to provide information about the external temperature of the cells as a function of the current values and how natural convection develops in the surrounding environment.

2. The Experimental Set-up

In this chapter we will introduce the experimental set-up used in this study. The test bench used was designed to ensure adequate data acquisition in safety, with the aim of maintaining a stable ambient temperature. The study, for reasons of statistical reliability, was conducted on 4 Li-ion cells electrically connected in parallel, NMC type and model N18650CL-29 of the company Zhengzhou BAK Battery Co., Ltd. This cell has a nominal voltage of 3.6 V, nominal capacity 2.9 Ah, with an internal impedance declared by the manufacturer ≤ 35 mΩ, and an energy density (at 1C) of 213 Wh/kg. These 4 cells have been arranged on a base designed to guarantee heat exchange with the air and avoid overheating. Figure 1 shows the batteries, the complete bench and the ITECH Battery Test System.

The cells' supports have been fixed on a wooden base, so as to ensure adequate electrical insulation. This base was closed with three walls of polystyrene, to ensure a stable temperature, and the fourth in plexiglass to ensure optical access. As it is possible to see, placed on a lower shelf, there is the entire data acquisition system including four K thermocouples to monitor the cell temperature, a PT100 thermocouple for the ambient temperature, four LEM current sensors – CTSR 1-P (characterized by a working range equal to +/- 1 A) and four LEM – CTSR 6-NP (characterized by a working range equal to +/- 6 A). For each cell there is one of these two different types of current sensors, so as to guarantee the acquisition accuracy for different current ranges.

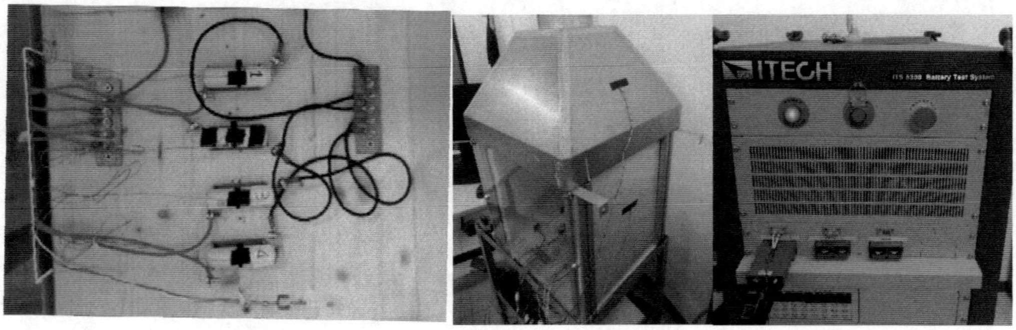

Figure 1. Test cells, complete bench and battery cycler.

These eight current sensors were first calibrated with a NI-Virtual Bench multimeter in the effective working range, and then managed by three acquisition cards, all then connected to a PC, and they are: the NI-USB 6001 which acquires voltage data, the NI-PCI-6221 which acquires current values, and the NI-USB-6008 for temperatures. The PC also communicates with the ITECH ITS 5300 "cycler" (e.g the system capable of absorbing/receiving current from the cells connected to it in parallel) via ITECH proprietary software. The ITECH battery test system consists of two IT-6015C bidirectional power supplies (with full scale 80 V and 360 A for one, and 500 V and 90 A for the other) [8]. The wiring diagram that summarizes this configuration can be seen in Figure 2:

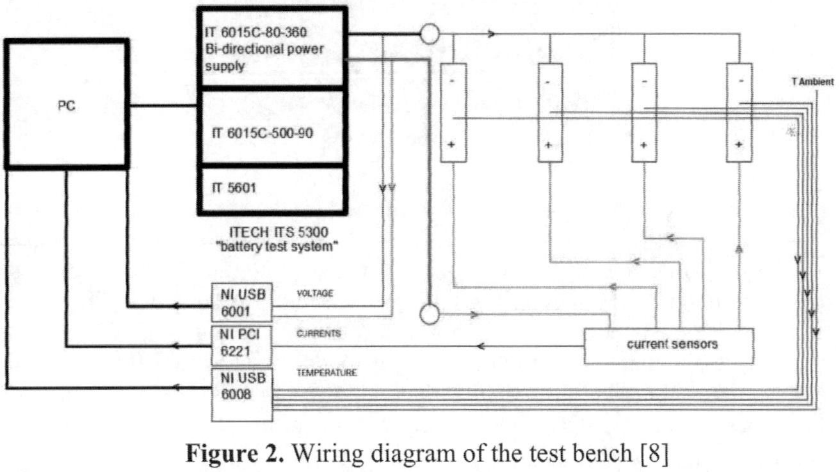

Figure 2. Wiring diagram of the test bench [8]

The management of the hardware used in the counter is entrusted to two control and acquisition systems. Respectively one, created internally in the LabVIEW environment, dedicated to data acquisition (currents, voltages and temperatures), the other implemented by ITECH which manages the charge and discharge cycles.

3. Implementation of the numerical model for the computational approach

CFD simulations are implemented by Ansys Fluent for thermal conditions and thermal exchange between cells and air. This part examines each step of simulation from creation of geometry to results of analysis (Figure 3).

As software, Solidworks and Autocad are used to design geometry of the simulation. Firstly, 3D test bench model is divided in four equal parts with the same conditions in the test bench. Only 2 batteries and air domain remain in the final 3D model.

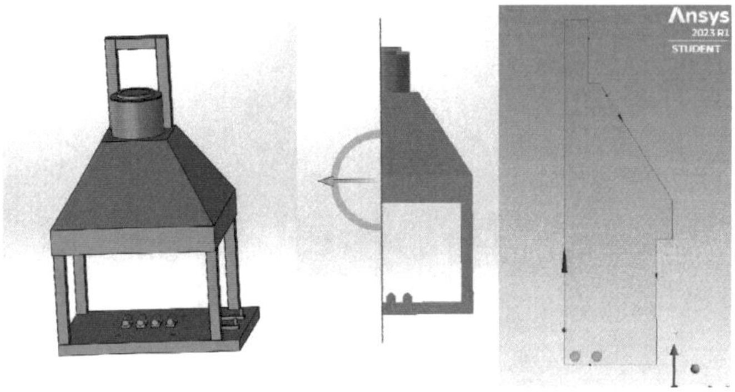

Figure 3. 3D Geometry created by Solidworks and 2D Geometry

Secondly, the 3D geometry is converted to a 2D geometry and imported to Ansys Workbench. The Fluent module is used to see the energy exchange between cells and air domain. "Mesh inflation" is applied inside (three layers) and outside (seven layers) of the cells and generated "First layer thickness", which equals 0.0005m. Triangular type mesh is selected for all the model. Symmetry edge, insulated edge, bottom, cap edges are applied. A total of 68,891 elements are generated after all these meshing selection.

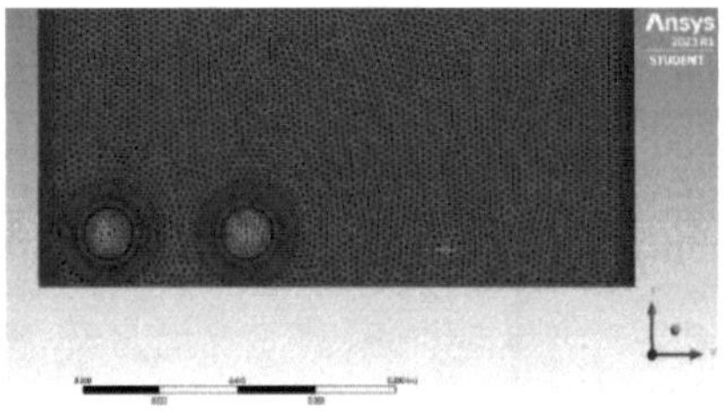

Figure 4. Meshing condition in batteries, air domain and walls

The next step for the CFD simulation is the Fluent-Setup section. Simulation conditions have been defined according to experimental test conditions. First of all, transient conditions are selected with 49 time steps. Each time step lasts 60 seconds with a maximum of 1,300 iteration. The chosen viscous model is "Laminar", pressure based conditions; gravity is obviously applied. Discharge is simulated in 2.75 A (C rate) for one cell. Each cell has 35 $m\Omega$ nominal internal resistance (assumed as measured in ambient temperature conditions, at SOC 100%), 2,360 $kg/m3$ density, 1,000 J/kgK specific heat and 1 W/mK thermal conductivity.

A.H. Mahmud *at al.* issued a study on the impact of SOC and temperature on internal resistance of Li-ion cells. Based on their research results, the battery internal resistance typically evidences a significant decrease with temperature, around 70 % for cells going from 10°C to 35 °C [12]. Considering the relevance of the energy source intensity for the thermal evolution of the cells, in our simulation the internal resistance has been supposed to assume different values, calculated accordingly the SoC values evaluated during the discharging phase. The assumed internal resistance vs. the SoC trend is shown in Figure 6.

The duration of the simulation steps in which different values of internal resistance are assumed have been selected after monitoring the SOC of batteries during the experimental test. Experimental test results gave the state of charge for each battery with the discharge duration. Figure 6 shows the voltage values vs. SOC percentage. Other researchers evidenced how the internal resistance as a function of SoC for Li-ion cells can have significantly different trends. In particular, in [9] experimental test results highlighted how that internal resistance was at maximum value when the battery capacity was empty.

The minimum value of the internal resistance was not noticeable at the time at 100% SOC, but at a point between 80% and 90%. Four batteries are used in their research and the results are shown as Figure 7. As it will be shown in the following section, assuming different trends for internal resistance can have dramatic effects on the cell temperature time-history.

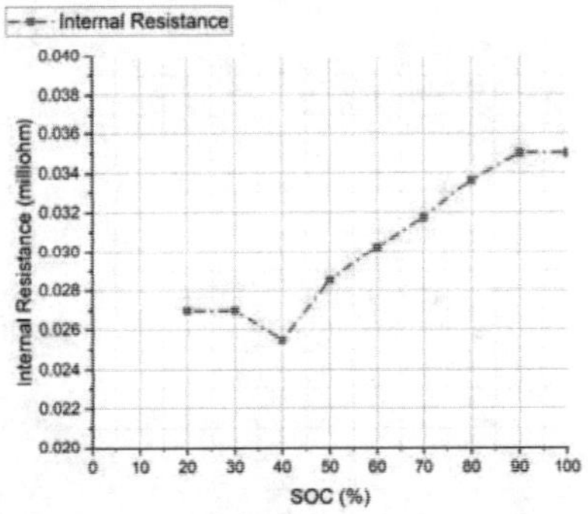

Figure 5. Internal resistance of test batteries vs. SoC, from [12]

% SoC	Voltage (V)
0	2.841
2.5	2.874
5	2.907
10	2.974
20	3.107
30	3.24
40	3.374
50	3.507
60	3.64
70	3.774
80	3.907
90	4.04
95	4.106
100	4.174

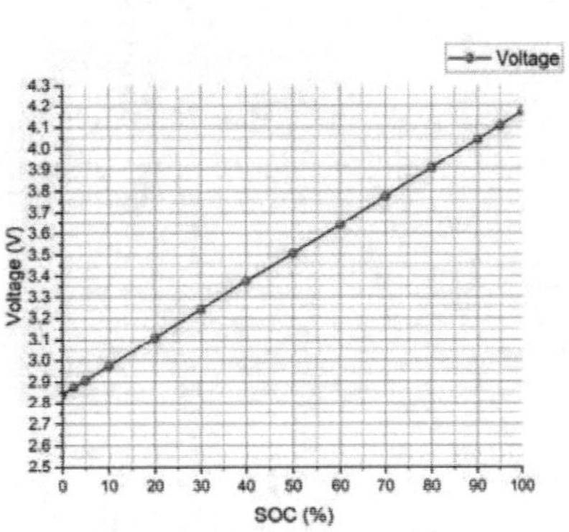

Figure 6. Voltage values which equals to SoC percentage

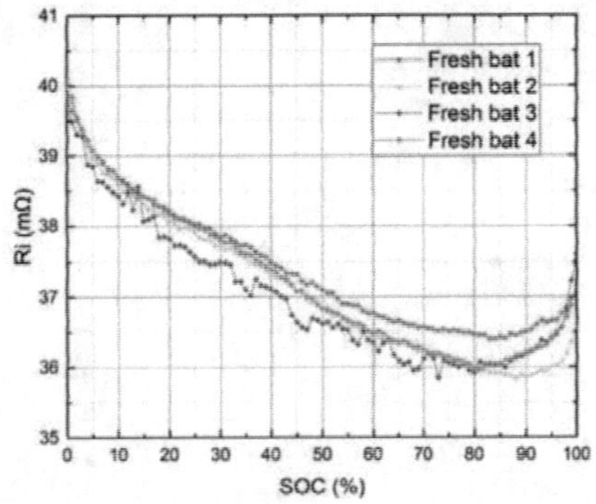

Figure 7. R_i-SOC plot for different batteries [9]

4. Experimental Test and Results

After the preparation of the experimental setup, the capacity test is started with the steps reported in Figure 8, while the test evolution is reported in Figure 9.

As can be seen in Figure 9, In step 1, the cell is simply following the ambient temperature evolution.

In step 2, a C/20 moderate current is flowing hence the cell temperature gradient is slightly increased in the first 2 hours in step 2. Later a moderate temperature increase is observed and then cells slowly reach a thermal equilibrium with ambient temperature; equilibrium is achieved about at time 40k seconds. Step 3 starts from a thermal equilibrium condition and immediately the temperature starts raising with a significand gradient due to intense current (1C rate discharge). In step 3, the temperature gradient follows the resistance trend in fig 6 i.e. the initial temperature gradient is higher, for higher resistance; as the cell is discharged, the gradient is progressively reduced.

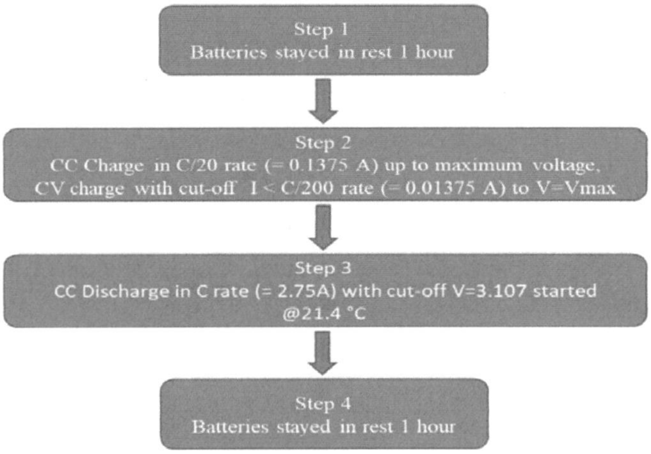

Figure 8. Sequence of the experimental setup steps

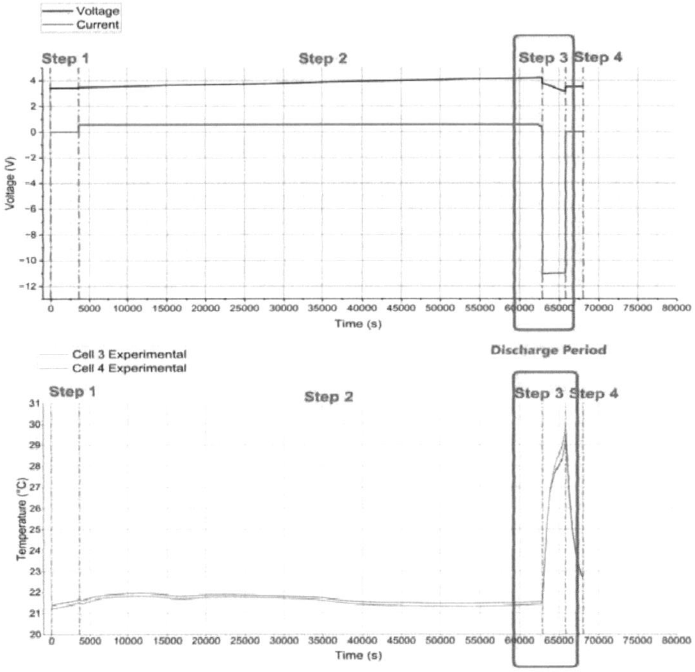

Figure 9. Capacity Test Voltage/Current-Time Graph and Cells-Temperature Graph

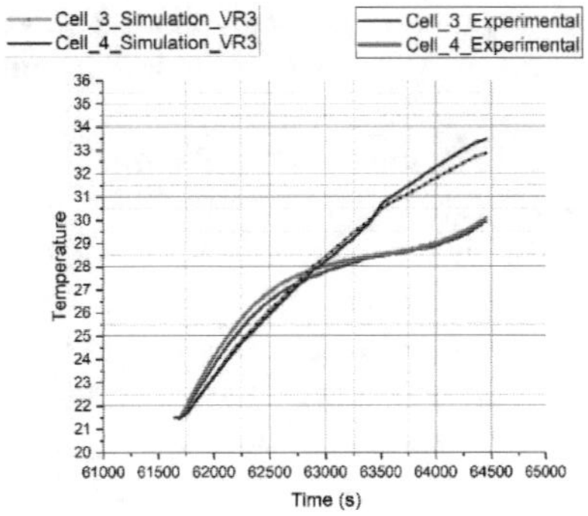

Figure 10. Comparison of experimental test and CFD simulation VR3 vs time.

Voltage and current values are shown in the top graph of Figure 9 during the charge and discharge conditions. Also each step is divided with red dash-dot lines. Current value is 0 during the resting time but OCV is 3.435V in the first step. In the second step, constant current is 0.55 A (for the 4 batteries). This section provides to the batteries to reach to maximum voltage level which is defined by manufacturer as 4.20V. After all batteries are fully charged, discharging duration is started in the step three. The discharge is applied with a constant current, -11 A for the 4 batteries to reach the OCV of batteries of 3.107 V. The discharging period goes for each battery from 100% of SoC to 20%. The resting period followed the discharging period and the resting time is 1 hour in the step four.

The comparison between tests and simulations are shown in Figure 10. Cell 3 and Cell 4 temperature time-profile are sketched during the discharge period, varying the resistance according to the assumption reported in Figure 5. Simulation_VR3 is run assuming seven different internal resistances values for each cell as the SoC was progressively decreased during the CC discharge Step 3. The temperatures of simulation results are almost 4 °C higher than the experimental ones.

In Simulation_VR2, the internal resistance *vs.* SoC dependence was assumed to be different, using three different internal resistance values. These internal resistances are 35 mΩ, 17.5 mΩ and 22.5 mΩ. The temperature time-profile resulting from Simulation_VR2 are reported in Figure 11; as can be seen, with this assumption of the temperature trend, the simulations are in very good agreement with the experimental data.

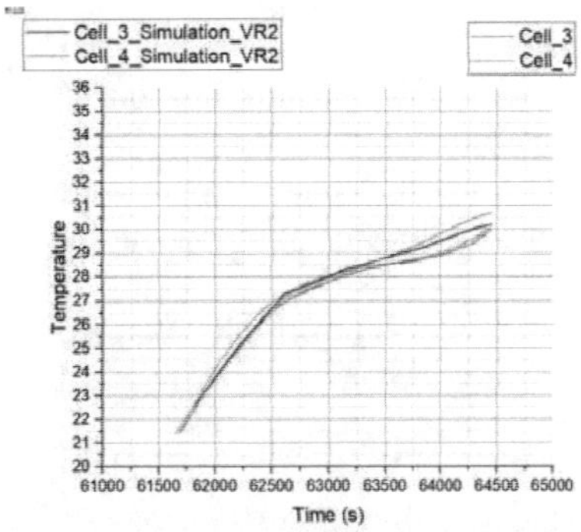

Figure 11. Comparison of experimental test and CFD simulation VR2 vs time.

5. Conclusions

In the present activity, an experimental test bench for the analysis of thermal behaviour of four Li-ion batteries was developed by using the NI-USB 6001 which acquires voltage data, the NI-PCI-6221 which acquires current values, the NI-USB-6008 for temperatures and the ITECH ITS 5300 which has battery test system consists of two IT-6015C bidirectional power supplies. Applied 4 steps are resting, CC-CV charging, CC discharging and finally resting step again, have been set. Batteries thermal condition are monitored, between from 100% SoC to 20% SoC of batteries, during only step 3 which is CC discharging.

A numerical methodology for the CFD simulation of thermal evolution of cells, thermal fields and thermal exchange with surrounding air. CFD simulation is run as 2D analysing. 2 cells, surrounding air and boundary conditions have been defined respecting to the experimental test conditions. Also in the simulation, only step 3, CC discharging, is analysed.

The relevant influence of internal resistance dependence on SoC was proved, with different assumptions resulting in significantly different thermal evolution of cells. Internal resistance is related to temperature, SoC and discharge rate and any changes of temperature, SoC or discharge rate can impact the value of internal resistance [10]. The study shows that internal resistance can be decrease almost half values in the different SoC but when it reaches minimum degree, it doesn't increase reasonable during the discharging.

6. References

[1] V. Ruiz, A. Pfrang, A. Kriston, N. Omar, P. van den Bossche, and L. Boon-Brett, "A review of international abuse testing standards and regulations for lithium-ion batteries in electric and hybrid electric vehicles," *Renewable and Sustainable Energy Reviews*, vol. **81**. Elsevier Ltd, pp. 1427–1452, Jan. 01, 2018. DOI: 10.1016/j.rser.2017.05.195.

[2] E. Vergori, F. Mocera, and A. Somà, "Battery modeling and simulation using a programmable testing equipment," *Computers*, vol. **7**, no. 2, 2018, DOI: 10.3390/computers7020020

[3] H. A. Gabbar, A. M. Othman, and M. R. Abdussami, "Review of Battery Management Systems (BMS) Development and Industrial Standards," *Technologies*, vol. **9**, no. 2, p. 28, Apr. 2021, DOI: 10.3390/technologies9020028.

[4] S. Singirikonda and Y. P. Obulesu, "Advanced SOC and SOH Estimation Methods for EV Batteries—A Review," in *Lecture Notes in Electrical Engineering*, 2021, vol. 700, pp. 1963–1977. DOI: 10.1007/978-981-15-8221-9_182.

[5] M. Thingvad, L. Calearo, A. Thingvad, R. Viskinde, and M. Marinelli, "Characterization of NMC Lithium-ion Battery Degradation for Improved Online State Estimation," Sep. 2020. DOI: 10.1109/UPEC49904.2020.9209879.

[6] D. Cittanti, A. Ferraris, A. Airale, S. Fiorot, S. Scavuzzo, and M. Carello, "Modeling Li-ion batteries for automotive application: A trade-off between accuracy and complexity," 2017. DOI: 10.23919/EETA.2017.7993213.

[7] D. I. Stroe, M. Swierczynski, A. I. Stroe, and S. K. Kær, "Generalized characterization methodology for performance modeling of lithium-ion batteries," *Batteries*, vol. **2**, no. 4, Dec. 2016, DOI: 10.3390/batteries2040037.

[8] E. Valentini, L. Postrioti, G. Baldinelli, "Analisi Sperimentale Delle Caratteristiche Elettriche E Di Scambio Termico Di Celle Li-Ion Per Impiego Automotive,"2022, Master Thesis, Perugia University.

[9] D. Wang, Y. Bao, J. Shi, "Online Lithium-Ion Battery Internal Resistance Measurement Application in State-of-Charge Estimation Using the Extended Kalman Filter," 2017, *Energy MDPI*, Basel, Switzerland.

[10] L. Chen, M. Zhang, Y. Ding, S. Wu, Y. Li, G. Liang, H. Li, H. Pan, "Estimation the internal resistance of lithium-ion-battery using a multi-factor dynamic internal resistance model with an error compensation strategy", 2021, *Electrochem. SoC.* Interface 21-61, Volume **7**, Pages 3050-3059.

[11] A. Amini, Ö. Ekici, T. Özdemir, S. Ç. Başlamışlı, M. Köksal, "A thermal model for Li-ion batteries operating under dynamic conditions", 2021, *Applied Thermal Engineering*, Vol **185**

[12] A.H. Mahmud, Z. H. C. Daud, Z. Asus, "The Impact Of Battery Operating Temperature And State Of Charge On The Lithium-Ion Battery Internal Resistance", 2017, *Jurnal Mekanikal*, Vol **40**, 01-08

Numerical validation of ice formation on a lattice structure evaporator

L Bernardini[1], A I Garivalis[1], B Marangolo[1], E Rezaei[1,*] and P Di Marco[1]

[1] DESTEC, University of Pisa, Largo Lucio Lazzarino 1, Pisa 56122, Italy

* Corresponding author: ehsan.rezaei@adm.unipi.it

Abstract. This paper presents a comparative analysis between experimental and numerical results of the ice formation process on the surface of the evaporator in a Sparkling Water Dispenser. The experimental setup comprises an R290 refrigeration cycle that is fully equipped with thermocouples, pressure transducers, a wattmeter, and a Coriolis-meter. The evaporator, which is specifically designed to enhance the ice formation speed, is situated within a water-filled tank, where the ice formation process takes place. Due to the phase change phenomenon, which involves the interface between two phases moving, solving transient heat transfer problems involving solidifications can be inherently challenging. Analytical solutions are only possible under certain simplified circumstances. In cases where exact solutions are not available, semi-analytic, approximate, and numerical methods can be utilized to address phase-change problems. A numerical model of the solidification process based on the energy equation and conjugate heat transfer was developed using COMSOL Multiphysics. The software was found to be effective in simulating the physical processes associated with heat transfer through conduction and convection, as well as the behaviour of phase change. The results showed a very good agreement between experimental and numerical results.

1. Introduction

Refrigeration systems are widely used in many industries and applications. Among the various applications of refrigeration systems, we also find water dispensers. These devices extract potable water from a source (usually from the water supply or a reservoir), cool it to the setpoint temperature (typically below 10 °C), and dispense it. Given the physical properties of water and the specific requirements of this type of device (which must be small and consume low electrical power), an auxiliary system is needed to handle intermittent peaks of thermal power. Therefore, these devices are generally associated with a thermal storage unit, which mitigates load peaks and prevents the refrigeration circuit from being oversized and activated whenever there is a demand for chilled water. Thermal storage plays a central role in the operation of a Sparkling Water Dispenser (SWD). Due to its availability, low cost, and high energy density (333 kJ/kg), ice water storage is the most widely used technology in SWDs.

Another critical component of a refrigeration system is the evaporator. In these devices, the evaporator is immersed inside the storage unit and charges the storage by solidifying the water. Ice formation can significantly reduce the heat transfer rate and increase the system's energy consumption. The ice formation process is complex and depends on various factors, such as the evaporator geometry, ambient and refrigerant temperatures, and the flow rate of the refrigerant. To better understand this process, researchers have developed various experimental and numerical models to investigate the ice formation mechanism and optimize the design of the heat exchanger [1-4]. Given objective to optimize the heat exchanger geometry, the utilization of a numerical model offers a significant advantage in a cost-effective and time-efficient manner. The phase change process has been studied for a variety of geometries, the most frequent of which being shell-and-tube systems, encapsulated systems, and ice-on-coil systems. Ice-on-coil energy storage tanks are widely used in phase change commercial cooling applications [5-9]. Numerous studies on ice formation on the coil heat exchangers have been conducted during the past decay using commercial CFD programs. Afsharpanah et al. [9] numerically examined a cuboid-shaped ice container with serpentine tubes and plates by evaluating various dimensionless parameters regarding the flow and geometric aspects. In numerical research, Hamzeh & Miansari [10] looked explored the quantity and configuration of refrigerated tubes as well as the impact of the fin dimensions on ice development. A novel ice-on-coil cold storage system with coil tube was investigated in experimental and numerical research by Mousavi Ajarostaghi et al., [11] by proposing a new evaporator concerning better cooling process and uniform ice production. In all mentioned research, Ansys Fluent was used for the numerical simulation of the ice formation process. COMSOL Multiphysics is one such software that also can be used for this purpose. However, it is important to validate the simulation results with experimental data to ensure their accuracy.

In this paper, we will compare the experimental data obtained from lattice structure evaporators with the simulation results obtained using COMSOL Multiphysics. We will also discuss the agreement between the experimental and simulated results.

2. Experimental Apparatus and Procedures

2.1. Refrigerant Loop

Figure 1 shows the sketch of the experimental apparatus. The facility consists of a propane (R290) refrigeration cycle composed of a reciprocating compressor (COM), a water-heated evaporator (EV), an air-cooled condenser (CO), and an expansion device (CT). An oil filter (OF) is located before the capillary tube. The evaporator, consisting of a copper finned-tube structure, is placed inside a tank of water kept in motion by an agitator. The temperature of the hot source is kept constant in the laboratory environment, while the cold source's temperature evolves during ice formation on the evaporator's surface.

Temperatures at each point of interest are measured with calibrated T-type thermocouples with a measurement uncertainty of ±0.2 °C. Thermocouples are located at the inlet of the capillary tube (T1), at the outlet of the capillary tube (T2), at the outlet of the evaporator (T3), at the outlet of the compressor (T4), and in front of the condenser fan (T5). Thermocouples T1, T2, T3, and T4 are placed inside the copper tube in direct contact with the refrigerant fluid. The pressure at the sides of each component is measured with absolute pressure transducers with an uncertainty of ±0.025 bar. The pressure transducers (KELLER-PAA-23 SY Ei) are placed at the inlet of the capillary tube (P1), at the outlet of the capillary tube (P2), at the outlet of the evaporator (P3) and the outlet of the compressor (P4). A Coriolimeter (Bronkhorst - MINI Cori-Flow, CF) placed at the condenser outlet measures the R290 mass flow rate with an uncertainty of ±0.2%. Finally, a digital energy meter (Socomec COUNTIS E03/E04, WM) measures the electrical power absorbed by the compressor.

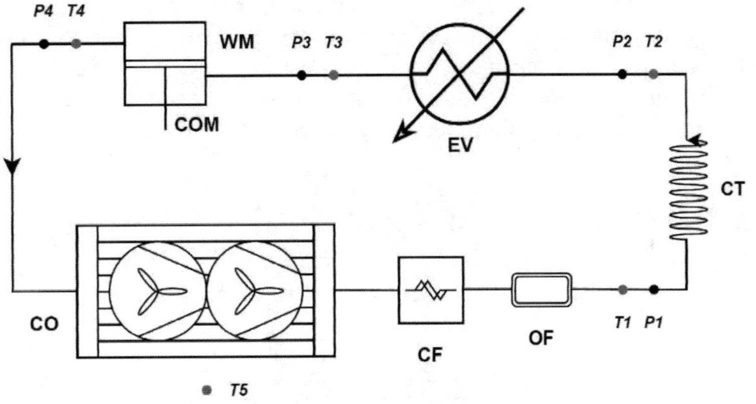

Figure 1. Schematic of the experimental setup

Temperatures at each point of interest are measured with thermocouples of T-type with an uncertainty of ±0.2 °C. Thermocouples are located at the inlet of the capillary tube (T1), at the outlet of the capillary tube (T2), at the outlet of the evaporator (T3), at the outlet of the compressor (T4), and finally in front of the condenser fan (T5). Thermocouples T1, T2, T3, and T4 are placed inside the copper tube in direct contact with the refrigerant fluid. The pressure at the sides of each component is measured with absolute pressure transducers with an uncertainty of ±0.025 bar. The pressure transducers (KELLER - PAA – 23 SY Ei) are placed at the inlet of the capillary tube (P1), at the outlet of the capillary tube (P2), at the outlet of the evaporator (P3) and the outlet of the compressor (P4). A Coriolimeter (Bronkhorst - MINI Cori-Flow, CF) placed at the condenser outlet measures the R290 mass flow rate with an uncertainty of ±0.2%. Finally, a digital energy meter (Socomec COUNTIS E03/E04, WM) measures the electrical power absorbed by the compressor. Table 1 summarizes the main characteristics and accuracy of the measuring devices.

Table 1. Characteristics and accuracy of the measurement elements.

Quantity	Variable	Type	Calibration range	Accuracy
5	Temperature	T-type thermocouple	-30 to 100 °C	±0.2 °C
2	Pressure	Pressure transducer KELLER - PAA – 23 SY Ei	0 to 25 bars	± 0.0625 bar
2	Pressure	Pressure transducer KELLER - PAA – 23 SY Ei	0 to 10 bars	± 0.025 bar
1	Mass flow rate	Coriolimeter Bronkhorst MINI Cori-Flow	0 to 10 kg/h	± 0.02 kg/h
1	Electric power	Single-phase Digital Energy meter – Socomec COUNTIS E03/E04		±0.5% lecture

2.2. Evaporator and Thermal Storage

The evaporator is made of a spiral-shaped copper tube with an 8 mm outer diameter installed on a square base with 19 cm on each side and a height of 21 cm. The heat exchanger tube has seven turns and a longitudinal copper fin that runs its entire length except for the corners. The fin has a thickness of 1 mm and a length of 24 mm. In addition, 80 copper bars with a 3 mm thickness, spanning the entire height of the heat exchanger, are positioned perpendicular to the fins. The thermal storage, a water-filled tank containing 5.4 litres of water, holds the evaporator. A 10 cm-long radial agitator is in the center. Six T-type thermocouples, arranged between the fourth and fifth turns starting from the top, are used to instrument the heat exchanger. These are positioned between two fins and two rods and installed into a 3D-printed support, as shown in figure 2. Thermocouples are calibrated and of the same type as those used in the refrigerant loop, with an uncertainty of 0.2 °C.

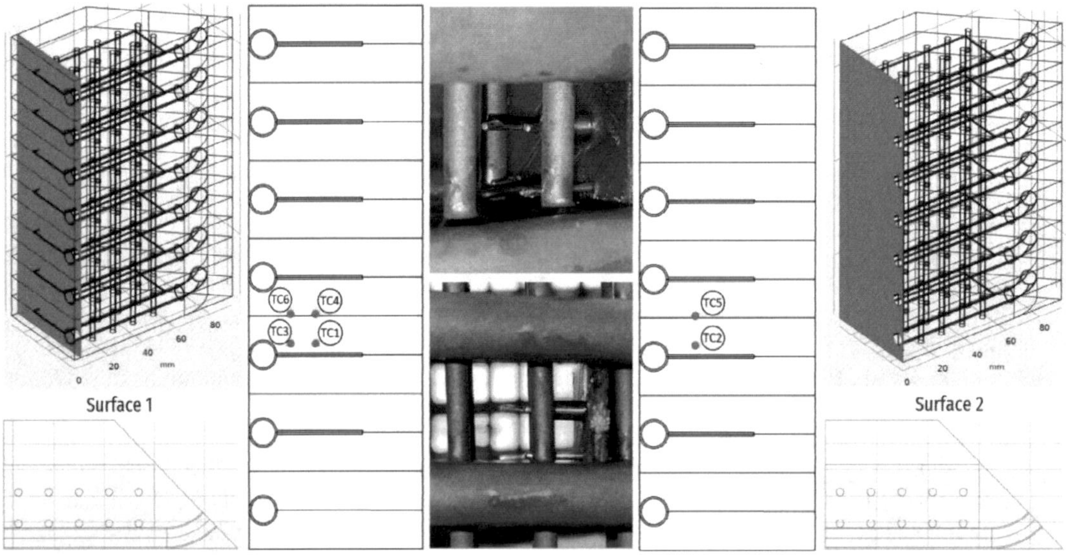

Figure 2. Six thermocouples location on two different surfaces.

2.3. Experimental Procedures

The experimental procedure unfolds as follows. The storage is at 27 °C, then, the refrigeration cycle and the agitator inside the storage are turned on, starting the sensible cooling phase. When the average temperatures measured by the thermocouples are close to 0°C, the agitator is switched off, and the ice formation begins. The test ends when 4 kg of ice has been formed. The water level and the energy balance on the evaporator assess the amount of ice. Three experiments were conducted one day apart making sure that the same environmental temperature of 22 °C was maintained

3. Numerical simulation

The most suitable method for simulating our case is the conjugated heat transfer approach, as it accurately represents the physics involved. This approach considers the heat transfer occurring at the interface between a copper tube carrying the heat transfer fluid (refrigerant) in motion and the stationary storage fluid (water), which undergoes a phase change in both processes. As shown in figure 3a, the study was carried out using a 3D geometry, and transient analysis under conduction and convection heat transfer was used to account for the time dependence of the problem as well as for the heat transfer and fluid flow in the entire system. The propane temperature was determined based on the pressure that was observed throughout the test.

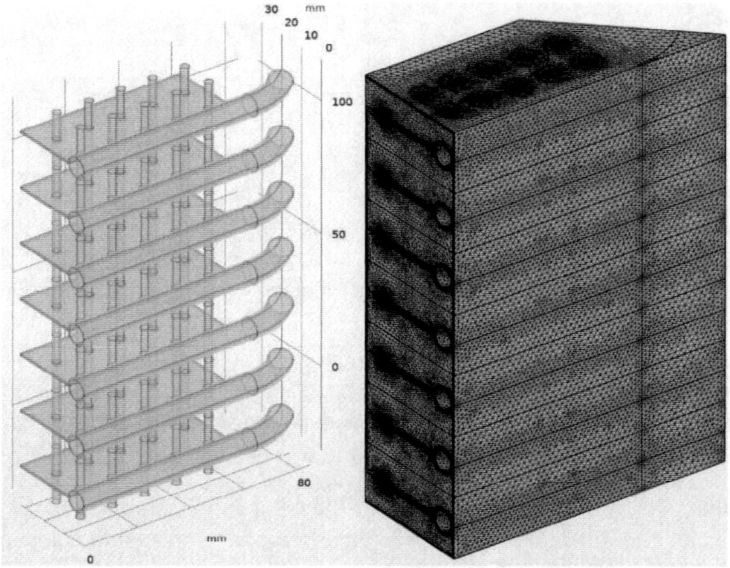

Figure 3. a) 3D geometry, b) Created grid mesh.

3.1. Grid Mesh

In accordance with the physics and geometry concerns, the computational grid was created using COMSOL Meshing (figure 3b). A sensitivity analysis was done to choose the best grid while taking accuracy and computing cost into account. To do this, the simulation was run with varying cell sizes *coarser* (1.9 million cells), *coarse* (3.4 million cells), *normal* (6.8 million cells), and *fine* (9.6 million cells). By comparing the temperature variations at five different times, the findings in the *normal* and *fine* grid meshes can be shown to have a small difference. As a result, all simulations were done using a *normal* grid mesh.

3.2. Governing Equations

The buoyancy effects are taken into account using the Boussinesq approximation:

$$\rho = \rho_0\left[1 - \beta\left(T - T_f\right)\right] \tag{1}$$

The continuity:

$$\nabla . \vec{V} = 0 \tag{2}$$

The momentum equation:

$$\rho \frac{\partial V}{\partial t} + \rho \vec{V} . \nabla \vec{V} = -\nabla P + \mu \nabla^2 \vec{V} + \rho g \beta \left(T - T_{Ref}\right) \tag{3}$$

The energy equation:

$$\frac{\partial h_{s+l}}{\partial t} + \nabla . \left(\vec{V} h_s\right) = \nabla . \left(\frac{k}{\rho C_p} \nabla h_s\right) \tag{4}$$

Where h_s and h_l are sensible and latent enthalpy. In these equations, ρ, C_p and k are calculated based on the properties of two phases:

$$\rho = \theta_1 \rho_1 + \theta_2 \rho_2$$
$$C_p = \frac{1}{\rho}\left(\theta_1 \rho_1 C_{p,1} + \theta_2 \rho_2 C_{p,2}\right) + L_{1 \to 2} \frac{\partial \alpha_m}{\partial T} \tag{5}$$

$$\alpha_m = \frac{1}{2}\frac{\theta_2\rho_2 - \theta_1\rho_1}{\theta_1\rho_1 + \theta_2\rho_2}$$
$$k = \theta_1 k_1 + \theta_2 k_2$$

Where, $L_{1\rightarrow 2}$ is the latent heat energy, and α is the thermal diffusivity.

3.3. Boundary Conditions, Initial Conditions

For boundary conditions, the following presumptions are made:

- The outer, top, and bottom boundaries are considered as perfectly insulated surfaces.
- Due to geometry symmetry, one-eighth of geometry was simulated, and two surfaces were considered as symmetric.
- No-slip conditions was considered on all walls.
- The temperature is based on the measured pressure and mass flow rate inside the evaporation tube used for the convective heat flux boundary condition for the inner surface of the tube. The temperature difference between tube rows was neglected.

The initial condition for simulations is a temperature field of 27 °C for all domains. All the properties of the material were used in simulations considered as temperature variants. Some parameters related to the simulation are presented in Table 2.

Table 2. Fixed simulation parameters.

Phase change temperature	0 [°C]
Reference temperature	20 [°C]
Latent heat	333 [kJ/kg]
Melting temperature range	3 [°C]

3.4. Solver Setting

Conjugated heat transfer, turbulent, and laminar flow models are employed as the physics for the simulation. With a total duration of 3500 seconds and a time step of 1 second, a time-dependent approach is used. Three variables, pressure (P), velocity (V), and temperature (T), are used to solve the physics. The solver employed by COMSOL is coupled direct solver PARDISO.

4. Results and Discussion

The simulation is implemented by dividing it into two stages (0-1250 and 1251-3500). The first stage involves modeling the turbulent flow physics interface, considering the experimental conditions with a working impeller. In the second stage, the simulation focuses on the laminar flow to replicate the conditions without the impeller. The results of the numerical modeling of the ice-formation process in the geometry under the study and its comparison with the experimental data are presented in this section. Figure 4a shows the comparison between numerical and experimental data for the six temperatures along the process and the absolute error between experimental and numerical results represented in figure 4b. The maximum absolute error for six thermocouples is below two degrees. It is clear that the simulation results are in good agreement with the experimental data.

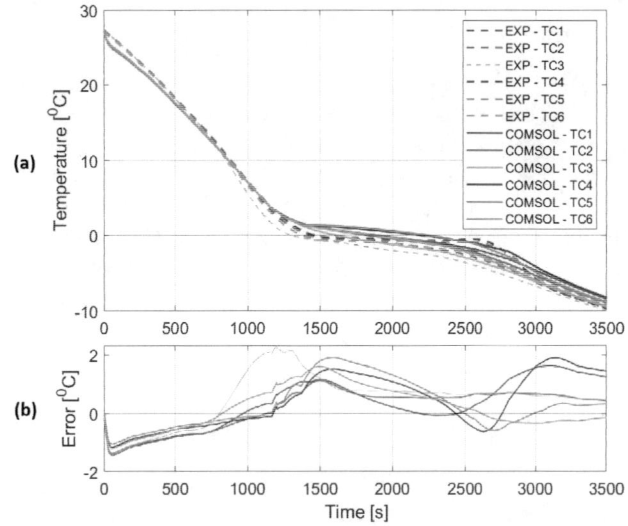

Figure 4. Comparison between experimental and numerical results for six thermocouples, a) temperature, b) absolute errors.

The velocity profile and streamlines at surface 1 are presented in figure 5a. The temperature contours for the surface one for different times have been illustrated in figure 5b. The observation reveals that as time passes, a growing region surrounding the heat exchanger attains the temperature required for solidification. Subsequently, this area solidifies, and the temperatures within these regions even fall below the phase change temperature. To examine how ice is distributed within the storage, one can utilize the contours of the liquid fraction. Figure 5c shows phase fraction contours at different times. These findings are in excellent accord with the outlines of the temperature shown in figure 5b.

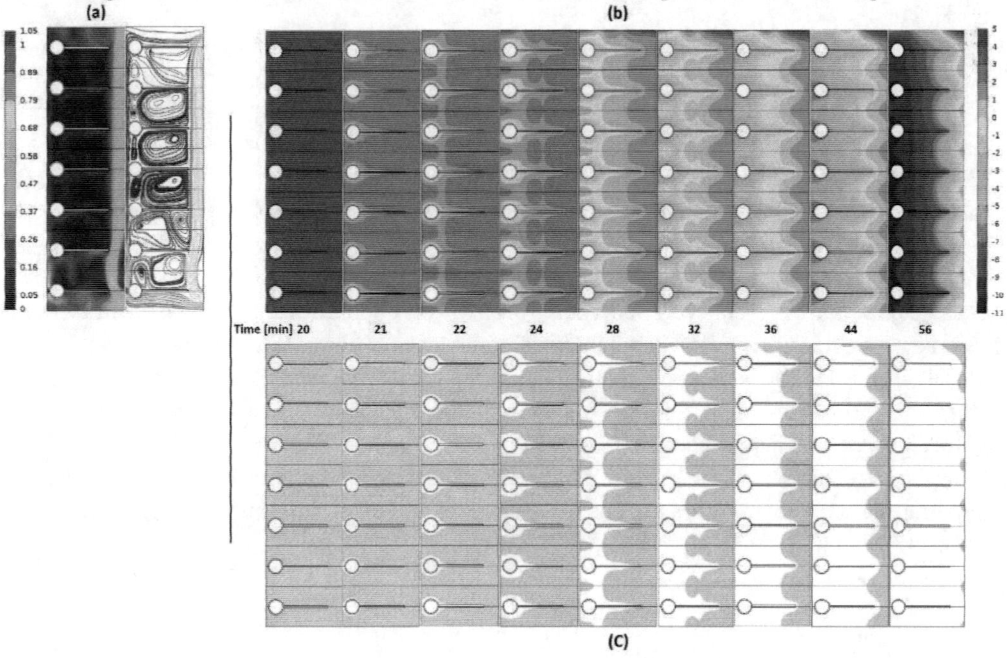

Figure 5. Numerical results for, a) velocity profile, b) temperature profiles at different times, and c) liquid fraction at different times (blue as liquid and white as solid).

5. Conclusion

This study used a 3D numerical simulation to examine the ability of COMSOL Multiphysics in simulating the ice formation process in a find-and-tube heat exchanger. The numerical results were validated with experimental results of an evaporator part of a refrigeration cycle. The findings of this study demonstrate that the various physical phenomena involved in water flow, heat transfer through conduction and convection, and phase change can be effectively simulated using COMSOL Multiphysics software. Future endeavors will involve investigating the complete refrigeration cycle, encompassing ice formation and ice melting using COMSOL Multiphysics and performing an experimental validation of the numerical results.

6. References

[1] Jannesari H and Abdollahi N 2017 Experimental and numerical study of thin ring and annular fin effects on improving the ice formation in ice-on-coil thermal storage systems Appl. Energy. **189** 369–384

[2] Mađerić D Čarija Z Pavković B and Delač B 2021 Experimental and numerical study on water ice forming on pipe columns in a limited-volume storage Appl. Therm. Eng. **194** 117080

[3] Li Y Yan Z Yang C Guo B Yuan H Zhao J and Mei N 2017 Study of a Coil Heat Exchanger with an Ice Storage System Energies **10** 1982

[4] Yang T Sun Q and Wennersten R 2018 The impact of refrigerant inlet temperature on the ice storage process in an ice-on-coil storage plate Energy. Procedia. **145** 82–87

[5] Saraceno L Boccardi G Celata G P Lazzarini R Trinchieri R 2011 Development of two heat transfer correlations for a scraped surface heat exchanger in an ice-cream machine Appl. Therm. Eng. **31** 4106-4112

[6] Wang Baolong, Li Xianting, Zhang Maoyong, Yang Xudong 2011 Experimental Investigation of Discharge Performance and Temperature Distribution of an External Melt Ice-on-Coil Ice Storage Tank HVAC&R Res. **9** 291-308

[7] Ezan M A Erek A 2012 Solidification and Melting Periods of an Ice-on-Coil Latent Heat Thermal Energy Storage System J. Heat. Transf. **134** 062301

[8] López-Navarro A Biosca-Taronger J Torregrosa-Jaime B Martínez-Galván I Corberán J M Esteban-Matías J C Payá J 2013 Experimental investigation of the temperatures and performance of a commercial ice-storage tank Int. J. Refrig. **36** 1310-1318

[9] Mousavi Ajarostaghi S S Poncet S Sedighi K Amiri L 2023 Solidification analysis in an ice-on-coil ice storage system: Experimental and numerical approaches J. Energy Storage **65** 107291

[10] Afsharpanah F Pakzad K Mousavi Ajarostaghi S S and Arıcı M 2022 Assessment of the charging performance in a cold thermal energy storage container with two rows of serpentine tubes and extended surfaces J. Energy. Storage. **51** 104464

[11] Hamzeh H A and Miansari M 2020 Numerical study of tube arrangement and fin effects on improving the ice formation in ice-on-coil thermal storage systems Int. Commun. Heat. Mass. Transf. **113** 104520

Acknowledgments

The PhD grant of Mr. Bernardini and Mr. Marangolo is sponsored by DRINKSTATION Inc., Los Angeles, USA.

Numerical simulation of shear driven film instability over heterogeneous surfaces via enhanced lubrication theory

Nicola Suzzi[1] **and Giulio Croce**[1]

[1]DPIA - Dipartimento Politecnico di Ingegneria e Architettura - Universitá di Udine - Via delle Scienze - 33100 - Udine (UD) - Italy

E-mail: nicola.suzzi@uniud.it, giulio.croce@uniud.it

Abstract. The prediction of the transition between continuous film, ensemble of rivulets and moving droplets is crucial in applications such as in-flight icing on airfoil wings or a number of chemical reactors. Here, lubrication theory is used to numerically investigate the stability of a continuous liquid film, driven by shear, over a heterogeneous surface. The disjoining pressure is used to model surface wettability, while the full implementation of the film curvature allows to investigate contact angles up to 60°. Different heterogeneous surface configurations occurring in real problems are investigated. An extended computational campaign records the transition from continuous film to rivulet regime and, if present, the further transition from rivulet to droplets at different flow conditions. A moving grid approach allows for accurate prediction of instability phenomena at low computational cost. The numerical results are successfully validated with experimental evidence in case of critical flow rate leading to a stable dry patch and compared with literature results involving the inherently multiscale in-flight icing phenomenon, providing useful statistical information, required to transfer the present detailed small-scale information into larger scale CFD computational approaches.

1. Introduction

The evolution of a liquid layer over a solid substrate is involved in several engineering applications. In in-flight icing phenomenon over airfoil wings, supercooled droplets from the clouds deposit on wing surface, forming a thin liquid layer, which evolution affects the ice accretion and, thus, the surface roughness, which in turn may influence the airfoil aerodynamics [1–3]. In CO_2 absorption through structured packing a liquid solvent flows down corrugated layers, while gas CO_2 flows up. Maximizing the liquid-gas interface area allows to enhance the absorption process [4, 5].

Film instability phenomenon has been largely investigated both numerically [4,6–12] and experimentally [13, 14]. In particular, finger instability of a film falling down an inclined plate with non-uniform contact angle, modeled through disjoining pressure in accordance with the precursor work of [15], was numerically investigated via lubrication theory in [6], where the effect of solid substrate on the film morphology was investigated. The effect of a solid occlusion on an advancing film front was numerically investigated by [16] via lubrication theory, with the liquid bulk that may cross the occlusion and wet the whole domain or form a stable dry-patch, depending on contact angle and flow condition. A dry-patch over a continuous film falling down an inclined plate was experimentally generated via a small perturbation in the liquid bulk and the critical flow rate leading to dry-patch shedding was traced under different flow conditions in [13, 14]. If gravity driven films, typical of chemical and heat transfer related applications, have been largely investigated, little information is available for shear driven films [17, 18], as those encountered in in-flight icing conditions. Nonetheless, the prediction of transition between film and rivulets in an icing environment affects the length of the potentially iced region, as well as the transient evolution of the accreted ice shape, and is thus an important feature to be considered. Here, in order to fill this gap, a shear driven film is numerically investigated under lubrication theory. The full implementation of capillary pressure allows to investigate contact angles up to $\theta = 60^\circ$ [9], while small-slope approximation leads to inaccurate results for $\theta \geq 30^\circ$ [9, 19]. Furthermore, a moving grid approach is adopted, in order to catch instabilities occurring far downstream with low computational cost. As the flow regime play a crucial role in both in-flight icing and CO_2 absorption, finger instability of an advancing film, driven by shear over an initially dry domain, is investigated, with the aim of identifying thresholds between continuous film and rivulet regimes. An application to the water film around an airfoil provides a first validation of the procedure. Furthermore, the stability of a continuous flat film, which initially wet the whole domain, is also investigated, introducing a small perturbation, that may induce a stable dry-patch or the de-wetting of the whole domain, with transition to droplet regime. The presented results can be used for developing simplified models, that does not go into the details of the film length scale.

2. Mathematical model

Consider a thin film flowing over a flat plate, driven by shear. Let $\boldsymbol{\tau}$ be the shear, applied at liquid-gas free surface, h_0 and u_0 the undisturbed film thickness and velocity, calculated according to Nusselt theory as $u_0 = \tau h_0/(2\mu)$. Lubrication theory allows to integrate the continuity equation along the plate normal direction. Assuming that the coordinate system moves with a constant velocity $\mathbf{u}^*$, retaining the viscous dissipation terms arising from velocity gradient across the film thickness and neglecting liquid inertia, the momentum equation gives a parabolic velocity profile. Averaging the film velocity and substituting in the continuity equation gives the well known lubrication equation for a shear driven film:

$$\frac{\partial h}{\partial t} - \mathbf{u}^* \cdot \nabla h + \nabla \cdot \left(\frac{\boldsymbol{\tau} h^2}{2\mu} - \frac{\nabla p}{3\mu} h^3 \right) = 0, \; p = -2\sigma\kappa - D, \quad (1)$$

where the liquid pressure p is given by the sum of capillary pressure and disjoining pressure. The free surface curvature is equal to,

$$2\kappa = \frac{\left[1 + \left(\frac{\partial h}{\partial y}\right)^2\right] \frac{\partial^2 h}{\partial x^2} + \left[1 + \left(\frac{\partial h}{\partial x}\right)^2\right] \frac{\partial^2 h}{\partial y^2} - 2\frac{\partial h}{\partial x}\frac{\partial h}{\partial y}\frac{\partial^2 h}{\partial x \partial y}}{\left[1 + \left(\frac{\partial h}{\partial x}\right)^2 + \left(\frac{\partial h}{\partial y}\right)^2\right]^{3/2}}, \quad (2)$$

while the disjoining pressure term is equal to [6],

$$D = B\left[\left(\frac{d}{h}\right)^n - \left(\frac{d}{h}\right)^m\right], B = \frac{\sigma}{d}\frac{(n-1)(m-1)}{n-m}(1-\cos\theta), \quad (3)$$

where $n > m > 1$ and $d \ll h_0$ is the precursor film thickness.

Assuming that shear is directed over y, which means $\boldsymbol{\tau} = \tau\hat{\mathbf{j}}$, introducing the capillary number of the undisturbed film as $\mathrm{Ca} = \mu u_0/\sigma$ and defining the following non-dimensional quantities,

$$H = \frac{h}{h_0}, \delta = \frac{d}{h_0}, \mathbf{X} = \frac{\mathbf{x}}{L_0}, \mathbf{U} = \frac{\mathbf{u}}{u_0}, T = \frac{t}{(L_0/u_0)},$$
$$P = \frac{p}{(\sigma h_0/L_0^2)}, 2K = \frac{2\kappa}{(h_0/L_0^2)}, \Pi = \frac{D}{(\sigma/h_0)}, \quad (4)$$

with L_0 being the characteristic length scale, $L_0 = h_0\,(3\mathrm{Ca})^{-1/3}$, the governing lubrication equations can be recast in non-dimensional form,

$$\frac{\partial H}{\partial T} - \mathbf{U}^* \cdot \nabla H + \nabla \cdot \left(H^2\hat{\mathbf{j}} - H^3\nabla P\right) = 0, \; P = -2K - \frac{\Pi}{(3\mathrm{Ca})^{2/3}}, \quad (5)$$

where the non-dimensional free surface curvature is:

$$2K = \frac{\left[1+(3\,\mathrm{Ca})^{2/3}\left(\frac{\partial H}{\partial Y}\right)^2\right]\frac{\partial^2 H}{\partial X^2}+\left[1+(3\,\mathrm{Ca})^{2/3}\left(\frac{\partial H}{\partial X}\right)^2\right]\frac{\partial^2 H}{\partial Y^2}-2\,(3\,\mathrm{Ca})^{2/3}\frac{\partial H}{\partial X}\frac{\partial H}{\partial Y}\frac{\partial^2 H}{\partial X\partial Y}}{\left[1+(3\,\mathrm{Ca})^{2/3}\left(\frac{\partial H}{\partial X}\right)^2+(3\,\mathrm{Ca})^{2/3}\left(\frac{\partial H}{\partial Y}\right)^2\right]^{3/2}}. \tag{6}$$

Eqs. (5) and (6) were numerically solved on an orthogonal, structured grid of $N_x \times N_y$ elements via Finite Volume Method (FVM), using an in-house code, previously developed in `FORTRAN` language and largely validated by the Authors [9,11]. Spatial discretization was achieved via the second order centered scheme of [20], while Alternating Direction Implicit (ADI) approximate factorization of [21] was implemented for marching in time. Parallelization for shared memory machines was implemented using `OpenMP` in order to speed up computations.

3. Problem statement

3.1. Non-wettable patch array

As a first case, we evaluated the effect of a localized perturbation on the stability of an advancing film. In particular, a film front driven by shear was perturbed by an array of small, non-wettable square patches of non-dimensional pitch $L_p = 60$ along X and size $L_\theta = 1$. Different configurations were considered, in order to evaluate whether the instability triggered by the patch will increase or will dampen as the film flows downstream. Computations were run on a domain of size $L_X \times L_Y = 60 \times 60$. The film was initialized with the solution of an advancing 1D shear driven film, with contact line front located at $Y \sim L_Y/2$ (to do so, a preliminary computation on a uniform substrate had to be run, in order to get the shape of the interface of a 1D advancing film). Periodic boundary conditions were applied at $X = 0, L_X$. Inlet condition was applied through $Y = 0$, while fully developed flow was imposed through $Y = L_Y$. The non-wettable patch of size $L_\theta = 1$, with a 10% increased value of the contact angle, was initialized in the dry region of the computational domain. In order to follow finger instability further downstream of the generating patch, we use a relative frame of reference, moving along Y with the advection speed of the undisturbed film contact line, equal to $U^* = \left(1-\delta^3\right)/\left(1-\delta\right)$. Time marching was stopped when moving film front is affected by boundary conditions imposed through $Y = 0, L_Y$ (i.e. receding front close to inlet at $Y = 0$ and outlet section at $Y = L_Y$ reached by advancing front).

3.2. Heterogeneous surface

A second example considers a shear driven advancing film front over an heterogeneous surface, representative of a real fouled surface as usually encountered in real world environment, in particular on airfoil surfaces naturally exposed to external contamination. The contact angle variation, which follows,

$$\theta(X,Y) = \sum_{k=-n}^{+n}\sum_{j=-m_k}^{+m_k} \Theta_{j,k}\cos\left(2\pi\frac{X\cdot j+Y\cdot k}{L_\pi}+\Psi_{j,k}\right),\, m_k = \left\lceil m\left(1-\frac{k^2}{n^2}\right)\right\rceil, \tag{7}$$

is periodic over $L_\pi = 60$ with $m, n = 120$ and ranges in $\theta_0 \pm \Delta\theta$, with $\Delta\theta = 0.1\,\theta_0$. Domain size, boundary and initial conditions are defined as in section 3.1.

3.3. Dry-patch generation

Finally, a flat, continuous film initially wetting the whole domain, $H(X,Y)|_{T=0} = 1$, was perturbed via an array of small, localized square dry-patches of pitch $L_p = 60$ and size $L_\theta = 1$, characterized by an increased value of the contact angle (+10% of the base contact angle) and initialized in the computational domain, of size $L_X \times L_Y = 60 \times 60$. Boundary conditions are the same as in sections 3.1 and 3.2, while the moving grid velocity was set to zero, $U^* = 0$. When a stationary dry-patch hanging to the small non-wettable patch was detected, the latter was removed, checking if the stationary dry-patch slowly shed away or remains stationary.

Two values of the reference contact angle, $\theta_0 = 30°, 45°$, were investigated for the three test cases, while the capillary number was ranged in $\mathrm{Ca} \in \left[10^{-3}, 10^{-1}\right]$. Following [6], the precursor film thickness was set to $\delta = 5 \times 10^{-2}$ for all the computations. After a grid dependency analysis, the spatial discretization steps were set to $\Delta X, \Delta Y \leq 5 \times 10^{-2}$.

4. Result

4.1. Non-wettable patch array

Parametric computations were run, in case of an array of equally spaced small non-wettable patches on a hydrophilic surface, at $\theta_0 = 30°, 45°$ and capillary number ranging from $\mathrm{Ca} = 0.1$ down to the critical value leading to rupture of the unperturbed 1D film front. Below Ca_{cr}, which depends on the substrate wettability, droplet regime is expected. Running preliminary computations of 1D shear driven film, it was found $\mathrm{Ca}_{cr} = 10^{-2}$ for $\theta = 45°$ and $\mathrm{Ca}_{cr} = 3 \times 10^{-2}$ for $\theta = 30°$. Then, the moving contact line was traced in the 2D computations, saving the curve (X_{cl}, Y_{cl}) at each time step. Both the advancing front and the receding front, computed as the topmost and the bottomest points of the contact line, were traced at each time step and used to estimate the receding and advancing contact line speed. Figures 1(a)

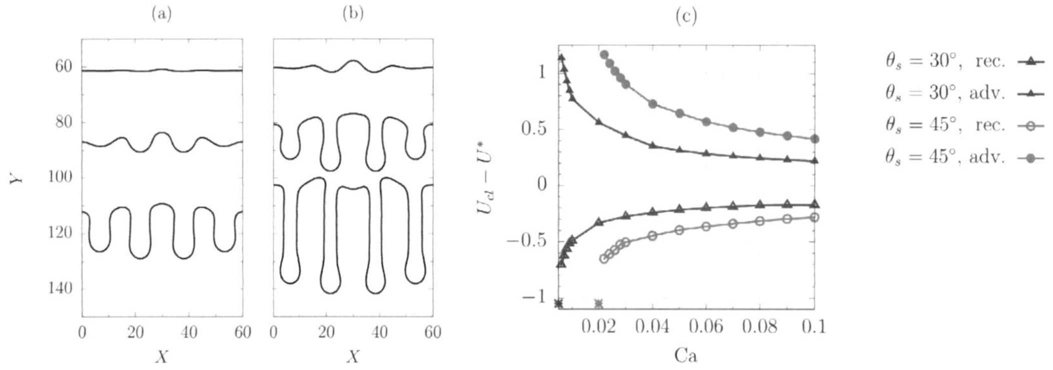

Figure 1. Moving contact line at different instants: $T = 25, 50, 80$. $\mathrm{Ca} = 10^{-1}$, $\theta_0 = 30°$ (a); $\mathrm{Ca} = 10^{-1}$, $\theta_0 = 45°$ (b). Relative velocity of advancing and receding fronts as a function of film capillary number, at $\theta_0 = 30°, 45°$ (c). Finger instability of the shear driven film is induced by an array of small non-wettable patches.

and 1(b) show the apparent contact line evolution after the film crossed the small non-wettable patch, at $\mathrm{Ca} = 10^{-1}$ and different contact angles of the base surface. Note that the rivulets form far from the non-wettable patch, located at $Y \sim 30$, meaning that the moving greed approach is required. As expected, rivulets grow faster and have a narrower shape at increasing contact angle. Indeed, the stabilized speed of both the receding and the advancing fronts, plotted in figure 1(c), are higher when $\theta_0 = 45°$. Figure 1(c) also shows that rivulets grow faster at decreasing capillary number. Furthermore, it was observed that the receding front may hang on the small non-wettable patch even at $\mathrm{Ca} > \mathrm{Ca}_{cr}$ (spare, star markers in figure 1(c)), demonstrating that the additional degrees of freedom of the 2D solution may anticipate the instabilities. However, we cannot derive general conclusions from such computations since the details of the non-wettable patch affects the final solution.

4.2. Heterogeneous surface

Parametric computations at $\theta_0 = 30°, 45°$ and $\mathrm{Ca} \in \left[\mathrm{Ca}_{cr}(\theta_0), 10^{-1}\right]$ were run in case of an heterogeneous surface, characterized by random variation of the contact angle. Figure 2(a) shows the evolution of the contact line over the heterogeneous surface for Ca $= 10^{-1}$, $\theta_0 = 45°$. It is important to point out that the shape of the growing rivulets is qualitatively similar to figure 1(b), at same θ_0 and Ca, but the contact line looks more irregular due to random contact angle variation. Again, the contact line

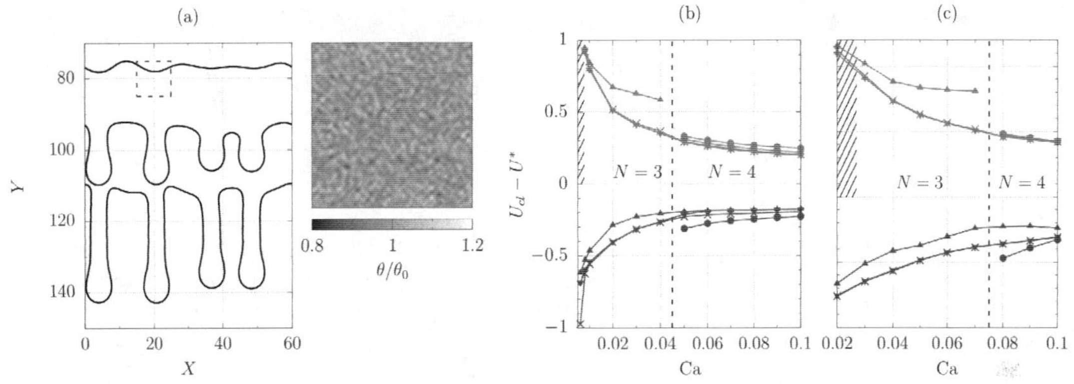

Figure 2. Moving contact line at $T = 42, 65, 88$ and zoom of the contact angle variation: Ca $= 10^{-1}$, $\theta_0 = 45°$ (a). Relative velocity of advancing (green) and receding (purple) fronts vs film capillary number (dotted region characterized by rupture of advancing front in droplets): $\theta_0 = 30°$ (b); $\theta_0 = 45°$ (c). Finger instability of the shear driven film is induced by an heterogeneous surface with $L_\pi = 60$.

was traced at each time step and the rivulets position tracked. The advancing speed of each rivulet front was computed as well as the receding speed of the valleys between neighbor rivulets, giving the velocity of the receding and the advancing fronts. Tracing only the topmost and the bottomest points would lead to scattered results, due to disturbance induced by a random contact angle variation. The rivulet advancing and receding speeds as a function of Ca are shown in figures 2(b) and 2(c). A bifurcation of the solution, with the number of stable rivulets passing from $N = 4$ to $N = 3$, can be observed: for $\theta_0 = 30°$ at Ca $\simeq 4 \times 10^{-2}$, figure 2(b); for $\theta_0 = 45°$ at Ca $\simeq 8 \times 10^{-2}$, figure 2(c). It is important to note that the number of stable rivulets is upper limited by $N \leq [L_\pi/\lambda]$, with $L_\pi \equiv L_X = 60$ and $\lambda = 9.4$ being the characteristic perturbance length from linear stability analysis [6]. Comparing figures 2(b) and 2(c), it can be observed that, at fixed Ca, a lower surface wettability leads to higher front speeds, meaning that rivulets grow faster at increasing θ_0. Receding fronts with $U_{rec} < 0$, which means that the film is not stable as the receding front goes upstream, were observed, for $\theta_0 = 30°$, at Ca $= 5 \times 10^{-3}$, still higher than $\mathrm{Ca}_{cr} = 3 \times 10^{-3}$. Thus, an unstable film with rivulet front moving upstream may be observed at values higher that the critical one, derived from 1D analysis. The results, presented in figures 2(b) and 2(c), are useful for practical applications such as in-flight icing phenomenon on airfoil wings, where knowing the dynamic flow pattern evolution over the whole wing surface is crucial [2, 3]. As a proof of this, the numerical results of [3] were replicated: the critical thickness of a continuous water film on NACA0012 airfoil, leading to film break with transition to rivulet pattern, was evaluated from the critical capillary number as $h_{cr} = (2\sigma/\tau)\,\mathrm{Ca}_{cr}$, with the local shear being estimated from the friction drag coefficient of the surrounding air flow, $\tau = \frac{1}{2}\rho_{air} c_f u_\infty{}^2$. The friction drag coefficient, figure 3(a), was derived from the CFD solution of a flow past NACA0012 airfoil at $\mathrm{Re} = \rho_{air} u_\infty l_c/\mu_{air} = 1.7 \times 10^5$ and zero angle of attack. Flow field was computed via open source code SU2 on a quadrangular mesh of 230336 nodes, using Spalart–Allmaras turbulence model. Figure 3(b) compares the computed critical film thickness as a function of chord position with literature data of [3], for an hydrophilic surface characterized by uniform contact angle $\theta = 30°$, showing good agreement.

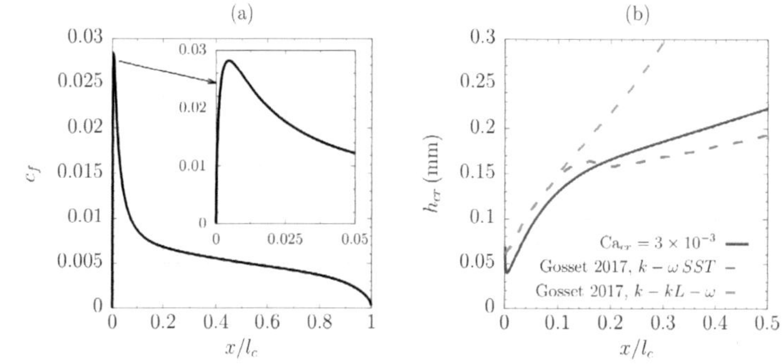

Figure 3. Friction drag coefficient as a function of chord position, air flow past NACA0012 airfoil at $\mathrm{Re} = 1.7 \times 10^5$ (a). Critical film thickness leading to film-rivulet transition: numerical vs literature data from [3] (b). $l_c = 101\,\mathrm{mm}$, $u_\infty = 25\,\mathrm{m\,s^{-1}}$, $\theta = 30°$. Air and water properties evaluated at atmospheric pressure and temperature 20°C.

4.3. Dry patch generation

Dry-patch generation from a continuous film was investigated, at $\theta_0 = 30°, 45°$. The capillary number was varied, looking for the two critical cases: (1) a receding, unstable dry-patch, growing into rivulets and drops, forms at low values of Ca; (2) the perturbation does not induce the generation of a stationary dry-patch, leading to fully wetted domain at high values of Ca. The thresholds of Ca are shown in table 1. The

	$\mathrm{Ca} = \mu\, u_0/\sigma$					
$\theta(°)$	4×10^{-3}	5×10^{-3}	1×10^{-2}	2×10^{-2}	3×10^{-2}	4×10^{-2}
30	$*$	•	•	⋆	⋆	⋆
45	$*$	$*$	$*$	•	•	⋆

Table 1. Flow regimes: rivulets and moving droplets, $*$; stationary dry-patch, •; fully wetted domain, ⋆. Small perturbation initialized in a flat, continuous film.

unstable dry-patch, growing into rivulets and droplets regime, was observed at $\theta_0 = 45°$ and $\mathrm{Ca} = 10^{-1}$, which is close to the critical value, leading to rupture of the 1D advancing film front. However, a receding contact line, with generation of rivulets and drops, was observed at $\mathrm{Ca} = 4 \times 10^{-3} > \mathrm{Ca}_{cr} = 3 \times 10^{-3}$ for $\theta_0 = 30°$. The threshold leading to fully wetted domain is not discussed, as it may be affected by the value of the increased contact angle, imposed inside the small perturbation. When a large, stationary dry-patch is generated, as in figure 4(a) for $\theta_0 = 45°$ and $\mathrm{Ca} = 3 \times 10^{-2}$, the small perturbation is removed, letting the dry-patch moving forward and slowly shed away, as shown in figure 4(b). In figure 5, referred to $\theta_0 = 45°$, $\mathrm{Ca} = 10^{-2}$, the evolution of the unstable dry-patch is followed: at the initial stage, a receding dry-patch is generated; however, after the moving contact line reaches the outlet section, finger instability is observed, with a rivulet forming in the core domain and breaking into droplets; then, further rivulets and droplets form, with the receding front of the main contact line approaching the inlet section.
The thresholds listed in table 1 can be compared with the experimental results of [13, 14]. In the experiments, a continuous film, driven by gravity, flowing down an inclined plate, which is fully wetted, was perturbed via a localized blow on the film, generating a stationary dry-patch. Then, the flow rate was slowly increased, until the dry-patch starts moving downstream. A critical flow rate per unit length Γ_{cr} was, thus, estimated in [13, 14]. Knowing fluid properties and applying Nusselt theory, $\Gamma_{cr} = \left(\rho\, g h_{cr}{}^3 \sin\alpha\right) / (3\mu)$ with α being the plate slope, the critical capillary number can be derived from Γ_{cr}. As clearly shown in table 2, the critical flow rates from [13, 14], experimentally derived from a similar test case and referred to contact angles that are close to the ones here investigated, fall inside the

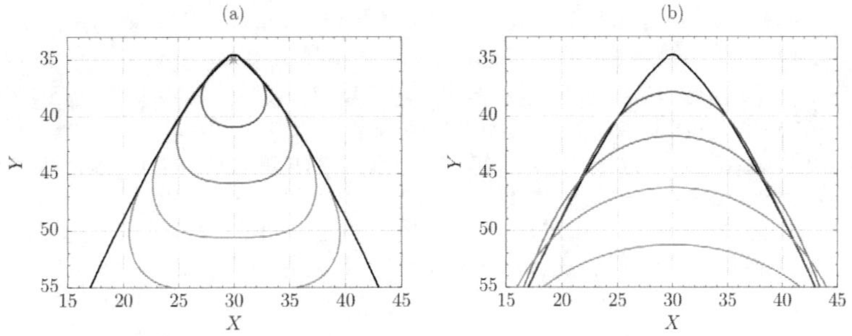

Figure 4. Contact line evolution: continuous film perturbed via localized non-wettable dry-patch, identified by a green marker, $T = 4, 8, 12, 16, 34$ (a); dry-patch shedding after non-wettable patch removal, $T = 34, 42, 49, 57, 64$ (b). $\theta_0 = 45°$, $\mathrm{Ca} = 3 \times 10^{-2}$.

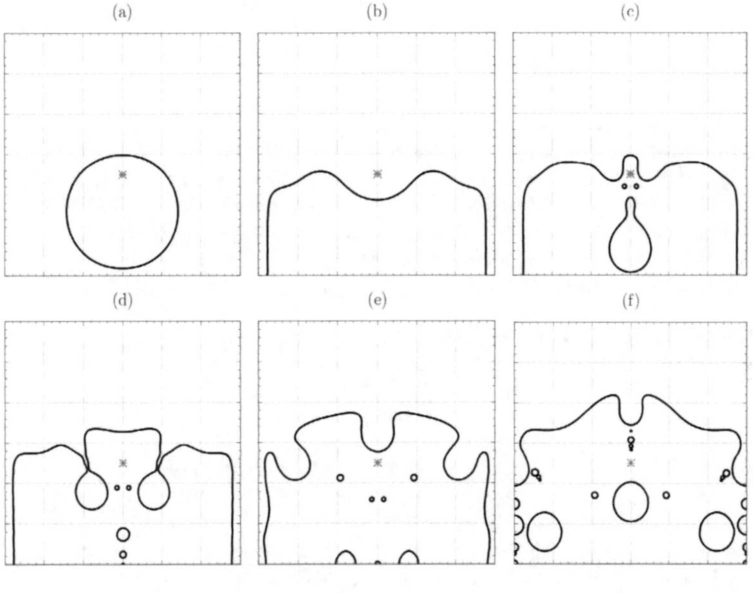

Figure 5. Contact line evolution at different instants: $T = 8$ (a); $T = 38$ (b); $T = 45$ (c); $T = 52$ (d); $T = 60$ (e); $T = 68$ (f). Continuous film perturbed via localized dry-patch, identified by a green marker. $\theta_0 = 45°$, $\mathrm{Ca} = 10^{-2}$, $L_X \times L_Y = 60 \times 60$.

α (deg)	θ (deg)	Γ_{cr} (m^2 s)	Ca_{cr}
90°	38° − 47°	3.5×10^{-6}	1.16×10^{-2}
90°	15° − 30°	$\sim 5 \times 10^{-7}$	3.18×10^{-3}

Table 2. Critical flow rate Γ_{cr} leading to shedding dry-patch in case of gravity driven film, from experimental results of [13, 14]. Silicone oil: $\rho = 950\,\mathrm{kg\,m^{-3}}$, $\mu = 0.019\,\mathrm{kg\,m^{-1}\,s^{-1}}$, $\sigma = 0.0206\,\mathrm{N\,m^{-1}}$.

numerical range of table 1.

5. Conclusion

Finger instability induced by non-uniform surface wettability and dry patch stability of a shear driven film were numerically investigated. The thresholds leading to transition between continuous film and de-wetting regime, with the receding film replaced by droplets and rivulets, were found, in terms of critical capillary number of the undisturbed film. Slightly different Ca_{cr} were observed, depending on the investigated test case (unperturbed 1D advancing film, heterogeneous surface, dry-patch generation). However, the stability thresholds appear coherent with literature experimental evidence [13, 14] and numerical literature simulations [3], the latter based on a different approach in case of a film along an aircraft airfoil. Furthermore, the moving grid approach allowed to catch finger instability occurring late after the perturbation (single patch, heterogeneous surface) was introduced. The speeds of the advancing and receding rivulet fronts were computed, leading to useful information about the evolving flow pattern, to be used in simplified modelization of in-flight icing. The dry-patch stability analysis, which results are in good agreement with experimental literature [13, 14], is useful in applications as CO_2 absorption, where the stationary flow solution is required rather that its evolution in time. As a future work, further computations to better identify the thresholds between flow regimes will be run and the effect of the surface characteristics will be analysed. Furthermore, contact angle up to 60° and the combined effect of shear and gravity will be investigated.

References

[1] Croce G, De Candido E, Habashi W G, Munzar J, Aubé M S, Baruzzi G S and Aliaga C 2010 *Journal of Aircraft* **47** 1283–1289
[2] Zhang K, Wei T and Hu H 2015 *Experiments in Fluids* **56** 173
[3] 2017 *Prediction of rivulet transition in anti-icing applications*
[4] Hoffmann A, Ausner I, Repke J U and Wozny G 2005 *Computers & Chemical Engineering* **29** 1433
[5] Subramanian K and Wozny G 2012 *International Journal of Chemical Engineering* **2012** 838965
[6] Zhao Y and Marshall J S 2006 *Journal of Fluid Mechanics* **559** 355–378
[7] Shkadov V Y and Sisoev G M 2005 *Computers & Fluids* **34** 151–168
[8] Thiele U, Brusch L, Bestehorn M and Bar M 2003 *The European Physical Journal E* **11** 255–271
[9] Suzzi N and Croce G 2021 *Fluids* **21** 405
[10] Suzzi N and Croce G 2021 *Journal of Physics: Conference Series* **1868** 012010
[11] Suzzi N and Croce G 2019 *Physics of Fluids* **31** 122106
[12] Suzzi N and Croce G 2017 *Journal of Physics: Conference Series* **796** 012038
[13] Podgorski T, Flesselles J M and Limat L 1999 *Physics of Fluids* **11** 845
[14] Rio E and Limat L 2006 *Physics of Fluids* **18** 032102
[15] Schwartz L and Eley R 1998 *Journal of Colloid and Interface Science* **202** 173–188
[16] Sellier M 2015 *International Journal of Multiphase Flow* **71** 66–73
[17] Abas S S and Yatim Y M 2014 *Physics of Fluids* **1605** 361–366
[18] Sullivan J M, Wilson S K and Duffy B R 2006 *Process in Industrial Mathematics at ECMI 2006* **12** 774–778
[19] Perazzo C A and Gratton J *Journal of Fluid Mechanics* **507** 367–379
[20] Diez J A and Kondic L 2002 *Journal of Computational Physics* **183** 274–306
[21] Witelski T P and Bowen M 2003 *Applied Numerical Mathematics* **45** 331

Acknowledgments

The authors are grateful to the European Commission for supporting the present work, performed within the IMPACT project, funded by the Clean Aviation Joint Undertaking under the European Union's Horizon 2020 research and innovation program (Grant Agreement no. 885052). This publication solely reflects the authors' view and neither the European Union, nor the funding Agency can be held responsible for the information it contains.

A New Concept of Direct Oil Cooling System for AFPM Machines, Numerical Analysis and Optimization

L. Pirillo[1], F. Nardecchia[2] and F. Bisegna[3]

[1] Ph.D. Student, Department of Mechanical and Aerospace Engineering, Via Eudossiana 18, Roma (RM), 00184, Italy.
[2] Ph.D., Department of Austronautic, Electric and Energy Engineering, Via Eudossiana 18, Rome (RM), 00184, Italy.
[3] Associate Professor, Department of Austronautic, Electric and Energy Engineering, Via Eudossiana 18, Rome (RM), 00184, Italy.

lorenzo.pirillo@uniroma1.it

Abstract. Given the strong growth of the electric drive industry, the automotive sector needs to increase the power density of its motors, leading to a size decrease and, consequently, to economic savings. The increase in power density brings with itself the consequence of increasing the Joule effect losses in the active conductors of the machine. For this reason, the problem of cooling turns out to be of utmost importance precisely because of its direct coupling with mechanical and electromagnetic design. This paper has the aim to perform a CFD analysis concerning the cooling system of a permanent magnet axial flux electric machine. The innovative cooling system studied is a direct one which takes advantage of oil to cool down the DC powered coils. The innovation brought by the present research lays in both the direct axial ejection and the use of oil as coolant. At the current state of art, the coolant flows tangentially and, moreover, water is usually employed instead of oil. The main aim of this analysis will be, after a validation of the numerical code by comparison of the literature data, to minimize the maximum temperature of the coil by means of the proper positioning of the nozzle.

1. Introduction

The state of the art in AFPM machines has been rapidly advancing in recent years, due to the growing demand for high-efficiency electric motors. Overall, the state of the art in AFPM machines is characterized by a continuous push for higher power density, improved efficiency, and lower cost [1]. As the demand for electric motors continues to grow, it is likely that we will see further advancements in AFPM technology in the coming years [2] The air gap flux permanent magnet (AFPM) machine is a promising technology that has gained significant attention in recent years due to its high torque density, compact size, and improved efficiency [3]. As the demand for higher power density and continuous operation under extreme conditions continues to rise, efficient cooling systems are crucial for managing the generated heat and ensuring optimal performance of AFPM machines. Nowadays, the methods adopted for heat removal are several. There are traditional cooling methods, such as natural convection or surface cooling techniques [4][5] and forced convection cooling methods, involving the use of external fans or blowers to enhance the airflow and heat transfer within the machine [6][7]. Moreover, liquid cooling methods, including direct liquid cooling (DLC) and indirect liquid cooling (ILC), have shown significant potential in efficiently managing the heat generated by

AFPM machines. The choice of coolant and optimization of the cooling channel design are critical factors in achieving effective liquid cooling in AFPM machines [8][9]. Hybrid cooling configurations may involve a combination of air and different liquid cooling strategies. The design and optimization of these complex systems require comprehensive analyses and multi-physics simulations to ensure effective heat transfer and thermal stability [10].

The aim of this work is to analyse and optimize, using a CFD code, a new direct oil cooling system, [11][12] which consists in a direct axial ejection of the coolant. The final purpose is to optimize the nozzle position and size in order to maximize heat removal and consequently, minimize the maximum coil temperature.

2. Physical problem and methodology

Of the many models of electric machines for wheeled transport that exist in the current state of the art, Axial Flux Permanent Magnet (AFPM) machines are particularly promising because they offer a smaller axial footprint for the same power density with respect to their radial counterpart [13]. Nowadays, the most used cooling system used for these machines consists in enclosing the stator copper strip coils in a metal jacket [14], which is then cooled down by water flowing in annular channels (see Figure *1* 1).

For the CFD simulations of the new concept cooling system, a 3-D CAD of the coil has been built and then partially enclosed in a fluid domain. In this way, the coolant can be injected in order to lap the upper surface of the coil, giving thus rise to a heat removal. The critical part of the coil, in terms of hotspot temperature, is the bottom one, which is the farthest from the heat sink.

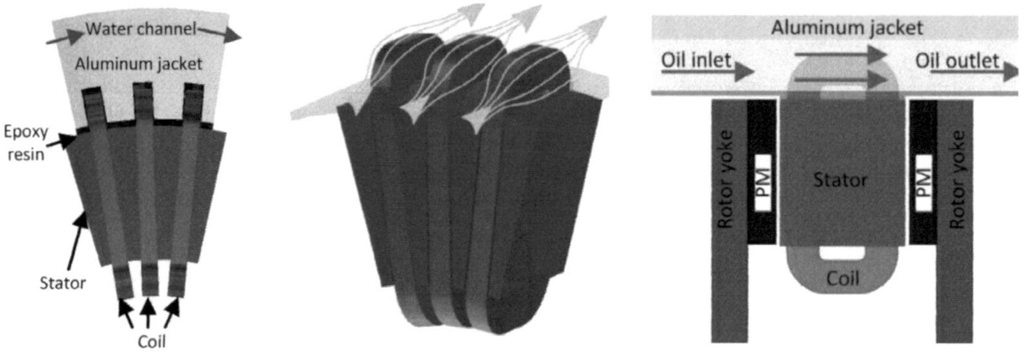

Figure 1: State of the art cooling system for AFPM Machines (left). New concept cooling system (centre and right).

3. Computational Setup

3.1. Test case description

The machine considered in this work has a stator pack and two rotor packs, coaxial to each other and arranged so that the stator, is placed between the two rotors and it is separated from them by two air gaps (left side of figure 2). The coils are located in the perimetral area of the stator core, and they are supplied with direct current (DC). The permanent magnets are housed on both rotors. The machine consists of 8 pole pairs, 48 stator slots in which each coil is housed, and it has as many nozzles responsible for the coolant ejection. The coolant is ejected axially, and it invests the coil as it is a fin [15]. It then descends by gravity into the area between the rotor and the coil and is then conveyed, together with all flows from the other nozzles, into a lower plenum. Then, a recirculating pump draws the fluid from the plenum and redistributes it as evenly as possible all over the machine's nozzles, thus

Journal of Physics: Conference Series **2685** (2024) 012020 doi:10.1088/1742-6596/2685/1/012020

closing the cycle. The coil housed on the stator consists of a 'ribbon' conductor of section 1mm x 5mm of enamel-insulated copper. It consists of eight windings in order to create a rectangular structure with smoothed corners. In the contest of this work, each winding of the coil has been 3-D modelled and then enclosed in a fluid domain. After that, a cylindrical nozzle has been obtained (right side of figure 2).

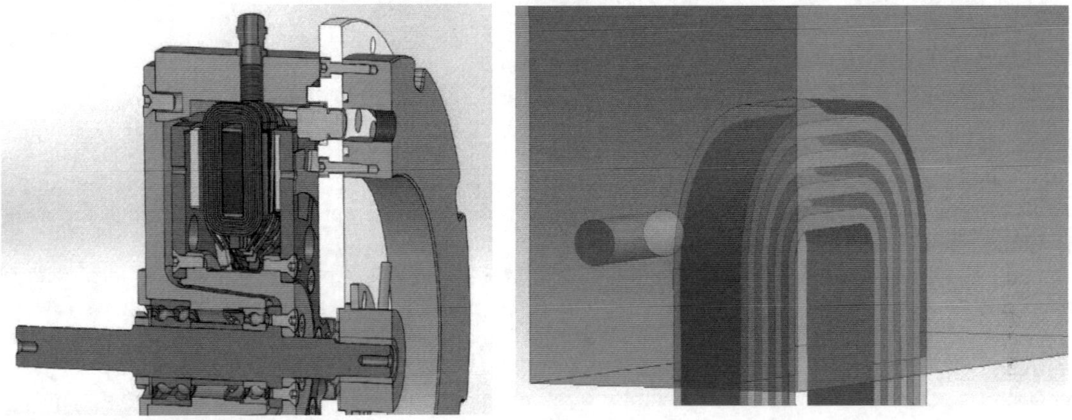

Figure 2: Detail of the machine (left). 3-D CAD of the eight strips of the coil, the fluid enclosure, and the nozzle (right).

3.2. Materials

The materials considered for this work are copper and enamel for the coil active material and insulator [15], while for the coolant a mineral transformer oil has been chosen. The physical parameters important for the analysis are reported in table 1.

Table 1: Thermophysical properties of the materials used in this work.

	Copper	Enamel	Oil
ρ [kg/m^3]	8978	2719	780
c_p [J/kg K]	381	871	3002
λ [W/m K]	386	0.93	0.24
μ [Pa s]	-	-	0.0195

3.3. Parametrization test Matrix.

In this paper, a campaign of simulations has been performed, with the purpose of finding an optimal configuration for both the nozzle diameter (D) and its position (L), evaluated from the bottom of the fluid enclosed domain. Starting from the baseline configuration (Sim 5) which reproduces the experimental setup of [15] both D and L have been changed in order to obtain a matrix of 9 test cases, summarized in Table **2** 2. Each case it runs with the same mass flow rate and inlet temperature in order to find the configuration which leads to a minimum hot spot temperature of the coil.

Table 2: Test matrix of simulations.

D – L	3mm	5mm	7mm
10mm	Sim 1	Sim 2	Sim 3
12mm	Sim 4	Sim 5	Sim 6
14mm	Sim 7	Sim 8	Sim 9

3.4. Numerical and Meshing Strategy

Given the axial symmetry of the machine, the analysis on which this work is based will be concentrated on an angular section equal to 1/48 of the revolution angle, i.e., on a single coil, to reduce the computational cost of the simulations, but still obtaining results representative of the behaviour of the entire machine.

The CFD simulation have been conducted with a pressure based coupled algorithm, they are steady state, one phase and viscous. The turbulence model adopted is the realizable k-ε with standard wall functions. A block-structured mesh [17] is chosen for the coil, in such a way that the layer of insulator between one winding and the next, which represents a lumped contact resistance is exactly halfway between two adjacent cells, which, by communicating, will take it into account in an optimal manner. In the fluid domain, the grid generated, consists of a tetrahedral structure in the bulk of the fluid and it contains a structured liner in the "near wall region". The structured liner consists of 30 prismatic layers, whose height grows following a geometric sequence with growth rate equal to 1.2, starting from a first cell height of 2.4 μm. in the downstream region of the fluid, the mesh size is increased in order to prevent pressure disturbances to rise back to the coil. A detail of the mesh is reported in figure 3.

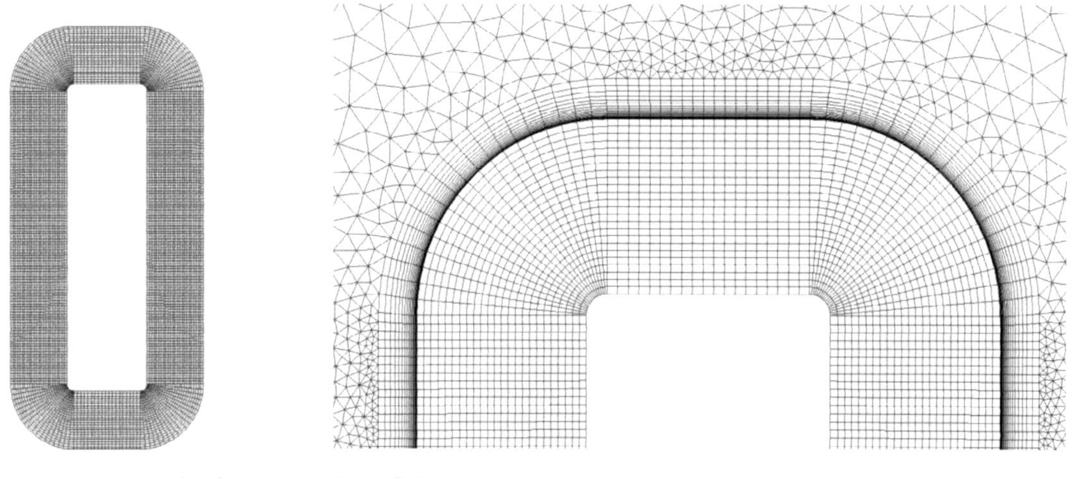

Figure 3: Mesh of the coil and the fluid, together with a detail of the near wall region refinement.

3.5. Initial and Boundary Conditions

The boundary conditions used for all the test cases investigated are the following:

- Mass flow inlet for the inlet section of the nozzle;
- Pressure outlet for the outlet plane;
- Wall with free slip for the fluid boundaries;
- Wall with no slip condition for the fluid-facing coil surfaces;
- Wall with thin layer and shell conduction for the coil-coil interfaces.

For each one of the cases under analysis it was assumed a thermal power generation inside the coil of 100 W [19]. The fluid field has been initialized by using constant parameters taken from the inflow.

4. Results

4.1. Validation

Both the numerical code and the mesh have been validated. A "mesh sensitivity analysis" was performed to obtain a convergent grid.

Four grids with a different degree of refinement have been built. They have been run with the same boundary conditions, and the results have been compared to each other. The features of the four grids are highlighted in table 3. The results, in terms of static temperature, have been extracted along the red line of figure 4, and they are reported together. Mesh 3 can be considered convergent, since the maximum difference between its results and Mesh 4 results is 1.92%, a value well below the limit for which mesh convergence can be declared.

After that, the software and the numerical method have been validated by comparison of the results with experimental data coming from literature. The experimental facility of [15] was reproduced in the simulation 5, in which the diameter of the nozzle is 5 mm and the distance of its centre, calculated form the bottom of the fluid domain (L) is 12 mm. the experimental data are well reproduced when the inlet velocity is 0.5 m/s and the inlet temperature is 25 °C, resulting in a hotspot temperature of the coil of 178°C.

Table 3: Features of the four different meshes created.

Mesh number	Number of cells	Average cell dimension [μm]	First cell height [μm]	Growth rate
1	650.000	500	2.4	1.2
2	980.000	97	2.4	1.2
3	1.270.000	35	2.4	1.2
4	1.550.000	1.1	2.4	1.2

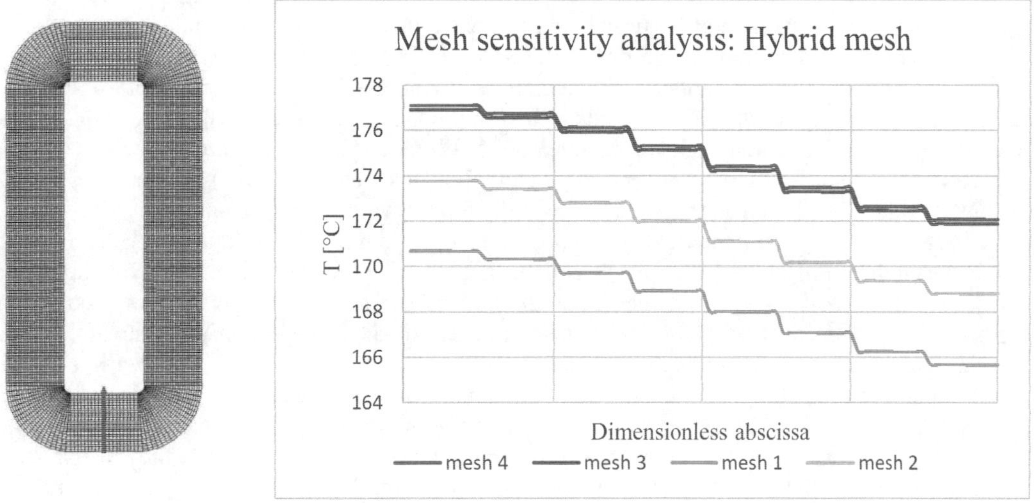

Figure 4: Temperature profile along the hot spot line for every mesh generated.

4.2. Optimal Configuration

After the validation is performed, the test matrix of simulations presented previously is implemented to find the situation in which the hotspot temperature of the coil is minimum. All the cases have been carried out by keeping the value of the mass flow rate as a constant, which results in an increase of the inlet velocity in simulations 1, 4 and 7, and a decrease of it in simulations 3, 6 and 9, due to the change in the value of the nozzle diameter. An inlet temperature of 25 °C has been assumed for every simulation. The results have been extracted, for each one of the 9 simulations, along the hotspot line and they are presented aside in figure 5.

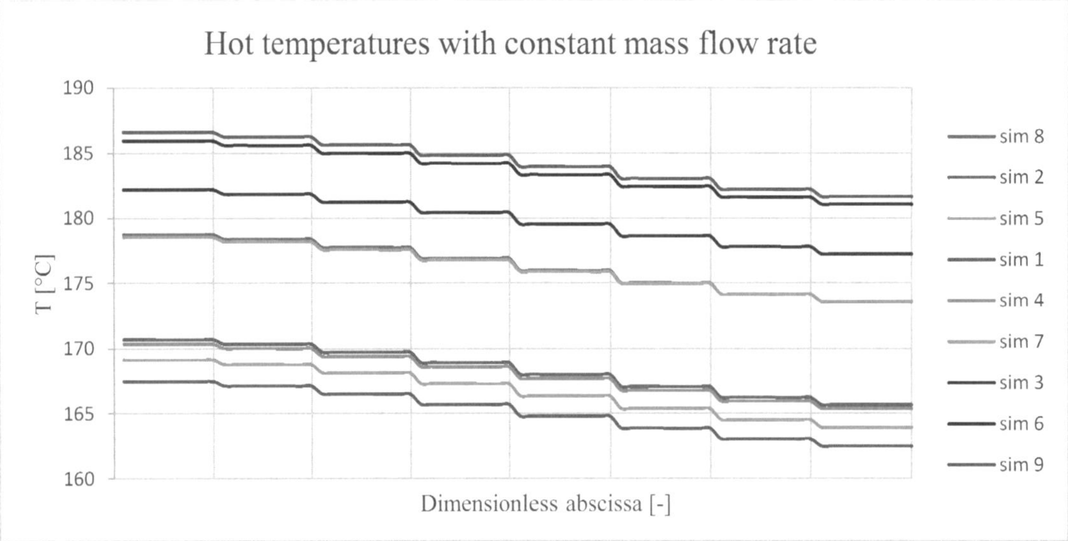

Figure 5: Hot temperatures of the coil for all the 9 simulations performed.

The simulation for which the minimum hot spot temperature of the coil is obtained is simulation 1, i.e. the one where the diameter is minimum and the nozzle height is maximum.

It can be seen that with the same diameter, for all 3 solutions, the best configuration is the one in which the nozzle is placed as high as possible. The nozzle placed at the top, although geometrically further away from the hot spot, is the one that results in the most effective cooling, no matter its diameter value, with an advantage that is more marked for intermediate diameters. The nozzle placed at the top is placed in a better position regarding the fluid motion established in the coil-fluid interface. In fact, given the curvature that characterizes the coil itself, injecting the fluid further down means that it interacts with the surface through a very large angle of attack (tending to 90°). In that case, the coil basically acts like a 'wall', and strongly slows down the fluid, severely penalizing the Reynolds number and the heat transfer. Injecting the fluid at the top, on the other hand, means adopting smaller angles of attack between the undisturbed flow and the coil surface. In this way, the reduction in velocity due to interaction with the coil is less pronounced, and the fluid is also able to effectively lap the coil in the downstream area, improving the amount of heat transferred. the three-dimensional streamlines are shown in figure 6 as a comparison between simulations 1 and 5.

For the purpose of heat removal, the most important comparison to be made is the value of the global heat transfer coefficient. Its behaviour is extracted along the midline of the coil's upper surface for all 9 simulations carried out and it is reported in figure 7. Simulation 1, which is the one that led to a lower value of the hot spot temperature, is the one with the highest values of the heat transfer coefficient. From these simulations, therefore, the solution to be adopted is to minimize the diameter of the nozzle and, secondly, to place it as high as possible according to geometric constraints.

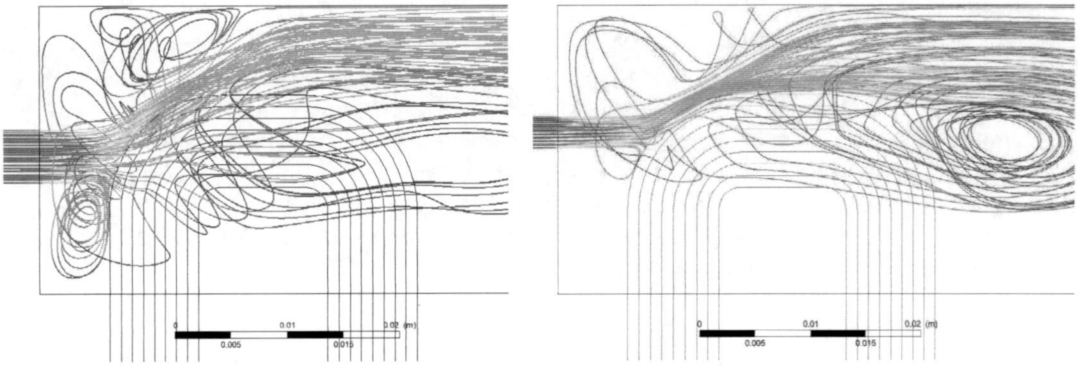

Figure 6: velocity streamlines coloured with velocity magnitude. Sim. 5 (left). Sim. 1 (right).

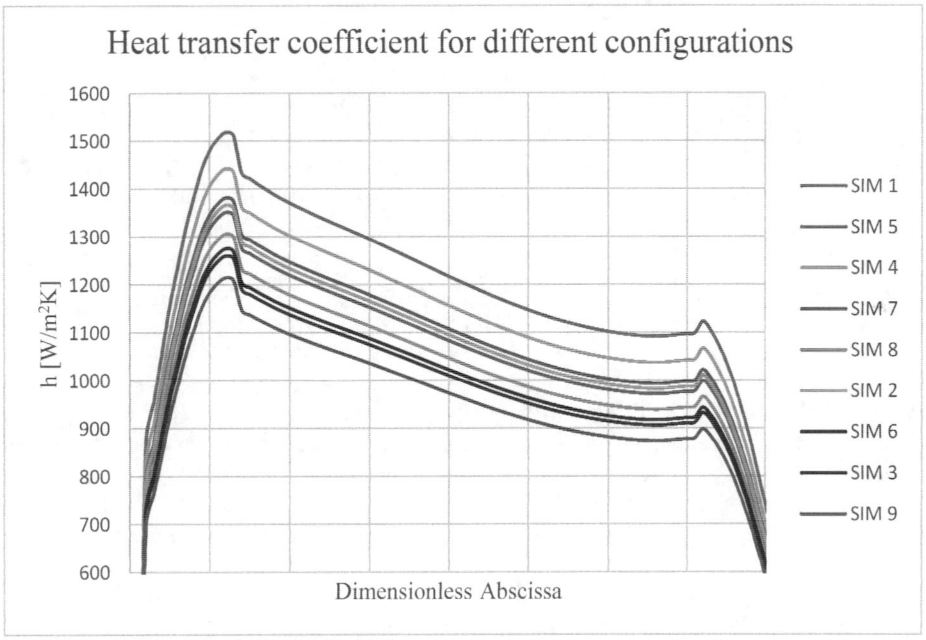

Figure 7: Heat transfer coefficient for each one of the simulations of the test matrix.

5. Conclusions

The analysis carried out during this paper highlighted the important improvements that a direct oil cooling system would bring to the state of the art. Moreover, by means of a CFD campaign, it was shown that the nozzle, responsible of the coolant injection, results in a more effective cooling of the coil when it is placed as high as possible (coherently with geometrical limits) and it is characterized by a small diameter, for the same mass flow rate. The reason for that is given by the strong decrease of velocity suffered from the impinging flow if the nozzle is located too low. Moreover, this work can be considered as a basis for further analysis and optimizations of AFPM cooling systems, and it can be extended also in other configurations in which the coils are not fixed in space or the fluid investing them is considered by means of a two species approach, together with the air permeating the domain.

6. References

[1] Bhagubai P P C, Bucho L F D, Fernandes J F P and Costa Branco P J 2022 *Optimal Design of an Interior Permanent Magnet Synchronous Motor with Cobalt Iron Core. Energies* **15**(8) 2882

[2] Hüner E and Toylan H 2022 *Design optimization with genetic algorithm of open slotted axial flux permanent magnet generator for wind turbines. Int. J. of Green Energy* **20**(4) pp 1-9

[3] Kelch F, Yang Y, Bilgin B and Emadi A 2018 *Investigation and design of an axial fluxpermanent magnet machine for a commercialmidsize aircraft electric taxiing system. IET Electrical Systems in Transportation* **8**(1) pp 52-60

[4] Wan Z, Sun B, Wang X, Wen W and Tang Y 2020 *Improvement on the heat dissipation of permanent magnet synchronous motor using heat pipe. Proceedings of the Institution of Mechanical Engineers, Part D: Journal of Automobile Engineering* **234**(5) pp 1249-1259

[5] Tikadar A, Johnston D, Kumar N, Joshi Y and Kumar S 2021 *Comparison of electro-thermal performance of advanced cooling techniques for electric vehicle motors. App. Th. Eng.* **183** (2) 116182

[6] Liu Y and Lei G 2018 *Thermal analysis of an air-cooled air gap flux permanent magnet machine for hybrid electric vehicle. Energies* **11**(4) 880

[7] Yao Z and Li S 2020 *Analysis and optimization of forced air cooling for air-gap flux permanent magnet machines. IEEE Transactions on Magnetics* **56**(1) pp 1-10

[8] Mao W and Yang G 2021 *A comprehensive review of cooling techniques for air-gap flux permanent magnet machines. Energies* **14**(5) 1277

[9] Deng W and Huang J 2020 *Thermal analysis and optimization of a direct liquid cooling system for an air-gap flux permanent magnet machine. IEEE Transactions on Magnetics* **56**(3) pp 1-10

[10] Chen X and Yang Y 2020 *Thermal design of a hybrid cooling system for air-gap flux permanent magnet machines. IEEE Transactions on Magnetics* **56**(6) pp 1-9

[11] Giulii Capponi F, De Donato G and Caricchi F 2012 *Recent advances in axial-flux permanent-magnet machine technology. IEEE Transactions on Industry Applications* **48**(6) pp 2190-2205

[12] Vainel G, Staton D A, Giulii Capponi F, De Donato G and Caricchi F 2013 *Thermal modelling of a fractional-slot concentrated-winding kaman type axial-flux permanent-magnet machine. 2013 IEEE Energy Conversion Congress and Exposition* pp 1505-1511

[13] Chai F, Bi Y and Chen L 2020 *A Comparison between Axial and Radial Flux Permanent Magnet In-Wheel Motors for Electric Vehicle. International Conference on Electrical Machines (ICEM)* pp 1685-1690

[14] Gai Y, Kimiabeigi M, Chong Y C, Widmer J D, Deng X, Popescu M, Goss J, Staon D A and Steven A 2019 *Cooling of Automotive Traction Motors: Schemes, Examples, and Computation Methods. IEEE Transactions on Industrial Electronics* **66**(3) pp 1681-1692

[15] Marcolini F, De Donato G and Caricchi F 2019 *Direct Oil cooling of End windings in Tour-Type Axial-Flux Permanent-Magnet Machines. IEEE Energy Conversion Congress and Exposition (ECCE)* pp 5645-5651

[16] Dombek G and Nadolny Z 2018 *Influence of paper type and liquid insulation on heat transfer in transformers. IEEE Transactions on Dielectrics and Electrical Insulation* **25**(5) pp 1863-1870

[17] Ning F, Wang Y, Yang L and Wang E 2020 *Research on Viscosity-Temperature Characteristics of PAO Oil and Ester Oil. Int. J. Of Eng. Res. & Tec. (IJERT)* **9**(6)

[18] Edelsbrunner H 2009 *Geometry and topology for mesh generation*, Cambridge University press

[19] Wang C, Qu R, Li J, Fan X, Li D and Lu Y 2017 *Rotor loss calculation and thermal analysis of a dual-stator axial-flux permanent magnet machine with combined rectangle-shaped magnets. 20th International Conference on Electrical Machines and Systems (ICEMS)* pp 5.

Influence of different heating systems on thermal comfort perception: a dynamic and CFD analysis

Vincenzo Ballerini, Eleonora Palka Bayard de Volo, Beatrice Pulvirenti, Eugenia Rossi di Schio*, Paolo Valdiserri and Paolo Guidorzi

Alma Mater Studiorum – University of Bologna, Department of Industrial Engineering DIN. Viale Risorgimento 2, I-40136 Bologna, Italy.
vincenzo.ballerini2@unibo.it, eleonora.palkabayar2@unibo.it, beatrice.pulvirenti@unibo.it, eugenia.rossidischio@unibo.it, paolo.valdiserri@unibo.it, paolo.guidorzi@unibo.it

* e-mail: eugenia.rossidischio@unibo.it

Abstract. In this paper, we investigate the influence of different heating systems on the thermal comfort indexes, Predicted Mean Vote (*PMV*) and Predicted Percentage of Dissatisfied (*PPD*), for a residential apartment located in Bologna (Italy). The apartment has an area of 40 m^2 and is located on the ground floor of 4 floors building. The envelop consists in horizontal perforated bricks with internal thermal insulation material and two windows. The analyses are performed employing Trnsys, a commercial dynamic simulation software and Simcenter STAR-CCM+, a multiphysics computational fluid dynamics (CFD) software. The CFD analysis regards a steady condition of a typical winter day in Bologna. Thermal comfort indexes and thermal energy demand are studied comparing two different heating generation systems existing in the considered apartment: a condensing gas boiler coupled with radiators as terminal emitters and an air-to-air heat pump. By crossing the results obtained by the dynamical approach and by the CFD simulations, a two-objective methodology where energy consumption is minimised while thermal comfort is obtained, is presented.

1. Introduction

Nowadays, an important part of greenhouse gas emissions comes from fossil fuels; in particular, in Europe natural gas is widely employed for residential heating [1]. In order to achieve the decarbonization goals [2], a possible solution is the use of electric heat pumps, instead of generators based on fossil fuels, for building heating. According to market data [3], in Europe the most widespread typology of heat pumps employed in residential sector for heating and cooling are air source heat pumps, ASHP. The prevalence of ASHP is related basically to the availability of the air as heat source and to the limited installation cost respect to other types of heat pumps as for example ground source heat pumps [4, 5]; air source heat pumps, on the other hand, are characterized by reduced performances with respect to ground source heat pumps, and their performance strongly depends on weather conditions and frost formation on the external evaporator, as suggested in many papers in the literature [6-8]. The transition towards more sustainable energy sources in building heating, however, should not overlook the importance of thermal comfort and indoor air quality. Ballerini et al. presented an economic analysis comparing gas boilers and heat pumps as generators for building heating [9]. The aim of the present paper is to complement the analysis presented in [9] by taking into account the comfort analysis. According to the ISO standard 7730 definition [10], thermal comfort is the "condition of mind which expresses satisfaction with the thermal environment". The predicted mean vote *PMV* and the associated predicted percentage of dissatisfied *PPD* indexes, developed by Fanger [11], are widely used to evaluate indoor thermal comfort in moderate environments. Aqilah N. et al. propose a review on thermal comfort in residential building [12]. Their findings show that an adaptive thermal comfort model, which will improve both the comfort requirements and the building energy performance, is still required. Several researchers focus their attention on the optimization of thermal comfort perception within residential environments and many of them uses CFD as a powerful tool to predict thermal comfort, as well as to simulate the behavior of screens on glass facades [13]. Concerning thermal comfort, Usman et al. [14], propose the CFD analysis of thermal comfort in a house under natural and mechanical ventilation. Results show how *PMV* values are improved when ceiling fan coil units are used. In the present paper the dynamic and CFD analyses concerning the comfort perception and building thermal energy demand are presented, employing an air-to-air heat pump and a gas boiler as heating generation system, considering a real residential apartment located in Bologna.

2. Setting of the analysis

2.1. *Dynamic simulations*

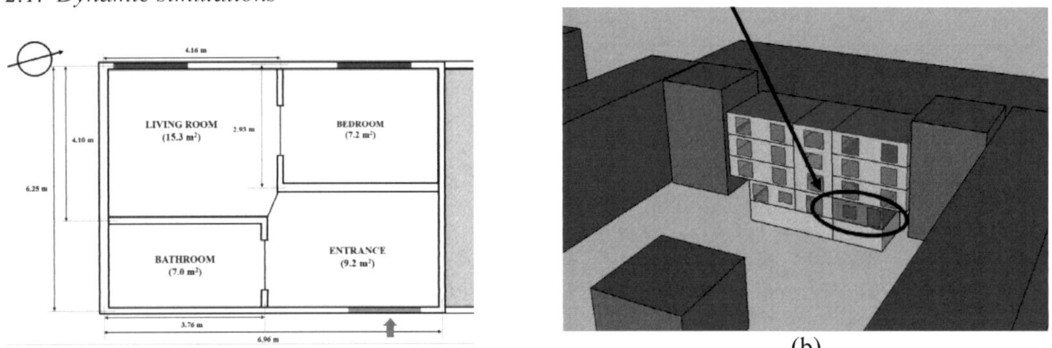

(a) (b)

Figure 1 – Layout of the apartment (a); 3D view of the building modeled with Google SketchUp (b).

Table 1 – Transmittances *U* and dimensions (thickness/area) of apartment envelope components.

Component	Thickness (m)	Area (m^2)	U ($W/(m^2K)$)
External walls	0.30	-	0.667
Dividing walls	0.10	-	2.047
Inter-floor	0.42	-	0.595
Entrance door	-	2.43	5.54
Window (bedroom)	-	1.89	1.69
Window (living room)	-	5.94	1.69

The apartment, modelled through Google Sketch-up [15] and imported in Trnsys [16, 17] employing Trnsys3d plugin [18], displays 4 thermal zones (Figure 1) rather than 3 zones as it really is: in particular the kitchen – living room has been split in two different thermal zones since all the thermal zones must be convex, in order to obtain a detailed calculation of the mean radiant temperature considering the view factors for each zone.

In Table 1 the dimensions and thermophysical characteristics of the main envelope elements are reported; the transmittance values of the main envelope elements (vertical walls and int-floor) are typical of the northern Italian buildings built in years 1970-1990, as reported in [9].

The building has been modelized employing type 56 (multizone building), while the generators have been modelized in two different ways: the gas boiler have been modelized employing type 122 coupled to radiators radiators (types 320 and 362 [19]), while the heat pump has modelized employing two interpolators, in order to determine step-by-step, the Coefficient of performance of the machine (*COP*) and the thermal power output (P_{th}), function of the external air temperature and of the inverter frequency. The performance maps of the *COP* and heat pump thermal power output are reported in Figure 2, for different inverter frequencies. The set-point temperature in the case of radiators is maintained employing thermostatic valves, while, when the heat pump is assumed as the heating generator system, the heat pump thermostat is modelized employing a *PID* (Proportional Integral Derivative) controller.

Four different dynamic simulations have been performed, as reported in Table 2, with the same boundary conditions. In detail, the air change rate of the thermal zones in the apartment is 0.6 h^{-1}, except for the bathroom, where it is 0.8 h^{-1}; in the garage the air change rate is 2 h^{-1}. The air change rate in the garages is higher than in all the other zones because they present large openings to the outside air, while the bathroom presents higher air change rate than other zones of the considered apartment since it is equipped with an extractor fan.

The temperature of the upper apartment, i.e., the only heated boundary zone, has been set to 20°C during the day (6:00 am – 11:00 pm) and 17°C during the night (11:00 pm – 6:00 am). The temperature of the garage and stairwells is determined by the dynamic simulation software: in fact, these two adjacent thermal zones are non-heated zones with respect to the considered apartment.

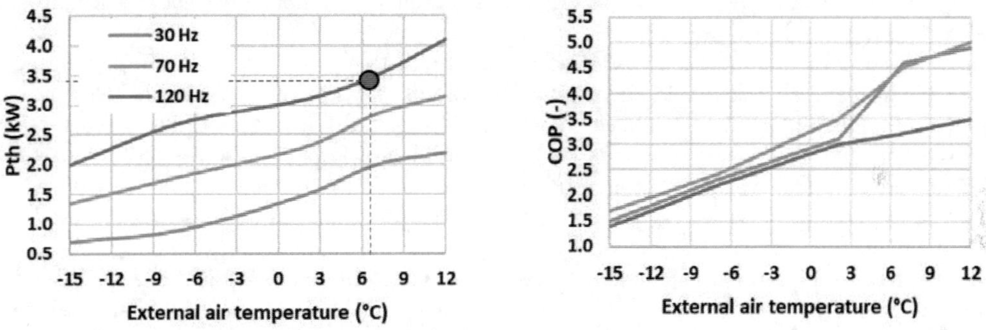

(a) (b)

Figure 2 – Thermal power output (a) and *COP* (b) of the heat pump considered, for three different inverter frequencies and for a fixed indoor air temperature of 20 °C. The red dot indicates the thermal power output at rated conditions (3.5 kW when the indoor air temperature is 20 °C, while the outdoor temperature is 7°C).

Table 2 – Dynamic simulations performed. In case DS1 the emitters are the radiators, while in case DS2, DS3 and DS4 the emitter is the internal unit of the air-to-air heat pump.

Case	Heating generation system	Daily set-point, 6:00 am – 11:00 pm (°C)	Night set-point, 6:00 am – 11:00 pm (°C)	Annual energy demand (kWh)
DS1	Gas boiler	20.5	17.5	3486
DS2	Heat pump	20.5	17.5	2669
DS3	Heat pump	21.5	17.5	2983
DS4	Heat pump	22.5	17.5	3303

In order to determine the ground temperature, typical thermophysical characteristics of the soil in Bologna (conductivity 1.8 W/(mK), density 2800 kg/m^3 and thermal capacity of 0.85 kJ/(kgK)) have been used [20].

The internal gains related to electric equipment and occupancy have been considered according to IAE Task 44 [21], considering a single occupant of the apartment. The latent internal gains (due to occupancy, cooking, equipment and wet surfaces) have been set to be 0.25 kg/h, as prescribed by UNI-TS 11300-1 standard for residential buildings [22]. The coupling air flow between the 4 zones (entrance, living – room, bedroom and bathroom) has been determined employing CONTAM [23], as function of the air density between the zones. The simulation time-step has been set to 15 s and, for all the considered cases, the analysis is restricted to the standard heating period for Bologna (15 October – 14 April).

2.2. *CFD analysis*

The commercial software STARCCM+ has been used to carry out a CFD analysis. The analysed governing equations consist of the mass (1), momentum (2), and energy (3) balance equations for steady state, incompressible, turbulent flows of an ideal gas, namely:

$$\nabla \cdot \vec{v} = 0 \tag{1}$$

$$\rho_0[(\vec{v} \cdot \nabla)\vec{v}] = -\nabla(\mathrm{p} + \rho_0 g z) + \rho_0 g \beta (T - T_0)\nabla \mathrm{z} + \nabla \cdot \tilde{\tau}_{eff} \tag{2}$$

$$\nabla \cdot \big(\vec{v}(\rho e + p)\big) = \nabla \cdot [(k + k_t)\nabla \mathrm{T}] + \tilde{\tau}_{eff} \cdot \vec{v} \tag{3}$$

where $\vec{v}$ is the velocity vector, p is the pressure, β is the thermal expansion coefficient, $\tilde{\tau}_{eff}$ is the effective stress tensor, e is the specific enthalpy, k is the thermal conductivity and k_t the turbulent thermal conductivity.

To accurately account for turbulence anisotropy, the Reynolds Stress Transport (RST) turbulence model, along with the Elliptic Blending near-wall Reynolds-stress turbulence closure, was utilized. The governing equations were discretized into a set of algebraic equations using a second-order upwind scheme, and the pressure-velocity coupling problem was solved using the semi-implicit method for pressure-linked equations (SIMPLE algorithm). The simulation was monitored for convergence by measuring residuals, velocity, and temperature values at various randomly selected points in the room.

A mesh sensitivity analysis was conducted to optimize the computational demand and quality of the results. The final mesh consisted of 9 x 10^6 elements with a base size of 4 cm. A boundary layer was modelled with 5 layers and a total thickness of 5% of the base size. Additionally, surface refinements were necessary for the heat pump internal unit, radiators, and internal furniture. The distribution of furniture within the apartment is shown in Figure 3(a), while Figure 3(b) illustrates the top view of the final mesh on the internal surfaces.

Two different steady state simulations have been performed to compare the two different heating systems equipped in the apartment during a typical winter day in Bologna. An external temperature of 6.9 °C has been prescribed. The ceiling, floor and internal wall (adjacent to the stairs hall) display a temperature of 20 °C, 10 °C and 12 °C respectively. The heat pump internal unit is composed by an inlet section where a fixed velocity inlet boundary condition is imposed of 2.4 m/s, according to experimental data, with a discharging angle of 45° with respect to the ceiling and an outlet section where a pressure outlet boundary condition is applied. Radiators, instead, are modelled as heat sources, obtaining a surface temperature of 60 °C. The parameters obtained by the simulations have been used to evaluate the global thermal comfort indexes *PMV* and associated *PPD* within the apartment.

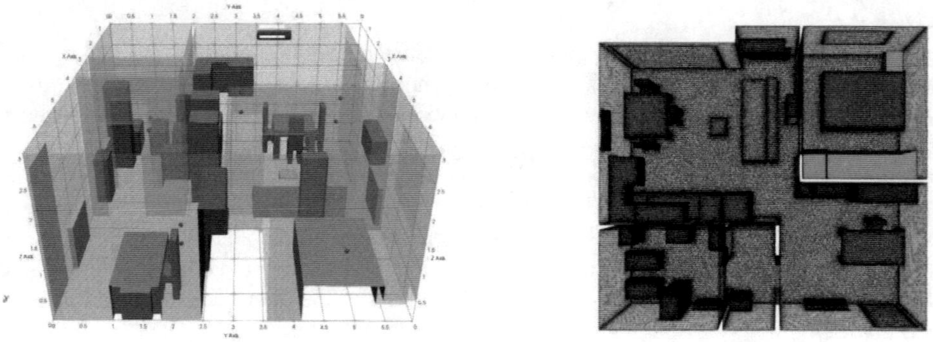

(a) (b)

Figure 3 – Furniture layout (a) and top view of the internal mesh (b).

3. Results

3.1 Dynamic analysis results

In the last column of Table 2 the results of the dynamic simulations for cases DS1 – DS4 are reported. The results clearly show the difference in thermal energy demand of the building in the case of radiators and heat pump. In Figure 4, the PMV and PPD are shown for a representative day of the heating season that presents a mean outdoor temperature is 6.9 °C. The comfort indexes have been calculated considering a fixed air velocity of 0.1 m/s, a clothing factor of 1.1 clo and a metabolic rate of 1.1 Met. In particular, in Figure 4 the red line refers to the case when the gas boiler coupled to radiators as heating system are considered, while the other three lines refers to the cases when the air-to-air heat pump is considered, with three different air set-points.

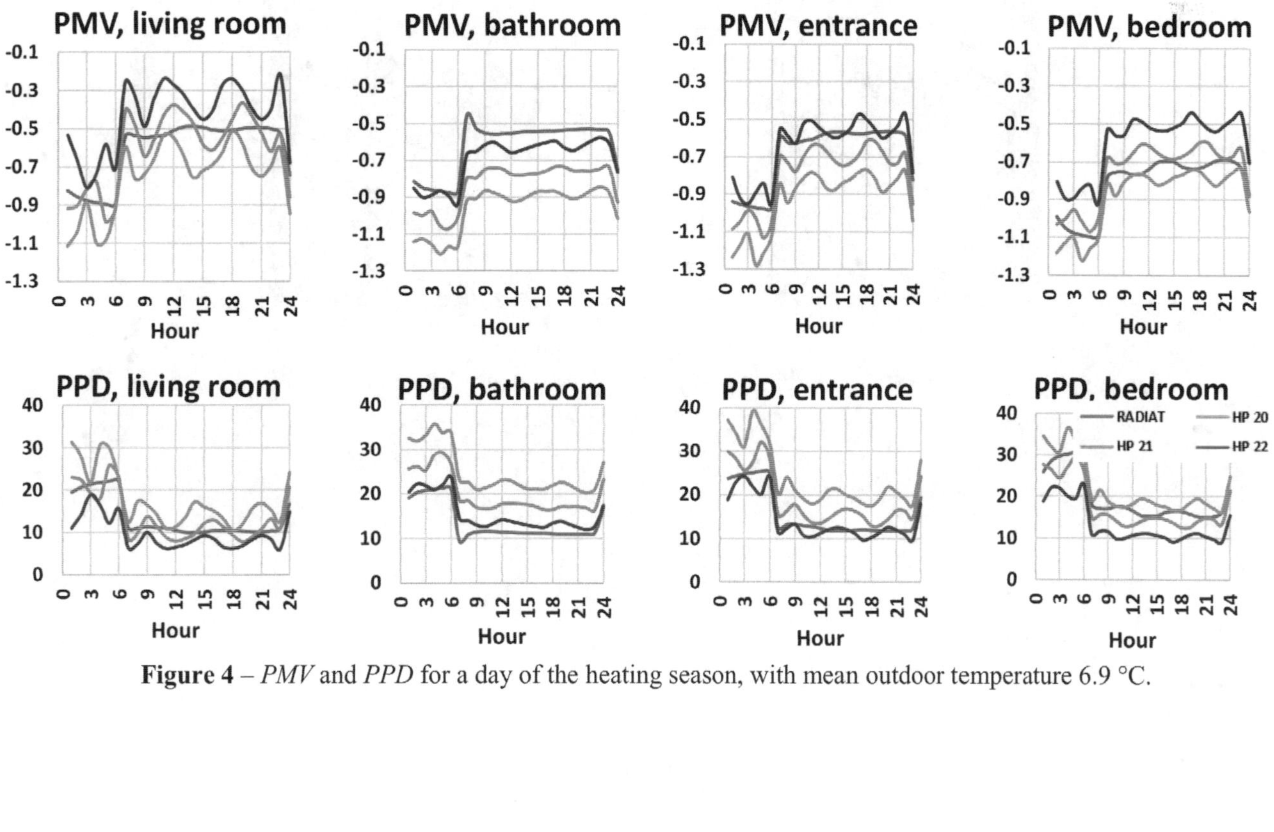

Figure 4 – *PMV* and *PPD* for a day of the heating season, with mean outdoor temperature 6.9 °C.

3.2 CFD analysis results
Figure 5 shows the *PMV* and *PPD* distribution obtained within the apartment through the numerical simulation for both heating systems. It can be noted how the choice of the system lead to different distribution of the thermal comfort parameters. As the matter of facts, the distribution of the relevant parameters for the evaluation of the PMV index significantly differs among the two configurations. As shown in Figure 6, the streamlines associated to the heat pump follows the discharging angle of 45° reaching the sofa zone, and then, under the buoyancy effect, tends to go upwards and enters the bedroom. As a result, less hot air is delivered to the bathroom and entrance, leading to lower degrees of comfort in these areas. The situation changes when using the radiators as emitters since they improve the mean radiant temperature distribution throughout the apartment, resulting in a different comfort situation.

4. Discussion
Table 2 presents an important reduction of thermal energy demand in the case of a heat pump as a generator, instead of a gas boiler: the reduction for case DS2 with respect to case DS1 is 23 %, while there is a similar energy demand in cases DS1 and DS4, but in this latter case the heat pump has a set-point temperature of 22 °C during the day. The different thermal energy demand between the two different heating systems is primarily due to the fact that the heat given by the heat pump is purely convective, while the heat coming from radiators is partly convective and party radiative.

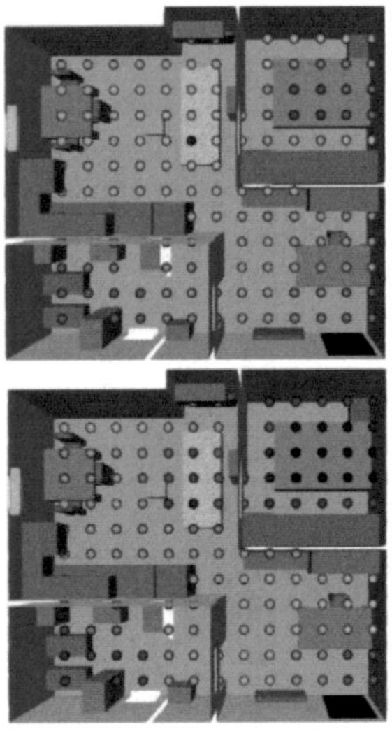

(a)

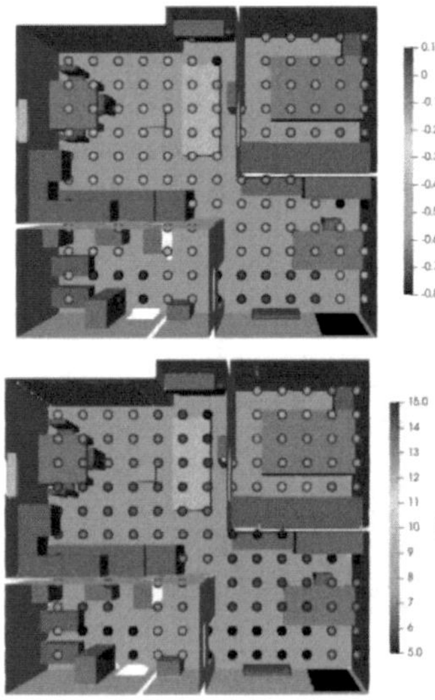

(b)

Figure 5 – Top view of *PMV* (upper figures) and *PPD* (lower figures) distribution for heat pump (a) and radiators (b).

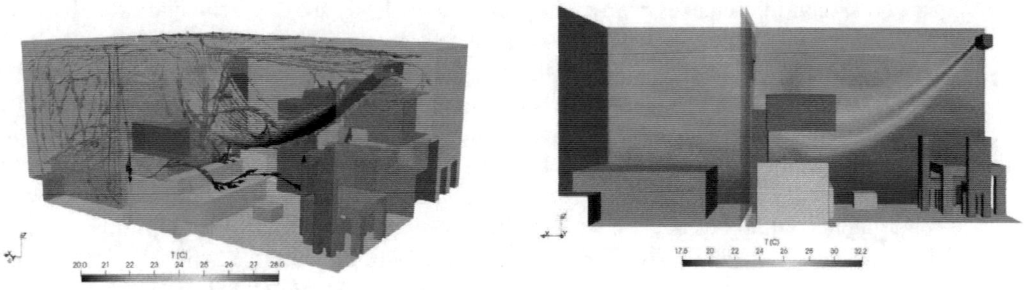

(a) **(b)**

Figure 6 – Streamlines of the heat pump discharged air (**a**) and temperature profile (**b**) along a section in correspondence of the heat pump internal unit.

The use of the heat pump instead of the gas boiler results advantageous in terms of energy savings, but affects the comfort in the apartment, as shown in Figure 4: in particular, it can be noticed that the *PPD* is higher for the case DS2 (heat pump with a set-point of 20 °C during the day) in all the zones except the bedroom that has no radiators. It can also be observed that *PPD* are comparable only in cases DS1 and DS4 in almost all the thermal zones: therefore, the heat pump can guarantee the comfort given by the gas boiler coupled with radiators only with a set-point of 22 °C. Moreover, the analysis on comfort indexed referring to the coldest day of the heating season, with mean outdoor temperature −4 °C, highlights an important reduction of comfort indexes, especially for the heat pump cases: for example, the *PPD* is up to 50 % in bedroom, entrance and bathroom for case DS1; in case DS4 the *PPD* results slightly lower (5-10 %) with respect to case DS1 only for bedroom and living room. In terms of comfort index, the most important difference between case DS1 and the other cases is the mean radiant temperature: in the representative day of the heating season characterized by a mean outdoor temperature of 6.9 °C, the difference in mean radiant temperature calculated in the thermal zones is about 1-2 K, i.e., in the case DS1 the mean radiant temperature is always higher than in the heat pump cases. This aspect become more relevant considering the coldest day of the heating season, when the mean radiant temperature for all the zones referring to case DS1 is higher up to 3 °C. The CFD analysis reveals differences in the levels of comfort achievable with the two heating systems. For instance, the use of the heat pump results in nearly zero *PMV* values in the bedroom, whereas low *PMV* values are observed in both the bathroom and the entrance (Figure 5(a)). Furthermore, the high velocity and temperature of the discharged air from the heat pump can create discomfort zones, with hot thermal plumes reaching the sofa, placed in the living room. In contrast, the use of radiators leads to a more uniform distribution of *PMV* throughout the apartment. This is due to the positioning of the terminals, which results in higher *PMV* values in both the bathroom and the entrance compared to the heat pump heating mode. However, this improvement comes at the expense of thermal comfort in the bedroom, where *PMV* values are significantly lower due to the absence of a radiator and the presence of a huge window.

5. Conclusion

In conclusion, the main results of the present analysis are the following:

- The air-to-air heat pump as alternative to gas boiler leads to important savings in terms of thermal energy demand of the building (up to 23 %)
- As determined by dynamic simulation, the comfort perception inside the considered apartment is worse in case of heat pump instead of gas boiler coupled with radiators, especially during the colder days of the heating season

- The poorer comfort indexes, particularly on colder days, are due principally to the reduced mean radiant temperature determined using the heat pump instead of the radiators.
- As shown by the CFD analysis, while the air-to-air heat pump system can provide satisfactory *PMV* values in the living room and the bedroom, it cannot meet the thermal requirements of the bathroom and entrance areas. Additionally, due to the high velocity and temperature of the discharged air, local discomfort may be experienced by occupants. To improve the overall thermal comfort throughout the apartment, the internal unit's position could be changed, or a supporting unit could be added to address the areas with lower *PMV* values.
- The CFD simulation shows radiators can provide a more comfortable experience for occupants due to the uniform distribution of internal thermal parameters within the apartment.

Further analyses will focus on the integration of the dynamic and CFD results with experimental measurements to evidence the influence of certain dynamic parameters on the simulation results.

6. References

[1] Pezzutto S, Croce S, Zambotti S, Kranzl L, Novelli A, Zambelli P 2019 *Energies* **12** 1760
[2] EU 2050 long term strategy. https://ec.europa.eu/clima/policies/strategies/2050_en
[3] European heat pump association (EHPA) 2022. https://www.ehpa.org/publications/heat-pump-sales-in-europe-2021-ehpas-2022-market-report/
[4] Rivoire M, Casasso A, Piga B, Sethi R 2018 *Energies* **11** 1941
[5] Ołtarzewska A, Krawczyk D.A 2021 *Energies* **14** 6701
[6] Ballerini V, Dongellini M, di Schio E.R, Valdiserri P 2022 *J. Phys. Conf. Ser.* 2177 012025
[7] Rossi di Schio E, Ballerini V, Dongellini M, Valdiserri P 2021 *Appl. Sci.* **11** 8003
[8] Vocale P, Morini G.L, Spiga M 2014 *Energy Procedia* **45** 653–662
[9] Ballerini V, Rossi di Schio E, Valdiserri P 2022 *Buildings* **12** 1936
[10] ISO 7730:2005. https://www.iso.org/standard/39155.html
[11] Fanger O 1970 *Thermal Comfort* (Copenhagen: Danish Technical Press)
[12] Aqilah N, Rijal H.B, Zaki S.A 2022 *Energies* **15** 9012
[13] Santolini E, Pulvirenti B., Guidorzi P, Bovo M, Torreggiani D, Tassinari P 2022 *Build Environ.* **211** 108691
[14] Usman F, Bakar A.R.A 2020 Proceedings of AICCE'19. AICCE 2019. *Lecture Notes in Civil Engineering* **53** Springer, Cham.
[15] Trimble Inc. 2015, https://www.sketchup.com/it
[16] Klein S.A et al, 2017 TRNSYS 18: A Transient System Simulation Program, Solar Energy Laboratory, University of Wisconsin, Madison, USA.
[17] Klein S.A, Duffie A.J, Mitchell J.C, Kummer J.P, Thornton J.W, Bradley D.E, Arias D.A, Beckman W.A 2010 University of Wisconsin: Madison, WI, USA.
[18] TRNSYS3D - plugin for SketchUp. http://www.trnsys.com/features/suite-of-tools.php.html
[19] Holst S 1996 http://trnsys.de/static/641c965bac4f593c3b4cc569839a51c2/Type_361_362_320_en.pdf
[20] Natale C, Naldi C, Dongellini M, Morini G.L 2022 *J. Phys. Conf. Ser.* **2385** 012090
[21] Dott R, Haller M, Ruschenburg J, Ochs F, Bony J 2013 http://www.taskx.iea-shc.org/data/sites/1/publications/T44A38_Rep_C1_B_ReferenceBuildingDescription_Final_Revised_130906.pdf
[22] UNI/TS 11300-part1 2014 ENERGY PERFORMANCE OF BUILDINGS - PART 1: EVALUATION OF ENERGY NEED FOR SPACE HEATING AND COOLING.
[23] Dols W, Polidoro B 2020 https://tsapps.nist.gov/publication/get_pdf.cfm?pub_id=930691

Microscale damper prototype: A preliminary study on suppressing air flow oscillations within microchannels

A Fichera, A Pagano and R Volpe*

Department of Electrical, Electronics and Computer Engineering, University of Catania, via Santa Sofia 64, 95125 Catania, Italy

* Corresponding author: rosaria.volpe@unict.it

Abstract. This research introduces a novel micro-damper designed to mitigate pressure and velocity oscillations from a piezoelectric micropump in microfluidic environments. Unlike existing research focusing on damping in incompressible liquid flows with methods like elastic films and PDMS membranes, this study proposes a novel micro-damper prototype. Integrated into a microdevice for particle granulometric separation and detection, the damper connects to a piezoelectric micropump outlet and to a focusing microchannel inlet, followed by a capacitive sensor for size-based particle counting. Preliminary analysis determined an optimal airflow velocity at w = 0.5 m/s for accurate focusing and counting under laminar conditions. The micro-damper, constrained by the piezoelectric pump's geometry, features a 27 μm high and 1000 μm wide cross section. Its outlet supports two potential focusing microchannel inlet configurations of 30 μm or 40 μm. Distinctively, it incorporates two symmetrical backward micro-channels connecting to the atmosphere, allowing direct piezometric contact between the main flow and an infinite compliant volume. OpenFOAM simulations confirm the damper's effectiveness in maintaining laminar outlet flow and suppressing micropump disturbances. Thus, the proposed micro-damper ensures optimal inlet conditions for subsequent microchannel processes, enabling stable particle separation and detection in controlled airflow samples.

1. Introduction

Microfluidic systems have shown vast potential for applications from biomedical diagnostics to chemical synthesis. Whitesides pioneered in recognizing their significance, setting the stage for a plethora of subsequent research [1].

A primary challenge in microfluidics is oscillations in incompressible fluid flows. Addressing this, Utz et al. [2] used an elastic film to dampen pressure oscillations within micro-channels, exploring how pressure perturbations propagate, and how a channel's damping response can be varied based on its geometry and film elasticity, provided a fundamental understanding of this phenomenon. Yang et al. [3] and Iyer et al. [4] evolved this concept using a PDMS membrane to separate liquid flow from compliant chambers, which compensated for pressure and flowrate perturbations. Their methods were in line with the discussion by Squires and Quake [5], who emphasized the intricacies inherent in microfluidic designs and how subtle alterations could lead to pronounced effects. For gas flows, Zhang et al. [6] compared a damper to an RC low-pass filter, efficiently damping frequencies over 5 Hz. Veenstra et al. [7] introduced a compliant capacitive element. An analytical model was developed and experimentally validated to describe the dynamic damping response. However, their methodology faced the intrinsic limitation of air's low compressibility, leading to the realization that damping capacity is contingent upon the available compliant volume. Merging diverse methodologies, Jiao et al. [8] presented an innovative damper encapsulating a gas volume within an elastic envelope. This approach eliminates the need for a membrane at the interface between the channel and the compliant volume. The dynamic damping response of this system is simulated using a numerical finite element model, complemented by an in-depth description of the chip fabrication process and the measurement instrumentation used in experimental tests. This research is in line with the discussions by Gervais et al. [9] and Tsao et al. [10], who explored flow-induced deformations. Delving into complex flow patterns, Dendukuri et al. [11] and Atencia et al. [12] highlighted the importance of controlled flow in microfluidics. This is paramount, especially when the applications involve tasks like precision particle detection or trapping.

In this context, this study proposes a damper to mitigate oscillations from piezoelectric micropumps in a novel microfluidic system for particulate matter classification and measurement, underlining the need for undisturbed laminar airflow.

2. Microscale damper design for the suppression of flow perturbations

The micro-scale damper described in this study is part of an advanced microscale particulate detection system, aiming to characterize airborne particulate matter within a controlled airflow sample. This process involves a focusing channel for spatial separation based on particle diameter and an array of capacitive sensors for precise particle counting. Preliminary analysis suggested an optimal average airflow of w = 0.5 m/s for the device's operation, highlighting the need for laminar and steady flow. The damper, placed between the piezoelectric micropump and the focusing microchannel, ensures laminarity at the inlet by suppressing flow disturbances. It has a transversal section height of 27 μm and an inlet of 1000 μm, aligning with the pump's outlet for consistent flow. The outlet section width accommodates two configurations of 30 μm and 40 μm. Multiple damper configurations have been developed with distinct features, as depicted in figure 1.

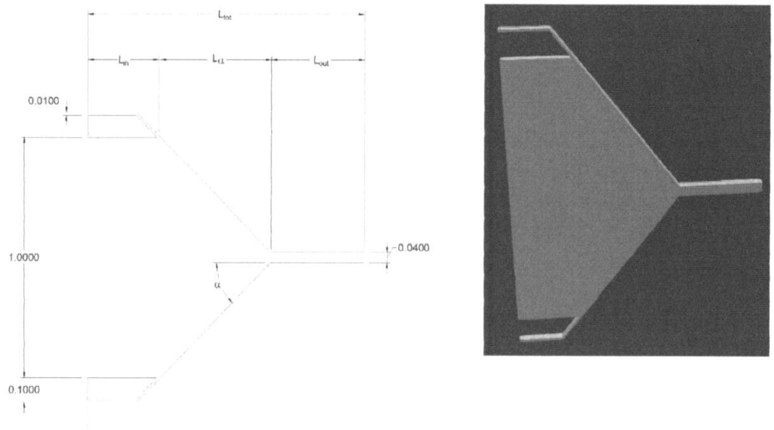

Figure 1. Figure with short caption (caption centred).

The entry region maintains a constant rectangular inlet cross-section of 1000 μm x 27 μm; a converging section following the entry region, with an axial length L_α defined by a convergence angle α of either 45° or 30°; the outlet micro-channel, interfacing with the external environment, can have cross-section dimensions of either 40 μm x 27 μm or 30 μm x 27 μm. This section has an axial length, L_{out}, strategically designed to be $L_{out} > 10\ L_{DH}$, i.e. ten times greater than the equivalent hydraulic diameter D_H, to ensure flow's re-laminarization before reaching the focusing channel; two backward micro-channels connect directly to the atmosphere, facilitating direct communication between the primary flow in the main micro-channel and an infinite compliant volume, crucial for damping perturbations from the piezoelectric micropump. Consequently, the micro-damper's design not only meets the operational constraints but also introduces novel features that actively dampen disturbances. This stabilization is critical for ensuring that downstream components, such as the focusing channel, receive a steady flow crucial for precise analysis. Further details on the various design configurations are provided in Table 1.

Table 1. Different design configuration

	Inlet Section	L_{in}	α	Outlet Section	D_H	L_α	L_{out}	L_{tot}
A1	1000μm x 27μm	360μm	45°	30μm x 27μm	28.4μm	284μm	305μm	1150μm
A2	1000μm x 27μm	431.5μm	30°	30μm x 27μm	28.4μm	284μm	305μm	1568μm
B1	1000μm x 27μm	320μm	45°	40μm x 27μm	32.2μm	322μm	400μm	1200μm
B2	1000μm x 27μm	391.5μm	30°	40μm x 27μm	32.2μm	322μm	400μm	1615μm

Table 1 presents four distinct damper configurations (A1, A2, B1, B2), each with unique convergence angles, outlet dimensions, and lengths, impacting fluid dynamics and performance characteristics. All configurations share a standardized inlet of 1000 μm x 27 μm for uniform flow entry. The axial length of the entry region L_{in} varies with A1 and B1 shorter length than A2 and B2, affecting initial flow distribution. The convergence angle α is steeper (45°) in A1 and B1 for rapid convergence, while A2 and B2 have an angle of 30° for gradual narrowing. Outlet dimensions differ; A1 and A2 have 30 μm x 27 μm outlets, and B1 and B2 have wider 40 μm x 27 μm outlets, influencing flow rate and distribution. The equivalent hydraulic diameter D_H is consistent with A1 and A2, and slightly larger in B1 and B2, potentially affecting flow rates and fluid dynamics. The axial length of the converging section, L_α, is uniform across configurations. L_{out}, the axial length of the outlet section, is consistent in both A and B configurations to ensure steady flow into subsequent channels. Finally, L_{tot} represents the total axial length of the damper, with variations across configurations.

3. Numerical modelling of the microscale damper

The fluid dynamics within the micro-damper were simulated using OpenFOAM.

3.1. Model assumptions

The micro-damper simulations assume incompressible and predominantly laminar flow, with the *icoFoam* solver selected for its appropriateness for incompressible fluids with mainly laminar motion. The pulsatile behavior of the piezoelectric micropump, manifesting as sinusoidal flow variations, is an essential factor considered in the simulations.

3.2. Geometry and mesh generation

The geometry of the micro-damper has been accurately meshed, as illustrated in figure 2, ensuring that the flow domain is precisely resolved.

3.3. Boundary conditions

The employment of the "*uniformTotalPressure*" inlet boundary condition has been chosen to maintain a consistent total pressure (encompassing both static and dynamic components) across the inlet. This selection is particularly appropriate for incompressible flows. Concurrently, the utilization of the "*pressureInletVelocity*" for velocity coupling at the inlet refers to a system where the velocity field is derived based on the prescribed total pressure and internal field metrics, ensuring an equilibrium in the flow dynamics representation. The "*pressureNormalInletOutletVelocity*" function, designated for the outlet, caters to scenarios with potential flow direction reversals. Essentially, this condition enforces a zero-gradient when the flow is outgoing and transitions to the "*pressureInletOutletVelocity*" paradigm upon ingoing flow.

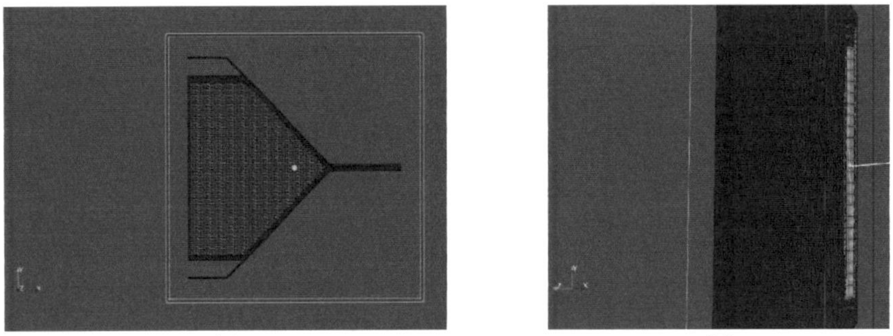

(a) (b)

Figure 2. Schematic of the mesh (a) plan view; (b) 3D view on a cross section.

3.4. Input parameters

The following specific pressure conditions at the inlet and outlet sections have been applied: (a) imposed normalized pressure at the input section; (b) imposed normalized pressure at the main outlet section, given as p_{out}/ρ_{air} = const: (c) imposed constant normalized (relative) atmospheric pressure at the outlets of the side compensation channels, i.e., $p_{atm}/\rho_{air} = 0$. One important aspect that must be considered is the dampening effect of flow perturbations caused by the pulsatile behavior of the piezoelectric micropump. These pulsations typically manifest as sinusoidally pulsating flow variations. Therefore, to simulate such a condition, a second set of simulations have been performed with the following imposed values:

1. at the inlet section:
   a) the normalized pressure is expressed as:

$$\frac{p_{in}(t)}{\rho_{air}} = \frac{p_{norm}}{\rho_{air}} \cdot [1 + \sin(2\pi f \cdot t)] \tag{1}$$

   Here, the pressure pulsates at a frequency f with both an oscillation amplitude and a mean pressure value equivalent to p_{nom}/r_{air};

   a) the velocity field is derived from the pressure-velocity coupling by imposing only inflow at the inlet section.
2. at the outlet of the main channel (intended for connection to the focusing channel):
   a) an imposed constant normalized pressure is defined as p_{out}/ρ_{air};
   b) the velocity field emerges from the pressure-velocity coupling, both outflow and inflow;
3. at the side channels' outlet (compensation channels):
   a) a set constant (relative) atmospheric pressure is established, denoted as $p_{atm}/\rho_{air} = 0$;

the same condition used for the main outlet section.

4. Preliminary results

To assess the damper's capability to ensure laminar flow at the focusing channel inlet, simulations were executed under steady flow conditions. The outcomes of these simulations are depicted in figure 3, highlighting the pressure field and of the axial component of the velocity field.

In the configuration with the angle α = 45°, although the flow remains laminar throughout, the pronounced narrowing induces a local detachment of the flow streamlines. Consequently, this leads to a transient generation of vorticity. Nevertheless, this localized vortex is quickly attenuated within the initial segment of the outlet channels. The rapid damping of this vorticity is further corroborated by the distribution of the axial velocity component presented in figure 3(c), taken at two-thirds the length of the outlet channel.

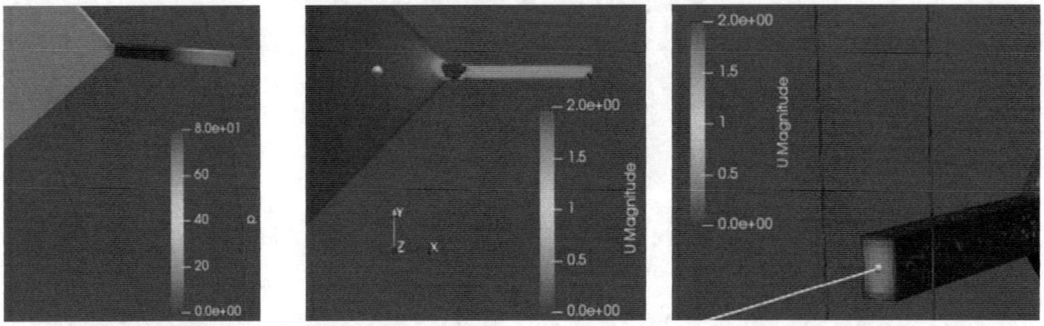

(a) (b) (c)

Figure 3. Simulation result for steady state flow supply condition, $p_{in}/\rho_{air} = 40\ m^2/s^2$: detail of the pressure field (a) and of the flow field (b) in the convergent section and in the outlet channel; (c) velocity field on a cross-section in the re-laminated region in proximity of the outlet

Hence, the criterion selected for determining the length of the straight outlet channel ensures that flow re-lamination is achieved before reaching the focusing channel. The laminarity of the flow at the outlet of the damper is ensured also when the pulsatile nature of the piezoelectric micropump is considered, i.e., the imposed inlet pressure is given as in equation (1).

This is demonstrated by the laminar structure of the pressure and velocity fields reported in figure 4 for $p_{in}/\rho_{air} = 40\ m^2/s^2$ and f = 10 Hz. For all the reported cases the detail of the velocity field is reported on a cross-section situated in the re-laminated zone, chosen at the abscissa that corresponds to the middle third of the outlet channel.

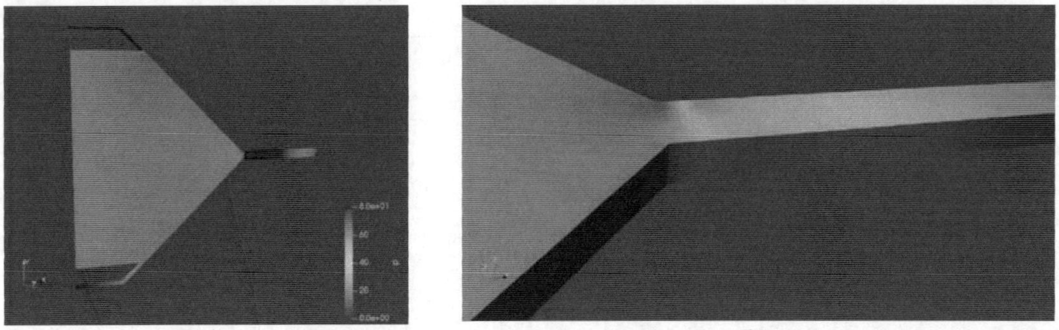

(a) (b)

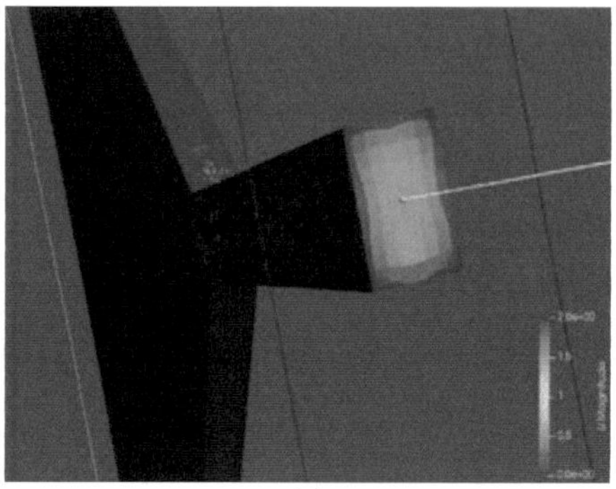

(c)

Figure 4. Inlet pressure with amplitude and average value both equal to $p_{in}/\rho_{air} = 40\ m^2/s^2$ and pulsation frequency 10 Hz

(a)

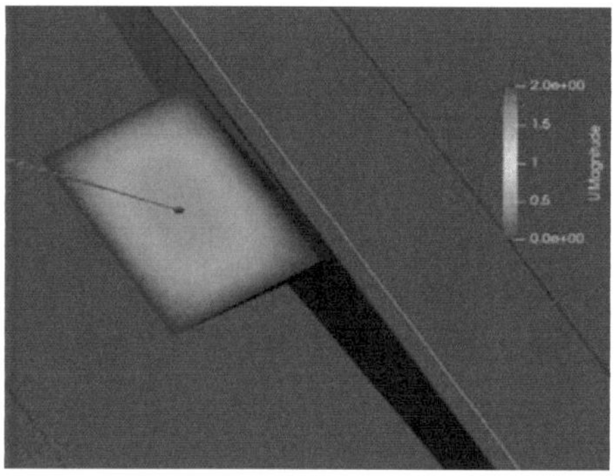

(b)

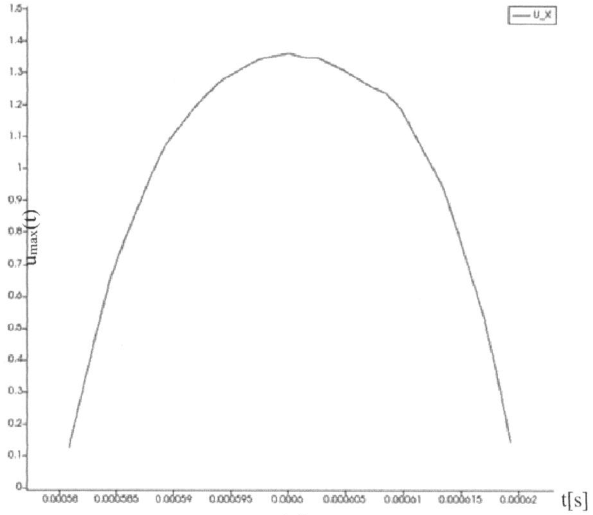

(c)

Figure 5. Inlet pressure with amplitude and average value equal to $p_{in}/\rho_{air} = 40\ m^2/s^2$ and pulsation frequency 100 Hz: (a) pressure field throughout the convergent section and the damper outlet channel; (b) velocity field on a cross section in the re-laminated region of the damper outlet channel; (c) parabolic velocity profile along the y-direction passing by the centroid of the cross section.

To characterize the behavior of the damper, it is not only crucial to consider the re-lamination of the flow, but also to gauge the amplitude of residual pulsations which might disturb the stationarity of the flow at the outlet section. Figure 6 provides insight into the residual velocity pulsations of the axial component u_{max}, evaluated at the center of a cross-section in the re-laminated zone. The simulations were based on an inlet pressure characterized by an amplitude and mean value both equivalent to $p_{in}/\rho_{air} = 40\ m^2/s^2$ at a pulsation frequency of 50 Hz. As they refer to the center of the cross section, the data reported correspond to the residual perturbation of the local maximum of the parabolic velocity profile. Leveraging the Poiseuille relation:

$$u_{av} = \frac{u_{max}}{2} \tag{2}$$

allows for the deviation of the average velocity u_{av} based on the maximum value u_{max} identified along the duct's axis. Consequently, the residual pulsation within the average velocity can be discerned from the residual pulsation in u_{max}.

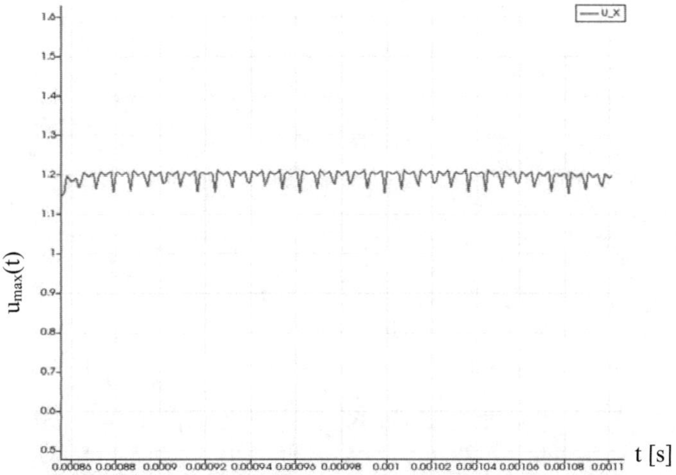

Figure 6. Residual velocity pulsation of the axial velocity component, umax, for an inlet pressure of amplitude and average value both equal to $p_{in}/\rho_{air} = 40\ m^2/s^2$ and pulsation frequency 50 Hz

Table 2 summarizes the average velocity values alongside the amplitude of the residual fluctuations at the cross section of the damper outlet channel for a triad of benchmark frequencies. It's imperative to note that, for comparative consistency across all conditions, the inlet pressure pulsation was standardized to an amplitude and mean value both equal to $p_{in}/\rho_{air}=40\ m^2/s^2$.

Table 2. Average axial velocity and maximum amplitude of residual perturbation

Frequency [Hz]	$\bar{u}_{av}$ [m/s]	$\frac{\Delta u_{av}}{\bar{u}_{av}}$ %
100	0.68	2.8%
50	0.58	2.4%
10	0.38	0.4%

Reported results highlight that across the spectrum of assessed pulsation frequencies, the residual pulsation remains sufficiently low. This confirms the efficacy of the proposed damper in ensuring optimal inlet conditions for the subsequent focusing channel.

5. Conclusions

The objective of this study was to conceptualize, simulate, and analyse a micro-scale damper that ensures the laminarity of the input flow to a focusing channel, specifically designed to detect particulate matter. Air is introduced inside the damper by a piezoelectric micropump, known to introduce time-dependent flow perturbations. By using OpenFOAM, a preliminary study delineated the behaviour of the fluid under various condition. In particular, several damper designs were evaluated in the context of their geometrical and operational constraints. The study demonstrated that, even under laminar conditions, certain damper configurations could introduce localized vortices, especially when a narrowing is encountered in the damper's geometry. However, these vortices were found to dissipate rapidly, ensuring a relatively smooth flow downstream. The criterion for selecting the length of the straight output channel proved effective in achieving flow re-laminarization before introducing the fluid to the subsequent focusing channel. In light of the discussed results, the proposed damper stands as a robust solution, proficient in mitigating the challenges posed by the piezoelectric micropump. It not only ensures the laminarity of the flow, but also minimizes the residual pulsations to a negligible extent. This accomplishment is of paramount importance for the subsequent focusing channel, which is integral to the accurate detection of particulate matter. Future research will consist in the validation of the obtained numerical results with a properly built experimental setup.

6. References

[1] Whitesides G M 2006 *Nature* **442** 368
[2] Utz M, Begley M R, Hariri H H 2011 *Lab on a Chip* **11** 3846
[3] Yang B, Lin Q 2009 *J. of Microelectromechanical System* **18** 539
[4] Yver V, Raj A, Annabattula R K, Sen A K 2015 *J. of Micromechanics and Microengineering* **25** 075003
[5] Squires T M, Quake S R 2005 *Reviews of Modern Physics* **77** 977
[6] Zhang X, Zhu Z, Xiang N, Ni Z 2016 *Biomicrofluidics* **28** 10
[7] Veenstra T T, Sharma N R, Forster F K, Gardeniers J C E, Elwenspoek M C, van den Berg A 2002 *Sensors and Actuators B: Chemical* **82** 2
[8] Jiao Z, Zhao J, Chao Z, You Z, Zhao J 2019 *Microfluidics and Nanofluidics* **23** 2
[9] Gervais T, El-Ali J, Gunther A 2006 *Lab on a Chip* **6** 4
[10] Tsao C W, DeVoe D L 2009 *Microfluidics and Nanofluidics* **6** 1
[11] Dendukuri D, Doyle P S 2009 *Advanced Materials* **21** 41
[12] Atencia J, Beebe D J 2005 *Nature* **437** 648

Acknowledgments
This research has been conducted under the activities of the project S.A.L.VO. – Realizzazione di nodo multisensore per il monitoraggio degli ambienti di lavoto – Prog. N. F/190012/01-03/X44.

Thermal-hydraulic study of the EU-DEMO Helium Cooled Pebble Bed Breeding Blanket Primary Heat Transport System

E Vallone[1,*], G Agnello[1], G Bongiovì[1], F M Castrovinci[1], S D'Amico[2], P A Di Maio[1], I Moscato[2], A Quartararo[1]

[1]Department of Engineering, University of Palermo, Viale delle Scienze, Ed. 6, 90128 Palermo, Italy
[2]EUROfusion Consortium, PPPT Department, Boltzmannstr. 2, Garching, Germany

* Corresponding author: eugenio.vallone@unipa.it

Abstract. The European DEMO reactor will work under normal operating conditions in accordance with a pulsed duty cycle. However unplanned and planned transients of plasma overpower may occur compromising the integrity of its plasma-facing components structures. Consequently, adopting appropriate tools is essential to accurately and consistently model the thermal-hydraulic response of the involved cooling systems under both normal operating conditions and accidental events. Given this background, the University of Palermo, in collaboration with EUROfusion, started a research work to study the thermal-hydraulic behaviour of the Primary Heat Transport System (PHTS) of the Helium Cooled Pebble Bed Breeding Blanket (HCPB BB) of the DEMO device under steady-state and transient conditions. The activity has been performed adopting a computational approach, employing the thermal-hydraulic system code TRACE version 5.0 patch 6. The key point of the work has been the code-to-code benchmark with the outcomes previously obtained with the RELAP5-3D code, to estimate the impact of the physical models, numerical resolution schemes and modelling techniques adopted on the predictive capabilities of the system codes considered. The models and the analysis results are presented and critically discussed herein.

1. Introduction

By the middle of this century, in agreement with the European roadmap, nuclear fusion is going to be part of the energy mix for electricity production [1]. To achieve this goal, the EUROfusion consortium is working on the development of a DEMOnstration fusion reactor (DEMO) following the ITER path in exploring fusion energy. Unlike ITER reactor, DEMO is conceived to provide electricity to the grid. Since the energy conversion and cooling systems of the tokamak ([2], [3]) have a fundamental role to the realization and the licensing of the plant, a detailed assessment of their features is needed.

Based on the tokamak design, the EU-DEMO operates according to a pulsed operating cycle. These operating conditions induce thermal and mechanical cycling and could potentially compromise the qualified lifetime of the relevant equipment. Furthermore, planned and unplanned overpower transients of the plasma could put the integrity of the plasma-facing components structures at risk [4]. Thus, it is crucial to have adequate computational tools which can accurately and consistently model the thermal-hydraulic features of the cooling systems of these kinds of machines, during both major operational and accidental transients.

Thermal-hydraulic system codes, like RELAP5-3D [5], developed by Idaho National Laboratories, and TRACE [6], developed by the U.S. Nuclear Regulatory Commission, are extensively used to characterize the thermal-hydraulics of operational and innovative nuclear fission power plants, and their use in this field is widely validated by the experience of nuclear power plants that operated in the second half of the 20th century and those that are currently in operation and by validation campaigns with experimental data on innovative reactors.

At present, the study of the predictive capabilities and, therefore, the verification of the potential use of thermal-hydraulic system codes in the field of fusion reactors is severely limited by the scarce availability of experimental data. A powerful tool currently available to the research community involved in Research and Development (R&D) on fusion technologies to assess the impact on the predictive capacities of a system code of the physical models, numerical resolution schemes and modelling techniques adopted is indeed the code-to-code benchmark.

Within this framework, University of Palermo started a thermal-hydraulic research programme to assess the Primary Heat Transport System (PHTS) in a DEMO device equipped with the Helium Cooled Pebble Bed (HCPB) Breeding Blanket (BB) ([7], [8]). The activities have been carried out in collaboration with EUROfusion. A computational methodology has been adopted to conduct the thermal-hydraulic analyses, and the thermal-hydraulic system code TRACE (version 5.0 patch 6) has been used to simulate both steady-state and transient conditions. First efforts have been focused on creating a TRACE nodalization which could realistically predict the response of the system with an acceptable computational time. Afterwards, the core point of the activity has been the code-to-code benchmark with results previously obtained with the RELAP5-3D code [9] to estimate the impact of the physical models, numerical resolution schemes, and modelling techniques adopted on the predictive capabilities of the system codes considered and to obtain a preliminary validation of their use in the field of fusion reactors. The models, assumptions and analysis results are herein presented and critically argued.

2. The design of the HCPB BB PHTS

The latest baseline design of the HCPB BB for the DEMO machine has 16 identical sectors. Each sector is composed of 5 BB segments, namely Left Inboard Blanket (LIB), Right Inboard Blanket (RIB), Left Outboard Blanket (LOB), Central Outboard Blanket (COB) and Right Outboard Blanket (ROB) [7]. In Figure 1, a general scheme of one of the 16 sectors is shown.

A EUROFER97 exterior box consisting of a First Wall (FW), bottom and top caps, sidewalls, and the backplate containing the Breeding Zone (BZ) makes up a typical HCPB BB section. An Advanced Ceramic Breeder (ACB) is present in the BZ and positioned in the fuel-breeder pins configuration. Its main function is the tritium production. The ACB is positioned within the inner and outer cladding of a pin, which is made up of two concentric tubes. The pins serve as structural components to prevent pressurisation in the event of a potential LOCA scenario. They are placed within pressure tubes that link the FW to the BZ backplate. The Be12Ti is positioned outside the pressure pipe holding the pins in the shape of hexagonal prismatic blocks and functions as a neutron multiplier [10]. Pressurized helium at 8 MPa and an inlet temperature of 300 °C cools the FW and BZ zones. First flowing from the FW inlet manifold to its cooling channels, helium is then delivered to the pins via the BZ inlet manifold. Inside the BZ inlet manifold it is chilled in a counterflow system to equalise the temperature distribution. It finally reaches the BZ pins at approximately 520 °C into the outlet plenum and sent to the BZ outlet manifold [7].

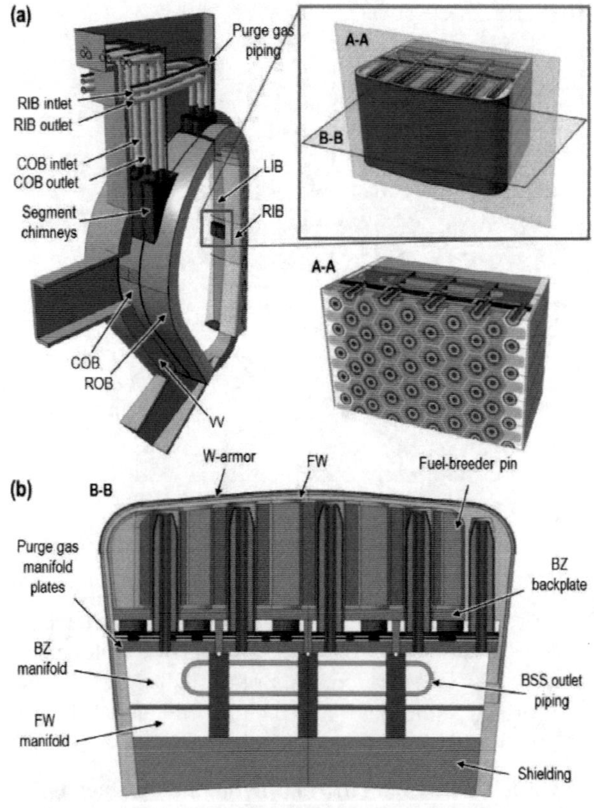

Figure 1. Typical HCPB BB Segment [1].

The thermal power removal is the main goal of the HCPB BB primary circuit. From the components inside the vessel, it is transferred via the Intermediate Heat eXchangers (IHXs) to an Intermediate Heat Transport System (IHTS). At an 8 MPa inlet pressure and operating temperatures of 300÷520 °C, it is intended to extract around 2029 MW of thermal power using helium [7]. Table 1 shows the principal data of the HCPB BB primary circuit.

Table 1. Primary circuit main parameters.

Parameter	Value
Coolant [-]	Helium
Inboard Blanket segments [-]	32
Outboard Blanket segments [-]	48
Inboard Blanket power [MW]	575.4
Outboard Blanket power [MW]	1453.7
I/O Blanket inlet pressure [MPa]	8.0
I/O Blanket inlet temperature [°C]	300.0
I/O Blanket outlet temperature [°C]	520.0
I/O Blanket pressure drop [kPa]	79.9

The HCPB BB PHTS is divided in eight distinct cooling loops in its current configuration [2]. Each loop feeds two blanket sectors and consists of In-vessel BB cooling circuits, an IHX, two circulators, and the piping that connects these sections. Figure 2 illustrates the HCPB BB PHTS [2].

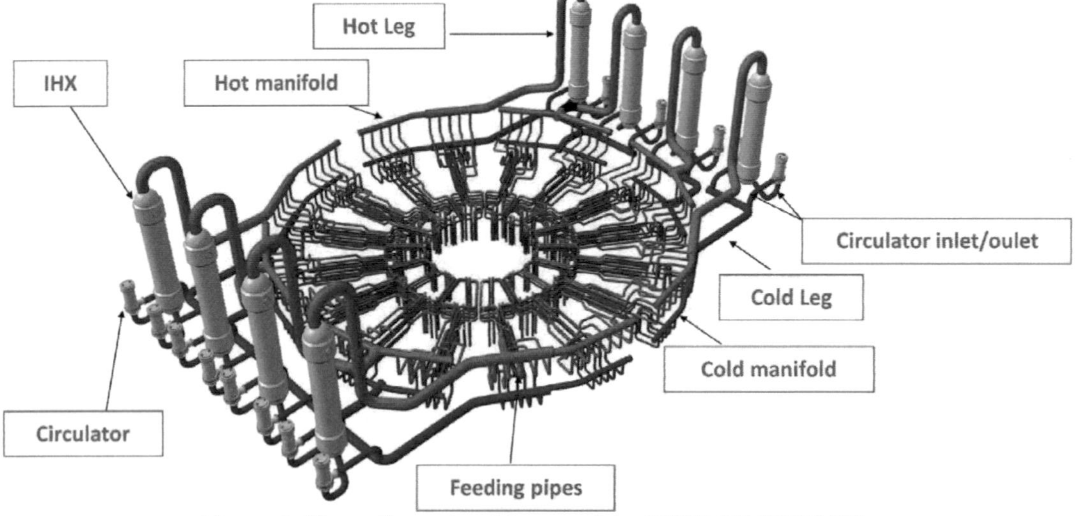

Figure 2. Three-dimensional layout of the HCPB BB PHTS [2].

In the tokamak cooling rooms, which are situated on the upper level of the Tokamak Building, the eight IHXs are placed on two opposing sides (see Figure 2). Every IHX is installed in a row and distributed equally at the same location [2]. The IHX thermally couples the HCPB BB PHTS to the IHTS. The main design option for the IHX consists of a shell and tube type [9], with the helium circulating on the tube side (primary side) and the HITEC® molten salt [11] flowing in the shell side (secondary side). In Table 2 the key thermal-hydraulic features of the latest IHX design are presented.

Table 2. IHX main thermal-hydraulic parameters.

Parameter	Helium	Molten salt
Inlet pressure [MPa]	7.89	0.8
Inlet temperature [°C]	520.0	270.0
Outlet temperature [°C]	290.9	465.0
Mass Flow Rate [kg/s]	222.2	867.9
Pressure Drop [kPa]	60.7	462.7

The IHXs lower head is not far from the helium circulators. Each circulator has two short pipes that connects it upstream to the IHX and downstream to the cold leg [2].

3. Model setup

As the HCPB BB PHTS is composed by eight independent circuits, attention has been focused on modelling only one loop of the HCPB BB PHTS in analogy to what has been done in [9] and simulating its thermal-hydraulic response under the transient conditions expected during the duty cycle.

The main goal has been to create an accurate finite-volume model representing all of the significant thermal-hydraulic properties that distinguish both components inside and outside the vessel, in compliance with the TRACE system code criteria. The finite volume model has been developed starting from the RELAP5-3D model described in [9] in order to obtain valuable indications from the benchmark between the two system codes. More in detail, the model is made of four different sub-models.

The cooling circuit architecture is approximated in quasi-2D by the geometrical sub-model. The flow domain nodalization has been done while maintaining the volume of every component, enabling an accurate modelling of the total amount of coolant. In addition, sub-volumes have been appropriately positioned and modelled to replicate their respective heights and positions. Therefore, concerning the geometrical factors involved in their evaluation, both gravitational effects and distributed pressure drops have been accurately represented. Additionally, the model nodalization has been created to accurately forecast the selected loop overall thermal-hydraulic behaviour while requiring allowable calculation time. Figure 3 shows the nodalization of the single loop of the HCPB BB PHTS.

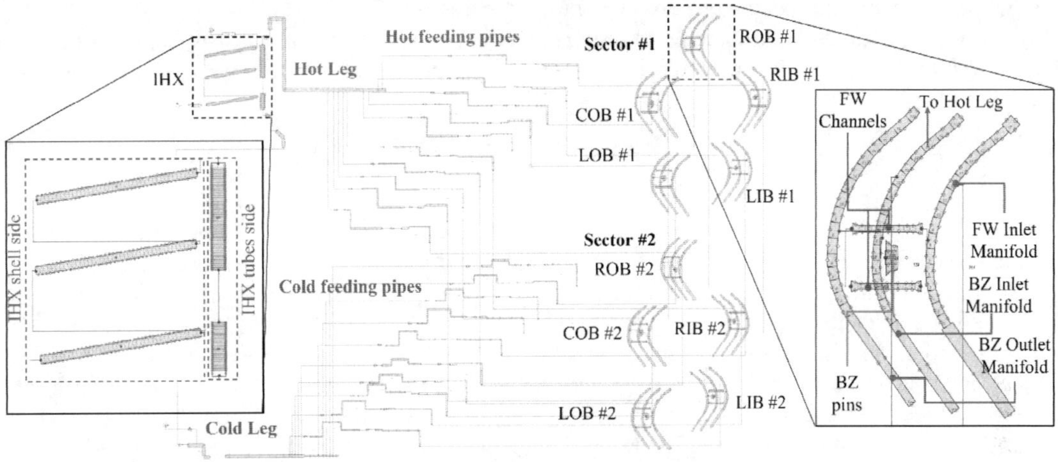

Figure 3. Nodalization of the HCPB BB PHTS cooling loop.

The constitutive sub-model explains how the thermophysical features of the adopted fluids depend on temperature and pressure and how the thermophysical properties of the structural materials present in the model depend on temperature. In this respect, as far as the helium is concerned, the properties dependence on pressure and temperature is based on TRACE libraries while, regarding the HITEC® molten salt, its thermophysical properties have been introduced within the TRACE code [11] as an external material library. Furthermore, several materials have been considered to accurately simulate the thermal response of structural elements inside and outside the vessel. Regarding the BB structural materials, their thermophysical properties, e.g. EUROFER and Tungsten ([12] and [13]), have been inserted in the TRACE code as tabular temperature functions. As far as the IHX is concerned, only the tube bundle structures, which are made of Alloy-800H [14], have been simulated.

The goal of the hydraulic sub-model is to replicate the fluid flow through the systems. The model described in [9] has been considered as a reference to assess the hydraulic performance of the helium in the PHTS side and the HITEC® on the shell-side of the IHX while obtaining valuable indications from the benchmark between the two system codes.

The aim of the thermal sub-model is to replicate the heat transport processes via the cooling systems as accurately as possible. All the structural components have been modelled in order to accurately simulate their heat capacity. The heat transfer phenomena have been replicated by using appropriate heat transfer models. Specifically, the widely used Gnielinski correlation, which is incorporated into the TRACE code, has been employed in conjunction with the Bell-Delaware procedure, a method that has been also used in [9], to replicate the heat transfer characteristics of the IHX. The TRACE model incorporates both the nuclear heating, and the FW heat fluxes in its thermal loads, which vary according to the DEMO duty cycle [9].

4. Thermal-hydraulic analyses

The benchmark between the RELAP5-3D analysis [9] and the purposely set-up TRACE model has been conducted considering the nominal DEMO duty cycle, as reported in [9].

It is important to note that the TRACE field equations have been developed with the idea that viscous shear stresses are negligible, at least in the first approximation, and therefore explicit turbulence modelling is not coupled to the conservation equations [6]. Although this assumption can generally be considered acceptable, it can lead to significant errors when the working fluid is a gas and one is dealing with complex circuits such as those envisaged for the DEMO device.

Aiming at assessing the effect of this assumption associated with the formulation of the field equations adopted in TRACE and at the same time overcoming its drawbacks, two models have been developed: a base model, called Model I, and an alternative model, called Model II, in which a fictitious thermal power source has been added to the helium domains.

In particular, in order to evaluate this additional power source, the helium coolant domain of the BB PHTS circuit has been divided into r regions. In each of these regions, the following heat balance may be applied:

$$\sum_{i=1}^{n} G_i h_i - \sum_{j=1}^{p} G_j h_j + \sum_{k=1}^{m} Q_k = W_r \tag{1}$$

where W_r is the additional power for the r-th region, G_i and h_i are respectively the mass flow rate and the fluid enthalpy of the i-th inlet section and G_j and h_j are respectively the mass flow rate and the fluid enthalpy of the j-th outlet section while Q_k is the heat transferred from the k-th heat structure of the r-th region to the helium coolant. Then, the additional energy source calculated for the r-th region is evenly distributed over the r-th region.

Furthermore, as far as the helium mass flow rate is concerned, it has been varied at a rate of 12.5 percent per minute throughout the ramp-up and ramp-down phases, from its largest value during the pulse period (see Table 2) to its minimum of 83.34 kg/s for the duration of the dwell time (Figure 4). The mass flow rate of the secondary side is calculated by the code by a control system adopting a proportional integral control. The aim is to keep the inlet temperature of the BB system at 300°C (see Figure 5).

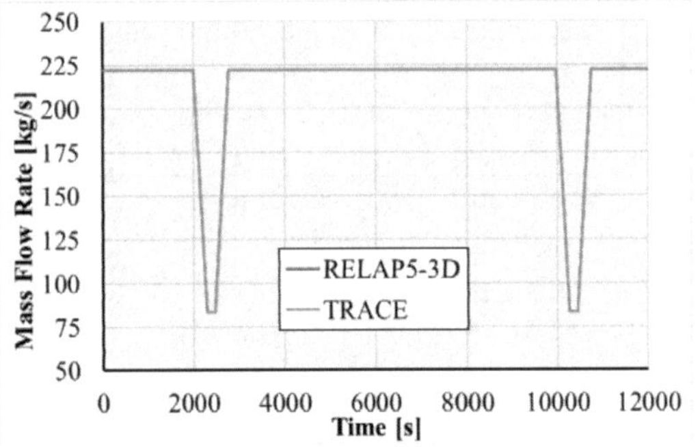

Figure 4. Primary side mass flow rates.

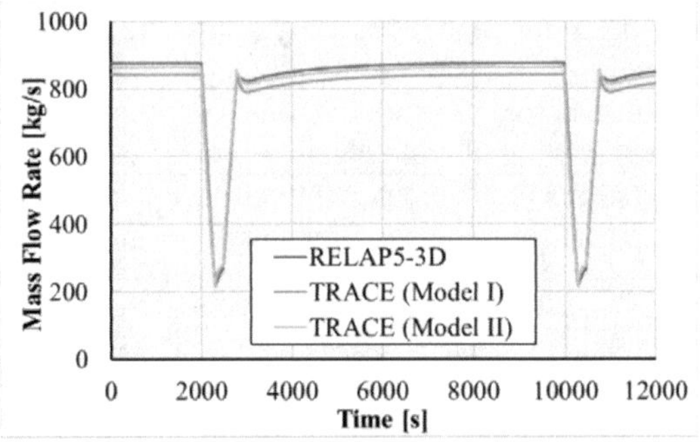

Figure 5. Secondary side mass flow rates.

The time evolution of the inlet pressure and inlet temperature of the cold leg and the inlet temperature of the hot leg are reported in figures from Figure 6 to Figure 8, respectively, while Figure 9 gives the temperature trends along the axis of the intermediate heat exchanger during the pulse period. Furthermore, Table 3 shows the main results for the pulse period compared with the design data.

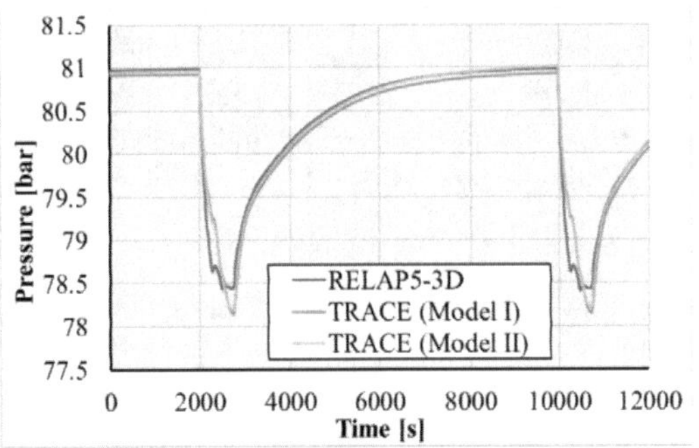

Figure 6. Cold leg inlet pressure.

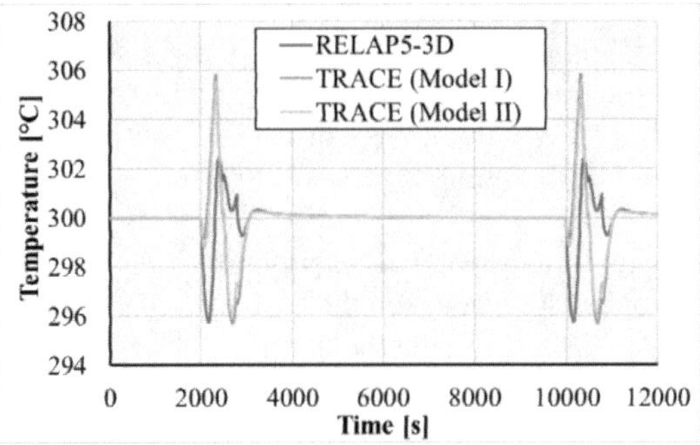

Figure 7. Cold leg inlet temperature.

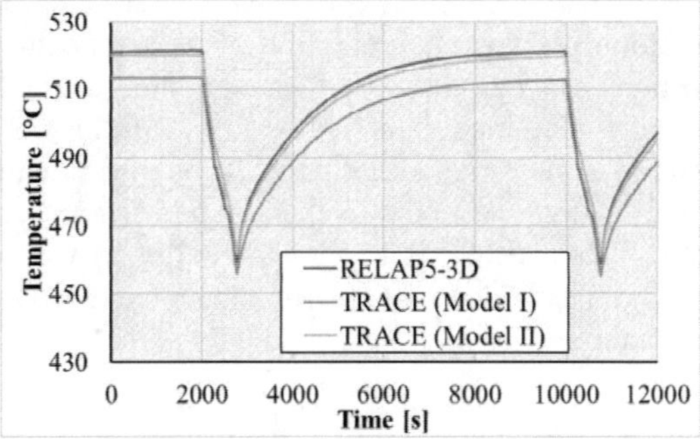

Figure 8. Hot leg inlet temperature.

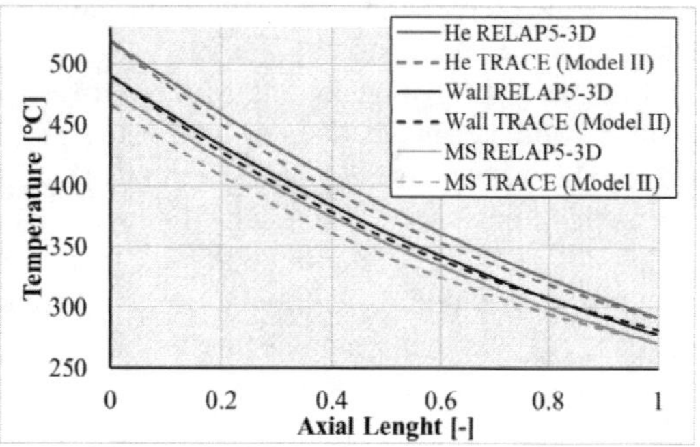

Figure 9. IHX temperature profiles.

Table 3. Comparison of main numerical results.

Parameter	Design	RELAP5-3D	TRACE (Model I)	TRACE (Model II)	$\varepsilon_{RELAP5\text{-}3D}$ [%]	$\varepsilon_{Model\ I}$ [%]	$\varepsilon_{Model\ II}$ [%]
G_{Helium} [kg/s]	222.2	222.2	222.2	222.2	0.00%	0.00%	0.00%
$G_{Molten\ Salt}$ [kg/s]	867.9	876.09	842.68	864.60	0.94%	-2.91%	-0.38%
$p_{Cold\ Leg}$ [bar]	78.29	80.98	80.91	80.96	3.44%	3.35%	3.41%
$T_{Cold\ Leg}$ [°C]	300.00	300.00	300.00	300.00	0.00%	0.00%	0.00%
$T_{Hot\ Leg}$ [°C]	520.00	521.43	513.45	520.25	0.27%	-1.26%	0.05%

From the analysis of the results, it can be inferred that the time trends of the main quantities examined are very close to those previously estimated by the RELAP5-3D calculation [9] with the exception of the temperature trend at the hot leg inlet for the Model I results. This last point is related to what has been mentioned above regarding the formulation of the field equations adopted in TRACE which translates in practice into a numerical heat sink distributed along the circuit. The introduction of a fictitious thermal power source in Model II significantly improves the quality of numerical results. This is further confirmed by the comparison with design values that refer to hypothetical steady-state conditions analogous to those occurring at the end of the pulse phase (see Table 3) and the temperature trend along the axis of the intermediate heat exchanger at the end of the pulse period (see Figure 9).

It is worth emphasising that the two codes examined have slightly different temperature and pressure trends at the inlet of the cold leg during the dwell period. This point could be connected with the slightly different definition of proportional-integral control between the two codes. Nonetheless, it can be seen from the results that this does not significantly affect the timing of the transient.

5. Conclusions

In accordance with the EUROfusion Consortium initiative, the University of Palermo initiated an investigation of the thermal-hydraulic response of the HCPB BB PHTS ([7], [8]) of the EU-DEMO reactor during steady-state and transient conditions. A computational approach has been adopted to perform the activity and the TRACE thermal-hydraulic system code has been adopted. The core point of the activity has been the code-to-code benchmark with the results previously obtained with the RELAP5-3D code [9] to estimate the impact of the physical models, numerical resolution schemes, and modelling techniques adopted on the predictive capabilities of the system codes considered and to obtain a preliminary validation of their use in the field of fusion reactors. In this context, to assess the effect of a potential weakness identified in the TRACE system code with respect to RELAP5-3D and associated with the particular formulation of the field equations adopted in TRACE, and at the same time overcome its drawbacks, two models have been developed: a base model, called Model I, and an alternative model, called Model II, in which a fictitious thermal power source has been added to the helium domains.

The results have highlighted that the time trends of the main quantities examined are very close to those previously estimated by the RELAP5-3D calculation [9] with the exception of the temperature trend at the hot leg inlet for the Model I results. In practice, the particular formulation of the field equations adopted in TRACE results in the presence of a numerical heat sink distributed along the circuit, which becomes evident when the working fluid is a gas and one is dealing with complex circuits such as those envisaged for the DEMO device. The introduction of a fictitious thermal power source in Model II significantly improves the quality of numerical results. This is further confirmed by the comparison with design values that refer to hypothetical steady-state conditions analogous to those occurring at the end of the pulse phase.

Acknowledgements

This work has been carried out within the framework of the EUROfusion Consortium, funded by the European Union via the Euratom Research and Training Programme (Grant Agreement No 101052200 – EUROfusion). Views and opinions expressed are however those of the author(s) only and do not necessarily reflect those of the European Union or the European Commission. Neither the European Union nor the European Commission can be held responsible for them.

References

[1] Donné T and Morris W 2018, *European Research Roadmap to the Realisation of Fusion Energy*, p. 74 ISBN: 978-3-00-061152-0
[2] Moscato I et al. 2022 Tokamak cooling systems and power conversion system options *Fus. Eng. and Des.* **178** 113093
[3] Barucca L 2022 Maturation of critical technologies for the DEMO balance of plant systems *Fus. Eng. and Des.* **179** 113096
[4] Maviglia F 2022 Integrated design strategy for EU-DEMO first wall protection from plasma transients *Fus. Eng. and Des.* **177** 113067
[5] Idaho National Laboratory 2018 *Models and Correlations (RELAP5-3D Code Manuals)*
[6] U.S. Nuclear Regulatory Commission 2010 *TRACE V5.0 Theory Manual*
[7] Spagnuolo G A et al. 2021 Integrated design of breeding blanket and ancillary systems related to the use of helium or water as a coolant and impact on the overall plant design *Fus. Eng. and Des.* **173** 112933
[8] Boccaccini L V et al. 2022 Status of maturation of critical technologies and systems design: Breeding blanket *Fus. Eng. and Des.* **179** 113116
[9] D'Amico S et al 2021 Preliminary thermal-hydraulic analysis of the EU-DEMO Helium-Cooled Pebble Bed fusion reactor by using the RELAP5-3D system code *Fus. Eng. and Des.* **162** 112111
[10] Hernàndez F A et al. 2019 Advancements in the helium cooled pebble bed breeding blanket for the EU DEMO: holistic design approach and lessons learned *Fus. Sci. and Tec.* **75** pp 352-364
[11] Coastal Chemical Co. L.L.C. *HITEC® Heat Transfer Salt*
[12] Gillemot F et al. 2016 *Material Property Handbook pilot project on EUROFER97 (MTA EK, KIT)* EUROfusion IDM Ref.:2MRP77
[13] *ITER Material Properties Handbook* ITER Document No. G74 MA 16
[14] ASME 2015 *ASME Boiler and Pressure Vessel Code - 2015* ASME BPVC - Section II - Subsection D ASME

Thermal performance of clay bricks with different fillings: a CFD analysis

F Merli[1,*], S Susta[2] and C Buratti[1]

[1] Engineering Department, University of Perugia, Via G Duranti 93, Perugia 06125, Italy
[2] F.B.M. – Fornaci Briziarelli Marsciano, Via XXIV Maggio I, Marsciano 06055, Italy

* Corresponding author: francesca.merli@unipg.it

Abstract. The thermal conductivity of 15 loadbearing and infill bricks was calculated through a Computational Fluid Dynamic (CFD) analysis. Clay bricks filled with air, expanded polystyrene EPS, and graphitized expanded polystyrene were selected from the catalogue of the manufacturer F.B.M. – Fornaci Briziarelli Marsciano. The company provided the geometric features of each block and the thermal characteristics of the clay mix and the filling. The thermal conductivity value of each air cavity was calculated in compliance with UNI EN ISO 6946. Each volume was discretized in triangular meshes. The equivalent thermal conductivity of each block was calculated starting from the simulation of the isotherms trend inside the sample and from the heat flux passing through it. The obtained results were discussed depending on the filling, the geometry, and the hole percentage. In general, thermal conductivity increases with hole percentage in the bricks with air. Similar geometries have similar thermal behaviour. EPS or graphitized EPS involve better thermal performance as the hole percentage increases. Finally, the influence of the different orientations of the blocks was investigated. Equivalent thermal conductivity values increase for horizontally-brick placed, especially in homogeneous and compact geometries.

1. Introduction

In recent years, there was a growing emphasis on energy efficiency and sustainable constructions, considering that buildings account for more than 40% of social energy consumption [1]. For this reason, building energy saving has a great relevance in order to reduce the energy crisis and protecting the environment [2]. Since heat loss in building envelopes accounts for 60-80% of the total loss [3], it is crucial to create a high level of comfort and to reduce the energy consumption by improving the thermal performance of the envelopes, particularly of the opaque walls. As a result, researchers and engineers are working to develop materials that offer enhanced thermal insulation properties while minimizing energy consumption. Among these, hollow bricks represent a promising standard solution able to provide excellent insulation, due to the hollow cores. However, further improvements in their thermal performance are still desirable.

Expanded polystyrene (EPS) is a widely used insulation material known for its low thermal conductivity (0.033 – 0.039 W/mK) and lightweight properties [4]. It was successfully employed in several applications to enhance the thermal resistance of building envelopes. Recently, researchers started to explore the potential of incorporating EPS into hollow bricks as a filling material to enhance their thermal insulation performance. The introduction of EPS as a filling material has the potential to significantly avoid the convective heat transfer within the holes, resulting in a high thermal resistance [5]. Moreover, the field of materials science has witnessed the advent of graphitized EPS, a novel derivative of EPS with enhanced thermal conductivity properties (λ=0.031 W/mK) [6]. It opens new possibilities for enhancing the thermal performance of hollow bricks by incorporating graphitized EPS as a filling material.

However, despite the potential benefits of using air, EPS and graphitized EPS in hollow bricks, a comprehensive and systematic understanding of their effects on the thermal performance of such building materials is still lacking. This knowledge gap underscores the need for systematic investigations that evaluate and compare the thermal characteristics of hollow bricks filled with EPS and graphitized EPS. Therefore, this study aims to address this research gap by investigating the effect of air, EPS and graphitized EPS filling on the thermal performance of hollow bricks. Equivalent thermal conductivity values of several bricks were calculated by means of CFD analysis and the results were discussed taking into account the geometry, the hole percentage, and the orientation of the blocks, in order to establish the correlations between the features of the block and its thermal performance.

2. Materials and Methods

Several clay bricks were selected from the catalogue of the manufacturer F.B.M. – Fornaci Briziarelli Marsciano [7] (Table 1, Figure 1). They are both load-bearing bricks (BTP) and partition clay blocks (BTT), some with joints (inc) and some smooth (lis) among the most marketed. The clay mixture used to produce the bricks is characterized by a thermal conductivity λ equal to 0.328 W/mK, a density ρ equal to 1800 kg/m^3, and a specific heat Υ of 937 J/kgK. Air, EPS (λ = 0.038 W/mK, ρ = 30 kg/m^3 and Υ = 1450 J/kgK), and graphitized EPS (λ = 0.031 W/mK, ρ = 20 kg/m^3 and Υ = 1450 J/kgK) were considered to fill the holes in each block. The geometric features of the bricks and those of the clay mixture and fillings were provided by the company. The dimensions are in the 45x30x18 cm^3 (brick 1) – 30x25x18 cm^3 (brick 15). Blocks 2, 5-9, 11, 12, 14, 15 are characterized by homogeneous holes (height in the 8.8 – 25.4 mm range), whereas holes of different sizes are present in the bricks 1, 3, 4, 10, 13. Hole percentages are in the 44.65 (brick 8) – 65.53 (brick 11).

Table 1. Features of the selected bricks.

N.	Description	Hole percentage [%]	Hole height [mm]
1	BTP 45x30x18	53.93	23.0/47.0[a]
2	BTT inc 42x30x18	62.97	14.3
3	BTT inc 42x30x18	62.27	14.3/72.1[a]
4	BTP (36+6)x25x18	49.37	9.3/57.7[a]
5	BTP 42x25x18	46.71	9.3
6	BTT 38x30x18	60.55	11.3
7	BTT inc 38x30x18	60.32	11.3
8	BTP 38x25x18	44.65	8.8
9	BTT inc 35x30x18	60.6	11.5
10	BTT inc 35x30x18	65.15	11.5/72.1[a]
11	BTT 30x25x25	65.53	25.4
12	BTT_10s 30x25x25	57.59	18.8
13	BTP 30x25x18	45.40	14.6/38.8[a]
14	BTT_14s inc 30x25x18	57.95	14.3
15	BTT_14s lis 30x25x18	53.86	12.5

[a] small/large hole height.

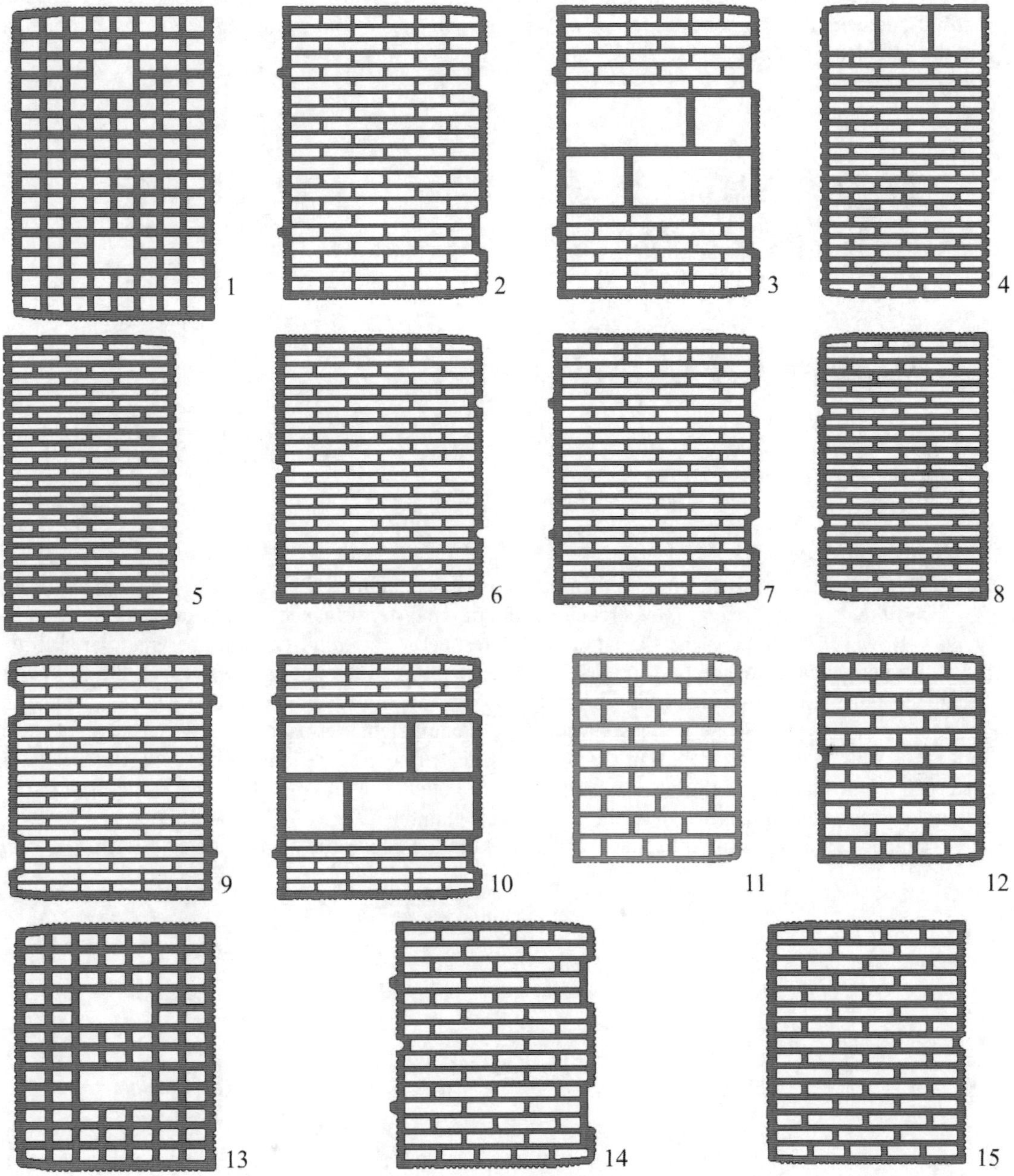

Figure 1. The investigated clay bricks.

Thermal resistance of the air gaps in each brick was calculated in compliance with UNI EN ISO 6946 [8]. The analyses were carried out by means of a finite volume approach. The 2D model from the external CAD software was imported in the software component called *Design Modeler*. Each continuum was discretized into elements of finite volume, choosing triangular meshes with an average element size of 0.001 m, as shown in Figure 2-a (block 14, as an example). The boundary conditions of the model (right and left adiabatic sides, convective conditions on the upper and lower sides) and the features of the materials (thermal conductivity, density, and specific heat of the clay, air/EPS/graphitized

EPS) were defined in the dedicated software. The thermal resistance and the corresponding equivalent thermal conductivity λ_{eq} of each block was obtained by means of the simulated isotherms (Figure 2-b) and the calculated heat flux.

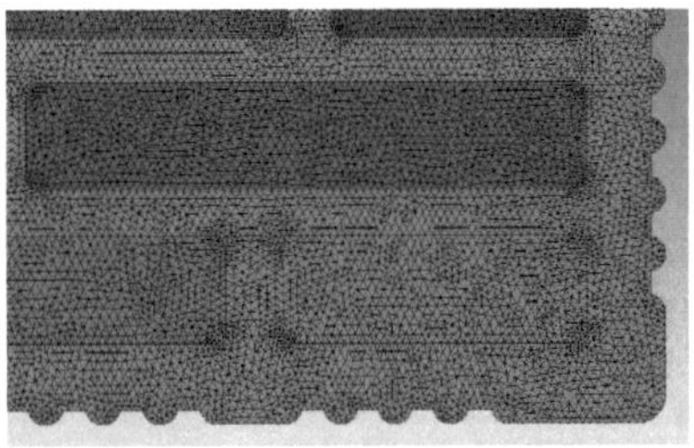
(a)

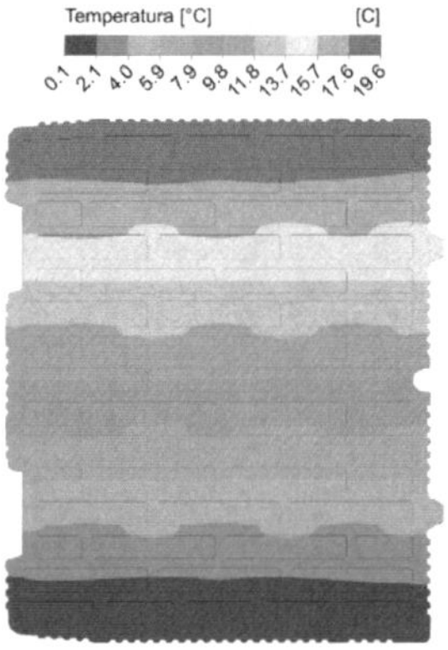

(b)

Figure 2. Example of meshes and isotherms.

3. Results and discussion

Figure 3 shows the equivalent thermal conductivity values obtained for the bricks with three different fillings of the gaps. Increasing hole percentage, the thermal performance of the bricks with air tend to worsen. However the trend is rather scattered with respect to the trend-line and the great variability is related to the geometry of the brick, especially the height of the holes, as discussed in the following. When the holes are filled with EPS or graphited EPS, an improvement in thermal performance is obtained as the hole percentage increases (reduction of equivalent thermal conductivity is in the 8 – 61% and 12 – 65% ranges with EPS and graphitized EPS, respectively). The trends with the insulation materials are practically parallel; lower λ_{eq}-values are obtained with graphitized EPS, according to the thermal performance of the materials; the reduction is about 4-12%.

A detailed discussion should be made with respect to the geometric characteristics (Figure 4).

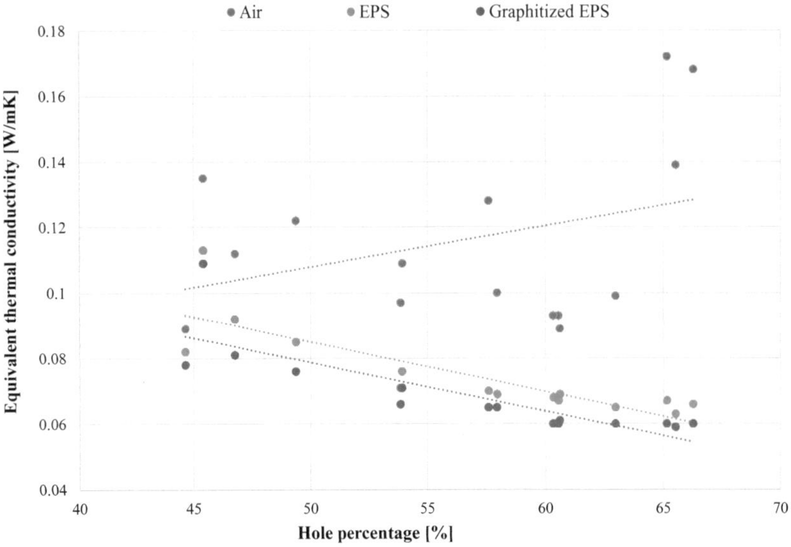

Figure 3. λ_{eq} vs. hole percentage: influence of the filling gaps.

Journal of Physics: Conference Series **2685** (2024) 012024 doi:10.1088/1742-6596/2685/1/012024

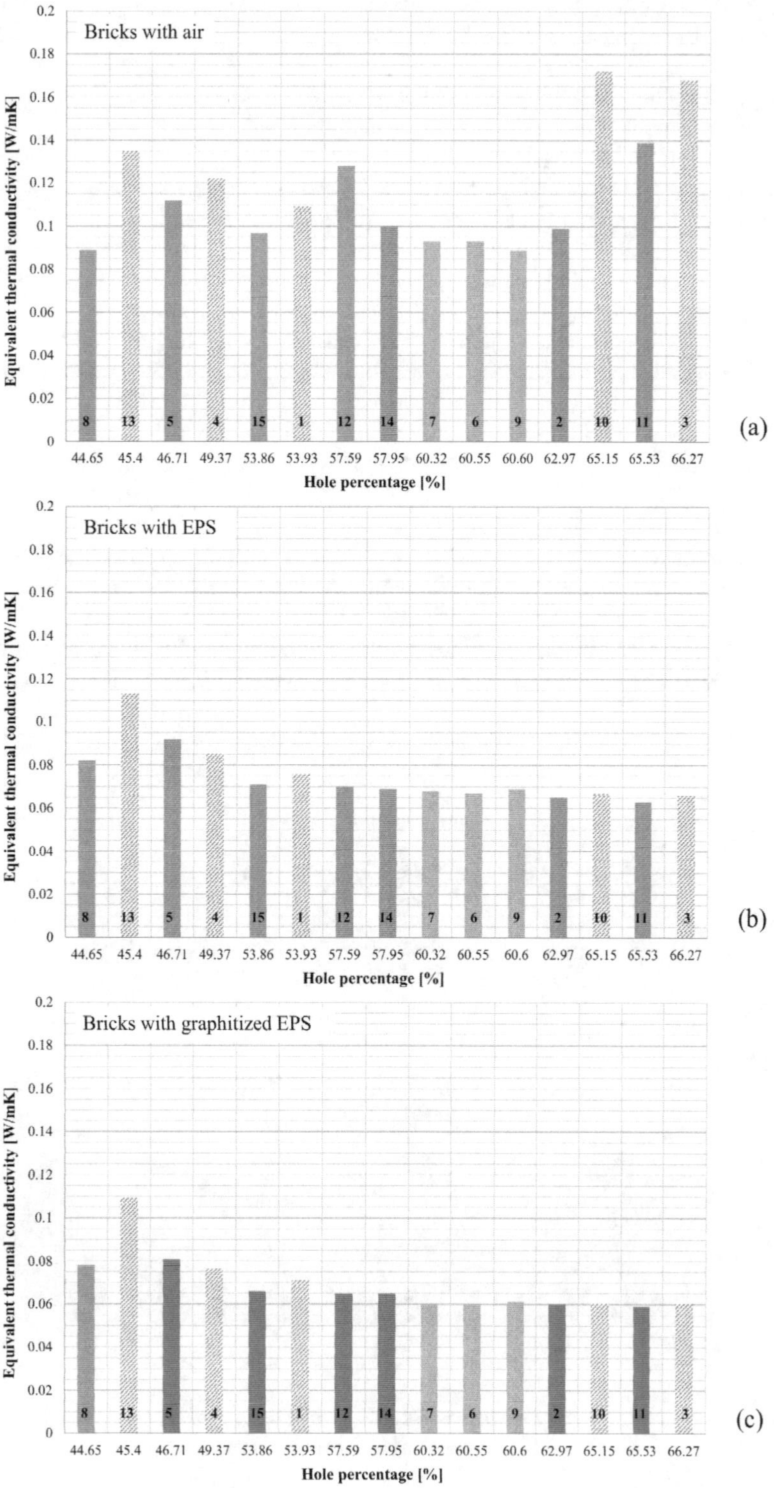

Figure 4. Thermal performance of the bricks: influence of filling and hole percentage.

Increasing the hole percentage in the air-filled bricks with very small holes (high hole of 8.8 and 9.3 mm in 8 and 5 bricks, respectively), λ_{eq} increases (Figure 4-a). When both the percentage and the height of the holes are about equal (yellow bars), similar thermal performance are obtained. For hole percentages in the 53.9 – 65.5 range, the higher holes (n. 15, 2, 14, 12, 11, blue bars) involve a worsening of the thermal performance of the brick. Conversely, the increase in drilling leads to a reduction in λ_{eq}-values, in geometries with non-homogeneous cavities (red dotted bars).

The increase in the percentage of both EPS and graphited EPS leads to an improvement in thermal performance, especially in bricks with non-homogeneous holes (λ_{eq} is reduced up to 42% and 45% with EPS and graphitized EPS from brick n. 13 to brick n. 3) (Figure 4-b and c, red dotted bars). A particular behavior is found in block 8 (hole percentage of 44.7%), which is characterized by the smallest gaps (8.8 mm): this behavior is probably due to a reduction in convection heat transfer. Similar geometry has comparable performance. The influence of the geometry is less marked as the performance of the material inside the cavities increases; with graphitized EPS, when the hole percentages is higher than 60%, the performance is approximately equal in all the bricks (λ_{eq}=0.059 – 0.06 W/mK) (Fig. 4-c).

The shaping does not affect the thermal properties with any filling (blocks 6 and 7 with equal size but one smooth and one with joints).

Furthermore, 4 blocks (n. 8, 13, 15, 1) which can be installed on site also according to a horizontal arrangement were selected. The comparison of the equivalent thermal conductivity values obtained with respect to the vertical position is shown in Figure 5. The horizontal position involves a worsen behavior in thermal performance, especially in the homogeneous and compact bricks (8 and 15); the thermal conductivity-value increases by 50 and 49% with air, 45 and 43% with EPS, and 47 and 45% with graphitized EPS for blocks number 8 and 15, respectively, regardless of geometry. In bricks with large holes (n. 13 and 1), the increase in λ_{eq} due to the horizontal position is much higher as the % of holes increases (from 26% to 40% with air, from 18% to 37% with EPS, from 19% to 26% with graphitized EPS).

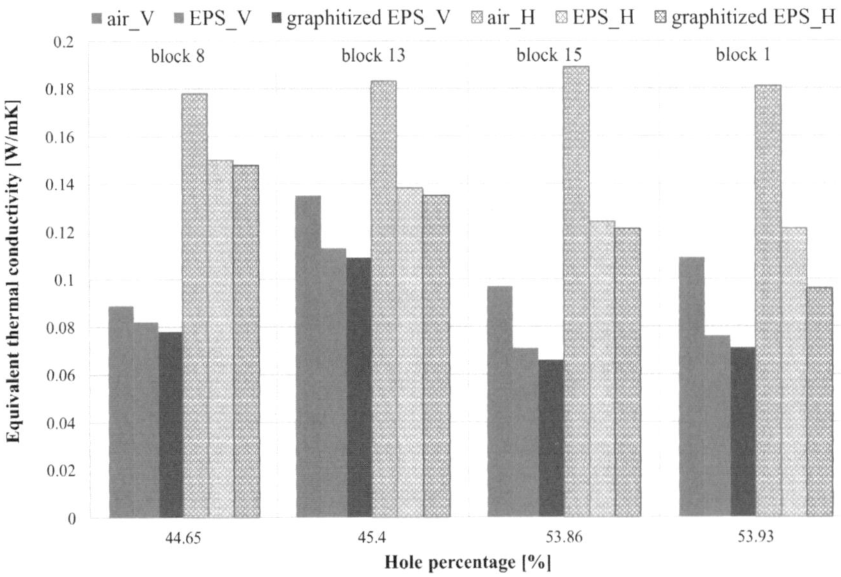

Figure 5. Comparison between λ_{eq} values obtained for blocks n. 8, 13, 15, and 1 in horizontal (H) and vertical (V) arrangment.

4. Conclusion

In the present study, thermal performance of 15 clay bricks in terms of equivalent thermal conductivity were calculated by means of CFD simulations. The brick-cavities are filled with air (thermal conductivity of each gap calculated according to UNI EN ISO 6946), EPS (λ=0.035 W/mK), or graphitized EPS (λ=0.031 W/mK). For each geometry, starting from the simulated trend of the isotherms and the calculated heat flux, the thermal resistance and the corresponding equivalent thermal conductivity values of the block were determined. A critical analysis of the results is presented in this paper and the following main conclusions are highlighted:

- increasing the hole percentage from about 45% to 66%, the thermal conductivity decreases when the cavities are filled with insulation materials (up to 44% and 46% with EPS and graphitized EPS, respectively), whereas it increases with air (up to 48%);
- in air-filled bricks, the height of the holes greatly influences the thermal performance: the higher holes result in a thermal deterioration (up to 30% with holes about 25-mm high);
- in geometries with non-homogeneous holes, the increase of the drilling with EPS and graphitized EPS involves a reduction of the λ_{eq}-values with three different fillings;
- similar geometry is characterized by similar thermal properties;
- bricks with holes greater than 60% filled with graphitized EPS have a similar thermal behavior;
- the horizontal positioning of the brick leads to an increase in thermal conductivity, especially in geometries with homogeneous and compact holes (increases up to 50, 45, and 47% with air, EPS, and graphitized EPS, respectively).

5. References

[1] Zhao H X and Magoulès F 2012 *A review on the prediction of building energy consumption. Renewable Sustainable Energy Rev.* **16(6)** 3586-3592.

[2] Ma G, Lin J, Li N and Zhou J 2017 *Cross-cultural assessment of the effectiveness of eco-feedback in building energy conservation. Energy Build.* **134** 329-338.

[3] Meng X, Luo T, Gao Y, Zhang L, Shen Q and Long E 2017 *A new simple method to measure wall thermal transmittance in situ and its adaptability analysis. Appl. Therm. Eng.* **122** 747-757.

[4] Hu W, Xia Y, Li F, Yu H, Hou C and Meng X 2021 *Effect of the filling position and filling rate of the insulation material on the insulation performance of the hollow block. Case Stud. Therm. Eng.* **26** 101023.

[5] Li J, Meng X, Gao Y, Mao W, Luo T and Zhang L 2018 *Effect of the insulation materials filling on the thermal performance of sintered hollow bricks. Case stud. Therm. Eng.* **11** 62-70.

[6] Lakatos Á 2023 *Thermal insulation capability of nanostructured insulations and their combination as hybrid insulation systems. Case Stud. Therm. Eng.* **41** 102630.

[7] Available online: https://www.fbm.it/.

[8] UNI EN ISO 6946: 2018. Building components and building elements - Thermal resistance and thermal transmittance - Calculation methods.

Acknowledgments

Authors wish to thank the F.B.M. – Fornaci Briziarelli Marsciano for allowing to conduct this research. Francesca Merli would like to thank the Italian funding program Fondo Sociale Europeo REACT EU – PON Research and Innovation 2014-2020 (D.M. n. 1062 of August 10, 2021) for supporting her research through project Messa a punto di una filiera per la produzione di pannelli isolanti eco-sostenibili mediante recupero degli scarti del legno utilizzato nella produzione di sistemi finestrati di un'azienda locale, in ottica di economia circolare.

Influence on energy demand of thickness, thermal conductivity, and volumetric heat capacity of ladle working lining in secondary steel-making process

M Neri[1,*], M Pilotelli[1] and A M Lezzi[1]

[1] Department of Mechanical and Industrial Engineering, University of Brescia, Brescia, Italy

[*] e-mail: manuela.neri@unibs.it

Abstract. The secondary steel-making process involves several steps during which steel is kept in a ladle, that is, a vessel made of an outer steel layer (carpentry), an intermediate refractory layer, and an internal refractory layer. Unlike the intermediate layer, the internal layer undergoes a progressive reduction in thickness and a periodic restoration. Traditionally, it is made of alumina or magnesite. During the process, the ladle undergoes unsteady heating and cooling; therefore, heat transfer depends on thermal conductivity and heat capacity. This study aims to identify the ladle internal layer characteristics that affect the energy demand. This analysis investigates the effect of the internal layer thickness S, volumetric heat capacity C, and thermal conductivity λ. Through the Design Of the Experiments (DOE), different scenarios have been selected and analyzed by means of numerical simulations performed on a numerical model defined in COMSOL Multiphysics. The energy demand as a function of the internal layer properties has been estimated, and it has emerged that low thermal conductivity and heat capacity values require a lower amount of energy.

Keywords: *Secondary steel making process, refractory materials, energy consumption, Design Of the Experiments, ladle.*

1. Introduction

1.1. The secondary steel-making process

The secondary steel-making process involves several steps during which steel is kept in a ladle as that schematized in figure 1. The ladle undergoes several production cycles which, according to figure 2, consist of several phases: preheating (only once at the beginning of a series of cycles), waiting while empty, filling with liquid steel poured from the electric arc furnace (EAF), heating using electrodes (LF), emptying in the continuous casting (CC), and cleaning. Along the process, the ladle undergoes heating and cooling which make the process unsteady and complex to be analyzed. During the steel-making process, several ladles are used in turns, and each of them undergoes between 70 and 90 cycles of use. In a standard cycle, before entering the production, gas burners placed inside the ladle preheat the internal surface to avoid thermal shocks and consequent breaks. If the wait between two cycles in a series is too long, the ladle is heated again, but this occurs rarely. After a certain period of waiting, the ladle is moved under the EAF furnace, where it is filled with steel and slag. Once filled, gas is continuously insufflated in the ladle to mix the steel and maintain its temperature as homogeneous as possible. Then, the ladle is sent to the LF station where ferroalloy is added to obtain the final composition, and energy (E_{LF}) is supplied using electrodes; since this is the last step where the steel temperature may be controlled, the energy supply must counteract energy losses due to heat transfer towards the ambient and the addition of materials at ambient temperature. Later, the ladle is closed with a lid to reduce energy losses and sent to the CC station where the steel is progressively drained and transformed into the final product. At this point, the ladle is sent to the cleaning station where slag and residuals of steel are removed. Finally, the ladle returns in front of the furnace to enter the next cycle.

In the literature, the secondary steel-making process was investigated from several points of view, with a particular interest in evaluating the heat transfer across the ladle surfaces [1-13]. Heat losses through the ladle were evaluated for the preheating phase with and without steel [3], for the tapping phase focusing on the effect of the slag characteristics (that is thickness, thermal properties, or by imposing a defined heat flux on the top surface) [4,5], and during the holding phase [6,7]. The study reported in [6] investigates the motion of particles in the liquid steel, and in [7] a simplified model for the heat loss from the molten steel to the refractory is proposed. Also, the thermal efficiency and mechanical strength of the ladle were analyzed [8,9]. The temperature distribution in the ladle was determined through a mathematical model in [10]. Other studies instead, analyzed the entire process, that is, from the tapping to the cleaning phases [11,12].

1.2. The ladle in the secondary steel-making process

Ladles are vessels made of an outer steel layer (carpentry), an intermediate refractory layer, and an internal refractory layer (figure 1). Unlike the intermediate layer, the working lining undergoes a progressive reduction in thickness and a periodic restoration. The intermediate layer is made of the so-called permanent lining, whereas the inner layer undergoes a progressive reduction in thickness and a periodic restoration: it is called either wear or working lining. Traditionally, the wear layer is made of alumina or magnesite, which differ in their thermophysical properties and manufacturing processes. The study presented in this paper analyses the energy demand E_{LF} in the secondary-steel-making process and, more precisely, how the characteristics of the working lining affect the energy demand in the LF facility. According to the study reported in [11], the energy demand in the LF depends on both thickness S and material used as working lining. However, the study was performed by considering real materials characterized by temperature-dependent properties, and this did not allow a full understanding of how single variables affect the energy demand. Given the complexity of the phenomenon, the numerical approach is the only viable method of investigation, since an experimental campaign would be prohibitively expensive, time-consuming, and influenced by production needs. Therefore, a simple but sound numerical model has been designed in COMSOL Multiphysics to simulate the process and focus on the working lining variables only. Experimental data have been used to validate the model as explained in [11]. The variables of the working lining have been analyzed using the Design Of the Experiments (DOE), a structured approach to collect data and to investigate the relationship between input and output variables [14].

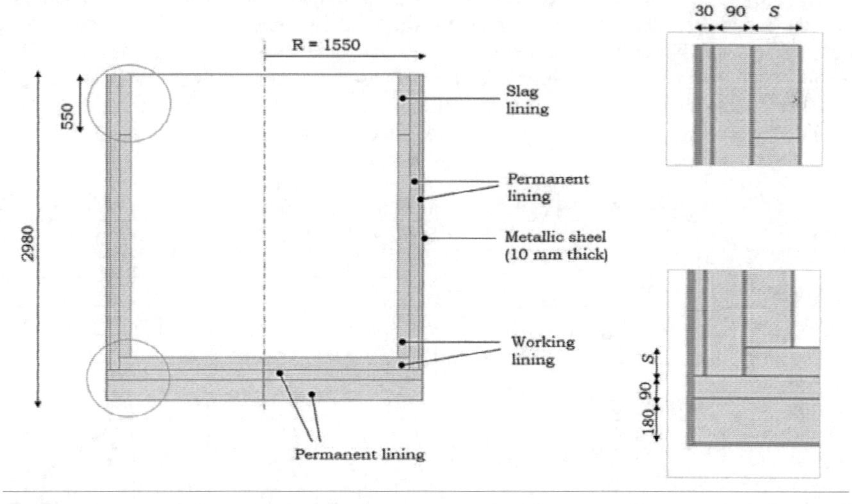

Figure 1. The geometry of the numerical model defined in COMSOL Multiphysics. Variations in the working lining properties and thickness S have been investigated. The dimensions are in millimeters.

2. Methods

The objective of this study is to provide manufacturers with suggestion on which material is the most efficient, that is, the materials that allow a greater energy savings. By means of numerical simulations, the effect of the working lining properties on the energy demand has been assessed. Several scenarios have been analyzed. They were identified through the Design Of the Experiments (DOE) technique, which allows to plan the tests to be performed. The DOE is generally used to plan experimental tests and it based on the principles of randomization, replication, and blocking

[15]; however, given that in this study data is obtained numerically, some assumptions could be neglected. Indeed, results obtained numerically are not affected by noise and sequence repetition.

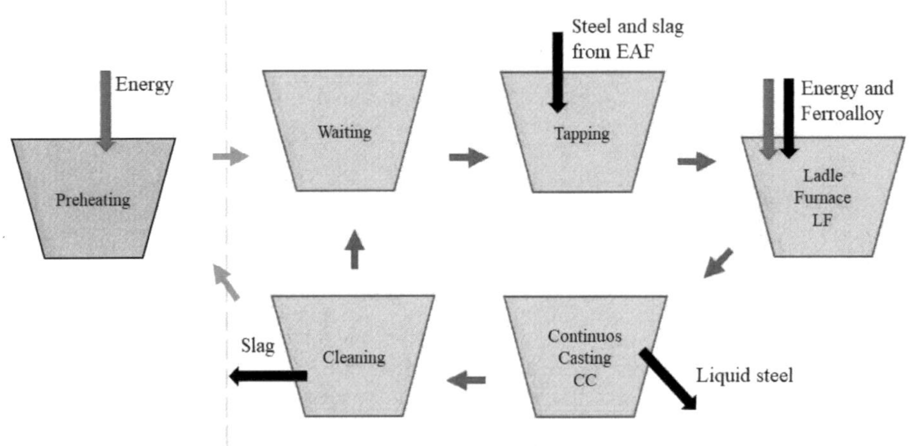

Figure 2. Phases that a ladle goes through in the secondary steelmaking process. After the cleaning phase, if necessary, the ladle can be resent to the preheating station.

2.1. Numerical model

Figure 1 shows the ladle geometry in which the working lining characteristics (thickness and thermal properties) varied while those of other layers remained constant. The process has been divided into cycles (*C*) and phases (*F*) as schematized in figure 3, and each phase has been modelled by means of one or more numerical simulations. To consider the variation of the liquid steel level along the process, the model geometry varied as shown in figure 3. The temperature detected at the end of each simulation has been set as initial conditions in the next one. The same initial volume of steel has been considered for the three thickness values considered, and this resulted in a higher steel free surface in the ladle with maximum thickness. The properties of the working lining have been identified as explained in section 2.2, while the properties of the other layers are reported in Table 1.

Table 1. Properties of the materials defined in the numerical model.

Material	ρ	c_p [kJ/(kg K)]	λ [W/(mK)]	ε [-]
Slag	2800 [13	0.52 [13]	0.48 [13]	0.6 [12]
Perm. lining (wall)	2700	1.05	2.3	-
Carpentry	7200 [13]	0.79 [13]	52 [13]	0.85
Perm. lining (bottom)	2100	1.05	2	-
Molten steel	6981	0.78	41	-
Lid internal	2390	1	1.35	0.45
Lid external	2390	1	1.35	0.52
Slag lining	3030	0.89 (200°C), 1.06 (700°C), 1.14 (1200°C)	8 (0°C), 6 (400°C), 4.8 (800°C), 4 (1200°C)	0.45

In the model, some simplifications have been adopted. The liquid steel convective motions in real ladles are forced by the insufflation of gas, but in the model the steel has been considered at uniform but time-dependent temperature by imposing an isothermal condition. Chemical composition effects have been neglected. On the ladle external surface, a mixed convective-radiative condition has been imposed (ambient temperature at 30 °C). The surface-to-surface radiation model is set on the internal surface of the ladle free from steel. Between steel and ladle, the perfect thermal contact condition has been imposed. Between the lid and the slag, the convective and radiative heat transfer is considered. The presence of the gas burners in the preheating phase *F1* has been modelled by imposing a temperature-time-dependent Dirichlet condition on the ladle's internal surface: the temperature *T* grows linearly from 30 °C to 1100 °C in 11 hours, while in the last 2 hours it is kept at 1100 °C. Then, the ladle waits ($F1_{bis}$) and in the filling phase (*F2*), it is full of steel and slag. In phase *F2*, the filling process has been considered instantaneous with an initial steel temperature of 1600 °C. Phase *F3* is representative of the process in the *LF* where energy E_{LF} is supplied to the steel. In real processes, E_{LF} varies depending on many variables. In our model it has been calculated in the following way.

Time duration of phase *F3* is fixed and equal to ΔT = 2400 s (40 minutes). During phase F3 a uniform and constant volumetric source term q''' is imposed to the steel. Through an iterative process, the value of q''' is determined as the value that allows the steel to reach the threshold temperature of 1595 °C in ΔT = 2400 s. The energy supplied is computed as $E_{LF} = q'''\cdot V\cdot \Delta t$, where V is the steel volume. In phase *F4*, the ladle is closed with a lid which is at an initial temperature of 250 °C. The emptying in the continuous casting CC (phase *F5*) has been analyzed through 11 numerical simulations in which the steel level has been progressively lowered by 200 mm from one simulation to the next one. The ladle in phases *F6* and *F7* is empty, and convection and radiation are set on all surfaces. To approach the steady condition, the simulation process has been repeated for several cycles. Therefore, for each case, 100 numerical simulations have been performed to simulate the steel making process in the first seven cycles: indeed, the first cycle *C1* counts eight phases (*F1-F7*) because of the preheating *F1* and waiting $F1_{bis}$ phases, while the following ones, *C2-C6,* only six phases (*F2-F7*).

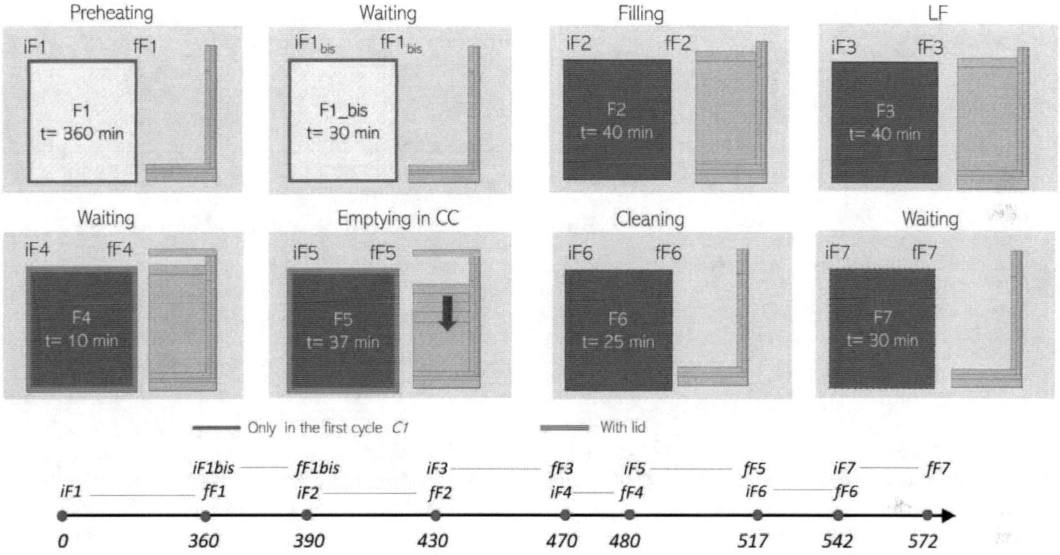

Figure 3. Representation of the phases in a cycle with their time duration. Phase *F5* has been divided into 11 sub-phases in which the level of the steel is progressively lowered. Phases *F1* and $F1_{bis}$ are present in the first cycle only, while the lid is present only in phases *F4* and *F5*.

2.2. Design Of the Experiments: selection of levels and factors

The Design Of the Experiments (DOE) is a technique used to systematically classify and quantify cause-and-effect relations between variables and outputs [14]. To apply this technique, it is necessary to identify the study objective, the variables affecting the phenomenon under analysis called *factors*, the range of the variables called *levels*, the experimental design type (fractional or full factorial design), and the type of experiments (experimental or numerical experiments). The objective of the study is to assess if and how the characteristics/properties of the working lining affect the energy demand in the secondary steel making process. In [11], it emerged that heat transfer through the ladle surface is influenced by thickness and working lining material. However, since real materials were considered and, therefore, temperature-dependent thermal properties, it is not understandable how density ρ, thermal conductivity λ, and specific heat c affect the energy demand. This type of analysis is performed in this paper, where each variable is analyzed separately. Through the Design Of the Experiments (DOE) technique, different scenarios have been identified. As can be seen in figure 4, three factors have been considered, that are, thickness S, thermal conductivity λ, and volumetric heat capacity C (obtained as the product of density ρ and specific heat c). Based on surveys of refractory material producers (who wish to remain anonymous), the range of each variable has been determined. Due to the possibility that the response to a factor variation may not be linear, three levels have been determined that can detect the possibility of a curvature in the response. The properties have been considered constant, and the identified levels for the thickness S are 80, 115 and 150 mm, for the thermal conductivity λ are 1, 5.5, and 10 W/mK, for the specific heat c are 800, 1150, and 1500 J/kg K, for the density ρ are 2000, 2750, and 3500 kg/m^3. Therefore, three levels and three factors have been considered, with a total of 3^3 cases to be analysed (figure 4). The intermediate value of the volumetric heat capacity C is the product of the intermediate level of ρ and c, with consequent values that are not the central one. By comparing the estimated energy-demand E_{LF} for each analyzed case, it has been possible to assess the influence of each factor as better explained in the following section. Referring to table 2, by comparing the results obtained for *Mat 1_1, Mat 2_1, Mat 3_1* the influence of the volumetric heat capacity is assessed. By comparing the results obtained for *Mat 1_1, Mat 1_2, Mat 1_3* the influence of the thermal conductivity is assessed. The initial idea was to test all the combinations but, successively, it has been decided to test only one scenario for the intermediate and the minimum thicknesses.

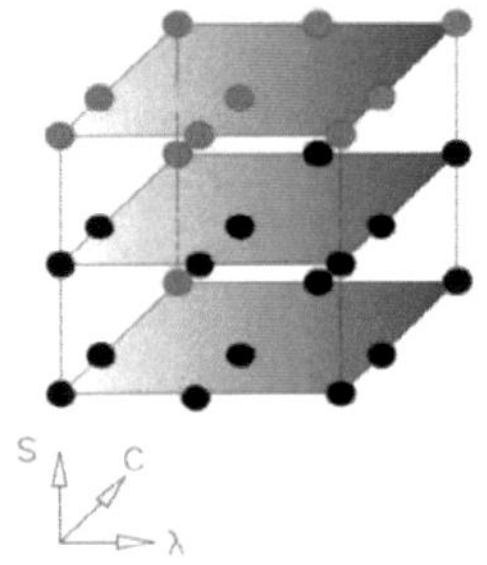

Figure 4. Representation of the factorial design considered in the study. Numerical experiments have been performed for the scenarios in red color only.

Table 2. Levels identified for the thermal conductivity λ and the volumetric heat capacity *C*. Materials resulting from this process are fictitious since they are not representative of real materials. In the nomenclature, the first number refers to the volumetric heat capacity.

	C [kJ/kg m^3] / λ [W/mK]	**Low**	**Int**	**High**
		1600	**3162.5**	**5250**
Low	**1**	Mat 1_1	Mat 2_1	Mat 3_1
Int	**5.5**	Mat 1_2	Mat 2_2	Mat 3_2
High	**10**	Mat 1_3	Mat 2_3	Mat 3_3

3. Results and discussion

Results of the numerical analysis reported in figures 5-10 are discussed to assess the influence of the working lining thermal conductivity *λ*, volumetric heat capacity *C*, and thickness *S*. According to figures 5 and 6, there is a greater difference between the energy demand in successive cycles at the beginning of the process, specifically between cycles *C1* and *C2*. In phase *F1*, present only in cycle *C1*, the ladle is heated with gas burners, and the temperature on the internal surface at the end of the process is 1100°C. Depending on the ladle characteristics, the temperature of the liquid steel varies between 1600°C and 1300°C throughout the cycle. Steel's initial temperature is higher than the temperature imposed at the end of preheating phase, so the ladle heats up when it comes in contact with it. In phase *F3* of the first cycle *C1*, the ladle has been in contact with liquid steel for 40 minutes only and, therefore, a higher quantity of energy is required to reach the threshold steel temperature. In the following cycles (*C2-C7*), less energy is required thanks to the longer contact with the liquid steel. In figure 5, at the beginning of the process, in cycle *C1,* the energy demand is lower for the minimum thickness but, after four cycles, the highest energy demand is related to the intermediate thickness (S_{int}). In the seventh cycle, the highest energy demand is estimated for the minimum (S_{min}) and intermediate (S_{int}) thicknesses. To better understand the causes of these differences, the temperature detected along the ladle wall for different thickness values (but same volumetric heat capacity and thermal conductivity) at the end of the preheating phase (*F1*) are reported in figure 7. The difference in energy demand is due to the opposite effect played by the working lining thickness: a greater thickness offers a greater thermal resistance but, at the same time, a greater heat capacity. On the other hand, the scenario with the greater thickness has a limited heat exchange surface. Figure 6 shows the energy demand E_{LF} required along the process as a function of the working lining thermal conductivity and of the volumetric heat capacity. The energy demand shows higher values at the beginning of the process, and the steady condition is approached depending on the scenario. A dependency on the material properties is also detected at the beginning of the process. Thermal conductivity *λ* strongly affects the energy demand since, for *C* equal to 5250 kJ/kg m^3, it ranges between 1 MWh (for λ=1 W/mK, *Mat 3_1*) and 2.4 MWh (for λ=10 W/mK, *Mat 3_3*). As regards the effect of the volumetric heat capacity, for *λ* equal to 10 W/mK, E_{LF} ranges between 1.7 MW (for C=1600 kJ/m^3 K, *Mat 1_3*) and 2.4 MW (for *C*=5250 kJ/m^3 K, *Mat 3_3*). Therefore, a correlation between the thermal conductivity and the volumetric heat capacity is detected but not

Journal of Physics: Conference Series **2685** (2024) 012025 doi:10.1088/1742-6596/2685/1/012025

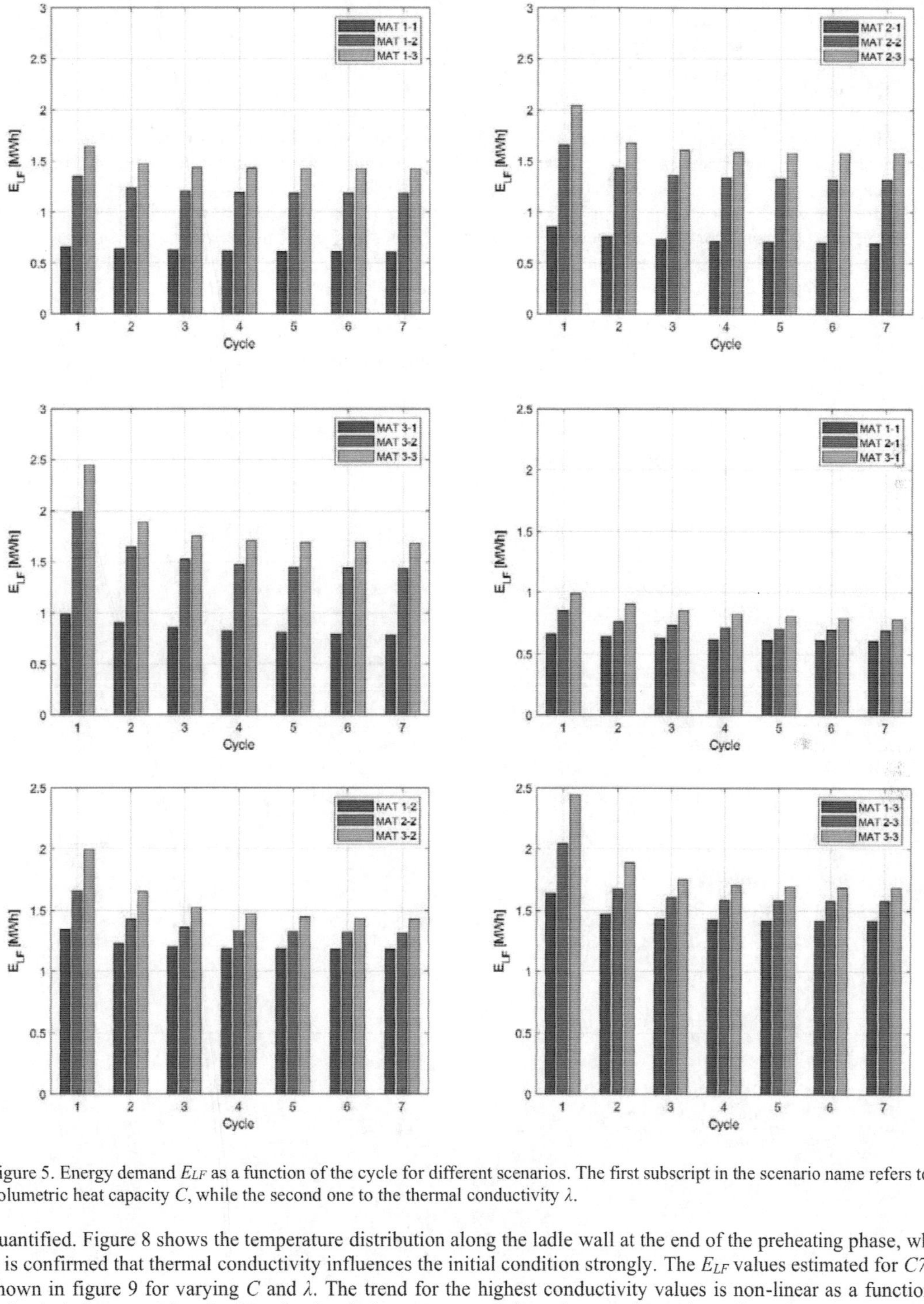

Figure 5. Energy demand E_{LF} as a function of the cycle for different scenarios. The first subscript in the scenario name refers to the volumetric heat capacity C, while the second one to the thermal conductivity λ.

quantified. Figure 8 shows the temperature distribution along the ladle wall at the end of the preheating phase, where it is confirmed that thermal conductivity influences the initial condition strongly. The E_{LF} values estimated for *C7* are shown in figure 9 for varying C and λ. The trend for the highest conductivity values is non-linear as a function of volumetric heat capacity. On the other hand, the energy trend depends strongly on thermal conductivity. In contrast, the variation as a function of thickness is limited and lower than 0.1 MWh (0.78 MWh for S=150 mm, 0.80 MWh for

S=115 mm, 0.80 MWh for S=80 mm), as also observed in [11]. Therefore, this factor can be neglected in the analysis to limit the number of experiments to be conducted.

4. Conclusions

The study analyzed the energy consumption in the secondary steelmaking process as a function of working lining characteristics. The study has been performed through numerical simulations performed with COMSOL Multiphysics and the analyzed scenarios have been selected through the DOE technique. Three factors and three levels have been identified for a total of 27 cases, of which only 11 have been analyzed. The identified factors are the working lining thickness, volumetric heat capacity, and thermal conductivity. A total of 1000 numerical simulations have been performed to determine the energy demand supplied in the electric arc furnace to guarantee a threshold temperature of the liquid steel. The results show that the most influencing factor is the thermal conductivity, with lower energy consumption for the material with the lower conductivity value. Heat capacity also affects energy consumption with lower values for lower values of heat capacity. Finally, thickness does not heavily affect the results, indicating that the heat transfer phenomenon can be treated by leaving out this variable, which makes the analysis much simpler.

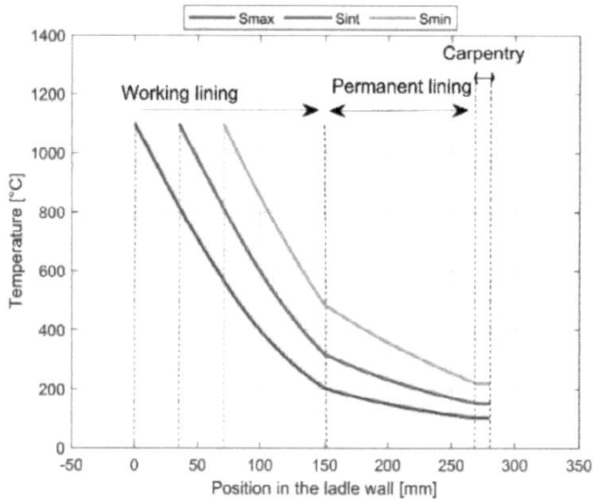

Figure 6. Temperature distribution in the ladle wall as a function of the working lining thickness *S*, at the end of the preheating phase *F1*.

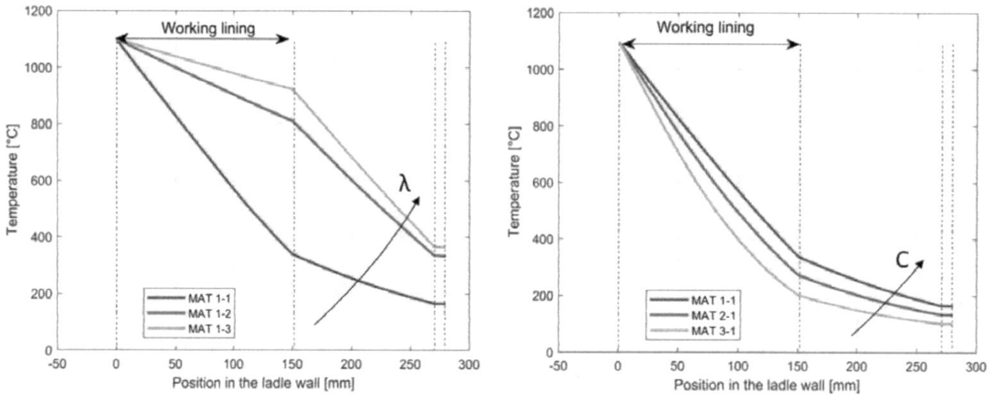

Figure 7. Temperature distribution along the ladle wall at the end of the preheating phase *F1* in the first cycle *C1* as a function of the thermal conductivity λ, and of the volumetric heat capacity *C*.

Journal of Physics: Conference Series **2685** (2024) 012025 doi:10.1088/1742-6596/2685/1/012025

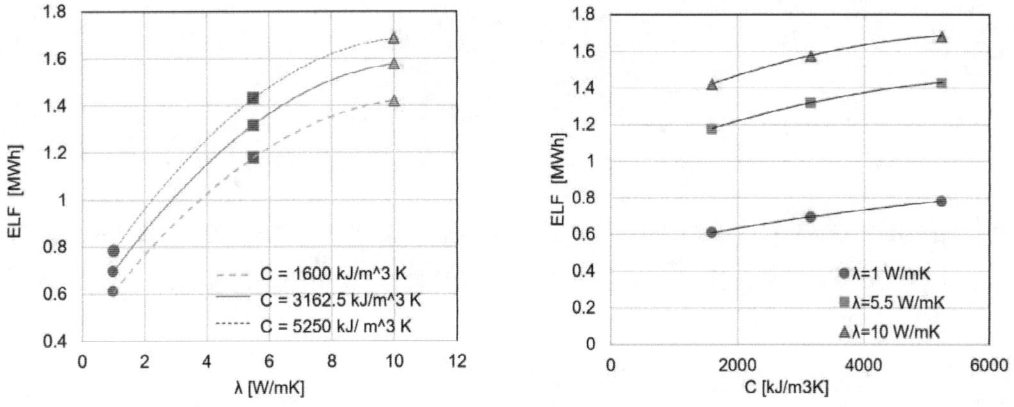

Figure 8. Energy demand E_{LF} as a function of the thermal conductivity λ and of the heat capacity C, estimated for S_{max} at the end of the process (cycle *C7).*

REFERENCES

[1] Santos M F et al 2018 Enhanced numerical tool to evaluate steel ladle thermal losses *Ceramics International* **44** pp 12831-12840

[2] Zimmer A 2008 Heat Transfer in Steelmaking Ladle *Journal of iron and steel research international* **15** pp 11-14.

[3] Glaser B, Görnerup M, Sichen D 2011 Thermal Modelling of the Ladle Preheating Process *Steel research int.* **82** pp 1425-1434

[4] Glaser B, Görnerup, M, Sichen D 2011 Fluid Flow and Heat Transfer in the Ladle during Teeming *Steel research international* **82** pp 827-835

[5] Sanjib C, Yogeshwar S 1992 Effect of slag cover on heat loss and liquid steel flow in ladles before and during teeming to a continuous casting tundish *Metallurgical Transactions B* **23** pp 135-151

[6] Haojian D, Ying R, Lifeng Z 2019 Thermal Stratification, and Inclusion Motion During Holding Period in Steel Ladles *Metallurgical and Materials Transactions B* **50** pp 1476-1489

[7] Xia J L, Ahokainen T 2001 Transient flow and heat transfer in a steelmaking ladle during the holding period, *Metallurgical and Materials Transactions B* **32** pp 733-741

[8] Aidong H, Shengli J, Harald H, Dietmar G 2018 A Method for Steel Ladle Lining Optimization Applying Thermomechanical Modeling and Taguchi Approaches *Journal of The Minerals, Metals & Materials Society* **70** pp 2449-2456

[9] Moreira M H, Pelissari P I B G B, Angelico R A, Sako E Y 2017 Steel ladle Energy saving by refractory lining design, Conference paper 201, UBITECR 2017, Santiago, Chile

[10] Fredman T P, Saxen H 1998 Model for temperature profile estimation in the refractory of a metallurgical ladle *Metallurgical and material transaction B* **28** pp 611-659

[11] Neri M, Lezzi A M 2022 Energy demand in secondary steel making process: numerical analysis to assess the influence of the ladle working lining properties 39th UIT International Heat Transfer Conference, Gaeta (Italy) 20-22 June 2022.

[12] Santos M F et al 2018 Enhanced numerical tool to evaluate steel ladle thermal losses *Ceramics International* **44** pp 12831-12840

[13] Farrera-Buenrostro J E et al 2019 Analysis of temperature losses of the liquid steel in a ladle furnace during desulfurization stage *Transactions of the Indian Institute of Metals* **72** pp 899-909

[14] Jankovic A, Chaudhary G, Goia F 2021 Designing the design of experiments (DOE) – An investigation on the influence of different factorial designs on the characterization of complex systems *Energy and Buildings* **250**

[15] Natoli C, Oimoen S 2019 Classical Designs: Full Factorial Designs *STAT COE-Report-35-2018* Available at *http://www.afit.edu/STAT,* accessed on the 10th May 2023

CFD analysis of thermal fluid dynamic parameters inside an organ pipe: influence of air temperature variations

P Domenighini[1,*] **and C Buratti**[1]

[1] Engineering Department, University of Perugia, Via G Duranti 93, Perugia 06125, ITA

* Corresponding author: piergiovanni.domenighini@studenti.unipg.it

Abstract. Organ pipes represent a fluid-dynamic challenge since the time of Bernoulli and Reynolds. A complete and reliable description of the fluid behaviour at the basement of the resonator named the "mouth" of the pipe is still absent. In the context of the design of an air-heating system for the attenuation of the inconsistencies between sounding frequencies of organ pipes, computing efforts were carried out focusing on the fluid-dynamic behaviour of the air at the mouth of the pipe at different temperature conditions. Describing the air dynamic and the heat transfer at the mouth level of the pipe, it would be possible to predict the heating conditions within the pipe resonator, which was the main subject of the project. Multiple points on the pipe organ air blowing scheme were considered for simulation results extraction. Temperature, pressure, and audio measurements were performed and no significant influence of the temperature on those parameters was raised from measurements, rising questions about the effective behaviour of the blowed air at the mouth of the pipe. Simulations were performed to predict both the fluid dynamic and the heat transfer. A finite volume approach in Fluent environment was chosen. The SST k-omega model was considered in reliability vs computation time balance. 3D simulations were performed based on a CAD reconstructed model of a real pipe. The results of the simulations confirmed experimental data, and they may prove the absence of a significant mass exchange between the in and outside of the mouth of the pipe. The oscillation of the air membrane at the mouth of the pipe may be cleared from instabilities and turbulence regime phenomena, as usually described in the literature.

Keywords: *CFD, organ pipe, heat transfer, SST model, k-omega.*

1. Introduction

Pipe organs, with their rich and resonant tones, have enthralled audiences for centuries with their majestic and awe-inspiring sound. The intricate design and craftsmanship of these instruments involve a delicate balance of mechanics, acoustics, and materials. However, pipe organs also face thermal challenges that can significantly impact their performance and acoustic tuning [1,2]. To address these challenges, the integration of Computational Fluid Dynamics (CFD) techniques has emerged as a state-of-the-art approach for understanding the physics behind sounding organ pipes [3-6].

Pipe organs are complex musical instruments composed of a multitude of elements, including pipes, wind reservoirs, bellows, keyboards, and mechanisms for controlling airflow. The sound produced by a pipe organ is intricately connected to the physical properties of its pipes, which are designed to generate specific pitches and timbres. Achieving optimal acoustic tuning requires meticulous attention to factors such as pipe dimensions, materials, voicing techniques, and wind pressure regulation. Temperature and humidity fluctuations can significantly impact the performance, stability, and longevity of pipe organs. Recently, studies conducted report a shift of the 6% in the pipe tuning of a real pipe organ in a church in the Mediterranean area of Perugia, Italy, when the indoor temperature and relative humidity are in the $6 - 30\ °C$ and $40 - 90\%$ ranges, respectively [2]. Understanding and managing these thermal challenges are crucial for maintaining the desired sound quality and overall functionality of pipe organs. In the literature, it is possible to find examples of heating system tested on church pipe organs. In St. Martin church in Oberesslingen, Germany, a fan coils heating system was experimented to fix beating effects between the different sections of the instrument [7]. The researchers identified a threshold temperature difference of $\Delta T = 0.5°C$ between the sections of the instrument, resulting in emitted frequency difference of $0.3\ Hz$, which generates a perceivable beating effect. Unfortunately, the designed system could limit the temperature variations only $\Delta T < 1.4°C$.

In this work, a different heating method and system was tested on a test-bench pipe organ provided by Pinchi srl at the Environmental Control Lab at the University of Perugia. The heating system is directly embedded into the blowing system of the pipe organ itself, providing hot air at different temperature stages. The influence of the changing air conditions of the blowing air into the pipe-resonant system were measured in terms of temperature, acoustic pressure, and frequency. The absence of a significant influence on the resulting sounding frequency by the temperature change was investigated by a CFD model of the measured pipe. A Reynolds Average Navier-Stokes (RANS) SST k-omega model was used for the simulations and the hot-air flux was analyzed and compared to measurements. Interestingly, results may prove the absence of turbulence at the mouth level of the pipe, conversely to the hypothesis of Reinholds and Rosenhead [8,9], and the modal excitation of the resonance tube may be guided by an oscillating air flux where no turbulence phenomena happen.

2. Materials and method

To investigate the influence of temperature changes on the blowing air through the pipe organ system on the resulting frequency, a heating system was installed on the laboratory pipe organ provided by Pinchi srl organbuilder at the Environmental and Control Laboratory at the Engineering department of the University of Perugia. Italy.

2.1. The heating system and the measurement campaign

The heating system design followed the scheme reported in Fig.1 and consisted in an electric heater with two temperature set points (333 and $383\ K$), which influence the air boosted to the windchest of the instrument by the fan-bellows system (Fig.2).

A F1 sounding pipe was chosen for measurements because its dimensions (reported in Table.1) are average for a pipe of an organ. Acoustic pressure and temperature were measured during the whole process, and the following measurement set-up was involved:

- 8 thermocouples managed by a central device-manager SpecView (± 0.02°C);
- 2 Tinytags TGU-1500 probes for room T and RH (± 0.2°C from 0°C to 50°C);
- 2 Endevco pressure transducers class 8507 C-1 / C-2 - 2 mm Ø (±50 mV/psi);
- 2 Bruel&Kjaer pressure transducer types 4938 and 4939 (± 2 dB in 4 Hz to 70 kHz range);
- NI-9234 cards for transducer acquisition with a $fs = 51.2\ kHz$.
- Zoom H2 audio recorder (accuracy combined with post processing ±0.01 Hz).
- Fluke 922 airflow meter for air velocity data acquisition (±1 m/s).

Temperature and pressure probes' locations are reported in Fig.3. The Endevco pressure probes were located at the rank-chest and the inner part of the mouth of the pipe, together with the thermocouples (Fig.4), whereas the Bruel&Kjaer pressure probes (bigger in size) were outside the mouth of the pipe and at the top (Fig.5). Air velocity was measured at the basement of the foot of the pipe, into the rank-chest valve.

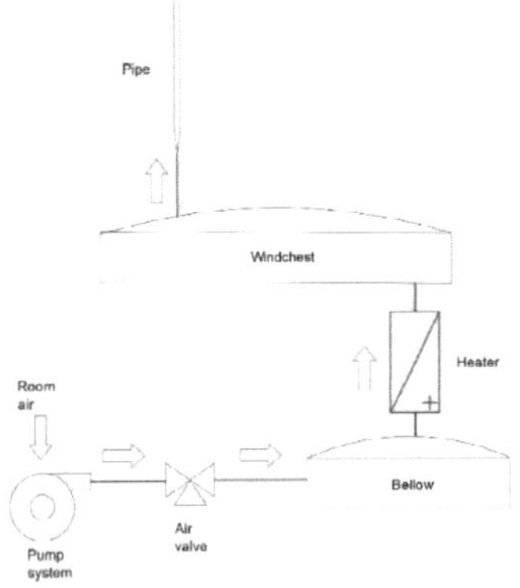

Figure 1. Scheme of the pipe organ heating system.

Figure 2. The heating system installed before the windchest stage of the laboratory organ at the Environmental and Control Lab at the University of Perugia.

Table 1. Geometrical characteristics of the F1 sounding pipe.

Fundamental frequency			192 Hz
Geometrical dimensions	General	H [cm]	108.80
		R [cm]	2.95
	Resonator	H [cm]	89.10
		R [cm]	2.95
		V [cm^3]	2435.97
	Mouth	H [cm]	1.30
		W [cm]	4.50
		A [cm^2]	5.85
	Foot	H [cm]	19.70
		R_{min} [cm]	0.45
		R_{max} [cm]	2.95
		V [cm^3]	269.29

2.2. The CFD model

The Computational Fluid-Dynamic CFD simulation was done in Fluent Ansys environment. The pipe was modeled together with the volume of the inner air in a 3D CAD file. The 3D model (Fig.6) was modelled based on metric measurement and photo relief, and represents a simplified geometrical model of the pipe especially at the mouth level, where craftmanship modifications to the geometry of the pipe made it difficult to be realistically reported into the virtual model. The 3D model was imported into Ansys mesher module and a detailed mesh of 1003107 nodes and 5349166 elements were generated (element size 1.2e-003 m [max 2.4e-003m, defeature 6e-006m], target Skewness=0.9). Further refinement was reserved to the area of the mouth, with an increase of the volume number of 6 times (Fig. 7).. Then inlet and outlet surfaces were defined (Fig. 6), and the 3D model was imported into the Fluent software environment for CFD calculation, where gravity acceleration was also considered in the y-vertical-direction. The fluid boundaries were set up with wall boundary condition with roughness constant of 0.5. Default Fluent converge criteria were used, 1e-03 for continuity, momentum and 1e-06 for energy equations.

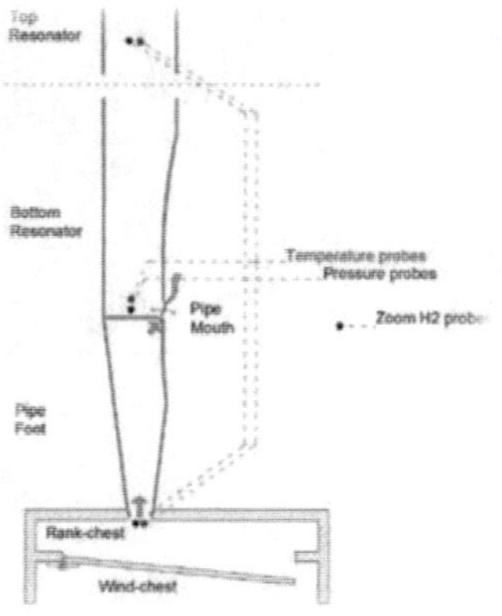

Figure 3-a. Positions of the probes during measurements inside the sounding pipe.

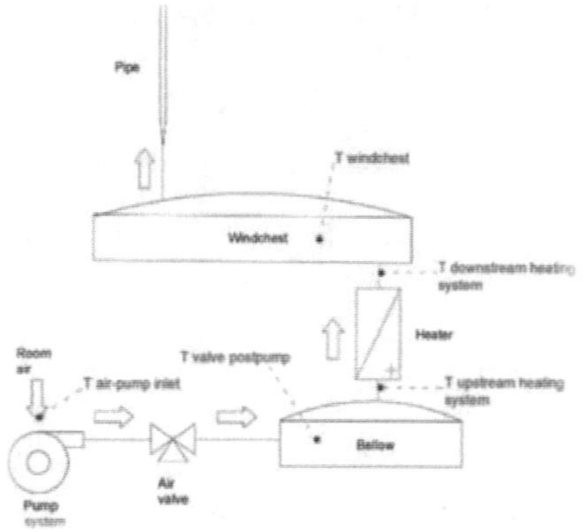

Figure 3-b. Positions of the probes during measurements inside pipe organ system.

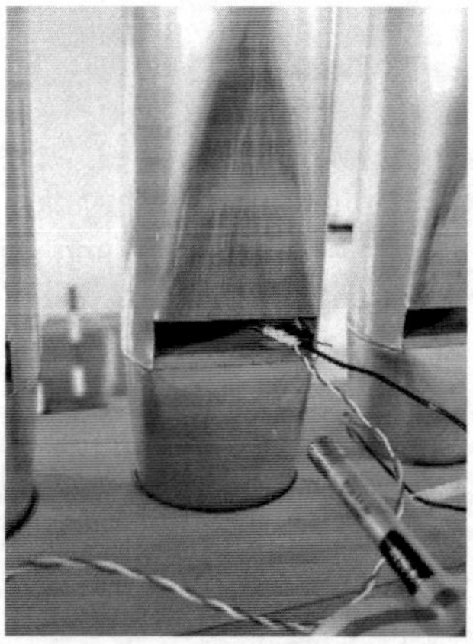

Figure 4. Pressure and temperature probes at the mouth of the pipe.

Figure 5. Pressure and temperature probes at the top end of the pipe.

Although it would be possible to analyse only half of the 3D Geometry of the pipe due to its symmetry around the y-longitudinal axis, the model was completely considered to study turbulence within the pipe. RANS analysis was considered because both the $k - \epsilon$ and $k - \omega$ models have complementary pros and cons in the simulation of a pipe, where the turbulent regime is not well defined, and the presence of the wall is essential, especially at the mouth level. It was chosen to perform a transient analysis using the SST k-ω model.
The model is based on the differential equation system as available in the literature [10,11].

SST k-ω model was chosen among the possible RANS model because:

- it combines the high reliability in predicting flows present in boundary layers due to gradients of contrary pressure and flow separation of the $k - \omega$ model, and the high accuracy of $k - \epsilon$ model solving turbulent flows;
- compared with the BSL model, it considers the shear stress transport due to the presence of turbulence. This may be important in the dynamic of the fluid at the mouth to evaluate the cause of the oscillation of the "air membrane", as introduced by the work of Elder [9].

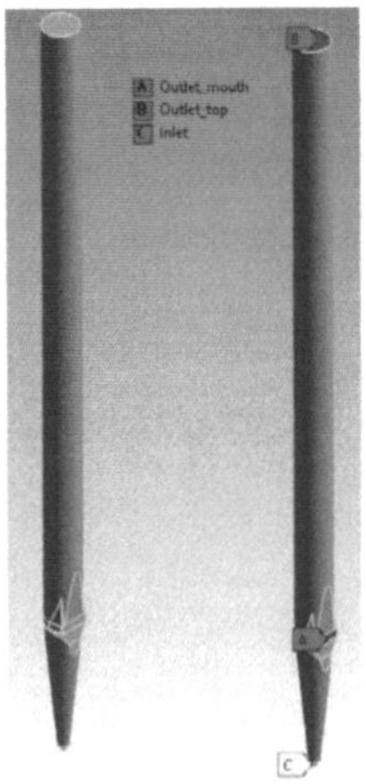

Figure 6. The 3d model of the pipe (on the left) and the Inlet-outlet configuration (on the right). In green, the wall boundary condition layer.

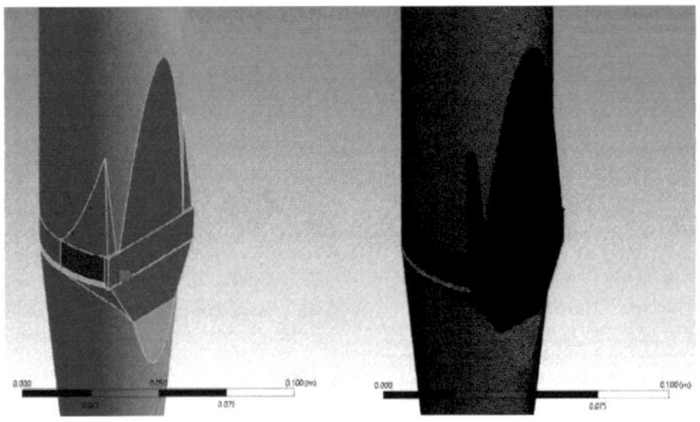

Figure 7. Detail of the 3D with the faces object of the mesh refinement process (in blue, on the left), and mesh representation (on the right).

3. Results and discussion

The results of the measurements on F1 pipe, at different working stages of the heater, are reported in Fig. 8-13 and Table 2.

Fig. 6-9 show the temperature trend over functioning time of the heater at the measurement points reported in Fig. 3, plus all the stages of the air blowing system. In particular, the three stages are highlighted by the temperature trends at the upstream measurement, from the off stage, the first stage appear between 1000 and $1500\ s$, while the second stage between 1500 and $2000\ s$. Fig. 3 highlights the significant thermal dissipation at the heating windchest stage, where more than $80\ K$ of temperature decrease are reported. Furthermore, temperature trends focused on the sounding pipe system show important thermal dissipation estimated in a $3-4\ K$ decrease between the mouth and the top of the pipe. However, although there is a thermal dissipation between the mouth and the top of the pipe, the most significant thermal dissipation is measured at the previous windchest-pipe foot stage, where temperature decreases by about $20\ K$. The thermal dissipation at that stage, and the consequent absence of a significant influence of the heating system influence on the air-flow temperature, can explain why variations on the pressure are visible only at the pipe-foot inlet level (Fig. 10, Table 2), and there are variations at the mouth and top of the pipe (Fig. 11-12, Table 2). Pressure magnitudes from measurements reported in Table 2 are trustworthy because the higher pressure is expected to be at the inlet of the pipe foot, where the section of the pipe is the smaller and the air velocity is the higher, while pressure at the mouth and the top of the pipe (the two extremities of the resonator) are according to the physics of the resonator of the pipe. Indeed, the closer to the extremities of the pipe resonator, the closer to zero is the average pressure by the oscillations. However, differences in terms of pressure in function of the temperature are reported only at the inlet stage. There, a higher temperature corresponds to a higher average pressure with comparable oscillations (Standard deviations St.d). No differences are reported at the other stages. Consistently to the absence of influence of temperature on the pressure, the sound of the pipe remained the same during the measurements, and the spectral content of its sound reports no significant differences (Fig.11).

The absence of influence of the temperature changes on the resonator acoustic features, pressure patterns, justifiable by the temperature drops between the foot and the basement of the resonator where the temperature probe was located, suggests that the air flow at the mouth of the pipe direct the warmer air outside the pipe rather than splitting in and out from the pipe as hypothesized in the literature [8,9].

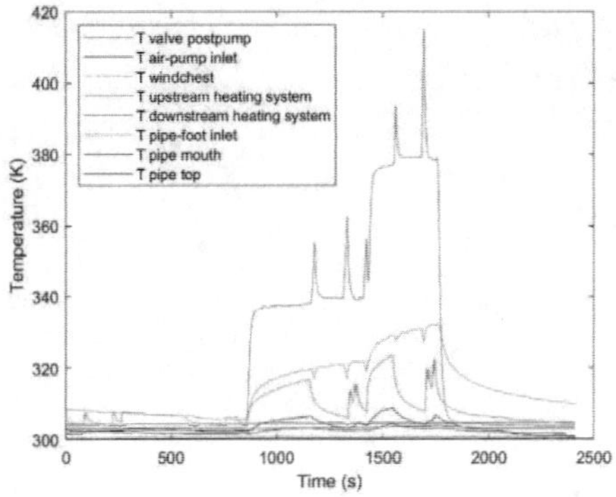

Figure 8. Temperature measurements results from the sensors at the different stages of the blowing system of the instrument.

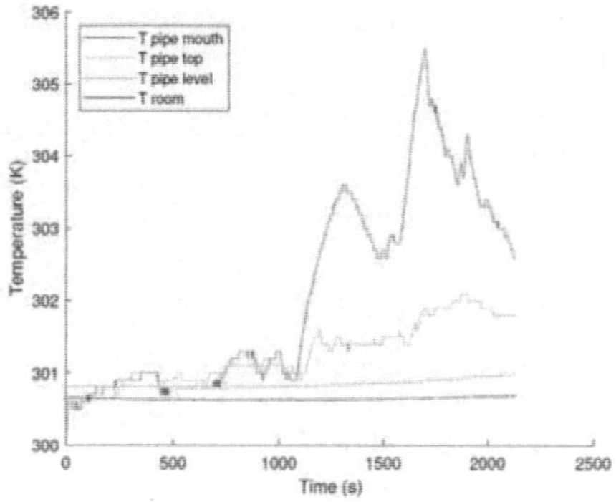

Figure 9. Temperature measurements results from the sensors at the three points of the pipe (as reported in Fig.3).

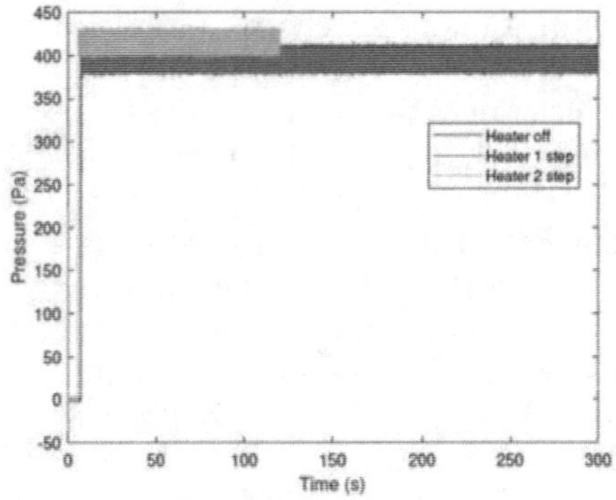

Figure 10. Pressure measurements at the foot of the pipe at the three different temperature stages of the heating system.

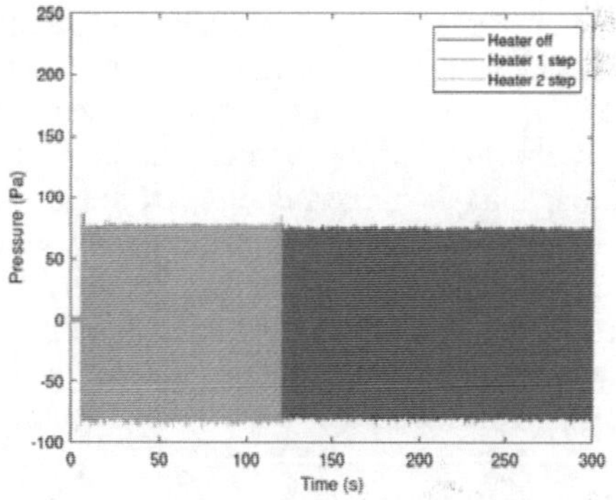

Figure 11. Pressure measurements at the mouth of the pipe at the three different temperature stages of the heating system.

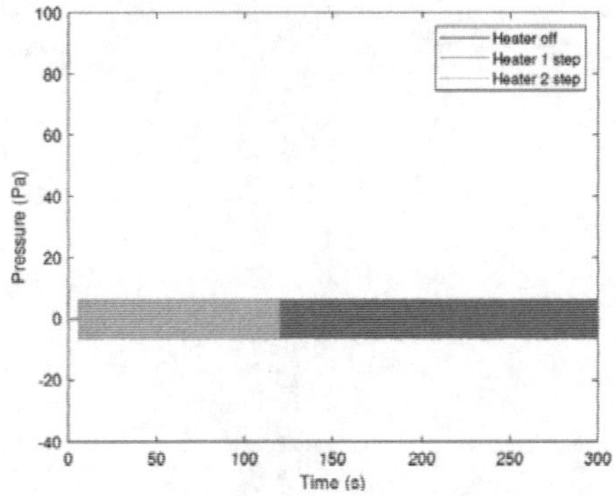

Figure 12. Pressure measurements at the top of the pipe at the three different temperature stages of the heating system.

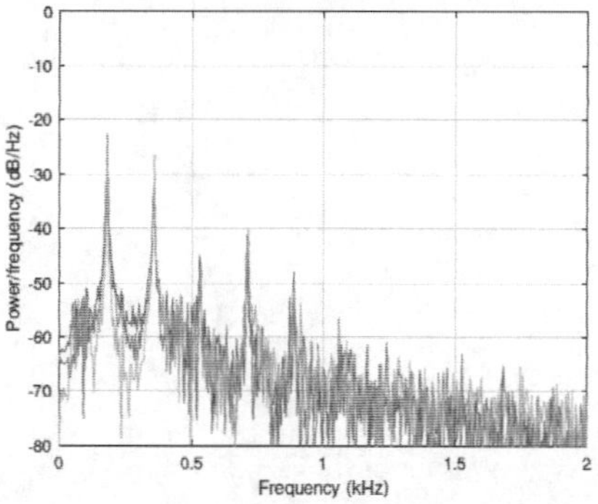

Figure 13. FFT of the sound recordings of the pipe at the three different temperature stages of the heating system (yellow-red-blue).

Table 2. Pressure means and standard deviations values at each heating step, at each measurement point, for both pipes.

Pipe	Measurement point	*Pressure at each Heating stage [Pa]*					
		off		1		2	
		mean	St.d	*mean*	St.d	*mean*	St.d
F1	Inlet	349.83	4.30	397.16	5.39	416.12	5.45
	Mouth	$-1.67e^{-4}$	46.57	$8.12e^{-5}$	46.79	$3.76e^{-4}$	47.76
	Top	$-1.30e^{-2}$	3.83	$4.80e^{-3}$	3.84	$2.19e^{-4}$	3.95

To visualize the path of the airflow from the foot of the pipe to the mouth, and then the resonator, the RANS SST k − ω model in Fluent was used to investigate the thermal dissipation. The air velocity measurement at the inlet of the pipe (air velocity $= 20 \pm 1\ m/s$) and the temperature measurements as reported in Fig.9 were used as input into the CFD model. Fig.14-15 graphically represent the results in function of time and not a specific frame of the simulation.

The results of the simulations (Fig.14-15) show the vector direction at the mouth level as outgoing the mouth of the pipe. In particular, the dominant direction is toward the external upper side of the mouth, where the largest magnitude is reported (Fig.14). In the same figure it is possible to observe the preserve of other vector directions, smaller in magnitude, which can be due to interactions between the airflow, the low and up borders of the mouth of the pipe. However, velocity vectors in different directions than the largest one report magnitude of several order smaller, and they may be neglected. To investigate the behaviour of the airflow at the mouth of the pipe and evaluate if the warmer air flow moved toward the resonator of the pipe or externally, the velocity vector component along the perpendicular direction to the plane of the mouth of pipe were isolated (Fig.15). There, it is possible to observe the largest magnitude of the vector outcoming the mouth of the pipe, only a small number of vectors is represented as incoming the mouth of the pipe and their magnitude is at least smaller for one order.

This may demonstrate that the airflow at the mouth of the pipe generates a driving oscillation which moves the air membrane at the mouth, but there is no significant mass exchange from outside the mouth as frequently hypnotized in the literature [8,9].

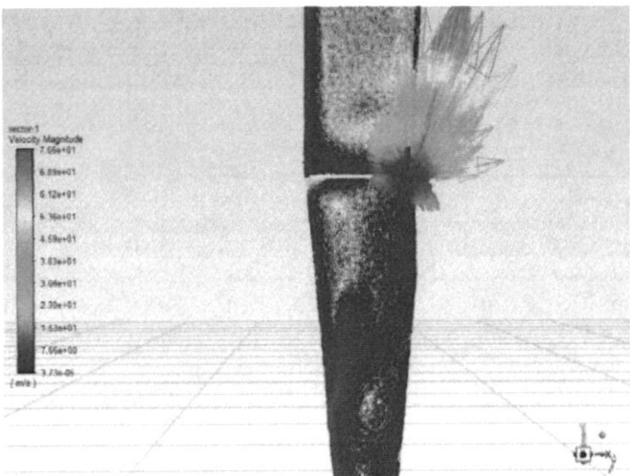

Figure 14. Vector velocity magnitude of the airflow from CFD simulations.

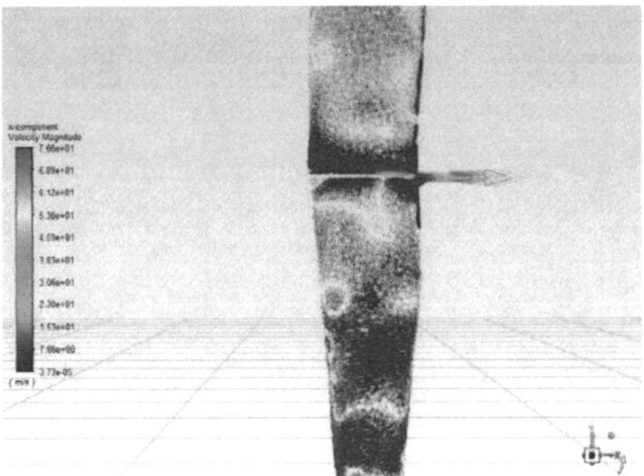

Figure 15. x-component of the vector velocity magnitude of the airflow from CFD simulations.

4. Conclusions and future perspectives

In the present work the influence of temperature changes on the airflow driving-resonant of a pipe organ is introduced and investigated. To evaluate this phenomena, temperature, pressure, and acoustic frequency measurements were carried out on a modified test-bench pipe organ provided by Pinchi srl organ builder at the Environmental Control Laboratory at the Department of Engineering, University of Perugia. The pipe organ was provided with an electrical heating system installed directly between the bellows and the windchest of the instrument. The measurement campaign highlighted significant thermal dissipation between the heater and the windchest, as well as between the foot of the pipe and the resonator, rising questions about the effective direction of the air flow from the foot, mainly hypothesized as equally distributed inside and outside the mouth of the pipe because of instabilities and turbulence phenomena. The absence of a significant change in the temperature of the air at the top of the pipe suggested the use of a CFD model in Fluent environment to confirm and better investigate the directions of the velocity vector of the air incoming from the foot of the pipe. RANS SST $k - \omega$ model was used to consider the transport of the share stress in the air flow. Results from simulations showed the outgoing direction as the dominant one, and the absence of contribution of the warmer air into the resonator of the pipe was proved by the absence of a significant magnitude of the air vector velocity inside the mouth of the pipe from the external part. Further investigations are required to improve the precision of the RANS simulation. For example, DNS simulation may provide data useful for the RANS approach, improving the numerical setup, and justify and prove the behaviour of the air at the mouth level.

5. References

[1] Pitsch S, Sedlbauer K and Angster J 2008 Einfluss der Lufttemperatur auf die Verstimmung von Orgelpfeifen *Bauphysik* **30** pp. 363–365 doi: 10.1002/bapi.200810046.

[2] Domenighini P and Buratti C 2023 The Influence of Thermal-Hygrometric Properties of Air on the Tune of Pipe-Organs *Applied Acoustics* **213** doi: 10.1016/j.apacoust.2023.109611.

[3] Angster J, Rucz P and Miklos A 2017 Acoustics of Organ Pipes and Future Trends in the Research *Acoustics Today* **13** pp. 12–20.

[4] Rucz P 2016 *Innovative methods for the sound design of organ pipes* Budapest: University of Technology and Economics.

[5] Rucz P, Angster J, and Miklos A 2022 Physics-based model of the sound generation by dynamic lingual organ pipes *Proceedings of DAGA2022* - **48** Jahrestagung fur Akustik. pp. 822–825.

[6] Nagy PT, Rucz P and Szabo 2022 A Examination of jet growth and jet-drive in the recorder by means of linearized numerical and lumped models *Journal of Sound and Vibration* **527** p. 116857. doi: https://doi.org/10.1016/j.jsv.2022.116857.

[7] Pitsch S, Holmberg S and Angster J 2010 Ventilation system design for a church pipe organ using numerical simulation and on-site measurement *Building and Environment* **45.12** pp. 2629–2643 doi: 10.1016/j.buildenv.2010.05.022.

[8] Rosenhead L 1931 The formation of vortices from a surface of discontinuity *Proceedings of the Royal Society of London* **Series A** Containing Papers of a Mathematical and Physical Character 134.823, pp. 170–192 doi: 10.1098/rspa.1931.0189

[9] Elder SA 1978 Self-excited depth-mode resonance for a wall-mounted cavity in turbulent flow". *The Journal of the Acoustical Society of America* **64.3** pp. 877–890. doi: 10.1121/1.382047.

[10] Kundu PK, Cohen IM and Dowling DR 2015 *Fluid mechanics* ed. Academic press isbn: 978-0-1240-5935-1.

[11] ANSYS Fluent 2020 *ANSYS Fluent Theory Guide, Release 2020 R2*

Acknowledgments

Authors wish to thank Pinchi srl organ builder for providing the test pipe organ, for the technical support, and for the help during the measurements.

Saccardo nozzle ventilation system: cfd analysis to upgrade the performance of the ventilation system

M Musto[1,*], **F Cascetta**[2], **N Bianco**[1] , **G Rotondo**[1]

[1] Department of Industrial Engineering, University of Naples “Federico II”, Piazz.le Tecchio 80 - 80125 Naples, Italy
[2] Department of Engineering, University of Naples “Vanvitelli”, Via Roma – 81031 Aversa (CE), Italy

* Corresponding author: marilena.musto@unina.it

Abstract. This paper presents the results of a 3D numerical investigation carried out on a Saccardo ventilation system operating in road tunnel, with the aim to improve the Performance of the Ventilation System, P_{VS}. 1D mathematical model, here presented, has highlighted that the pitch angle and length of inflow opening are the main parameters that influence the P_{VS}. The results have shown that the pitch angle of the Saccardo nozzle, influences moderately the P_{vs}, instead the opening inflow lengths influence considerably the transferring of the momentum, increasing the mass flow rate. P_{vs} trend vs opening inflow length shows an improvement up to 47% with respect to the reference opening inflow length.

1. Introduction

In one-way road tunnel, the longitudinal ventilation systems are adopted to maintain acceptable air quality and thermal comfort. Tunnel ventilation system, typically, is designed to provide the optimal environmental condition for both normal and congested traffic scenarios, both in ordinary and emergency cases. The main systems adopted to provide impulse ventilation in tunnels include the Saccardo nozzles sometimes coupled with impulse exhaust dampers, traditional jet fan and its optimization such as the Banana Jet® and MoJet©. Tarada F. and Brandt R. [1] have presented a brief overview of impulse ventilation for tunnels, as guidance regarding their advantages and drawbacks. The verification of the goodness of the design choices or the activity of optimization for the ventilation systems is, usually, achieved by full scale or reduced scale experiments, both of which are expansive in terms of cost and time. Instead, the CFD analysis enables to carried out, regardless of the complexity of geometrical characteristics of boundary (spatial and time) conditions, a quite easy prediction of velocity and temperature fields both in ordinary situations and in emergency (in case of fire or traffic jam). Therefore, the CFD analysis, once validated on experimental data, is the cheapest way to optimize the performance of the fluid dynamic systems with respect to full or reduced scale experiments. Alston J. et al. [2] studied the retrofitting feasibility in a transverse ventilation system with a Saccardo nozzle placed at the entrance of a road tunnel. Sturm P. et al. [3] proposed in their research a simple calculation method for 1-D approach, derived from the conservation equation of momentum in order to calculate the Saccardo efficiency coefficient. Waymel, F. et al. [4] presented a methodology based on combination of 1D and 3D modelling used for the design and specific uses related to the Saccardo system design development. Hang, J.G. [5] et al. optimized nozzle design reaching a significant reduction in the overall

maximum tunnel air temperature and proposed an alternative nozzle design. In particular, they present a detailed study of the performance of an optimized Saccardo Nozzle design using a CFD model combined with a mono-dimensional temperature prediction model to evaluate the carry-over effects in the tunnel. Ganjiazad, R. et al. [6] evaluated the influence of volumetric flow rate and inclination angle of air jet on the Saccardo ventilation system performance and inclination angle range to optimize the performance in fire scenario. Alston J. et al [7] have studied retrofitting of an existing transverse ventilation system to a longitudinal system using a Saccardo Nozzle using 3D CFD. Chammem T. et al [8] in their paper suggested a new longitudinal ventilation technique based on supplying external air into the tunnel by two inclined jets, combining the advantages of Banana Jet and Saccardo nozzles. Sturm P. [8] discussed the methods of fire ventilation with a focus on the requirements for sensors for Saccardo ventilation system.
In this work the authors present a 3D CFD analysis of a longitudinal ventilation system equipped with Saccardo nozzle with the aim to improve the Performance of the Ventilation System, Pvs. A preliminary 1D mathematical model highlights in the pitch angle and length of inflow opening the main parameters that influence the Pvs. The CFD results show that the pitch angle of the Saccardo nozzle influences moderately the Pvs, instead the different opening inflow lengths influence considerably the mean air velocity in the tunnel and mass flow rate. In particular, a Pvs trend vs opening inflow length shows an improvement up to 47% with respect to the reference opening inflow length. In order to validate the CFD simulation, this paper refers to tunnel investigated to Hang et al. [5].

2. Details of the Tunnel and Ventilation System models

The physical domain under investigation is shown in figure 1 (not to scale); it represents a road tunnel of 1 km length, 14.6 m of width (with 3 lanes) and 6 m of height equipped with longitudinal ventilation system. The air flow is induced by Saccardo nozzles that provides fresh air, while the dumpers remove the exhausts. The three Saccardo nozzles and exhaust dumpers (as shown in figure 1 on the right side) are arranged in side-by-side configuration, both placed in the middle of the tunnel at a distance of 500 m from inlet portal. The arrangement and geometrical dimensions of the ventilation system are shown in figure 2 and table 1; they are the same of that investigated experimentally by Hang et al. [5] and here used with the aim to validate CFD model. Both exhaust dampers and Saccardo nozzles engine are built by HVAC Vents connected to each other by ducts as shown in figure 2. The ducts represent a continuous flow path, instead the nodes attached to vents surface represent the interface between the 1D-model and 3D-CFD regions.

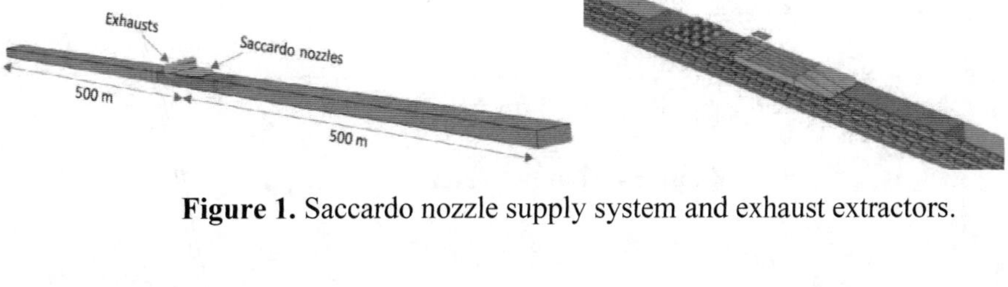

Figure 1. Saccardo nozzle supply system and exhaust extractors.

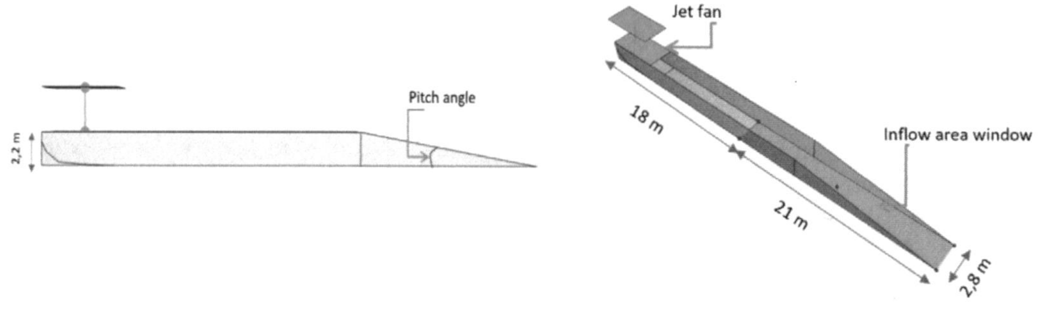

Figure 2. Saccardo nozzle dimensions details.

Table 1. Saccardo Nozzle and HVAC Vent dimensions.

	Parameters	values
w_S	Width of inflow duct of Saccardo nozzle	2.8 m
H_{duct}	High of inflow duct of Saccardo nozzle	2.2 m
L_{duct}	Length of duct of Saccardo nozzle	18 m
L_{inflow}	Length of inflow area of Saccardo nozzle	21 m
α	Pitch angle	10°
$A_{exhaust}$	Exhaust dampers area	9.0 m^2
$D_{S\text{-}d}$	Distance between Saccardo Nozzle and Exhaust dampers	16 m
$\dot{m}_S$	Saccardo mass flow rate	kg/s
$\dot{m}_{exh}$	Exhaust mass flow rate	kg/s

2.1. 1D Mathematical model of the Ventilation System.

A 1D mathematical model to investigate the Saccardo system performance is presented. This model, even if simplified, can lead to recognize the main parameters to be investigated. Starting from the momentum balance (equation 1) referred to the volume control presented in figure 3,

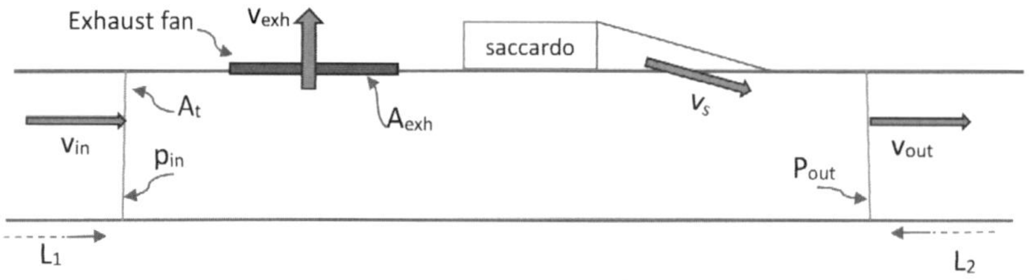

Figure 3. Control volume.

$$(p_{in} - p_{out})A_t = \dot{m}_{out}v_{out} - \dot{m}_{in}v_{in} + \dot{m}_{exh}v_{exh} - \dot{m}_S v_S \cos(\alpha) \quad (1)$$

where $\dot{m}_{out}$and $\dot{m}_{in}$ are the mass flow rates at the outlet and inlet tunnel portal, respectively, and $\dot{m}_S$ *and* $\dot{m}_{exh}$ are the the mass inflow and outflow provided by Saccardo and exhaust respectively, and taking into account the pressure drops to the portals (equations 2 and 3):

$$p_{portal} - p_{in} = \frac{1}{2}\rho v_{in}^2 + \frac{1}{2} f_{total} \frac{\rho L_1}{D_H} v_{in}^2 \quad (2)$$

$$p_{out} - p_{portal} = \frac{1}{2}\rho v_{out}^2 + \frac{1}{2} f_{total} \frac{\rho L_2}{D_H} v_{out}^2 \quad (3)$$

the equation (1) becomes the equation (4):

$$\left(-\frac{1}{2}\rho v_{in}^2 - \frac{1}{2}f_{total}\frac{\rho L_1}{D_H}v_{in}^2 - \frac{1}{2}\rho v_{out}^2 - \frac{1}{2}f_{total}\frac{\rho L_2}{D_H}v_{out}^2\right)A_t = \dot{m}_{out}v_{out} - \dot{m}_{in}v_{in} + \dot{m}_{exh}v_{exh} + -\dot{m}_S v_S\cos(\alpha) \tag{4}$$

where f_{total} is the total friction losses coefficient and L_1, L_2 tunnel are lengths upstream and downstream the control volume, respectively. This coefficient sums the friction due to the tunnel walls and the friction due to cars and the losses due to vorticity among cars [9,10].
Considering, also, the continuity equation (5):

$$\rho v_{out}A_t = \rho v_{in}A_t \tag{5}$$

the equation (4) can be rewritten as equation (6):

$$v_{out}^2 = \frac{1}{A_T(1 - f_{total}\frac{L}{D_H})}(-\dot{m}_{exh}v_{exh} + \dot{m}_S v_S\cos(\alpha)) \tag{6}$$

The equation (6) shows that the velocity in the tunnel (v_{out}) depends only on v_S and α, while $\dot{m}_S$, $\dot{m}_{exh}$ and f_{total} were design parameters. Through the CFD investigation, the influence of the pitch angle α and the jet velocity v_S are investigated in terms of Performance of the Ventilation System "P_{vs}", defined as ratio between the mass flow rate at the outlet tunnel portal ($\dot{m}_{out}$) and the mass inflow provided by Saccardo ($\dot{m}_S$), as reported in equation (7):

$$P_{vs} = \frac{\dot{m}_{out}}{\dot{m}_S} \tag{7}$$

3. CFD model

3.1. Mathematical model description

The time Averaged mass Navier–Stokes combined with LES Standard model equations (8-10) describe the unsteady and turbulent states and three-dimensional flux, in the fluid region. These equations model the motion of viscous fluid substances:

$$\text{the mass conservation } \frac{\partial \rho}{\partial t} = \nabla \cdot (\rho \vec{v}\ ) = 0 \tag{8}$$

$$\text{the momentum conservation } \frac{\partial}{\partial t}(\rho \vec{v}) + \nabla \cdot (\rho \vec{v}\vec{v}) = -\nabla p + \nabla \cdot \bar{\bar{\tau}} + \rho \vec{g} + \bar{S}_{fan} \tag{9}$$

$$\text{the energy conservation } \frac{\partial}{\partial t}\left[\rho\left(u + \frac{v^2}{2} + |\vec{g}|z\right)\right] + \nabla \cdot \left[\rho \vec{v}\left(h + \frac{v^2}{2} + |\vec{g}|z\right)\right] - \nabla \cdot (\bar{\bar{\tau}} \cdot \vec{v}) + \nabla \cdot \vec{q} = 0 \tag{10}$$

3.1.1. Boundary conditions

In this numerical CFD analysis, carried out by FDS simulator [11], the following boundary conditions were imposed:

- the jet fans installed in Saccardo ventilation system are simulated as simple momentum source able to obtain:
  - total inflow mass flow rate $\dot{m}_s$ up to 202 kg/s;
  - total outflow mass flow rate of exhausts $\dot{m}_{exh}$ up to 202 kg/s;
- the air is considered an ideal gas with properties valuated at the ambient temperature (20 °C);
- the ceiling and floor roughness height of tunnel, modelled by "log wall model", are set equal to 0.005 m;
- the pitch angle of the Saccardo system is set equal to 10° [5];
- the pressure at inlet and outlet portals is set equal to 101325 Pa.

3.2. Sensor positions

Three types of sensors were placed over of the simulation domain: velocity magnitude (v), axial velocity (v_x), static pressure (p). The figure 4 shows how the sensors are arranged vertically and transversely in the tunnel.

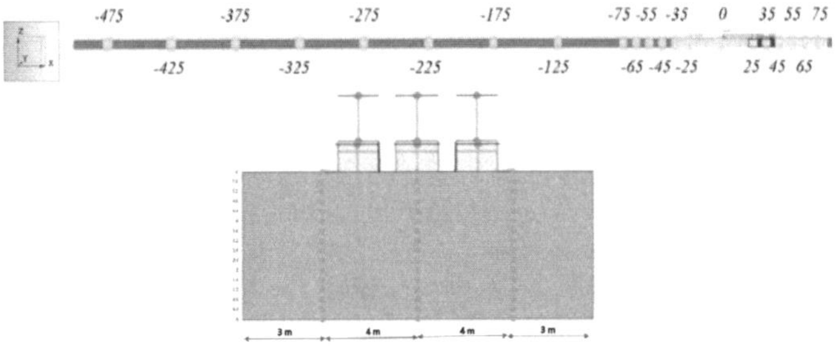

Figure 4. Sensor's position, on each point in the sketch, 3 thermo-fluid-dynamic sensors are placed: velocity (v), axial velocity (v_x), static pressure (p).

3.3. Mesh choice and model validation

The domain is divided in two mesh zones, the mesh type A (0.5 m) covers the computational domain longitudinally (x-axis) from -500 m to -38.5 m and from 81.5 m to 500 m (see figure 5); the mesh type B (0.25 m) covers the zone where Saccardo nozzle and exhaust dampers have placed as shown in figure 5. The mesh types are the same (0.5 m and 2.5 m) in both congested and empty tunnel traffic. The mesh choice was carried out by comparing the CFD results and experimental ones (reference model) provided by Hang et al. [5], in terms of air flow velocity and total static pressure. The figure 6 shows the good agreement between the CFD simulation and CFD-JG Hang (validated by experimental results) in terms of air velocity and gauge pressure. This result allows the CFD model validation. Figure 7 shows the air average velocity and mass flow trends over time; one note that the stationary condition is reached after about 300 s. During the simulation, only one Saccardo fan was kept running, the one closest to the left wall of the tunnel, with the corresponding extraction module active [5].

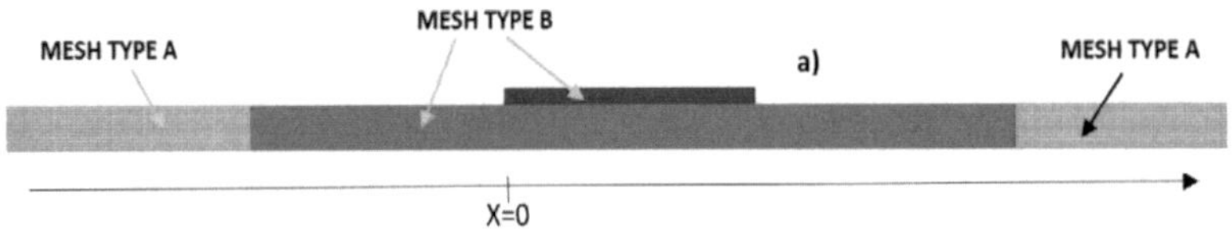

Figure 5. Mesh size and connection of the two types of mesh adopted.

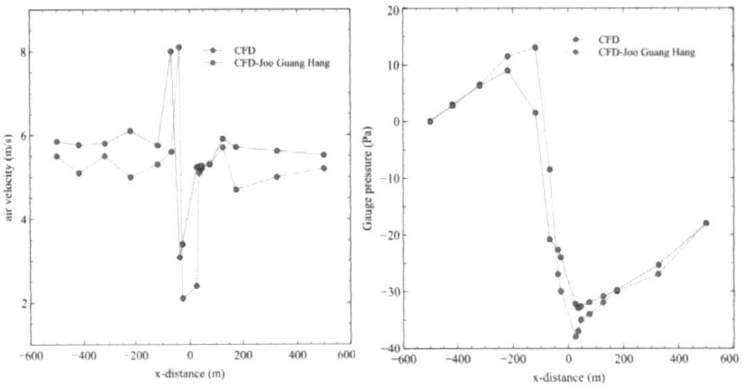

Figure 6. CFD model validation.

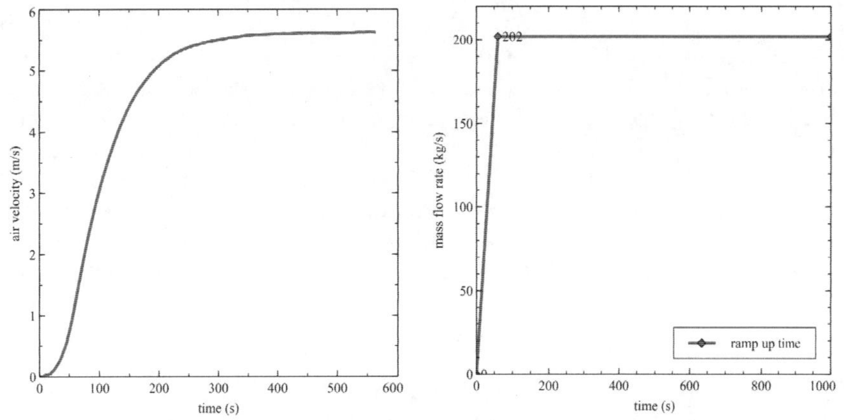

Figure 7. At exit portal of the tunnel: left side) average air velocity vs time, right side) mass flow rate vs time.

4. CFD INVESTIGATION

In this paragraph, the Performance of the Ventilation System "P_{vs}" versus the magnitude and direction (pitch angle) of the inflow velocity v_s is discussed. The P_{vs} is evaluated for different inflow air direction (pitch angle) in congested traffic condition and velocity magnitude in empty tunnel condition. About the first case, the cars are modelled by two overlapped parallelepipeds. The vehicle dimensions and their positioning inside the tunnel are reported in the table 2. In the second case, different velocity magnitude is achieved just reducing the inflow opening, with constant mass flow rate.

Table 2. Car dimensions and positioning inside the tunnel

	unit/m
Total high of cars	1.5
Bottom parallelepiped high	1.0
Top parallelepiped high	0.5
Top parallelepiped length	2.0
Car width	2.0
Car length	4.5
Distance between the wall of tunnel and cars	2.3
Distance between two consecutive cars	1.0
Distance between two lateral cars	2.0

4.1. Pitch angle influence analysis

Three different pitch angles, 8°, 10° and 12°, were investigated numerically to evaluate the interaction between the air flux direction and cars with the consequently pressure losses. The pitch angle has not been further reduced (less than 8°) because the high jet velocity tends to deflect the jet towards the tunnel ceil due to the 'Coanda effect [8] increasing the friction losses and it has not been further increased (greater than 12°) because the pressure drops tends to increase due to the air flow of the Saccardo system directed on the cars.

The figure 8 (left side) shows that the mass flow rate $\dot{m}_{out}$ is not influenced significantly by the pitch angle and consequently the P_{vs}. The figure 8 (right side) shows that the pitch angle 8° degree seems improve slightly the ventilation in the road tunnel. Therefore, in the following a pitch angle of 8° is adopted.

4.2. Influence of inflow opening length of the Saccardo ventilation system

Having fixed the angle of inclination equal to 8°, we have investigated the influence of the inflow opening length reduction on Saccardo system performance. The different configurations investigated consist in reduction of the opening length of n x 1.8 m, for n=1 to 5, starting from the base configuration of 21 m. The results are provided in terms of P_{vs} vs opening length reduction (figure 9, left side) and Saccardo outflow velocity vs opening length reduction (figure 9, right side). We can note that as the opening length decreases the P_{vs} and Saccardo outflow velocity increase. The table 3 shows the influence of different inflow opening lengths on the mass flow rate at the outlet tunnel portal $\dot{m}_{out}$ and consequently P_{vs}.

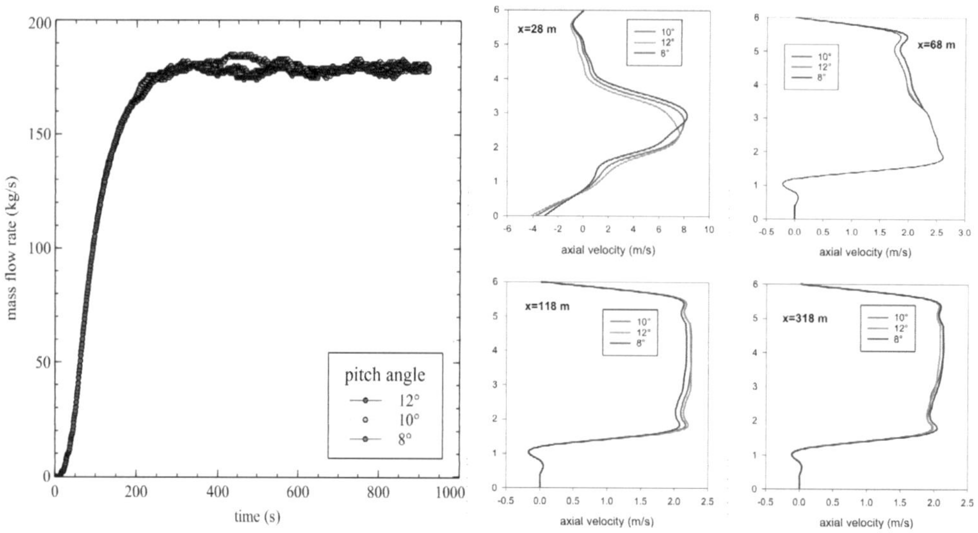

Figure 8. Left side: Mass flowrate vs time with pitch angle as parameter, velocity profile of air in the tunnel at right side: different 'x' distance from Saccardo inflow.

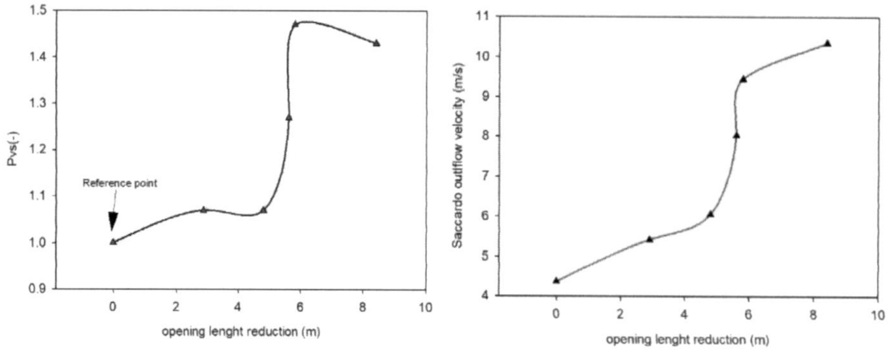

Figure 9. Left side: P_{vs} vs opening length reduction (m); right side: outflow velocity (m/s) vs opening length reduction (m).

Table 3. Inflow opening lengths vs mass flow rate and P_{vs} fixed the $\dot{m}_s$(=202 kg/s)

L_{inflow} (m)	$\dot{m}_{out}$(kg/s)	P_{vs}
21.0	201.6	1.00
18.1	216.0	1.07
16.2	216.0	1.07
14.4	254.4	1.27
14.2	295.2	1.47
12.6	286.8	1.43
21.0	201.6	1.00
18.1	216.0	1.07
16.2	216.0	1.07

5. Conclusion

In this work the authors have presented the results of a 3D numerical investigation carried out on a Saccardo ventilation system operating in road tunnel. The boundary conditions and the geometrical dimensions of the ventilation system were the same of that investigated experimentally by Hang et al. [5] used to validate the CFD model. Through the CFD investigation, the influence of the pitch angle $\boldsymbol{\alpha}$ and the jet velocity $\boldsymbol{v_S}$ were investigated in terms of Performance of the Ventilation System "P_{vs}", defined as ratio between the mass flow rate at the outlet tunnel portal ($\dot{\boldsymbol{m}}_{out}$) and the mass inflow provided by Saccardo ($\dot{\boldsymbol{m}}_S$). The results have shown that the pitch angle of the Saccardo nozzle influences moderately the P_{vs}, instead different opening inflow lengths influence considerably the mean air velocity in the tunnel and mass flow rate. A P_{vs} trend vs opening inflow length has shown an improvement up to 47% for an inflow length of 14.2 m (6.8 m inflow length reduction), this allows to obtain a significant energy saving.

6. References

[1] Tarada F and Brandt R 2009 *Proc. ISAVFT* **1** 95-107
[2] Alston J, Kandra D and Potter R 2013 *Proc. ISAVFT* **1** 89-104
[3] Sturm P, Beyer M and Bacher M 2013 *Proc. ISAVFT* **1** 105-119
[4] Waymel F, Béraud E, Lorenz M and Rathat L 2019 *Proc. ISAVFT* **1** 223-239
[5] Hang JG, An H, Wang PC, Yu SCM, Chan S, Thong M, Lim LW, Cheong MK and Louis G 2021 *TUST* **118**,104185
[6] Ganjiazad R, Kazemipour A, Afshin H and Farhanieh B 2013 *Appl. Mech. Mater.* **527**: 146
[7] Alston J and Richard Potter 2014 *ASHRAE Transactions*, **120**: 00012505
[8] Chammem T,Vauquelin O and Mhiri H 2014 *TUST* **41**: 53
[9] Casetta F, Musto M, Rotondo G, D'Alessandro C and De Maio D 2019 *TUST* **86**: 165
[10] Casetta F, Musto M, Rotondo G, D'Alessandro C and De Maio D 2020 *TUST* **96**:103213
[11] FDS User's Guide, January 18, 2017 FDS Version 6.5.3

Journal of Physics: Conference Series **2685** (2024) 012028 doi:10.1088/1742-6596/2685/1/012028

Critical assessment of thermal conductivity models for Miscibility Gap Alloy-based composite Phase Change Materials for high temperature Thermal Energy Storage

M. Molteni[1], I. M. Carraretto[2], P. Bassani[3], E. Gariboldi[1], A. Lucchini[2] and L. P. M. Colombo[2,*]

[1] Dipartimento di Meccanica, Politecnico di Milano, via Giuseppe La Masa 1, 20156 Milano (MI), Italy
[2] Dipartimento di Energia, Politecnico di Milano, via Lambruschini 4a, 20156 Milano (MI), Italy
[3] CNR-ICMATE, via Previati 1/E, 23900 Lecco (LC), Italy

* Corresponding author: luigi.colombo@polimi.it

Abstract Miscibility Gap Alloys (MGAs), such as Al-Sn-based systems, provide a viable solution for the development of composite Phase Change Materials (PCMs) for Thermal Energy Storage (TES) purposes. Their successful production depends on the cooling rate imposed to the melt. Finite Element Analyses (FEA), which relies also on thermal conductivity values, represent a powerful tool for the design of the production process. Thermal conductivity, which depends on the arrangement of the phases in the system, also affects the thermal response of the alloy. In the view of evaluating the impact of the phase morphology, the authors adapted some of the models developed for composites and solutions to Al-Sn and Al-Sn-Si-Mg alloys, characterized by broad solidification ranges in terms of composition and temperature and by significantly different phases thermal conductivity. In the fully-liquid range, Filippov and Novoselova model was selected for the description of both alloys. Models that consider sphere-like dispersions give values quite close to the theoretical upper Wiener bound when the high-melting phase is solid. The phase morphology impact is relevant when the solidification range is considered. The resulting arrangement-related thermal conductivity curves are compared to those supplied by CALPHAD-based software and to available literature data.

1. Introduction

In the last decades, Phase Change Materials (PCMs) have increasingly attracted the attention of the scientific community as a promising response to the incoming energy hunger for their capability of absorbing/releasing heat at constant temperature while undergoing phase transition. In general, the most exploited transformation is the solid-liquid one. This perspective makes PCMs very appealing for Thermal Energy Storage (TES) and Thermal Management (TEM) purposes [1]. Recently, the research has been focused on metallic PCMs [2] due to their moderate to high melting temperature, combined with high thermal conductivity and latent heat of fusion per unit volume. However, the difficulty of managing metals in the molten state, especially for their corrosivity and reactivity [3], represents a strong limitation to their use as PCMs. An approach to overcome this issue is the adoption of Miscibility Gap Alloys (MGAs) [4], i.e. systems composed of two different elements that do not interact in both solid and liquid states [5]. Their microstructure resembles that of two-phase composites, in which the compositions of solid phases are close to the pure starting elements. The Al-Sn binary system represents a possible solution [4]. The melting onset temperature of the alloy is also close to the one of the pure Sn (232°C, highlighted by the red line in Figure 1(a)), i.e., the active PCM, and can be modified adding other elements, such as Si [6,7]. While producing the PCM material, the arrangement of phases has to be carefully controlled for the achievement of TES/TEM purposes. In particular, the active phase, i.e., PCM, which undergoes solid-liquid transition, should be well-dispersed in the thermally stable and highly thermally conductive passive phase (Figure 1(b)).

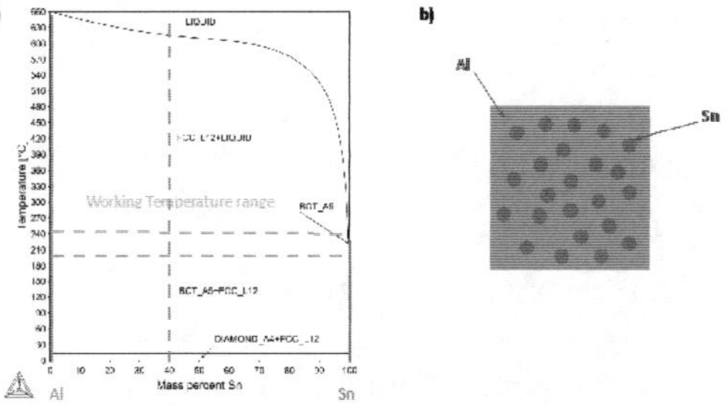

Figure 1. Al-Sn binary system showing the investigated Al-40Sn as vertical green dashed line (a) and microstructural sketch for the target MGA microstructure (b).

This microstructure could be obtained by mixing the phases in solid state [8], or by the molten metal route, if high solidification and cooling rates are considered [7,9]. In this view, the simulation of the solidification process based on Finite Element Analysis (FEA) represents a valuable tool for the correct design of these materials when produced by casting techniques. FEA simulations rely on the knowledge of the material thermophysical properties, i.e., density, specific heat capacity at constant pressure and thermal conductivity. Among them, thermal conductivity depends on the arrangement of the active phase within the matrix [10]. In addition, thermal conductivity determines the thermal response of the system. However, literature offers few data on the thermal conductivity of Al-Sn-based systems. On the other hand, the commercially available software Thermo-Calc [11] is commonly used to determine the properties of a multicomponent system, among which thermal conductivity. Its calculations rely on the CALPHAD approach, a semi-empirical approach based on experimentally obtained thermodynamic property databases. These latter allow the derivation of polynomial expressions, which describe the thermal conductivity temperature dependency. The composition influence instead, is estimated with Redlich-Kister expression [12]. It is worth mentioning that only phase arrangement-independent results are supplied by the software. Hence, the aim of this work consists in evaluating the impact of second-phase arrangement on the thermal conductivity of MGA alloys with the means of simplified analytical models. These latter were adapted from literature-granted models and developed in three different temperature-steps: fully solid, fully liquid and solidification ranges. The investigated MGA alloys are obtained by adding of 40mass% Sn to both pure Al and Al-7Si-0.4Mg alloy. The resulting systems solidify over a broad composition range and are composed of phases with different thermal conductivity. Cooling rate changes can result in a different phase-arrangement, that may lead to a consistent thermal conductivity data scatter. These latter influences both the final microstructure and the thermal response of the system. In this view, the obtained results are further compared to those directly supplied by Thermo-Calc software [11] and to available literature data.

2. Materials and methods

The study focuses on a binary Al-Sn alloy and a more complex system also containing Si and Mg. For both, the nominal mass content of Sn is 40%, as shown in Table 1, where also the reference names used in the paper are given. Figure 1 shows the presence of a namely pure Al phase (light blue line in Figure 1(a)) during and at the completion of the solidification of Al-40Sn alloy. Similarly, previous studies for the Al-7Si-0.4Mg-40Sn, revealed that in most of the solidification range and at the end of it, there is a solid phase namely matching the chemical composition of the starting Al-7Si-0.4Mg [7]. The above situations suggested the possibility to consider Sn-bearing MGAs in their solid stage as composites made of a pure Sn and an Al/Al-7Si-0.4Mg phases. For this reason, thermal conductivity models were developed and compared on the bases of the thermal conductivity and amount/compositions of the two pure phases.

Table 1. Chemical compositions of the investigated alloys, expressed in weight percentages, simulated with Thermo-Calc.

	Al	Sn	Si	Mg	Fe	Ti
Al-40Sn	60	40	-	-	-	-
Al-7Si-0.4Mg-40Sn*	55.315	39.87	4.427	0.237	0.082	0.069

*The composition is calculated from Al-7Si-0.4Mg, considered in [13].

2.1. Thermo-Calc simulations

The thermodynamic equilibrium calculator of Thermo-Calc software [11] was adopted for the estimation of thermal conductivity, mass-normalized composition and amounts of liquid and solid phases of Sn-bearing alloys in Table 1 at temperatures from 20 to 700°C. At boundary temperatures, all of them were in fully solid and fully liquid state, respectively. A temperature step of 10°C, which automatically gets refined in proximity of phase transitions, was selected. The same has been done for Al, Sn and Al-7Si-0.4Mg to create a consistent dataset of reference phase properties to estimate the arrangement effect.

2.2. Analytical models for Sn-bearing alloys thermal conductivity

The thermal conductivity models for liquid phase (k_c) consider it as a solution. Their analytical expressions (Table 4) include the thermal conductivity of the solvent (k_1) and solutes (k_i, $i \geq 2$), and the corresponding mass (atomic) fractions n_i (x_i). The solvent is considered as the phase with volume fraction ≥0.5. For the physical meaning of the models, including $\phi_{k,ij}$, y and w parameters, the reader can refer to [14] and [15].

Table 2. Models adopted for the estimation of the thermal conductivity of the liquid phase.

Model	Equation
Gas mixtures [14]	$k_c = \sum_{i=1}^{n} \frac{x_i k_i}{\sum_{j=1}^{n} x_j \phi_{k,ij}}$
Dul'nev & Zarichnyak [15]	$k_c = k_1 \left[y^2 + w(1-y)^2 + \frac{2wy(1-y)}{wy + (1-y)} \right]$
Filippov & Novoselova [15]	$k_c = k_1(1-n_2) + k_2 n_2 - 0.72(k_2 - k_1) n_2 (1-n_2)$

In the fully solid and solidification temperature ranges, when at least one of the two phases is solid, the analytical models designed for two-phases composite materials were adopted for the description of the Sn-bearing system. The models consider second phases dispersed in a matrix. The matrix is the phase with volume fraction ≥0.5. Al/Al-7Si-0.4Mg and Sn were considered for the fully solid range, whereas solid Al/Al-7Si-0.4Mg and Sn-rich liquid for the solidification range. k and v stand for the thermal conductivity [W m^{-1}K^{-1}] and the volume fraction of phases, respectively. The subscripts 1 and 2 refer to the high melting phase matrix and the second phases.

The analytical models considered for the above temperature ranges consider either a spherical dispersion or a fibre-like geometry of the same phases (Table 3). For exhaustive understanding of the former models and parameters A, B and ψ in Lewis-Nielsen model the reader is referred to [10]. The values of A, B and ψ depends on the second phases arrangement (BCC, FCC and SC for spheres and hexagonal, SC, 1D and 3D random for the fibres). The detailed explanations of models and parameters ζ and γ, listed in Table 4, can be found in [10] and [16]. Another particular arrangement considered by analytical thermal conductivity models is the one here presented in Figure 2. The equivalent thermal network for the control volume is described by German in [17] (Figure 2).

Table 3. Models adopted in the fully solid and solidification ranges for the description of composites thermal conductivity with spherical dispersion.

Model	Equation
Maxwell [10]	$k_c = k_1\left\{1 + 3v_2\left[\left(\frac{k_2 - 2k_1}{k_2 - k_1}\right) - v_2\right]^{-1}\right\}$
Rayleigh [10]	$k_c = k_1\left\{1 + 3v_2\left[\left(\frac{k_2 - 2k_1}{k_2 - k_1}\right) - v_2 + 1.569\left(\frac{k_1 - k_2}{3k_2 - 4k_1}\right){v_2}^{10/3}\right]^{-1}\right\}$
Lewis-Nielsen [10]	$k_c = \frac{1 + AB\psi}{1 - B\psi v_2}$
Bruggeman [10]	$(1 - v_2)\frac{k_1 - k_c}{k_1 + 2k_c} + v_2\frac{k_2 - k_c}{k_2 + 2k_c} = 0$

Table 4. Models adopted in the fully solid and solidification ranges for the description of composite thermal with fibrous dispersion.

Model	Equation
Rayleigh$_{par}$ [10]	$k_{c,\parallel} = k_1\left[1 + \left(\frac{k_2 - k_1}{k_1}\right)v_2\right]$
Rayleigh$_{per}$ [10]	$k_{c,\perp} = k_1\left[1 + \frac{2v_2}{C_1 - v_2 + C_2(0.30584{v_2}^4 + 0.013363{v_2}^8)}\right]$
Halpin-Tsai [16]	$k_c = \frac{2}{3}\left(\frac{1 + \zeta\gamma v_2}{1 - \gamma v_2}\right) + \frac{1}{3}(v_2k_2 + v_1k_1)$

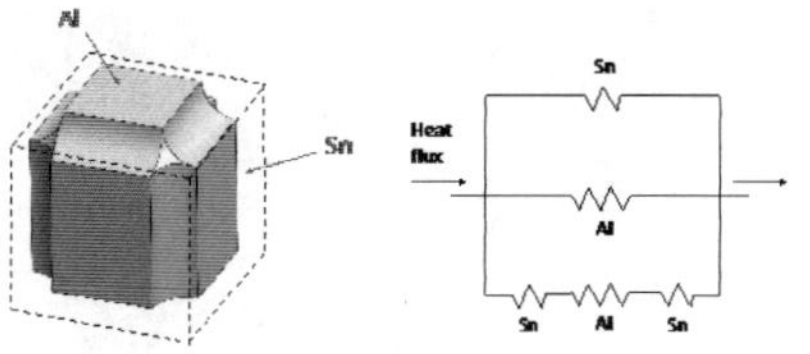

Figure 2. Sketch of the control volume and equivalent thermal resistance calculation for German model [17].

Finally, the above models are compared to the theoretical upper ($k_{c,u}$) and the lower ($k_{c,l}$) Wiener bounds for composite thermal conductivity, which corresponds to two-phase composites with second phases arranged perpendicularly or parallel to heat flux are described by eqs. (1) and (2), respectively [18]:

$$k_{c,u} = k_1v_1 + k_2v_1 \quad (1)$$

$$k_{c,l} = \frac{k_1k_2}{k_1v_2+k_2v_1} \quad (2)$$

3. Results and discussion

3.1. Modelled thermal conductivity vs. temperature curves

The comparison among the thermal conductivity vs temperature curves of the investigated MGAs, derived from phase arrangement-related analytical models and those directly obtained by Thermo-Calc simulations, is presented and commented here with reference to the various solidification steps.

3.1.1. Fully liquid range

The comparison of thermal conductivity curves for the two alloys shown in Figure 3 suggests that predictions can significantly vary with the adopted model. Further, the thermal conductivity of the quaternary alloy, mainly differing in weight by 4.42%Si from the binary Al-40Sn, shows significant differences with respect to this latter. "Gas mixture" model provides the highest conductivities for the alloys, among the models proposed. The conductivity change induced by the presence of Si and Mg is particularly relevant for Dul'nev and Zarichnyak model, mining its reliability. The semi-empirical Filippov and Novoselova equation displays thermal conductivity curves close to those directly simulated by Thermo-Calc. Table 5 quantifies the predictions by different models with respect to simulations at the lowest and highest temperatures for fully liquid range. For these reasons, the authors adopted Filippov and Novoselova model for the description of the liquid phase in the solidification range.

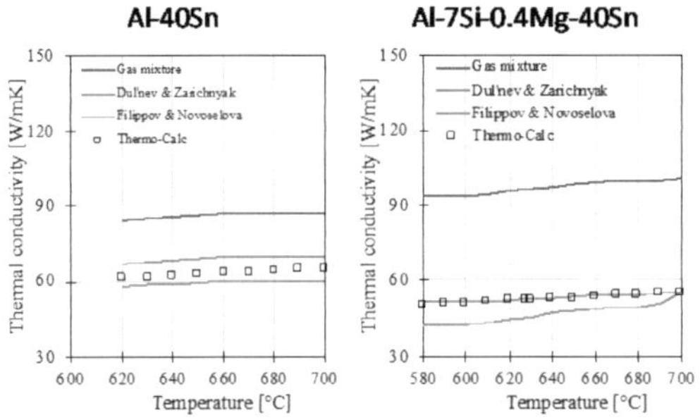

Figure 3. Thermal conductivity predicted by analytical models and by Thermo-Calc in fully liquid range.

Table 5. Percentage differences in the thermal conductivity at the lowest and highest temperatures in the fully liquid range among the models proposed and Thermo-Calc data, with respect to the latter.

Relative difference [%]	Al-40Sn liquid		Al-7Si-0.4Mg-40Sn liquid	
Model	Lowest T	Highest T	Lowest T	Highest T
Gas mixture	-37.768	-34.429	-87.222	-83.352
Dul'Nev & Zarichnyak	-10.128	-7.473	14.518	-0.235
Filippov & Novoselova	5.020	7.263	-2.477	-0.223

3.1.2. Fully solid range

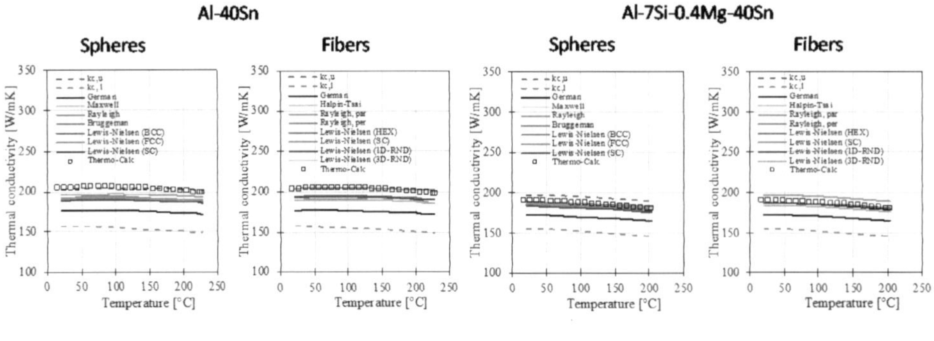

Figure 4. Thermal conductivity predicted by analytical models and Thermo-Calc in fully solid range.

In both the MGAs considered, the thermal conductivity curves predicted by most models are close to the upper Wiener bound (Figure 4), specifically the ones obtained with a spherical dispersion of second phases, described by Maxwell, Rayleigh and Bruggeman. The thermal conductivity decreases as the dispersion arrangement changes from that of regularly arranged spheres (Lewis-Nielsen models) to the 3D arrangement of German model. The thermal conductivity data directly simulated are close to the theoretical upper Wiener bound in the case of Al-40Sn and close to spherical dispersion models for Al-7Si-0.4Mg-40Sn alloy, as quantified in Table 6. Notwithstanding the crude approximation of considering the Al-7Si-0.4Mg as a unique phase, the differences between the simulated data and spherical arrangement analytical models is lower than 4%. In general, minimal changes in the fully solid range, where phase compositions remain almost constant along the whole range, were confirmed by simulation derived curves. At least part of decrement in thermal conductivity as the temperature increases can be related to the minimal increment in solubility of Al/Mg in Sn-rich phase and vice-versa. This effect is taken into account in Thermo-Calc simulations but not in the models, simply based on the thermal conductivity of pure Sn and Al/Al-7Si-0.4Mg.

Table 6. Percentage differences in the thermal conductivity among the proposed models with respect to Thermo-Calc data at the lowest and highest temperatures of both the fully solid and solidification ranges.

Relative difference [%]	Al-40Sn solid		Al-7Si-0.4Mg-40Sn solid		Al-40Sn solidification		Al-7Si-0.4Mg-40Sn solidification	
Model	Lowest T	Highest T	Lowest T	Highest T	Lowest T	Highest T	Lowest T	Highest T
$\mathbf{k_{c,u}}$	-0.171	-2.230	-3.211	-5.224	-0.399	4.128	-18.357	-24.577
$\mathbf{k_{c,l}}$	22.528	24.575	18.376	18.784	47.284	4.175	7.142	-24.564
German	12.961	12.985	9.34	8.575	24.784	4.044	-8.829	-24.567
Maxwell	3.788	2.212	0.592	-1.119	6.022	4.144	-16.037	-24.573
Rayleigh	3.798	2.223	0.601	-1.109	6.042	4.144	-15.401	-24.573
Bruggeman	3.598	5.211	1.252	0.057	6.801	4.143	-16.887	-24.570
L.N. (BCC)	6.461	4.750	3.023	1.460	15.118	4.602	-11.840	-24.666
L.N. (FCC)	5.474	4.078	2.201	0.603	11.074	4.342	-13.827	-24.612
L.N. (SC)	7.295	5.622	3.814	2.283	18.964	4.872	-9.840	-24.722
Halpin-Tsai	6.571	5.392	3.256	1.788	10.920	4.151	-14.104	-24.57
$\mathbf{Rayleigh_{par}}$	-0.171	-2.230	-3.211	-5.224	-0.399	4.129	-18.357	-24.577
$\mathbf{Rayleigh_{per}}$	6.568	5.388	1.504	-0.196	10.911	4.151	-14.104	-24.570
L.N. (HEX)	4.769	3.316	1.531	-0.104	8.22	4.191	-15.138	-24.581
L.N. (SC)	5.083	3.648	1.829	0.206	9.789	4.242	-14.412	-24.601
L.N. (1D-RND)	4.975	3.534	1.727	0.010	11.210	4.253	-14.661	-24.594
L.N. (3D-RND)	6.936	5.610	3.590	2.041	18.241	4.849	-10.206	-24.718

3.1.3. Solidification range

The thermal conductivity vs. temperature curves shown in Figure 5 are quite spread apart, both for spherical and fibres models, suggesting a strong effect of phase arrangement. This influence is particularly evident in the intermediate temperature region, where, as the temperature decreases, the liquid phase enriches in Sn reducing its thermal conductivity. Besides, all the models predict the strong thermal conductivity reduction close to 600°C corresponding to the melting range of the Al-rich phase and the abrupt change related to Sn-rich phases solidification (200-230°C). The analytically-derived thermal conductivity curves are close to the upper bound along the whole solidification range and to the simulation-calculated up to about 350°C for Al-40Sn. At higher temperature simulated thermal conductivity values decrease more rapidly but in a smoother way than those analytically modelled. For the quaternary alloy instead, Thermo-Calc-derived curves are coherent with those derived from analytical spheres-based models along the solidification range, but they display steeper changes at its extreme portions. For both alloys, the rapid changes in the solid and the liquid compositions, possibly lead to the different trends shown in the graphs.

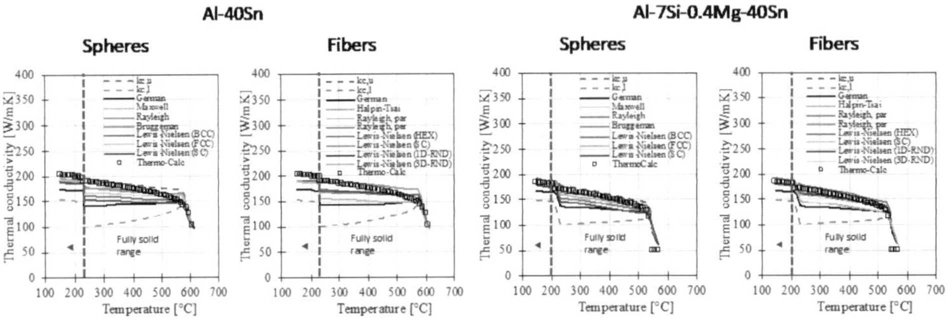

Figure 5. Thermal conductivity predicted by analytical models and Thermo-Calc in the solidification range.

3.2. *Comparison with literature data*

The modelled thermal conductivity curves for the Sn-containing systems, in particular those obtained by Bruggeman model and simulations, are compared to gathered literature data in order to evaluate the accuracy of the calculations. However, to the authors' best knowledge, there is a lack of experimental results regarding the investigated alloys. In the case of Al-Sn binary alloys, Meydaneri [19] investigated the thermal conductivity of a set of alloys including Al-50wt.%Sn and Al-25wt.%Sn. Their thermal conductivity well match those obtained by Li [20] by applying the Lattice Monte Carlo method to their microstructure considering the phases as pure Sn and Al. These data are presented in Figure 6(a). The best-fitting of thermal conductivity vs. Sn volume content data suggest coherent values and trends with both the Bruggeman and the Thermo-Calc curves (Figure 6). Lattice Monte Carlo method has also been applied by Rawson [18] to a Al-40Sn alloy obtained with a different production technique, leading to different microstructure. The author considered other values for the conductivity of the phases, which lead to significantly lower thermal conductivity curve (Figure 6(a)). This result highlights the importance of both the microstructure and the thermal conductivity data. No data instead was found for the quaternary Al-7Si-0.4Mg-40Sn alloy. A comparison with Al-25Sn-5Si [6] was attempted considering a mixture of 25 mass% of Sn phase and 75% of an Al-6.6%massSi phases, close Al-7Si-0.4Mg. The agreement between thermal conductivity curves of the Al-7Si-0.4Mg-40Sn and Al-25Sn-5Si alloy, displayed in Figure 6(b), was considered satisfactory. However, no indication were found in [6] regarding the temperature at which the data of solid and liquid were measured. In the graph below, they were plotted at 20°C and 600°C, respectively.

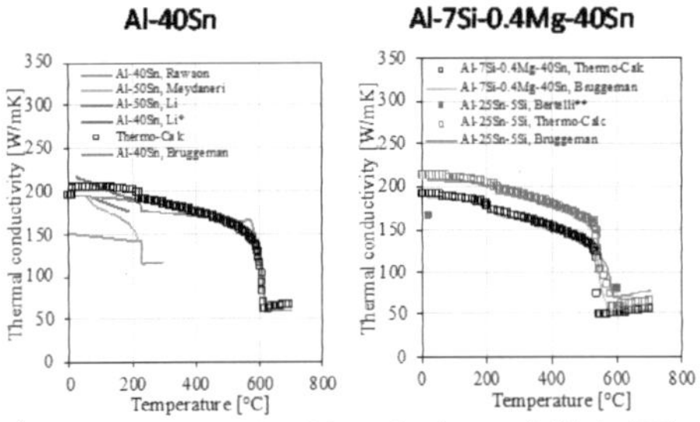

*The curve is interpolated from the data available in [20]

**The temperature related to the thermal conductivity data, here plotted at 20°C and 600°C, is not specified in [6]

Figure 6. Comparison among the experimental or modelled thermal conductivity curves for a) Al-40Sn with also data on Al-50Sn); b) Al-7Si-0.4Mg-40Sn (with also data on Al-25Sn-5Si).

4. Conclusions

This work presents a critical comparison of temperature dependent thermal conductivity data for Al/Al alloy-Sn-based MGAs. These are obtained by classical analytical models for composites/solutions combined with temperature-dependent Al, Sn, Al-7Si-0.4Mg thermal conductivity, liquid phase amount and composition derived from Thermo-Calc software. Thermal conductivity modelled data are compared to those directly obtained from the software for the Sn-based MGAs and to available literature data in the fully solid, fully liquid and solidification ranges. The results suggest that the arrangement of the Sn/Sn-rich liquid phase is particularly important in determining the thermal conductivity, especially in the solidification range. Here, the significant liquid compositional change limits the effectiveness of the models. In particular, the models tend to diverge close to the Sn transition, a temperature range crucial for the PCM applications. The analytically derived curves with the spherical dispersion hypothesis approach the theoretical upper bound, when one of the Al-phase is solid. Moreover, these latter values are quite close with the simulation-calculated ones. Filippov and Novoselova model for solutions was adopted in the fully liquid range. The comparison with the literature is limited by the scarcity of available data, a priority for evaluating the reliability of the modelling tools considered.

5. References

[1] I Sarbu C Sebarchievici 2018 A comprehensive review of thermal energy storage *Sustain. 10*
[2] PJ Shamberger NM Bruno 2020 Review of metallic phase change materials for high heat flux transient thermal management applications *Appl Energy 258 113955.*
[3] A Rawson C Villada M Kolbe V Stahl F Kargl 2021 Suitability of aluminium copper silicon eutectic as a phase change material for thermal storage applications: Thermophysical properties and compatibility *Energy Storage. 1–15.*
[4] H Sugo E Kisi D Cuskelly 2013 Miscibility gap alloys with inverse microstructures and high thermal conductivity for high energy density thermal storage applications *Appl Therm Eng 51 1345–1350.*
[5] JZ Zhao T Ahmed HX Jiang J He Q Sun 2017 Solidification of immiscible alloys: A review *Acta Metall. Sin. (English Lett. 30) 1–28.*
[6] F Bertelli N Cheung IL Ferreira A Garcia 2016 Evaluation of thermophysical properties of Al-Sn-Si alloys based on computational thermodynamics and validation by numerical and experimental simulation of solidification *J. Chem. Thermodyn. 98 9–20.*
[7] P Bassani M Molteni E Gariboldi 2023 Microstructural features and thermal response of granulated Al and A356 alloy with relevant Sn additions *Mater. Des. 229 111879.*
[8] C Confalonieri M Perrin E Gariboldi 2020 Combined powder metallurgy routes to improve thermal and mechanical response of Al–Sn composite phase change materials *Trans. Nonferrous Met. Soc. China. 30 3226–3239.*
[9] C Confalonieri R Casati E Gariboldi 2022 Effect of Process Parameters on Laser Powder Bed Fusion of Al-Sn Miscibility Gap Alloy *Quantum Beam Sci. 6.*
[10] W. Karol 2011 A review of models for effective thermal conductivity of composite materials *J. Power Technol. 95 14–24.*
[11] Thermo-Calc Documentation Set *Thermo-Calc Version 2023a, (n.d.).*
[12] O Redlich AT Kister 1948 Algebraic Representation of Thermodynamic Properties and the Classification of Solutions *Ind. Eng. Chem. 40 345–348.*
[13] L Ceschini A Morri A Morri 2013 Effects of the delay between quenching and aging on hardness and tensile properties of A356 aluminum alloy *J. Mater. Eng. Perform. 22 200–205.*
[14] I Setiawan M Nohtomi M Katsuta 2015 Critical temperature differences of a standing wave thermoacoustic prime mover with various helium-based binary mixture working gases *J. Phys. Conf. Ser. 622.*
[15] GN Dul'nev YP Zarichnyak 1966 Thermal Conductivity of liquid mixtures *Inzhenerno-Fizicheskii Zhurnal. 11 747–750.*
[16] AJ Rawson E Kisi C Wensrich 2015 Microstructural efficiency: Structured morphologies *Int. J. Heat Mass Transf. 81 820–828.*
[17] RM German 1993 A model for the thermal properties of liquid-phase sintered composites *Metall. Trans. A. 24 1745–1752.*
[18] A Rawson E Kisi H Sugo T Fiedler 2014 Effective conductivity of Cu-Fe and Sn-Al miscibility gap alloys *Int. J. Heat Mass Transf. 77 395–405.*
[19] F Meydaneri T Buket 2016 Structural and thermo-electrical properties of Sn – Al alloys *Appl. Phys. A Mater. Sci. Process. 122:906.*
[20] Z Li C Confalonieri E Gariboldi 2021 Numerical and experimental evaluation of thermal conductivity: an application to Al-Sn alloys *Metals 2021, 11, 650.*

Journal of Physics: Conference Series **2685** (2024) 012029 doi:10.1088/1742-6596/2685/1/012029

Evaluating the potential of persistent luminescence in counteracting urban overheating

C Chiatti[1], C Fabiani[1], E Bou-Zeid[2] and AL Pisello[1,3,*]

[1] Engineering Department, University of Perugia, Via G. Duranti, 93 – 06125 Perugia (Italy)

[2] Department of Civil and Environmental Engineering, Princeton University, Princeton, 08544, NJ (United States)

[3] CIRIAF - Interuniversity Research Centre, Via G. Duranti, 67 - 06125 Perugia (Italy)

* Corresponding author: anna.pisello@unipg.it

Abstract. In recent years, the use of numerical simulations to model real atmospheric conditions over cities has become increasingly popular. One of the primary objectives of these models is to assess the efficacy of various strategies for mitigating the Urban Heat Island (UHI) phenomenon. At the same time, researchers have developed and studied new adaptive materials for building applications that aim to reduce buildings' energy consumption and improve urban microclimate conditions, while performing radiative cooling. Among the new generation of passive cooling solutions, persistent luminescent (PL) materials have emerged as a cutting-edge option for energy-saving purposes, owing to their ability to reject the incident solar radiation through both reflection and light emission. Here, the Princeton Urban Canopy Model (PUCM) is used to evaluate the potential of an advanced PL roof coating to counteract urban overheating. The phenomenon of persistent luminescence is modeled for the first time, taking advantage of experimentally obtained parameters coming from previous studies. Results demonstrate how persistent luminescence can effectively mitigate surface overheating reducing the roof's surface temperature and net shortwave radiation up to 1.15 °C and 35 W/m^2 respectively, with consequent benefits to the overall energy balance of the envelope. Such results may be further increased with the optimization of PL materials for engineering solutions.

1. Introduction

The Urban Heat Island (UHI) phenomenon refers to the fact that cities are hotter than nearby rural areas, and it is caused by factors such as the use of artificial materials that retain heat and have low cooling capacity [1]. With urban expansion and population growth, UHI has become a threat to the livability of cities and poses risks to human health, especially for vulnerable populations [2]. UHI also has implications for building energy consumption, leading to increased cooling loads and energy costs. To address UHI, sustainable urban planning and design are crucial, including the use of green infrastructure and cool materials [3]. Sohaili et al. [4] have shown that green roofs can reduce surface temperatures by 4.3 °C compared to traditional dark roofs. Additionally, green roofs can lead to a 3.7 °C reduction in average air temperature and an 8.46% reduction in relative humidity compared to standard bitumen roofs [5]. Despite the promising results, implementing such applications requires extensive economic and practical efforts. For this reason, the attention of the scientific community has focused also on the development of innovative roof applications, characterized by optimized solar reflectance and thermal emittance capability, i.e., cool roof solutions [6]. As a matter of fact, surface treatments are not only cost-effective, but also easy to implement. Moreover, they require significantly less maintenance than traditional green roofs, while still providing comparable temperature reductions to vegetated roofs. Roof paintings have been extensively studied by Synnefa et al. [7]. They found that certain coatings made of transparent polymers and white pigments can significantly reduce surface temperature during the day and night under summer conditions. Levinson et al. [8] found that using NIR-reflective roof tile coatings can result in a reduction in surface temperature and heat flux. More recently, thermochromic materials, which change their optical properties based on temperature, have been investigated to reduce winter heat loss [9]. Fabiani et al. [10] demonstrated that an innovative thermochromic roof configuration can reduce cooling load by 6.59% per square meter, slightly less than a more common cool application, but with no negative effects in winter.

Among the latest generation of cool materials for the built environment, photoluminescent solutions have emerged owing to their ability to combine light reflection and emission [11]. Despite the numerous studies suggesting the suitability of photoluminescence for several UHI mitigation solutions, most of the research available in literature is limited to laboratory-scale investigations focused on chemical compounds [12]. In the present study, the Princeton Urban Canopy Model (PUCM) has been employed to explore the potential application of photoluminescent roof coating as a UHI mitigation strategy. The study aims to investigate the advantages of optimized photoluminescent paintings over purely cool roofs at the inter-building scale. The numerical analyses have been critically evaluated to assess the expected benefits of introducing photoluminescence in the built environment.

2. Materials and methods

2.1. Materials

Various commercially available pigments were selected in the frame of an innovative research and development project and considered as starting-point materials in this work. These pigments are typically dispersed into different solutions to create more complex prototypes that glow in the dark, such as glasses, ceramics, and coatings. The aim of encompassing different photoluminescent emission colors, spanning the entire visible spectrum, led to the selection of the seven compounds showed in figure 1. Table 1 provides a summary of the pigments' composition, including their hosting matrix and doping ions responsible for the specific color of emission, as already characterized in a previous research work [13]. Additionally, the quantum yield (QY), i.e., the ratio between the emitted over the absorbed photons, and the lifetime decay constant (τ) were taken into consideration. The latter was derived from the fitting curve describing the luminance decay of the photoluminescent emission over time ($L_v(t)$, equation 1) after the exposure to solar radiation has ended.

$$L_v(t) = L_v(t_0) \cdot e^{-\frac{1}{\tau}t} \tag{1}$$

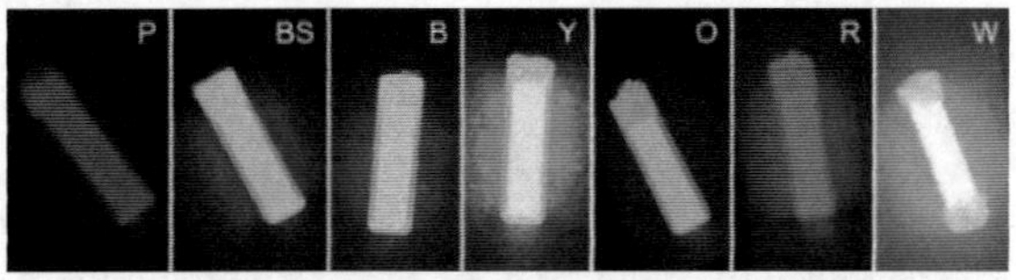

Figure 1. Investigated photoluminescent pigments.

Table 1. Photoluminescent compounds and their main properties, i.e., the wavelength corresponding to the peak quantum yield (QY) and lifetime decay constant (T).

Code	Emission color	Matrix	Doping ions	λ_{PL}[nm]	λ_{AE}[nm]	QY	T
Y	Yellow	$SrAl_2O_4$	Eu^{+2}, Dy^{+3}	520	390	0.44	0.7792
B	Blue	$Sr_4Al_{14}O_{25}$	Eu^{+2}, Dy^{+3}, B^{+3}	490	370	0.88	1.2263
BS	Blue sky	$Sr_2MgSi_2O_7$	Eu, Dy	645	465	0.29	0.9138
P	Purple	$CaAl_2O_4$	Eu^{+2}, Nd^{+3}, La^{+3}	440	326	0.34	0.8607

R	Red	CaS	Eu	468	360	0.62	0.6269
O	Orange	Y_2O_2S	Eu	468, 626*	310	0.18	0.6350
W	White	$YSr_2Al_7O_{14}$	Eu, Dy, B	445, 626*	330	0.30	0.7794

* The first wavelength corresponds to the first peak in the fluorescence profile, while the second corresponds to the maximum intensity of the emission.

2.2. *Methodology*

This study explores the potential of photoluminescent roof coating as a novel strategy for mitigating urban heat island (UHI) effects. To achieve this, we utilized the urban canopy model (PUCM) developed by Wang et al. [14] at Princeton University. The PUCM represents a single-layer urban canyon and incorporates a detailed surface exchange scheme that considers energy transport and hydrological processes over natural and/or engineering surfaces. The PUCM was validated for different climate conditions through in-field monitoring campaigns performed in different towns, in a stand-alone mode driven by 5 to 30-minute averaged meteorological data. In the model, the one-dimensional *interfacial energy balance* for an infinitesimally thin layer of each considered roof, wall, or ground surface, at the surface-air boundary, is defined as:

$$R_{net} = H + LE + G \tag{2}$$

where R_{net} is the net radiation, i.e., the sum of all net shortwave and longwave components, H is the convective sensible heat flux, LE is the latent heat flux, and G is the heat conducted into the wall, ground, or roof. The PUCM treats the canyon air as a separate thermal mass and any anthropogenic emissions are directly released into that mass (except for emissions due to space heating, which are indirectly represented by maintaining a fixed indoor temperature in the winter). Additionally, the model allows for detailed simulation of the thermo-optical properties of surfaces in a canyon environment, thanks to the ability of dividing surfaces into subfacets and accounting for the effects of trees. The model setup, including the specific properties of ground surfaces, building walls, and roofs is similar to those used by Ryu et al. [15] and enable the simulation of the effects of trees on the canyon environment. Table 2 summarizes the main properties of the materials involved in the model.

Table 2. Input parameters used in the PUCM for materials' properties.

Thermo-optic parameters	Value	Thermo-optic parameters	Value
Roof albedo (cool / dark)	0.66/0.16	Roof conductivity	0.94 W/(mK)
Wall albedo	0.25	Wall conductivity	0.94 W/(mK)
Ground albedo (asphalt/concrete)	0.10/0.20	Ground conductivity (asphalt/concrete)	1.20/2.00 W/(mK)
Roof emissivity	0.95	Roof heat capacity	1.40 MJ/(m^3K)
Wall emissivity	0.95	Wall heat capacity	1.40 MJ/(m^3K)
Ground emissivity (asphalt/concrete)	0.95/0.93	Ground heat capacity (asphalt/concrete)	1.80/1.30 MJ/(m^3K)

2.2.1. *Mathematical description of photoluminescence.*

A traditional building envelope surface of finite depth absorbs a net flux of shortwave and longwave radiation ($R_{SW,net} + R_{LW,net} = R_{net}$), and releases the resulting energy into the air as H and LE, and into the solid surface as G at the inner boundary of the surface. The finite depth also allows some storage of energy in the layer. For photoluminescent surfaces, an additional contribution to the rejection of the incident shortwave radiation must be introduced, which is a different energy dissipation term (Q_{PL}) due to radiation emission. Therefore, the surface energy balance of equation 1, with a photoluminescent layer and storage capacity, should be modified to the following equation:

$$R_{SW,net} + R_{LW,net} - LE - H - G - Q_{PL} = mc\frac{dT_{surf}}{dt} \tag{3}$$

where $mc\frac{dT_{surf}}{dt}$ is the amount of heat stored by the surface in time per unit area (it can be considered equal to zero for an infinitesimally thin layer to recover a form similar to equation 1). $R_{SW,net}$ is directly proportional to the incoming shortwave radiation (I_{SW}) and the shortwave solar absorbance (a). On the other hand, $R_{LW,net}$ is related to the incoming longwave from the atmosphere (given as observed input to PUCM) and the outgoing longwave, which depends on the longwave surface reflectivity, the thermal emissivity (ε), the Stefan-Boltzmann constant (σ) and the fourth power of surface temperature (T_{surf}^4). Since the photoluminescent emission of the pigments investigated in this study occurs in the visible range of the electromagnetic spectrum (380-780 nm), representing the effect in PUCM simply requires the modification of the equation of the net shortwave radiation ($R_{SW,net}$) of a light-emitting surface to the following:

$$R_{SW,net} = I_{SW}(1-\alpha) - Q_{PL} \quad (4)$$

where α is the albedo of the surface, determined by the material's capability of reflecting the incident solar radiation.

In recent years, Garshasbi et al. [13] proposed an analytical model to investigate the effect of fluorescent and optical properties on the cooling potential of photoluminescent materials. According to their work, Q_{PL} can be calculated as:

$$Q_{PL} = \frac{h \cdot c}{\lambda_{PL}} \int_{300nm}^{\lambda_{AE}} QY(\lambda) \frac{a(\lambda) \cdot I_{SW} \cdot \varphi(\lambda)}{\frac{h \cdot c}{\lambda}} d\lambda = \left(\frac{1}{\lambda_{PL}} \int_{300nm}^{\lambda_{AE}} QY(\lambda) \cdot a(\lambda) \cdot \varphi(\lambda) \cdot \lambda \cdot d\lambda \right) \cdot I_{SW} \quad (5)$$

where h is the Planck constant; c is the speed of light; λ_{PL} and λ_{AE} correspond to the material's peak of emission and absorption respectively; a(λ) is the material's solar absorption coefficient; I_{SW} is the incoming shortwave radiation; φ(λ) is the spectral distribution of the global solar radiation according to the standard spectrum AM 1.5g. However, equation 5 does not give any information about the decay of the photoluminescent emission when the incoming energizing radiation is turned off; i.e., it describes a purely fluorescent material that is receiving radiation. For this reason, in this work, a model for the description of phosphorescence is introduced for the first time, moving from the equation of Grashasbi et al.:

$$Q_{PL} = \Psi(\lambda_{PL}, \lambda_{AE}, QY, a) \cdot I_{SW} - \tau \frac{dQ_{PL}}{dt} \quad (6)$$

where Ψ is a hypothetical dimensionless term including all the fluorescent properties (a, QY, λ_{PL} and λ_{AE}) and τ is the lifetime decay constant of the photoluminescent compound. Once the last term of the equation is discretized in time using a forward Euler method in PUCM or an equivalent model, the last equation is written as:

$$Q_{PL}^{t} = \frac{\Psi(\lambda_{PL}, \lambda_{AE}, QY, a)}{1+\frac{\tau}{\Delta t}} \cdot I_{SW} + \frac{\tau}{1+\frac{\tau}{\Delta t}} \frac{Q_{PL}^{t-\Delta t}}{\Delta t} \quad (7)$$

where $Q_{PL}^{t=0} = \Psi(\lambda_{PL}, \lambda_{AE}, QY, a)$ and Δt is a 10-second time step.

2.2.2. Numerical experiments design. Three different analyses were carried out in this work. The first analysis involved assessing the impact of photoluminescent roof coatings with different afterglow colors and benchmarking them against a traditional dark (α = 0.16) and cool (α = 0.66) non-luminescent roof. Simulations were performed for three sunny days (3-5 July 2011), to identify the most promising option for passive cooling purposes. The input parameters for the photoluminescence equation were obtained from in-lab experimental data presented in Table 1. The second analysis involved a deeper investigation of the best performing photoluminescent coating through monthly-averaged simulations during the hottest (July-August 2011) and the coldest months of the year (January-February 2012). This analysis aimed to assess the coating's impact on the overall energy balance of the roof. Finally, an ideal photoluminescent material with optimized properties (QY, a, and τ) was simulated to quantify the maximal potential of the coating for UHI mitigation. More in details, QY and a were considered in the maximum value they could achieve, i.e., 100% and 1 respectively. The simulations were driven by in-situ observations collected in Princeton, NJ, from a dedicated weather station with a 30-minute sampling rate. The urban canyon considered in this study had an aspect ratio (building height to canyon width) of 0.80 and an impervious area fraction of 0.84. The area was thus densely urbanized, but its spatial extent was limited, which minimized the impact of the urban surface fluxes on the atmospheric variables and made it suitable for offline UCM applications and testing. The observations are obtained from a dedicated weather station, with a 5-minute sampling rate [14]. The atmospheric variables required for the simulations included air temperature, specific humidity, atmospheric pressure, wind speed, shortwave and longwave solar radiation, and precipitation data. All these variables were interpolated from the original weather file to fit the simulation time interval of 10s.

3. Results

3.1. Comparison between photoluminescent coatings with different afterglow color

Figures 2 and 3 show the differences in surface temperature (ΔT_{surf}) and net shortwave radiation ($\Delta R_{SW,net}$) of a typical dark roof and cool roof, both non-luminescent, relative to their identical counterparts with a photoluminescent coating applied on the outer skin. In particular, three sunny days were simulated (3-5 July 2011) to detect the maximum impact of photoluminescence in the best-case scenario, i.e., under favorable weather conditions, with a good level of solar radiation and no clouds.

Results show how the red-emitting pigments (R) had the greatest cooling potential, providing for a maximum surface temperature reduction of around 0.75 °C if applied to a dark roof ($\alpha = 0.16$), and around 1.15 °C in the case of a cool roof ($\alpha = 0.66$). The net shortwave radiation was also reduced by about 35 W/m^2 in both cases. Blue- (B) and yellow- (Y) emitting coating resulted in less than half the cooling benefit in both ΔT_{surf} and $\Delta R_{SW,net}$, even if they showed good potential for lighting purposes in other studies [16, 17]. Blue sky (BS), white (W) and purple (P) pigments followed in order, while orange (O) afterglow did not provide a significant change in surface temperature and net shortwave radiation if compared to a non-luminescent cool or dark roof. For this reason, the R-emitting coating was investigated deeper through the hottest (July-August) and coldest (January-February) months of the year.

3.2. Best-performing roof coating and its possible optimization

Figures 4 and 5 represent the contribution of the red-emitting coating, selected in section 3.1 as the best-performing for UHI mitigation purposes, in comparison to the behavior of the non-luminescent roof with the same thermo-optical cool properties. The monthly-averaged profiles refer to the hottest (July-August, figure 4) and the coldest (January-February, figure 5) period of the year. The roof surface temperature (T_{roof}) and the net shortwave radiation ($R_{SW,net}$) were taken as reference parameters to compare the performance of the non-luminescent and red-emitting roof.

The behavior of the cool roof was generally good in terms of UHI mitigation potential during the hottest months of the year, when the envelope was positively affected by reduced radiative absorption of incoming solar radiation. However, the presence of a photoluminescent coating on the outer layer of the roof further reduced heat gains, with peak reductions in net shortwave radiation of 27.94 W/m^2, and roof temperature of 0.77 °C. The impact of photoluminescence during the coldest months, instead, inevitably negated part of the winter-time benefits of the UHI phenomenon, but its magnitude was significantly lower than the summer impact (max $\Delta R_{SW,net}$ = 16.85 W/m^2), resulting in a surface temperature reduction of only 0.38 °C if compared to the non-luminescent scenario.

The capability of the red-emitting roof coating to dampen the heat gains during summer with negligible effects for the winter period, paved the way for the investigation on the maximum benefits that could be obtained by applying an optimized photoluminescent material with ideal fluorescent properties. Three scenarios were hypothesized, modeling a photoluminescent coating with the following characteristics: (i) QY = 100% and a = 1 ($R_{opt(QY,a)}$); (ii) $\tau_{id} = \tau \times 10^4$ ($R_{opt(\tau \times 10^4)}$); (iii) QY = 100%, a = 1 and $\tau_{id} = \tau \times 10^4$ ($R_{opt(QY,a,\, \tau \times 10^4)}$). Scenario (i) maximizes the optical potential of the material, while scenario (ii) extends its storage capacity and its re-emission time scale τ. Specifically, the effect on the net shortwave radiation and surface temperature reduction due to the optimized red-emitting coating were quantified and compared to the performance of a non-luminescent roof, during three sunny summer days (3-5 July 2011).

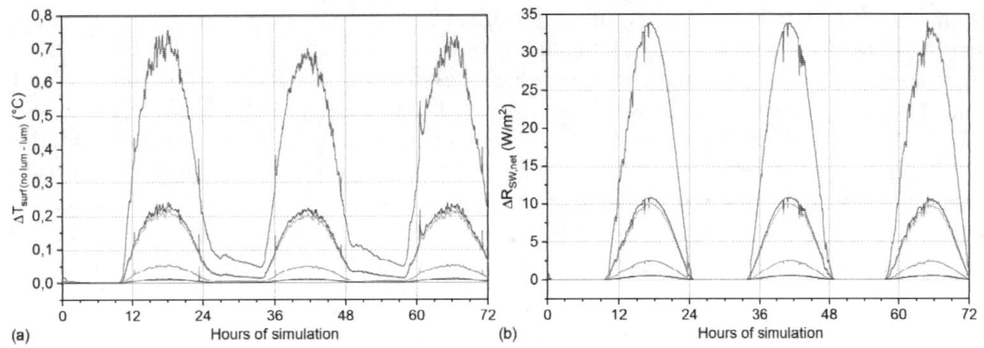

Figure 2. Differences in (a) superficial temperature (ΔT_{surf}) and (c) net SW radiation ($\Delta R_{SW,net}$) of the dark roof in its photoluminescent configuration and its non-luminescent version, according to different emission colors (Y-yellow, B-blue, BS-blue sky, P-purple, R-red, O-orange, W-white).

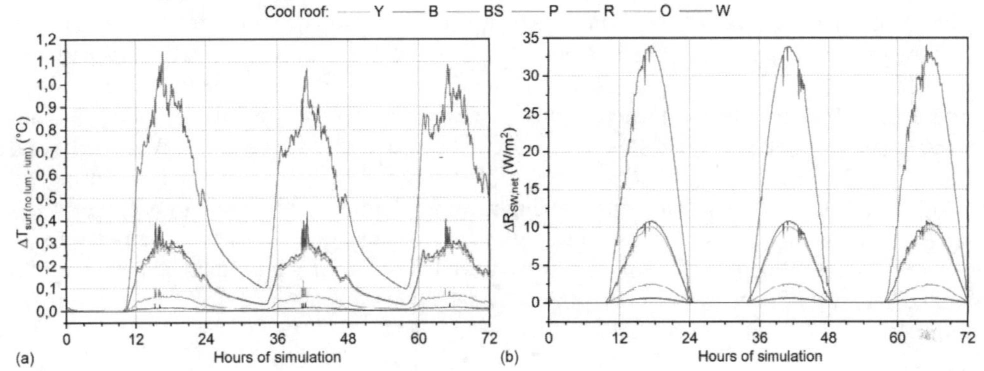

Figure 3. Differences in (a) superficial temperature (ΔT_{surf}) and (c) net SW radiation ($\Delta R_{SW,net}$) of the cool roof in its photoluminescent configuration and its non-luminescent version, according to different emission colors (Y-yellow, B-blue, BS-blue sky, P-purple, R-red, O-orange, W-white).

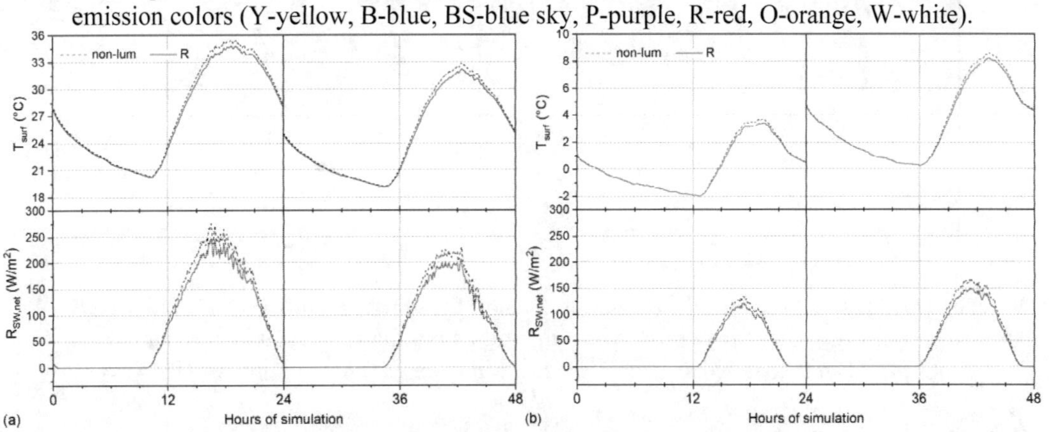

Figure 4. Comparison between superficial temperature (T_{surf}) and net SW radiation ($R_{SW,net}$) of the non-luminescent (non-lum) and red-emitting (R) roof. Monthly-averaged data for: (a) the hottest and (b) the coldest period of the year.

Figure 5 shows how, by maximizing the material's ability to absorb solar radiation and emit light ($R_{opt(QY,a)}$), a more than doubled net shortwave radiation (≈77 W/m^2) and surface temperature (≈2.4 °C) reduction could be obtained with respect to the non-luminescent case, if compared to the impact of a real R-emitting material. Such significant results could be further improved if a more persistent photoluminescent materials can be developed for urban applications. Indeed, when modelling a red-emitting material with a more persistent photoluminescence ($R_{opt(\tau \times 10^4)}$), the impact of the phenomenon itself beomes more evident. The net shortwave radiation of $R_{opt(\tau \times 10^4)}$ and $R_{opt(QY,a, \tau \times 10^4)}$ profiles of figure 5c demonstrate how negative values of $R_{SW,net}$ occurred during the night period, indicating that the radiation emitted from the surface, i.e., the photoluminescence, was greater than the incident radiation. A 3-hour emission corresponded to a lifetime constant of the material equal to $\tau \times 10^4$, which is in line with data from literature on similar R-emitting materials. A longer photoluminescence decay also resulted in a slightly smoother and right-shifted $\Delta R_{SW,net}$ and $\Delta \tau_{roof}$ profiles, meaning that the peak in both net shortwave radiation and surface temperature of the roof were delayed during the day.

4. Conclusions

This study developed a numerical model to simulate the performance of a photoluminescent coating for roofs in mitigating the urban heat island effect. The model compared different photoluminescent compounds with various afterglow colors on both dark and cool roofs. The cooling effect of the compounds, in terms of temperature reduction and shortwave radiation reduction, was measured to identify the best-performing compound. The latter was deeper investigated during the hottest and coldest periods of the year, comparing the impact of photoluminescence on the roof's energy balance to the non-luminescent counterpart. The study found that the photoluminescent coating effectively reduced surface temperature and shortwave radiation when applied to a cool roof, with red-emitting coating showing the highest reduction in both temperature (≈ -1.2 °C) and net shortwave radiation (≈ -35 W/m^2). The effects were more pronounced during the hottest period, while negligible effects were observed during winter. Additionally, the study suggests that by optimizing the photoluminescent material's properties, such as increasing absorption and emission capabilities, even greater reductions in surface temperature (≈ -2.4 °C) and shortwave radiation (≈ -77 W/m^2) can be achieved. The findings highlight the potential of using photoluminescent materials to mitigate the urban heat island effect.

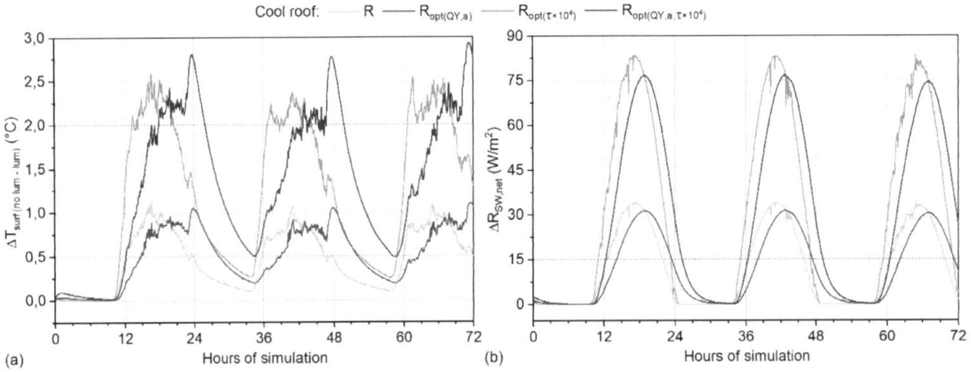

Figure 5. Comparison between differences in (a) superficial temperature (ΔT_{surf}) and (b) net SW radiation ($\Delta R_{SW,net}$) of the cool roof in its red-emitting configuration and its non-luminescent counterpart. The R coating has been considered in its real (R) and optimized (R_{opt}) status.

References

[1] A. G. Silva e M. C. Torres, «Proposing an effective and inexpensive tool to detect urban surface temperature changes associated with urbanization processes in small cities,» *Building and Environment,* vol. 192, p. 107634, 2021.

[2] H. L. Macintyre, C. Heaviside, X. Cai e R. Phalkey, «The winter urban heat island: Impacts on cold-related mortality in a highly urbanized European region for present and future climate,» *Environment International,* p. 106530, 09 2021.

[3] H. Lu, A. Gaur, E. Scott Krayenhoff, Z. Jandaghian, M. Lacasse e T. Moore, «Thermal effects of cool roofs and urban vegetation during extreme heat events in three Canadian regions,» *Sustainable Cities and Society,* vol. 99, p. 104925, 2023.

[4] J. Sohaili, L. Kar Yan, S. K. Muniyandi e S. S. Mohamad, «Urban heat island mitigation using green roof approach,» *Jurnal Teknologi,* vol. 80, n. 3, 2018.

[5] M. Moghbel e R. Erfanian Salim, «Environmental benefits of green roofs on microclimate of Tehran with specific focus on air temperature, humidity and CO2 content,» *Urban Climate,* vol. 20, pp. 46-58, 2017.

[6] M. Rawat e R. Singh, «A study on the comparative review of cool roof thermal performance in various regions,» *Energy and Built Environment,* vol. 3, n. 3, pp. 327-347, 2022.

[7] A. Synnefa, M. Santamouris e K. Apostolakis, «On the development, optical properties and thermal performance of cool colored coatings for the urban environment,» *Solar Energy,* vol. 81, n. 4, pp. 488-497, 2007.

[8] R. Levinson, H. Akbari e J. C. Reilly, «Cooler tile-roofed buildings with near-infrared-reflective non-white coatings,» *Building and Environment,* vol. 42, n. 7, pp. 2591-2605, 2007.

[9] Z. Zhang, L. Zhang, Y. Zhou, Y. Cui, Z. Chen, Y. Liu, J. Li, Y. Long e Y. Gao, «Thermochromic Energy Efficient Windows: Fundamentals, Recent Advances, and Perspectives,» *Chemical Reviews,* vol. 123, n. 11, pp. 7025-7080, 2023.

[10] C. Fabiani, A. Pisello, E. Bou-Zeid, J. Yang e F. Cotana, «Adaptive measures for mitigating urban heat islands: The potential of thermochromic materials to control roofing energy balance,» *Applied Energy,* vol. 247, pp. 155-170, 2019.

[11] C. Chiatti, I. Kousis, C. Fabiani e A. Pisello, «Luminescence for the built environment: from lighting to urban heat island mitigation purposes,» in *Global Urban Heat Island Mitigation*, Elsevier, 2022, pp. 47-69.

[12] C. Chiatti, C. Fabiani e A. Pisello, «Long Persistent Luminescence: A Road Map Toward Promising Future Developments in Energy and Environmental Science,» *Annual Review of Materials Research,* vol. 51, pp. 409-433, 2021.

[13] C. Fabiani, M. Gambucci, C. Chiatti, G. Zampini, L. Latterini e A. Pisello, «Towards field implementation of photoluminescence in the built environment for passive cooling and lighting energy efficiency,» *Applied Energy,* vol. 324, p. 119687, 2022.

[14] Z. Wang, E. Bou-Zeid e J. Smith, «A coupled energy transport and hydrological model for urban canopies evaluated using a wireless sensor network,» *Quarterly Journal of the Royal Meteorological Society,* vol. 139, n. 675, pp. 1643-1657, 2012.

[15] Y. Ryu, E. Bou-Zeid, Z. Wang e J. Smith, «Realistic Representation of Trees in an Urban Canopy Model,» *Boundary-Layer Meteorology,* vol. 159, pp. 193-220, 2016.

[16] C. Chiatti, C. Fabiani e A. Pisello, «Toward the energy optimization of smart lighting systems through the luminous potential of photoluminescence,» *Energy,* vol. 266, p. 126346, 2023.

[17] C. Chiatti, C. Fabiani e P. A.L., «Optimization of photoluminescent materials for lighting energy saving in the built environment,» *Journal of Physics: Conference Series,* vol. 2042, p. 012178, 2021.

Acknowledgments
Funded by the European Research Council within the framework of Horizon Europe Programme (ERC, HELIOS, G.A. 101041255). Views and opinions expressed are however those of the authors only and do not necessarily reflect those of the European Union or the European Research Council. Neither the European Union nor the granting authority can be held responsible for them. EBZ is funded by the Army Research Office under contract W911NF2010216.

Investigating the relationship between surface roughness and reflectance properties of building materials

F Marchini[1,2,*], C Chiatti[2], C Fabiani[2,3], L Latterini[1] and AL Pisello[2,3]

[1] Department of Chemistry, Biology and Biotechnology, University of Perugia, Via Elce di sotto 8, Perugia, 06123, Italy
[2] CIRIAF - Interuniversity Research Center, University of Perugia, Via G. Duranti 67, Perugia, 06125, Italy
[3] Department of Engineering, University of Perugia, Via G. Duranti 93, Perugia, Italy

* Corresponding author: francesco.marchini@studenti.unipg.it

Abstract. Materials used in the exterior envelope of buildings and open urban surfaces, in general, strongly affect the urban thermal balance, determining the general magnitude of urban overheating. The surface temperature of reflective materials varies as a function of physical and geometrical properties. Quantifying the influence of surface roughness on reflectance properties has crucial relevance since reflectance can significantly affect the reduction of the absorbed solar radiation and, in turn, the energy demand for cooling. Through an experimental and statistical investigation, this research aims to analytically assess the impact of surface roughness on the reflectance and thermal performances of building materials and, in turn, the role that roughness could play towards urban cooling and mitigation. Results show that the surface's different roughness affects the sample's reflectance coefficient, leaving it basically unchanged in the Ultraviolet and Visible ranges but with appreciable differences in the Near-Infrared wavelengths. This outcome confirms the correlation between the surface roughness and the optical and thermal characteristics of the building materials, making evident the importance of the study of superficial topography towards the mitigation of Urban Heat Island phenomena.

1. Introduction

The thermal balance of urban areas is significantly influenced by the choice of materials used in the built environment, such as the exterior envelopes of buildings and other open urban surfaces [1,2]. Due to the absorption of the incident solar radiation, the closely packed buildings and paved surfaces that make up our cities trap and amplify heat, exhibiting a high surface temperature and a high release of sensible heat to the atmosphere, increasing the Urban Heat Island (UHI) phenomenon [3,4]. In recent years, scientific researchers have focused on developing mitigation technologies that increase the albedo of cities by reflecting solar radiation [5,6]. For example, Synnefa et al. compared the thermal performances of reflective coatings, showing that they can reduce surface temperature of white concrete tiles even under hot summer conditions [7]. Hernández-Pérez presented a review on the state of the art of the application of reflective materials on buildings' walls and roofs that highlighted a consistent reduction of the daily heat gains and indoor air temperatures, and thus of the daily cooling energy consumptions [8]. However, even highly reflecting white materials can exhibit significant variability in measured daily surface temperatures, despite being of the same colour [9]. Indeed, the surface temperature of reflective white materials varies as a function of several parameters, such as physical (solar reflectance, emissivity, and thermal capacity) and geometrical (surface roughness, texture, size, and thickness) properties.

In particular, superficial roughness can play a relevant role in determining an object's interaction with the environment, affecting the functional properties of its surfaces [9,10]. Irregularities can affect reflectance in several ways, including scattering light in many directions and altering the average slope of the surface [11]. Generally, a rougher surface will have a lower reflectance than a smoother surface because the roughness causes light to scatter in many directions and eventually causes multiple reflections and, consequently, absorption phenomena [12]. However, the relationship between reflectance and superficial properties is only sometimes straightforward, depending on the specific wavelength and incidence angle of light and the observation scale [13]. Different studies demonstrated the impact of the material's roughness on thermal characteristics, showing that smooth and flat surfaces are of lower surface temperature than similar rough [14]. However, surface roughness is usually considered only from a qualitative and macroscopic point of view. Quantifying the influence of surface roughness on reflectance properties has crucial relevance since reflectance can significantly affect the reduction of the absorbed solar radiation and, in turn, the energy demand for cooling.

2. Materials and Methods

2.1. Samples production

In this work, eighteen outdoor porcelain stonewares (80 mm × 80 mm × 20(h) mm) for flooring in outdoor environments are selected to investigate a possible correlation between superficial roughness and reflectance properties. Samples' finishes differ in colour and surface texture, guaranteeing heightened anti-slip characteristics thanks to the different superficial bumpinesses. A checkered pattern is produced on the samples dividing the surfaces into four squares (40 mm × 40 mm) and applying opaque spray paint in three different colours (white, grey, black), as shown in Figure 1. In this way, the contribution of colour became uniform between the treated samples, preserving the differences in the surface roughness characteristics.

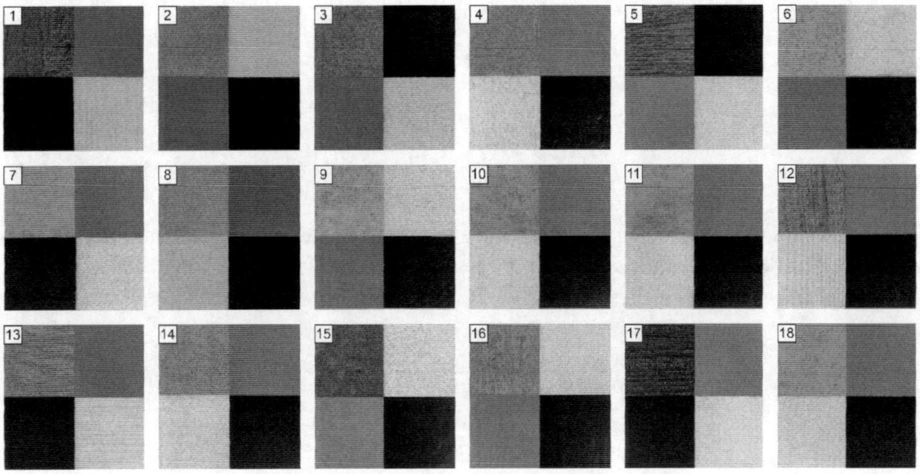

Figure 1. Samples investigated in the experimental campaign.

2.2. Experimental methodology

2.2.1. Optical characterization. For the first type of analysis, reflectance measurements (300-2500 nm) were performed using a Lambda 1050+ UV/Vis/NIR spectrophotometer equipped with a 150 mm integrating sphere. For each sample, the average value and standard deviation upon three different tests were considered to obtain the material's reflection coefficients and expected deviation, according to the ASTM E903 standard [17].

2.2.2. Surface characterization. Surface roughness was analysed through a Nanovea JR25 optical profilometer designed with Chromatic Confocal technology examining a 1x1 cm area for each sample and colour (acquisition rate 2000 Hz). The instrument was equipped with a Sentech STIL optical pen consisting of a CL4 chromatic lens and an MG35 magnifier with a maximum range of 3.0 mm. According to the evaluation frameworks provided by ISO 25178 standard [18], data obtained were analysed using MountainsMap (Digital Surf) software. The roughness component was isolated by applying S- and L-filters, which eliminated microroughness and waviness, respectively. Height, spatial and hybrid parameters, calculated on the entire surface, were used to quantify surface texture properties and evaluate their functional performances.

2.2.3. Statistical analysis. Analysis of variance (ANOVA) is a collection of statistical models used to determine whether there are significant differences among groups. ANOVA can be applied to various types of data with different levels of complexity, e.g. Repeated-Measures ANCOVAs compare means across one or more variables based on repeated observations while controlling for up to ten covariate factors. In ANCOVA RM models, the F statistic tests whether the difference in means between the groups is significant after controlling for the effect of one or more continuous covariates. The p-value indicates the probability of obtaining an F value as extreme as the one observed, assuming the null hypothesis is true. A p-value less than the significance level (usually 0.05) indicates a significant difference between the means of the groups after controlling for the effect of the covariates.

In this work, the experimental results of the coloured samples (B, G, W) represented the input for a Repeated-Measures ANCOVA statistical analysis investigating the correlation between the solar reflectance properties and the roughness parameters. Among these parameters, Mean surface roughness (Sa), representing the height difference of each point compared to the arithmetical mean of the surface as defined in Equation 1, was assumed as the predictor variable.

$$Sa = \frac{1}{A} \iint_A |Z(x,y)| \, dx \, dy \tag{1}$$

For each sample, reflectance coefficients in the three regions of the solar spectrum (UV, Visible and NIR) were calculated, obtaining the three dependent variables. The other topographic parameters collected by the profilometric analysis (Table 1) were used as the covariates in order to control for the effect of slope and complexity of the surfaces.

Table 1. Surface parameters defined by ISO 25178 [18].

Group	Parameter	Definition	Description
Height	Sa	Mean surface roughness	Arithmetic mean height
	Sq	Root-mean-square height	Standard deviation for the amplitudes
	Ssk	Skewness	Symmetry of the height distribution
	Sku	Kurtosis	Flatness of the height distribution
	Sp	Maximum peak height	Height between highest peak and mean plane.
	Sv	Maximum pit height	Depth between mean plane and deepest valley.

	Sz	Maximum height	Height between highest peak and deepest valley.
Spatial	Ssw	Dominant spatial wavelength	Size of the dominant surface irregularities
Hybrid	Sdq	Root-mean-square slope	Root-mean-square slope of the surface.
	Sdr	Developed interfacial area ratio	Complexity of the surface

3. Results and Discussions

This research seeks to analyse the impact of surface roughness on the reflectance and thermal performance of building materials. By conducting an experimental and statistical investigation, we aim to determine the role that roughness plays in urban cooling and mitigation. Reflectance is influenced by various factors, including colour and surface roughness. Colour relates to which wavelengths of light are absorbed and which are reflected, while surface roughness affects the amount and angle of light reflection. To isolate the contribution of superficial properties, eighteen outdoor porcelain stonewares, differing in colour and surface textures, were painted in three different colours (white, grey, and black).

The application of paint has resulted in significant changes in the optical properties of the samples. The reflectance curves of the untreated samples exhibit high variations in the entire solar spectrum, whereas the curves of the coloured samples tend to be more uniform in the UV and Visible ranges but with appreciable differences in the Near-Infrared wavelengths (Figure 2). In the case of white samples, significant variations can also be seen in the Visible wavelengths, probably due to the higher reflectance values associated with the white colour. For each sample, reflection coefficients in three regions of the solar spectrum were obtained as the average value of three different tests.

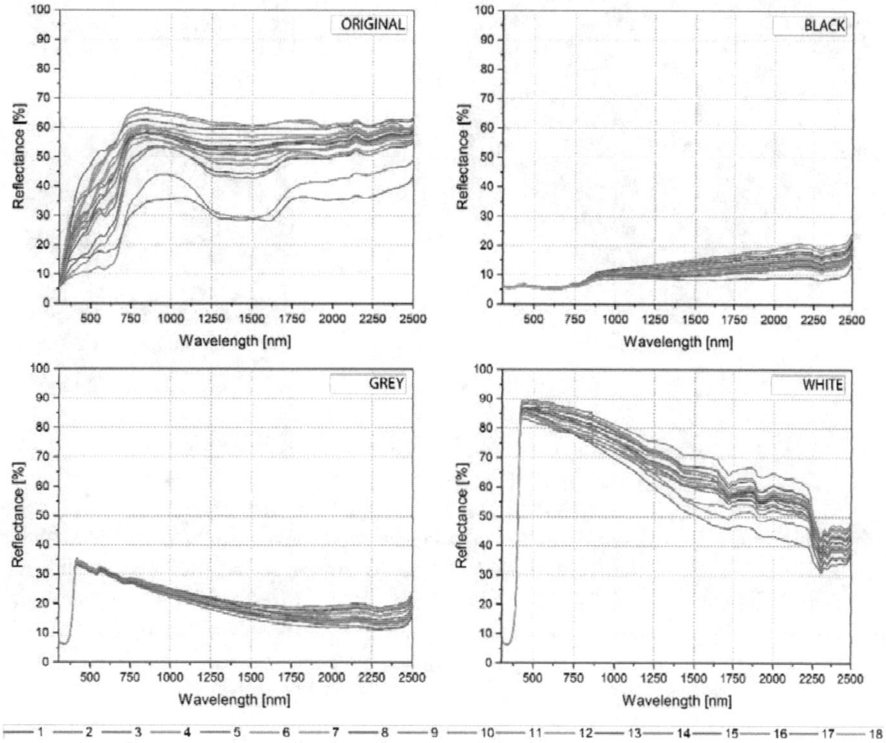

Figure 2. Reflectance curves of the investigated samples varying the superficial colour finishing.

A thorough surface characterization was performed using ISO 25178, the first international standard that specifies and measures surface properties in three dimensions. This standard introduced non-contact measurement methods, which allowed us to investigate surface properties of 54 treated samples using optical profilometry, as shown in Figure 3. Our analysis of roughness went beyond simply measuring the distance of points from an ideal smooth surface. We also considered the inclinations and the general complexity of the surface, including curvature, symmetry, and texture.

To accurately describe the surface properties, various classes of parameters were taken into account. Height parameters were used to measure the Z-axis values perpendicular to the mean plane of the surface, including *Sq, Ssk, Sku, Sz, and Sa.* Spatial parameters were also considered, which use spectral analysis to describe the topographic features and measure the lateral information on the X- and Y- axes (*Ssw*). Additionally, hybrid parameters measure the information on all three axes (X, Y, and Z), including criteria that depend on both amplitude and spacing, such as slopes and curvatures (*Sdq, Sdr*).

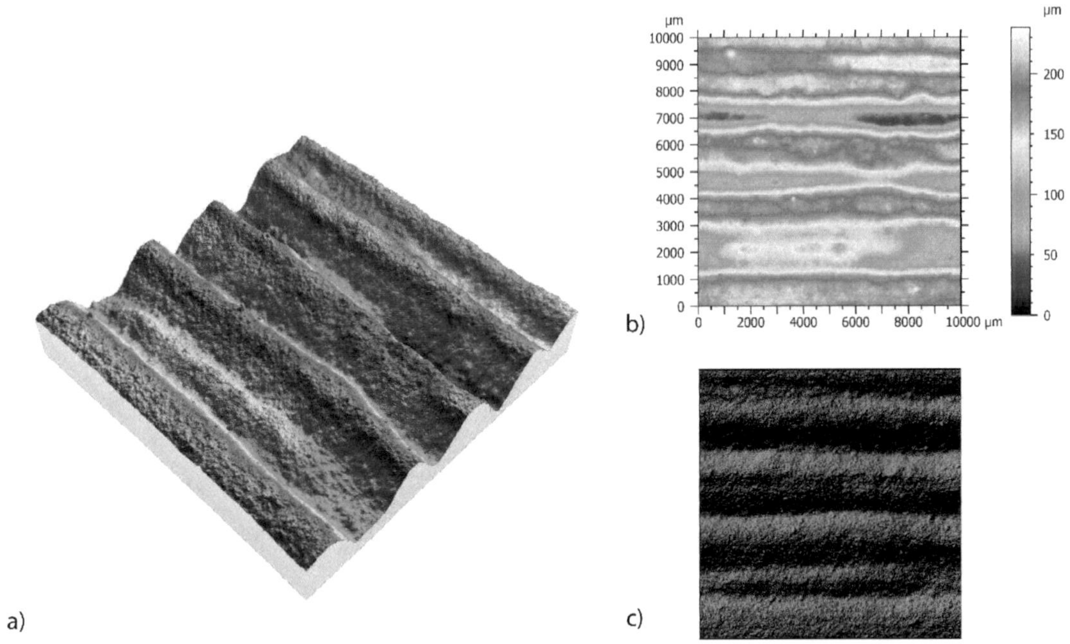

Figure 3. Surface analysis of 13W sample: a) 3D view, b) false-colour view, c) photo-simulation.

As a simple and widely used parameter to describe the roughness of a surface, the arithmetic mean roughness (*Sa)* was considered the most representative parameter to be included as the independent variable of the statistical analysis. Since ANCOVA analysis requires categorical data for the independent variables, *Sa* collected by the profilometer were divided into five classes with increasing value, from 1 (very low roughness) to 5 (very high roughness), as shown in Table 2.

Table 2. Surface characterization: Mean Surface Roughness of the investigated samples.

	Black		Grey		White	
Sample	Sa [μm]	Class	Sa [μm]	Class	Sa [μm]	Class
1	5,39	2	5,7	2	4,74	1
2	4,73	1	6,22	3	6,17	3
3	6,57	3	7,1	4	6,31	3
4	6,96	3	9,89	5	9,39	5
5	4,22	1	4,67	1	5,58	2
6	8,32	5	5,81	2	6,9	3
7	7,8	4	6,85	3	6,76	3
8	8,37	5	7,44	4	6,72	3
9	7,53	4	7,41	4	13,39	5
10	5,58	2	5,6	2	6,01	3
11	5,39	2	5,58	2	5,56	2
12	5,22	2	6,06	3	5,72	2
13	6,86	3	5,53	2	6,92	3
14	5,33	2	6,72	3	6,19	3
15	5,63	2	7,61	4	8,58	5
16	4,36	1	5,05	2	4,21	1
17	4,8	1	4,66	1	4,92	1
18	5,8	2	5,56	2	5,38	2

After verifying the meeting of assumptions, an RM Ancova analysis has been carried out to investigate the effects of Mean surface roughness on reflectance properties in the solar spectrum, while controlling for other topographic parameters. Table 3 resumes the results of the RM ANCOVA analysis. Mean surface roughness significantly influences the three dependent variables – the variations of reflectance properties in the entire solar spectrum – with a resulting p-value of 0.005, i.e., ten times smaller than the assumed confidence level, after controlling for the effects of covariates.

Table 3. Repeated Measures ANCOVA - Test of Between-Subjects Effects

Source	Type III Sum of Squares	df	Mean Square	F	Sig.	Partial Eta Squared
Intercept	1691.034	1	1691.34	2.991	.041	.066
Cov - Sq	1184.573	1	1184.573	2.095	.155	.048
Cov - Sku	5066.685	1	5066.685	8.961	.005	.176
Cov - Ssk	700.224	1	700.224	1.238	.272	.029
Cov - Sdq	7278.682	1	7278.682	12.873	<.001	.235
Cov - Sdr	7976.540	1	7976.540	14.108	<.001	.251
Cov - Ssw	4.066	1	4.066	.007	.933	.000
Cov - Sz	354.197	1	354.197	.626	.433	.015
Sa	9830.038	4	2457.510	4.346	.005	.293
Error	23747.181	42	565.409			

Depedent Variables: UV Reflectance, VIS Reflectance, NIR Reflectance

If a covariate in ANCOVA has statistical significance, it suggests a relationship between the covariate and the dependent variable beyond the influence of the independent variable(s) studied. In this analysis, three covariates showed a significant effect on the dependent variable: (i) *Sku* - Kurtosis of the height distribution, qualifying the flatness of the surface (Sig. = 0.005); (ii) *Sdq* - Root-mean-square slope of the surface (Sig. < 0.001); and (iii) *Sdr* - Developed interfacial area ratio, indicating the complexity of the surface respect to a projected horizontal surface (Sig. < 0.001). Conversely, other topographic parameters did not demonstrate any additional variance in the dependent variable beyond what was already explained by the predictor variable.

Estimated Marginal Means (EMMs) are statistical outputs in ANCOVA repeated measures that evaluate the mean response of each condition while accounting for the covariates' influences. Visual representations of EMMs aid the interpretation of the results by providing a clearer picture of the differences between the groups when comparing the effects of different treatments. Figure 4 shows EMMs of Reflectance in the entire solar spectrum for each class of *Sa* while considering covariates at their mean values. The mean response in reflectance tends to decrease as the mean surface roughness grows. This result is consistent with the literature [19], confirming that surface roughness can negatively influence reflectance properties.

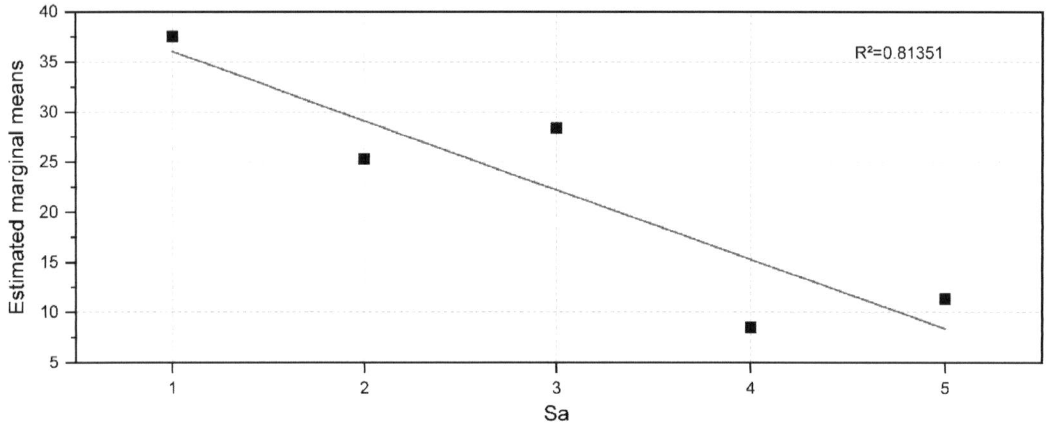

Figure 4. Estimated marginal means of Reflectance. Covariates are considered at their mean values.

4. Conclusions

This study delved into the connection between surface roughness and reflectance properties of building materials. Samples, differing in colour and surface textures, were painted in three different colours to isolate the contribution of superficial properties. Results of an RM ANCOVA statistical analysis showed that mean surface roughness significantly influences the sample's reflectance coefficients. Indeed, a rougher surface was associated with minor reflectance properties, with appreciable differences in the Near-Infrared wavelengths. On the other hand, UV reflectance remained more uniform, while noticeable variations could be seen in the Visible wavelengths for white samples. This outcome confirms the correlation between the surface roughness and the optical and thermal characteristics of the building materials, making evident the importance of the study of superficial topography towards the reduction of solar radiation absorption and the mitigation of UHI phenomena. Moreover, the analysis suggested that there is a relationship between other surface parameters, acting as covariates, and the dependent variables, i.e. the reflectance coefficients. This relationship needs to be further examined.

5. References

[1] Mirrahimi S, Mohamed M F, Lim, C H, Nik Ibrahim N L, Yusoff, W and Aflaki, A 2016 *Renewable and Sustainable Energy Reviews* **53** 1508-1519
[2] Su M, Ngarambe J, Santamouris M and Yun G 2021 *iScience* **24** 102495
[3] Santamouris M 2014 *Energy and Buildings* **82** 100–113.
[4] Yang X and Li Y 2015 *Building and Environment* **90** 146–156
[5] Falasca S, Ciancio V, Salata F, Golasi I, Rosso F and Curci G 2019 *Building and Environment* **163** 106242
[6] Kleerekoper L, van Esch M and Salcedo T B 2012 *Resources, Conservation and Recycling* **64** 30–38
[7] Synnefa A, Santamouris M and Livada I 2006 *Solar Energy* **80** 968–981
[8] Hernández-Pérez I, Álvarez G, Xamán J, Zavala-Guillén I, Arce J and Simá E 2014 *Energy and Buildings* **80** 81–105
[9] Synnefa A, Santamouris M 2013 *Advances in the Development of Cool Materials for the Built Environment* 33–71
[10] Grzesik W 2016 *Journal of Materials Engineering and Performance* 4460–4468
[11] Xu M, Zhang Q, Yang X, Wang Z, Liu J and Li Z 2016 *The Journal of Supercritical Fluids* **107** 286–297
[12] de la Rie E R, Delaney J K, Morales K M, Maines C A and Sung L P 2010 *Studies in Conservation* **55** 134–143
[13] Nayar S, Ikeuchi K and Kanade T 1991 *IEEE Transactions on Pattern Analysis and Machine Intelligence* **13** 611–634
[14] Gujrati A, Khanal S R, Pastewka L and Jacobs T D B 2018 *ACS Applied Materials & Interfaces* **10** (34) 29169–29178
[15] Murrell K A, Ghetu C C and Dorman F 2018 *Anal. Methods* **10** 4909–4916
[16] Ogunjemiyo S, Parker G, Roberts D 2005 *IEEE Geoscience and Remote Sensing Letters* **2** 90-93
[17] ASTM E903-12, Standard Test Method for Solar Absorptance, Reflectance, and Transmittance of Materials Using Integrating Spheres (2012)
[18] ISO 25178-2:2021, Geometrical Product Specifications (GPS) - Surface Texture: Areal - Part 2: Terms, Definitions and Surface Texture Parameters., International Organization for Standardization: Geneva, Switzerland, (2021)
[19] Castaldo V L, Coccia V, Cotana F, Pignatta G, Pisello A L, Rossi F 2015 *Urban Climate* **14** (2) 301-314

Acknowledgements

Funded by the European Research Council within the framework of Horizon Europe Programme (ERC, HELIOS, G.A. 101041255). Views and opinions expressed are however those of the authors only and do not necessarily reflect those of the European Union or the European Research Council. Neither the European Union nor the granting authority can be held responsible for them. F.M.'s acknowledgements are due to the PhD school in Photoinduced Processes and Technologies from the University of Perugia.

Experimental investigation of moisture influence on biochar and biochar-soil blends thermophysical properties

N Morselli[1], M Puglia[1]*, F Ottani[1], S Pedrazzi[1], G Allesina[1], A Muscio[1], P Tartarini[1]

[1] Università di Modena e Reggio Emilia – Dipartimento di Ingegneria "Enzo Ferrari", Via Pietro Vivarelli, 10-41125 Modena, Italy;

* Corresponding author: marco.puglia@unimore.it

Abstract. Biochar is a carbonaceous and porous material obtained through pyrolysis or gasification. It can be extremely valuable as soil amendment since it increases the organic matter content and fertility, the microbial activity, the water retention, and the crop yields. Moreover, biochar soil application has the potential for long-term carbon sequestration which makes its application to soil interesting even outside agricultural crops. In recent years, the study of the variation of the thermophysical properties of the soil induced by mixing with biochar has attracted interest.

In this work, the effect of the water content on thermal conductivity of biochar was investigated by means of the guarded hot plate apparatus λ-Meter EP500e. The same procedure was applied to various mixtures of biochar and soil. Furthermore, the specific heat was measured in order to obtain the thermal diffusivity in the various conditions through a calorimeter. Solar reflectance was also measured following the ASTM C1549 using a solar spectrum reflectometer SSR-ER. The obtained thermophysical properties can be used for the evaluation of the temperature trend of soil at different depths during the seasonal variations.

Keywords: *Thermophysical properties, biochar, conductivity, reflectance, specific heat*

1. Introduction

In line with the objectives of the 2030 agenda, with the Sustainable Development Goal 7 "Affordable and clean energy", the scientific community and institutions are focusing their efforts on reducing energy consumption, making energy production systems and energy usage more efficient and identifying new technologies, materials and/or combinations of these to achieve the common goals set by the 2030 Agenda of the United Nations [1].

One of the main targets is to "increase the share of renewable energy in the global energy mix, by 2030". In addition, the European Union has set as an important objective the zeroing of CO_2 emissions by 2050 [2]. In this context of energy sustainability and energy efficiency there must be space for technologies and materials that allow for a significant reduction in global GHG emissions and among them the atmospheric carbon capture and storage technologies find a prominent position. In particular, bioenergy with carbon capture and storage (BECCS) is one of the most promising strategies and it is based on the production of biochar through pyrolysis and gasification. Biochar is a vegetable carbon composed of up to 80% of biogenic carbon: it is a highly porous charcoal like an activated carbon with high recalcitrance that allows it to remain stable for decades when it is buried into the soil.

When biochar fulfils specific requirement it can be used as soil improver [3]. Its soil application can increase crop productivity, organic matter, water retention and microbial activity, thus it decreases the need for fertilizers and erosion [4,5,6]. The scientific community recognizes this material as a long-term storage medium for atmospheric CO_2 [7, 8, 9], in fact, 100 kg of biochar prevent the emission of about 150 kg of equivalent CO_2 [10]. In the last two decades several scientific studies have proven the benefit of the direct application of biochar in soil, but nowadays the attention of the scientific community has been shifting to more complex applications such as water filtering, microbial remediation of contaminated soil, sludge treatments, electronic and electric application and building technology like new materials and bricks [11, 12, 13, 14].The scientific community has moved a long way to understand the interactions between biochar and its retention capacity [15], and several studies have already confirmed the high capacity to retain large quantities of water, aiming to obtain more fertile soils and water saving [16]. Unlike agricultural applications, there are still few works that investigate the exploitation of the water holding capacity to improve existing technological processes in the energy sector. Most of them report the characterization of thermal properties of biochar as a function of the water content [17, 18, 19], but these studies focus on the characteristics of the biochar or soil-biochar matrices to characterize the material.

This work instead focuses on the thermal characterization of biochar, soil and soil-biochar mixtures related to the use of biochar as a geothermal backfill material, in particular on specific heat, thermal conductivity and solar reflectance of the previously mentioned solid matrices. Measurements were performed on three samples: a matrix composed of soil only (SOIL) and matrix composed of 80% soil and 20% biochar (SOIL-CHAR) and a matrix composed of char only (CHAR). The objective of these measurements was to compare a standard soil and the thesis (addition of biochar) to quantify the improvement obtained with the application of biochar and the positive or negative contribution in moist soil conditions. Biochar, in fact, with its water retention capacity improves the thermal conductivity conditions of the soil [18] in which it is applied as it causes the soil to remain moist for a longer time.

In addition, solar reflectance was investigated to study the effect on albedo that would occur when biochar or soil-biochar mixture is applied on the surface of a geothermal field rather than deep into the soil. The aim of this measure is to investigate the best way to apply these mixtures, also taking into consideration any effects of variation of solar reflection of earth surface.

2. Materials and methods

A laboratory analysis campaign for the evaluation of the thermophysical was carried out.

The properties of the biochar are not well-known, aside from a few properties such as the hydrophilicity of the biochar if obtained under specific conditions [3, 17, 18]. Since biochar has a pronounced ability to make the soil saturated with water, an increase of soil heat transfer performance is expected. Moreover, biochar can be obtained from different raw materials and with different methodologies, and this could heavily impact the resulting thermophysical properties, thus a final product can be better than others (as perlite or bentonite) [6, 20]. During the measurement campaign three different samples were taken into consideration under different humidity content. SOIL and SOIL-CHAR matrices were tested at three different degrees of humidity: 0% (DRY), 20% w/w wet basis (20) and 30% w/w (30) wet basis. In addition to these, SOIL-CHAR and CHAR were also tested at higher moisture content. SOIL-CHAR_36 and CHAR_67 samples were made by adding water to the matrices until they can retain water to study the thermal characteristics in saturation conditions. Table 1 summarizes and explains the composition of all the samples.

Table 1. Samples name, composition and humidity content.

Sample	Solid matrix composition	*Moisture content* kg_{water}/kg_{tot}
SOIL_DRY	Soil 100%	0%
SOIL_20	Soil 100%	20%
SOIL_30	Soil 100%	30%

SOIL-CHAR_DRY	Soil 90% + biochar 10%	0%
SOIL-CHAR_20	Soil 90% + biochar 10%	20%
SOIL-CHAR_30	Soil 90% + biochar 10%	30%
SOIL-CHAR_36	Soil 90% + biochar 10%	36%
CHAR_DRY	Biochar 100%	0%
CHAR_50	Biochar 100%	50%
CHAR_67	Biochar 100%	67%

2.1 Biochar

The biochar used during the experimental campaign is a commercial biochar purchased from the Italian company Bio-Esperia Srl and produced through an updraft gasifier with a nominal power of 170 kW.

The biochar production capacity of the gasifier is 900 m^3 per year, obtained from quality biomass deriving from secondary cuts of the forest in non-humanized areas. The producer declares that from 1000 kg of dry biomass, it is possible to produce 250-300 kg of biochar. This biochar is certified as a soil improver, according to Italian legislation, and can therefore be applied to the soil. Table 2 summarizes the characteristics of the BioDea White Onyx biochar used in this research.

Table 2 BioDea White Onyx biochar characteristics declared by the company [21].

Parameter	Value
Total Nitrogen (N)	< 0.5%
Carbon from carbonates (C)	< 0.1 %
Maximum water retention	115 %
Total Carbon of biological origin - dry basis	70 %
Salinity	110 mS/m
pH	9.85
Ash content	4.6 % dry basis
H/C - mole fraction	0.2
Fraction of grain size < 0.5 mm	1 %
Fraction of grain size < 2.0 mm	1 %
Fraction of grain size < 5.0 mm	60 %

The BioDea biochar was then minced using a mixer to make the material uniform: all the grains reached a size of less than 1 mm.

2.2 Thermal conductivity measurements

2.2.1 Sample preparation

The thermal conductivity measurement was performed by means of a guarded hot plate apparatus, λ-Meter EP500e.

The instrument consists of a fixed plate on which the specimen is placed and a motorized plate which is positioned on the upper face of the specimen to be analysed. The use of 0.5×0.5 m specimens is recommended to obtain more precise data. The minimum thickness is 0.02 m and the maximum 0.2 m. In addition to the set-point temperature at which the measurement is performed and the temperature difference between the two plates, it is possible to choose the pressure with which to compress the specimen, in a range between 250 Pa and 2500 Pa.

The tests during the experimental campaign were performed at a set point temperature of $T_{set} = 20°C$ and a temperature difference between the two plates equal to $\Delta T = 15$ K. The plate maintains the horizontal centre line of the sample at the setpoint temperature and the two horizontal surfaces (upper and lower) to the difference ΔT. The pressure was set to 2500 Pa. The instrument ended the measurement when it converged to a unique conductivity value λ *[W/ (m K)]* with variations lower than 1% around the identified value. In addition to the λ value, the instrument returned the error on the measurement performed *S [W/m K]* and the thickness of the sample analysed as $L \pm 0.01$ *[mm]*. As the

SOIL and CHAR samples are composed of solid but not compact matrices, it was necessary to place them in a box. An OSB wooden box open on the upper side was used, having dimensions of 0.5×0.5×0.07 m and walls with a thickness of 0.012 m, waterproofed with epoxy resin. A 0.005 m thick neoprene tape was placed along the entire upper perimeter to facilitate compression in the hot plate.

For each test, the procedure was the following: the desired sample was prepared according to Table 1, it was placed in the box and levelled, it was uniformly compressed using a weight of 50 kg and finally the whole box was coated with a layer of transparent film to preserve the instrument from water. Finally, the box was placed in the hot plate.

The samples were made of the same soil which had previously been dried and homogenized in terms of grain size. 50 kg of soil were divided into two equal parts, one destined for SOIL samples and one destined for SOIL-CHAR samples. This second was finally mixed with a quantity of biochar equal to 10% of the sample final weight, Eq. 1. In this way SOIL-CHAR_DRY samples were obtained.

$$M_{biochar} = \frac{M_{soil} \times 10}{100-10} \; [kg] \tag{1}$$

Once the tests with the DRY samples were completed, water was added to create the wet samples at 20% w/w wet basis, following Eq. 2, where M_{sample_dry} represents the considered sample (SOIL_DRY, SOIL-CHAR_DRY and CHAR_DRY). Similarly, water was added for testing the other wet basis samples.

$$M_{water} = \frac{M_{sample\_dry} \times 20}{100-20} \; [kg] \tag{2}$$

2.2.2 Experimental measurements campaign

For each sample two measurements were performed (rotating 90° the samples).

In addition to the measurements on the SOIL, SOIL-CHAR and CHAR samples, two measurements of the thermal conductivity of the wooden base of the waterproofed box were also carried out. This data is necessary to find the thermal conductivity of the sample contained in the box.

All the variables used in every test for the calculation of the thermal conductivity are summarized in Table 3. Figure 1 shows the layers of the sample.

Table 3. Variables for thermal conductivity calculation.

Variable	Symbol	Error	*Notes*
Thickness of the wood base	L_{wood}	$1 \times 10^{-5}\ m$	Data from hot plate
Sample thickness (box base + unknown sample)	L_{sample}	$1 \times 10^{-5}\ m$	Data from hot plate
Thickness of the unknown sample	L_x[a]	ε_{L_x}[a]	Data from calculation
Wood base thermal conductivity	λ_{wood}	ε_{wood}	Data from hot plate
Sample thermal conductivity (box base + unknown sample)	λ_{sample}	ε_{sample}	Data from hot plate
Unknown sample thermal conductivity	λ_x[a]	ε_{l_x}[a]	Data from calculation
Wood base thermal resistance	R_{wood}[b]	ε_{wood}	Data from calculation
Sample thermal resistance (box base + unknown sample)	R_{sample}[b]	ε_{sample}	Data from calculation
Unknown sample thermal resistance	R_x[a]	ε_{R_x}[a]	Data from calculation

[a]Values of the unknown sample, "x" generically represents the SOIL, SOIL-CHAR and CHAR samples.
[b]The thermal resistance is considered "specific thermal resistance", in fact the surface area is not considered since it is the same for all the samples

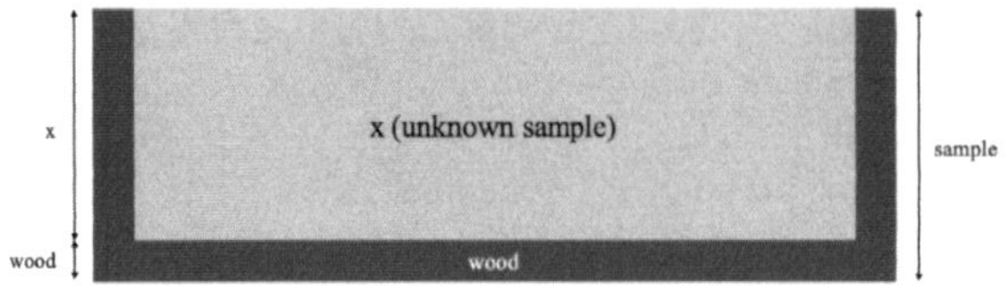

Figure 1: Simplified scheme of the sample layers and related nomenclature.

The equations used for the calculation of λ_{SOIL}, $\lambda_{SOIL-CHAR}$ and λ_{CHAR} are described in Eq 3., 4. and 5.

$$R_{wood} = \frac{L_{wood}}{\lambda_{wood}} \left[\frac{m^2 K}{W}\right] \quad (3)$$

$$R_{sample} = \frac{L_{sample}}{\lambda_{sample}} \left[\frac{m^2 K}{W}\right] \quad (4)$$

$$\lambda_x = \frac{(L_{sample} - L_{wood})}{(R_{sample} - R_{wood})} \left[\frac{W}{m\,K}\right] \quad (5)$$

Where $\boldsymbol{\lambda_x}$ represents the thermal conductivity of the unknown sample. The thermal conductivity value $\boldsymbol{\lambda_x}$ for each sample was calculated as the average value of the two tests performed.

$$\lambda_x = \frac{\lambda_{x_1} + \lambda_{x_2}}{2} \quad (6)$$

The measurement errors were calculated according to the propagation of uncertainties proposed by Kline McClintock [22], Eq 7.

$$\varepsilon_f = \left[\left(\frac{\partial f}{\partial x_1} * \varepsilon_{x_1}\right)^2 + \left(\frac{\partial f}{\partial x_2} * \varepsilon_{x_2}\right)^2 + \cdots + \left(\frac{\partial f}{\partial x_n} * \varepsilon_{x_n}\right)^2\right]^{1/2} \quad (7)$$

where $f = f(x_1, x_2, \ldots, f_n)$ is a generic equation with generic variables $x_1, x_2, \ldots, x_n$.

2.3 Specific heat measurements.

2.3.1 Sample preparation

Only two samples were characterized: biochar and soil.

Once the specific heat of both elements has been obtained, it is possible to proceed with the combination in different percentages to calculate the composition of different matrices.

2.3.2 Experimental measurements campaign

The experimental apparatus was composed of a 2-liter glass beaker which was placed in a specially created extruded polystyrene box with a minimum thickness of 0.10 m.

The beaker was filled with water at room temperature which was constantly mixed by a stirrer placed on the lid of the box. A thermocouple measured the water temperature T_{H_2O} (Figure 2).

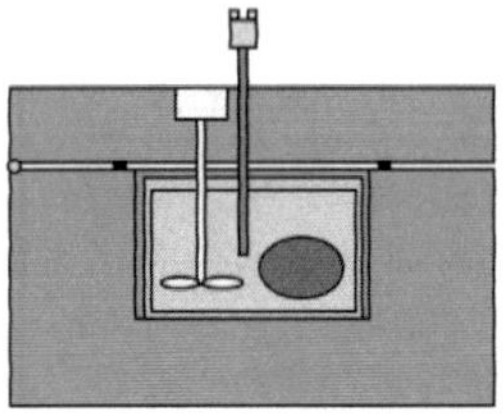

Figure 2: Simplified scheme of the apparatus for specific heat measurement

The analysed sample, closed in an airtight bag, was immersed for three hours in a water bath maintained at the temperature of $T_{\text{bath}} = 30°C$ through a Julabo ME thermostatic bath. In this way the sample was completely brought to 30°C. The sample was then extracted from the thermostatic bath, immersed in the beaker water and the stirrer was activated. The temperatures T_{H_2O} and T_{air} were logged using a Pico-Technology TC08 data logger. The test was interrupted when the water temperature stabilized at an intermediate value between the ambient temperature T_{H_2O} and T_{bath}.

Two blank tests of the insulating box, without samples, were carried out to evaluate the dispersion through the walls. The computation of the heating value of the unknown sample starts from the Eq. 8.

$$m_{H_2O} \times Cp_{H_2O} \times (T_{H_2O\_end} - T_{H_2O\_start}) = m_{sample} \times Cp_{sample} \times (T_{sample\_start} - T_{sample\_end}) \quad (8)$$

Where $\text{Cp}_{\text{sample}}$ is the specific heat (the unknown value of the equation). $T_{sample\_start} = 30\ °C$ due to thermal conditioning in the thermostatic bath and $\text{T}_{\text{sample\_end}} = \text{T}_{\text{H}_2\text{O\_end}}$ as the equilibrium has been reached.

2.4 Measurements of solar reflectance

The solar reflectance of the samples was measured through a solar spectrum reflectometer, (Model SSR-ER version 6) following the ASTM C1549.

For each sample five measurements of solar reflectance were made in five different areas of the surface of the 0.5×0.5 m samples. The final value was calculated as the average value.

3. Results and discussion

Thermal conductivities measured for every sample are summarized in Table 4.

Table 4. Samples thickness and thermal conductivity

	Sample thickness[mm]	Thickness uncertainty [mm]	Thermal conductivity λ_x [W/(mK)]	Thermal conductivity uncertainty ε_{λ_x} [W/(mK)]
WOOD	11.37	0.01	102.54×10^{-3}	1×10^{-3}
SOIL_DRY	40.00	0.01	228.50×10^{-3}	0.05×10^{-3}
SOIL_20	84.16	0.01	380.66×10^{-3}	0.22×10^{-3}
SOIL_30	84.36	0.01	1200.00×10^{-3}	1.11×10^{-3}
SOIL-CHAR_DRY	83.89	0.01	221.86×10^{-3}	0.06×10^{-3}
SOIL-CHAR_20	84.98	0.01	376.24×10^{-3}	0.12×10^{-3}
SOIL-CHAR_30	86.83	0.01	1132.24×10^{-3}	0.90×10^{-3}
SOIL-CHAR_36	32.28	0.01	1867.93×10^{-3}	0.1×10^{-3}
CHAR_DRY	19.40	0.01	70.00×10^{-3}	0.02×10^{-3}
CHAR_50	22.2	0.01	500.00×10^{-3}	0.02×10^{-3}
CHAR_67	21.7	0.01	2480.20×10^{-3}	0.03×10^{-3}

Literature values for biochar thermal conductivity are between 0.10 to 0.13 W m^{-1} K^{-1}, depending on the temperature of the process which is responsible for the microstructure, mineralogy, and physicochemical properties [20]. Biochar thermal conductivity is lower compared to soil thermal conductivity. The addition of biochar to soil does not significantly affect its thermal conductivity in dry conditions, while the effect of humidity is dominant. This effect is confirmed also by SOIL-CHAR_36 sample whose thermal conductivity value is the highest for the soil – char matrices. The CHAR_50 and CHAR_67 enable the attribution of a significant role to water in the thermal conductivity value of this material. In fact, an increase in humidity of only 17% produces a significant increase in the λ_x value.

Table 5. Solar reflectance and specific heat

	Solar reflectance [-]	Specific Heat [J/(kg K)]
SOIL_DRY	0.283	930
SOIL_20	0.201	-
SOIL_30	0.240	-
SOIL-CHAR_DRY	0.134	-
SOIL-CHAR_20	0.136	-
SOIL-CHAR_30	0.139	-
CHAR_DRY	0.109	1310
CHAR_20	0.098	
CHAR_30	0.095	

The specific heat of the mixture can be calculated as weighed average of the specific heat of the components of the matrices, for this reason the measurements were performed only on soil and biochar and not on their mixtures. The specific heat of pure soil was in line with literature values that range from 1.17 to 2.25 kJ kg^{-1} $°C^{-1}$ for clay and from 0.83 to 1.67 kJ kg^{-1} $°C^{-1}$ for sand (moisture content 20/25%) and density of 1300 kg m^{-3} [23]. Specific heat of biochar was similar to soil (slightly higher). Therefore, the addition of biochar could alter the specific heat of the mixture primarily because of its higher water retention capacity. This, in turn, increases the maximum soil moisture and extends its duration, while the contribution of the specific heat of the carbonized material can be considered negligible. Concerning solar reflectance, it was lower for biochar than soil and adding biochar to soil is reduces its solar reflectance. From Table 5 it can be seen that the humidity seems to reduce the solar reflectance of the matrix, however the relationship humidity-solar reflectance is not straightforward.

4. Conclusion

Biochar and soil thermal properties measured in this work are in line with scientific literature for the agronomic field.

Soil thermal conductivity can be properly measured by means of a hot plate instrument. This method, compared to the most common one consisting in the dual needles probes, allows a fast measurement of different soil matrices with different compositions (lamoy soil, clay soil, sandy soil) and also of different biochar/soil mixtures with different humidity even with little amount of available material. It would be also possible to simulate different soil depths by compressing the soil with the proper pressure. All this data can be extremely useful for studying possible new soil applications such as low enthalpy geothermal heating and cooling systems that combine energy efficiency and carbon sequestration and storage. The addition of biochar to soil does not significantly modify its thermal conductivity in dry condition, but through biochar water retention capability a higher conductivity can be reached at high moisture content. Biochar reduces solar reflectance of soil.

5. References

[1] UN General Assembly, Transforming our world: the 2030 Agenda for Sustainable Development, 21 October 2015
[2] Net Zero by 2050: A Roadmap for the Global Energy Sector. International Energy Agency. 2021.
[3] G Ravenni, Z Sárossy, J Ahrenfeldt, U B Henriksen, Activity of chars and activated carbons for

removal and decomposition of tar model compounds – A review, *Renewable and Sustainable Energy Reviews*, Volume 94, 2018, pp 1044-1056

[4] J Lehmann, M C Rillig, J Thies, C A Masiello, W C Hockaday, D Crowley, Biochar effects on soil biota – A review, *Soil Biology and Biochemistry*, Volume 43, Issue 9, 2011, pp 1812-1836

[5] S Baronti, F P Vaccari, F Miglietta, C Calzolari, E Lugato, S Orlandini, R Pini, C Zulian, L Genesio, Impact of biochar application on plant water relations in Vitis vinifera (L.), *European Journal of Agronomy*, Volume 53, 2014, pp 38-44

[6] G Allesina, S Pedrazzi, F Allegretti, N Morselli, M Puglia, G Santunione, P Tartarini, Gasification of cotton crop residues for combined power and biochar production in Mozambique, *Applied Thermal Engineering*, Volume 139, 2018, pp 387-394

[7] Woolf D, Amonette J, Street-Perrott F, Lehmann J, Joseph S, Sustainable biochar to mitigate global climate change. *Nature Communications* 1, 56 (2010)

[8] S Pedrazzi, G Santunione, M Mustone, G Cannazza, C Citti, E Francia, G Allesina, Techno-economic study of a small scale gasifier applied to an indoor hemp farm: From energy savings to biochar effects on productivity, *Energy Conversion and Management*, Volume 228, 2021, 113645

[9] N. Morselli, M. Puglia, S. Pedrazzi, A. Muscio, P. Tartarini, G. Allesina, Energy, environmental and feasibility evaluation of tractor-mounted biomass gasifier for flame weeding, *Sustainable Energy Technologies and Assessments*, Volume 50, 2022, 101823

[10] Smith P (2016). Soil carbon sequestration and biochar as negative emission technologies. *Global Change Biology*, 22(3), 1315–1324

[11] P Godlewska, H Peter Schmidt, Y Sik Ok, P Oleszczuk, Biochar for composting improvement and contaminants reduction. A review, *Bioresource Technology*, Volume 246, 2017, pp 193-202

[12] P Guan, S O Prasher, M T Afzal, S George, J Ronholm, J Dhiman, R M Patel, Removal of Escherichia coli from lake water in a biochar-amended biosand filtering system, *Ecological Engineering*, Volume 150, 2020, 105819

[13] M Mamera, J J van Tol, M P Aghoghovwia, Treatment of faecal sludge and sewage effluent by pinewood biochar to reduce wastewater bacteria and inorganic contaminants leaching, *Water Research*, Volume 221, 2022, 118775

[14] G Bruno, O Alves, B Rijo, G Lourinho, and C Nobre. 2022, Biochar: Production, Applications, and Market Prospects in Portugal, *Environments*, 9, no. 8: 95

[15] F Razzaghi, P Bilson Obour, E Arthur, Does biochar improve soil water retention? A systematic review and meta-analysis, *Geoderma*, Volume 361, 2020, 114055

[16] B M C Fischer, S Manzoni, L Morillas, M Garcia, M S Johnson, S W Lyon, Improving agricultural water use efficiency with biochar – A synthesis of biochar effects on water storage and fluxes across scales, *Science of The Total Environment*, Volume 657, 2019, pp 853-862

[17] S Adhikari, W Timms, M A Parvez Mahmud, Optimising water holding capacity and hydrophobicity of biochar for soil amendment – A review, *Science of The Total Environment*, Volume 851, Part 1, 2022, 158043, ISSN 0048-9697

[18] B Usowicz, J Lipiec, M Łukowski, Z Bis, J Usowicz, A E Latawiec, Impact of biochar addition on soil thermal properties: Modelling approach, *Geoderm*a, Volume 376, 2020, 114574

[19] Z Liu, J Xu, X Li, J Wang, Mechanisms of biochar effects on thermal properties of red soil in south China, *Geoderma*, Volume 323, 2018, pp 41-51

[20] D Patwa, U Bordoloi, A Aishwarya Dubey, K Ravi, S Sekharan, P Kalita, Energy-efficient biochar production for thermal backfill applications, *Science of The Total Environment*, Volume 833, 2022, 155253

[21] Biodea site: https://biodea.bio/prodotto/biochar-white-onyx-scaglie-5-a-7-mm

[22] Kline S J and McClintock F A, Describing uncertainties in single-sample experiments, *Mechanical Engineering*, 75 (1) (1953), pp. 3-8

[23] Nidal H Abu-Hamdeh, Thermal Properties of Soils as affected by Density and Water Content, *Biosystems Engineering*, Volume 86, Issue 1, 2003, Pages 97-102, ISSN 1537-5110

Acknowledgement
The authors express their gratitude to Antonio Giovannelli for his assistance in data collection.

Numerical investigation of radiative transfer during oxyfuel combustion of hydrogen and hydrogen enriched natural gas in the container glass industry

F Ott[1, *], J Losacker[1], N Schmitz[1] and H Pfeifer[1]

[1] Department for Industrial Furnaces and Heat Engineering, RWTH Aachen University, Kopernikusstr. 10, 52074 Aachen, Germany

* Corresponding author, e-mail: ott@iob.rwth-aachen.de

Abstract. The substitution of natural gas with hydrogen is one way to eliminate direct CO_2 emissions. However, oxyfuel combustion of hydrogen or hydrogen enriched natural gas leads to different exhaust gas properties due to a changed composition compared to conventional combustion. In combustion simulation, the emissivity of a gas mixture is usually approximated using a Weighted Sum of Gray Gases (WSGG) model. Most of the existing WSGG models have been validated for natural gas combustion with air or oxyfuel and are therefore not applicable to hydrogen-oxyfuel combustion. CFD simulations showed, that none of the investigated WSGG models is able to predict the radiative heat transfer for all considered combustion scenarios with appropriate accuracy. In addition, in container glass manufacturing more than 95% of the heat flux to the glass surface is transferred by radiation because of the high process temperatures. Due to the changed gray gas emissivity, the high content of water vapor leads to a different emission spectrum of the exhaust gas. The influence of the changed emission spectrum on radiative heat transfer and the penetration depth of radiation in the glass melt is investigated using a simulation model of a pilot plant and non-gray modelling of the radiation transport. The CFD simulations show slightly enhanced radiative heat transfer to the glass and a slightly deeper penetration depth especially for wavelength below 2.2 μm for hydrogen-oxyfuel combustion.

1. Introduction

With the European Green Deal, the European Union established the goal of Europe becoming the first climate-neutral continent. In order to reach this goal, especially the energy-intensive industries need to eliminate their CO_2 emissions [1]. In container glass production, about 85% of the required energy is used for the melting process where soda, sand, glass shards and other ingredients are heated up to about 1450°C, refined and afterwards cooled to a temperature suitable for forming of glass containers. This is usually done in a natural gas-powered end fired glass furnace with air combustion and up to 20% electrical boosting. For a CO_2-neutral container glass production different approaches are being followed. One is an all-electric furnace with a cold top, another possibility is to substitute natural gas with green hydrogen or hybrid solutions [2]. Both ways eliminate the energy-related CO_2 emissions. In addition, the raw-material related emissions need to be considered as well. These emissions can locally be avoided with a carbonate free batch composition. This will be studied in a pilot hybrid glass melting furnace which is being built as part of the BMWK founded ZeroCO₂Glas project. This paper focuses on the substitution of natural gas with hydrogen oxyfuel with a conventional batch composition and the effects coming with it regarding heat transfer.

In many technical plants, radiative heat transfer is the dominant heat transfer mechanism due to the high process temperatures [3]. Since experiments evaluating the radiative heat transfer between surfaces through a scattering, absorbing and emitting gaseous media are often hard to realize, the radiation modeling is an important alternative. Due to the high dependencies of the radiative properties on the wavelength, temperature and spatial direction, the accurate calculation of heat transfer can easily cause large computational costs. The recent literature regarding the accuracy of CFD simulation of combustion systems mainly focuses on the influence of turbulence-chemistry interaction. Radiation is often modeled using a gray approach or an optical thin simplification. These approaches can lead to an underprediction of up to 100°C of temperature depending on the conditions or inaccurate calculation due to spectral variations of the radiative properties [4]. In addition, many gray models were only developed and validated for natural gas-air combustion [5].

2. Hydrogen combustion

Due to the different gas properties of hydrogen compared to natural gas, many combustion properties differ as well. Table 1 shows an overview of the important properties for hydrogen and methane, the main component of natural gas.

Table 1. Gas properties of Hydrogen and Methane

	H_2	CH_4
Density in kg/m³ [6]	0,09	0,72
Gross calorific value in kWh/m³ [6]	3,54	11,06
Upper Wobbe-Index in kWh/m³ [6]	11,36	14,85
Adiabatic flame temperature in °C [7]	2380	2222
Minimum air demand in $m^3_{air}/m^3_{fuel\ gas}$	0,95	0,79
Minimum flue gas volume (wet) $m^3_{offgas}/m^3_{fuel\ gas}$	2,88	10,64
Laminar flame speed in m/s [7]	3,46	0,43

Apart from the higher adiabatic flame temperature, the main differences in the radiative heat transfer are caused by the flue gas composition, see figure 1 [8]. The flue gas of combustion mainly consists of the radiating gases CO_2 and H_2O and the non-radiating gases N_2 and O_2 [3].

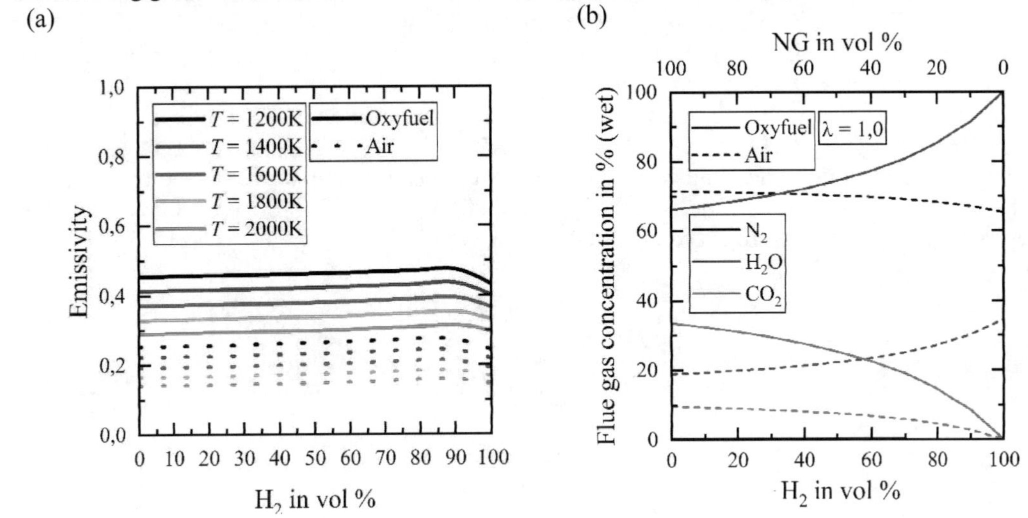

Figure 1. Gray gas emissivities [8] (a) and flue gas composition (b) for hydrogen-natural gas mixtures

The water vapor content in the flue gas for oxyfuel combustion with an air ratio of λ=1 rises with the hydrogen content in the fuel gas up to 100% while the CO_2 proportion decreases until 0%. For combustion with air the maximum H_2O content of 34.7% is reached with pure hydrogen as fuel gas. The emission spectrum of water and carbon dioxide differ, which leads to a change in the gray gas emissivity calculated according to Alberti [8].

3. Radiative transfer and the WSGG model

The Weighted Sum of Gray Gases model is a global approach for calculating the gray emissivity ε of gases according to formula (1). Where T is the Temperature, κ is the absorptivity, p the pressure and s the optical path length [5]. Hottel and Sarafim [9] proposed the approximation in 1970 and many authors modified the model and parameter sets since then. The gray emissivity of gases or gas mixtures is thereby calculated as the sum of partial intensities using usually four or five (j) gray gases plus one transparent one. The radiative properties of these gray gases are assumed constant over the whole spectrum [5].

$$\varepsilon = \sum_{i=0}^{I} a_{\varepsilon,i}(T)(1 - e^{-\kappa_i p s}) \tag{1}$$

The temperature dependence of factor $a_{\varepsilon,I}$ is normally approximated by

$$a_{\varepsilon,i} = \sum_{j=0}^{J} b_{\varepsilon,i,j} T^{j-1} \tag{2}$$

but it is also possible to use any other function. The values for $b_{\varepsilon,i,j}$ and κ_i are nowadays usually obtained by fitting the emissivity to Line-by-line calculations. In earlier papers different methods like exponential wide band modelling or spatial integration of the radiative transfer equation (RTE) were used to calculate an emissivity as reference [5]. In a participating medium like the flue gas of a combustion absorption, emission and scattering take place. The spectral radiative intensity I_η is then described by the following RTE [4]:

$$\frac{dI_\eta}{ds} = \kappa_\eta I_{b\eta} - (\kappa_\eta + \sigma_{s\eta})I_\eta + \frac{\sigma_{s\eta}}{4\pi} \int_0^{4\pi} I_\eta(\hat{s}_i)\Phi_\eta(\hat{s}_i, \hat{s})d\Omega_i \tag{3}$$

When it comes to computational costs the advantage of using the WSGG model is, that the integration over the spectrum happens before solving the RTE, so the RTE only needs to be solved once for every direction needed to determine the radiation field. The LBL method on the other hand computes an emissivity for every line of the spectrum resulting in the need of solving the RTE for each of the lines for every direction [10]. For modelling of complex industrial applications, the WSSG model often is chosen and delivers accurate estimates [11]. Most WSGG formulations offer different sets of parameters depending on the gas composition in the cell. The composition is generally characterized by the molar ratio of the partial pressure of H_2O to the partial pressure of CO_2, figure 2 shows the molar ratio of the flue gas as a function of the hydrogen content in the fuel gas [5].

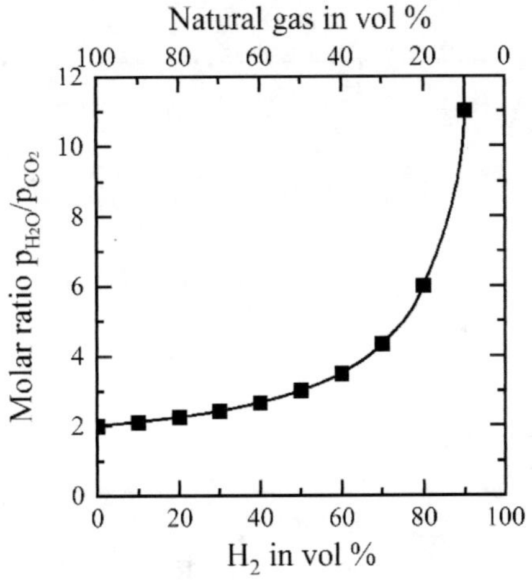

Figure 2. Molar ratio of H_2O to CO_2 for the flue gas of combustion of natural gas with different hydrogen admixtures

Many WSGG authors proposed WSGG formulations with improved accuracy for natural gas-oxyfuel combustion or operating points of special interest. A choice of relevant WSGG models are listed in table 2. ANSYS Fluent, as used for the calculations in this paper, computes the emissivity with the WSGG coefficients proposed by Smith [12] or Coppalle [13] depending on the Temperature [14].

Table 2. Overview of recent or commonly used WSGG Models

Author	Year	Condition	Molar Ratios	Temperature [K]	Reference
Smith	1982	Air-fuel	1, 2	600-2400	[12]
Coppalle	1983	Air-fuel	1, 2	2000-3000	[13]
Yin	2010	Oxyfuel & Air-fuel	0.125, 0.25, 0.5, 0.75, 1, 2, 3, 4	500-3000	[15]
Kangwanpongpan	2012	Oxyfuel	0.125, 0.25, 0.5, 0.75, 1, 2, 4	400-2500	[16]
Yin	2013	Air-fuel	0.05, 1, 2	500-3000	[17]
Bordbar	2014	Oxyfuel	0.1-4	300-2400	[18]
Guo	2015	Oxyfuel	0.05-2	600-2500	[19]
Bordbar	2020	all	all (0-10^8)	300-2400	[20]

Due to the varying molar ratio considered in Bordbar's 2014 formulation and the consideration of all molar ratios in his 2020 formulation, these are the two models which will be compared to the commonly used Smith/Coppalle and the LBL based calculation by Alberti in the following.

4. Radiative properties of glass melts

The probably most significant property of glass is its permeability to visible light. When it comes to industrial glass production, the rapid change of the absorption over the wavelength has a significant influence on the overall process. Figure 3 shows the spectral absorption over the spectrum obtained with the transmission method for a flint glass with 0.02 wt% Fe_{2+}. The radiative properties of a flint glass are mainly determined by the Fe_{2+} absorption bands since it nearly does not contain Cr_2O_3 [21].

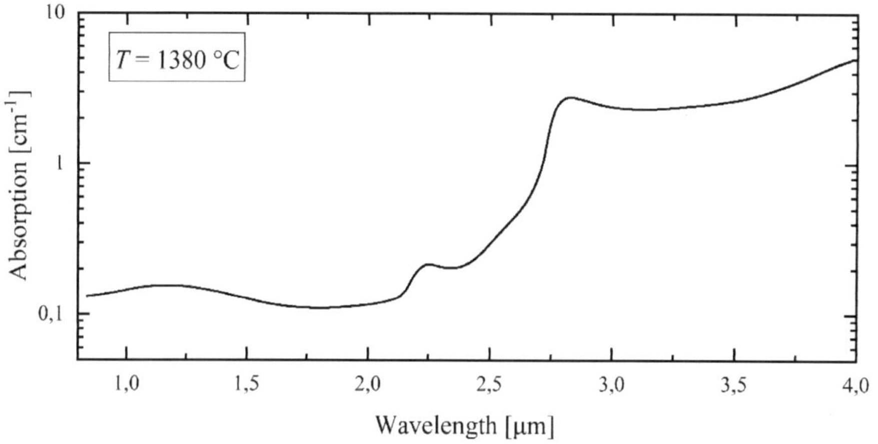

Figure 3. Spectrum of flint production glass at 1380 °C [21]

In Figure 3, the sudden increase in absorption at a wavelength of about 2.75 μm is particularly noticeable. The glass melt is therefore transmissive for lower wavelengths and more impervious for radiation of longer wavelengths. Water vapor has more bands at lower wavelength and emits more radiation in the transmissive band of the glass. The typical mean free path l_m of photons in a clear silicate flint container glass melt for the more transparent spectral region is about 20 cm [22]. The refractive index of the glass melt is assumed to be 1.5 according to Rubin [23].

5. Description of the test cases

In the following section the CFD test cases are described.

5.1. Test case for WSGG Model testing

For comparison of the selected WSGG models a test case was set up. For the 3D CFD simulation, the geometry and the refractory material properties of a pilot glass melting furnace were used. The ceiling is located 1 m above the glass bath and has a fixed temperature of 1773 K and an emissivity of 0.5. The glass surface is 1700 K and the fuel gas in between has a fixed temperature of 2000 K. The fuel gas emissivities were calculated using the WSGG parameters of the selected models for a mean beam length s of 0.899 m, according to the definition in ANSYS Fluents Theory Guide [14] where V is the volume considered and A is the area of the surrounding walls, see formula (4). The volume and absorbing area were obtained from the geometry of a planned pilot glass melting furnace.

$$s = 3.6\frac{V}{A} \tag{4}$$

The target value of the calculation is the heat flux density on the glass surface.

5.2. Test case for radiative transfer in the glass melt

For evaluating the radiative transfer in the glass melt the CFD model was extended to include the glass melt. The depth of the glass bath is 0.75 m as in the planned pilot melting furnace. The radiation transport in the glass melt was modelled with the non-gray Discrete Ordinates Method. To consider the variation of the absorption by the glass melt over the wavelength, three non-gray bands were implemented, as recommended in the ANSYS Fluent Users Guide [24]. The absorption coefficients of the gas for the three bands were obtained from Line-by-Line calculations of the HITEMP 2010 [25] database and are listed in table 3. Due to the high computational costs of these calculations in this paper only natural gas-oxyfuel and hydrogen-oxyfuel were considered for the comparison of the radiation transport in the glass melt. The glass surface was treated as Coupled Wall being semi-transparent for radiation.

Table 3. Emissivity calculated by Line-by-Line calculations for the three gray bands for an optical path length of 0.899 m

	Emissivity Band 0 0 – 2.2 µm	Emissivity Band 1 2.2 - 2.75 µm	Emissivity Band 2 > 2.75 µm
H_2-Oxyfuel	0.120	0.437	0.545
CH_4-Oxyfuel	0.088	0.395	0.583

6. Results and discussion

The following section discusses the results of the CFD simulations.

6.1. Results of WSGG model testing

Figure 4 shows the radiation heat flux received by the glass surface for different WSGG formulations. When comparing figure 4 (a) and (b) it becomes clear that oxyfuel combustion results in a higher heat flux to the glass bath. This is due to the higher content of radiating gas components. However, the lower exhaust gas volume is not considered in this calculation. For both, oxyfuel and air combustion, the WSGG model by Smith/Coppalle shows a deviation of at least 10 to 22 kW. This is equal to 5-12% for oxyfuel and 7-14% for combustion with air. The heat flux is thereby underestimated for oxyfuel and overestimated for air fuel.

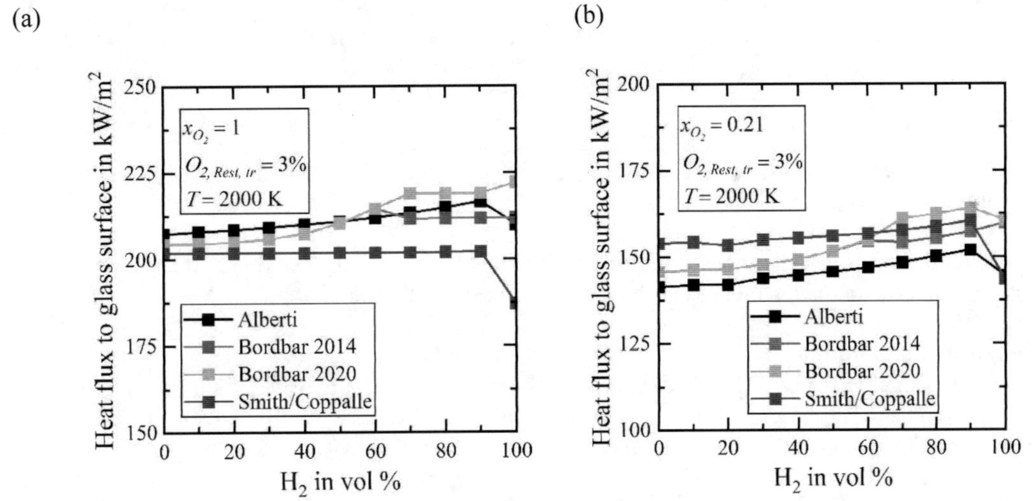

Figure 4. Radiative heat flux received by the glass surface calculated with different WSGG formulations for different H_2-contents in the fuel gas for (a) oxyfuel and (b) combustion with air

The WSGG formulations proposed by Bordbar et al. in 2014 and 2020 match the heat flux calculated with Albertis emissivity with a deviation of less than 5 kW (2.3% for oxyfuel and 3,1% for combustion with air) up to a hydrogen content of about 50% for air combustion and 90% for oxyfuel. For higher hydrogen admixtures in air fuel combustion the 2014 formulation keeps its accuracy up to 90% hydrogen content, while the 2020 formulation exceeds the Alberti calculation by up to 25 kW or about 17%. For pure hydrogen the commonly used coefficients by Smith and Coppalle fit the reference heat flux, whereas both Bordbar models overpredict the heat flux by about 25 kW. The drop in the heat flux at 100% hydrogen is not shown by Bordbar's 2014 nor by his 2020 formulation, only the coefficients proposed by Alberti and Coppalle display the drop qualitatively. The 2014 formulation was developed explicitly for oxyfuel, while the 2020 formulation is intended to cover all scenarios. Nevertheless, despite the great dependence of the emissivity on the water vapour partial pressure, only coefficients for

pure water vapour at 1 atm were given. This explains the great deviation of the predicted heat flux for combustion with air.
None of the tested models can be safely used for all considered combustion scenarios.

6.2. Results of radiation transport in the glass melt
In total, 15.9 kW/m^3 more radiation is absorbed by the glass melt when switching to hydrogen-oxyfuel combustion, while just considering the different flue gas composition at the same temperature. To evaluate the origin of the increased heat flux, the radiation received by a horizontal area in dependence of the glass bath depth is shown in figure 5. It becomes clear that the biggest difference in the heat flux occurs in Band 0, where for hydrogen-oxyfuel at the surface 21 kW/m^2 more are absorbed by the glass melt. In addition, the radiation penetrates the glass slightly deeper for fuel gas composition of hydrogen-oxyfuel combustion. The difference in radiation heat flux is significantly lower for Bands 1 and 2.

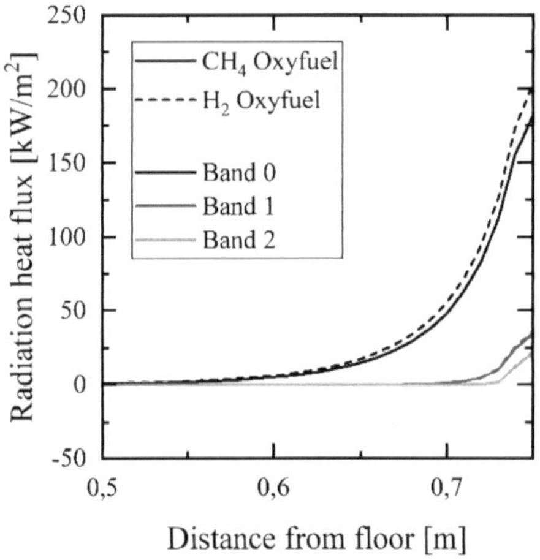

Figure 5. Comparison of incident radiation heat flux near the glass surface due to the gas radiation
This is consistent with the differences in the emissivity shown in table 3.

7. Conclusion and outlook
None of the evaluated WSGG Models delivers suitable results for all investigated operating points. To enable all combustion scenarios to be calculated with one model, new coefficients need to be fitted, especially for water vapor at different partial pressures, or different WSGG Models need to be combined. The simulation of the radiation depth with the three gray bands in the glass melt shows a slightly increased heat transfer for hydrogen-oxyfuel compared to natural gas oxyfuel. This is mainly caused by radiation at wavelengths below 2.2 μm. The obtained difference needs to be evaluated further regarding its impact on the formation of the flow in the glass melt.

8. References
[1] European Comission 2019 Communication from the comission to the european pariliament, the european council, the european economic and social committee and the committee of the regions *The European Green Deal*
[2] Matthias Leisin 2019 Energiewende in der Industrie: Potentiale und Wechselwirkungen mit dem Energiesektor, Branchensteckbrief der Glasindustrie *Bericht an Bundesministerium für Wirtschaft und Energie*
[3] Wang P, Fan F, Li Q 2014 Accuracy evaluation of the gray gas radiation model in CFD simulation *Case Studies in Thermal Engineering* vol 3
[4] Moest M F, Haworth D C 2016 *Radiative Heat Transfer in Turbulent Combustion Systems* (Springer) ed F A Kulacki

[5] Fraga G C, Silva F R, Zanin L G B, da Fonseca R J C, França F H R, Centeno F R 2021 A comprehensive evaluation of the WSGG model for air- and oxy-fuel combustion conditions through three-dimensional calculations *Fire Safety Journal* vol 125
[6] Gas und Wärme-Institut Essen e.V. 2013 gwi-Arbeitsblätter: Verbrennungskennwerte, Gaseigenschaften, Berechnungen (Vulkan Verlag)
[7] Joos F 2006 *Technische Verbrennung* (Springer)
[8] Alberti M, Weber R, Mncini M 2018 Gray gas emissivities for H_2O-CO_2-CO-N_2 mixtures *Journal of Quantitiative Spectroscopy & Radiative Transfer* vol 219
[9] Hottel H C, Sarofim A F 1970 Models of radiative transfer in furnaces *Journal of engineering physics* vol 19
[10] Yadav R, Balaji C, Venkateshan S P 2023 *Radiative Heat Transfer in Participating Media* (Springer)
[11] André F, Coelho F, Consalvi J-L, França F H R, Galtier M, Nmira F, Solovjov V, Webb B 2019 Accuracy of engineering methods for radiative transfer in CO_2-H_2O mixtures at high temperature *9th International Symposium on Radiative Transfer* pp.407-414
[12] Smith T F, Shen Z F, Friedmann J N 1982 Evaluation of coefficients for the weighted sum of gray gases model *Journal of Heat Transfer* vol 104 (4)
[13] Coppalle A, Vervisch P 1983 The total emissivities of high-temperature flames *Combustion and flame* vol 49
[14] ANSYS Fluent Theory Guide 2022 R2 (Canonsburg, PA)
[15] Yin C, Johansen L C R, Rosendahl L A, Kær S 2010 New weighted sum of gray gases model applicable to computational fluid dynamics (CFD) modeling of oxy-fuel combustion: Derivation, validaiton and implementation *Energy & Fuels* vol 24 (12)
[16] Kwanpongpan T, França F H R, Corrêa da Silva, Schneider P S, Krautz H J 2012 New correlations for the weighted-sum-of-gray-gases model in oxy-fuel conditions based on HITEMP 2010 database *International Journal of Heat and Mass Transfer* vol 55
[17] Yin C 2013 Refined weighted sum of gray gases model for air-fuel combustion and its impacts *Energy & Fuels* vol 27
[18] Bordbar M H, Weçel G, Hyppänen T 2014 A line by line basted weighted sum of gray gases model for inhomogeneous CO_2-H_2O mixture in oxy-fired combustion *Combustion and Flame* vol 161
[19] Gou J, Li X, Huang X, Liu Z, Zheng C 2015 A full spectrum *k*-distribution based weighted-sum-of-gray-gases model for oxy-fuel combustion *International Journal of Heat and Mass Transfer* vol 90
[20] Bordbar M H, Fraga G C, Hostikka S 2020 An extended weighted-sum-of-gray-gases model to account for all CO_2 - H_2O molar fraction ratios in thermal radiation *International Communications in Heat and Mass Transfer* vol 110
[21] Faber A J, Rongen M, Lankhorst A, De D, Meneses S 2020 Characterization of high temperature optical spectra of glass melts and modeling of thermal radiation conductivity *International Journal of Applied Glass Science* vol 11 (3)
[22] Faber A J 2002 Optical properties and redox state of silicate glass melts *Comptes Rendus Chimie* vol 5 (11)
[23] Rubin M 1985 Optical properties of soda lime silica glasses *Solar Energy Materials* vol 12
[24] ANSYS Fluent Users Guide 2022 R2 (Canonsburg, PA)
[25] Rothman L S, Gordon I E, Barber R J, Dothe H, Gamache R R, Goldman A, Perevalov V, Tashkun S A, Tennyson J 2010 HITEMP, the high-temperature molecular spectroscopic database *Journal of Quantitative Spectroscopy and Radiative Transfer* vol 111 (15)

Acknowledgments
The work was funded within the framework of Project (FKZ: 03EN2075B) by the German Federal Ministry for Economic Affairs and Climate Action.

Experimental analysis of insulating materials using Guarded Hot Box – Preliminary results

T de Rubeis[1,*], A Ciccozzi[2], D Paoletti[2] and D Ambrosini[2]

[1] Department of Civil, Construction-Architectural and Environmental Engineering, University of L'Aquila, 67100 L'Aquila, IT
[2] Department of Industrial and Information Engineering and Economics, University of L'Aquila, 67100 L'Aquila, IT

* Corresponding author: tullio.derubeis@univaq.it

Abstract. The containment of energy consumption in the construction sector strongly depends on the envelope, which is mainly responsible for heat loss in buildings. Thus, great attention should be paid to the selection of thermally-performing materials. In this work, preliminary results of three different configurations of walls in real size have been compared, conducting the analyses with the heat flow meter method inside the Guarded Hot Box apparatus, capable of guaranteeing repeatable and controlled conditions. The analyses were carried out with two types of heat flow meters, characterized by different sensitivity. The side of the wall facing the hot chamber has been insulated with rock wool for all the experiments, while the other side has been insulated first with expanded polystyrene (EPS) with graphite, then with hemp, and finally with cork. The results showed that the sample with the best thermal behaviour is the first one, i.e., the wall with EPS with graphite, characterized by a transmittance value between 0.148 W/m^{2}K and 0.153 W/m^{2}K. The other two configurations, characterized by the use of natural materials, showed worse performance with conductance values about 20% higher than EPS. The percentage differences between the two heat flux sensors for the experimental campaigns ranged from 2.8% to 4.4%.

1. Introduction

Energy use in buildings increased from 115 EJ in 2010 to almost 135 EJ in 2021, representing 30% of global final energy consumption. As emerges from these data from the International Energy Agency's September 2022 report [1], it is extremely important to invest in energy efficiency interventions in the building sector to align with the Net Zero Emissions scenario by 2050 [2]. The continuous attention paid to building energy efficiency has led to the development of increasingly complex construction solutions capable of satisfying the requirements of current regulations in force. As the complexity of building components increases, it is necessary to grow the investigation criteria and develop measurement and analysis systems as accurately as possible. In this context, therefore, research and experimentation assume a particular and strategic importance.

In the last years, the Guarded Hot Box apparatus is attracting a lot of attention among researchers. This instrument allows the measurement of the transmittance of building components by recreating controlled thermal conditions inside its chambers. Various studies have been conducted on the applicability of the GHB method for calculating thermal parameters in construction.

Chowdhury et al. [3] used a Guarded Hot Box to explore the thermal characteristics of walls and roofs commonly used in Indian residential construction. Bai et al. [4] integrated the numerical analyses carried out with the finite element method with experimental analyzes conducted in Hot Box to study the thermal properties of a wall made of hollow shale blocks. Trgala et al. [5] have experimentally analyzed various building envelopes potentially suitable for thermal insulation. The measurements were carried out under dynamic conditions in the cold chamber of the Hot Box. The study shows that the accuracy of the results seems to increase with the length of the measurements. Ferrari et al. [6] tested the thermal performance of four different walls. For the study, a single climatic chamber (considering the laboratory room where the chamber was placed as the indoor space) was used. The measurements were conducted under dynamic conditions. de Rubeis et al. [7] presented a novel integrated measuring and control system for hot box experiments. In this work, the thermal parameters of a doubly insulated X-lam wall sample were measured by keeping the thermal conditions inside the chambers under control, thanks to the use of 135 probes which constantly recorded the temperature values during the 72 hours of measurements. The results obtained experimentally with the heat flow meter method (HFM) showed a difference of less than 20% from the theoretical values. In another study, Nardi et al. [8] evaluated the thermal parameters of a sample wall placed in a Guarded Hot Box. The results obtained through a thermographic investigation were compared with heat flow meter measurements and theoretical values.

The aim of the present work is to compare the thermal performance of three wall samples with real dimensions of 300 cm x 300 cm, using two different heat flow meters (Hukseflux HFP01 and HFP03) inside the Guarded Hot Box, under repeatable and controlled conditions. The side of the wall facing the hot chamber has been insulated with rock wool for all the experiments carried out, while the other side has been insulated first with expanded polystyrene (EPS) with graphite, then with hemp, and finally with cork. Thanks to the Hot Box, known thermal conditions were imposed, with the aim of having a temperature difference that can give rise to an appreciable thermal flow. In this specific case, stationary analyses were carried out to understand which of the three wall configurations was the most performing.

2. Materials and Methods

In this work, experimental analyses were conducted to compare three wall configurations characterized by the use of different insulating materials on the outer surface of the wall: i) EPS, ii) hemp, iii) cork.

2.1 Methodology

The present study has a twofold objective: i) to analyze the thermal characteristics of the insulating materials under examination, ii) to compare the results obtained with two different heat flow meters.

The methodology used for this study can be divided into two macro-phases.

The first phase focuses on the creation of the three wall samples (S1, S2, S3), after the insulating materials selection through a scientific literature review.

The second phase, i.e., the analysis phase, focuses on the study of the thermal behaviour of the three wall configurations through theoretical and experimental analyses conducted in Hot Box using the HFM method.

3. Experimental setup

3.1 Insulating materials selection

After a scientific literature review [9- 25], the insulating materials chosen for the realization of the various wall configurations are rock wool, EPS mixed with graphite, hemp, and cork.

3.2 Creation of the wall samples

The specimen has actual dimensions of 300 cm x 300 cm and is made up of a double-insulated X-lam load-bearing element known as CLT (Cross Laminated Timber). For the first experiment, the internal surface of the wall was insulated with rock wool, while the external surface was insulated with EPS mixed with graphite. The wall was sandwiched between double layers of plasterboard (Figure 1a). For the second experiment, EPS with graphite was replaced with hemp panels, keeping the thickness unchanged (Figure 1b). In the last cycle of measurements, the insulating layer on the cold side was replaced with cork panels, keeping the thickness unchanged (Figure 1c).

(a) (b) (c)

Figure 1. Samples' creation. (a) First configuration: EPS and graphite. (b) Second configuration: hemp. (c) Third configuration: cork.

4. Analysis phase
The analyses of the three samples were carried out via theoretical and experimental approaches.

4.1 Theoretical approach
The theoretical approach was carried out under steady and one-dimensional conditions.

The calculation of the total resistance (R_{tot}) has been carried out through Ohm's law, expressed by the Equation (1), based on the analogous electrical circuit:

$$R_{tot} = R_{s,i} + \sum_1^n R_{cond,n} + R_{s,e} \quad [\text{m}^2\text{K/W}] \tag{1}$$

where:
- $R_{s,i}$ and $R_{s,e}$ are the internal and external surface resistances [m^2K/W], equal to 0.13 m^2K/W and 0.04 m^2K/W respectively, as indicated by EN ISO 6946 standard [26];
- $R_{cond,n}$ is the conductive thermal resistance of each layer [m^2K/W].

Table 1 shows the various layers of the three samples with their thermal properties.

Table 1. Properties of the samples' layers from cold chamber to hot chamber.

Description	Thickness [cm]	Density [kg/m^3]	Conductivity [W/mK]	Resistance [m^{2}K/W]
Plasterboard (double layer)	2.50	900	0.210	0.12
EPS with graphite (S1) **or** Hemp panel (S2) **or** Cork panel (S3)	10.00	32 (S1) **or** 50 (S2) **or** 150 (S3)	0.031 (S1) **or** 0.040 (S2) **or** 0.041 (S3)	3.23 (S1) **or** 2.50 (S2) **or** 2.44 (S3)
X-lam panel	10.00	470	0.130	0.77
Rock wool	5.00	135	0.039	1.28
Plasterboard (double layer)	2.50	900	0.210	0.12

Therefore, based on Equation (1) and on the thermal properties of the samples' layers, the total thermal resistance calculated for the samples is 5.69 m^2K/W for (S1), 4.96 m^2K/W for (S2) and 4.90 m^2K/W for (S3).

Known the total resistances of the walls, the thermal transmittance was calculated using Equation (2):

$$U = \frac{1}{R_{tot}} \quad [\text{W/m}^2\text{K}] \tag{2}$$

Finally, by imposing a temperature difference of 20 °C between the inner and the outer surfaces of the walls, the heat flux passing through each sample was calculated using Equation (3).

$$q = U \cdot \Delta\text{T} \quad [\text{W/m}^2] \tag{3}$$

Table 2 shows the thermal transmittance and the heat flux values calculated for the three samples.

Table 2. Theoretical transmittance and heat flux values of the three samples.

Samples	U [W/m^{2}K]	q [W/m^2]
S1	0.176	3.52
S2	0.202	4.04
S3	0.204	4.08

4.2 Experimental approach
The Hot-Box used for the experimental analyses is the GHB (Guarded Hot-Box) type. In the Hot-Box with a guard chamber, the metering box is surrounded by a chamber that has the control function (guard box). A detailed description of the apparatus can be found in previous works [8, 27, 28]. By using the GHB approach, it is possible to set and control the temperatures within the chambers, guaranteeing repeatable and controlled

conditions. At the same time, it allows the evaluation of the influence of boundary conditions on the thermal behaviour of the analyzed building component. Through the use of electrical resistances and the introduction of thermal energy into the hot chamber, the temperature was brought up to the set point value of 20 °C. In the cold chamber, on the other hand, the thermal conditions were regulated by a refrigeration unit which allowed the set point value of 0 °C to be reached. During the experimental phase, the HFM method was used. The analyses were carried out using two different heat flux probes, installed on the internal surface of the specimen, two temperature probes, one of which applied to the cold side wall and the other to the hot side wall, and a data logger. The recommendations provided by ISO 9869 are followed [29]. The measurements were carried out using Hukseflux HFP01 and HFP03 heat flow meters, whose characteristics are summarized in Figure 2. Hukseflux HFP01 is one of the most commonly used heat flow meters, while the HFP03 can be considered the higher sensitivity version of the HFP01 probe. Although based on the same technology, the two sensors differ mainly in sensitivity, accuracy, and measurement area (equal to 8 x 10^{-4} m^2 for HFP01 and 64 x 10^{-4} m^2 for HFP02).

The schematization of the experimental setup and the main characteristics of the measuring instruments are shown in Figure 2.

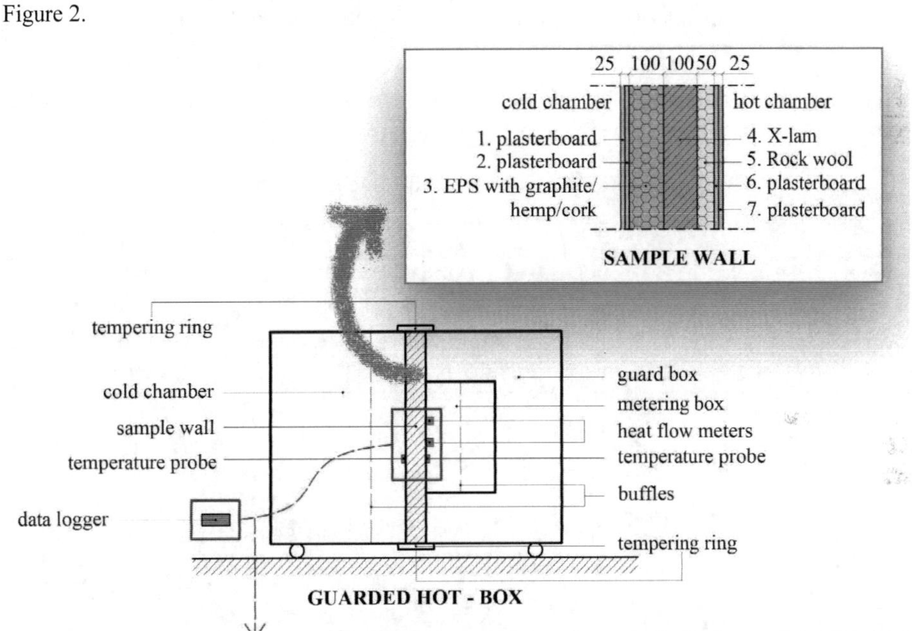

Technical specifications of the measuring instruments used for the experimental analyses.

Instrument	Type	Mesurement ranges	Sensitivity	Accuracy
Heat flow meter	Hukseflux HFP01	From -2000 to 2000 W/m²	60 μV/W/m⁻²	± 3%
Heat flow meter	Hukseflux HFP03	From -2000 to 2000 W/m²	500 μV/W/m⁻²	± 6%
Temperature probe	LSI Lastem DLE 124	From - 50 to 80 °C	0.01 °C	± 0.15 °C
Data logger	LSI Lastem E-Log ELO3305	From -300 to 1200 mV	40 μV	± 100 μV

Figure 2. Experimental setup's configuration, sample wall's detail (dimensions in millimeters), and technical specifications of the measuring instruments.

The measurements had a total duration of three weeks. They were carried out from 7 March 2022 at 1:00 p.m. to 14 March at 10:00 a.m. for the first sample (S1), from 16 March at 1:00 p.m. until 23 March at 11:30 a.m. for the second (S2) and from 28 March at 1:30 p.m. to 4 April at 8:50 a.m. for the last sample (S3). The data logger acquired the values with a sampling interval of 10 minutes. The progressive average method, specified in ISO 9869 standard [29], was utilized to process the data previously collected. In detail, the conductance (Λ) and thermal resistance (R) values have been obtained utilizing Equations (4) and (5), respectively:

$$\Lambda = \frac{\sum_{j=1}^{n} q_j}{\sum_{j=1}^{n}(T_{s,in,j} - T_{s,out,j})} \quad [\mathrm{W/m^2K}] \tag{4}$$

$$R = \frac{\sum_{j=1}^{n}(T_{s,in,j} - T_{s,out,j})}{\sum_{j=1}^{n} q_j} \quad [\mathrm{m^2K/W}] \tag{5}$$

where:
- $\sum_{j=1}^{n}(T_{s,in,j} - T_{s,out,j})$ is the progressive sum of the differences between the internal and external surface temperatures [°C];
- $\sum_{j=1}^{n} q_j$ is the progressive sum of the density of the heat flux [W/m^2].

Finally, the transmittance value (U) was calculated according to Equation (2).

5. Results

Two tests were conducted on each wall configuration: the first test was performed using the HFP01, the second using the HFP03. Table 3 shows the values obtained from the experimental analyses.

Table 3. Results of the experimental campaigns.

	Test 1 – HFP01			Test 2 – HFP03		
Sample	**q [W/m²]**	**Λ [W/m²K]**	**U [W/m²K]**	**q [W/m²]**	**Λ [W/m²K]**	**U [W/m²K]**
S1	3.08 ± 0.09	0.157 ± 0.004	0.153 ± 0.004	2.98 ± 0.18	0.151 ± 0.008	0.148 ± 0.008
S2	3.65 ± 0.11	0.187 ± 0.005	0.182 ± 0.005	3.49 ± 0.21	0.179 ± 0.009	0.174 ± 0.009
S3	3.61 ± 0.11	0.185 ± 0.005	0.180 ± 0.005	3.44 ± 0.21	0.177 ± 0.009	0.172 ± 0.009

The uncertainty analysis and propagation of uncertainty in the conductance values were carried out following Holman's method [30], using Equation (6):

$$W_\Lambda = \left[\left(\frac{\partial \Lambda}{\partial q} w_q\right)^2 + \left(\frac{\partial \Lambda}{\partial T_{s,i}} w_{T_{s,i}}\right)^2 + \left(\frac{\partial \Lambda}{\partial T_{s,e}} w_{T_{s,e}}\right)^2\right]^{1/2} \tag{6}$$

where:
- W_Λ is the uncertainty in the conductance value;
- w_q, $w_{T_{s,i}}$ and $w_{T_{s,e}}$ are the uncertainties in the independent variables q, $T_{s,i}$ and $T_{s,e}$.

Figure 3 shows the results obtained using the two different heat flow meters for each of the wall configurations. The graphs contain the measurements made in the last 72 hours of each measurement cycle.

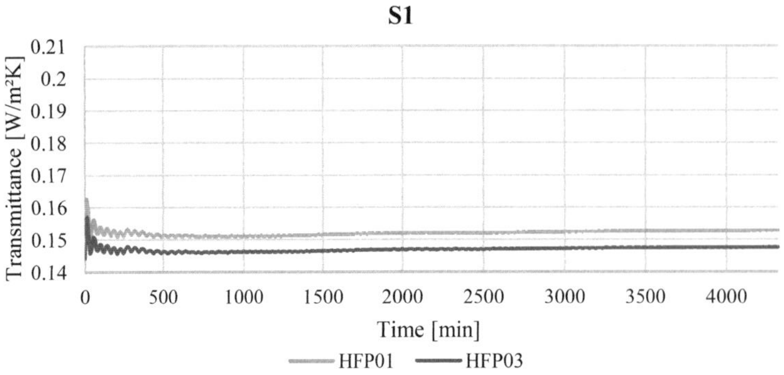

(a)

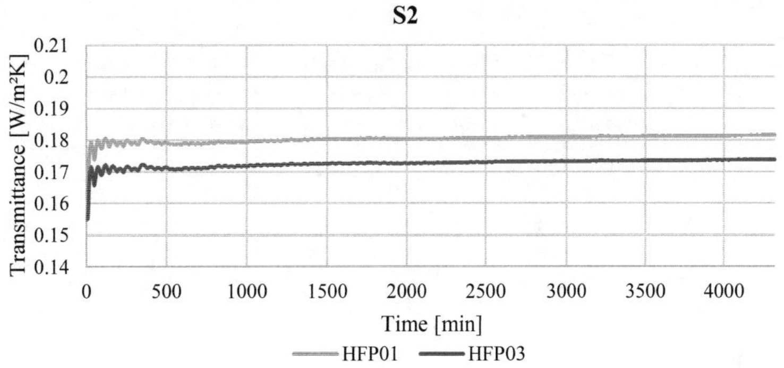

(b)

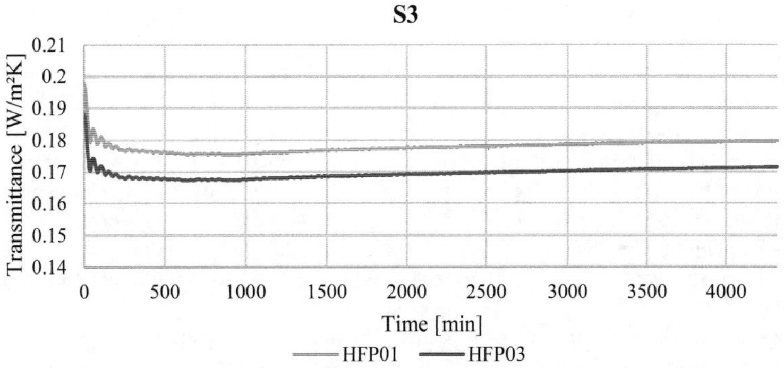

(c)

Figure 3. Thermal transmittance experimental results. (a) First sample - S1. (b) Second sample – S2. (c) Third sample – S3.

A comparison of the transmittance values obtained with the two HFM for each sample shows that HFP03 always gives slightly lower values than HFP01, with percentage differences ranging from 2.8% to 4.4%, as shown in Figure 4.

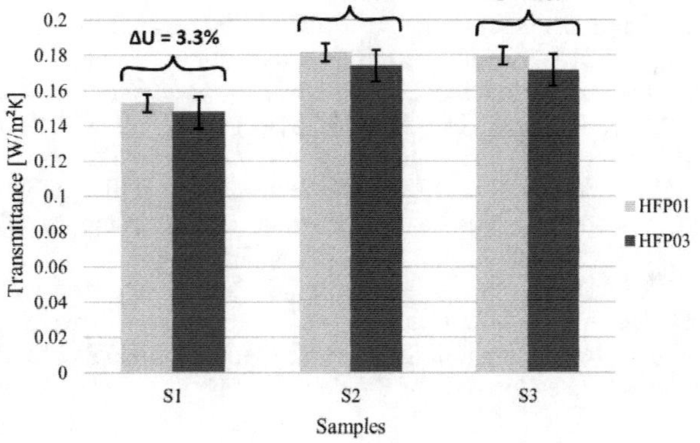

Figure 4. Transmittance values obtained with the HFP01 and HFP03 and relative percentage difference.

Table 4 illustrates the percentage differences between the theoretical and experimental transmittance values for the tests conducted.

Table 4. Comparison between theoretical and experimental transmittance values (values in [W/m^{2}K]).

Sample	*Test*	*Theoretical U*	*Experimental U*	*Percentage difference*
S1	**Test 1 – HFP01**	0.176	0.153 ± 0.004	13.1 %
	Test 2 – HFP03	0.176	0.148 ± 0.008	15.9 %
S2	**Test 1 – HFP01**	0.202	0.182 ± 0.005	9.9 %
	Test 2 – HFP03	0.202	0.174 ± 0.009	13.9 %
S3	**Test 1 – HFP01**	0.204	0.177 ± 0.005	13.2 %
	Test 2 – HFP03	0.204	0.172 ± 0.009	15.7 %

6. Conclusions

The aim of the present work was to compare the thermal performance of three wall samples with real dimensions of 300 cm x 300 cm, characterized by the presence of different insulating materials: i) EPS with graphite, ii) hemp, and iii) cork.

The analyses were conducted using two heat flow meters (Hukseflux HFP01 and HFP03) in laboratory conditions thanks to the Guarded Hot Box apparatus. The starting point of the study was the knowledge of the materials' characteristics, the instruments' technical specifications, and the theoretical values, which made possible a comparison with the results obtained through the experimental analyses.

The sample that showed the best thermal behaviour is the first one, i.e., EPS with graphite. However, the results obtained for the other two wall configurations, i.e., hemp and cork, are promising, also considering that these materials are natural, recyclable and with low environmental impact. Moreover, the comparison of the two measuring instruments, characterized by different measurement area, sensitivity, and accuracy, showed that the more sensitive instrument (i.e., HFP03) always provided slightly lower conductance values than those obtained with HFP01.

This preliminary study offers several research points. The analyses of insulating materials already present on the market are an excellent basis for undertaking a line of research aimed at discovering innovative solutions that also consider unusual insulating materials under development, if they demonstrate that they have the necessary requirements to make a significant gap in the energy field. Furthermore, the measurement system could also be deepened and improved. The next step will be integrating the analyses already carried out in this work with dynamic regime analyses, to simulate the energy behaviour of the building as close as possible to the real behaviour.

References

[1] Buildings – Analysis – IEA. Available online: https://www.iea.org/reports/buildings (accessed on 14 November 2022).
[2] DECRETO LEGISLATIVO 10 giugno 2020, n. 48. Available online: https://www.gazzettaufficiale.it/eli/id/2020/06/10/20G00066/sg (accessed on 3 May 2023)
[3] Chowdhury D, Neogi S 2019 Thermal performance evaluation of traditional walls and roof used in tropical climate using guarded hot box *Construction and Building Materials* **218** 73-89
[4] Bai G, Du N, Xu Y, Qin C 2017 Study on the thermal properties of hollow shale blocks as self-insulating wall materials *Advances in Materials Science and Engineering* 9432145.
[5] Trgala K, Pavelek M, Wimmer R 2019 Energy performance of five different building envelope structures using a modified Guarded Hot Box apparatus—Comparative analysis *Energy and Buildings* **195** 116-125
[6] Ferrari S, Zanotto V 2013 The thermal performance of walls under actual service conditions: evaluating the results of climatic chamber tests *Constr Build Mater* **43** 309–16
[7] de Rubeis T, Muttillo M, Nardi I, Pantoli L, Stornelli V, Ambrosini D **2019** Integrated Measuring and Control System for Thermal Analysis of Buildings Components in Hot Box Experiments *Energies* **12** 2053
[8] Nardi I, Paoletti D, Ambrosini D, de Rubeis T, Sfarra S 2016 U-value assessment by infrared thermography: A comparison of different calculation methods in a Guarded Hot Box *Energy and Buildings* **122** 211-221
[9] Schiavoni S, D'Alessandro F, Bianchi F, Asdrubali F 2016 Insulation materials for the building sector:

a review and comparative analysis *Renewable and Sustainable Energy Reviews* **62** 988-1011
[10] Fangueiro R 2011 Fibrous and composite materials for civil engineering applications (Cambridge: Woodhead Publishing Limited)
[11] Technical datasheet. Available online: http://www.rockwool.com/ (accessed on 20 December 2022).
[12] Karamanos A, Hadiarakou S, Papadopoulos AM 2008 The impact of temperature and moisture on the thermal performance of stone wool *Energy Build* **40** 1402-1411
[13] Sulong NHR, Mustapa SASM, Rashid MKA 2019 Application of expanded polystyrene (EPS) in buildings and constructions: A review *Journal of Applied Polymer Science* **136** 47529
[14] Dalmay P, Smith A, Chotard T, Sahay-Turner P, Gloaguen V, Krausz P 2009 Properties of cellulosic fibre reinforced plaster: Influence of hemp or flax fibres on the properties of set gypsum *Journal of Materials Science* **45** 793– 803
[15] Colinart T, Glouannec P, Chauvelon P 2012 Influence of the setting process and the formulation on the drying of hemp concrete *Construction and Building Materials* **30** 372– 380
[16] Nozahic V, Amziane S, Torrent G, Saïdi K, De Baynast H 2012 Design of green concrete made of plant-derived aggregates and a pumice–lime binder *Cement & Concrete Composites* **34** 231– 241
[17] Magniont C, Escadeillas G, Coutand M, Oms-Multon C 2012 Use of plant aggregates in building ecomaterials *European Journal of Environmental and Civil Engineering* **16** S17– S33
[18] Zampori L, Dotelli G, Vernelli V 2013 Life cycle assessment of hemp cultivation and use of hemp-based thermal insulator materials in buildings *Environmental Science & Technology* **47** 7413–7420
[19] Pervais M, Sain MM 2003 Carbon storage potential in natural fiber composites *Resources, Conservation and Recycling* **39** 325– 340
[20] Linkosalmi L, Husgafvel R, Fomkin A, Junnikkala H, Witikkala T, Kairi M, Dahl O 2015 Main factors influencing greenhouse gas emissions of wood-based furniture industry in Finland *Journal of Cleaner Production* **113** 596 – 605
[21] Martínez-Alonso C, Berdasco L 2015 Carbon footprint of sawn timber products of Castanea sativa Mill. in the north of Spain *Journal of Cleaner Production* **102** 127-135
[22] Dias A, Arroja L, Capela I 2012 Carbon storage in harvested wood products: implications of different methodological procedures and input data - a case study for Portugal *European Journal of Forest Research* **131** 109-117
[23] Dias A, Arroja L 2014 A model for estimating carbon accumulation in cork products *Forest Systems* **23** 236-246
[24] Demertzi M, Sierra-Pérez J, Paulo JA, Arroja L, Dias AC 2017 Environmental performance of expanded cork slab and granules through life cycle assessment *Journal of Cleaner Production* **145** 294–302
[25] Sierra-Pérez J, Boschmonart-Rives J, Gabarrell X 2015 Production and trade analysis in the Iberian cork sector: economic characterization of a forest industry *Resources, Conservation and Recycling* **98** 55-66
[26] *EN ISO 6946:2017*; Building Components and Building Elements—Thermal Resistance and Thermal Transmittance—Calculation Methods. European Committee for Standardization: Brussels, Belgium, 2017.
[27] De Rubeis T, Nardi I, Muttillo M 2017 Development of a low-cost temperature data monitoring. An upgrade for hot box apparatus *J. Phys. Conf. Ser.* **923** 012039
[28] Nardi I, Paoletti D, Ambrosini D, de Rubeis T, Sfarra S 2015 Validation of quantitative IR thermography for estimating the U-value by a hot box apparatus *J. Phys. Conf. Ser.* **655** 012006
[29] *EN ISO 9869-1:2014*; Thermal Insulation—Building Elements—In-Situ Measurement of Thermal Resistance and Thermal Transmittance, Part 1: Heat Flow Meter Method. European Committee for Standardization: Brussels, Belgium, 2014.
[30] Holman JP 2012 *Experimental Methods for Engineers* 8th ed. (New York: McGraw-Hill Series in Mechanical Engineering)

Improving the energy performance of a 3D-printed wall using recycled material

M Neri[1,*], L Licciardello[2], A Reggia[2], M Pilotelli[1] and A M Lezzi[1]

[1] Department of Mechanical and Industrial Engineering, University of Brescia, Brescia, Italy
[2] Department Civil, Environmental, Architectural Engineering and Mathematics, University of Brescia, Italy

[*] Corresponding author: manuela.neri@unibs.it

Abstract. In this paper, we present the results of a preliminary experimental campaign conducted on a 3D-printed wall 40 cm thick made of three concrete walls, connected by metal pins and concrete curbs to form three types of cavity - one rectangular and two triangular. The tests were performed in a climatic chamber at the *Pietro Pisa Laboratory* of the University of Brescia. The wall thermal performance has been evaluated by examining four scenarios in which the cavities have been filled with insulating material made of cellulose-based recycled flakes characterized by a declared thermal conductivity of 0.038 W/mK. The wall thermal transmittance U is measured based on the temperatures and heat fluxes measured through the structure. Based on the degree of filling, the wall thermal transmittance ranges between 1.58 W/m^{2}K for the wall without insulation, and 0.28 W/m^{2}K for the configuration with all cavities completely filled.

1. Introduction

The 3D-Printing Concrete (3DPC) technology allows the conversion of digital models into real ones through, as the name suggests, a 3D printer that deposits a concrete mixture layer-by-layer. This technology has gained the interest of the scientific community since it can reduce construction waste by 60%, production time by 70%, and labor costs by 80% [1]. However, the lack of regulation prevents the widespread of this technology with few exceptions. The geometrical printability and standardization of 3D printing concrete components are high-priority concerns that need to be addressed by codes and regulations [2,3].

3D manufacturing allows flexibility in the geometry design, and it can be either implemented to realize prefabricated elements, or in situ. As a result of the movement of a mechanical arm, 3D walls are not typically chunk elements but rather have cavities. Thus, only the indispensable elements are created, namely the frames that must guarantee the structure's stability and strength.

Several studies have analyzed the properties of 3D-printed walls from different points of view: fire resistance [4], mechanical resistance [5,6], and thermal performance [7-9]. For example, the numerical analysis in [7,8] analyzed how 3D-printed wall configuration affects the thermal transmittance U, and it was

Journal of Physics: Conference Series **2685** (2024) 012034 doi:10.1088/1742-6596/2685/1/012034

stated that thermal performance improvement can be achieved by filling the cavities of the walls. In [9], the mechanical and thermal performance of seven 30 cm-wide walls was investigated as a function of the concrete design. Other investigations compared the effect of introducing parallel cavities and of increasing cavities size: it was shown that the first option is more effective. Based on the climate zone and the compactness ratio of the building, the Italian regulations [10,11] define for buildings a maximum allowable value of the mean overall heat transfer coefficient (by thermal transmission). Typical values are reported in Table 1. Due to their superior thermal performance, vertical opaque walls should have a thermal transmittance U lower than those in Table 1, where the maximum heat transfer coefficient is expressed as a function of $X=A_{env}/V_g$ where A_{env} is the building surface envelope and V_q is the building volume.

In this paper, thermal aspects are the focus of the research. This preliminary study evaluates the thermal performance of the 3D-printed wall shown in Figure 1 and identifies ways to reduce its thermal transmittance. As 3D printing allows for a wide range of wall profiles, it is possible to incorporate cavities to be filled with insulating material, thereby enhancing the thermal resistance of the wall. The aim of this study, outlined in this paper, is to quantify the increase in thermal resistance based on the level of wall insulation. The results obtained can assist designers in predicting varying insulation levels.

Table 1. Maximum overall heat transfer coefficient in W/m^{2}K by the thermal transmission for new buildings realized in the Italian area [10,11].

	ZONES A and B	**ZONE C**	**ZONE D**	**ZONE E**	**ZONE F**
$X < 0.4$ m^{-1}	0.80	0.80	0.80	0.75	0.70
0.4 m$^{-1} \leq X \leq 0.7$ m^{-1}	0.63	0.60	0.58	0.55	0.53
$X > 0.7$ m^{-1}	0.58	0.55	0.53	0.50	0.48

Figure 1. The 3D-printed wall analyzed at the *Pietro Pisa Laboratory* of the University of Brescia (Italy). The left picture shows the wall realized for structural analysis, and the right picture shows the details of the wall interior.

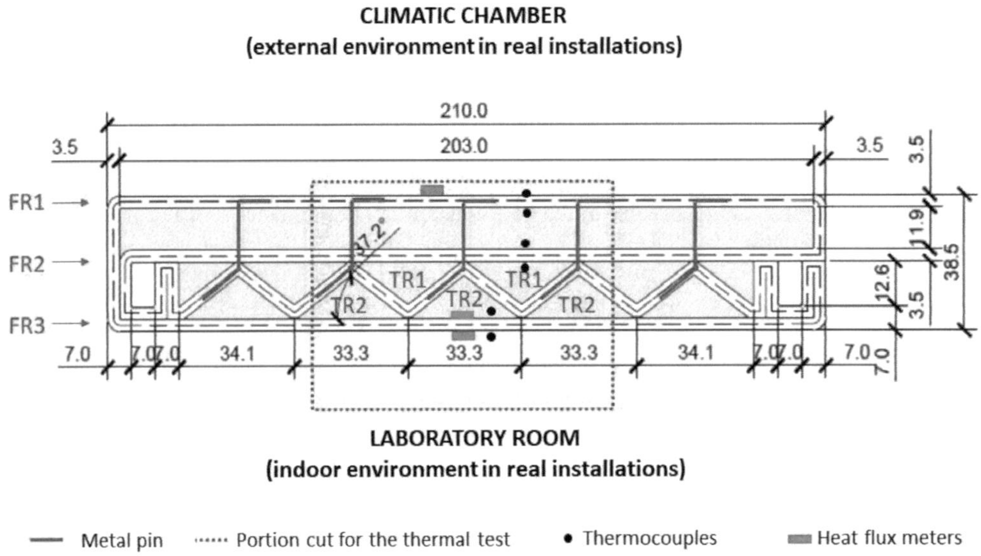

Figure 2. Wall cross section and dimensions (in centimetres). A portion of the wall (framed in the red dotted line) 1 m x 1m x 0.39 m has been cut and installed in the climatic chamber.

2. Methods and results

The aim of the experiments described in this paper is the preliminary evaluation of the thermal transmittance U of the 3D-printed wall shown in Figures 1-3. The experimental tests have been performed at the *Pisa Laboratory* at the University of Brescia. A real-size wall has been realized for structural analysis, and a square portion with a side length of 1 m has been cut to be tested in a climatic chamber. According to Figures 1 and 2, the analyzed structure consists of three walls connected through curbs and metal pins, forming cavities of different shapes - one rectangular and two triangular. The wall cavities have been filled with recycled material made of cellulose flakes obtained from newspapers, with a declared thermal conductivity of 0.038 W/mK [12]. Several degrees of filling have been analyzed, and results have been compared to the reference case represented by the empty wall. In this way, it has been assessed how different cavities filling degrees affect the wall thermal performance. The wall is made of a pre-mixed concrete product designed for 3D extrusion technologies: it consists of polymer fibers[a], and sand with a maximum diameter of 2 mm. The mortar ensures self-support during the printing phase, maintaining the shape envisaged by the starting 3D model, while at the same time ensuring excellent workability and progressive development of mechanical strength.

The structure analyzed is a portion of a larger wall realized for structural analysis. The wall has been installed in a climatic chamber, and thermocouples and heat flux meters have been installed in the wall as schematized in Figure 2. The wall is made of three parallel frames connected through curbs and metal pins to form three main types of cavity: in relation to the climatic chamber position, the two external cavities have a triangular shape, while the internal one is rectangular.

[a] Diameter of aggregates 0-2 mm. Bulk density of premixed powder 1.15 ± 0.15 kg/dm^3 [12].

Figure 3. **a**) The wall has been mounted in a metal frame (in blue colour), and XPS panels (in pink colour) have been installed to limit lateral heat losses, **b**) the rectangular cavity filled with recycled cellulose flakes, **c**) each cavity has been sealed with impermeable plastic sheets to limit air circulation between adjacent cavities.

Table 2. Description of the tests performed in the climatic chamber.

Test name	Test description	Weight of insulation	Climatic chamber temperature	Laboratory ambient temperature
EMPTY	No insulating material in the cavities	0 kg	60.3±0.1°C	27.9±1.2°C
BIG	Insulating material in the rectangular cavity	4.2 kg in the rectangular cavity	60.5±0.2°C	24.6±0.8°C
TR_1	Insulating material also in the triangular cavities *TR1*	+ 2.9 kg in the *TR1* cavities	60.5±0.1°C	24.7±0.3°C
FULL	Insulating material in all the cavities.	+ 3.5 kg in the *TR2* cavities	60.5±0.1°C	25.5±1.0°C

As can be seen in Figure 2, the structure is 1 m x 1 m x 0.38 m in dimensions and consists of three parallel continuous walls – denoted as *FR1*, *FR2* and *FR3*. In real constructions, wall *FR3* faces the indoor environment. Walls *FR1* and *FR2* are connected using small metal pivots to form a rectangular cavity. Walls *FR2* and *FR3* are connected by inclined curbs which form two series of triangles denoted as *TR1* and *TR2*: the base of triangles *TR1* is located on wall *FR2*, whereas the base of triangles *TR2* is located on wall *FR3*. The metal pivots are B450C construction bars 6 mm in diameter. To limit lateral heat flux, XPS panels have been installed along the wall perimeter (Figures 3 a-b) and, to reduce air circulation between adjacent cavities, each of them has been sealed employing impermeable plastic sheets (Figure 3 c). The temperature measured along the wall is reported in Figure 4. During the tests, the temperature in the climatic chamber has been set at 60°C, while the external temperature (laboratory temperature) was maintained at about 25°C, with limited fluctuations that could not be prevented.

According to Table 2, four tests have been performed in which the cavities have been filled with insulating material made of flakes obtained from pure newsprint cellulose with a declared thermal conductivity λ of 0.038 W/mK [12]. In the test denoted as EMPTY, no filling was used so air could flow freely; however, the plastic sheets positioned at the top and the bottom of the structure reduced the air exchange among the cavities, even if it cannot be prevented completely. In test BIG, the rectangular cavity bounded by *FR3* and *FR2* was filled with 4.2 kg of insulating material (Figure 3 b). In test TR_1, the triangular cavities *TR1* were filled with 2.9 kg of insulating material, while cavities *TR2* were left empty (Figure 3 c). Finally, in test FULL, 3.5 kg of insulating material was added to *TR2* cavities. The different insulation weights have ensured the same material density in all the cavities. The structure was equipped with three thermal flow meters and K-type thermocouples, and two thermocouples have been used to measure the temperature in the climatic chamber and in the laboratory room respectively, as schematized in Figure 2. At each considered point, two thermocouples have been installed to determine the temperature.

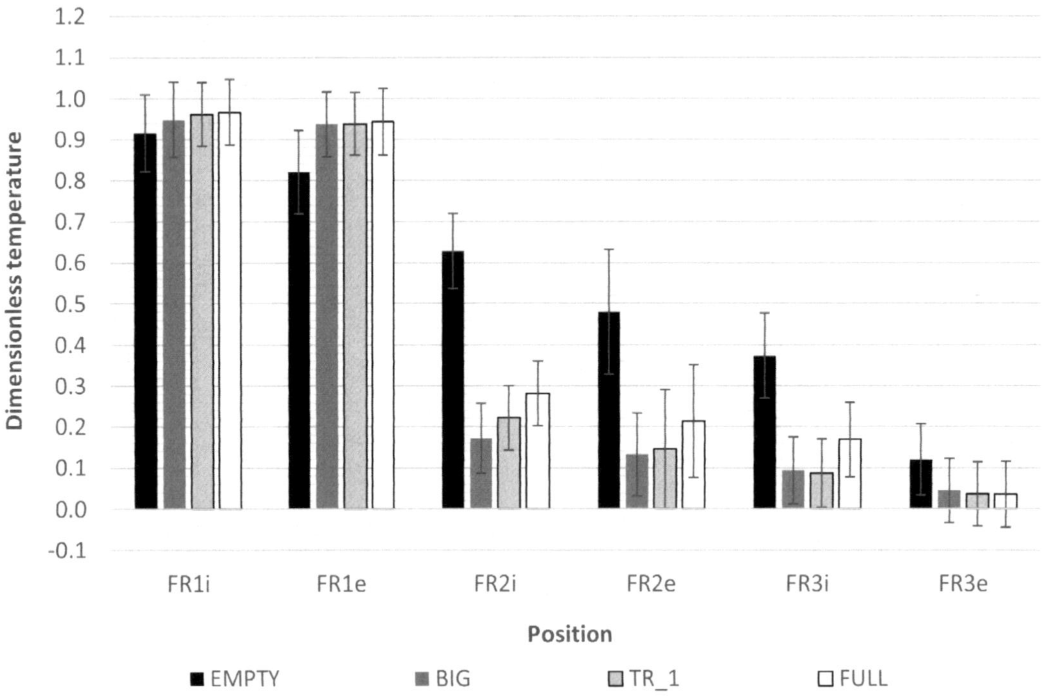

Figure 4. Temperature measured on the surfaces of the wall. The subscript '*i*' refers to the side oriented towards the climatic chamber, while the subscript '*e*' refers to the side oriented toward the laboratory. $FR1_i$ and $FR1_e$ have been measured on wall *FR1*, $FR2_i$ and $FR2_e$ on wall *FR2*, $FR3_i$ and $FR3_e$ on wall *FR3*.

The thermal transmittance U has been obtained as $U=q''/\Delta T$ where q'' is the specific heat flux measured by the heat flow meters, and ΔT is the difference between the temperature measured in the climatic chamber and the laboratory respectively. The specific heat flux q'' and the wall thermal transmittance U are reported in Table 3.

Table 3. Specific heat flux q'' and thermal transmittance U measured for the different wall configurations.

CONFIGURATION	q'' [W/m²]	U [W/m²K]
EMPTY	54.04 ± 1.42%	1.58 ± 0.03
BIG	12.16 ± 6.31%	0.34 ± 0.02
TR1	11.35 ± 6.76 %	0.32 ± 0.02
FULL	10.01 ± 7.67%	0.28 ± 0.02

The worst situation is the one with all cavities empty, as it entails the greatest thermal transmittance U. On the other hand, filling the largest cavity, reduces the thermal transmittance significantly, from 1.58 W/m^2K to values lower than 0.34 W/m^2K. This shows that preventing convective motion inside the rectangular cavity, that is, the largest one, is critical. By comparing the thermal transmittance U measured for configurations BIG, TR_1 and FULL, increasing the filling degree does not change the thermal performance significantly, since triangular cavities do not guarantee continuous and thick insulation. Moreover, as concrete is more conductive than insulating material, the heat flux through the curbs accounts for most of the heat transfer between *FR2* and *FR3*. By comparing the heat flux measured for case EMPTY with the others, the percentage reduction in the specific heat flux is 77.5% for case BIG, 79.0% for case TR1, and 81.5% for case FULL.

By comparing the measured thermal transmittance U with those imposed by the Italian regulation [10] and reported in Table 1, it emerges that filling the rectangular cavity with insulating material makes the wall suitable to be used in the Italian area, but other considerations need to be done to assess whether the overall heat transfer coefficient accomplishes the regulations. However, leaving the wall empty does not satisfy any of the climatic zones' requirements.

To better understand the influence of the filling on the wall's thermal performance, temperatures measured along the wall and shown in Figure 4 are analyzed. It is important to note that, as specified in Table 2, the laboratory temperature was not the same for all the tests, and this could have slightly affected the detected values. By analysing the temperature diagram, the greatest temperature variations through the wall occurred within the larger cavity ($FR1_e$ and $FR2_i$), and the greatest variation is detected for configuration BIG, as expected. Indeed, the rectangular cavity is the thicker one, and it is crossed by the metal pins only. For the configuration EMPTY, the temperature decreases along the thickness almost homogeneously. The temperature in the laboratory was comparable for the configurations BIG, TR1 and FULL, and the temperature measured on $FR3_e$ is almost the same. By analyzing the temperature detected at the position denoted as $FR2_i$, higher values are measured for a higher degree of filling. By comparing the temperatures detected for cases BIG and TR_1, emerges that filling the triangular cavity *TR1* does not affect the temperatures on the more external walls; however, given the lower thermal transmittance U detected, the filling is beneficial.

Heat transfer takes place by conduction, convection, and radiation. Convective motions may occur in empty cavities with a consequent increase of the heat transfer depending on cavity size and shape. Therefore, from a thermal point of view, a correct design must consider this aspect and ensure that convective motions are limited if not negligible, for example, by working on the cavities' shape and dimensions.

Therefore, 3D-printed walls need to be adequately insulated to meet the thermal insulation requirements that, however, depend on the considered location. The curb design should be done both from the standpoint of structural resistance and energy saving. For example, shapes that lengthen the path would reduce the thermal bridge effect. The thermal transmittance U can be reduced by substituting the metal pivots with plastic ones, to limit their thermal bridge effect. To improve the thermal performance also in the summer period characterized by important temperature fluctuations along the day, it would be advisable to consider heavy insulating materials to increase the temperature wave phase shift. As the cavities are closed, it is possible to use end-of-life materials as filling materials [14,15] that do not provide adequate fire protection if used individually, but which do not pose a problem if confined within the cavities.

3. Conclusions and future works

The paper presents the results of an experimental campaign performed on a 3D-printed wall that presents cavities that have been filled with recycled insulating material. In line with what is expected, the greatest degree of filling determines the lowest thermal transmittance value; however, the filling of more external cavities does not affect the thermal performance strongly. Results have shown that to meet the values of thermal transmittance U prescribed on the Italian territory, the cavities of the structure cannot be left empty, but a certain degree of filling is required. Since 3D-printed walls are light structures, to improve their thermal performance also in the summer period, when external ambient conditions are not stationary, it would be advisable to consider insulating material with high thermal capacity to reduce the periodic thermal transmittance. In this case, it will also be necessary to make assessments from a structural point of view. In the next phase of the project, experimental data will be compared with numerical results to validate numerical models capable of simulating the wall thermal behaviour.

4. References

[1] Panda B, Paul S C, Hui L J, Tay Y W D, Tan M J 2017 Additive manufacturing of geopolymer for sustainable built environment *Journal of Cleaner Production* **167** pp 281-288

[2] Alabbasi M, Agkathidis A, Chen H 2023 Robotic 3D printing of concrete building components for residential buildings in Saudi Arabia *Automation in Construction* **148** 104751

[3] Buswell R A, Da Silva W R L, Bos F P, Schipper H R, Lowke D, Hack N, Kloft H, Mechtcherine V, Wangler T, Roussel N 2020 A process classification framework for defining and describing digital fabrication with concrete *Cement and Concrete Resources* **134** 106068

[4] Suntharalingam T, Gatheeshgar P, Upasiri I, Poologanathan K, Nagaratnam B, Corradi M, Nuwanthika D 2021 Fire performance of innovative 3D printed concrete composite wall panels – A Numerical Study *Case Studies in Construction Materials* **15** e00586

[5] Le T T, Austin S A, Lim S, Buswell R A, Law R, Gibb A G F, Thorpe T 2012 Hardened properties of high-performance printing concrete *Cement and Concrete Research* **42** 3 pp 558-566

[6] Wang L, Jiang H, Li Z, Ma G 2020 Mechanical behaviors of 3D printed lightweight concrete structure with hollow section *Archives of Civil and Mechanical Engineering* **20** pp 1-17

[7] Suntharalingam, T, Upasiri I, Gatheeshgar P, Poologanathan K, Nagaratnam B, Santos P, Rajanayagam H 2021 Energy Performance of 3D-Printed Concrete Walls: A Numerical Study *Buildings* **11** p 432

[8] Alkhalidi A, Hatuqay D 2020 Energy efficient 3D printed buildings: Material and techniques selection worldwide study *Journal of Building Engineering* **30** 101286

[9] Cuevas K, Strzałkowski J, Kim J S, Ehm C, Glotz T, Chougan M, Ghaffar S H, Stephan D, Sikora P 2023 Towards development of sustainable lightweight 3D printed wall building envelopes – Experimental and numerical studies *Case Studies in Construction Materials* **18** e01945

[10] Italian Republic. Decree of the Ministry of Economic Development 26/06/2015 "Applicazione delle metodologie di calcolo delle prestazioni energetiche e definizione delle prescrizioni e dei requisiti minimi degli edifici" *Official Journal of the Italian Republic* 15 July 2015 (in Italian)

[11] Corrado V, Ballarini I, Dirutigliano D, Murano G (2016) Verification of the New Ministerial Decree about Minimum Requirements for the Energy Performance of Buildings, *Energy Procedia* **101** pp 200-207

[12] Climacell data sheet. Available at *https://climacell.it/wp-content/uploads/scheda-tecnica-climacell-S_cam-2.pdf*, Accessed on the 6th February 2023.

[13] EN 1097-3:1998 Tests for mechanical and physical properties of aggregates - Part 3: Determination of loose bulk density and voids

[14] Neri M 2022 Thermal and Acoustic Characterization of Innovative and Unconventional Panels Made of Reused Materials *Atmosphere* **13** 1825

[15] Neri M, Cuerva E, Levi E, Pujadas P, Müller E, Guardo A 2023 Thermal, acoustic, and fire performance characterization of textile face mask waste for use as low-cost building insulation material *Developments in the Built Environment* **14** 100164

Numerical performance investigation of High Vacuum Flat Plate Hybrid Photovoltaic-Thermal devices.

P Strazzullo[1,3*], D De Luca[2,3], A Caldarelli[1,3], E Gaudino[1,3], M Musto[1,3], Emiliano Di Gennaro[2,3], R Russo[3]

[1] Industrial Engineering Department, University of Napoli "Federico II", Piazzale Vincenzo Tecchio, 80, Napoli, 80125, Italy
[2] Physics Department, University of Napoli "Federico II", Via Cinthia, 21, Napoli, 80126, Italy
[3] Institute of Applied Sciences and Intelligent Systems, National Research Council of Italy, Via Pietro Castellino 111, Napoli, 80131, Italy

* e-mail: paolo.strazzullo@unina.it

Abstract. We propose an innovative flat plate hybrid Photovoltaic-Thermal system under high vacuum (HV PV-T) optimized for solar-to-thermal energy conversion. It consists of a glass cover, metallic vessel, and the actual PV-T device, which englobes a low-emissive Transparent Conductive Oxide (TCO), a perovskite-based PV cell, a Solar Absorber, and a copper substrate. We investigate, through a 1-D model developed in MATLAB, the performances of the proposed PV-T system, still mined by radiative losses, varying the operating temperature (T_{op}) and the emittance of the TCO (ε_{TCO}) in the ranges of (25÷175) °C and (0.05÷0.45) respectively. The annual thermal and electrical productions are evaluated considering the Typical Meteorological Year of Naples, Italy. Specific annual costs and emission savings are evaluated and compared with the ones assured by commercial High Vacuum Flat Plate Solar-Thermal (HVFP ST) and PV collectors. Results indicate that the proposed HV PV-T increases the annual cost savings by 34% and 11% when compared to HVFP ST and PV collectors, respectively. Moreover, the presented HV PV-T increases the annual CO_2 emissions savings by 7% and 48% when compared to HVFP ST and PV collectors, respectively.

1. Introduction

Renewable energies have a key role in the fight against climate change. Since the Earth's surface receives from the Sun enough power to satisfy the demands of the whole world [1,2], solar energy could provide the most relevant contribution, but its true potential is still unexplored. Currently, solar energy is predominantly converted into electrical or thermal energy thanks to PV and Solar-Thermal (ST) collectors: different devices with different conversion efficiencies and different performance decay with increasing operating temperature. ST collectors can reach efficiency values up to 80%, while, on the other hand, the highest PV electrical efficiency value recorded is 29.1% for single-junction cells made of GaAs [3,4]. Hybrid Photovoltaic-Thermal (PV-T) devices combine a solar cell and a thermal collector, exploiting more solar radiation than a typical PV collector due to the recovery of the fraction of incident power not converted into electricity. Several PV-T architectures and layouts have been studied since the 1970s [5,6,7]. Concentrating PV-T devices reach the highest operating temperatures, but since they require reflectors, lenses, and a sunlight-tracking mechanism, the flat-plate layout is preferred due to its affordability. Classical PV-T applications aim to increase the electrical efficiency of PV panels preventing their heating [8]: these systems are optimized for electrical production, while the thermal output is a secondary product, which can meet loads at temperatures equal to or lower than 50 °C. However, the Heating and Cooling sector accounts for 51% of Total Final Energy Consumption, against the 17% related to the Power sector [9]. Moreover, thermal loads require frequently supply temperatures higher than 50°C.

Mellor *et al.* [10] and Hu *et al.* [11] proposed Silicon-based PV-T encapsulated in vacuum, demonstrating the thermal efficiency enhancement due to the abatement of convective losses, but the lamination of Silicon Cells on copper doesn't extend the spectral exploitation of the device: copper becomes highly reflective long before the wavelength corresponding to the bandgap of Silicon (1.1 eV = 1029 nm). The effectiveness of high vacuum encapsulation in flat plate architectures has been validated for both Solar-Thermal (ST) [12,13,14] and PV-T devices [10,11,15]. For the PV-Ts, the abatement of convective losses can double the thermal efficiency against the non-evacuated architecture [10]. Materials with energy bandgap higher than that of Si, like GaAs, exhibit a weaker dependence on the temperature and let a higher portion of the spectrum shine through the PV cell. GaAs-based PV modules are already adopted in space applications, which are similar to terrestrial applications under high-vacuum due to the high operating temperatures and absence of convective cooling. GaAs-based PV cell electrical efficiency experimental values, in the operating temperature range (25÷175) °C [16], will support our modelling. Unfortunately, GaAs PV modules are far more expensive than the ones adopted for terrestrial applications, so they are not suited for industrial-scale applications. However, cheaper materials like Perovskites are emerging in the PV sector. Perovskite PV cells (PSC) reached 20% electrical efficiency even in the inverted configuration [17] and Romano *et al.* and Hoang *et al.* [18,19] contemplate the tandem PSC/CIGS cells eligibility in space applications. Considering their positive trend in the last years [20], is safe to assume that PSCs will soon reach the same thermal stability as GaAs-based PV cells. The adoption of low-emitting TCO contributes to the minimization of the predominant loss mechanism which still mines the thermal performance of our device, i.e., radiative losses. Thermal coupling of the PV cell and SA is another means at our disposal to expand solar spectrum exploitation. Thanks to their spectral absorptivity higher than 0.95 in the wavelength range (350÷1500) nm, SAs like the ones proposed by D'Alessandro *et al.* [14] can enhance the thermal absorption above the energy bandgap of the PV cell, allowing the collection of almost all the radiation transpired through the semiconductor. In this sense, the PV cell acts as a spectral splitter. De Luca et. al [15] performed the simulations of two spectral splitting layouts: thermally decoupled and thermally coupled. In the thermally decoupled layout, the PV cell is placed below the glass cover so that it does not experience an efficiency decay due to high operating temperatures. In the thermally coupled layout, the PV cell is placed directly above the SA so that almost all the radiation that has been absorbed by the PV cell but not converted into electricity (thermalization and extraction losses) is conducted to the SA, providing an additional availability for the solar-to-thermal conversion. They demonstrated that the thermally coupled layout outperforms the decoupled one in terms of thermal and total efficiency.

In this work, we propose a novel HV PV-T layout optimized for solar-to-thermal-conversion which ensures a non-negligible amount of electrical output. This layout is based on the following main concepts: a) High-Vacuum encapsulation (p<0.1 Pa); b) adoption of high bandgap PV cells; c) adoption of low-emitting TCOs; d) thermal coupling of the PV cell on a Solar Absorber (SA) (figure 1). The performances of the proposed device are evaluated through a 1-D model developed in Matlab. Section 2 describes the thermodynamic model and the performance parameters investigated through a parametrical analysis, which is carried out by varying those parameters that govern the radiative losses, i.e., operating temperature (T_{op}) and TCO emittance (ε_{TCO}). The performance parameters are evaluated considering both standard (subsection *2.1*) and hourly (subsection *2.2*) global irradiation in the city of Naples, Italy. An economic and environmental impact analysis, contextualized in the Neapolitan climate, is also carried out comparing PV, HVFP ST, and HV PV-T devices as proposed systems (subsection *2.3*). In **Results and discussion**, we expose the results of the analysis, while in **Conclusions** the main achievements are highlighted.

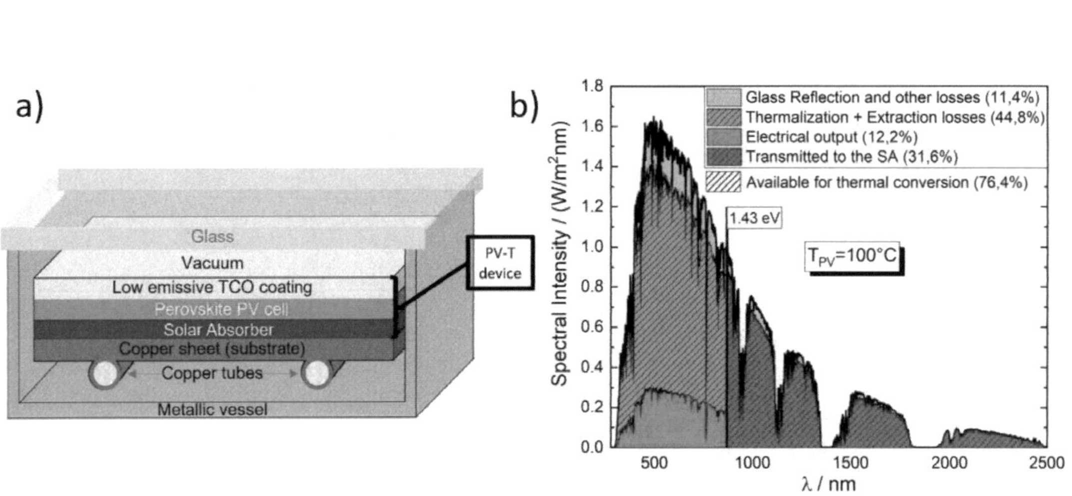

Figure 1. (a) Schematics of the proposed HV PV-T; **(b)** Exploitation of the solar spectrum enabled by thermal coupling of high bandgap PV cell and SA.

2. Methods

This section describes the thermodynamic model implemented in Matlab to evaluate the standard and annual energetic performances of PV, HVFP ST, and HV PV-T collectors. Moreover, the parameters and the equations adopted to translate the energetic performances in terms of specific annual costs and emission savings are exposed.

2.1. Standard Performance

Standard Performance indicates the performance that the three types of collector exhibit when exposed to the standard AM1.5G Spectrum (I = $1000\frac{W}{m^2}$). For the PV collector, we assume the electrical efficiency is independent of the incoming radiation and equal to the value obtained in the Standard Test Condition (STC) [21]. Instead, as regards the HVFP ST and HV PV-T collectors, the performances have been simulated through a 1-D model implemented in MATLAB. Equations (1-3) describe the model concerning the HV PV-T, which is based on the power balances related to the glass cover (equation (1)), the PV cell (equation (2)), and the Solar Absorber (equation (3)):

$$\alpha_g I + h_w(T_{amb} - T_g) + \varepsilon_{g-TCO}\sigma_{SB}(T_{PV}^4 - T_g^4)\alpha_g^{IR} + \varepsilon_g\sigma_{SB}(T_{amb}^4 - T_g^4) = 0 \quad (1)$$

$$\tau_g \int_0^{+\infty} \alpha_{PV} I_\lambda d\lambda + \frac{k_{SA-PV}}{t}(T_{SA} - T_{PV}) + \varepsilon_{g-TCO}\sigma_{SB}(T_g^4 - T_{PV}^4)\varepsilon_g - \eta_{PV}(T_{PV})\tau_g I = 0 \quad (2)$$

$$\tau_g\alpha_{SA} \int_0^{+\infty} \tau_{PV} I_\lambda d\lambda + \frac{k_{SA-PV}}{t}(T_{PV} - T_{SA}) + \varepsilon_{s-v}\sigma_{SB}(T_v^4 - T_{SA}^4) - \dot{Q}_{ABS} = 0 \quad (3)$$

where $\varepsilon_{g-TCO} = 1/(1/\varepsilon_g + 1/\varepsilon_{TCO} - 1)$ is the equivalent emittance of the two parallel plane surfaces, i.e., glass cover and TCO. The same goes for $\varepsilon_{s-v} = 1/(1/\varepsilon_s + 1/\varepsilon_v - 1)$. All the parameters are described in detail in Table (1). T_{PV} and ε_{TCO} are the input variable of the model and the numerical analysis is performed by varying them in the ranges of (25÷175) °C and (0.05÷0.45) respectively. For each combination of T_{op} and ε_{TCO}, our model evaluates the performance parameters of the HV PV-T device in terms of thermal, electrical, and total (thermal + electrical) output, i.e., $\dot{Q}_{ABS}, \dot{P}_{el}$, and $\dot{P}_{tot}$ in $\frac{W}{m^2}$. The corresponding standard efficiencies, i.e., $\eta_{th}^{PV-T}, \eta_{el}^{PV-T}$, and η_{tot}^{PV-T}, are easily evaluated by normalizing the output with respect to global irradiation. The electrical efficiency, as well as the electrical output, is evaluated according to the experimental data provided by Maros *et al.* [16]:

$$\dot{P}_{el}^{PV-T}(T_{PV}) = \eta_{el}(T_{PV})\tau_g I \left(\frac{W}{m^2}\right) \quad (4)$$

Table 1. Geometric, thermal, and optical parameters of the proposed PV-T device

Parameter	Name	*Value*	*Unit*	Parameter	Name	*Value*	*Unit*
α_g	Glass Absorptance	0.02		I	Global Solar irradiation	1000	Wm^{-2}
τ_g	Glass Transmittance	0.915		T_{amb}	Ambient Temperature	25	°C
ε_g	Glass Emittance	0.89		v_{wind}	Wind Velocity	1	ms^{-1}
t	PV cell Thickness	1.50E-6	m	h_w	Convective Heat transfer coefficient	4.5+2.9v_{wind}	Wm^{-2}K^{-1}
α_{PV}	PV cell Absorptivity	0 for E<E_{BG} 0.95 for E>E_{BG}		T_v	Vessel Temperature	25	°C
τ_{PV}	PV cell Transmissivity	1 for E<E_{BG} 0 for E>E_{BG}		ε_v	Metallic Vessel emittance	0.15	
α_{SA}	SA Absorptance	0.95		ε_s	Copper substrate emittance	0.02	
σ_{SB}	Stefan-Boltzmann constant	5.67×10^{-8}	Wm^{-2}K^{-4}	k_{PV-SA}	Equivalent SA-PV cell thermal conductivity	100	Wm^{-1}K^{-1}

Results show that the temperatures of the PV cell and SA differ by less than 1/100 °C, therefore, from now on, $T_{op} = T_{PV} = T_{TCO} = T_{SA}$ (as was easily expected due to the low thickness of the layers composing the absorbent surface).

The model implemented for the HVFP ST collector is like the set of equations 1-3, except for the absence of the PV cell and the electrical output. In this case, the absorbent surface is a Selective Solar Absorber (SSA) deposited on an aluminium substrate and its thermal properties have been experimentally evaluated in [13].

2.2. *Annual Performance*

Annual performances have been evaluated considering the hourly variation of the incoming Global Irradiation in the city of Naples [22]. Hourly thermal efficiencies are calculated according to equation (5) and equation (6) for HV PV-T and HVFP ST:

$$\eta_{th}^{PV-T}(h) = \tau_g\alpha_{ABS} - \tau_g\eta_{el}(T_{op}) - \frac{\varepsilon_{g-TCO}\sigma_{SB}(T_{op}^4-T_g^4)+\varepsilon_{s-v}\sigma_{SB}(T_{op}^4-T_v^4)}{I(h)} \quad (5)$$

$$\eta_{th}^{ST}(h) = \tau_g\alpha_{SSA} - \frac{\varepsilon_{g-SSA}\sigma_{SB}(T_{op}^4-T_g^4)+\varepsilon_{s-v}\sigma_{SB}(T_{op}^4-T_v^4)}{I(h)} \quad (6)$$

where α_{ABS} is the mean absorption coefficient of the whole PV-T absorbent surface (PV + SSA). Radiative losses towards the glass and the vessel are governed solely by the T_{op} and the emittance of the surface, therefore they have a different weight on the thermal efficiency depending on the hourly global irradiation I(h).

On the contrary, the electrical efficiencies of both HV PV-T and PV collectors can be considered independent of the hourly irradiation because the electrical output is directly proportional to I(h). Hourly thermal and electrical output in $\frac{W}{m^2}$ are evaluated by simply multiplying the hourly efficiencies by the I(h). Their values eventually allow the evaluation of annual thermal and electrical output in terms of $\frac{kWh}{m^2 year}$, i.e., Q_{ABS}, E_{el}, and E_{tot}, and annual efficiency ($\bar{\eta}$).

2.3. *Specific Annual Costs and Emissions Saving evaluation.*

In the industrial sector, the reference solution to satisfy the annual thermal demand is burning Natural Gas with furnaces in loco. Instead, the electrical demand is satisfied by linking to the national electric grid. In this section, systems based on PV, HVFP ST, and HV PV-T solar collectors are proposed to partially satisfy thermal and electrical demands and contain running costs and emissions. Thermal and electrical demands are contextualized in the climate of Naples. Specific Annual Costs Saving (SACS, $\frac{€}{m^2 year}$) and Specific Annual Emissions Saving (SAES, $\frac{t_{CO_2}}{m^2 year}$) are predicted starting from the annual energetic performances of the named solar devices. A system based on PV collectors produces $E_{el}^{PV}\ \frac{kWh}{m^2 year}$, avoiding, an annual cost of

$$\mathrm{SACS(PV)} = E_{el}^{PV} c_{el}\ \frac{€}{m^2 year} \tag{7}$$

and an annual emission of

$$\mathrm{SAES(PV)} = \frac{E_{el}^{PV} f_{NTEP}}{1000}\ \frac{t_{CO_2}}{m^2 year} \tag{8}$$

both related to the link with the national electric grid, which is characterized by the specific cost of electricity $c_{el} = 0.292\frac{€}{kWh}$ and the emission factor of the national thermos-electric park $f_{NTEP} = 0.483\frac{kg_{CO_2}}{kWh}$.

For both proposed systems based on HVFP ST and HV PV-T collectors, we assume:

- The integration of a thermal storage system large enough to fully decouple thermal production and demand, which are supposed to be at $T_{op} = 100\ °C$;
- A heat Removal Factor (HRF) of 0.82, was evaluated from a commercial HVFP ST collector datasheet [23].

A system based on HVFP ST collectors produces effectively $Q_{ABS}^{ST} \cdot HRF\ \frac{kWh}{m^2 year}$, obtaining:

$$\mathrm{SACS(ST)} = \frac{Q_{ABS}^{ST} \cdot HRF}{\eta_b LHW} c_{NG}\ \frac{€}{m^2 year} \tag{9}$$

where η_b is the combustion efficiency of industrial furnaces (assumed equal to 0.8), LHW= 9.59 $\frac{kWh}{m^3}$ is the Lower Heating Value of natural gas, and c_{NG}=0.7$\frac{€}{m^3}$ is its specific cost. Considering a natural gas emission factor $f_{NG} = 0.202\ \frac{kg_{CO_2}}{kWh}$ (referred to as primary energy), we obtain:

$$\mathrm{SAES(ST)} = \frac{Q_{ABS}^{ST} \cdot HRF}{\eta_b 1000} \cdot f_{NG}\ \frac{t_{CO_2}}{m^2 year} \tag{10}$$

Similarly, a system based on HV PV-T collectors provides the following annual specific savings:

$$\mathrm{SACS(PV-T)} = \frac{Q_{ABS}^{PV-T} \cdot HRF}{\eta_b LHW} c_{NG} + E_{el}^{PV-T} c_{el}\ \frac{€}{m^2 year} \tag{11}$$

$$\mathrm{SAES(PV-T)} = \frac{Q_{ABS}^{PV-T} \cdot HRF}{\eta_b 1000} \cdot f_{NG} + \frac{E_{el}^{PV-T}}{1000} \cdot f_{NTEP}\ \frac{t_{CO_2}}{m^2 year} \tag{12}$$

3. RESULTS AND DISCUSSION

The colourmap in figure 2(a) shows, for the proposed HV PV-T device, the standard thermal output (subsection 2.1) dependence on ε_{TCO} and T_{op}. The higher these parameters, the higher the thermal losses, and the lower the thermal output. The vertical black dotted line in figure 2(a) represents the case where the PV-T absorbent surface is coated with the low-emissive TCO proposed by Alonso-Alvarez *et al.* ($\varepsilon_{TCO} = 0.21$) [24]. This case is highlighted in figure 2(b), where the standard energetic performances of the proposed HV PV-T and commercial PV and HVFP ST collectors are compared. Increasing the operating temperature, the thermal production of the HV PV-T decreases faster than that of the HVFP ST due to the electrical power extraction and an ε_{TCO} three times greater than the emittance of the HVFP ST collector. With $T_{op} > 25\ °C$, the HV PV-T device has a lower total output than that of an HVFP ST collector, but always higher than that of a PV collector.

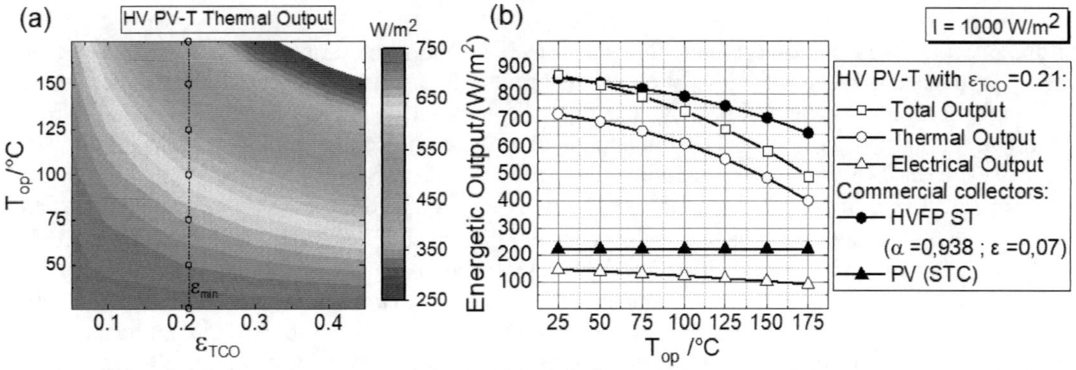

Figure 2. Standard performances under I=1000W/m². **(a)** Colourmap of the HV PV-T standard thermal output in dependence of ε_{TCO} and T_{op}. The vertical black dotted line (ε_{min}) represents the case where the PV-T absorbent surface is coated with the low-emissive TCO proposed by Alonso-Alvarez *et al.* ($\varepsilon_{TCO} = 0.21$). **(b)** Comparison among standard energetic performances of HV PV-T, HVFP ST, and PV collectors. The PV collector is supposed to work in STC.

This trend is also observed on an annual basis, as shown in figure 3(a), where the annual energetic producibilities are compared. In Naples, $T_{op} = 100\ °C$, which is a common temperature for several industrial processes, leads to interesting results since the HV PV-T collector provides a thermal output of 941 $\frac{kWh}{m^2 year}$: this amount corresponds to 68% of the annual production of the commercial HVFP ST at the same temperature (green and orange columns in figure 3(a)). Moreover, the HV PV-T electrical output of 235 $\frac{kWh}{m^2 year}$ (yellow column, figure 3(a)) leads to $\bar{\eta}_{tot}^{PV-T}(100\ °C,\ \varepsilon_{min}) = 0.61$ (empty squares, figure 3(a)). $\bar{\eta}_{tot}^{PV-T}$ drops to 0.37 with $T_{op} = 150\ °C$: higher temperatures mean higher thermal losses and a lower number of hours of effective thermal production (equation (5)). In Naples, with $T_{op} = 100°C$ and $\varepsilon_{TCO} = 0.21$, the annual energetic performance of the proposed HV PV-T device led to promising values of SACS and SAES. Figure 3(b) shows the comparison among solar field types in terms of SACS and SAES:

- SACS (HV PV-T) = 1.34 SACS (HVFP ST)
- SACS (HV PV-T) = 1.11 SACS (PV)
- SAES (HV PV-T) = 1.07 SAES (HVFP ST)
- SAES (HV PV-T) = 1.48 SAES (PV)

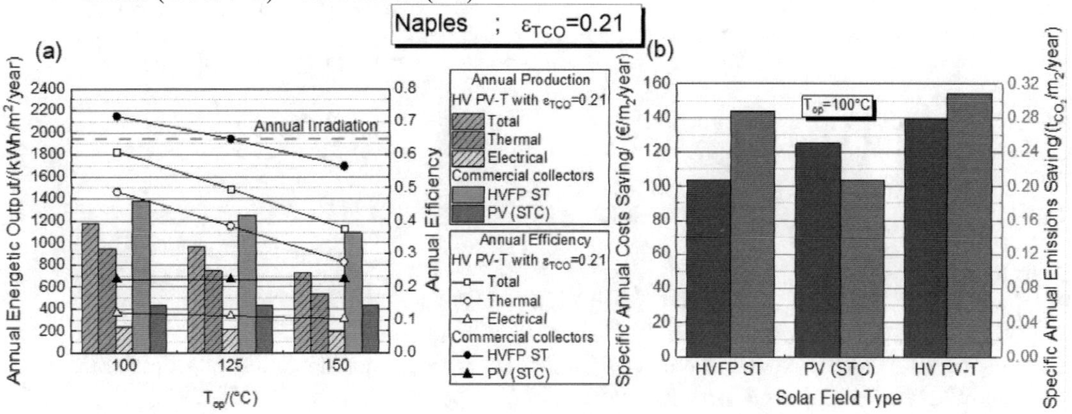

Figure 3. (a) Comparison among annual energetic performances of HV PV-T with $\varepsilon_{TCO} = 0.21$, HVFP ST, and PV collectors. The PV collector is supposed to work in STC **(b)** SACS and SAES parameters evaluated for the three solar field types.

4. Conclusions

In this paper, authors proposed a novel layout of an HV PV-T device optimized for solar-to-thermal energy conversion based on the following concepts: high-vacuum encapsulation, adoption of perovskite PV cells, low-emitting TCO coating, and thermal coupling of the PV cell on a Solar Absorber. Performances have been simulated under standard global irradiation varying the TCO emittance (0.05÷0.45) and the operating temperature (25-175) °C. Referring to the Typical Meteorological year of Naples, fixing the ε_{TCO} at 0.21 and T_{op} at 100°C, the specific annual cost and emissions saving parameters have been calculated for all three collectors. HV PV-T collector increases the specific annual cost saving by 34% and 11% when compared to HVFP ST and PV collectors, respectively. Moreover, it increases the annual CO_2 emissions saving by 7% and 48% when compared to HVFP ST and PV collectors, respectively. Future research on how to reduce TCO emittance would lead to further improvements in the performance of our device. In conclusion, the results suggest that HV PV-T collectors could play a key role in the renewable energy market.

NOMENCLATURE

c_{NG}	Natural Gas specific cost, €/m^3	LHV	Natural Gas Lower Heating Value, kWh/m^3
c_{el}	Specific cost of electrical energy, €/kWh	NG	Natural Gas
E_{el}	Annual Electrical production per m^2, kWh/m^{2}year	$\dot{P}_{el}$	Electrical power per square meter, W/m^2
f_{NG}	standard natural gas emission factor, kg_{CO_2}/kWh	$\dot{Q}$	Thermal power per square meter, W/m^2
f_{NTEP}	standard national thermo-electric park emission factor, kg_{CO_2}/kWh_{el}	Q	Annual Thermal production per square meter, kWh/m^2 year
h	hour	SACS	Specific Annual Costs Saving, €/m^{2}year
HRF	Heat Removal Factor	SAES	Specific Annual Emissions Saving, t_{CO2}/m^{2}year
HVFP ST	High-Vacuum flat plate solar-thermal	SSA	Selective Solar Absorber
HV PV-T	High-Vacuum flat plate photovoltaic-thermal	t	thickness, m
I	Global Irradiation, W/m^2	T	Temperature, °C
I_λ	Spectral Irradiance, W/m^{2}nm	TCO	Transparent Conductive Oxide

Greek Letters

α	absorptance	λ	wavelength, nm
ε	emittance	σ	Stefan-Boltzmann constant, W/m^2 K^4
η	efficiency	τ	transmittance
η_b	combustion efficiency		

Subscripts

ABS	absorbent surface	s	substrate
amb	ambient	SA	Solar Absorber
el	electrical	SB	Stefan-Boltzmann
g	glass	ST	relative to the HVFP ST collector/solar field
NOCT	Normal Operating Conditions Temperature	th	thermal
op	operating	tot	total (thermal + electrical)
PV	PV cell	v	vessel
$PV-T$	relative to the HV PV-T collector/solar field	w	wind

Superscripts

$PV-T$	relative to the HV PV-T collector/solar field
ST	relative to the HVFP ST collector/solar field

5. References

[1] Lauterbach C, Schmitt B, Jordan U, Vajen K 2012 The potential of solar heat for industrial processes in Germany *Renewable and Sustainable Energy Reviews* 16 5121–5130

[2] Shahsavari A, Akbari M 2018 Potential of solar energy in developing countries for reducing energy-related emissions *Renewable and Sustainable Energy Reviews*, 90:275–291

[3] Green M, Dunlop E, Hohl-Ebinger J, Yoshita M, Kopidakis N, Hao X Solar cell efficiency tables (version 57) *Progress in Photovoltaics: Research and Applications* **29** (1) 3–15

[4] Kayes B M, Nie H, Twist R, Spruytte S G, Reinhardt F, Kizilyalli I C, Higashi G S 2011 27.6% conversion efficiency, a new record for single-junction solar cells under 1 sun illumination 37th *IEEE Photovoltaic Specialists Conference*
[5] Wolf M 1976 Performance analyses of combined heating and photovoltaic power systems for residences *Energy Conversion* 16:79–90
[6] Florschuetz L 1979 Extension of the Hottel-Whillier model to the analysis of combined photovoltaic/thermal flat plate collectors *Solar Energy* 22:361–366
[7] Hendrie S D 1982 Photovoltaic/thermal collector development program. Final report
[8] Joshi S S, Dhoble A S 2018 Photovoltaic -Thermal systems (PVT): Technology review and future trends *Renewable and Sustainable Energy Reviews* 92:848-882
[9] REN21 2022 Renewables 2022 Global Status Report https://www.ren21.net/wp-content/uploads/2019/05/GSR2022_Full_Report.pdf
[10] Mellor A *et al* 2018 Roadmap for the next-generation of hybrid photovoltaic-thermal solar energy collectors *Solar Energy* 174:386–398
[11] Hu M, Guo C, Zhao B, X Ao, Suhendri, Cao J, Wang Q, Riffat S, Su Y, Pei G 2021 A parametric study on the performance characteristics of an evacuated flat-plate photovoltaic/thermal (PV/T) collector *Renewable Energy* 167: 884–898
[12] D'Alessandro C, De Maio D, Musto M, De Luca D, Di Gennaro E, Bermel P, Russo R 2021, Performance analysis of evacuated solar thermal panels with an infrared mirror *Applied Energy* 288: 116603
[13] De Maio D, D'Alessandro C, Caldarelli A, De Luca D, Di Gennaro E, Russo R, Musto M 2021 A Selective Solar Absorber for Unconcentrated Solar Thermal Panels *Energies* 14, 900
[14] De Maio D, D'Alessandro C, Caldarelli A, Musto M, Russo R 2022 Solar selective coatings for evacuated flat plate collectors: Optimisation and efficiency robustness analysis *Solar Energy Materials and Solar Cells* 111749
[15] De Luca D, Caldarelli A, Gaudino E, Di Gennaro E, Musto M, Russo R 2023 Modeling of energy and exergy efficiencies in high vacuum flat plate photovoltaic thermal (PV-T) collectors *Energy Reports* 9:1044–1055
[16] Maros A, Gangam S, Fang Y, Smith J, Vasileska D, Goodnick S, Bertoni M I, Honsberg C B 2015 High-temperature characterization of GaAs single junction solar cells *IEEE 42nd Photovoltaic Specialist Conference (PVSC)* June 14-19 New Orleans, LA, USA
[17] Haider M I, Hu H, Seewald T, Ahmed S, Sultan M, Schmidt-Mende L, Fakharuddin A 2023 Ethylenediamine Vapors-Assisted Surface Passivation of Perovskite Films for Efficient Inverted Solar Cells *Solar RRL* 2201092
[18] Romano V, Agresti A, Verduci R, D'Angelo G 2022 Advances in perovskites for photovoltaic applications in space *ACS Energy Letters* 7 (8) 2490–2514
[19] Hoang M T, Yang Y, Tuten B, Wang H 2022 Are metal halide perovskite solar cells ready for space applications? *The Journal of Physical Chemistry Letters* 13:2908–2920
[20] https://www.nrel.gov/pv/cell-efficiency.html
[21] SUNPOWER, https://us.sunpower.com/sites/default/files/media-library/data-sheets/sunpower-x-series-residential-solar-panels-x22-360-datasheet-514618-revc.pdf
[22] PVGIS Online Tool, https://re.jrc.ec.europa.eu/pvg_tools/en/
[23] MT-Power DataSheet v4 SK, https://www.tvpsolar.com/attach/MT-Power%20Datasheet%20(v4%20SK).pdf
[24] Alonso-Alvarez D, Llin L F, Mellor A, Paul D J, Ekins-Daukes N J 2017 ITO and AZO films for low emissivity coatings in hybrid photovoltaic-thermal applications Solar Energy 155:82–92.

Experimental study on a Dual-Source Heat Pump in ground mode to assess the soil thermal response by means of a Distributed Temperature Sensing system

C Natale*, M Dongellini, C Naldi and G L Morini

Department of Industrial Engineering, Alma Mater Studiorum – University of Bologna, Viale del Risorgimento 2, 40136, Bologna, Italy

* Corresponding author: christian.natale3@unibo.it

Abstract. In this paper, a Dual-Source Heat Pump (DSHP), able to exploit both aerothermal and geothermal energy sources, has been tested in ground mode to evaluate experimentally the soil thermal response in presence of an undersized Borehole Heat Exchanger (BHE) field. The field is instrumented with a Distributed Temperature Sensing (DTS) system, by which the geothermal fluid temperature can be measured over the entire length of the boreholes during the heat pump operation. The DSHP has been tested to reproduce the working profile of a heat generator coupled to a reference building, which has been numerically simulated by means of ALMABuild, a Matlab-Simulink tool. Three operating profiles have been identified within the simulation results to define three typical days of the heating season, characterized by different required loads. The results show that a DSHP operated in ground-mode and coupled to a borefield 50% undersized can meet completely the heating needs of a typical winter day, whilst higher building loads must be satisfied exploiting both air and ground sources. In this case, 80% of the undisturbed temperature of the soil can be recovered in an hour when aerothermal energy is extracted, thus the unit efficiency remains high and the investment cost is strongly reduced.

1. Introduction

In the recent decades, the increase of greenhouse gas emissions has encouraged the adoption of electric heat pumps to replace fossil fuel boilers in HVAC systems. Nonetheless, both Air-Source Heat Pumps (ASHPs) and Ground-Coupled Heat Pumps (GCHPs) present some limits: ASHPs operate with low performance when the outdoor temperature drops, due to thermodynamic reasons and the deposition of an ice layer on the external heat exchanger, while GCHPs need high investment cost to install the Borehole Heat Exchanger (BHE) field. In order to overcome these problems, Dual-Source Heat Pumps (DSHPs) able to exploit both aerothermal and geothermal energy sources and coupled to an undersized BHE field have been introduced [1-3]. In parallel, a correct design of the borefield is necessary to guarantee proper performance of the system and limit the investment cost linked to BHEs installation [4]. Although the traditional design procedures, such as the ASHRAE method, provide rapid and simple guidelines to calculate the total length of the boreholes, with these methodologies the borefield size is typically oversized and, consequently, the capital expenses increase. In order to limit this problem, a deeper knowledge of the ground properties and its thermal response can be achieved by means of Distributed Temperature Sensing (DTS) systems, typically based on laser measurements in optical fibres inserted within the BHEs or installed in the soil nearby [5-6].

With the aim to evaluate the benefits of the above-mentioned technologies, in this paper the prototype of an innovative DSHP has been tested in ground-source mode for different operating conditions. The prototype is coupled to a borefield made of 4 in-line vertical double-U tube BHEs having different lengths. The size of the field can be varied during each test with respect to its total length (i.e., 300 m) since each borehole can be intercepted individually. In order to evaluate dynamically the performance of the borefield for several operating conditions, the geothermal fluid temperature over the entire length of 2 BHEs has been measured experimentally by a DTS system during the heat pump operation, assessing the recovery time and the heat transfer rate.

2. Experimental test rig and Dual Source Heat Pump

The experimental test rig employed in this paper is shown in figure 1. The innovative DSHP prototype under test, able to exploit both aerothermal and geothermal energy sources, is placed within an underground climate chamber and is coupled to a hydronic loop, by means of which the building heating/cooling load can be emulated according to the Hardware-in-the-Loop (HiL) approach. A borefield made of 4 in-line vertical double-U tube BHEs having different lengths (i.e., two are 55 m long, namely BHE1 and BHE2, and two are 95 m long, namely BHE3 and BHE4) is adopted to reject/absorb heat to/from the ground when the DSHP operates in ground source mode. It is worth mentioning that the overall length of the borefield can be varied to check the influence of the BHE field size on the heat pump performance. In fact, each borehole can be intercepted independently by means of a shut-off valve. Additionally, as pointed out by figure 1, groundwater movement can be between 25-30 m and 36-50 m of depth.

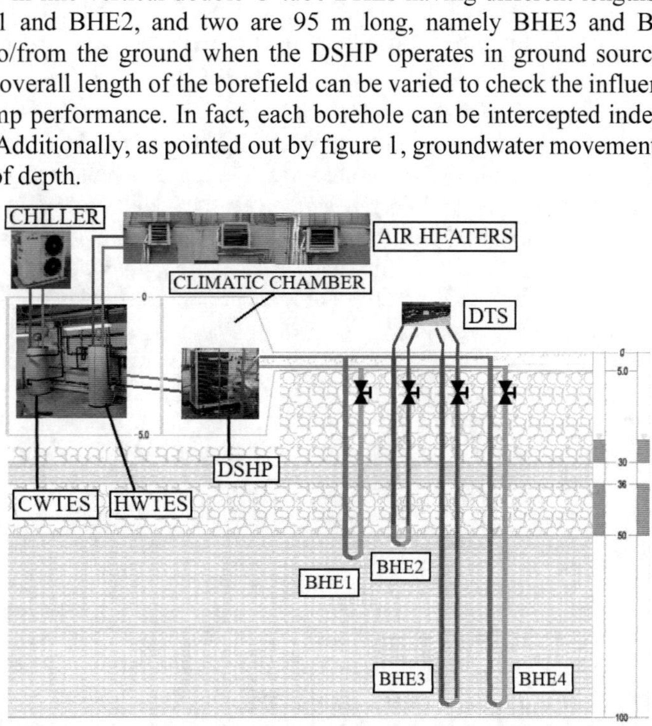

Figure 1. Experimental test rig.

A Distributed Temperature Sensing (DTS) system is installed in the BHE field to measure dynamically the fluid temperature distribution along different boreholes when the DSHP is tested in ground-source mode. More in detail, fibre optical cables are placed within supply and return pipes of a single U-tube of BHE2 and BHE3, having a length of 55 m and 95 m, respectively. The installed DTS system (SMARTEC, DiTemp Light Reading Unit) has a minimum spatial resolution of 2 m, a measurement time higher than 10 s and a resolution of ±0.1 K.

The heat pump tested in this work is a variable-speed dual-source prototype using R-410A as refrigerant. The rated heating capacity of the unit operating in ground source mode is 12.5 kW, evaluated in the following conditions: borehole fluid temperature at the inlet of the heat pump evaporator equal to 10 °C, load water inlet and outlet temperature equal to 35 °C and 40 °C, respectively. In figure 2a and 2b, values of thermal capacity (P_h) and Coefficient of Performance (COP) given by the manufacturer are reported for both operating modes (i.e., air source and ground source) as functions of the external source temperature at the heat pump inlet, for different inverter frequencies.

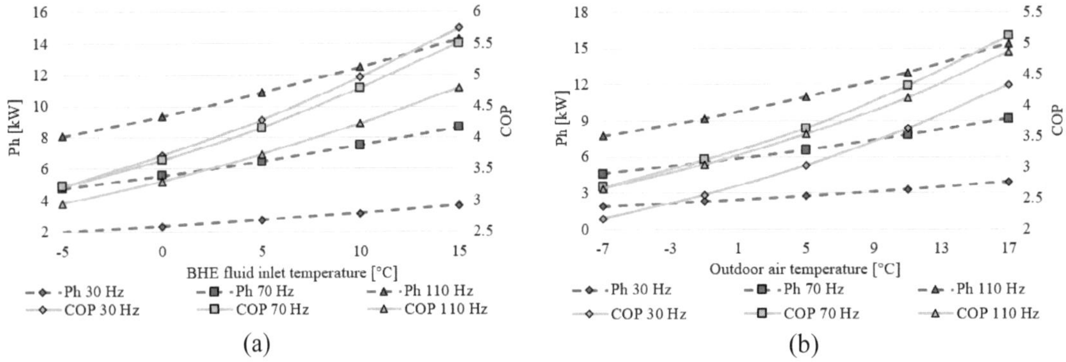

Figure 2. Heating capacity (P_h) and *COP* of the DSHP as functions of the BHE fluid inlet temperature (a) and outdoor air temperature (b) for different inverter frequencies (water temperature at the inlet/outlet of the heat pump condenser of 35/40 °C).

In order to guarantee proper operations of the DSHP, a hot water thermal energy storage (HWTES in figure 1) tank is placed in the hydronic loop (load side). The tank has a volume of 0.5 m^3 and the hot water produced by the heat pump flows in a coiled heat exchanger, immersed within the tank. The HWTES is connected to a second thermal energy storage tank of cold water (CWTES, see figure 1), having a volume of 0.5 m^3. An external chiller coupled to the CWTES, and three external air heaters linked to the HWTES (see figure 1) allow to dissipate heat to the ambient air (when needed). Further details on the experimental test rig, the sizing of its components and the typical operating conditions can be found in [7] and are not reported here for sake of brevity.

Additionally, temperature sensors (RTD Pt100, range from −5 °C to 50 °C, accuracy class 1/10 DIN, uncertainty ±0.15 K), electromagnetic flow meters (Siemens, SITRANS F M MAG 1100 as sensor and MAG 6000 as transmitter, range 0–5000 l h^{-1}, overall accuracy 0.2% ± 1.77 l h^{-1}) are installed in the hydronic loop to check the operation of the system and evaluate the energy performance of the heat pump under test. A power meter is also used during experimental tests (Fluke 1735 three-phase power quality logger, accuracy ± 1%) to measure the electric power input of the tested heat pump and assess its *COP* experimentally.

3. Methods

In the first phase of the research, the DSHP has been operated continuously up to 8 hours in heating mode exploiting the ground as external heat source. Different configurations of the BHE field (i.e., total BHE length ranging between 55 m and 300 m) were considered to assess the temperature distribution along the boreholes when heat is extracted from the soil (i.e., during the heat pump operation) and during the recharging process (i.e., the recovery of the undisturbed ground temperature when the DSHP is switched off). The heat transfer rate between BHE fluid and soil, as well as the Recharging Time (RT, namely the time interval needed by the BHE fluid to achieve 95% of its initial temperature) have been evaluated at different depths. Since water is used as BHE fluid to improve heat transfer coefficient, a threshold temperature value has been defined to avoid freezing. For borefield configurations characterized by a reduced length, the heat pump operation is stopped when the BHE fluid temperature at the heat pump outlet drops below 5 °C.

A second series of experimental tests has been carried out to evaluate the thermal response of the soil when the heat pump is operated with an intermittent working profile, typical of heat generators dedicated to the climatization of buildings. The detached residential building described in [3], located in Bologna and coupled to a GCHP having the same size of the tested DSHP, has been considered. Building heating and cooling loads and the heat pump behavior have been simulated by means of ALMABuild [8], a freely-available Matlab-Simulink tool for the dynamic modelling of buildings and HVAC systems. In figure 3 the monthly energy demand of the building is reported. It is evident how the heating demand during the cold season is 85% higher than the cooling energy need, making the building loads strongly unbalanced. To size the length of the BHE field coupled to the GCHP of the reference building, and to evaluate the linear thermal resistance of the boreholes, the methodology reported by ASHRAE [9] and by [10] have been followed, respectively, by considering the main characteristics of the borefield (diameter D, conductivity k, volume flow rate per tube $\dot{V}_{tube}$, shank spacing s, specific thermal capacity c_p, density ρ, thermal diffusivity α, undisturbed ground temperature T), reported in table 1. Also, no groundwater movement has been considered in the numerical model of the BHE field [4].

Table 1. Borefield main characteristics; subscript: internal i, external e, grout gt.

Pipe: PE100-RC			Borehole: double U-tube			
k [W m^{-1} K^{-1}]	D_i [m]	D_e [m]	D [m]	$\dot{V}_{tube}$[l min^{-1}]	k_{gt} [W m^{-1} K^{-1}]	s [m]
0.36	0.0262	0.032	0.152	17.6	1.6	0.083
Fluid: water			**Ground: clay and gravel**			
c_p^1 [J kg^{-1} K^{-1}]	ρ^1 [kg m^{-3}]	k^1 [W m^{-1} K^{-1}]	α [m^2 s^{-1}]	k [W m^{-1} K^{-1}]	T [°C]	
4203	999.8	0.56	7.99E-07	1.53	14.2	

According to the standard [9], the nominal length of the borefield is 198 m. To perform experimental tests with a well-sized borefield coupled to the DSHP, BHE 1, 2 and 3 have been selected for a total length of 205 m. This BHE length has been also adopted for the numerical simulation considering 2 boreholes (2×102.5). Within the numerical results, three typical days of the heating season, characterized by severe (S), average (A) and mild (M) outdoor air conditions and, thus, heating loads, have been identified to characterize the entire season. For each day, the heat pump operating profile has been evaluated with ALMABuild and has been reproduced in the test rig. The selection of different building loads allows to understand if an undersized borefield can be exploited in some periods of the heating season. In figure 4 the heat pump operating profiles and the outdoor air temperature for the selected days are illustrated. For each selected day, experimental tests have been performed by varying the overall length of the BHE field between 55 m and 300 m to evaluate the thermal response of the soil and assess the heat pump performance with undersized or oversized BHE fields.

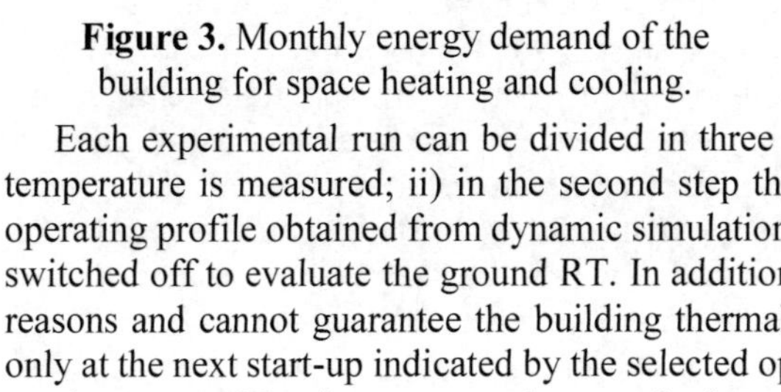

Figure 3. Monthly energy demand of the building for space heating and cooling.

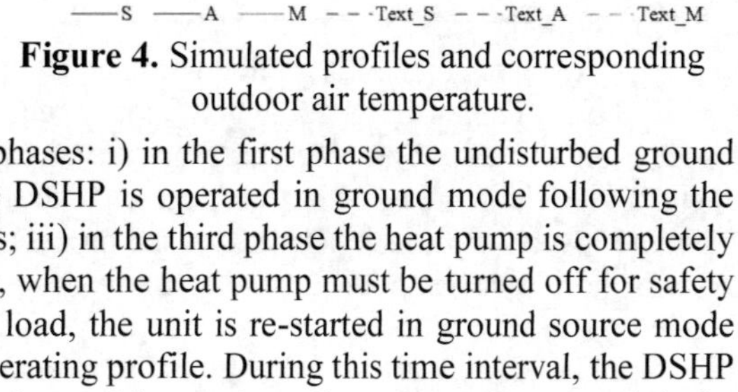

Figure 4. Simulated profiles and corresponding outdoor air temperature.

Each experimental run can be divided in three phases: i) in the first phase the undisturbed ground temperature is measured; ii) in the second step the DSHP is operated in ground mode following the operating profile obtained from dynamic simulations; iii) in the third phase the heat pump is completely switched off to evaluate the ground RT. In addition, when the heat pump must be turned off for safety reasons and cannot guarantee the building thermal load, the unit is re-started in ground source mode only at the next start-up indicated by the selected operating profile. During this time interval, the DSHP can be operated in air source mode even for low values of the outdoor air temperature to exploit the capability of the ground to recover its initial thermal level. It is important to highlight that in this circumstance, the BHE field configuration coupled to the DSHP operating in ground mode only is not proper and a longer field would be needed to match the load. Five configurations of the borefield have been considered and are reported in table 2 as a function of the selected typical days. As evidenced in table 2, all configurations have been tested in the severe day (S1, S2, S3, S4, S5), three configurations in the average day (A1, A2, A3) and one in the mild day (M1). During the operation of the DSHP, in all the described experimental tests, the water temperature at the inlet/outlet of the heat pump condenser has been maintained equal to 35 °C and 40 °C, respectively, whilst the inverter frequency of the compressor has been set to the maximum value (110 Hz).

Table 2. Configurations of the BHE field considered in the second series of tests with corresponding percentage of under- and over-sizing.

Configuration	Mild day (M)	Average day (A)	Severe day (S)
1 (-73%)	M1) 1×55 m	A1) 1×55 m	S1) 1×55 m
2 (-54%)		A2) 1×95 m	S2) 1×95 m
3 (-27%)		A3) 1×55 m + 1×95 m	S3) 1×55 m + 1×95 m
4 (0%)			S4) 2×55 m + 1×95 m
5 (+46%)			S5) 2×55 m + 2×95 m

4. Results

In the first part of this work, the DSHP has been tested in geothermal mode with different configurations of the borefield (i.e., configurations 1, 2 and 5) to analyze the heat transfer rate between fluid and soil and the RT at different depths. It is worth mentioning that when the BHE field is significantly undersized, the heat pump operation had to be interrupted before the estimated timeline (i.e., 8 hours) due to the overcoming of the threshold temperature value linked to the freezing risk. Figure 5 shows the Charge Percentage (CP) of the BHE fluid (namely the ratio between the mean water temperature within the BHE and its undisturbed value) during the recharging period as a function of the borefield configuration. For sake of clarity, it must be declared that the described configurations have been tested in different days. As pointed out by figure 5, the geothermal fluid recovers 95% of its initial undisturbed temperature in less than 7 hours (around 400 min) after the heat pump shut down in every BHE configuration. Results point out also that, in the complete BHE field configuration (configuration 5), the longer borehole (black curve in figure 5) presents a larger CP (82% at the initial time) with respect to the shorter one (75% at the initial time, red curve). Indeed, although the water temperature at the borefield inlet is the same throughout the period of operation, the temperature distribution along the boreholes changes according to their length: the longer borehole shows a mean BHE fluid temperature higher due to a larger heat transfer area. When configurations 1 and 2 are considered (single borehole active, green and blue curves in figure 5), it is interesting to note that in the first 60 minutes of the recharging period the heat transfer rate between ground and BHE2 and BHE3 employed singularly is significantly higher (0.08 K min^{-1}) compared to that of the same boreholes (red and black curves) tested in the complete borefield configuration 5 (0.03 K min^{-1}). In fact, when the boreholes are utilized alone in an undersized field, the fluid temperature decreases faster along the test and a bigger temperature gradient between fluid and soil is present at the heat pump shut down. Furthermore, in a single borehole configuration, the fluid within the shortest BHE takes slightly less time, on average, than that in the longer one to recover its initial temperature, equal to the undisturbed soil temperature. This phenomenon can be better explained analyzing the data reported in figure 6, in which the fluid temperature distribution along the boreholes, measured by the DTS system, is illustrated as a function of time and depth. When the recharging period begins (time 0 min) the fluid temperature distribution along both boreholes overlaps. In the first stage of the process, up to 40 min, the water stored in the shorter BHE2 absorbs more heat from the surrounding soil compared to that within the longer BHE3 (about +1 K in the first 50 meters). It is interesting to highlight that the water temperature difference between BHE2 and BHE3 in deeper layers of the ground (20-50 m) remains almost constant between 40 and 220 min after the beginning of the recharging phase and vanishes completely after 24 h.

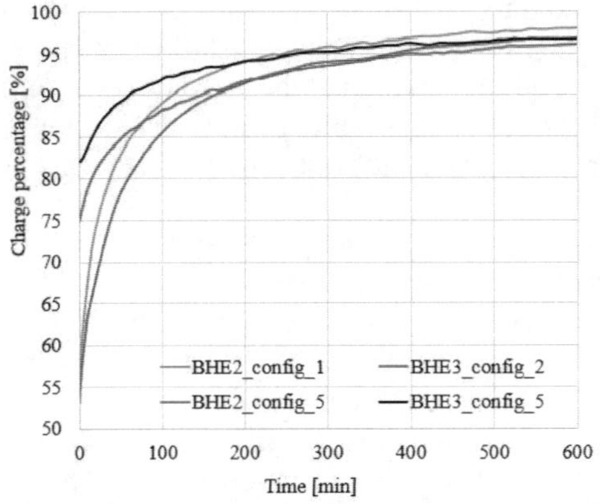

Figure 5. Charge percentage of the BHE fluid during the recharging period.

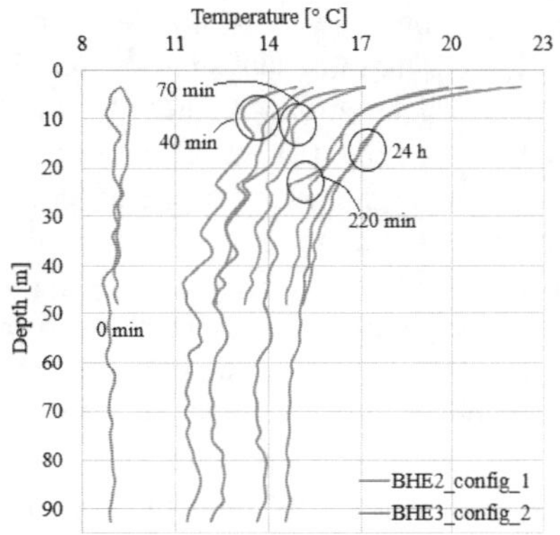

Figure 6. Vertical temperature profile within BHE2 and BHE3 as function of time during the recharging period.

As can be seen in figure 5, for a single borehole configuration the recharging process is fast: the ground recovers 80% of its undisturbed temperature in about one hour. For this reason, the analysis of the ground thermal response under an intermittent operation of the DSHP coupled to an undersized borefield has been performed. In this way, the capability of the ground to restore rapidly its thermal level between two consecutive start-ups of the heat pump can be investigated to figure out if the building energy demand during the winter season can be met by adopting an undersized BHE field. This second series of experimental tests has been carried out considering the three operating profiles of the heat pump reported in figure 4, reproduced by the DSHP operated in geothermal mode and coupled to the borefield configurations of table 2. Figure 7 shows the water temperature distribution within the inlet pipe of BHE3 during case study A2 (i.e., operating profile of the average day with borefield 54% undersized) as a function of time and depth. Results point out that sharp variations of the fluid temperature within the BHE can be observed in correspondence of the heat pump start-ups and shut-downs.

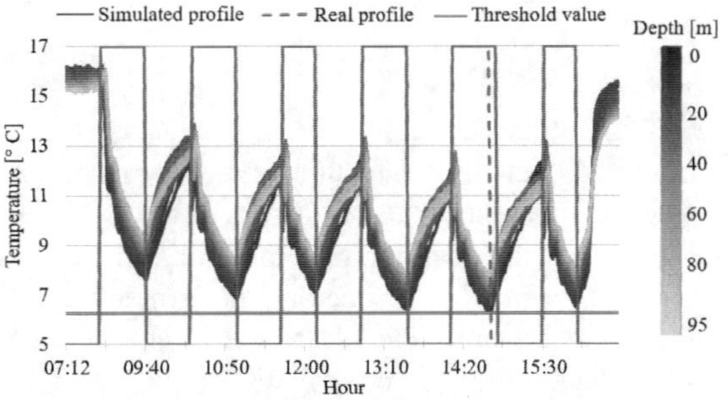

Figure 7. Water temperature distribution within the inlet pipe of BHE3 (case study A2).

For example, in the first start-up in figure 7 the mean water temperature drops of 4 °C in 6 minutes. Moreover, data shown in figure 7 indicate that the required building thermal load with average outdoor conditions can be satisfied by BHE3 only, thus employing a borefield of reduced length (54% undersizing). Indeed, the effective heat pump operating profile (dashed red line in figure 7) overlaps the simulated profile (continuous red line) for almost the whole day. The heat pump must be switched off to prevent freezing only for a limited period during the fifth start-up, for about 5 minutes. Coherently, during the heat pump operation the highest fluid temperature is obtained in the inlet pipe at the borehole bottom, as pointed out by the light blue colour of the chromatic scale. Additionally, it can be noticed that the minimum value of the fluid temperature achieved at the borehole head (dark blue colour) does not vary from the fourth start-up. It is possible to conclude that steady-state conditions were achieved:

heat extracted from the fluid at the heat pump evaporator and thermal energy gained from the soil have the same value. Therefore, the analysed configuration of the BHE field, characterized by an undersizing of about 50%, can be adopted to meet the building heating load for most part of the season. It is important to stress that, adopting a glycol solution instead of water as geothermal fluid, the minimum achievable temperature would be much lower and, consequently, no heat pump stops due to freezing risk would have been necessary in this case.

In order to fully characterize the heat pump performance along the experimental tests carried out in this work, table 3 reports for each test the average values of the Coefficient of Performance (namely, the ratio between heating power supplied and electric power absorbed by the heat pump) in ground-mode (COP_g), air-mode (COP_a) and in dual-source mode (COP_d). Regarding COP_d, both sources must be exploited (due to freezing risk). According to table 3, the thermal energy demand of the building in the severe day test can be satisfied in ground-mode only by configurations S3 (27% undersizing), S4 (nominal length of borefield) and S5 (46% oversizing) in which the operation time matches the scheduled time. On the contrary, with configurations S1 and S2 (undersizing of 73% and 54%, respectively) the heat pump cannot guarantee the whole building load during the most severe part of the season due to the rapid drop of water temperature and the consequent shut-down of the heat pump for freezing risk. However, the adoption of a DSHP permits to install a borefield strongly undersized, such as that in S1 and S2 configurations, even for the most critical outdoor conditions thanks to the capability to exploit also the aerothermal energy source. As pointed out in the previous part of the paper, the temperature of the soil around the boreholes can increase sharply when the heat pump works in air-source mode, and so, the ground thermal level can be restored.

Table 3. Main experimental results.

Case study	Under/over- sizing [%]	COP_g	COP_d	COP_a	Operation time [min]	Scheduled time [min]
M1	-73	4.29	4.31	4.89	115	120
A1	-73	4.23	3.87	3.39	150	220
A2	-54	4.29	4.25	3.39	215	220
A3	-27	4.25	4.25	3.39	220	220
S1	-73	4.32	3.40	2.85	153	320
S2	-54	4.32	3.69	2.85	221	320
S3	-27	4.17	4.17	2.85	320	320
S4	0	4.40	4.40	2.85	320	320
S5	+46	4.39	4.39	2.85	320	320

Furthermore, it can be observed that the heat pump experimental performance in ground-mode (COP_g) is only slightly influenced by the borefield configuration. This mainly depends on the temperature of the water at the inlet of the heat pump evaporator, which does not change significantly over the tests. On the other hand, for a fixed reference day, the time operated by the DSHP in ground-mode varies strongly as a function of the borefield size. As an example, the minimum period operated by the heat pump in ground-mode in the most severe day test ranges between 153 min (case S1) to 320 min (cases S3, S4 and S5, building load completed satisfied). This implicates that the less the ground is exploited, the more the DSHP works in air-mode with a consequent reduction of the overall heat pump efficiency (COP_d in table 3), especially in the most severe part of the season when the average performance of the unit mainly depends on the external air temperature and on the BHE field size. Indeed, when the air-source must be exploited, the heat pump performance drops by 21% and 15% for cases S1 and S2, respectively, whilst a smaller decrease of 9% and 1% is noticed for cases A1 and A2, respectively. However, the DSHP performance (COP_d) remains very high (+19% and +29% for cases S1 and S2, respectively) if compared to that of the air unit (COP_a). Regarding the mild day test (case study M1), results indicate that a single borehole 55 m long (undersizing of 73%) can be employed for 96% of the operating time. However, the operation in ground mode of a DSHP coupled to a strongly undersized borefield during the hotter part of the heating season is not suggested, even if technically feasible, since the heat pump performance in air-mode (4.89) could be higher (+13%) with respect to that achieved in dual-mode (4.31).

5. Conclusions

In this paper the prototype of an innovative DSHP, able to exploit both ground and air as external heat source, has been tested in ground-source mode with different configurations of the coupled BHE field which is composed by 4 in-line vertical boreholes having different lengths. Each BHE can be intercepted to reduce the total length of the field. Finally, two BHEs are instrumented by a DTS system by means of which the geothermal fluid temperature over the entire length of the borehole can be measured during the heat pump operation. Several experimental tests have been performed by varying the operating profile (continuous and intermittent operations) of the DSHP. In particular, the operating profiles of the heat pump, coupled to a reference residential building, during three typical days of the heating season characterized by mild, average and severe climatic conditions, have been calculated numerically. By the experimental campaign it has been demonstrated that the DSHP operated in ground-source mode and coupled to a borefield 50% undersized is able to meet completely the heating needs for most part of the season. On the contrary, the DSHP coupled to the shortest BHE field configuration (under-sizing of about 70%) is able to meet in ground-source mode only the thermal energy demand of the milder part of the season. However, experimental results show that the shortest BHE field configuration can be adopted with the tested DSHP even in the most severe part of the season. In fact, the DSHP can operate in geothermal mode until the BHE fluid temperature drops below a threshold value (linked to freezing risk or to an efficiency limit), below which the aerothermal energy source can be exploited. In this way, when the external air is used as source, 80% of the undisturbed temperature of the soil can be recovered in an hour. Therefore, the efficiency of the unit can be maintained high, whereas the investment cost is strongly reduced. Additional tests will be performed in further research activities to fully characterize the performance of the DSHP described in this paper.

6. References

[1] Corberán J M et al. 2018 Dual source heat pump, a high efficiency and cost-effective alternative for heating, cooling and DHW production *Int. J. Low Carbon Technol.* **13** pp 161–176
[2] Grossi I et al. 2018 Dynamic modelling and energy performance analysis of an innovative dual-source heat pump system *Appl. Therm. Eng.* **142** pp 745–759
[3] Natale C et al. 2022 Dynamic modelling of a dual-source heat pump system through a Simulink tool *J. Phys.: Conf. Ser.* **2385** 012090
[4] Naldi C et al. 2019 A new numerical method to determine isothermal g-functions of borehole heat exchanger fields *Geothermics* **77** pp 278–287
[5] Fujii H et al. 2009 An improved thermal response test for U-tube ground heat exchanger based on optical fiber thermometers *Geothermics* **38** pp 399–406
[6] Aranzabal N et al. 2020 Novel instruments and methods to estimate depth-specific thermal properties in borehole heat exchangers *Geothermics* **86** 101813
[7] Dongellini M et al. 2023 A new climate chamber for air-source and ground-source heat pump testing based on the Hardware-in-the Loop approach: Design and cross validation *J. Build. Eng.* **64** 105661
[8] Campana JP et al. 2019 BESTEST and EN ISO 52016 Benchmarking of ALMABuild, a New Open-Source Simulink Tool for Dynamic Energy Modelling of Buildings *Energies* **12** 2938
[9] ASHRAE Handbook 2007 – HVAC applications (ASHRAE, Atlanta, GA) chapter 32
[10] Conti P et al. 2016 Revised heat transfer modeling of Double U vertical ground-coupled heat exchangers *Appl. Therm. Eng.* **106** pp 1257–1267

Acknowledgments
The research leading to these results has received funding from the Italian Ministry of University and Research (MUR) within the framework of the PRIN2017 project «The energy FLEXibility of enhanced HEAT pumps for the next generation of sustainable buildings (FLEXHEAT)», grant 2017KAAECT. Authors wish to acknowledge the special work by Corrado Camponeschi and Nicolò Mazza and the assistance from Vincenzo Ballerini.

A novel Ground Source heat exchanger in an underground metro tunnel

F Ahmed*, N Massarotti, A Mauro and G Normino

Department of Engineering, University of Naples "Parthenope," Isola C4, Centro Direzionale, 80143 Naples, Italy

* Corresponding author: Fawad.ahmed001@studenti.uniparthenope.it

Abstract. Using Ground Coupled Heat Pump (GCHP) systems in urban areas can be particularly difficult due to space or legislative constraints, other than excessive costs due to drilling. To overcome these problems, the authors proposed to use Artificial Ground Freezing (AGF) probes, used for tunnel excavation, as Ground Heat Exchangers (GHE). It is a widespread practice to seal the probes in the tunnel after the completion. The conversion of existing AGF probes into GHE for GCHP allows us to avoid additional drilling costs, and other space or legislative constraints associated with the use of GCHP systems in urban areas. Such systems could be developed to use underground urban transportation tunnels for heating and cooling in smart cities. These systems were tested after the construction of two tunnels, as part of the GeoGRID project, in Piazza Municipio, Naples, Italy. The data obtained from the experimental setup have been analysed and used for validation of a finite element model developed by the authors to simulate heat transfer between the probes and the surrounding ground. A simplified pipe flow model was introduced, in combination with a 3D model of the ground, to reduce the complexity and computational effort to solve the discretized equations. The simulation results have been compared with experimental data, showing a good agreement, and observing differences in the range of 2 – 5 %. The model can therefore be used as a predictive tool for the development of this type of innovative heat exchanger.

Keywords: *Ground Source Heat Pump, Decarbonization, 3D numerical model, Sustainable Cooling, Underground metro tunnel*

1. Introduction

The goals set for decarbonization by the European Green Deal (EGD) aim to achieve carbon neutrality by 2050, and before that to reduce Greenhouse gas (GHG) emissions by 30% by 2030 compared to 1990 levels[1]. 30% of the GHG emissions are due to the use of fossil fuels for energy consumption across the world[2]. Including renewable energy is necessary to reduce the dependence on fossil fuels, which is not in good proportion compared to the energy supplied by conventional fuels. Heating is crucial for the European energy sector as it is the major contributing sector in energy consumption. The energy source mix for heat production in the EU states is shown in figure 1 including renewables contribution [3].

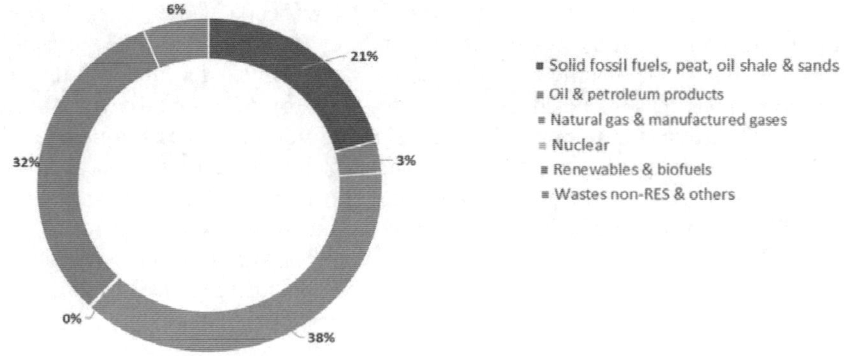

Figure 1: Energy mix for gross heat production in EU countries

Figure 1 shows that around 32% of the heat production is obtained by the renewables while the main contributor is the natural gas. If we specifically investigate the gross heat production mix of Italy which is shown in figure 2, we can see that the heat production in Italy highly depends on the supply of gas and then on solid fuels, which is an important cause of GHG emissions. It is worth mentioning that Italy is importing that gas from different countries and prices of the gas will likely increase under the current geopolitical situation between the Russia and the European countries. The energy transition is a gradual process that must include on one side more low carbon energy sources and on the other side technologies with higher efficiencies to reduce GHG emissions while reducing the energy consumption.

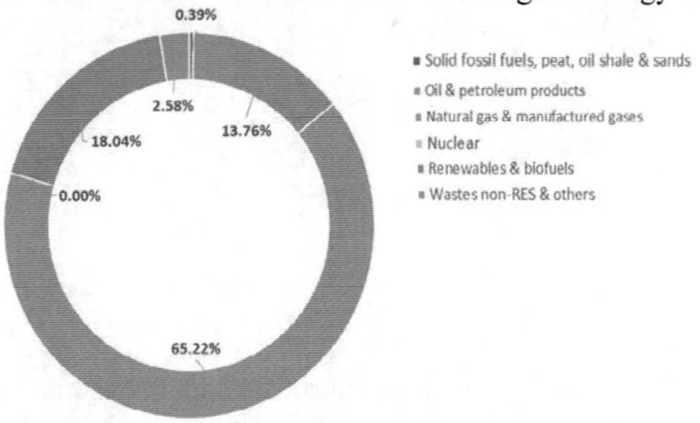

Figure 2: Energy mix of Italy for gross heat production

Heat pumps being an energy efficient and economically viable technology in terms of energy consumption can be a good fit for the purpose. An advantage of heat pump is its versatile nature of compatibility with diverse renewable energy sources. Recently, ground source heat pump (GSHP) systems attracted many researchers because of their sustainable nature in terms of low energy consumption and minimal impact on environment in terms of GHG emissions[4], [5], moreover it can be used to exploit the low enthalpy sub-surface sources which can be used for heating/ cooling[6], [7]. Among all GSHP types, one of the most convenient type of system is the Ground Coupled Heat Pump (GCHP) system which can further be categorized into a vertical and horizontal system based on the configuration of the loops laid into the ground[8]. However, more than 50% of the cost associated with GCHP is due to ground loop installation, due to the drilling installation costs for the loops[9]. These costs can be minimized by avoiding the expensive procedures of drilling.

Figure 3 shows the number of tunnels across Europe, many of those tunnels were built by using the Artificial Ground Freezing (AGF) technique[10]. The advantageous aspect about underground tunnels constructed with AGF is the presence of probes used during the construction of tunnels [11], [12]. The probes are then buried in the ground after the tunnel is completed. Conversion of these probes into ground heat exchangers (GHE) for coupling heat pumps and refrigeration system with the ground for heating/ cooling operations can reduce the drilling cost, but it is not easy.

This novel idea was implemented under the GeoGRID project[13] developed in the region of Campania in southern Italy, with the support of EU funding. The idea was to convert available AGF probes into GHE with no or little modification to maximize the benefits associated with their use. During the project, experiments were conducted by converting some of AGF probes with GHEs[14]. The numerical model validated in this work has been developed to evaluate the performance of the GHE with different conditions, to assess the potential for heat exchanges during the heating and cooling processes. The GHE has been numerically simulated to see the behaviour of the surrounding ground with the GHE [15]. This paper presents a validation of the model with the experiments conducted at Piazza Municipio metro lines in the city of Naples, Italy. The developed model is based on the coupling of heat transfer in solid module with non-isothermal pipe flow module to reduce the computational time and model complexity during simulation.

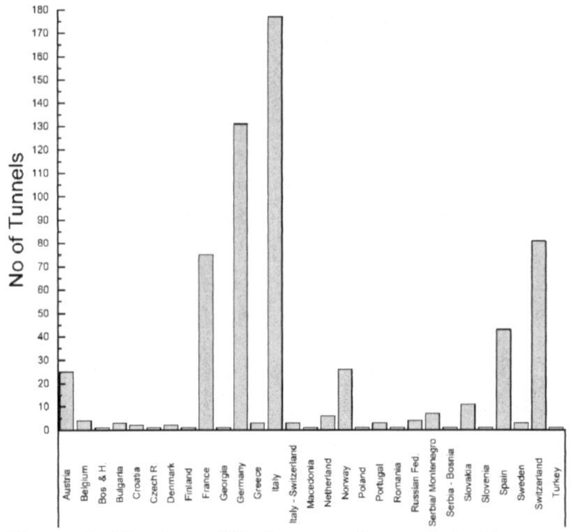

Figure 3: Number of Underground tunnels in Europe

2. Model & Equations

The numerical model developed in this work has been solved with commercial software, to simulate the effects of heat transfer between the fluid in the probes and the ground. The tunnel geometry has been obtained from the final construction and surroundings. After the tunnel geometry's construction, this was imported in COMSOL environment, where a 1D pipe flow model was combined with a 3D model of the ground and tunnel. The introduction of the pipe flow model reduces the overall complexity of the model and the computational time required for the simulations. Figure 4 shows the geometry details together with a picture of the activated tunnel and probes used in the experimental campaign.

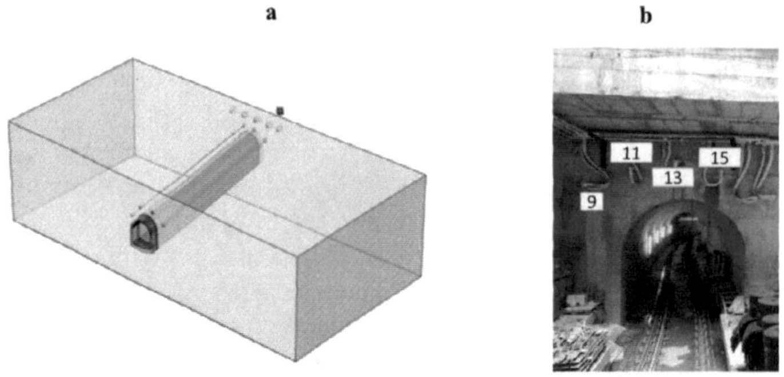

Figure 4: (a) 3D computational model, (b) Actual tunnel with probes 9,11,13 and 15 converted into GHEs (Ground Heat Exchangers)

Model Equations

To have a complete solution, two different physics are coupled in this model. Figure 4a presents the detailed model showing the heat transfer domain and the pipes. The thermo-fluid dynamics of the working fluid flowing inside the probes is modelled with the non-isothermal pipe flow (*nipfl*) one-dimensional module, while the interaction of pipes with the surrounding ground is modelled as a 3D heat conduction module. The equation used for the solution of non-isothermal one-dimensional model is the energy conservation equation for an incompressible fluid flowing in a pipe shown as equation 1[16].

$$\rho_f A C_{pf}\frac{\partial T_f}{\partial t} + \rho_f A C_{pf}\boldsymbol{u}\cdot\nabla T_f = \nabla\cdot\left(Ak_f\nabla T_f\right) + \frac{1}{2}f_D\frac{\rho A}{d_h}|\boldsymbol{u}^3| + Q_{wall} \quad (1)$$

Where, ρ_f (kg/m^3) is the fluid density, A (m^2) is the cross-sectional area of the pipe, C_{pf} (J/(kg.K)) is the specific heat of the fluid at constant pressure, $\boldsymbol{u}$ (m/s) is the velocity of the fluid in the pipe, T (K) is the temperature, k_f (W/(m.K)) is thermal conductivity of the fluid, the second term on right hand side of the equation corresponds to the friction heat dissipation, and Q_{wall} (W/m) takes into account the heat transfer from the pipe walls to the surrounding ground. A few assumptions are considered during the solution:

i. The flow through the pipes is incompressible, fully developed and one dimensional for the model.
ii. Frictional pressure drops are considered.
iii. All velocity components normal to the pipe axis are assumed to be zero.

The energy conservation equation used to model the heat transfer in the ground is shown as equation 2.

$$\rho_s C_{ps}\frac{\partial T}{\partial t} + \rho_s C_{ps}\boldsymbol{u}\cdot\nabla T - k_s\nabla T = Q_{wall} \quad (2)$$

Where ρ_s (kg/m^3) is the density of solid, C_{ps} (J/(kg.K)) is the specific heat of solid at constant pressure, T (K) is the temperature, $\boldsymbol{u}$ (m/s) is the velocity field which is assumed to be zero near the probes because of the reduced permeability due to freezing operation during the excavation of the tunnel, k (W/(m.K)) is the thermal conductivity of the solid, and Q_{wall} is the heat, which is exchanged with the pipes. In this model, pipe flow module was coupled with solid heat transfer module through temperature and the Q_{wall} term in the above equations.

Initial and Boundary Conditions for the Model

The initial and boundary conditions considered in this work have been derived from the experimental setup, to have a realistic comparison of simulated and experimental data. The initial and boundary conditions employed in the simulations are shown in figure 5. The boundary conditions on both sides are assumed to be adiabatic because of the infinite domain, therefore:

$$\frac{\partial T}{\partial n} = 0 \quad (3)$$

The undisturbed ground temperature, measured experimentally, is imposed to be constant and applied at the bottom of the domain.

$$T_{grnd} = 18.5\ ^{\circ}C \quad (4)$$

The top surface of the domain is exposed to the ambient environment, therefore a convective boundary conditions, with assumption of natural convection, is assumed which is exchanging heat with the domain and can be written by the following equation.

$$q_k = h\,(T - T_\theta) \quad (5)$$

In this equation, T represents the temperature of the domain while T_θ is the ambient temperature, whereas 'h' represents convective heat transfer coefficient for natural convection and the value of 6 W/m^{2}K is used during the calculation[17]. The front and back of the tunnel are assumed to be thermally, insulated so there is no heat transfer in the direction normal to these surfaces.

Initial conditions for the fluid flow are:

$$T_{in} = T(t) \quad (6)$$

$$u_{in} = u(t) \quad (7)$$

T_{in} and u_{in} are the inlet temperature and velocity of the Heat Transfer Fluid flowing through the pipe respectively. The temperature and velocity are time dependent and, are obtained from the experimental data.

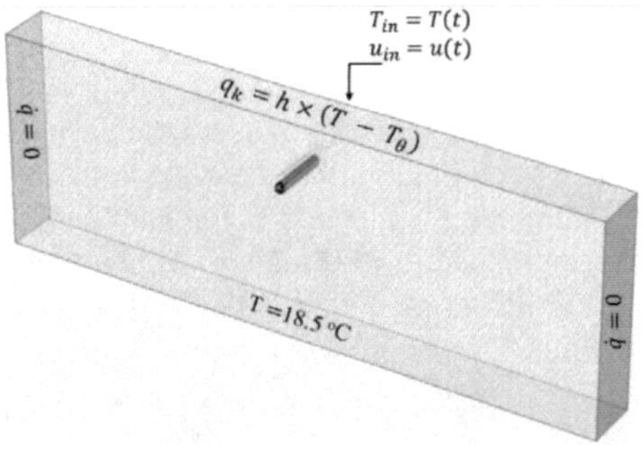

Figure 5: Boundary conditions on domain

Furthermore, the material properties have been chosen considering the actual conditions, the properties of the tunnel material are shown in Table 1[18].

Table 1. Ground properties adjacent to tunnel

Property	**Pozzolana**
Porosity Ɛ	0.51
Dry density ρ_d(kg/m^3)	1172
Thermal conductivity k (W/(mK))	1.28
Heat capacity C_p (kJ/(m^{3}K))	3150
Permeability k_f (m/s)	10^{-6}

Ground is considered to be homogeneous and material properties are considered to be constant during the simulation.

Mesh Characteristics

The mesh sensitivity analysis has shown that the best balance between computational time and independence of the results from the grid is the one presented in Figure 6 and Table 2.

Table 2. Details of the mesh elements of the model

Description	No of Elements
Tetrahedra	88389
Triangles	19260
Edge elements	5352
Vertex elements	410

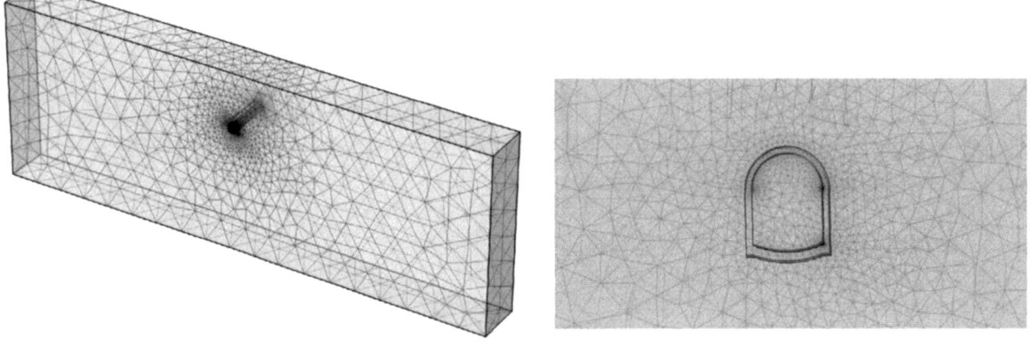

Figure 6: Distribution of elements on a particular mesh

Figure 6 shows the overall positioning and shape of the *Mesh* elements in the model and shows the area where two physics models are interacting, the size of the elements is much finer as compared to the areas which are far from the probes.

The model has been simulated for the month of July, with the above-mentioned boundary conditions, to observe the behaviour of the model. The experimental data were processed to be used as boundary conditions for the simulation model, and to obtain experimental data at the probes outlet to be used for validation of model.

3. Results & Discussions

In this section of the paper, results in terms of temperature field in the ground and transient thermal behaviour of the tunnel are discussed. In figure 7 (a), the fluid temperature in the probes is shown. The heat is transferred as the fluid flows through the pipe from the inlet towards the outlet. The inlet fluid temperature was 27 °C and at outlet the temperature dropped down to 26°C. The temperature reduces along the pipe as heat is transferred to the ground via AGF probes converted into the ground heat exchangers (GHE). In figure 7(b), the temperature distribution in the ground adjacent to the tunnel is shown throughout the computational domain. It is evident from the plot that the heat transfer occurs outwards from the tunnel. Temperatures near the tunnel is high and it reaches the ground temperature as the distance increases from the tunnel.

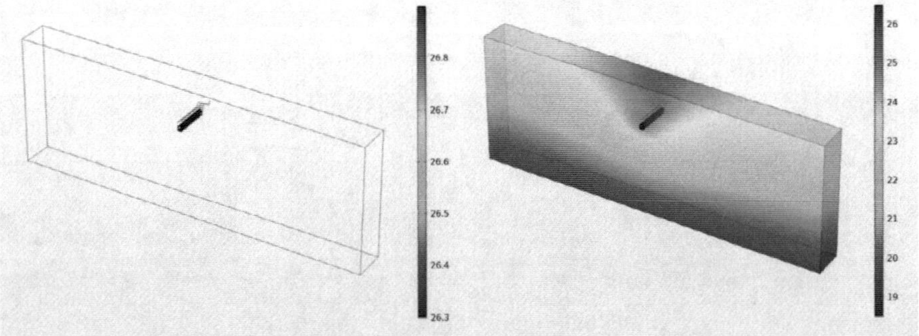

Figure 7: (a) Fluid temperatures in the probes, (b) Temperature distribution in the computational domain

Figure 8 shows the isotherms at the inlet and outlet of the tunnel. The average difference between each iso-surface is around 0.5 °C, and the temperature distribution can be seen through the computational domain. The surface adjacent to the pipes shows a temperature around 23 °C, while away from the tunnel, the temperature reduces to the average ground temperature of 18.5 °C. The main parameter employed to compare simulations with experiments is the outlet temperature of the working fluid from the GHE probes. Figure 9 shows the comparison between experimental data and numerical results in terms of outlet temperatures over time for different probes.

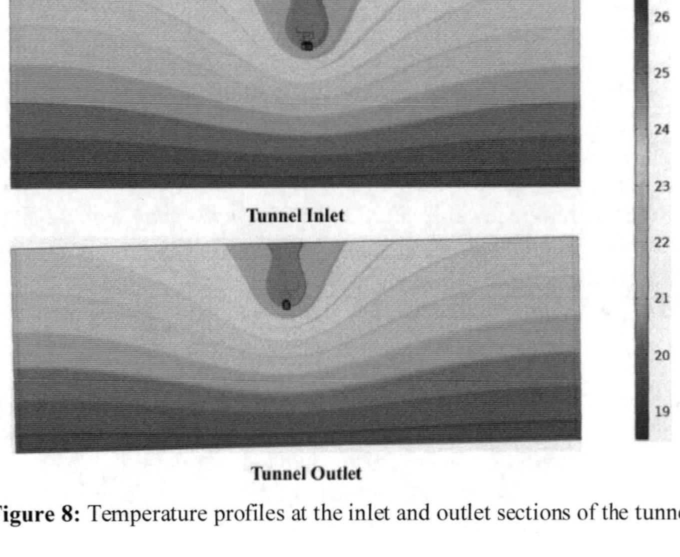

Figure 8: Temperature profiles at the inlet and outlet sections of the tunnel

The black dots represent the average daily outlet temperature obtained from the tunnel during the experiments. The uncertainties for each daily value were calculated within the confidence level of 95.5%. The red continuous line shows the outlet temperature obtained from the simulations. The red dot shows the location of the GHE at the tunnel inlet position. At the start of the month the temperature lines are around 21°C, which gradually increased to 27°C on 7th day of the month and after that it remained constant. The numerical model and experimental values are in good agreement.

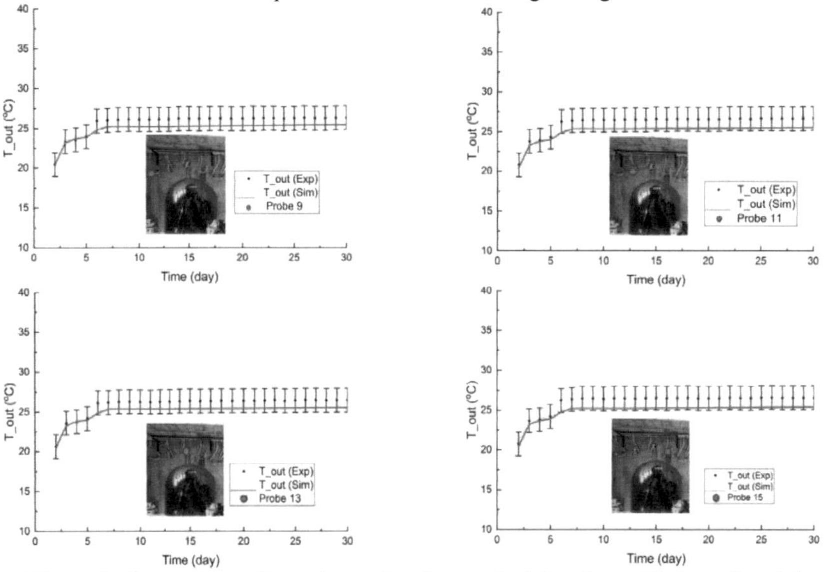

Figure 9: Comparison of experimental and numerical data for validation of model

The trend outlet temperature trend for all the probes is the same against the variation of the flow rate which shows that irrespective of the position of the probes the temperature may vary with variation in the flow rate of the HTF (Heat Transfer Fluid). It is due to the fact that the probes were fed in parallele.

4. Conclusions

The numerical model presented here has been validated with the experimental results obtained in correspondence of the probes installed in the tunnels at Piazza Municipio, Naples, Italy. A good agreement between the numerical and experimental results is observed. The model can be used to predict the actual behavior of the probes when used for actual building. The present ground heat exchanger system has economic benefits, related to the re-use of the freezing probes due to which the inflated cost of drilling can be minimized for such systems. Moreover, the model will be used to simulate the operation of GSHP coupled with the ground through the tunnel.

5. References

[1] European Parliament, "Directive of the European parliament and of the council as regards the promotion of energy from renewable sources," *Eur. Comissmion*, vol. 0218, pp. 5–24, 2021.

[2] P. Friedlingstein *et al.*, "Global Carbon Budget 2022," *Earth Syst. Sci. Data*, vol. 14, no. 11, pp. 4811–4900, Nov. 2022, doi: 10.5194/ESSD-14-4811-2022.

[3] *EU Energy in Figures, Statistical pocket book*. Publications office of the European Union, 2022.

[4] A. D. Carvalho, P. Moura, G. C. Vaz, and A. T. De Almeida, "Ground source heat pumps as high efficient solutions for building space conditioning and for integration in smart grids," *Energy Convers. Manag.*, vol. 103, pp. 991–1007, 2015, doi: 10.1016/j.enconman.2015.07.032.

[5] A. D. Carvalho, D. Mendrinos, and A. T. De Almeida, "Ground source heat pump carbon emissions and primary energy reduction potential for heating in buildings in Europe - Results of a case study in Portugal," *Renew. Sustain. Energy Rev.*, vol. 45, pp. 755–768, 2015, doi:

10.1016/j.rser.2015.02.034.
[6] A. Mauro and J. C. Grossman, "Street-heat: Controlling road temperature via low enthalpy geothermal energy," *Appl. Therm. Eng.*, vol. 110, pp. 1653–1658, 2017, doi: 10.1016/j.applthermaleng.2016.08.180.
[7] M. Iorio *et al.*, "Low Enthalpy Geothermal Systems in Structural Controlled Areas: A Sustainability Analysis of Geothermal Resource for Heating Plant (The Mondragone Case in Southern Appennines, Italy)," *Energies 2020, Vol. 13, Page 1237*, vol. 13, no. 5, p. 1237, Mar. 2020, doi: 10.3390/EN13051237.
[8] I. Sarbu and C. Sebarchievici, *Ground-Source Heat Pumps: Fundamentals, Experiments and Applications*. joe hayton, 2015.
[9] Q. Lu, G. A. Narsilio, G. R. Aditya, and I. W. Johnston, "Economic analysis of vertical ground source heat pump systems in Melbourne," vol. 125, 2017, doi: 10.1016/j.energy.2017.02.082.
[10] "Economic and Social: RAILWAY TUNNELS IN EUROPE AND NORTH AMERICA," vol. 7, no. 1,May, pp. 27–28, 2002, [Online]. Available: https://rosap.ntl.bts.gov/view/dot/5494%0A.
[11] G. Colombo, P. Lunardi, B. Cavagna, G. Cassani, and V. Manassero, "The artificial ground freezing technique application for the Naples underground," *Proc. Word Tunn. Congr. 2008 Undergr. Facil. Better Environ. Saf.*, pp. 910–921, 2008.
[12] A. Mauro, G. Normino, F. Cavuoto, P. Marotta, and N. Massarotti, "Modeling Artificial Ground Freezing for Construction of Two Tunnels of a Metro Station in Napoli (Italy)," *Energies 2020, Vol. 13, Page 1272*, vol. 13, no. 5, p. 1272, Mar. 2020, doi: 10.3390/EN13051272.
[13] "GeoGrid - Innovative technologies and systems for the sustainable use of geothermal energy." https://www.geogrid.it/#Progetto (accessed Mar. 06, 2023).
[14] "BIM Technology for low enthalpy geothermal energy plants | Request PDF." https://www.researchgate.net/publication/328335656_BIM_Technology_for_low_enthalpy_geothermal_energy_plants (accessed Jul. 08, 2022).
[15] A. Di Meglio, N. Massarotti, A. Mauro, and V. Trombetta, "Coupled Geothermal Energy Simulations," *J. Phys. Conf. Ser.*, vol. 2177, no. 1, 2022, doi: 10.1088/1742-6596/2177/1/012004.
[16] G. A. (George A. Olah, A. Goeppert, and G. K. S. Prakash, *Modeling of Oil Product and Gas Pipeline*. 2006.
[17] P. Kosky, R. Balmer, W. Keat, and G. Wise, "Chapter 14 - Mechanical Engineering," *Explor. Eng. (Fifth Ed.*, pp. 317–340, 2021, doi: 10.1016/B978-0-12-415891-7.00012-1.
[18] E. Pimentel, S. Papakonstantinou, and G. Anagnostou, "Numerical interpretation of temperature distributions from three ground freezing applications in urban tunnelling," *Tunn. Undergr. Sp. Technol.*, vol. 28, no. 1, pp. 57–69, 2012, doi: 10.1016/j.tust.2011.09.005.

Acknowledgments
The authors acknowledge the funding for green thematic doctoral research program under the projects of GEOGRID reference number CUP B43D18000230007 and GEODE (Geothermal energy for Decarbonization) reference number REACT-EU CUP I65F21001390001.

Development and validation of a model for an air-to-air air conditioner

W Ferretto[1,*], L Molinaroli[1], F Inzoli[1] and S De Antonellis[1]

[1] Politecnico di Milano, Department of Energy, 20156 Milano (MI), IT
[*] e-mail: william.ferretto@polimi.it

Abstract. Air conditioning units are responsible for a significant amount of global warming, with demand predicted to triple by 2050. Mathematical models aid in designing energy efficient air conditioners that can contribute to climate change mitigation. This study presents a steady-state model for a air-to-air air conditioner. The proposed model can assist in reducing the number of experiments and optimizing the efficiency of air conditioning systems. It utilizes a bottom-up approach to solve for compressor, condenser, capillary tube, and evaporator sub-models. The compressor is modelled using polynomial equations to determine refrigerant mass flow rate and power consumption based on operating temperatures. Heat exchanger models are solved using finite volume approach and reliable correlations are used for void fraction, friction factor, and heat transfer coefficient calculations. The closing equations are the mass conservation, i.e. constant refrigerant charge, and the equivalence between the mass flow rate sucked by the compressor and the one that flows through the capillary tube, while the evaporating and condensing pressures are the independent variables. The capillary tube is modelled using semiempirical correlations. The model is validated against experimental data at various operating conditions and shows a ±7% agreement for predicted cooling capacity.

Keywords: *Air conditioner, Numerical model, Energy efficiency*

1. Introduction

Air conditioning systems are essential for maintaining comfortable indoor environments in various contexts, including residential, commercial, and industrial buildings. The modelling of these systems is crucial to provide the customers and designers with the possibility of comparing different systems and machine configurations in terms of performance.

A significant amount of work on heat pumps and air conditioner modelling is available in the literature and can be divided into three main groups.

1. Equation-fit models: they describe the system as a black-box and the operation is based on equation parameters derived from performance maps.
2. General thermodynamic approaches: they describe the heat pump performances according to EN standards.
3. Analytical and semi-analytical models: they describe each component through physical analysis. Analytical and semi-analytical models differ from each other as the latter combine empirical relation for the description of the processes in subsystems while the first employs full physical description of each component.

In this work a semi-analytical model using the elemental volume approach is presented. Heat exchangers are fully described in an analytical way while the compressor and the expansion device are described through an empirical approach. The proposed approach allows the numerical comparison between the machine performance with different heat exchanger configurations.

Journal of Physics: Conference Series **2685** (2024) 012038 doi:10.1088/1742-6596/2685/1/012038

2. Model Description

2.1. Heat exchangers

The model of the fin-and-tube heat exchangers is developed using elemental volume approach. Each heat exchanger is divided into elemental volumes, sufficiently small to provide good computational accuracy. As shown in Figure 1, three different indexes corresponding to tube, volume in the tube and row are assigned to each volume. To simplify notation in the present work, a single number called *index w* is assigned to each elemental volume: the index value increases with the refrigerant circuitation, i.e. with w = 1 at the first tube first volume, w = T·V·R at the last tube last volume. The elemental volume shape is rectangular for an in-line tube arrangement and a hexagonal prism for staggered arrangement. Mass, momentum and energy equations are used in the solution of each elemental volume applying the following assumptions:(i) Steady-state operation regime. (ii) Superheated condensation and subcooled boiling are neglected. (iii) Kinetic and potential energy contributions are negligible. (iv) Return bends, joints, splits, headers are adiabatic. (v) Merging and splitting components are at constant pressure. (vi) Refrigerant properties are provided by RefProp 10.0 [1]

2.1.1. Refrigerant-side balances

Refrigerant-side mass, momentum and energy balances are reported in equations (1) – (3). Splitting and merging components (i.e. bends and headers) can be handled through appropriate balances.

$$\dot{m}_{REF,w} = \dot{m}_{REF,w+1} \tag{1}$$

$$P_{REF,w} = P_{REF,w+1} + \Delta P_{REF,F,w} + \Delta P_{REF,G,w} + \Delta P_{REF,A,w} \tag{2}$$

$$\dot{m}_{REF,w} h_{REF,w} = \dot{Q}_w + \dot{m}_{REF,w+1} h_{REF,w+1} \tag{3}$$

Pressure drops $\Delta P_{REF,F,w}$, $\Delta P_{REF,G,w}$ and $\Delta P_{REF,A,w}$ are respectively the frictional, gravitational and accelerative pressure drop over the w elemental volume. Blasius correlation is used to estimate friction factor in smooth tubes; Afroz and Miyara [2] correlation is used to estimate friction factor in microfin tubes. In two-phase flow, during condensation, the friction factor in the microfin tube is estimated according to Cavallini [3] while for the smooth tube the Kim Mudawar [4] is used. During evaporation, the friction factor in the microfin tube is estimated according to Diani [5] while for the smooth tube the Zhang – Webb [6] is used. Moreover, frictional pressure drop in refrigerant bends is taken into account through the Paliwoda [7] correlation.

2.1.2. Air-side equations

On air-side, different equations are used depending on tube arrangement: indeed, the shape of elemental volume changes between squared, for the in-line arrangement, and hexagonal, for the staggered arrangement. Generally, for a w elemental volume mass and energy balances take the following form:

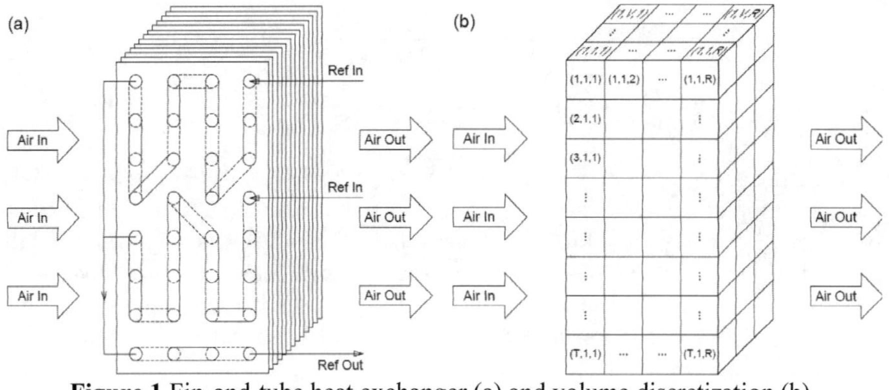

Figure 1 Fin-and-tube heat exchanger (a) and volume discretization (b)
(T = number of tubes per row, V = number of elemental volumes per tube, R = number of rows).

Table 1. Correlations for heat transfer coefficient in single-phase and two-phase flow (condensation and evaporation).

Tube type		Condensation	Evaporation
	Single - phase	Two - phase	Two - phase
Smooth	Gnielinski [11]	Cavallini [12]	Kim-Mudawar [13]
Microfinned	Copetti [14]	Cavallini [15]	Rollmann [16]

Table 2. Correlations for air side heat transfer coefficient.

Fin Type	Correlation
Flat	Wang [10]
Wavy	Wang [10]
Louvered	Wang [17]

$$\dot{m}_{AIR,IN,w} = \dot{m}_{AIR,OUT,w} \tag{4}$$

$$\dot{m}_{AIR,IN,w}\bar{h}_{AIR,IN,w} + \dot{Q}_w = \dot{m}_{AIR,OUT,w}\bar{h}_{AIR,OUT,w} \tag{5}$$

Air-side momentum balance is solved separately from the previous balances since the influence of pressure variation on heat transfer is negligible: the frictional pressure drop is computed, as reported in equation (6).

$$\Delta P_{AIR,F} = \frac{G^2}{2 \cdot \rho_{AIR,IN}} \cdot \left[f \cdot \frac{A_{TOT}}{A_{MIN}} \cdot \frac{\rho_{AIR,IN}}{\rho_{AIR,AVE}} + (1 + \sigma^2) \cdot \left(\frac{\rho_{AIR,IN}}{\rho_{AIR,OUT}} - 1 \right) \right] \tag{6}$$

The friction factor is computed according to correlations, depending on fin shape, tube arrangement and heat exchanger characteristics.

2.1.3. Heat transfer in elemental volume

In this work, each elemental volume is composed by the refrigerant, the tube, the fins and the air and is analysed as a crossflow heat exchanger in which the air is the unmixed stream and the refrigerant is the mixed one. Two different set of equations are developed to account for the dehumidification process that may occur on the evaporating section of the machine.

The heat transfer of an elemental volume in the condensing section is given by equation (7):

$$\dot{Q}_w = \varepsilon_w \cdot \left(\dot{m}c_p\right)_{MIN,w} \cdot \left(T_{REF,w} - T_{AIR,w}\right) \tag{7}$$

The effectiveness ε_w is computed through specific formulation for single and two-phase flow [8].

On the evaporator, firstly the occurrence of dehumidification at elemental volume scale is assessed by comparing the air dew temperature with the average temperature of the finned surface. If the latter exceeds the dew temperature, the elemental volume is modeled according to the methodology outlined for the condenser. In contrast, if dehumidification is present, it is accounted for using the local energy balance. The total heat transferred to the refrigerant is due to the latent heat ($d\dot{Q}_{LAT}$) and the sensible heat ($d\dot{Q}_{SENS}$), mathematically represented by equations (8) – (10).

$$d\dot{Q} = \frac{d\left(T_{WALL,EXT} - T_{REF}\right)}{R_{WALL} + R_{REF}} \tag{8}$$

$$d\dot{Q}_{SENS} = \frac{d\left(T_{AIR} - T_{WALL,EXT}\right)}{R_{AIR,DRY}} \tag{9}$$

$$d\dot{Q}_{LAT} = \frac{d\left(X_{AIR} - X_{WALL,EXT}\right)}{R_M} \cdot H_{LG}(T_{AIR}) \tag{10}$$

The humidity ratio $X_{WALL,EXT}$ refers to saturated moist air at $T_{WALL,EXT}$ temperature. The latent heat is expressed as function of the mass transfer resistance, which is computed through the Lewis analogy, assuming a constant Lewis number of 0.845 [9]. Once the unknown $T_{WALL,EXT}$ is obtained through an iterative approach, equations (8) – (10) are integrated over the elemental volume and the energy balance is applied. The air temperature and humidity at the outlet of each elemental volume w are computed and transferred as input values to the following volume. Table 2 and 3 summarize correlations used for heat transfer coefficients in the present work.

On the evaporator, to account for efficiency reduction due to the presence of condensed water, the Hong and Webb [18] correction is applied.

2.2. Compressor

The compressor is modelled using a polynomial function provided by the manufacturer. The compressor model provides the refrigerant flow rate and the power required by the machine as a function of T_{EVA} and T_{COND}.

From the energy balance on the compressor, it is also possible to compute the enthalpy of the refrigerant at compressor outlet. However, to retrieve the correct outlet temperature, oil circulation inside the compressor is considered. The energy balance to retrieve refrigerant outlet enthalpy is expressed in equation (11).

$$\dot{W}_{COMP} = \dot{m}_{REF,COMP} \cdot \left[h_{REF,OUT}(T_{REF,OUT}, P_{COND}) - h_{REF,IN}(T_{REF,IN}, P_{EVA})\right] + \dot{m}_{REF,COMP} \cdot \frac{RC}{1-RC} \cdot \bar{c}_{p,OIL} \cdot (T_{REF,OUT} - T_{REF,IN}) \quad (11)$$

Equation (11) is then solved in the only unknown $T_{REF,OUT}$. The recirculation rate RC is provided by literature analysis and, in this work, is set at a 1% value.

2.3. Expansion device

The expansion device considered in the model is an adiabatic capillary tube, therefore, an isenthalpic process is considered. The Shimizu [19] model has been selected in this work.

2.4. Refrigerant charge

2.4.1. Charge estimation in heat exchangers

The charge is estimated at the elemental volume scale as shown in equations (12) – (13).

$$M_{REF} = \sum_{i=1}^{T \cdot R \cdot V} \rho_i \cdot U_i \quad (12)$$

$$\rho_i = \begin{cases} \rho_L(T, P) \rightarrow \text{subcooled liquid} \\ \rho_G(T, P) \rightarrow \text{superheated gas} \\ \alpha \cdot \rho_{VS}(T) + (1-\alpha) \cdot \rho_{LS}(T) \rightarrow \text{two phase} \end{cases} \quad (13)$$

Void fraction is estimated through the model of Rouhani and Axelsson [20].

2.4.2. Charge estimation solved in compressor oil

Studies available in the literature [21,22] have shown that propane solubility with a mineral oil can reach values up to 40 %, hence it is necessary to account it for. Refrigerant solubility in lubricant oil is considered in this work through experimental data presented in the work of [21], which reports oil solubility at different pressure and temperature. The amount of oil contained in the machine and the oil sump positioning, which determines the oil temperature and pressure, is provided by the manufacturer.

2.5. Model structure

The simulation model presented in this work follows the iterative scheme proposed by Domanski and Didion [23]. The objective of the model is to characterize the steady state condition at each operating condition. The model is developed using pressure and enthalpy as state parameters, applying three different balances: mass, momentum and energy. A schematic of the cycle is reported in Figure 2.

The simulation process starts with a guessed refrigerant condensing and evaporating pressure and a compressor inlet superheat. Refrigerant mass flow rate and condenser inlet conditions are obtained through the compressor model. The condenser and capillary are simulated, and capillary mass flow rate is obtained: this value is compared with the one obtained in the compressor; if they differ more than the tolerance, the iterative cycle starts from the initial point with a different evaporating pressure. In case of mass flow rate through a capillary tube is higher than mass flow rate through the compressor, evaporating pressure is increased; if it is lower, evaporating pressure is decreased. Once convergence on mass flow rate is reached, the evaporator is solved using as input the data from the capillary tube. Outlet refrigerant temperature is compared with the one found in the previous iteration and is used as input in the following iteration. The entire cycle is simulated again up until convergence on the refrigerant superheat at compressor inlet. Then, the mass inventory is computed and compared with the machine value. If the difference is higher than the tolerance, the condensing pressure is changed, and the cycle is simulated again. In case of computed charge higher than the actual value, condensing pressure is decreased; if it is lower, condensing pressure is increased.

2.6. *Submodel structure*

The model structure for simulating the heat exchangers is the same for the evaporator and the condenser.

First, an initialization of the air temperature profile and all the elemental volume properties is required. Then, the procedure starts from the refrigerant inlet volume. For each volume, balances are solved, and output properties are obtained and transferred to the following volume. Once the thermal profile is solved, the hydraulic one is computed in terms of pressure drop and a check is set up to find convergence on pressure drop between the current and previous iteration. If convergence is not found, pressure profile is updated, and mass flow rate redistributed to find the same pressure drop across parallel circuits.

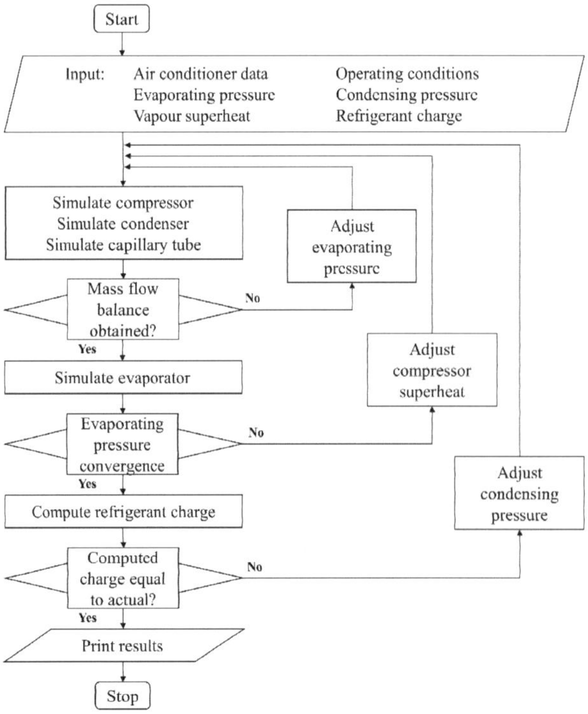

Figure 2. Schematic of the iterative model that simulates the air conditioner. At each iteration, the condensing and evaporating pressure are adjusted through the bisection method; the compressor superheat is computed and substituted in the following iteration.

Journal of Physics: Conference Series **2685** (2024) 012038 doi:10.1088/1742-6596/2685/1/012038

3. Model Validation

A set of experimental data was available from in-house experiments of a R290 prototype air conditioner carried out at the Department of Energy of Politecnico di Milano. The experiments are in the steady state condition. Refrigerant temperature was measured through T-type thermocouples while pressure was measured at condenser outlet and evaporator outlet, through Huba ® 520 pressure transducer. Air temperature and humidity were measured in 4 points, i.e., both heat exchangers inlet and outlet, through Vaisala ® HMT330 sensors. Air velocity at heat exchangers outlet was measured through a VelociCalc ® 5725 fan anemometer. Nineteen different operating conditions were analysed. The air temperature is in the range [20 – 38] °C whereas the air humidity is in the range [4.5 - 14.1] g/kg. The geometrical parameters and the refrigerant charge are retrieved from the machine technical data and are used as model input. The model is run at each operating condition and, once convergence is reached, results are collected. Two parameters were considered and reported in Table 3: the Mean Absolute Relative Deviation (MARD) and the Mean Relative Deviation (MRD), expressed for a quantity X as in equations (14.i) – (14.ii).

$$MARD = \frac{1}{n}\sum_{i}^{n}\frac{|X_{MODEL,i} - X_{EXP,i}|}{X_{EXP,i}} \quad \text{(i)} \qquad MRD = \frac{1}{n}\sum_{i}^{n}\frac{(X_{MODEL,i} - X_{EXP,i})}{X_{EXP,i}} \quad \text{(ii)} \qquad (14)$$

Results show that the model performed well in predicting both parameters. However, concerning the condenser, most of the data points were overestimated, as demonstrated by the parity plot Figure 3 (a) and the MRD positive value almost equal to MARD one.

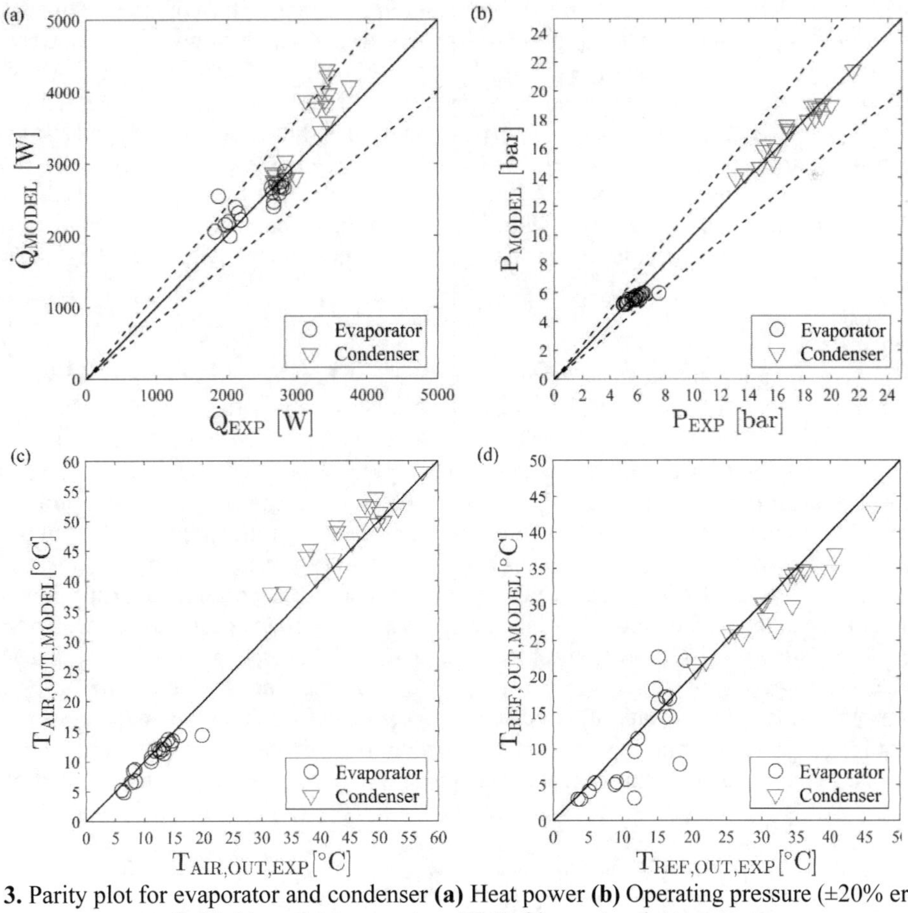

Figure 3. Parity plot for evaporator and condenser **(a)** Heat power **(b)** Operating pressure (±20% error is reported) **(c)** Air outlet temperature **(d)** Refrigerant outlet temperature

Table 3. Heat exchange, operating pressure and air conditions MARD and MRD

	$\dot{Q}_{COND}$	$\dot{Q}_{EVA}$	P_{COND}	P_{EVA}	$X_{OUT,EVA}$
MARD [%]	10.3	6.8	3.2	5.1	7.8
MRD [%]	9.3	2.1	0.7	-3.4	5.2

On the evaporator, the model performed better, with higher accuracy in predicting both pressure and heat transfer rate than the condenser. However, the evaporator capacity was overestimated due to the higher subcooling degree found at the condenser outlet, as shown in Figure 3 (d), which resulted in a lower enthalpy inlet at the evaporator. Since heat transfer performance at low vapor quality is better than at high vapor quality, the cooling power is overestimated. Additionally, the evaporator outlet refrigerant conditions in Figure 3 (d) showed a smaller superheat with respect to experimental data, supporting the explanation of a lower enthalpy inlet in the evaporator.

Despite the underestimation of evaporating pressure value, as shown in Figure 3 (b), the outlet evaporating absolute humidity was overestimated, indicating an underestimation of the dehumidification process. To improve accuracy, different models of water condensation that have been tested on a wider dataset could be implemented in future works. However, the overestimation of heat power implies the low influence of dehumidification on these experimental data.

As shown in Figure 3 (c), the evaporator outlet air temperature is correctly estimated.

The prediction performance on pressures suggests that the model correctly estimated the machine mass inventory and refrigerant flow rate. On the flow rate, it is related only to the compressor behavior, which is not simulated but described through empirical relations, and to the compressor inlet superheat, which is generally correctly estimated as previously explained. Hence, the flow rate and the evaporating pressure, which is the variable associated to the flow rate closure relation, are correctly estimated. On the mass inventory, it is related to the condensing pressure and the operating conditions at which the oil is retained, which at last depend on the oil sump positioning. The estimation on the condensing pressure suggests that the Rouhani Axelsson model for void fraction provides good accuracy. Nevertheless, the computed refrigerant solved in the oil is about 20% of the total mass inventory in all the simulated points, suggesting that a significant error takes place when this is not considered. Future work should always consider this element and develop a wide dataset considering the largest number of refrigerants and oils.

4. Conclusions

A model for the simulation of an air-to-air air conditioner has been developed in this work. The model is based on a semi-analytical approach. The approach provides the possibility of a preliminary analysis on innovative machine configuration performance without the burden of prototyping. The heat exchangers are described through the elemental volume approach and ε – NTU method. The compressor is described through the manufacturer polynomial curves and the capillary tube is described through semiempirical correlations. Mass inventory is computed considering refrigerant dissolved in the oil. The model iteratively solves the components using mass flow rate and mass inventory as closing relations for convergence. The model is validated against experimental data on a air-to-air air conditioner at different air temperature and humidity. Good agreement is found on the heat transfer, with a MARD of 6.8% on the evaporator and 10.3 % on the condenser, and on the operating pressures, with a MARD of 2.1% on the evaporating pressure and 3.2% on the condensing one. Probably, the heat transfer is overestimated due to air flow rate measurement underestimation, while operating pressures show a correct estimation on refrigerant flow rate and mass inventory estimation. Future work should assess the model with a wider database of experimental data with a more accurate air flow rate estimation.

5. References

[1] Lemmon EW, Bell IH, Huber ML and McLinden MO, 2018.
[2] Afroz HMM and Miyara A, 2007 *Int. J. Refrig.* **30** pp 1187–1194.
[3] Cavallini A, Del Col D, Doretti L, Longo GA and Rossetto L, 2000 *Int. J. Refrig.* **23** pp 4–25.

[4] Kim SM and Mudawar I, 2012 *Int. J. Heat Mass Transf.* **55** pp 3246–3261.
[5] Diani A, Mancin S and Rossetto L, 2014 *Int. J. Refrig.* **47** pp 105–119.
[6] Zhang M and Webb RL, 2001 *Exp. Therm. Fluid Sci.* **25** pp 131–139.
[7] Paliwoda A, 1992 *Int. J. Refrig.* **15** pp 119–125.
[8] ESDU 98005, Design and Performance Evaluation of Heat Exchangers: The Effectiveness and NTU Method, 2014.
[9] ASHRAE, Handbook Heating Ventilating and Air-Conditioning Applications, ASHRAE, 2019.
[10] Wang C-C, Chi K-Y and Chang C-J, 2000 *Int. J. Heat Mass Transf.* **43** pp 2693–2700.
[11] Gnielinski V, 1976 *Int Chem Eng* **16** pp 359–368.
[12] Cavallini A, Col DD, Doretti L, Matkovic M, Rossetto L, Zilio C and Censi G, 2006 *Heat Transf. Eng.* **27** pp 31–38.
[13] Kim SM and Mudawar I, 2013 *Int. J. Heat Mass Transf.* **64** pp 1239–1256.
[14] Copetti JB, Macagnan MH, De Souza D and De Césaro Oliveski R, 2004 *Int. J. Refrig.* **27** pp 876–883.
[15] Cavallini A, Del Col D, Mancin S and Rossetto L, 2009 *Int. J. Refrig.* **32** pp 162–174.
[16] Rollmann P and Spindler K, 2016 *Int. J. Therm. Sci.* **103** pp 57–66.
[17] Wang C-C, Lee C-J, Chang C-T and Lin S-P, 1999 *Int. J. Heat Mass Transf.* **42** pp 1945–1956.
[18] Hong KT and Webb R, 1996 *HVACR Res.* **2** pp 27–41.
[19] Shimizu GKK, Colmanetti ARA, Gardenghi ÁR, Cabezas-Gómez L and Tibiriçá CB, 2021 *J. Braz. Soc. Mech. Sci. Eng.* **43**.
[20] Rouhani SZ and Axelsson E, 1970 *Int. J. Heat Mass Transf.* **13** pp 383–393.
[21] Wu J, Chen Z, Lin J and Li J, 2018 *Int. J. Refrig.* **94** pp 24–32.
[22] Wang X, Jia X and Wang D, 2021 *Int. J. Refrig.* **124** pp 13–19.
[23] Domanski P and Didion D, 1984 *Int. J. Refrig.* **7** pp 249–255.

6. Nomenclature

Latin symbols

w	Elemental volume	[-]	RC	Oil recirculation rate	[-]
T	Number of tubes	[-]	$\dot{W}$	Power	[W]
V	Number of volumes per row	[-]	$\dot{Q}$	Heat power	[W]
R	Number of rows	[-]	X	Specific humidity	[g/g]
$\dot{m}$	Mass flow rate	[kg/s]	R	Resistance	[K/W]
h	Enthalpy	[J/kg]	M	Mass	[kg]
G	Mass flux	[kg/s/m^2]	U	Volume	[m^3]
F	Friction factor	[-]	A	Area	[m^2]
c_p	Specific heat at constant pressure	[J/kg/K]	H_{LG}	Liquid gas latent heat	[J/kg]
T	Temperature	[K]	MARD	Mean absolute relative deviation	[%]
P	Pressure	[Pa]	MRD	Mean relative deviation	[%]

Greek symbols

ρ	Density	[kg/m^3]	ε	Heat transfer effectiveness	[-]
σ	Contraction ratio of cross sectional area	[-]	α	Void fraction	[-]

Subscripts

REF	Refrigerant	WALL,EXT	External wall surface
F	Frictional	COMP	Compressor
G	Gravitational	OIL	Oil
A	Accelerative	EVA	Evaporator
IN	Inlet	COND	Condenser
OUT	Outlet	VS	Saturated vapor
AIR	Air	LS	Saturated liquid
AVE	Averaged on the rows	TOT	Total
SENS	Sensible	MIN	Minimum
LAT	Latent		

Development of lumped-parameters models for the thermal evaluation and air quality in aircrafts

G Tognon[1,*], **P Biasibetti**[1] **and A Zarrella**[1]

[1] Department of Industrial Engineering – Applied Physics Section, University of Padova, Via Venezia 1 – 35131, Padova, Italy

* Corresponding author: giacomo.tognon.2@studenti.unipd.it

Abstract. Aircraft cabins are a challenging category when dealing with thermal comfort and air quality inside means of transport. Two simplified dynamic models are developed. The first one is a lumped resistance-capacitance model for assessing the cabin thermal behaviour during the cruise phase. The fuselage is discretised into several slices and each one is represented through an RC network consisting of eleven nodes, thirteen resistances and three capacities. A thermal balance equation is set for each node and the linear system is solved to calculate the air and surfaces' temperatures. The model is validated by comparison with literature experimental data from ten flights, showing that the predicted temperatures agree well with the measured ones, presenting an RMSE of 1.5, 1.9 and 1.3 °C for cabin air, floor and cabin internal surface temperatures, respectively. A sensitivity analysis is conducted, for which the internal air temperature increases linearly with occupancy rate and decreases with cruising altitude. Secondly, an air quality model is proposed to evaluate the presence of pollutants inside the cabin, based on a simple concentration balance equation. Ventilation flow rates recommended from standards and a recirculation rate below 50-60% should be set to maintain acceptable CO_2 levels.

Keywords: *Lumped resistance-capacitance models, Aircrafts, Indoor Air Quality, Simplified dynamic models.*

1. Introduction

People spend most of their time in indoor spaces enhancing the necessity of providing adequate levels of thermal comfort to safeguard the occupants' well-being. Moreover, the outbreak of COVID-19 pandemic has raised a growing interest in air quality issues. By now, there is strong evidence that the airborne transmission of coronavirus is possible, through virus-laden microdroplets that remain suspended in the air and travel long distances transported by indoor airflows. For this reason, the importance of indoor ventilation for the infection risk mitigation has been increasingly highlighted [1].

Buildings and transport sectors were responsible for about the 40% and 30% of overall energy consumptions in Europe in 2018, leading to relevant GHG emissions [2]. Research is often oriented towards buildings, where the energy use is mainly aimed at maintaining good indoor environmental quality. However, people also use transport to move from one place to another daily. Energy use is primarily related to the vehicle motion, but a significant share is also required to guarantee a comfortable journey for passengers. From this point of view, the aircrafts represent a challenging category. Indeed, flights typically occur at an altitude between 9000 and 12000 m in the troposphere, where the external temperature, pressure and air composition are much different from those at sea level [3]. Additionally, aircrafts are characterized by high occupancy density over a contained volume, which means relevant heat gains and potential increase of CO_2 and other pollutants concentration. The cabin environment has a significant influence on passengers' well-being and health outcomes [4]. Mboreha et al. investigated the relation between COVID-19 spread and ventilation system in aircraft cabin, proposing control strategies regarding the seat arrangement [5]. The air-conditioning system must be well designed; ASHRAE gives guidelines for the environmental control system (ECS) in relation to the fluid dynamics and heat transfer phenomena occurring at high altitude [6].

Thermal comfort and indoor air quality levels achieved inside the passenger cabin can be evaluated experimentally or numerically. Environmental parameters were measured by Cui et al. on ten aircrafts in China; they concluded that thermal environment presents differences between middle, front and back of the cabin [7]. Mboreha et al. studied numerically temperature and airflow fields in a cabin equipped with distributed inlets above the seat rows for ventilation and compared it to conventional mixing system, showing that the former provides more uniformity and less re-circulation zones, where air quality could worsen [8]. Adopting the same approach, another paper investigates the performance of six innovative personalised ventilation systems in terms of thermal comfort [9]. Konstantinov et al. showed the importance of modelling in a proper way the clothing layer covering passengers' bodies for a realistic prediction of the radiative heat transfer [10]. Allen et al. studied the performance of 30 pilots on 21 manoeuvres in a flight simulator when exposed to different CO_2 concentration, revealing that the odds of passing increase by a factor of 1.69 going from 2500 to 700 ppm [11]. Cao et al. monitored CO_2 concentration in 179 U.S. domestic flights to verify that adequate outside airflow rates were supplied, detecting that only 73% met the recommended minimum of 4.7 l/s per person [12].

Most numerical simulation are based on CFD analysis, resulting computationally expensive. To overcome this issue, simplified modelling techniques are needed. In the case of buildings, several simplified dynamic modelling methods have been conceived, such as lumped parameters models: the building behaviour in terms of heat transfer and accumulation, occurring at the structural elements, is simulated through a resistance-capacitance (RC) network [13]. On the other hand, contaminant dispersion can be suitably evaluated employing a zonal modelling method, setting a simple mass balance within each control volume.

In this paper, two simplified dynamic models are developed. The first one is an RC model of the aircraft for the prediction of air and surface temperatures inside the passenger cabin. Its effectiveness is verified by comparison with data obtained from other literature works. The second one is constructed to assess the trend of pollutants concentration inside the cabin (e.g., CO_2) and the effect of ventilation in keeping acceptable air quality. A sensitivity analysis is carried out with both models to simulate the aircraft cabin behaviour varying different parameters, e.g., occupancy percentage.

2. Methodology

In this section the development and mathematical framework of two simplified models for the simulation of aircraft cabin environment are presented.

2.1. RC model of the cabin

The aircraft cabin is modelled as a resistance-capacitance network, exploiting the thermal-electrical analogy for heat transfer processes. Aircraft thermal characteristics are lumped into parameters that could describe its dynamic behaviour. The thermal resistances represent the branches of the circuit crossed by heat fluxes, whereas the thermal capacities are inserted to include the thermal inertia of massive structures and air volume. The model is thought to be employed for aircrafts of different sizes, from short to long-range flights. In the simulations, only the cruise phase of the flight is considered. From a geometrical point of view, the fuselage is modelled as a cylinder with a constant circular cross section, in relation to the airplane size. It is assumed that cabin rows consist of the same number of seats and are uniformly distributed along the aircraft, without distinguishing between different classes' arrangements. The occupancy density is considered as uniform, excluding the possibility of having cabin areas more crowded than the others. The fuselage internal volume is divided into two distinct spaces by the floor: the passenger cabin (upper part) and the hold (lower part or belly).
Once geometrically defined, the cabin is divided into several slices through parallel cross-sectional planes and the RC network is constructed for each one. The transparent surfaces are considered as evenly distributed along the airplane, dividing the total window area by the number of slices. The network consists of 11 nodes (node temperature denoted with θ), 13 resistances (R) and 3 capacitances (C), as shown in figure 1. Thermal resistances are conductive, convective or radiative based on the type of heat flux they model. Figure 1(a) reports the radiative thermal links between the nodes of the internal surfaces, whereas the convective branches connected to cabin air node are depicted in figure 1(b). The thermal capacities simulate the thermal inertia of fuselage and internal air volume. The former is differentiated between the thermal mass of the upper envelope portion surrounding the

passenger compartment, and the lower part containing the hold, both concentrated at the external nodes. The node state represents the average temperature at specific fuselage surfaces or air volumes. Among the temperature at nodes, θ_{inf} and θ_s are known since they represent the undisturbed external air temperature and the set supply air temperature from the air-conditioning system, respectively. θ_{ext} corresponds to the temperature near the fuselage external surface, and it is directly calculated from θ_{inf} through equation (1), which expresses the rise of air temperature caused by an isentropic compression due to airplane motion at high velocities. In this sense, the parameter R_p is only a fictitious resistance.

$$\theta_{ext} = \left(1 + \frac{\gamma - 1}{2} M^2\right) \theta_{inf} = R_p \theta_{inf} \tag{1}$$

In the equation, γ is the specific heats ratio of air (equal to 1.4) and M is the Mach number determined by the ratio between the aircraft speed (true air speed, TAS) and sound velocity in an ideal gas.

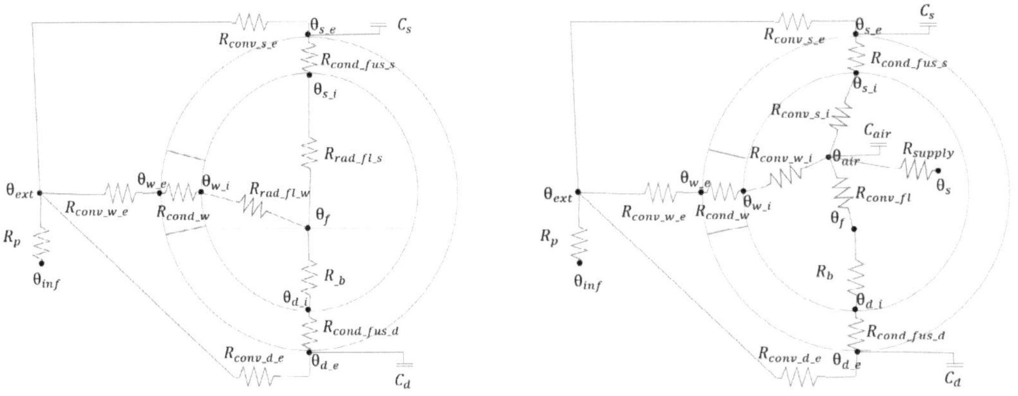

Figure 1. RC network for a fuselage slice, with internal radiative (a) and convective (b) heat exchanges.

The energy balance equations are set for the other eight nodes and the corresponding temperatures θ are determined by solving the resulting linear system at each simulation time step (equation (2) to (9)).

Floor node θ_f

$$\frac{\theta_{d,i} - \theta_f}{R_b} + \frac{\theta_{w,i} - \theta_f}{R_{rad,f,w}} + \frac{\theta_{air} - \theta_f}{R_{conv,f}} + \frac{\theta_{s,i} - \theta_f}{R_{rad,f,s}} + \varphi_{sol,f} + \sum_j (1 - f_{conv,j}) \eta_f \varphi_j = 0 \tag{2}$$

Window internal node $\theta_{w,i}$

$$\frac{\theta_{w,e} - \theta_{w,i}}{R_{cond,w}} + \frac{\theta_{f,i} - \theta_{w,i}}{R_{rad,f,w}} + \frac{\theta_{air} - \theta_{w,i}}{R_{conv,w,i}} = 0 \tag{3}$$

Window external node $\theta_{w,e}$

$$\frac{\theta_{ext} - \theta_{w,e}}{R_{conv,w,e}} + \frac{\theta_{w,i} - \theta_{w,e}}{R_{cond,w}} = 0 \tag{4}$$

Hold internal node $\theta_{d,i}$

$$\frac{\theta_f - \theta_{d,i}}{R_b} + \frac{\theta_{d,e} - \theta_{d,i}}{R_{cond,fus,d}} = 0 \tag{5}$$

Hold external node $\theta_{d,e}$

$$\frac{\theta_{ext} - \theta_{d,e}}{R_{conv,d,e}} + \frac{\theta_{d,i} - \theta_{d,e}}{R_{cond,fus,d}} - \frac{C_d}{\Delta\tau}\left(\theta_{d,e}{}^{\tau} - \theta_{d,e}{}^{\tau-\Delta\tau}\right) + \varphi_{sol,fus,d} = 0 \tag{6}$$

Ceiling internal node $\theta_{s,i}$

$$\frac{\theta_{air} - \theta_{s,i}}{R_{conv,s,i}} + \frac{\theta_{s,e} - \theta_{s,i}}{R_{cond,fus,s}} + \frac{\theta_f - \theta_{s,i}}{R_{rad,f,s}} + \varphi_{sol,s,i} + \sum_j (1 - f_{conv,j}) \eta_s \varphi_j = 0 \tag{7}$$

Ceiling external node $\theta_{s,e}$

$$\frac{\theta_{ext} - \theta_{s,e}}{R_{conv,s,e}} + \frac{\theta_{s,i} - \theta_{s,e}}{R_{cond,fus,s}} - \frac{C_s}{\Delta\tau}\left(\theta_{s,e}{}^{\tau} - \theta_{s,e}{}^{\tau-\Delta\tau}\right) + \varphi_{sol,fus,s} = 0 \tag{8}$$

Cabin air node θ_{air}

$$\frac{\theta_{w,i} - \theta_{air}}{R_{conv,w,i}} + \frac{\theta_{s,i} - \theta_{air}}{R_{conv,s,i}} + \frac{\theta_f - \theta_{air}}{R_{conv,f}} + \frac{\theta_s - \theta_{air}}{R_{supply}} - \frac{C_{air}}{\Delta\tau}\left(\theta_{air}{}^{\tau} - \theta_{air}{}^{\tau-\Delta\tau}\right) + \sum_j f_{conv,j} \varphi_j = 0 \tag{9}$$

In the above equations, θ are the node temperatures [K], R the thermal resistances [K W^{-1}], C the thermal capacitances [J/K], φ_j are the internal heat gains [W], φ_{sol} is the solar radiation load [W], f_{conv} represents the convective fraction of the internal heat gains, and $\Delta\tau$ is the time step length [s]. Infrared and solar radiation are uniformly distributed inside the cabin in proportion to surface area of the node element out of the total (expressed by floor and ceiling area ratio, η_f and η_s). The entering solar radiation share depends on the solar heat gain coefficient (SHGC) of the windows.
The kernel part of the model is the assessment of the lumped parameters. The thermal capacitance of both air and fuselage are obtained through equation (10).

$$C = \rho c_p V \quad (10)$$

where ρ is material density [kg m^{-3}], c_p is the specific heat [J kg^{-1} K^{-1}] and V is the volume of the element. The cabin air density must not be inferior to that of external air at 2440 m above sea level, in compliance with [14]. This minimum value is assumed to calculate air thermal capacity.
The conductive thermal resistances are related to fuselage envelope and window. They can be expressed by equation (11), in a general form for multi-layer structures.

$$R_{cond} = \frac{1}{A_i}\left(\sum_j \frac{s_j}{\lambda_j} + \sum_k r_{cavity,k}\right) \quad (11)$$

Where A_i is the surface area of the element [m^2] s and λ are the thickness [m] and thermal conductivity [W m^{-1} K^{-1}] of each solid layer, and r_{cavity} is the thermal resistance of intermediate cavities [m^2 K W^{-1}]. Convective thermal resistances are calculated through equation (12).

$$R_{conv} = \frac{1}{h_i A_i} \quad (12)$$

where h_i is the convective heat transfer coefficient [W m^{-2} K^{-1}], that has to be estimated based on flow regime and geometry. During the cruise phase, fuselage is affected by forced convection and the coefficient can be determined through equation (13).

$$h_{conv,ext} = \frac{0.74 Re^{-0.2} \rho c_p v}{3.52\, Pr^{0.4}} \quad (13)$$

where Re is Reynolds number, Pr is Prandtl number and v is the aircraft speed [m s^{-1}]. The radiative thermal resistances connecting two surface nodes are determined through equation (14), applying a linearization of the radiative heat exchange.

$$R_{rad,i,j} = \frac{1}{4\sigma A_i F_{ij} \theta_m{}^3} \quad (14)$$

where σ is Stefan–Boltzmann constant equal to 5.67 x 10^{-8} W m^{-2} K^{-4}, F_{ij} is the view factor between i-th and j-th surface, θ_m is the average of the two surfaces temperatures [K]. During the dynamic simulations, in the current calculation time step, T_m is obtained considering the surfaces temperatures of the previous one. The parameter R_b is calculated referring to ISO 6946 [15], considering the hold as a large cavity. Finally, the ventilation-related thermal resistance (R_{supply}) is defined in equation (15).

$$R_{supply} = \frac{1}{\dot{m}_s c_{p,air}} \quad (15)$$

where $\dot{m}_s$ is the supply mass airflow rate [kg s^{-1}], and $c_{p,air}$ is the specific heat of air (1004 J kg^{-1} K^{-1}). The internal heat gains are related to lights, avionics and passengers, who sit down at resting conditions. As reported in equations (2)-(9), they contribute to the heat balance of both air and internal surface nodes, in relation to the convective (f_{conv}) and radiative fraction (1-f_{conv}) of the emitted energy. For the interaction between fuselage and surrounding atmosphere, the radiative exchange with sky is neglected since the forced convective contribution induced by cruise speed is predominant. The undisturbed external air temperature is determined employing the "International Standard Atmosphere" model [3]: it decreases linearly from the sea level (S.L.) up to 11000 m (bottom of tropopause layer), then it is constant until 20000 m (lower boundary of stratosphere), as expressed by equation (16). It is assumed that it remains steady at a given altitude.

$$\theta_{inf}(h) = \begin{cases} \theta_0 + grad\ h & \text{if S.L.} < h < 11000\ \text{m} \\ \theta_{inf}(11000\ \text{m}) & \text{if } h \geq 11000\ \text{m} \end{cases} \tag{16}$$

where h is the aircraft altitude in the cruise phase [m], θ_0 is the reference temperature at sea level and 45° latitude, equal to 288.15 K, and *grad* is a descent gradient of -0.0065 K m^{-1}. The absorbed solar radiation by the i-th fuselage portion is calculated through equation (17), assuming that solar radiation impacts on the whole external surface, neglecting the relative orientation between aircraft and sun.

$$\varphi_{sol,fus,i} = A_{fus,i}\ \alpha\ C_y\ I \tag{17}$$

where α is the absorbance of the aluminum, I is the irradiance on a surface at sea level accounting for latitude and hour of day [W m^{-2}], and C_y is a correction factor for solar radiation based on altitude, which accounts for the attenuation effect of the air mass layer crossed by solar rays.
Theoretically, one slice interacts with the adjacent ones, but they are considered as independent, without resistances linking the respective nodes.

2.2. Air quality model

Air quality in the cabin is evaluated through a simplified dynamic model. The cylindrical fuselage is subdivided into slices where a pollutant mass balance is set (e.g., CO_2). It is assumed that each slice has its own inlet and outlet vents for ventilation and does not exchange airflows with the adjacent ones. The contaminant concentration balance, shown in equation (18), is solved at each simulation time step, after the aircraft took off.

$$V \frac{D^{\tau} - D^{\tau-\Delta\tau}}{\Delta\tau} = S + Q_{supply} D_{supply}{}^{\tau} - Q_{out} D^{\tau} \tag{18}$$

Where V is the slice volume [m^3], D is the contaminant concentration in the cabin [ppm], Q_{supply} is the supply flow rate [m^3 h^{-1}], Q_{out} is the exhaust flow rate [m^3 h^{-1}], equal to Q_{supply}, D_{supply} is the contaminant concentration in the fresh air [ppm], S is the contaminant emission rate from occupants [m^3 h^{-1}]. The last parameter depends on person's age and activity level [16], that can be assumed sedentary in the case of an aircraft cabin. The air-conditioning systems in the aircraft typically operate recirculating part of the exhaust air, which mixes with the air from the outside, as outlined in figure 2.

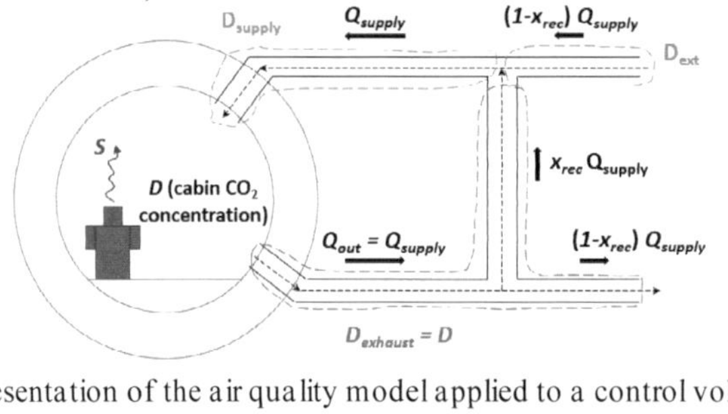

Figure 2. Schematic representation of the air quality model applied to a control volume.

Therefore, the inlet pollutant concentration is determined accordingly through equation (19).

$$D_{supply} = x_{rec} D + (1 - x_{rec}) D_{ext} \tag{19}$$

where D and D_{ext} are the cabin exhaust and external pollutant concentrations [ppm], respectively, and x_{rec} is the fraction of recirculated air, ranging from 0 to 1.

3. Applications and results

3.1. RC model validation

The RC model accuracy has been verified comparing its predictions with experimental data from [7]. The tests are conducted on ten flights on a Boeing 737-800, measuring air temperature and surface temperatures on floor, ceiling and side wall; the results are differentiated between front, middle and back part of the aircraft, so the mean values over the entire cabin length are calculated. Ceiling and side wall are represented by a single node in the RC model, thus the average temperature between the two measured values is employed. The obtained results for cabin air node are shown in figure 3.

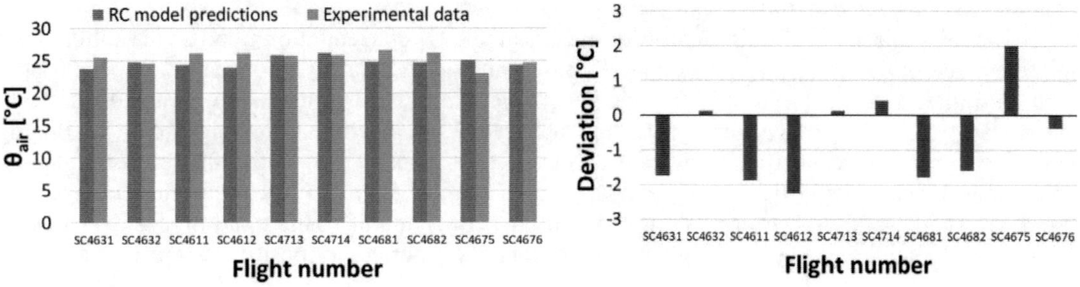

(a) (b)

Figure 3. Comparison (a) and deviations (b) between simulated and measured air temperatures.

From the deviations reported in figure 3(b), the RMSE is calculated through equation (20).

$$RMSE = \sqrt{\frac{\sum_{i=1}^{n}\left(\theta_{sim,i} - \theta_{meas,i}\right)^2}{n}} \tag{20}$$

where θ_{sim} and θ_{meas} are the simulated and measured temperatures [K], respectively, and n is the number of data. The obtained values for the RMSE related to air temperature, floor and cabin internal surface temperatures are 1.5, 1.9 and 1.3 °C, respectively. The larger deviation is observed for the floor node; however, the RC model is sufficiently accurate, and the errors present acceptable values.

3.2. RC model sensitivity analysis

After validation, a sensitivity analysis is conducted employing the RC model to evaluate the thermal behaviour of the cabin varying some parameters. Firstly, the influence of cruising altitude is studied for a Boeing 737-800 at full occupancy and with air conditioning system supplying 9,4 l/s per person at 23°C; the results are shown in figure 4(a) for the cabin air temperature. Three flight heights are chosen, namely 12947 (typical value), 9497 and 7497 m. The temperature decreases with height, but the trend is not linear due to a double aspect: the external temperature decreases linearly only up to 11000 m, then it remains constant, and the air mass layer becomes thinner, leading to lesser weakening of the solar radiation.

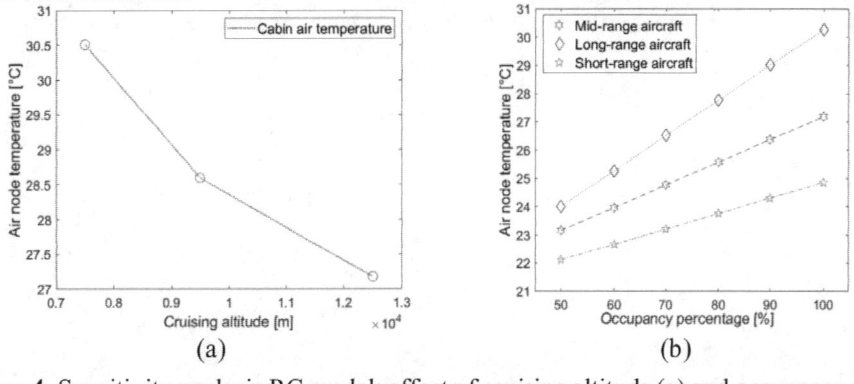

(a) (b)

Figure 4. Sensitivity analysis RC model: effect of cruising altitude (a) and occupancy percentage (b).

In the second instance, the influence of the occupancy percentage on the cabin air temperature, is investigated for a short, a mid and a long-range aircraft, all with similar cruising altitudes. It can be observed from figure 4(b) that the temperature increases linearly with occupancy rate; the long-range aircraft has higher internal air temperatures, whose progressive increase with cabin filling up is larger compared to the other two, as indicated by the slope of the straight lines.

3.3. Air quality model sensitivity analysis

The model for the air quality is employed to assess the CO_2 concentration trend in the cabin under different scenarios during the cruise phase, considering an external CO_2 concentration of 385 ppm and an initial internal of 400 ppm. The average generation rate of one passenger is 0.0032 l/s. In figure 5(a) the effect of occupancy varying between 50 and 100% is highlighted for a Boeing 737-800, where the ventilation flow rate is set at 9.4 l/s per person with 50% of recirculation, according to standards [14], [17]. As expected, the CO_2 concentration at equilibrium is higher for more crowded cabin, and steady state is reached later. In figure 5(b), the trend of steady-state CO_2 concentration as a function of occupancy percentage is reported. A positive linear correlation is observed.

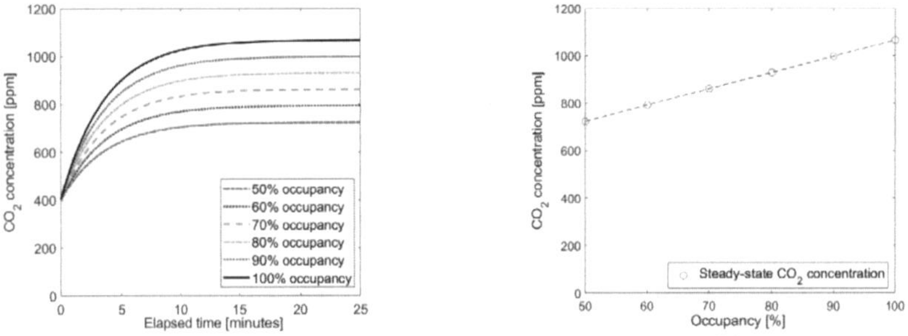

(a) (b)

Figure 5. Effect of occupancy percentage CO_2 concentration trends (a) and steady-state values (b).

The importance of ventilation rate and the effect of recirculation share are also investigated at 100% occupancy for the same aircraft. In the first case, the supply flow rate is varied between 1.4 and 11.4 l/s per passenger, with 50% of recirculation. From the CO_2 trends shown in figure 6(a), it is clear that higher dilution of contaminants is obtained increasing the ventilation flow rate. The time required to reach steady state increases as well; in the worst case it is not reached even after 1 h of flight. In figure 6(b) the effect of recirculation is illustrated, considering a standard supply flow rate of 9.4 l/s per passenger. Energy consumptions for air conditioning can be reduced increasing the recirculation rate, anyway, it should not exceed 50-60% to preserve the air quality in the cabin. At 100% recirculation, the CO_2 concentration increases linearly because the generation is not offset by removal.

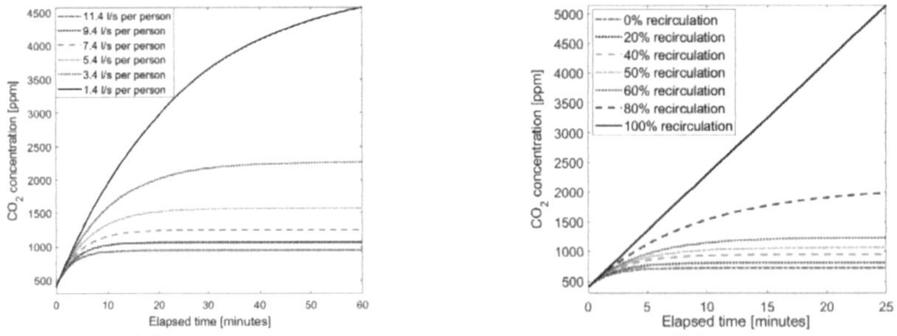

(a) (b)

Figure 6. Effect of ventilation flow rate (a) and percentage of recirculated air (b) on CO_2 balance.

4. Conclusions

In this paper, two simplified dynamic model has been developed to investigate the behaviour of an aircraft cabin, namely a lumped resistance-capacitance (RC) model to simulate the thermal balance of the cabin during the cruise flight, and another for the evaluation of the air quality in the cabin, based on a simple concentration balance for a given contaminant, e.g., CO_2. The validation of the RC model is firstly carried out, then the two models are applied to basic sensitivity analyses to study the influence of different parameters on the cabin environment. These outcomes emerged from the study:

- The comparison with experimental data from literature on a Boeing 737-800 shows good agreement with measurements; RMSEs of 1.5, 1.9 and 1.3 °C are obtained for the cabin air, the floor and the cabin internal surface temperatures, respectively.
- The increase of cruising altitude leads to a non-linear decrease in the cabin air temperature due to the double effect of external temperature drop and reduction of the air mass layer attenuating the solar radiation; instead, air temperature increases linearly with occupancy rate.
- The CO_2 steady concentration increases linearly and is reached later with higher occupancy percentages; the ventilation flow rate should be compliant with values recommended by the standards and the recirculation rate should be kept below 50-60% to preserve cabin air quality.

The proposed models present some limitations and need further improvements. At the moment, the passengers are supposed to uniformly distribute inside the cabin. Moreover, the slices into which the fuselage is discretised are considered as independent, whereas it would be interesting to include link branches between air nodes of adjacent cells, enabling the air movement across different sub-volumes.

5. References

[1] L. Morawska *et al.*, "How can airborne transmission of COVID-19 indoors be minimised?," *Environment International*, vol. 142. Elsevier Ltd, Sep. 01, 2020.

[2] European Commission. Statistical Office of the European Union (Eurostat), *Energy data 2020 edition*. 2020.

[3] NOAA, NASA, and USAF, *U.S. Standard Atmosphere*. Washington, D.C.: U.S. Government Printing Office, 1976, 1976.

[4] H. Hinninghofen and P. Enck, "Passenger well-being in airplanes," *Auton Neurosci*, vol. 129, no. 1–2, pp. 80–85, Oct. 2006.

[5] C. A. Mboreha and G. Kumar, "RISK AND PREVENTING OF COVID-19 IN A COMMERCIAL AIRCRAFT CABIN: AN OVERVIEW," *International Journal of Engineering Applied Sciences and Technology*, vol. 5, 2020.

[6] ASHRAE, "ASHRAE Handbook 2019 - HVAC Applications," 2019.

[7] W. Cui, Q. Ouyang, and Y. Zhu, "Field study of thermal environment spatial distribution and passenger local thermal comfort in aircraft cabin," *Build Environ*, vol. 80, pp. 213–220, 2014.

[8] C. A. Mboreha *et al.*, "Numerical simulations of the flow fields and temperature distribution in a section of a Boeing 767-300 aircraft cabin," *Mater Today Proc*, vol. 47, pp. 4098–4106, 2021.

[9] C. A. Mboreha *et al.*, "Investigation of thermal comfort on innovative personalized ventilation systems for aircraft cabins: A numerical study with computational fluid dynamics," *Thermal Science and Engineering Progress*, vol. 26, Dec. 2021.

[10] M. Konstantinov *et al.*, "Numerical simulation of the air flow and thermal comfort in aircraft cabins," *Notes on Numerical Fluid Mechanics and Multidisciplinary Design*, vol. 124, pp. 293–301, 2014.

[11] J. G. Allen *et al.*, "Airplane pilot flight performance on 21 manoeuvres in a flight simulator under varying carbon dioxide concentrations," *J Expo Sci Environ Epidemiol*, vol. 29, no. 4, pp. 457–468, Jun. 2019.

[12] X. Cao *et al.*, "The on-board carbon dioxide concentrations and ventilation performance in passenger cabins of US domestic flights," *Indoor and Built Environment*, vol. 28, no. 6, pp. 761–771, Jul. 2019.

[13] European Committee for Standardization-CEN, *EN ISO 13790:2008, Energy performance of buildings-Calculation of energy use for space heating and cooling*. 2008.

[14] Federal Aviation Administration (FAA), *FAR 14 CFR 25.831. Part 25-Airworthiness standards: transport category airplanes.* Washington, DC, 2013.

[15] European Committee for Standardization-CEN, *EN ISO 6946:2017, Building components and building elements - Thermal resistance and thermal transmittance - Calculation methods*. 2017.

[16] A. Persily and L. de Jonge, "Carbon dioxide generation rates for building occupants," *Indoor Air*, vol. 27, no. 5, pp. 868–879, Sep. 2017.

[17] ANSI/ASHRAE, *Standard 161-2007 Air quality within commercial aircrafts*. Atlanta, GA, 2007.

Journal of Physics: Conference Series 2685 (2024) 012040 doi:10.1088/1742-6596/2685/1/012040

Switching criteria analysis for a condenser moving boundary model

Michael Giovannini, Marco Lorenzini
Università di Bologna, DIN – Dipartimento di Ingegneria Industriale, Via Fontanelle 40, I-47121, Forlì (FC) Italy
E-mail: marco.lorenzini@unibo.it

Abstract. The switching moving boundary (SMB) scheme is widely used for numerical simulation of heat exchangers, especially in vapour compression refrigeration system. The method divides the heat exchanger into up to three zones (superheated vapour, two-phase fluid and subcooled liquid), depending on the operating conditions and treats each as a lumped-parameter system. During transients the number of zones may vary, and suitable switching criteria and related threshold values must be given, for this change to occur. Also, variables for the zones momentarily phased out need to be tracked, to allow smooth operation of the model. In this work, a switching moving boundary model of a brazed plate condenser in counter-flow arrangement is studied to obtain some useful guidelines for the selection of the optimal switching thresholds. It is shown that these values influence both the accuracy and the numerical robustness of the model; moreover an appropriate selection of these parameters may allow the usage of a fixed-step solver, which is more suitable for real-time simulations.

1. Introduction

Air conditioning and refrigeration are responsible for about 20% of the global energy demand, and this figure is expected to more than double by 2050 [1]. Vapour compression systems (VCSs) are the most widespread technology in Heating, Ventilation, Air-Conditioning and Refrigeration (HVAC&R). In these systems a working fluid (refrigerant) is recirculated between two sections in which it exchanges thermal energy with two environments at different temperatures; these sections are kept at two different pressure levels through the action of a compressor and an expansion valve. In order to contain energy consumption and greenhouse gases (GHG) emissions, HVAC&R stakeholders strive, among others, to improve the energy efficiency of VCSs, through better system design and to implement suitable control strategies. Numerical modelling of VCSs can be the best way to achieve these goals: models can be used in design, system analysis, control tuning and commissioning phases, but they can also be applied as software-in-the-loop controls for performance optimization or for automatic fault detection and diagnosis (AFDD) [2].

In VCS dynamic modelling, the largest effort must be devoted to the heat exchangers for two reasons:

- the dynamic behaviour of the VCSs strongly depends on the thermal transients, indeed the dominant time scale of the process is related to the heat transfer [3];
- the working fluid undergoes a non-instantaneous phase transition, in both heat exchangers for subcritical systems, or in the evaporator only for supercritical ones; this implies that the heat transfer rates and the pressure levels are coupled.

According to [4], heat exchanger models can categorised as lumped parameter, finite volume and moving boundary. In the first approach, the process is described employing a single control volume, for which the balance equations, for mass, energy and momentum, are written; heat transfer is therefore based on the physical properties averaged over the control volume. This assumption is liable to introduce non-negligible inaccuracies since physical properties of the working fluid vary significantly during the transport process, especially in the two-phase region, and so does the heat transfer coefficient. At the other end, the finite volume technique divides the heat exchanger volume into a given number of cells, for each of which the same balance equations are written; in this way, the properties of the fluids can be evaluated more precisely, reaching an almost local estimate when the grid becomes fine enough, and capturing the spatial gradient of the process variable; the drawback of this approach is that it requires an increasingly demanding computational effort if discretisation needs high degrees of grid refinement.

In the last decades, the moving boundary (MB) approach has been developed, with the aim of combining the accuracy and flexibility of the FV scheme and the low computational cost of a lumped parameter approach. In the early realizations of the MB scheme [5], the heat exchanger volume is divided into a maximum of three lumped systems, depending on the phase of the working fluid (superheated vapour, two-phase mixture or subcooled liquid). The same balance equations can be used for each subsystem, but in this approach, it is possible to deal with the discontinuities associated with phase transition; indeed, the physical properties are averaged over a volume of fluid under uniform conditions and a suitable integral correlation to evaluate heat transfer is employed for each zone. The drawback is that, during large transients, the conditions of the working fluid may vary sharply, causing zones to disappear or reappear, thus leading to numerical issues during computation, [2]. To avoid such complications, the MB scheme has been improved, allowing the interchange between different *representations*, [6], or *modes*, [7], defined on the basis of the number of active zones [8]. Transition is controlled by the definition of some switching criteria, which require the choice of a monitored variable and of its corresponding threshold; moreover, when one zone disappears, the corresponding equations are no longer available, so they must be substituted with some *pseudo state equations* [6], in order to avoid discontinuities at the zone's reappearance. The resulting scheme, called switching moving boundary, (SMB), has performed well when employed in the modelling of finned-tube condensers and evaporators for the simulation of on-off transient of a complete VCSs [6, 7].

The selection of the switching criterion has considerable influence on the accuracy and the stability of the model, as demonstrated in [9] in the case of an evaporator, yet no such analysis exists for condensers, nor for types of heat exchangers other than the fin-and-tube. In the present work, the effects of the activation and deactivation thresholds are numerically investigated for the case of a brazed-plate (BP) condenser; among the various formulations of the SMB scheme, the one based on the mean void fraction is chosen, in accordance with [6, 7]. The aim is to investigate the effects of the switching criteria on the accuracy of the model, analysing the behaviour of the predicted values of some variables of interest for the process as the thresholds vary, and on the stability and robustness of the model, exploring the numerical performance in terms of simulation time and minimum integration step size. Some useful guidelines are provided to select the optimal settings for the switching thresholds, so as to restrain the reduction in the integration step caused by the increase of both the stiffness and the ill conditioning of the system of differential equations, which occurs especially close to mode transition. The ultimate goal is to propose an SMB scheme suitable for fixed-step integration with optimal settings for the switching thresholds, to allow heat exchanges modelling to be included into the larger framework of the simulation of energy systems for control design purposes.

2. Model definition

In the MB description, a maximum of three control volumes can be defined in the heat exchanger, depending on the local thermodynamic state of the working fluid. Each control volume includes three elements, the working fluid, the secondary fluid and a portion of the wall of the heat exchanger, for each of which the balance equations may be written.

Consistently with previous works on SMB [5, 6, 8], some simplifying assumptions are made:

- one-dimensional, compressible and unsteady flow for the working fluid;
- one-dimensional, incompressible steady flow for the secondary fluid;
- cross-sectional area constant for both fluids;
- negligible pressure drops for both fluids;
- negligible axial conduction;
- negligible wall thermal resistance;
- thermal storage in the secondary fluid is neglected.

As a consequence of the assumptions above, the momentum equation is unnecessary for both fluids; moreover, since flow and heat transfer to the secondary fluid are stationary, no differential equation is needed to describe its behaviour, the algebraic equation for heat transfer in each control volume suffices.

In the following, the integral form of the balance equations applied to each control volume for the derivation of the MB formulation are presented.

The refrigerant mass balance equation:

$$\frac{d(\rho V)}{d\tau} = \dot{m}_{in} - \dot{m}_{out} \tag{1}$$

where ρ is the density of the working fluid, V is volume occupied by the fluid, and $\dot{m}_{in}$ e $\dot{m}_{out}$ are the inlet and outlet mass flow rates respectively.

The working fluid internal energy balance:

$$\frac{d(\rho u V)}{d\tau} = \dot{m}_{in} h_{in} - \dot{m}_{out} h_{out} + \dot{Q}_r \tag{2}$$

where u is the specific internal energy of the fluid, h_{in} and h_{out} are the specific enthalpies of the fluid at the inlet and at the outlet and $\dot{Q}_r$ is the heat transfer rate between the heat exchanger wall and the refrigerant.

The wall energy balance

$$\frac{d(C_w T_w)}{d\tau} = \dot{Q}_s - \dot{Q}_r \tag{3}$$

where C_w is the thermal capacity of the wall in the control volume, T_w is the wall temperature and $\dot{Q}_s$ is the heat transfer rate between the secondary fluid and the wall.

Selection of the state variable has a key role in the realization of the model; although this choice is arbitrary, given the redundant nature of thermodynamic properties [10], it may have consequences on the switching criteria employed and on mass conservation during the transients, [11].

2.1. SMB equations

In this work, the SMB model of a BP condenser is obtained using the mean void fraction scheme; the detailed procedure for the derivation of the governing equation, from the balance equations, eqs. (1) to (3), can be found in [12]. Expanding the balance equations for each zone, a system of nine differential equations is obtained. The working fluid is characterized by two state variables for each zone; the pressure, assumed as constant throughout the heat exchanger, must be added to these six state variables. Finally the condenser wall is described by its lumped temperature in each zone. Given that the ten time derivatives of the state variables are the unknown, another equation must be integrated to close the system. According to [6], it is convenient to add a tracking equation for the mean void fraction:

$$\frac{d\bar{\gamma}}{d\tau} = \frac{\partial\bar{\gamma}}{\partial P}\frac{dP}{d\tau} - K_{\bar{\gamma}}(\bar{\gamma} - \bar{\gamma}_{tot}) \tag{4}$$

where $\bar{\gamma}_{tot}$ represents the mean void fraction corresponding to a complete phase change, complete condensation in this case, and $K_{\bar{\gamma}}$ is a tunable relaxation gain. The mean void fraction is calculated employing the Zivi correlation [13].

The general form of the system of differential equations is $Z(x, u)\dot{x} = f(x, u)$, where the state vector is $x = [P, \zeta_1, h_1, \zeta_2, \bar{\gamma}, \rho_3, h_3, T_{w1}, T_{w2}, T_{w3}]$.

2.2. *Refrigerant heat transfer rate*
The heat transfer rate $\dot{Q}_{r,j}$ for the working fluid in each zone is calculated through:

$$\dot{Q}_{r,j} = \alpha_{r,j}\zeta_j A_r(T_{w,j} - T_{r,j}) \tag{5}$$

where $\alpha_{r,j}$ is the heat transfer coefficient for the jth zone, and A_r is the area of the heat transfer surface in the refrigerant side.

The physical properties of the working fluid averaged for each zone are evaluated employing a suitable table compiled using the CoolProp library [14]. The heat transfer coefficient for the refrigerant side is calculated for each zone employing a set of two correlations depending on its phase: for single phase in BP heat exchangers the correlation suggested by Martin, [15], is employed, while for condensation in BP heat exchangers the correlation proposed by Longo et al., [16] is adopted.

2.3. *Secondary fluid heat transfer rate*
Considering that the two fluids involved in the process are counter-flowing, the heat transfer rate $\dot{Q}_s$ of the secondary fluid in each zone can be calculated with:

$$\dot{Q}_{s,j} = \dot{m}_s c_{ps}(T_{si,j} - T_{w,j})(1 - e^{-NTU_{s,j}}) \tag{6}$$

where $\dot{m}_s$ is the mass flow rate of the secondary fluid, c_{ps} is its specific thermal capacity at constant pressure, $T_{si,j}$ is the temperature of the fluid at the inlet of the control volume of the jth zone , $T_{w,j}$ is the mean temperature of the wall, and $NTU_{s,j}$ is the number of transfer units.

The physical properties of the secondary fluid are evaluated at the inlet temperature. The heat transfer coefficient is calculated with the correlation recommended by Martin, [15].

2.4. *Model implementation*
The SMB model of a BP model presented in this work is realized in the Simulink® environment, employing the *S-function* block, that is a powerful mechanism for extending the capabilities of this system implementing code blocks written in Matlab® .

The core of the block is constituted by the system of equations which includes the ten differential equations and the algebraic heat transfer equations, eqs. (5) and (6). This system takes different forms depending on the number of active zones, i.e. on the phase of the working fluid.

The boundary conditions for the system of equations constitute the inputs to be provided to the S-function block; these are the refrigerant mass flowrate at the inlet and outlet of the condenser, the refrigerant temperature at the inlet and the stationary mass flow rate and the inlet temperature of the secondary fluid. Further details can be found in [12].

3. Analysis of the switching criteria

The system of ten differential equations, is employed only when all zones are active, that is, in the heat exchanger the working fluid is present as superheated vapour, two phase mixture and subcooled liquid. This representation or mode is called V-TP-L, consistently with the convection adopted in [11, 17].

In this mode the matrix $Z(x, u)$ is full rank only if $\zeta_j > 0$ for each zone, otherwise it is no longer invertible, [2]. Moreover, when a the normalized length tends to zero $\zeta \to 0$, the condition number of the matrix $Z(x, u)$ skyrockets, causing numerical issues in the solution. The SMB scheme overcomes these problems, allowing the disappearance and the reappearance of a zone by changing the set of state equations employed according to same criteria.

In order to avoid singularities, disappearance of a zone is triggered when the corresponding normalized length reaches a minimum value, ζ_{min}, rather than allowing the length to reach zero. The first switching criterion can be thus expressed as $\zeta_j < \zeta_{min}$.

From mode V-TP-L, the model can switch to V-TP mode if the subcooled liquid zone is extinguished; during normal operation of the condenser, disappearance of the super-heated vapour zone is unphysical, since the heat exchanger is directly fed by the compressor. When the subcooled liquid zone disappears, the corresponding differential equations are no longer available, and eq. (4) is no longer needed. In this case, the three state variables (ζ_3, h_2, T_{w3}) corresponding to the subcooled liquid zone must be tracked in some way in order to avoid discontinuities when the region reappears, as pointed out in [6, 7, 8]. The normalized length ζ_3 tracks the threshold value for the minimum length ζ_{min}, the enthalpy h_3 tracks the reference value corresponding to the saturated liquid state h_l and wall temperature T_{w3} tracks the wall temperature in the two-phase zone, as shown in eq. (7).

$$\frac{d\zeta_3}{d\tau} = K_\zeta(\zeta_{min} - \zeta_3) \qquad \frac{dh_3}{d\tau} = K_h(h_l - h_3) \qquad \frac{dT_{w3}}{d\tau} = K_{T_w}(T_{w2} - T_{w3}) \tag{7}$$

When operating conditions lead to the reappearance of the sub-cooled liquid zone, the system of state equations must be changed. This is achieved again using another switching criterion, based on the mean void fraction in the two-phase zone. Indeed, if complete condensation occurs in the TP zone, and sub-cooled liquid starts to accumulate towards the outlet of the heat exchanger, the mean void fraction in the TP region is less than the value corresponding the complete phase transition $\bar{\gamma}_{tot}$. So the sub-cooled zone formation can be identified by $\bar{\gamma} < \bar{\gamma}_{tot}$.

To make the two criteria more consistent, it is possible to convert comparison of the mean void fraction into comparison of the normalized lengths, since the same condition may be formulated defining a maximum value of normalized length ζ_{max} corresponding to the volume occupied by the subcooled liquid inside the two-phase region. In this way the second criterion can be written as $\zeta_2(\bar{\gamma} - \bar{\gamma}_{tot}) > \zeta_{max}$.

3.1. Analysis

The effects of the thresholds have been studied considering the accuracy in the evaluation of the process variables and the numerical performance of the model. The numerical test consists of a mass flow rate transient: all boundary conditions are kept constant, except the inlet mass flowrate of the refrigerant, which is subject to an increase of 10% of its initial value and then to an equal decrease before being brought back to its initial value. The input values used for the variables are $\dot{m}_{r,in,ref} = 0.05\, kg \cdot s^{-1}$, $\dot{m}_{r,out} = 0.05\, kg \cdot s^{-1}$, $\dot{m}_s = 0.28\, kg \cdot s^{-1}$, ,$T_{f,in} = 36\,°C$ while the refrigerant mass flow rate at the condenser's inlet varies by 10% around the equilibrium value ($\dot{m}_{r,in,ref}$). To the Authors' best knowledge, no experimental data are available for direct

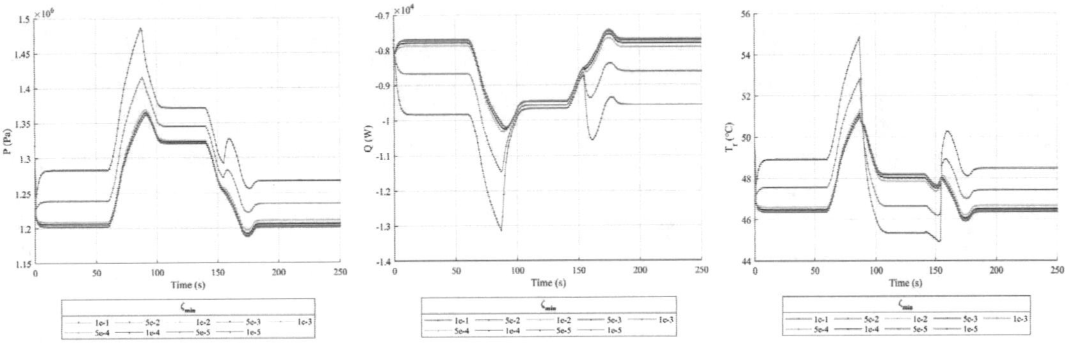

(a) (b) (c)

Figure 1: Time plots of the monitored variables for varying ζ_{min} threshold: pressure, (a), total heat transfer rate, (b), refrigerant outlet temperature, (c).

comparison, since imposing the mass flowrate would imply to be able to test the heat exchanger alone, on one hand and the evaporator should also allow optical access to determine the extent of the regions (although this could be bypassed by e.g. thermographic mapping of outer surfaces), for similar reasons no experimental comaprison was carried out in [9]. What can be compared is the influence of correctly assessing the switching threshold has on the overall simulation of a compression vapour system, which is beyond the scope of the paper.

The total heat transfer rate, the pressure of the working fluid, and the outlet temperatures of the refrigerant have been chosen as monitored variables in order to evaluate the accuracy.

The numerical integration is carried out via a variable-step solver available in Matlab®, the commonly used *ode45* with a relative tolerance of 10^{-3}.

On the other hand, a variable-step solver should not be employed in a control-oriented application involving several subsystems, which may require different integration step-size at any given time. Also, these methods are often unsuitable for use in a real-time environment due to requirements on sample inputs and update outputs at precise time intervals. This issue may be overcome using a fixed-step solver: in this case an appropriate integration step-size must be chosen to guarantee the accuracy required, stability of the solution and compatibility with larger frameworks [18].

For these reasons, the numerical performance of the model when the thresholds vary is evaluated in terms of simulation time and minimum step size; the latter is automatically chosen by the solver in order to achieve the prescribed accuracy, and it depends mostly on the condition number and the stiffness of the matrix $Z(x, u)$ at certain point of the simulation [19].

4. Results ans discussion

4.1. Effect of thresholds on the accuracy of the model

Reduction in thresholds' value produces a convergence of the predicted values; in fig. 1 the convergence of the dynamic behaviour of the monitored variables, pressure, heat transfer rate, outlet temperature of refrigerant when ζ_{min} drops from 10^{-1} to 10^{-5} can be appreciated qualitatively.

The normalized root mean squared error (NRMSE) is chosen as a metric for convergence. The NRMSE for each monitored variable (x) is calculated with eq. (8) using as reference signal (x_{ref}), corresponding to the minimum value of both thresholds (10^{-5}) with the relative tolerance set to 10^{-4}.

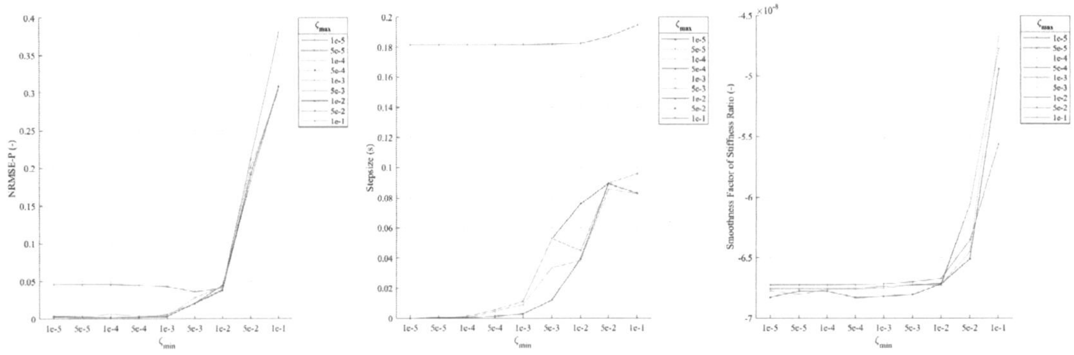

(a) (b) (c)

Figure 2: NRMSE for pressure, (a), minimum step-size, (b), and smoothness factor of the stiffness ratio, (c), versus ζ_{min} for different values of ζ_{max}.

$$NRMSE = \frac{\sqrt{\frac{\sum\left(x_i - x_{ref,i}\right)^2}{N}}}{x_{max} - x_{min}} \tag{8}$$

The threshold ζ_{min} is demonstrated to be more influential on convergence for pressure than ζ_{max}. A good level of of accuracy is reached for ζ_{min} and ζ_{max} lower than 0.5×10^{-2}.

4.2. Effect of thresholds on the numerical performance of the model

The numerical performance of the model was evaluated through the time needed to complete the simulation and the minimum integration step-size required by the solver. The variable-step solvers adapt the integration step-size to achieve the prescribed accuracy, indeed absolute or relative tolerance must be included in their settings. If the matrix of the system $Z(x, u)$ is ill-conditioned, accuracy of the solution deteriorates, due to numerical reasons; the solver tries to compensate for this effect reducing the integration step-size. Because of this, the minimum step-size used by a variable-step solver is a reliable indicator of the ill-conditioning and of the stiffness of the numerical problem at any time during the simulation. The stiffness of the system of differential equations $Z(x, u)$ is evaluated through a stiffness ratio (SR), i.e. the ratio between the absolute values of the largest and the smallest eigenvalues, in modulus, λ.

This is done at each time step: the SR is quite large for the whole duration of the simulation because the SMB system involves dynamics with very different time scales, which correspond to eigenvalues liable to vary of orders of magnitude. Moreover, when the normalized zone length tends to zero, the SR skyrockets, as does the condition number. The presence of such peaks in the SR time signal is evaluated using as smoothness factor (SF) the non-dimensional jerk metric defined as follows [20]:

$$SF = -\frac{\Delta\tau^3}{x_{max}^2} \int \left(\frac{d^2}{d\tau^2} x_i\right)^2 \tag{9}$$

Higher values of SF correspond to smoother signals. Using this metric, the presence of peaks close to the switching points can be conveniently detected.

When the thresholds become higher, the step size increases; this behavior is shown in fig. 2 (b). For larger values of ζ_{max}, the model is unable to capture the transition in mode; in these cases, the larger step-size corresponds to an non-physical description of the state of the working fluid in the condenser. For $\zeta_{max} < 0.5 \times 10^{-3}$, the minimum step-size shows a jump for larger values of ζ_{min}. This behaviour corresponds to a reduction of the peaks in the stiffness factor as shown in fig. 2 (c); indeed the smoothness factor of the SR increases for large values of ζ_{min}.

4.3. Optimal thresholds

The analysis above evidences two opposite trends. The accuracy is enhanced by the reduction of the thresholds ζ_{min} and ζ_{max}, yet, the stiffness of the system increases, and a smaller integration step-size is needed to satisfy accuracy requirements. A variable-step solver can do the task, while keeping simulation time within reasonable bonds. Yet, to include the SMB model into a more complex energy system framework, a fixed-step solver is preferable. The choice of the switching thresholds determines the accuracy in the computed values and the stiffness of the system of differential equations and, thus, the optimal integration step-size. This work proves that a value of the switching thresholds ζ_{min} and ζ_{max} between 0.5×10^{-2} and 10^{-3}, is a satisfactory compromise; the model yields a solution with a relative tolerance of 10^{-3} implementing a fixed step-size in the order of $0.05s$, which appears suitable for control design purposes.

5. Conclusions

In this work a numerical analysis of a SMB model based on the mean void fraction scheme for a BP condenser in counter-flow arrangement is carried out. The switching thresholds are checked with regard to the accuracy and the numerical performance of the simulation, studying a transient in the mass flow rate of the refrigerant. The general trend is that the smaller the thresholds, the higher the accuracy in the estimation of the process variables. On the other hand, if the threshold becomes too small, the system of differential equations becomes stiffer and may show ill conditioning. Some guidelines for the selection of the optimal thresholds are provided, with the ultimate goal to support the usage of a fixed-step solver, which is more suitable to include this condenser model in a real-time simulation framework for control design purposes.

6. Referecens

[1] Dupont, Domanski, Lebrun and Ziegler 38th Note on Refrigeration Technologies: The Role of Refrigeration in the Global Economy (2019)
[2] Rasmussen B P and Shenoy B 2012 *HVAC&R Research* **18** 956–973
[3] Bendapudi S and Braun J E 2002 A review of literature on dynamic models of vapor compression equipment
[4] Rasmussen B P 2012 *HVAC&R Research* **18** 934–955
[5] Grald E W and MacArthur J W 1992 *International Journal of Heat and Fluid Flow* **13** 266–272
[6] McKinley T L and Alleyne A G 2008 *International Journal of refrigeration* **31** 1253–1264
[7] Li B and Alleyne A G 2010 *International Journal of refrigeration* **33** 538–552
[8] Willatzen M and Pettit Ploug-Sørensen L 1998 *International Journal of refrigeration* **21** 398–403
[9] Rodriguez E and Rasmussen B 2017 *Applied Thermal Engineering* **112** 1326–1342
[10] Rasmussen B P and Alleyne A G 2004 *Journal of Dynamic Systems, Measurement, and Control* **126** 54–64 ISSN 0022-0434, 1528-9028
[11] Cecchinato L and Mancini F 2012 *International journal of refrigeration* **35** 349–364
[12] Lucchi M 2020 *Dynamic Models for the Analysis of Vapour-Compression Refrigerating Machines* Ph.D. thesis Alma Mater Studiorum Università di Bologna / alma
[13] Zivi S 1964 *Journal of Heat Transfer* **86** 247–251
[14] Bell I H, Wronski J, Quoilin S and Lemort V 2014 *Industrial & Engineering Chemistry Research* **53** 2498–2508 ISSN 0888-5885, 1520-5045
[15] Martin H 1996 *Chemical Engineering and Processing: Process Intensification* **35** 301–310 ISSN 0255-2701
[16] Longo G A, Righetti G and Zilio C 2015 *International Journal of Heat and Mass Transfer* **82** 530–536
[17] Zhang W J and Zhang C L 2006 *International Journal of Refrigeration* **29** 1119–1127 ISSN 0140-7007
[18] Murray-Smith D J 2012 *Modelling and Simulation of Integrated Systems in Engineering: Issues of Methodology, Quality, Testing and Application* (Elsevier)
[19] Quarteroni A, Sacco R and Saleri F 2010 *Numerical Mathematics* Texts in Applied Mathematics (Springer) ISBN 978-3-540-49809-4
[20] Balasubramanian S, Melendez-Calderon A and Burdet E 2012 *IEEE Transactions on Biomedical Engineering* **59** 2126–2136 ISSN 1558-2531

Thermally adaptive wall to enhance indoor comfort and energy performance

F Bianchi[1,*], G Baldinelli[2], J A Schnotale[3], A A Lechowska[3], A Presciutti[4]

[1] INFN sezione di Perugia, University of Perugia, Via Pascoli - 06123 Perugia – Italy

[2] Department of Engineering - University of Perugia, Via Duranti 67 - 06125 Perugia - Italy

[3] Department of Environmental Engineering and Energy - Cracow University of Technology. ul. Warszawska 24 31-155 Cracow – Poland

[4] Department of Economy, Universitas Mercatorum, Piazza Mattei 10 - 00153 Rome - Italy

* Corresponding author: francesco.bianchi@pg.infn.it

Abstract. In the present work, a new environment-adaptive wall is proposed, based on an inside and outside radiant panel with pipes drowned on the panels themselves and hosting a heat-carrying liquid pushed by a pump. The purpose consists of transferring heat across the thickness of the wall, in the direction required for energy saving and enhancement of inner indoor thermal comfort. A Computational Fluid Dynamic (CFD) analysis is reported and the results are applied on a single zone box-shaped building, where a whole-year study is implemented by means of the transient simulation tool TRNSYS. The efficiency of this solution respect to the state-of the-art static walls is finally discussed. A concept of a laboratory setup of an adoptive wall along with first result are presented and discussed.

1. Introduction

Active systems designed to increase the energy performance of the building envelope represent one of the research frontiers in the construction sector [1]. Thermally adaptive walls for building applications have been divided into the following types: PCM integrated walls, ventilated walls, dynamic insulation materials integrated walls, switchable insulations systems integrated walls, PCM integrated dynamic insulation walls [2]. Within this framework, the focus on the cooling requirements is relatively more recent respect to the massive investigations developed along the last decades on materials insulation properties [3,4].

Transparent surfaces knew a flourishing of solutions directed towards an active behaviour respect to the environmental conditions, such as photochromic, thermochromic or electrochromic glasses [5], aimed at defining a system that changes ideally its optic-energy properties as required by the users.

On the other hand, apart from phase change materials, a gap seems existing for the opaque fraction of the envelope respect to the glazing, as only a few components have been proposed in this direction such as, for instance, walls with modifiable air gaps, which effect on the overall varying thermal insulation properties is limited and difficult to control [6].

A system for adjusting the temperature of a building is described in [7], where two series of pipes are positioned at the two faces of a wall, i.e., straddling the insulating wall. Considering the summertime behavior of the walls, the circulation of the fluid is promoted at night by bringing heat from the inner portion of the building to the outside.

Another solution for internal walls was presented in [8], where concrete wall with embedded PEX pipes inside (Thermal Energy Storage System) was combined with copper pipes and rigid foam on both boundaries (Active Insulation System). Concrete with pipes inside cooperating with a chiller works as a thermal storage unit. When the stored energy is needed, it is transferred by the water flow though the copper pipes of the active insulation on both sides of the wall to reduce HVAC loads.

In the work of Bailey [9] a heat exchanger panel comprising an insulating layer mounted adjacently to a solid wall of a building is described. The setup comprises ducts connected by means of a fluid circuit to solar panels located on the roof. The heat exchanger panel can be part of an air conditioning system for a building consisting of a combination of a radiant heating or cooling system to best keep the desired temperature in the building. The system can include solar panels and radiant cooling plates.

The solution described in the present work consists of two radiant panels embedded in the outer and inner surface of any insulated wall [10]. The functioning concept is based on transferring the heat from and to the inner environment, according to energy and comfort requirements, controlling the pump that moves (or not) the fluid inside the circuit. The next paragraph describes in more the detail the working strategy.

2. Environment adaptive wall: the rationale

The environment adaptive wall core is obtained through the immersion in any horizontal, vertical or oblique wall made of any material, of pipes made of plastic, metal material, or other materials usually used for transporting liquids. A heat transfer fluid flows, moved by a thrust member such as a pump to transport the heat in the direction of the thickness of the wall, from the inside to the outside, and vice versa.

The control of the system is associated with two temperature sensors (Te and Ti), one located in the outdoor environment, and one located indoors: their differential value guides the on and off switching of the pump (fig. 1).

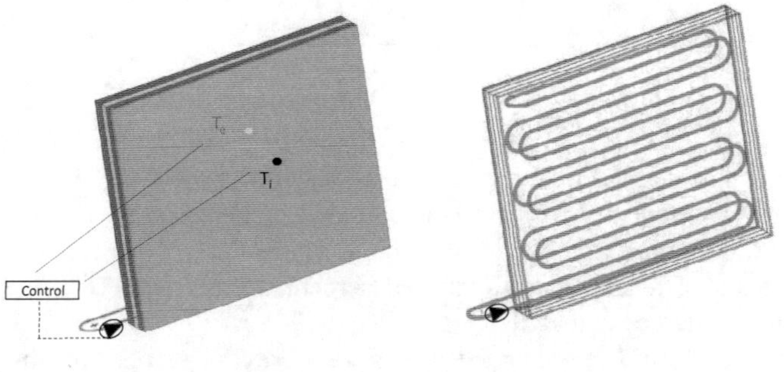

Figure 1. The proposed adaptive wall principle.

In wintertime, if the walls are properly insulated, the system does not work. The solar radiation in the late fall and early days of spring can heat the outer surfaces of the building envelope and this heat can be immediately transported into the buildings, by means of the fluid flow.

During the hot season, the outdoor temperature is lower than the indoor thermal wellbeing setpoint for a non-negligible number of hours in the day (especially the night-time hours). The transmission of heat (from the inside to the outside) in such periods is facilitated by the proposed device, thanks to the significant reduction of the wall thermal insulation and thermal inertia.

The adaptive wall further includes a control unit which acquires the data measured by the inner and outer temperature probes, compares them, and controls the pump. The central layers of the wall could be made of any material, while the two border layers where the pipes are embedded are realised with high thermal conductivity solutions, such as, for instance, plasterboard, cement, and other construction materials with similar characteristics (fig. 2).

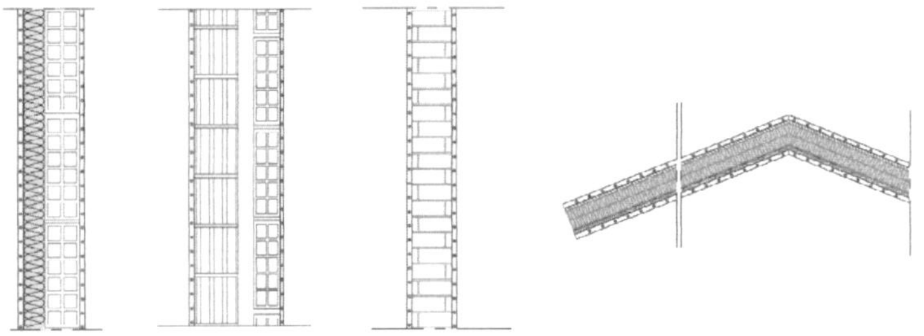

Figure 2. Application of the adaptive wall to various envelopes.

The indoor and outdoor temperature probes should be positioned in the middle of the wall, away from heat bridges and away from heat sources, embedded in the inner outer layers respectively.

The adaptive wall could be divided into several parts, each crossed by an independent trajectory of duct, and there could be an independent control of each circuit (fig. 3).

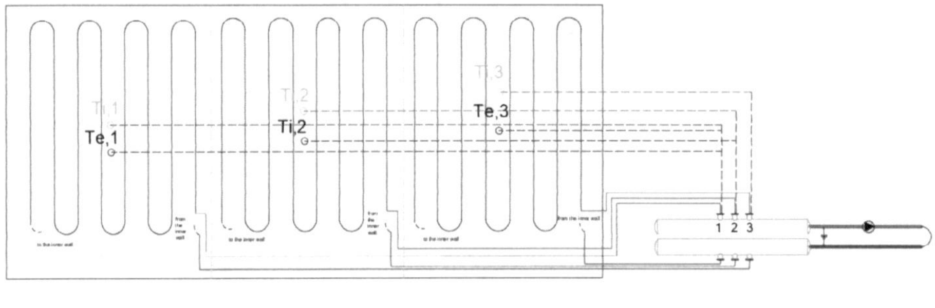

Figure 3. The proposed adaptive wall principle.

Preferably, the pipes forming the duct are made of a material selected from plastic, copper or iron, but it has not particular requirements respect to the common materials used in radiant panels. The heat transfer fluid is selected from pure water and water with the addition of other substances, such as glycol, which inhibits the freezing. The additive has varying percentages between 10% and 40% according to the climate zone where the device is installed.

The layout of the pipes could be designed in different ways. Figures 4b, 4b and 4c show three examples of pipe paths:

1) Sparse roundtrip pipes: the duct moves forward and backward in one face before going to the opposite side;
2) Sparse serpentine pipes: the duct moves only in one direction in one face before going to the opposite side;
3) Dense roundtrip pipes: the duct moves forward and backward in one face before going to the opposite side, the step of the path is lower respect to the first configuration.

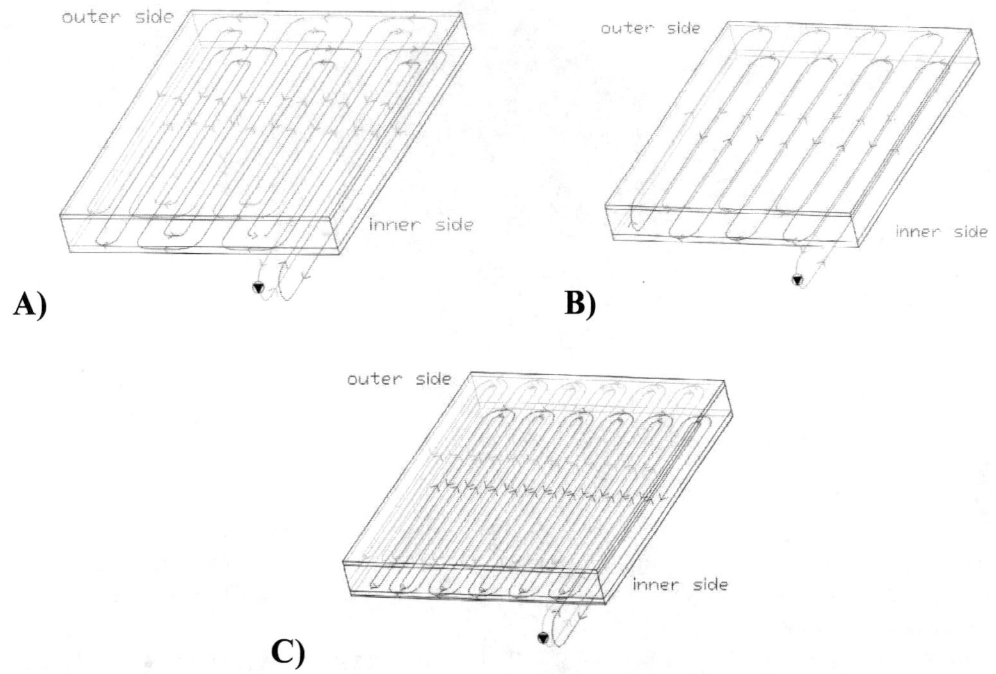

Figure 4. **A)** Sparse roundtrip configuration of the adaptive wall. **B)** Sparse-serpentine configuration of the adaptive wall. **C)** Dense roundtrip configuration of the adaptive wall.

3. CFD simulations of various configuration

The proposed adaptive wall has been simulated with a finite volumes code [11], following what will be tested in an experimental setup.

The model is a multi-layer highly insulated vertical wall, sized 1.23 m x 1.48 m, with a total thickness of 0.17 m. Starting from the inner environment, the first layer is made of gypsum and it hosts the first pipes layout. Then, an insulating layer of wood wool (with high thermal inertia) and another one of polystyrene are inserted, to finish with gypsum again (hosting the second path of the pipes) and a protective final layer of plaster on the outside. The thermophysical properties of each material are retrieved from the Standard ISO 10456 [12].

The inner diameter of the pipes is 8 millimetres, as this results one of the commercially available size for ducts used in radiant floor or walls.

The meshing process needed a strong effort to find a compromise between number of cells and their quality, because of the different granularities among the extension of the domain: the wall layers and the pipes. Figure 5 shows the grid for the dense roundtrip configuration.

The domain is meshed with tetrahedral cells; the total number of nodes are 2,223,396, with 10,583,055 elements. The mesh geometric characteristics has been checked with the element quality indicator, which reached its minimum value at 0.021; thus, it can be concluded that the wall is satisfactorily transformed in the computational domain. A mesh independence control was executed with a coarser grid, obtaining a heat flow rate that differed less than 1% respect to the chosen number of cells.

Figure 5. Grid of the computational domain for the dense roundtrip configuration.

All the domain is solid, except for the liquid inside the pipe: a mixture of water (80%) and glycol (20%) is considered. The boundary conditions in the border faces are taken from the Standard ISO 6946 [13], e.g. 7.69 W/m^{2}K in the inner surface and 25.00 W/m^{2}K in the outer one. The other lateral surfaces are considered adiabatic.

The fluid speed is defined by setting the pressure drop guaranteed by the pump that pushes liquid, therefore, the flow pattern depends on the pressure drop itself.

The steady-state simulations demonstrated that the wall global thermal transmittance may be raised by one order of magnitude when the fluid is in motion respect to the case with the fluid still [14]; furthermore, from the stationary analysis it emerged that a good compromise between the system, efficiency and the energy needed for the pumping is obtained with a fluid speed of 0.20 m/s. A simulation with a fluid speed of 0.40 m/s has been also conducted, showing negligible differences in terms of heat transfer respect to the case of 0.20 m/s. Since instead, the pressure drops increases of a factor close to 3, the speed of 0.20 m/s has been chosen.

Starting from these results, transient simulations have been implemented. The hot season case is following described, but the procedure and the results may be extended also to other periods of the year.

The inner temperature is fixed at a summer setpoint (26°C), while the outer temperature is taken from the weather data for the city of Perugia (Italy), taking also into account of the solar radiation for a south-facing wall with an absorption coefficient of 0.60 (sol-air temperature). Hypothesizing that the trend of the sol-air temperature is repeated for three days, the condition of stabilised periodic regime is fulfilled. Therefore, the transient simulation is repeated for three days with the same boundary conditions, but only the last 24 hours are analysed, when the initial transitory phase has disappeared.

In fig. 6 the thermal field of the analysed wall is reported during the first 8 hours of the days under analysis.

It is clear that the wall guarantees a free colling effect in the opaque wall as long as the outer temperature is lower than the inner setpoint temperature. When the boundaries changes (at 08:00), the pump stops and the wall turns back to be an insulating envelope.

The following graph (fig. 7) reports the results of two control strategies: one with the pump always on and the other with the pump that switches on only if the outer temperature is lower than the inner one. The heat flow entering the building is also reported, compared with the wall without the pipes.

In the configuration without the pipes, the wall cools the inner environment with an average value of 0.9 W/m$^{2;}$ the cooling effect is enhanced to 5.8 W/m^2 with the pump always on, to arrive at 6.6 W/m^2 when the pump controlled by the boundary conditions. No significant variations have been found varying the speed of the liquid.

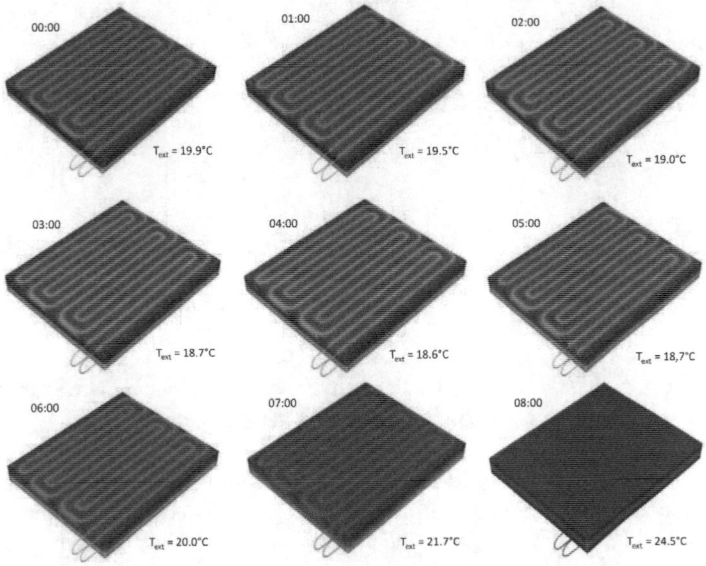

Figure 6. Trend of the thermal field created by the system during night-time in a summer typical day.

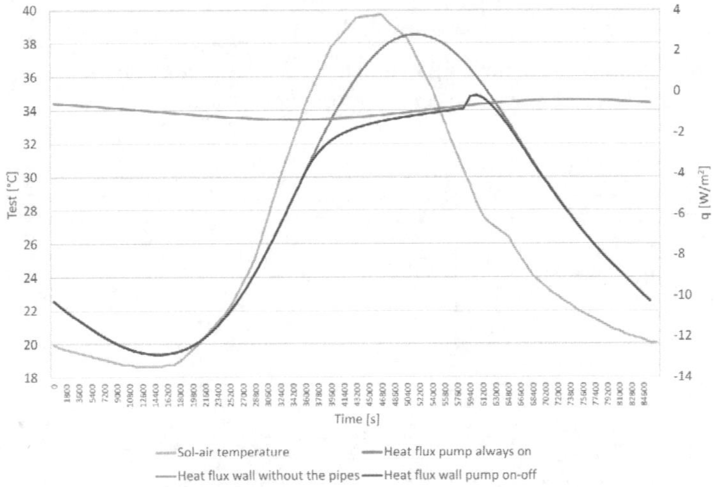

Figure 7. Sol-air temperature and specific heat fluxes entering the indoor environment for the different walls for a summertime day.

4. Description of the laboratory setup for the CFD validation

The performance of the proposed system will be also validated, beyond the simulations, through a hot box system. This device is made of a calorimetric chamber consisting of two rooms separated with a high thermal resistance wall, called surround panel, into which a specimen is mounted. The measurement setup is given in fig. 8.

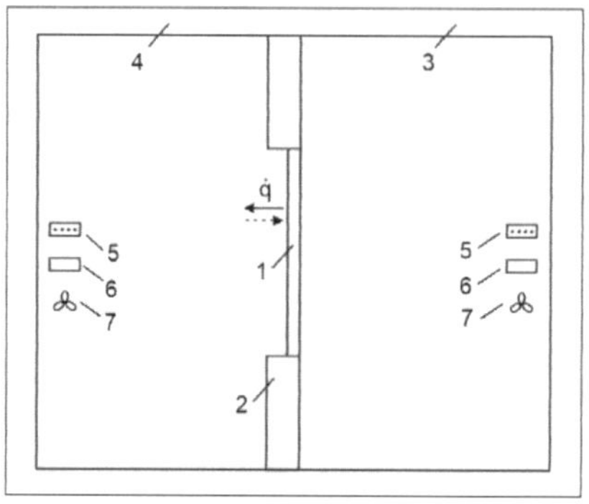

Figure 8. Measurement stand: 1 – tested specimen, 2 – surround panel, 3- warm chamber, 4 – cold chamber, 5 – cooling device, 6 – heating device, 7 - fan.

The two rooms simulate internal and external air conditions. The chamber is equipped with the measurement and control system which allows to set, control and measure the following data: air temperatures on both sides of the specimen, mean air velocities as well as surface temperatures on the demanded faces of the specimen. Within the chamber, it is possible to set steady-state normative conditions to perform measurements of thermal transmittance of the specimen. On the other hand, the conditions can be set as transient, simulating dynamic conditions, for example hourly external or internal air changes. The external air temperatures can be set as sol-air temperatures. The temperature sensors are T-thermocouples with the accuracy of 0.1 K. All the equipment is connected to the Ahlborn Almemo control and acquisition program with PID controller.
For the specific measurements, a sandwich wall with a coil embedded in the outer layers of the partition will be prepared. The size of the sample is 1230 mm x 1490 mm. The arrangement of the coil is shown in figg. 9a and 9b, while the cross-section of the test sample is shown in fig. 9c.

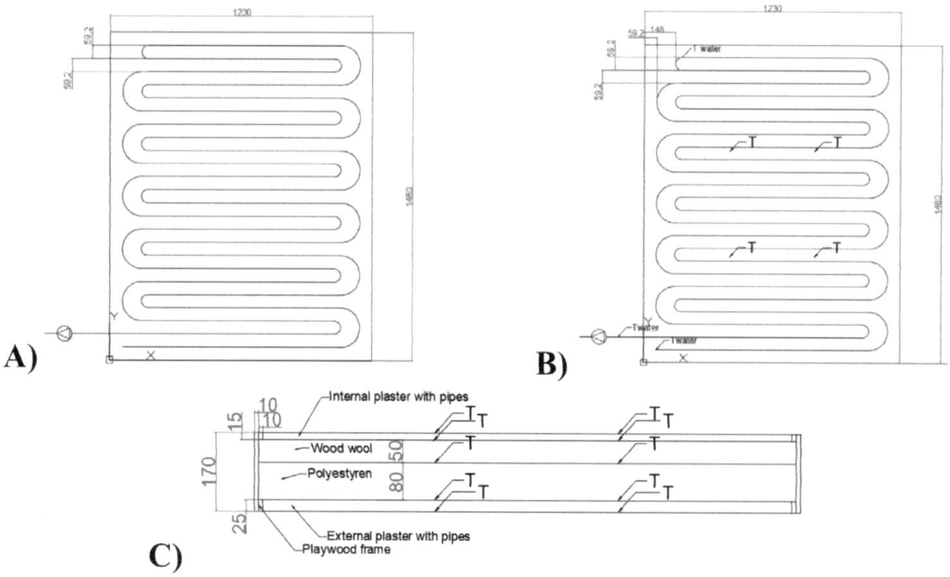

Figure 9. A) Pipe layout. **B)** Front view of thermocouples placement. **C)** Cross section with thermocouples placement.

The dimensions of the sample are determined by the size of the hole for mounting the measuring samples in the calorimetric chamber. The circulation pump with flow meter will be mounted in the lower part of the tested wall on the warm side. A pump with a motor isolated from the pumped liquid will be used to avoid the thermal effect of heat from the engine on the pumped liquid.

5. Conclusions and future work

The proposed solution provides the possibility of obtaining variable insulation of the building insulated external wall in the range of one order of magnitude. This gives the possibility of adjusting the heat flux flowing through the wall to the thermal conditions prevailing at a given moment, which allows both heat transfer from the external environment to the building and in warm periods to give off heat from the building to the external environment. The change in the thermal properties of the partition can be very quickly and easily achieved by controlling the operation of the pump forcing the flow of fluid in coils embedded in the walls. Future research will focus on measurements at a laboratory set up, which will give the opportunity to practically assess the accuracy of the developed CFD model and will allow to establish guidelines for the practical implementation of such systems in construction.

6. Acknowledgments

This work has been funded by the European Union - NextGenerationEU under the Italian Ministry of University and Research (MUR) National Innovation Ecosystem grant ECS00000041 - VITALITY. We acknowledge Università degli Studi di Perugia and MUR for support within the project Vitality.

7. References

[1] Xu L, Dai L, Yin L, Sun X, Xu W, Yang R, Wang X and Zhang Y 2020, *Energ. Buildings* **226** 110398.
[2] Zhang Y, Ma G, Wu G, Liu S and Gao L 2022, , *Energ. Buildings* **271** 112314.
[3] Aditya L, Mahlia T M I, Rismanchi B, Ng H M, Hasan M H, Metselaar H S C, Murazaf O, Aditiya H B 2017 *Renew. Sust. Energ. Rev.* **73** 1352-1365.
[4] Wang Z, Qiao Y, Liu Y, Bao J, Gao Q, Chen J, Yao H and Yang L 2021, *Build. Environ.* **194** 107709.
[5] Favoino F, Overend M, Jin Q 2014 *Appl. Energ.* **156** 1-15.
[6] Park B, Srubar III W V, Krarti M 2015 *Energ. Buildings* **103** 317-325.
[7] Lehmann M, “System for controlling the temperature of a building and method for controlling the temperature of a building using such a system”, Patent n. WO 2017/158100 A1, 21 September 2017.
[8] Iffa E, Hun D, Salonvaara M, Shrestha S, Lapsa M 2022, *Journal of Energy Storage* **46** 103815.
[9] Bailey S G, “Solar exterior cladding”, Patent n. GB 2482650, 8 March 2010.
[10] Bianchi F, Baldinelli G, "Variable thermal transmittance wall", Patent n. WO 2022/18270 A1, 09 September 2022.
[11] Ansys Fluent version 2021 R2.
[12] ISO 10456:2007 - Building materials and products — Hygrothermal properties — Tabulated design values and procedures for determining declared and design thermal values.
[13] ISO 6946:2017 - Building components and building elements - Thermal resistance and thermal transmittance - Calculation methods.
[14] Baldinelli G, Schnotale J, Lechowska A A, Presciutti A, Bianchi F 2022 *Energ. Buildings* **268** 112209.

Local heat-transfer coefficient estimation in cross-helix corrugated tubes under turbulent regime

Muhammad Waheed Azam[1,*], M Malavasi[1], L Cattani[1], F Bozzoli[1] and S Rainieri[1]

[1] Department of Engineering and Architecture, University of Parma, Parma, Italy
[*] Corresponding author: muhammadwaheed.azam@unipr.it

Abstract. In a wide variety of engineering applications, convection heat transfer enhancement plays a major role in reducing the heat exchanger's size and costs and increasing their efficiency. The passive methods for increasing convective heat transfer are the most interesting since they don't require any external power to achieve the desired enhancement. For this reason, this approach is preferred in the industrial sector of heat exchangers. For the enhancement of the performances of these apparatuses, in particular, the tubular ones used in the pharmaceutical, food, and chemical, the most adopted passive solution is the corrugation of their tube walls. Two pipes characterized by cross-helix corrugation have been experimentally analyzed both investigating the global heat transfer performance and studying the local heat transfer coefficient distribution to profoundly examine the thermal performance enhancement mechanisms of this passive technique. It has been observed that at Re = 16,000, the heat transfer rate in the smaller pitch tube is 18.18% higher than that in the larger pitch tube. The inverse heat conduction problem in the tube wall is addressed to achieve the local coefficient distribution method by using temperature readings collected from an infrared camera on the outer surface of the tube as the initial data.

1. Introduction

Convective heat transfer enhancement is a challenge for engineers and researchers to minimize the cost, reduce the size, enhance the efficiency, and effectiveness of heat exchangers. In this respect, for improving the heat transfer performances of heat exchangers it is generally possible to identify two different categories of techniques: passive ones and active ones. If there is no external power required, the approach is identified as passive otherwise active [1]. Normally passive techniques are preferred since they do not require external power, they are cheaper and easier to implement in many devices. Several passive approaches are used to enhance the heat transfer, such as treated and rough surfaces [2], swirl-flow and surface-tension devices [3], coiled tubes [4], or flow additives [5]. In many industrial sectors, the mostly adopted technique is represented by the wall pipe corrugation and among the different corrugation profiles adopted the cross-helical one demonstrated promising thermal performance [6].

This technique showed to be cost-effective and easy to manufacture, making it appealing for food industry applications. It is possible to manufacture this type of profiles easily and continuously by creating two helical corrugations in opposite directions, while also meeting the standards of hygienic design. Wall corrugation on a pipe's inner surface has several effects that increase heat transfer. It interrupts the thermal boundary layer, it enhances the heat transfer surface area by increasing the contact area between the fluid and the pipe wall, and it promotes the early onset of the transition to turbulent flow [7,8]. Bozzoli et al. [9] conducted an experimental study to investigate the performance of cross-helix corrugated geometries in terms of heat transfer enhancement and pressure drop for food thermal process. In this study, six different types of corrugated pipes, characterized by different corrugation pitch and depth values, were tested. The result revealed that the capacity of heat transfer enhancement of cross helix geometries was significantly higher than single helix and the performance of smaller pitch size pipes was higher than larger pitch size. Another research carried out the application of corrugated cross helix geometries in gas flow in turbulent regime [10]. To match a common range of gas-liquid heat exchangers, they analysed Reynolds number range was 5000–23000. They found the highest heat transfer with cross-helix tube having corrugation depth 1.86mm, but the drawbacks were higher friction factor.

Based on the literature review, it is concluded that previous studies on cross-helix corrugated pipes have only examined their performance in terms of average heat transfer. The investigation of the local convective heat transfer coefficient has significance for many different industrial procedures, such as food sterilization and pasteurization treatments. These processes use corrugated walls, and any non-uniform temperature distribution on the inside may compromise their overall performance. As a result, understanding and assessing the convective heat transfer coefficient at the local level is critical to ensuring the efficiency and effectiveness of these industrial applications. To the best of our knowledge, no research has been conducted on the local thermal performance of cross-helix corrugated pipes. In current research this gap starts to be filled with experimental investigation of cross helix local thermal performances by solving the Inverse Heat Conduction Problem (IHCP) [11] in the pipe wall employed by adopting as starting data the temperature measurements acquired by an infrared camera on the outer tube wall. The procedure approach is widely acknowledged to have some complications. This is mainly due to the ill-posed nature of the IHCP problem, which makes it extremely sensitive to even minor changes in the input data.

2. Pipe geometry and Experimental setup

In the present work, two cross-helix corrugated stainless-steel type AISI 304 pipes were tested. In figure 1 it is presented a sketch of the corrugation profile, while in figure 2 there are shown the pictures of the two analysed tubes and in (table 1) the dimensions of the corrugation profile are reported.

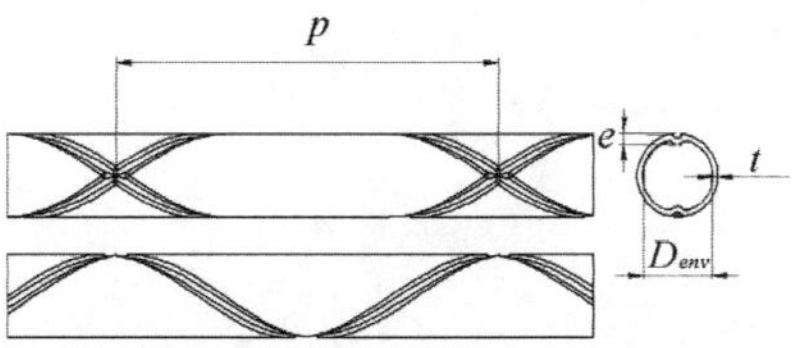

Figure 1. Visual representation of the wall corrugation pipes

Table 1. Dimensions of corrugated profile

Pipes	p (mm)	e (mm)	D_{env} (mm)	t (mm)
T1	16	1	14	1
T2	32	1	14	1

(a) (b)

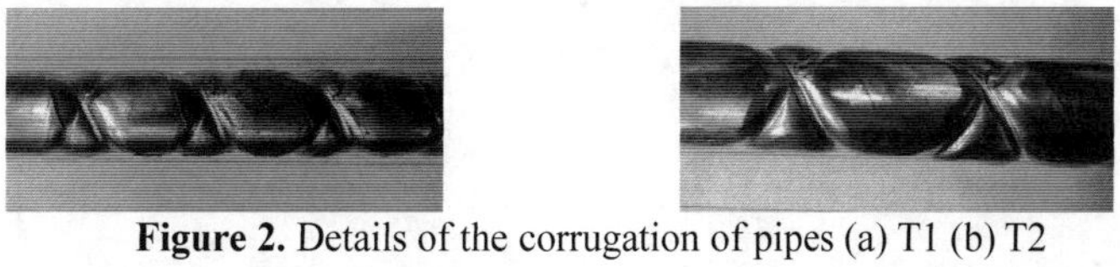

Figure 2. Details of the corrugation of pipes (a) T1 (b) T2

The pipes were connected to a test rig comprising of a volumetric pump that delivers the water from the storage tank to the pipe. Water was used as working fluid: before reaching the corrugated pipe, it flows inside a heat exchanger fed with tap water in order to keep constant the temperature at the inlet of the tested pipe. When the working fluid leaves the corrugated pipe, it is returned to the storage tank to restart the cycle.

The entire length of both pipes has three sections such as inlet, heating, and outlet section. The inlet section has a one-meter length, that allows the working fluid to achieve a developed hydrodynamic flow while in the heating section (1.84m) two steel fin electrodes are welded and connected to a power supply (HP 6671A) that operates in the range 0-220V and 0-8A. Joule heating effect was used to obtain the condition of uniform heat generated along the heat transfer section. The pipes were insulated with expanded polyurethane to minimize heat losses. At the inlet and outlet sections, T-type thermocouples were used to measure the temperature of the tube wall as well as the fluid. These thermocouples were precisely calibrated and linked to a multichannel ice point reference, especially the KAYE K170-50°C, ensuring exact and accurate temperature measurements throughout the experiment.

The graphic illustration of the experimental setup is shown in figure 3. A small portion of insulation was removed from corrugated pipes to be accessible to a thermal imaging camera. Then, that portion was uniformly painted with a thin layer of black paint with known emissivity (0.95). The effective emissivity of paints was estimated at different know temperatures. To measure the wall temperature six images are taken with the help of the system reported in figure 3. The FLIR SC7000 camera was characterized by a 640 x 512 pixels sensor with a sensitivity of 20 mK at 303 K and an accuracy of ±1 K. The complexity of the image processing procedure stemmed from the non-flat nature of the observed target surface. The image processing technique employed in [12] enabled the correction of optical distortions in the captured images caused by surface curvature. The obtained images were subsequently processed to create a continuous temperature map on the tube wall, due to suitable position references established on the tube wall.

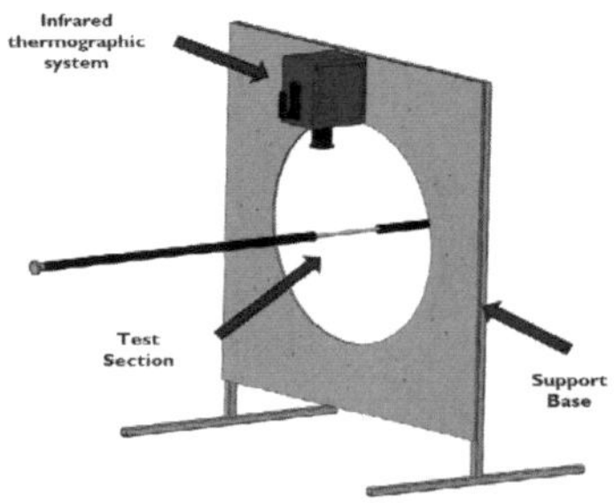

Figure 3. Graphic illustration of local experimental setup

3. Image processing and estimation procedure:

The inverse heat conduction problem (IHCP) for the wall surface was solved by calculating the local convective heat transfer coefficient. This was made possible with the help of the temperature distribution determined by infrared camera measurements. The given physical problem deals with the heat transfer through the wall of a cylindrical pipe (as shown in figure 4a). Moreover, the steady-state energy balance equation governing heat conduction within the domain of the cylindrical pipe's wall is expressed as follows:

$$k\nabla^2 T + q_g = 0 \tag{1}$$

The following two equations defined the boundary conditions.

$$-k\frac{\partial T}{\partial r} = \frac{(T - T_{amb})}{R_{env}}, \qquad r = r_{ext} \tag{2}$$

$$-k\frac{\partial T}{\partial r} = h_{int}(T - T_b), \qquad r = r_{int} \tag{3}$$

Where T_b is the bulk temperature of fluid flowing inside the tube, T_{amb} is the ambient temperature and R_{env} is the overall heat transfer resistance between the ambient and pipe wall.

a) (b)

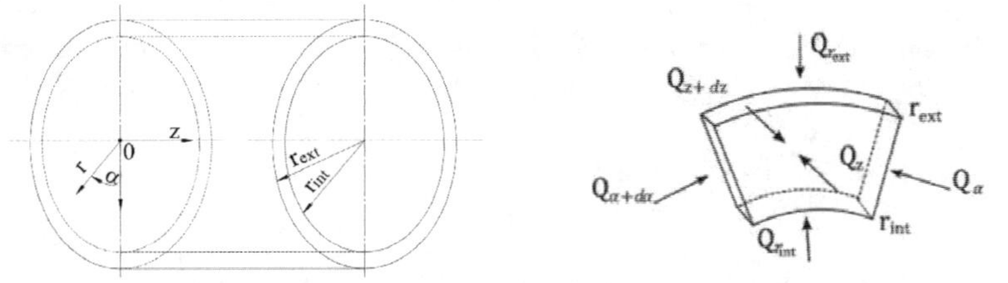

Figure 4. (a) Geometrical domain and b) test section representation.

The temperature on the external surface was assumed to be equal to the temperature on the internal surface under the thin-wall approximation.

$$T(\alpha, r, z) \cong T(\alpha, r_{int}, z) \cong T\,(\alpha, r_{ext}, z) \tag{4}$$

The Biot number is a dimensionless number that defines as the convective heat transfer coefficient multiplied by tube thickness and tube wall thermal conductivity. When the Biot number is less than 0.1, the thin-wall approximation is valid, suggesting that the conditions are adequate for simplified analysis. [13]. The thin-wall approximation simplifies the analysis in this context by assuming insignificant temperature gradients within the wall, which is generally true for systems with low Biot numbers. Figure 4b depicts the steady-state local energy balance equation with reference to the infinitesimal cylindrical section.

$$Q_{\alpha+d\alpha} + Q_{\alpha} + Q_{r_{ext}} + Q_{r_{int}} + Q_{z+dz} + Q_z + Q_g = 0 \tag{5}$$

By expressing all the terms in Eq. (5) and after some simplifications Eq. (6), one can arrive at the expression for the determination of the local convection coefficient at the internal wall-fluid interface:

$$h_{int} = \frac{k \cdot \ln\left(\frac{r_{ext}}{r_{int}}\right) \cdot \frac{\partial^2 T}{\partial \alpha^2} + \frac{k}{2}\left(r_{ext}^2 - r_{int}^2\right) \cdot \frac{\partial^2 T}{\partial z^2} - \frac{r_{ext}}{R_{env}}(T - T_{env}) + \frac{q_g}{2}\left(r_{ext}^2 - r_{int}^2\right)}{r_{int} \cdot (T - T_b)} \tag{6}$$

When dealing with discrete noisy data, Eq. (6) produces incorrect results due to the second-order derivative operator's extraordinary sensitivity to even small noise disturbances [11]. The accuracy and stability of the results may be compromised in such instances due to noise amplification during the differentiation process. A suitable method to solve this problem which was found in Gaussian filtering that allows to remove the noise in raw temperature data. Various Researchers [14] evaluated the performance of the Gaussian kernel using this particular method. In a 2-D frequency domain, the transfer function of this sort of filter can be written as follows.

$$H(u, v) = e^{-\left(\frac{u^2+v^2}{2u_c^2}\right)} \tag{7}$$

It provides valuable insights into the filter's frequency response, enabling the analysis of its behavior and performance in processing signals or images. Where u_c is the cut-off frequency, assuming that equally along the u and v coordinates. The cut-off frequency for this type of filter is unknown in the real application, so in order to successfully apply this method, a benchmark must be chosen. In the present work, this benchmark was given by the discrepancy principle as formulated by Morozov [15].

Journal of Physics: Conference Series **2685** (2024) 012042 doi:10.1088/1742-6596/2685/1/012042

4. Result and Discussion

In (figure. 5a) the Darcy friction factor, defined by following equation (8) for the two corrugated tubes under test is reported against the Reynolds number together with the results obtained for a smooth tube.

$$f = \frac{\Delta p}{p}\frac{D}{L}\frac{2}{w^2} \tag{8}$$

It can be observed that in (figure 5a) the T2 corrugated pipe, when compared to the T1 corrugated pipe, experiences a lower pressure drop in the range of $Re = 4 \times 10^3$-16×10^3. Moreover, values of f for both corrugated pipes were higher than those for the smooth tube. To evaluate the global efficiency (η) of both pipes by considering the thermal enhancement effect and the pressure drops caused by the corrugation is calculated by the following equation.

$$\eta = \frac{Nu/0.023\, Pr^{0.4} Re^{0.8}}{(f/f_0)^{1/3}} \tag{9}$$

where the subscript $_0$ refers to the smooth pipe. In (figure 5b) it can be seen that in both corrugated pipes, efficiency is decreases with the increase of Reynold number. Moreover, it is possible to observe that in terms of efficiency η, both T1 and T2 showing same efficiency at lower $Re = 10^4$, but with the increase of Reynolds number T1 pipe performs better than the T2.

a) (b)

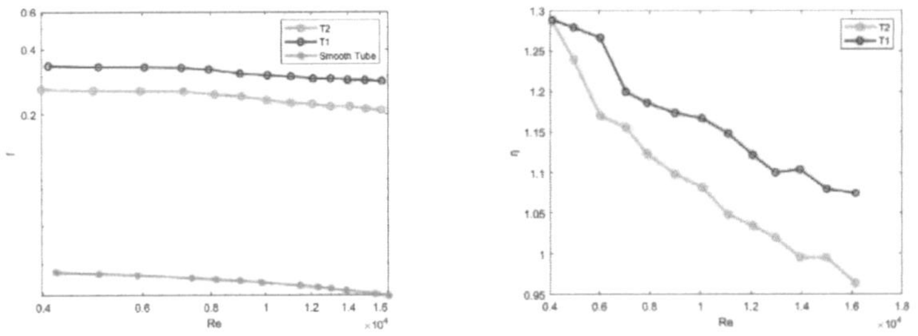

Figure 5. (a) Darcy friction factor vs. Reynolds number b) Efficiency (η) vs. Reynolds number

In this research, the performance of two cross-helix corrugated pipes is experimentally analysed in turbulent regime flow (Re values in the range 4×10^3 - 16×10^3). For a better understanding of the reader, schematic visualization of the corrugation position and flow direction is shown in figure 6.

a) *b)*

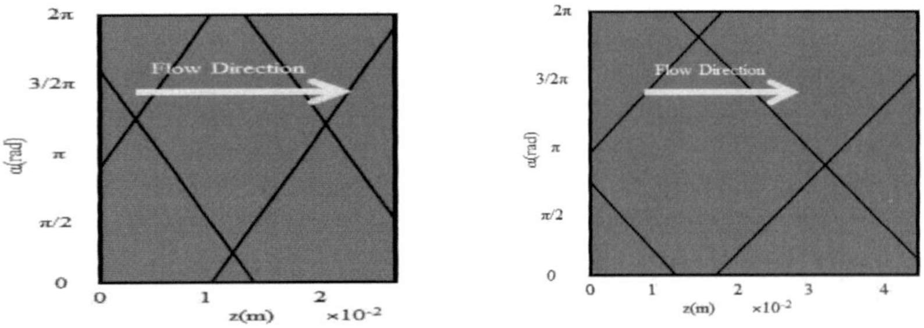

Figure 6. Flow direction and corrugation position in both cross-helix pipes (a) T1 (b) T2

The local heat transfer coefficient distribution at the internal wall to the liquid interface can be restored by starting from the temperature distribution on the outer pipe surface and using the IHCP approach explained in section 3. The filtered temperature maps, obtained by applying the Gaussian filter equation 7 on rough temperature distributions at Re = 4×10^3 of both tested pipes is reported in (figure 7), it is possible to observe that immediately after the corrugations, the wall temperature increases in both pipes, reaching a local maximum.

a) *b)*

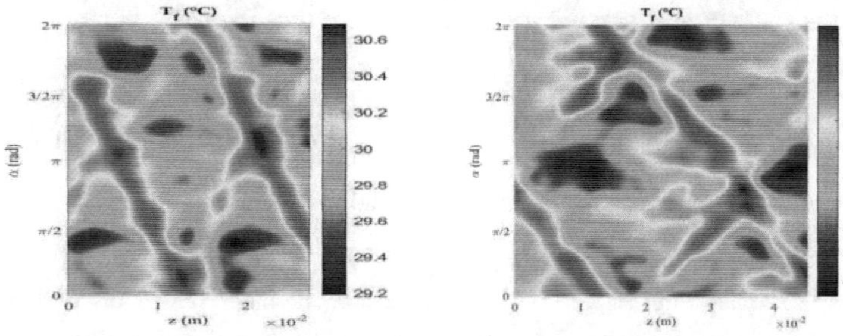

Figure 7. Filtered temperature distributions at Re=4000 a) T1 b) T2

The performance of both cross-helix corrugated pipes was investigated by analysing the internal convective heat transfer coefficient distributions for three different Reynolds numbers Re = 4×10^3, 10^4, and 1.6×10^3 were presented in figure 8,9 and 10. The connective heat transfer coefficient (h) confirmed the behavior of the temperature distribution, which had a minimum value after the corrugation and then reached a maximum as the fluid stream followed. The local heat transfer coefficient distribution at Re=4×10^3 is more uniform in the case of T1 pipe, promoting a higher heat transfer enhancement. Considering the T1 pipes having smaller pitch size, when compared to the T2 pipe, there is a considerable decrease in the h value between corrugations. The described trend holds true for situations with Re = 10^4 and Re = 16×10^3.

a) *b)*

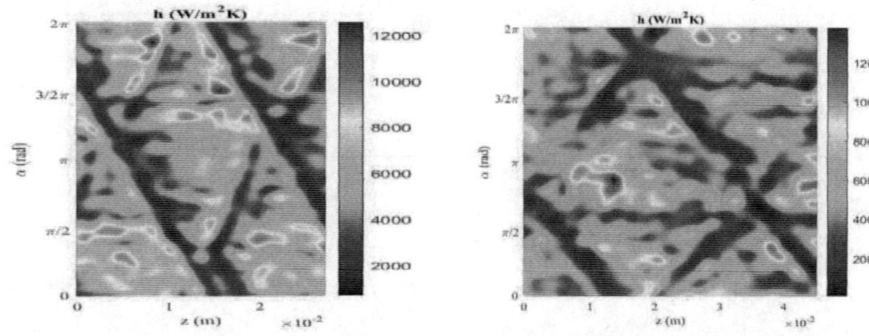

Figure 8. Convective heat transfer distribution (h) at Re= 4000 a) T1 b) T2

a) *b)*

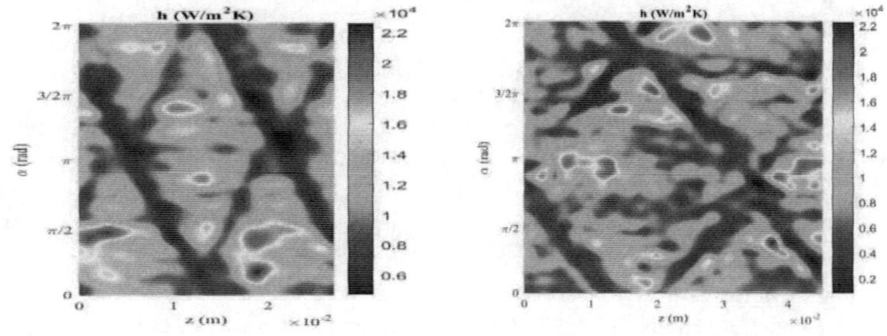

Figure 9. Convective heat transfer distribution (h) at Re= 10000 a) T1 b) T2

Journal of Physics: Conference Series **2685** (2024) 012042 doi:10.1088/1742-6596/2685/1/012042

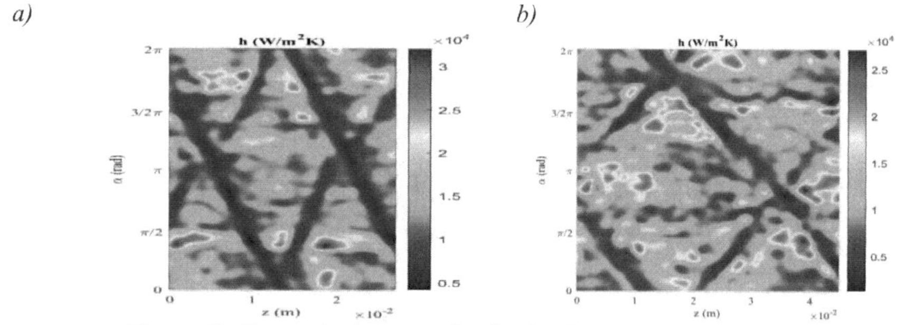

Figure 10. Convective heat transfer distribution (h) at Re= 16000 a) T1 b) T2

Moreover, it is feasible to declare that, in those regions, the heat exchange between the fluid and wall decreases. This type of behaviour could be found in the boundary layers disruption effect caused by the fluid confronting surface roughness along the direction of flow. Furthermore, after leaving the corrugation temperature gradually decreases until the next corrugation. This process is repeated when the fluid passes through from each corrugation.

Concluding, in figure 11 the Nusselt distribution for the radial coordinate $\alpha=\pi$ is represented. Also in this case, the behaviour of the internal convective heat transfer coefficient is confirmed: when the fluid flows after the corrugation, Nusselt number immediately decreases and then starts to increase until the next corrugation. This behaviour can be found also in the local convective heat transfer coefficient (*h*) as shown in previous figures 9,10 and 11. This phenomenon is repeated at each corrugation of the pipes. Moreover, it can be observed that the performance in terms of Nusselt number of T1 pipe is higher when compared to T2 pipe.

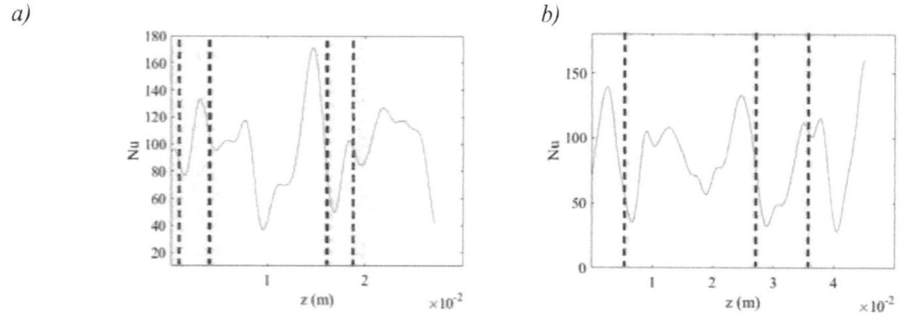

Figure 11. Local Nusselt number at $\alpha=\pi$ at Re=4000 a) T1 f) T2

5. Conclusions

In the present research work, the thermal performance of two cross-helix corrugated pipes are investigated from local experimental points of view in a turbulent flow regime. Both pipes have two different corrugation pitch values, i.e., 16 and 32 mm. The analysis of the global performance highlighted for both pipe and T1 pipe obtained higher efficiency as compared T2 pipe. With the increase of Reynold number efficiency decreases of both corrugated cross helix corrugated pipes. However, the drawback of this enhancement in both tested pipes is found that higher pressure drops as compared to reference geometry. Moreover, T1 pipe is the best choice when the aim is to obtain the highest value of convective heat transfer coefficient (*h*). The results from the measurements conducted locally indicate that using corrugation in both T1 and T2 geometries can effectively enhance the heat transfer capability of heat exchangers in turbulent conditions. Furthermore, another important result of the present research is the execution and testing of an effective experimental method that could be successfully applied for the optimization of wall corrugation in critical thermal processes.

References:

[1] R. L. Webb, "Principles of enhanced heat transfer," 1994.
[2] L. Ventola *et al.*, "Rough surfaces with enhanced heat transfer for electronics cooling by direct metal laser sintering," *Int. J. Heat Mass Transf.*, vol. 75, pp. 58–74, 2014, doi: 10.1016/j.ijheatmasstransfer.2014.03.037.
[3] M. Sheikholeslami, M. Gorji-Bandpy, and D. D. Ganji, "Review of heat transfer enhancement methods: Focus on passive methods using swirl flow devices," *Renew. Sustain. Energy Rev.*, vol. 49, pp. 444–469, 2015, doi: 10.1016/j.rser.2015.04.113.
[4] F. Bozzoli, L. Cattani, and S. Rainieri, "Effect of wall corrugation on local convective heat transfer in coiled tubes," *Int. J. Heat Mass Transf.*, vol. 101, pp. 76–90, 2016, doi: 10.1016/j.ijheatmasstransfer.2016.04.106.
[5] P. Freymuth, "Flow visualization in fluid mechanics," *Rev. Sci. Instrum.*, vol. 64, no. 1, pp. 1–18, 1993, doi: 10.1063/1.1144433.
[6] Z. S. Kareem, M. N. Mohd Jaafar, T. M. Lazim, S. Abdullah, and A. F. Abdulwahid, "Passive heat transfer enhancement review in corrugation," *Exp. Therm. Fluid Sci.*, vol. 68, pp. 22–38, 2015, doi: 10.1016/j.expthermflusci.2015.04.012.
[7] N. Kurtulmuş and B. Sahin, "A review of hydrodynamics and heat transfer through corrugated channels," *Int. Commun. Heat Mass Transf.*, vol. 108, p. 104307, 2019, doi: 10.1016/j.icheatmasstransfer.2019.104307.
[8] W. T. Ji, A. M. Jacobi, Y. L. He, and W. Q. Tao, "Summary and evaluation on single-phase heat transfer enhancement techniques of liquid laminar and turbulent pipe flow," *Int. J. Heat Mass Transf.*, vol. 88, pp. 735–754, 2015, doi: 10.1016/j.ijheatmasstransfer.2015.04.008.
[9] F. Bozzoli, L. Cattani, and S. Rainieri, "Cross-helix corrugation: The optimal geometry for effective food thermal processing," *Int. J. Heat Mass Transf.*, vol. 147, p. 118874, 2020, doi: 10.1016/j.ijheatmasstransfer.2019.118874.
[10] A. Harleß, E. Franz, and M. Breuer, "Heat transfer and friction characteristics of fully developed gas flow in cross-corrugated tubes," *Int. J. Heat Mass Transf.*, vol. 107, pp. 1076–1084, 2017, doi: 10.1016/j.ijheatmasstransfer.2016.10.129.
[11] J.Beck, "Inverse Heat Conduction Ill-posed problems," pp. 137-157,179-180,198-211, 2011.
[12] F. Bozzoli, L. Cattani, and S. Rainieri, "Effect of wall corrugation on local convective heat transfer in coiled tubes," *Int. J. Heat Mass Transf.*, vol. 101, pp. 76–90, 2016, doi: 10.1016/j.ijheatmasstransfer.2016.04.106.
[13] D. F.P. and De Witt, *Fundamentals of Heat and Mass Transfer*, vol. 4, no. 1. 2002.
[14] F. Bozzoli, L. Cattani, G. Pagliarini, and S. Rainieri, "Infrared image filtering applied to the restoration of the convective heat transfer coefficient distribution in coiled tubes," *Opto-Electronics Rev.*, vol. 23, no. 1, pp. 107–115, 2015, doi: 10.1515/oere-2015-0004.
[15] V. A. Morozov, *Methods for Solving Incorrectly Posed Problems*. New york. [Online]. Available: https://www.ptonline.com/articles/how-to-get-better-mfi-results

Acknowledgments: The work was funded under the National Recovery and Resilience Plan (NRRP), Mission 4 Component 2 Investment 3.1 - Call for tender No. 3264 of 28/12/2021 of Italian Ministry of University and Research funded by the European Union – NextGenerationEU. Award Number: Project code IR0000033, Concession Decree No. 120 of 21/06/2022 adopted by the Italian Ministry of University and Research, CUP I83C22001040006, Project title "Strengthening of the Italian Research Infrastructure for Metrology and Open Access Data in support to the Agrifood" (METROFOOD-IT).
The authors would also like to thank MBS S.r.l. for its support in the realization of the experimental setup.

Towards a more realistic MELCOR model for a dry cask for spent nuclear fuel. Part I: sensitivity studies.

M Angelucci[1,*] and S Paci[1]

[1] Department of Industrial and Civil Engineering, University of Pisa, Largo Lucio Lazzarino, Pisa, Italy

* Corresponding author: michela.angelucci@ing.unipi.it

Abstract. The United States (US) Department Of Energy (DOE) has addressed the thermal analysis of the Spent Nuclear Fuel (SNF) stored within a dry cask system as a matter of high priority. In this regard, it is of utmost importance that simulation tools effectively reproduce the general thermal behavior of the modelled cask, including heat exchange and removal. Temperature distribution in the different components of the system is usually the focus of performed thermal analyses. In particular, attention is paid to the maximum temperature reached in the fuel cladding, namely the Peak Cladding Temperature (PCT). Within this framework, the present paper is the first of a two-paper series aimed at developing a more accurate model for the HI-STORM 100S cask. The dry cask in question is modelled and its behavior is simulated by means of the MELCOR code (version 2.2.18019). Stressing the need for a more realistic model rather than a conservative one, this paper reports the efforts undertaken to evaluate the influence of some specific modelling choices on the PCT. The study of the cask performance is therefore conducted taking into consideration three main factors: the axial power distribution in the Fuel Assembly (FA), the flow losses in the air gap between the internal canister and the external overpack, and the conductivity of the overpack concrete.

Keywords: *Dry cask, spent nuclear fuel, MELCOR, Sensitivity, safety.*

1. Introduction

In response to Spent Fuel Pools (SFPs) reaching their maximum capacity and considering the delays in the construction of final repositories, dry casks have gained popularity worldwide as a valid alternative for the safe storage of Spent Nuclear Fuel (SNF). In this regard, several dry cask designs have been proposed and licensed: a list of dry storage systems approved by U.S. NRC is provided in [1]. Regardless of their design characteristics, compliance with regulatory requirements as reported in 10CFR72 [2] must be guaranteed. In particular, the maximum temperature reached in the fuel cladding, namely the Peak Cladding Temperature (PCT), must not exceed the temperature limits (673.15 K and 843.15 K for normal and off-normal/accidental conditions, respectively) in order to ensure the safety of the system [3]. In this regard, the thermal analysis of SNF stored within a dry cask system is addressed as high priority by U.S. Department Of Energy (DOE) [4].

Within this framework, many studies have been carried out in the past 40 years to thermally characterize various types of dry casks. Different classes of computer codes have been employed, with Computational Fluid Dynamic (CFD) tools being the most adopted ones nowadays [5–9]. However, previous studies proved that the MELCOR code [10,11] is able to reproduce the general behavior of the modelled cask in both normal operating condition and also when considering more challenging boundary conditions.

Continuing along this path, the present paper is the first of a two-paper series aimed at creating a more accurate model for the largely used HI-STORM 100S system [12]. In particular, pointing to a more realistic model rather than a conservative one, the influence of some modelling choices on the PCT is evaluated, taking into account three main factors: the axial power distribution in the FA, the flow losses in the air gap between the internal metallic canister and the external concrete-based overpack, and the conductivity of the concrete itself. As a first step, a comparison is made between the results, in terms of both PCT and its location, obtained considering a uniform axial power distribution and an axial power profile typical of Pressurized Water Reactor (PWR) fuel assemblies. Secondly, attention is paid to a more precise characterization of the flow losses in three different locations: air inlet, annular air channel, and air outlet. In particular, the complexity of inlet and outlet geometry seems to indicate the need for an improved definition of the flow losses, and consequently of the air mass flow rate, in order to reproduce the correct heat extraction capacity. Finally, a sensitivity study on the conductivity of the concrete is reported. Considering the multitude of plain concretes available and their different composition, an attempt is made to quantify the influence of different values of concrete conductivity on the PCT value.

2. Sensitivity Studies

As said before, the system addressed in this work is the HI-STORM 100S, the detailed description of which is reported in [12]. However, for the sake of clarity, few information are reported hereafter.

As it can be seen in Figure 1, the SNF is stored within a sealed metallic Multi-Purpose Canister (MPC) filled with 99.995% pure Helium. The canister is, in turn, housed within a concrete-based overpack. The MPC ensures confinement, whilst the overpack acts as radiation shielding. The cooling of the SNF is, instead, achieved through the combination of three mechanisms: the thermosiphon effect of the Helium inside the MPC, the natural circulation of the external air in the channel between the MPC and the overpack, and the radiative exchange from the MPC to the overpack walls. The external wall of the overpack is, in turn, cooled down by convective heat exchange with the environment.

The MPC can contain a different number of Fuel Assemblies (FAs) depending on the design. The one considered in the present analysis is the MPC-32, that houses 32 PWR 17x17 FAs, with a decay power of 30.0 kW.

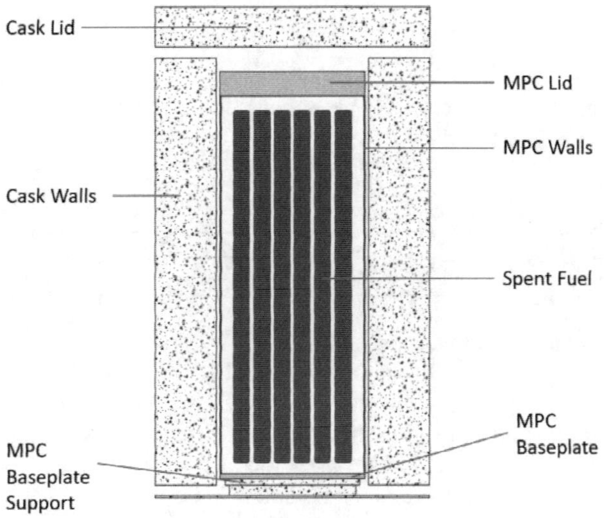

Figure 1. HI-STORM 100S system.

The entire system has been modelled with the MELCOR code (version 2.2.18019), that is originally developed for the simulation of Severe Accident (SA) scenarios. The model is described in [13,14]; however, few details will be reported in the following in relation to the performed sensitivity studies.

2.1. Axial Power Profile

The "core" region (representing the spent FAs within the MPC) is depicted in MELCOR through radial rings and axial levels, the combination of which results in the definition of cells. For each cell the amounts of fuel, cladding, and other components' masses have to be defined together with their hydraulic diameters, flow areas, boundary surfaces, etc, as requested in [10].

In the original model, a uniform axial power profile was employed, as a first approach, for the 8 (out of 12) axial levels containing the fuel. However, for a more realistic model, a non-uniform axial power profile has to be employed to take into account fuel operating history.

For this reason, the first sensitivity study is devoted to the evaluation of the influence of a non-uniform axial profile for the "core" region in the MPC. The considered profile has a shape peculiar to the majority of PWR FAs: relatively flat in the central section (with peak burnup of 1.1/1.2 times the FA average burnup) and significantly steep at the ends (with burnup of 0.5/0.6 times the assembly average), as it can be seen from Figure 2. More specifically, the profile (as proposed in [15]) has been exploited to obtain the "relative power density" for each axial level in the MELCOR model, as reported in Table 1.

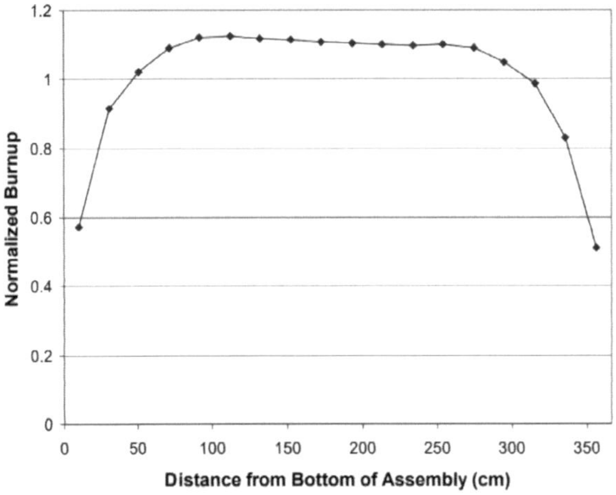

Figure 2. Normalized PWR Axial Burnup Distribution [15].

Table 1. Relative Power Density for each axial level in MELCOR.

Axial Level	4	5	6	7	8	9	10	11
Relative Power Density	0.68	1.0425	1.1225	1.1175	1.1075	1.075	0.85	0.5125

2.2. Loss Coefficients in the air channel

As shown in Figure 3, the annular air channel is modelled in MELCOR with 8 Control Volumes (CVs), connected by means of 7 vertical Flow Paths (FPs). Two additional horizontal FPs are meant to represent the flow of air entering or leaving the channel. For each FP, a loss coefficient, K, is defined and then used by the code to determine the pressure drop, across the FP, due to form losses.

In the original model, it was decided to keep default values, and loss coefficients for the FPs in the air channel were imposed as 1.0. However, a better characterization of the loss coefficients is needed to take into account the geometry of the system, with particular attention to the slightly more complex shape of inlet and outlet ducts connecting the environment with the annular space.

Considering what said, the second sensitivity study deals with a more realistic definition of the form losses in the air channel, with particular focus on three locations: air inlet, vertical (annular) channel, and air outlet.

Bearing in mind that form losses are usually linked to bends, changes in the flow area, valves, etc., a relatively low loss coefficient (K=0.1) has been selected for the FPs associated with the annular channel. On the contrary, higher loss coefficients have been considered for the FPs linked to the air inlet and outlet. More precisely, considering the presence, in the real cask, of 4 inlet and 4 outlet openings, and taking into account that, for sharp entrances or exits, the loss coefficients are between 0.5 and 1.0 [16], K values of 2.0 and 4.0 have been deemed appropriate for the analysis.

A summary of the values employed for each sensitivity study is reported in Table 2.

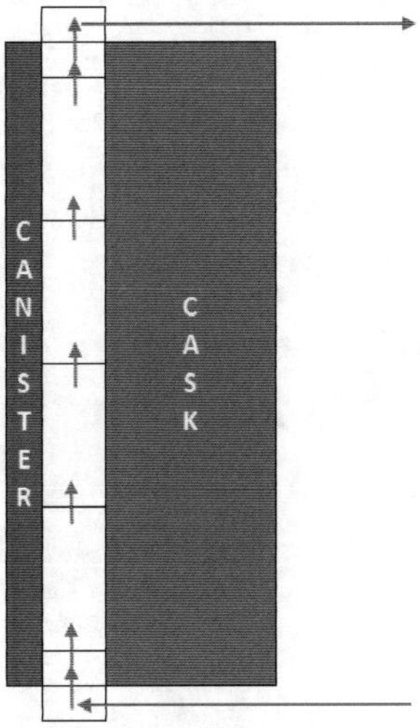

Figure 3. Sketch of the Air Channel Nodalization.

Table 2. Sensitivity Cases - Loss Coefficients values.

Case Name	Loss Coefficients		
	Vertical FPs	Inlet	Outlet
BaseCase	1.0	1.0	1.0
FlowLosses1	0.1	2.0	2.0
FlowLosses2	0.1	1.0	1.0
FlowLosses3	0.1	4.0	4.0

2.3. Concrete Conductivity

The overpack presents a sandwich structure composed of a thick core region made of plain concrete and two thin external layers of carbon steel. This means that the heat reaching the inner metallic wall of the overpack (by radiative exchange from the MPC or by convective exchange with the air in the channel) is almost completely transferred to the concrete. At this point, the heat transfer within the concrete wall is driven by two main factors: the thickness of the wall (that is 0.7 m) and the concrete thermal conductivity (that in the original model is imposed equal to 1.4 W/m-K, as in [6]). However, considering the multitude of plain concretes available and taking into account that the conductivity is dependent, among others, on the composition of the concrete itself, it is worth to make an attempt to evaluate the influence (if any) of different values of concrete conductivity on the PCT and, more generally, on thermal behavior of the system.

For this sensitivity study, after a literature review [17–19], two "extreme" values have been taken into account: a lower bound of 0.7 W/m-K, and upper bound of 2.1 W/m-K. Higher values have not been considered, since they are usually attributed to reinforced concrete.

3. Results

Results from the performed sensitivity studies are reported in the following subsections, one for each considered factor: power profile, loss coefficients in the air channel and concrete conductivity. Simulations outcomes will be presented mainly in terms of PCT; however, axial temperature profiles will be displayed and discussed, when needed.

3.1. Axial Power Profile

For what concerns the maximum cladding temperature, the use of the axial power profile in the "core" region results in a modest reduction of the PCT value: in fact, a decrease of 2 K is calculated (630.5 K for the base case, 628.5 K when considering the non-uniform axial power profile). The location of the PCT is the same i.e., the upper part (11th axial level) of the central FAs. However, as it can be seen in Figure 4, the addition of the axial power profile has a localized effect: following the shape of the burnup distribution, the temperatures in the upper and lower ends are slightly lower, whilst higher temperatures are calculated for the central part of FAs.

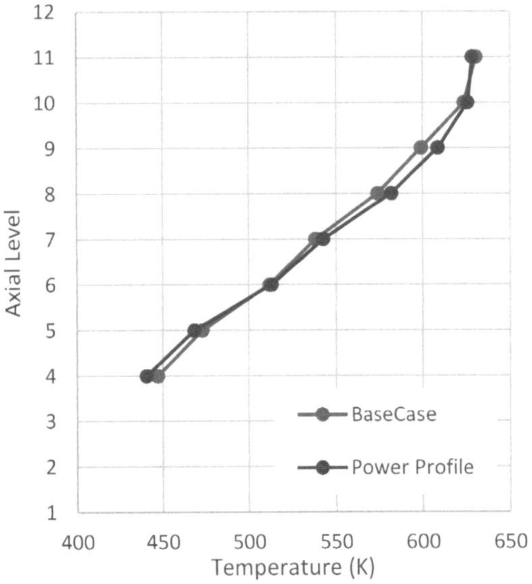

Figure 4. Axial Temperature Profile - hottest FAs.

3.2. Flow Losses in the air channel

The variation of loss coefficients in the FPs associated with the air channel causes a change in the mass flow rates. As it is reported in Table 3, the highest mass flow rate is associated as expected with the case "FlowLosses 2", in which the loss coefficients are minimized. Case "FlowLosses 3", on the other hand, presents a mass flow rate similar to the "BaseCase", even with a quite different assessment of loss coefficients. Mass flow rate for case "FlowLosses 1" is, instead, in between the other two cases.

The change in the mass flow rates is reflected in the temperatures of both the air channel and the fuel, since the ability to extract heat from the canister changes in turn. As it can be seen in Figure 5, trends are the same, but lower temperatures are calculated in all three cases. Observing the profile for the fuel, the ΔT with respect to "BaseCase" is at its minimum in the lower part, and it increases moving upward towards the hottest section of the FAs. The same behavior can be noticed for the air in the annular channel; however, maximum ΔT is a bit more pronounced.

As expected, the lowest temperatures are connected with the highest mass flow rate ("FlowLosses 2"), for which differences of 15 K and 20 K (with respect to the "BaseCase") are calculated for the PCT and the maximum air temperature, respectively. On the contrary, temperatures for case "FlowLosses 3" are almost the same as in the "BaseCase".

Table 3. Mass Flow Rates.

Case Name	Mass Flow Rate (kg/s)
BaseCase	0.307
FlowLosses 1	0.373
FlowLosses 2	0.423
FlowLosses 3	0.316

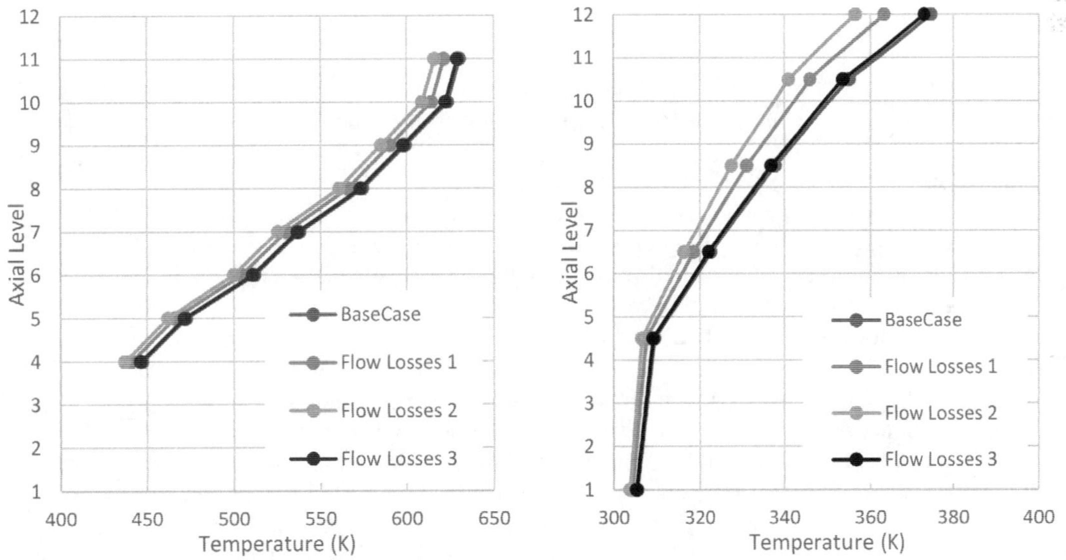

Figure 5. Axial Temperature Profiles for Fuel (left) and Air (right).

3.3. *Concrete Conductivity*

As it can be seen in Figure 6 (left), the thermal conductivity of the concrete has almost no influence on the PCT or even on the temperature, at any axial level. This confirms that the main mechanisms for the extraction of the heat from the canister are the convective heat exchange in the air channel and, to a less extent, the radiative heat transfer from the canister wall to the overpack wall.

However, concrete conductivity has an impact on the temperatures of the concrete overpack itself. As it can be observed in Figure 6 (right), a smaller ΔT between internal (hotter) and external (colder) concrete walls is linked to a higher thermal conductivity. On the contrary, the case with the lowest thermal conductivity shows a higher ΔT. This is consistent with the physical meaning of the thermal conductivity, that is the rate at which the material can transfer heat, either by receiving it from a hotter material or by passing it to a colder one.

The case with the lowest thermal conductivity is also the one with the highest temperature. However, temperatures are still below the limit as reported in [12], that is 422 K for normal operation.

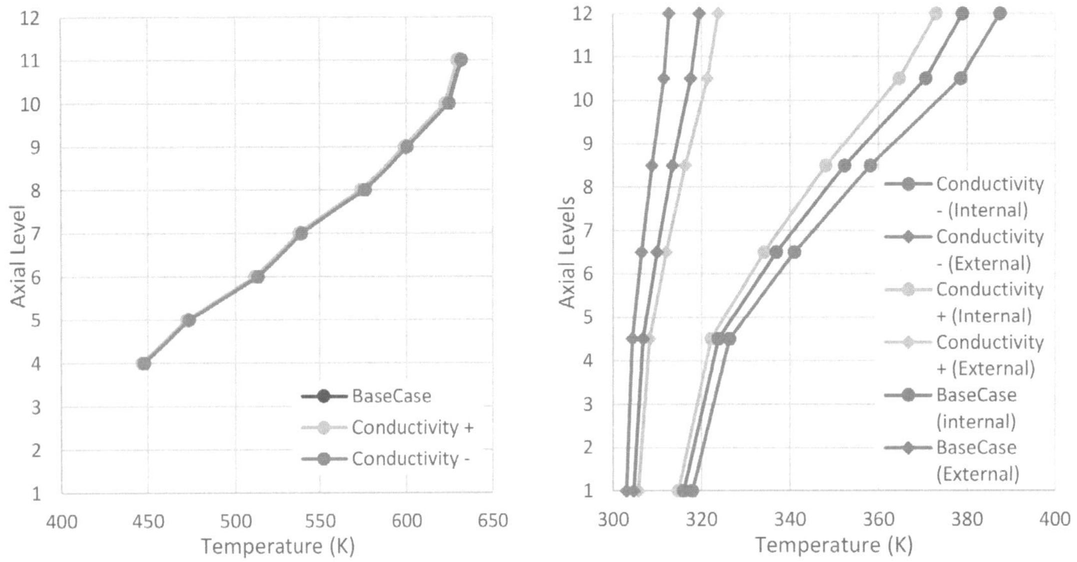

Figure 6. Axial Temperature Profiles for Fuel (left) and Concrete (right).

4. Conclusions

The aim of the work is the creation of a more accurate and realistic model for the HI-STORM 100S system. In this framework, the present paper reports the efforts made in the evaluation of the influence of some modelling choices on the thermal response of the system, and particularly on the PCT, being a key parameter for the safe storage of SNF within a dry cask.

From the performed sensitivity studies, some insights can be drawn:

- The employment of a non-uniform axial power profile has a localized effect: in fact, it affects mainly the fuel. Whilst there is no major impact on the PCT (only a 2K reduction), a change, even if limited, can be observed in the shape of the axial temperature profile: in fact, following the shape of the burnup distribution, the temperatures in the upper and lower ends are slightly lower, whilst a bit higher temperatures are calculated for the central part of FAs.
- The sensitivity study shows that a reduction of 15 K (with respect to the "BaseCase") is foreseen for the PCT when considering low loss coefficients in the FPs associated to the air channel. As a consequence, a proper definition of the loss coefficients in the air channel is of utmost importance for the correct reproduction of the heat extraction from the fuel.
- The thermal conductivity of the concrete has almost no influence on the PCT. This corroborate the idea that the main heat removal mechanisms from the canister are the convective heat exchange in the air channel and, to a less extent, the radiative heat transfer from the canister wall to the overpack wall. However, as expected, temperatures in the concrete layer of the overpack are influenced by the concrete conductivity values.

In short, chosen parameters, taken individually, seem to have an influence on the thermal behavior of the cask. However, a further step needs to be done to investigate their conjoint influence. This will be proposed in Part II, where findings from the presented sensitivity studies will be leveraged to set up a more realistic MELCOR model for the HI-STORM 100S cask.

5. References

[1] U.S. NRC Dry Spent Fuel Storage Designs: NRC Approved for General Use

[2] U.S. NRC *PART 72—LICENSING REQUIREMENTS FOR THE INDEPENDENT STORAGE OF SPENT NUCLEAR FUEL, HIGH-LEVEL RADIOACTIVE WASTE, AND REACTOR-RELATED GREATER THAN CLASS C WASTE | NRC.gov*

[3] U.S. NRC 2003 *Interim Staff Guidance-11, Revision 3 Issue: Cladding Considerations for the Transportation and Storage of Spent Fuel*

[4] U.S. NRC 2020 *Standard Review Plan for Spent Fuel Dry Storage Systems and Facilities Final Report Office of Nuclear Material Safety and Safeguards, NUREG-2215*

[5] Benavides J, Jimenez G, Galbán M and Lloret M 2019 Methodology for thermal analysis of spent nuclear fuel dry cask using CFD codes *Ann. Nucl. Energy* **133** 257–74

[6] Herranz L E, Penalva J and Feria F 2015 CFD analysis of a cask for spent fuel dry storage: Model fundamentals and sensitivity studies *Ann. Nucl. Energy* **76** 54–62

[7] Li J and Liu Y Y 2016 Thermal modeling of a vertical dry storage cask for used nuclear fuel *Nucl. Eng. Des.* **301** 74–88

[8] Brewster R A, Baglietto E, Volpenhein E and Bajwa C S 2013 CFD Analyses of the TN-24P PWR Spent Fuel Storage Cask *Am. Soc. Mech. Eng. Press. Vessel. Pip. Div. PVP* **3** 17–25

[9] Wang Y H, Lin Y P, Chang H Y and Lai C M 2021 Ventilation pattern and heat dissipation characteristics of a vertical dry storage cask for spent nuclear fuel: Wind tunnel experiments and CFD simulations *Ann. Nucl. Energy* **160** 108364

[10] Humphries L L, Beeny B A, Gelbard F, Louie D L, Phillips J, Schmidt R C and Bixler N E 2021 MELCOR Computer Code Manuals, Vol.1: Primer and Users' Guide, Version 2.2.19018

[11] Humphries L L, Beeny B A, Gelbard F, Haskin T, Louie D L, Phillips J, Schmidt R C and Bixler N E 2021 MELCOR Computer Code Manuals, Vol. 2: Reference Manual, Version 2.2.18019

[12] HOLTEC International 2010 *Final Safety Analysis Report (FSAR) for the HI-STORM 100 Cask System*

[13] Angelucci M, Herranz L E and Paci S 2021 Thermal analysis of HI-STORM 100S dry cask with the MELCOR code *J. Phys. Conf. Ser.* **1868**

[14] Angelucci M, Paci S, Feria F and Herranz L E 2022 Uncertainty and Sensitivity Analysis of a Dry Cask for Spent Nuclear Fuel *Energies*

[15] Wagner J C and Dehart M D 2000 Review of axial burnup distribution considerations for burnup credit calculations (ORNL/TM-1999/246)

[16] Idelchik I E, Ginevskiy A S, Kolesnikov A V, Malyavskaya G R and Shveyeva N K 2008 *Handbook of Hydraulic Resistance, 4th Edition Revised and Augmented* (Begell House)

[17] Yun T S, Jeong Y J and Youm K S 2014 Effect of surrogate aggregates on the thermal conductivity of concrete at ambient and elevated temperatures *Sci. World J.* **2014**

[18] Asadi I, Shafigh P, Abu Hassan Z F Bin and Mahyuddin N B 2018 Thermal conductivity of concrete – A review *J. Build. Eng.* **20** 81–93

[19] Cavalline T L, Gallegos J, Castrodale R W, Freeman C, Liner J and Wall J 2021 Influence of Lightweight Aggregate Concrete Materials on Building Energy Performance *Build. 2021, Vol. 11, Page 94* **11** 94

Towards a more realistic MELCOR model for a dry cask for spent nuclear fuel. Part II: application.

M Angelucci[1,*] **and S Paci**[1]

[1] Department of Industrial and Civil Engineering, University of Pisa, Largo Lucio Lazzarino, Pisa, Italy

* Corresponding author: michela.angelucci@ing.unipi.it

Abstract. Nowadays, a great deal of attention is devoted to the development of best-estimate models able to produce more realistic outcomes. This is also the case for system codes, such as MELCOR, that are being mostly used in a conservative way especially when dealing with the licensing process. The above-mentioned need for more realistic results is at the core of this two-paper series related to the creation of a more accurate MELCOR model for the HI-STORM 100S dry cask. The findings obtained from the sensitivity studies carried out in the Part I are leveraged to set up an improved MELCOR model, the characteristics of which are consistent with the typical features of Spent Nuclear Fuel (SNF), and with geometrical and material properties of the cask itself. The addition of an axial power profile in the Fuel Assembly (FA), the better characterization of the flow losses in the air gap between internal metallic canister and external concrete-based overpack, and the choice of an appropriate value for the concrete thermal conductivity, are taken into account conjointly in this Part II. The outcomes from the improved MELCOR simulation are reported mainly in terms of the Peak Cladding Temperature (PCT), being the variable under regulatory surveillance. However, in addition to PCT, calculated temperature profiles are displayed and compared against the ones resulting from the previous model.

Keywords: *Dry cask, spent nuclear fuel, MELCOR, sensitivity, safety.*

1. Introduction

In the early days, highly conservative design and safety analyses were performed to compensate the lack of adequate knowledge and the limitations in the modelling of some physical phenomena. However, conservative assumptions and conservative computer codes do not always lead to conservative results: indeed, over-conservative analysis might produce misleading or even unrealistic results [1]. For this reason, later on, with the development of improved simulation tools and with research programs filling in some knowledge gaps, a shift towards less conservative approaches was carried out. More realistic evaluations were foreseen also in the safety analysis of Emergency Core Cooling Systems (ECCS) [2].

Following this path, a number of studies were performed, during the past decades, to promote Best-Estimate (BE) analyses rather than conservative ones. In particular, Best Estimate Plus Uncertainty (BEPU) methodologies and their application in different areas, from thermal-hydraulics to Severe Accidents, have been largely investigated [3–11]. Furthermore, another option beyond BEPU has been introduced in [12], where a realistic approach with BE computer codes, BE assumptions and BE initial and boundary conditions (with no request for an uncertainty assessment) is foreseen.

Considering what said above, the present paper (that is the second of a two-paper series) pursues the need for more realistic results, when evaluating the thermal performance of a dry storage system for Spent Nuclear Fuel (SNF). More precisely, an improved model for the HI-STORM 100S system [13] is set up starting from a previously developed model [14], created employing the MELCOR code [15,16] and verified against a Computational Fluid Dynamic (CFD) model of the same system [17].

The improved model has been obtained exploiting the findings of the sensitivity studies carried out in the Part I [18] of this two-paper series. In particular, the three addressed factors (namely the axial power profile for the Fuel Assembly (FA), the loss coefficients in the air gap between the internal metallic canister and the external concrete-based overpack, and the thermal conductivity of the concrete itself) have been taken into account conjointly in this Part II. Considering that the objective is to have a more accurate and realistic model, the three factors mentioned above have been characterized as to have consistency with the typical features of the SNF, and to be coherent with geometrical and material properties of the system itself.

Results will be presented in comparison with outcomes from the previous model. Being the thermal analysis of utmost importance for the safety of the system [19], temperature profiles of the main components of the cask will be reported, and the new safety margins addressed.

2. Modelling

The modelling choices for the improved MELCOR model are described in the following as well as the rationale behind each decision made. However, before moving to the modelling, a summary of the findings from the sensitivity studies performed in Part I are reported.

2.1. Recap of Part I

As said before, the sensitivity studies addressed in Part I involved three main factors:

- the axial power profile for the FA;
- the loss coefficients in the air channel between the internal metallic canister and the external concrete-based overpack;
- the thermal conductivity of the concrete itself.

As for the axial power profile, a non-uniform profile with a shape peculiar to the majority of the Pressurized Water Reactor (PWR) FAs were considered to take into account the fuel operating history. As a result, a slight variation in the axial temperature distribution of the fuel is observed, with a 10 K rise in temperature in the central zone of the FA and a little reduction at the extremities. As a consequence, the calculated maximum temperature reached in the fuel cladding, namely the Peak Cladding Temperature (PCT), presents a minor decrement (2 K).

For what concerns the loss coefficients in the air channel, three different cases were run: for all three of them, the loss coefficients for the Flow Paths (FPs) defined in the vertical (annular) channel were imposed as 0.1, whereas for inlet and outlet FPs three values were selected, one for each case (1.0, 2.0 and 4.0). The variation of the loss coefficients produces a change in the heat extraction capability: when the loss coefficients are at their minimum, mass flow rate in the air channel rises, and lower temperatures are calculated for both air in the channel and fuel. On the other hand, when considering high loss coefficients, results are almost the same as in the original model. With regard to the PCT, a 15 K reduction is shown for the case with the lower loss coefficients, as a consequence of an enhanced heat removal from the canister.

As regards the thermal conductivity of the concrete in the overpack, two "extreme" values (0.7 W/m-K and 2.1 W/m-K) were analyzed following a literature review [20–22] for plain concrete. As expected, simulation outcomes confirmed that the thermal conductivity of the concrete mainly affect the temperature in the overpack, whilst it has almost no impact on the PCT.

2.2. Modelling Choices

Starting from the model presented in [14], an improved MELCOR model for the HI-STORM 100S system is implemented. Considering the outcomes of the sensitivity studies in Part I and having in mind the creation of a model as consistent to the reality as possible, modelling choices were made as presented below:

- Despite the impact of the addition of the axial power profile for the FA being localized at the fuel zone and limited in terms of temperature difference, it was decided in favour of it. The decision is based on the idea of having a model as realistic as possible. In fact, the fuel axial power profile is non-uniform along the entire fuel life. In particular, for a PWR, the initial quasi-cosine shape for the axial power profile is eventually substituted by a flattened curve with sharp ends, as an indication of a fuel more burned in the central zone and less burned at the extremities. Moreover, a slightly higher axial power is observed in the lower part of the FA than at the top, due to differences in the moderator (water) density.
  In Table 1, the axial power values as entered in the MELCOR model are reported for each axial level containing fuel. Values are in agreement with what enounced above.

Table 1. Axial Power Profile for the Fuel as imposed in MELCOR.

Axial Level	4	5	6	7	8	9	10	11
Relative Power Density	0.68	1.0425	1.1225	1.1175	1.1075	1.075	0.85	0.5125

- Having a lower PCT would be optimal in terms of increasing the safety margins. However, the case with the smallest values for the loss coefficients in the FPs (and, consequently, with the lowest PCT) was not chosen for the improved model: the loss coefficients selected for such case are deemed inadequate to represent the geometry of the channel and of the inlet and outlet ducts, especially. In fact, whilst a value of 0.1 is appropriate for a correct reproduction of the vertical flow of the air in the vertical annular gap between the external canister wall and the internal wall of the overpack, a value of 1.0 is not suitable for the proper representation of the form losses due to the combination of entrance/exit effects plus the direction changes. This is particularly true considering that the inlet and outlet FPs are meant to represent the actual 4 inlets and 4 outlets of the real cask.
  For the above mentioned reason, the loss coefficients for inlet and outlet FPs have been imposed as 2.0.
  Table 2 summarizes the loss coefficients chosen for the improved model.

Table 2. Loss Coefficients in the Air Channel.

	Air Inlet	Vertical Flow Paths	Air Outlet
Loss Coefficients	2.0	0.1	2.0

- Notwithstanding the fact that concrete thermal conductivity seems to have almost no influence on the PCT, it still plays a role in the heat transfer within the overpack. However, since the only requirements for the plain concrete in the overpack are its density and its compressive strength [13], it was decided to not to employ any extreme value for its thermal conductivity and to leave it as in the original case (1.4 W/m-K).

3. Results

As said before, the thermal analysis is considered a fundamental part in the safety assessment of a dry storage for SNF. Therefore, in the following, results will be presented mainly in terms of temperatures for the various components, taking into account the regulatory limits as reported in [13,23]. Comparisons between the original and improved MELCOR models outcomes will be displayed.

In view of the safety importance of the maximum temperature in the fuel, the PCT is reported first. As it can be seen in Table 3, the two models agree on the location of the PCT, that is in the upper part (11th axial level) of the central FAs. However, with the improved model, a reduction of around 11 K is foreseen, mostly due to the increased heat extraction by the air in the channel (that, in turn, is due to values chosen for the more realistic characterization of the loss coefficients in the air channel). This reduction in the PCT points to a broadening of the safety margins, with the maximum temperature being now more than 50 K lower than the regulatory limit for normal operating conditions (673.15 K).

Some differences can be observed also in the axial temperature profile for the central (hotter) FAs (Figure 1), where the influence of the use of the axial power profile is visible in the shape of the axial temperature profile, with extremities presenting a reduction with respect to the original model. Even so, the effect of the axial power profile in the central zone of the FAs is counterbalanced by the effect of a higher heat extraction from the canister, and temperatures are almost identical to the original model.

Table 3. PCT - comparison.

	BaseCase	New Model
PCT (K)	630.5	619.3
Axial Location	11th (central FAs)	11th (central FAs)

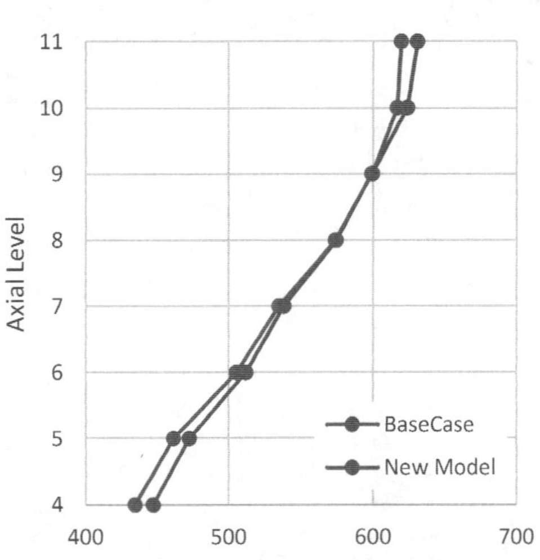

Figure 1. Axial Temperature Profile - Fuel Cladding - comparison.

The effect of the larger heat removal capability of the air in the channel can be observed also in the axial temperature profiles for the canister wall and for the air channel itself. As it can be seen in Figure 2, lower temperatures are obtained with the improved model in all axial locations, with the ΔT between outcomes by the original model and the improved one increasing along the axial direction. In particular, a ΔT of almost 15 K is reported for the upper part of the steel canister, with the predicted maximum temperature (402.2 K) being well below the maximum allowed temperature for the stainless stell in the canister in normal conditions (658.15 K [13]). As for the air in the channel, instead, the air temperature rises moving upwards toward the hotter part of the canister, with the maximum reached temperature being more than 10 K below the one calculated in the original model.

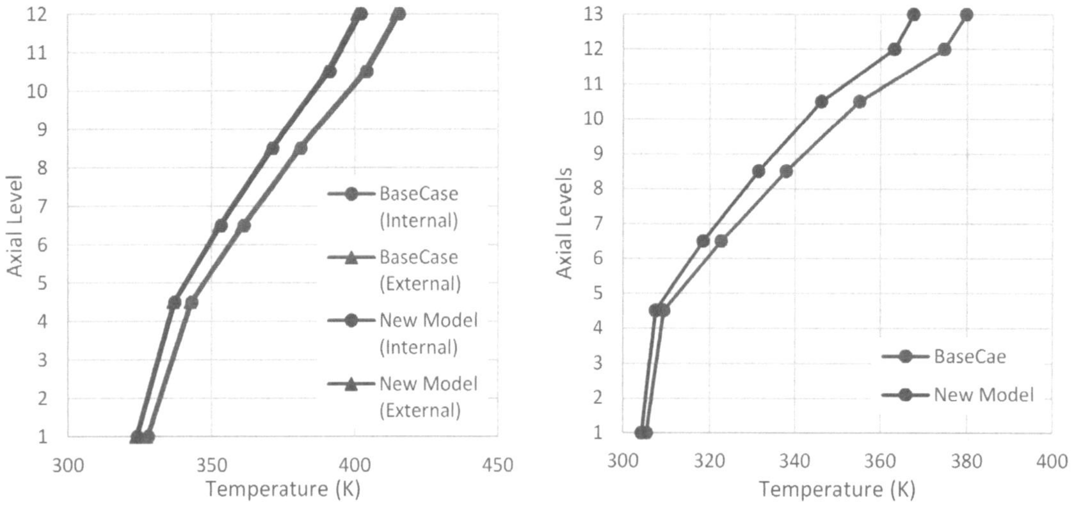

Figure 2. Axial Temperature Profile – Canister (left) and Air (right) - comparison.

Finally, to go a step further in the description of the performance of the system, attention is paid to the thermal behavior of the overpack. In Figure 3, the axial temperature profile of the concrete is displayed, for both the inner (hotter) and outer (colder) surfaces, together with the radial temperature profile within the overpack at the axial level corresponding to the PCT (11th axial level). As it can be seen from the radial profile, there is almost no temperature drop within the thin carbon steel liners surrounding the thick concrete wall. On the other hand, a ΔT of 50 K and 60 K can be observed within the concrete for the improved and original models, respectively. The difference in the temperature profile is not due to the thermal conductivity of the concrete, that is the same for both simulations, but rather to the lower predicted temperatures at the inner side of the overpack as a result of the higher heat removal by natural convection in the air channel.

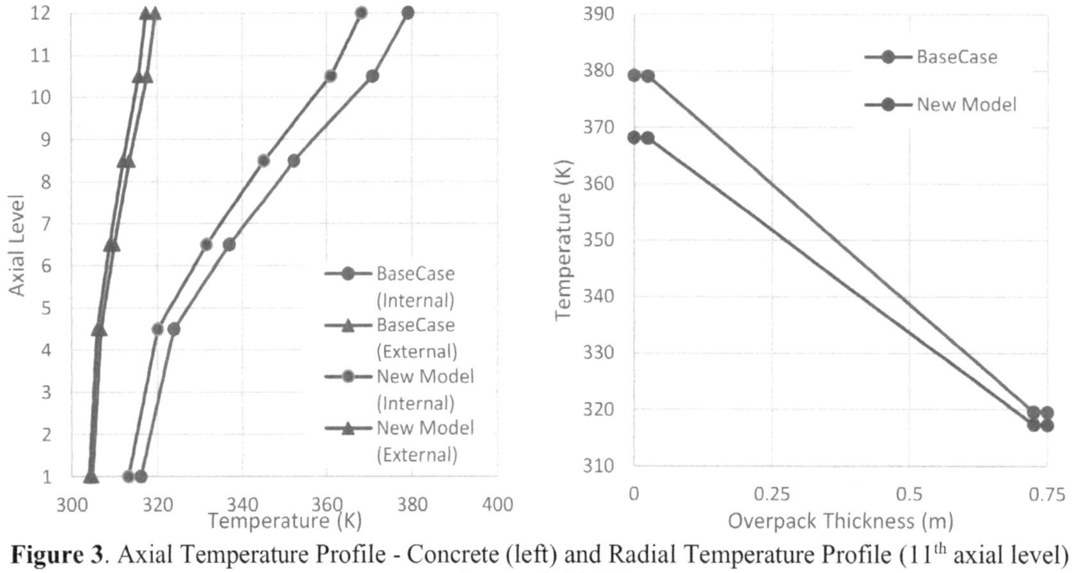

Figure 3. Axial Temperature Profile - Concrete (left) and Radial Temperature Profile (11th axial level) - Overpack (right).

4. Conclusions

The present paper carries on with the work started in Part I, where a number of sensitivity studies was performed as a first step in the creation of a more realistic model for the HI-STORM 100S storage system for SNF. In this framework, the findings from Part I are leveraged to set up an improved MELCOR model, the essential trait of which is to be as representative as possible of the typical characteristics of the SNF and of the geometry of the system.

To do so, a non-uniform axial power profile typical of a PWR FA has been implemented into the model, and the FPs associated with the air channel have been better characterized by imposing more suitable pressure loss coefficients.

Being temperatures of utmost importance for the safety of the cask system, the results have been analyzed mainly in terms of temperature profiles, paying attention to the limits imposed in the regulations.

A number of observations can be derived from the calculation outcomes:

- A PCT of 619.3 K is obtained, with a safety margin of around 50 K with respect to the regulatory limit for normal conditions. The PCT value is 11 K lower than the one from the original calculation. The improved (and more realistic) BE model results in less conservative outcomes.
- The axial temperature profile for the hotter FA is the expression of the combined effect of the presence of the axial power profile for the fuel and the increased heat removal from the air. Temperatures at the extremities are now lower than in the original case, as can be also inferred from the lower value of the PCT.
- Trends in the axial temperature profile for the different components are the same as in the original model, but lower temperatures are foreseen. This behavior is, again, due to the higher heat removed through natural circulation of air in the channel.
- The characterization of the loss coefficients in the air channel seems to play a major role in the thermal performance of the system. As mentioned above, results show that their values have a non-negligible influence on the capacity of the system to extract heat from the cask and, consequently, on the temperatures of any of the components of the system itself.
- Maximum calculated temperatures for fuel cladding, stainless steel in the canister and concrete in the overpack are all well below the regulatory limits.

Summarizing, it can be said that an improved (and more realistic) MELCOR model was implemented for the selected dry storage system. Being the proposed results consistent with the expected behavior of the system, the model seems suitable for the reproduction of its thermal performance, especially in cases in which no detailed fluid-dynamics analysis is needed.

5. References

[1] Jamali K 2015 Achieving reasonable conservatism in nuclear safety analyses *Reliab. Eng. Syst. Saf.* **137** 112–9

[2] U.S. NRC 1988 Compendium of ECCS [Emergency Core Cooling Systems] research for realistic LOCA [loss-of-coolant accidents] analysis: Final report

[3] Bucalossi A 2008 *Current use of Best Estimate plus Uncertainty methods on operational procedures addressing normal and emergency conditions (JRC Report)*

[4] D'Auria F, Camargo C and Mazzantini O 2012 The Best Estimate Plus Uncertainty (BEPU) approach in licensing of current nuclear reactors *Nucl. Eng. Des.* **248** 317–28

[5] D'Auria F 2019 Best Estimate Plus Uncertainty (BEPU): Status and perspectives *Nucl. Eng. Des.* **352** 110190

[6] Rohatgi U S and Kaizer J S 2020 Historical perspectives of BEPU research in US *Nucl. Eng. Des.* **358** 110430

[7] Martin R P and Petruzzi A 2021 Progress in international best estimate plus uncertainty analysis methodologies *Nucl. Eng. Des.* **374** 111033

[8] Zhang J and Schneidesch C 2023 Application of the BEPU safety analysis method to quantify margins in nuclear power plants *Nucl. Eng. Des.* **406** 112233

[9] Mazgaj P, Darnowski P, Kaszko A, Hortal J, Dusic M, Mendizábal R and Pelayo F 2022 Demonstration of the E-BEPU methodology for SL-LOCA in a Gen-III PWR reactor *Reliab. Eng. Syst. Saf.* **226** 108707

[10] Herranz L E, Beck S, Sánchez-Espinoza V H, Mascari F, Brumm S, Coindreau O and Paci S 2021 The EC MUSA project on management and uncertainty of severe accidents: Main pillars and status *Energies* **14** 1–11

[11] Angelucci M, Paci S, Feria F and Herranz L E 2022 Uncertainty and Sensitivity Analysis of a Dry Cask for Spent Nuclear Fuel *Energies*

[12] IAEA 2019 *Deterministic Safety Analysis for Nuclear Power Plants - IAEA SSG-2 Rev. 1*

[13] HOLTEC International 2010 *Final Safety Analysis Report (FSAR) for the HI-STORM 100 Cask System*

[14] Angelucci M, Herranz L E and Paci S 2021 Thermal analysis of HI-STORM 100S dry cask with the MELCOR code *J. Phys. Conf. Ser.* **1868**

[15] Humphries L L, Beeny B A, Gelbard F, Louie D L, Phillips J, Schmidt R C and Bixler N E 2021 MELCOR Computer Code Manuals, Vol.1: Primer and Users' Guide, Version 2.2.19018

[16] Humphries L L, Beeny B A, Gelbard F, Haskin T, Louie D L, Phillips J, Schmidt R C and Bixler N E 2021 MELCOR Computer Code Manuals, Vol. 2: Reference Manual, Version 2.2.18019

[17] Herranz L E, Penalva J and Feria F 2015 CFD analysis of a cask for spent fuel dry storage: Model fundamentals and sensitivity studies *Ann. Nucl. Energy* **76** 54–62

[18] Angelucci M and Paci S 2023 Towards a more realistic MELCOR model for a dry cask for spent nuclear fuel. Part I: sensitivity studies. *UIT2023*

[19] U.S. NRC 2010 *Standard Review Plan for Spent Fuel Dry Storage Systems at a General License Facility Final Report Office of Nuclear Material Safety and Safeguards, NUREG-1536*

[20] Cavalline T L, Gallegos J, Castrodale R W, Freeman C, Liner J and Wall J 2021 Influence of Lightweight Aggregate Concrete Materials on Building Energy Performance *Build. 2021, Vol. 11, Page 94* **11** 94

[21] Asadi I, Shafigh P, Abu Hassan Z F Bin and Mahyuddin N B 2018 Thermal conductivity of concrete – A review *J. Build. Eng.* **20** 81–93

[22] Yun T S, Jeong Y J and Youm K S 2014 Effect of surrogate aggregates on the thermal conductivity of concrete at ambient and elevated temperatures *Sci. World J.* **2014**

[23] U.S. NRC 2003 *Interim Staff Guidance-11, Revision 3 Issue: Cladding Considerations for the Transportation and Storage of Spent Fuel*

Comprehensive modeling of ventilation systems for Nearly Zero Energy Buildings

R Sedoni[1,*], G Cannistraci[1], P E Santangelo[1,2], D Angeli[1,3], M Romani[4], L Fioravanti[4]

[1] DISMI - Dipartimento di Scienze e Metodi dell'Ingegneria, Università degli studi di Modena e Reggio Emilia, Reggio Emilia (Italy)
[2] Centro interdipartimentale per la ricerca InterMech – MO.RE., Modena (Italy)
[3] Centro Interdipartimentale En&Tech, Reggio Emilia (Italy)
[4] Zehnder Group – Competence Center Campogalliano, Campogalliano (Italy)

* Corresponding author: 225791@studenti.unimore.it

Abstract. In the present work, a lumped-parameter model of a multifunctional ventilation unit for residential applications was developed in the Simulink environment, also relying on the Simscape toolbox with Moist Air and Two-Phase fluid libraries. A simple, yet effective method to analyze and optimize the efficiency of the combined HVAC – air distribution system is proposed. To investigate the impact of boundary conditions on system performance, a parametric study of different installation conditions was also carried out, including outdoor air and indoor air temperature, humidity, static pressure, filter fouling, pressure drop in the intake and distribution ducts. The model highlights a strong decrease in the useful cooling/heating heat flow rate produced by the system as the installation and maintenance conditions become more challenging.

Keywords: lumped-parameter model, multifunctional ventilation unit, moisture transport, ventilation mode, filter

1. Introduction

Nowadays HVAC systems play a crucial role in maintaining indoor air quality, thermal comfort, and energy efficiency in buildings. With the increasing focus on sustainability and energy efficiency, there is a growing need to improve HVAC systems and their operation. HVAC systems consume a significant amount of energy in buildings, and their inefficient operation can result in wasted energy, increased carbon emissions, and increased energy costs [1]. Therefore, improving the efficiency of HVAC systems has become an important research topic.

In the past, numerous efforts have been made by researchers for the development of simple analytical models to study some fundamental components of air treatment systems, including the most important one: the enthalpy recovery exchanger [2,3], and the diffusivity characteristics of its permeable membrane [4,5].

Studies have also been conducted to analyze the effect of specific boundary conditions on the performance of standard ventilation units. Manz et al. [6] investigated the performance of a single room ventilation unit with recuperative or regenerative heat recovery exchangers. Filis

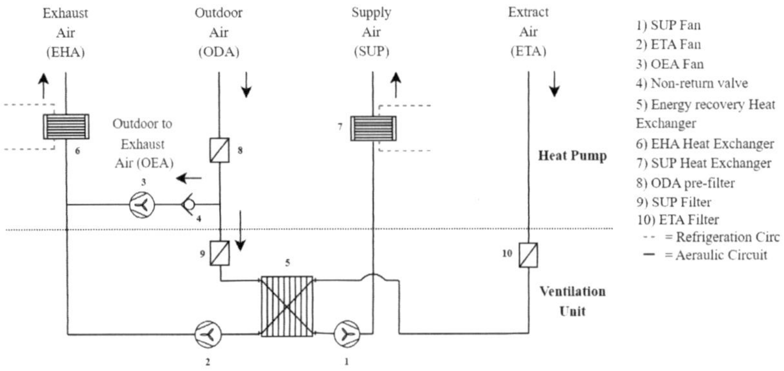

Figure 1. Full schematic of the HVAC unit investigated

et al. [7] investigated the effects of wind pressure and stack effect on total sensible heat recovery and supply air temperatures when using different fan types, airflow rates, and fan operating points.

This paper presents a comprehensive analysis of an innovative HVAC commercial unit for nZEB buildings, using a simple approach that can take into account all the possible effects to which the system may be subjected. For this purpose a new Simscape block for the Moist Air library was developed to model the enthalpy heat exchangers. The study specifically focuses on the critical aspects linked to the inclusion of an additional fan aimed at improving the unit performance.

2. System description

The system under consideration combines a balanced mass/volume flow ventilation unit with an air-to-air heat pump to optimize energy consumption and provide thermal comfort in residential buildings. This system works differently from standard products, in that the ventilation unit operates at a flow rate ranging between 0 and 450 m^3h^{-1}, while the heat pump unit operates with a balanced mass/volume flow rate ranging between 0 and 600 m^3h^{-1}, thanks to the contribution of a third fan (OEA) in the heat pump unit. Outdoor air (ODA) is extracted from the environment by the ventilation unit fan, and then conveyed through the heat pump ducts, preheater, energy recovery ventilation (ERV) or heat recovery ventilation (HRV) heat exchanger, and the evaporator or condenser heat exchanger of the heat pump. The resulting supply air (SUP) mass/volume flow is then delivered to the building where the system is installed. Another part of the circuit deals with extracted air (ETA) from the internal environment, which is discharged to the outdoor, called exhaust air (EHA), after passing through the ERV or HRV and heat exchanger of the HP. To increase the heat transfer rate at the evaporator (in heating mode) or condenser (in cooling mode), a third fan has been introduced in the heat pump. This fan draws air from the environment through the supply duct and drives it directly into the EHA duct upstream of the heat exchanger through a bypass line, thus increasing the EHA mass/volume flow rate. It is important to note that the OEA fan has specific operating limits, such as a maximum revolution speed of 2500 rpm, and a limited dimension. These limits were implemented to meet the design requirement of reducing and controlling noise levels, ultimately ensuring the acoustic comfort of the occupants. A full schematic view of the aeraulic circuit system is shown in Figure 1.

3. Model development

3.1. Aeraulic circuit

Regarding the aeraulic side of the circuit, the pressure drops of the individual components that make up the circuit (such as air ducts, filters, check valve, ERV/HRV, and T-junctions) are calculated using the known relationship:

$$\Delta p = R\dot{m}^2 \tag{1}$$

Where R is the specific resistance of the component, and $\dot{m}$ is the mass flow through the component. The values of R are commonly obtained from the technical data sheets of individual components. The 'Fan (MA)' [8] block is used to model the fans operating curves and the 'Condenser-Evaporator (2P-MA)' [8] block is used to model the two exchangers of the heat pump (EHA and SUP).

3.2. ERV/HRV modeling

ERV/HRV heat exchangers typically consist of a series of rectangular channels that primarily operate in counterflow. However, there is a small portion of the exchanger in cross-flow configuration. Despite of this portion, considering only the counterflow part of the heat exchanger for determining the exchange characteristics does not lead to a significant error.

A pre-defined type of heat exchanger similar to this one, currently, is not available in Simscape, so it was necessary to reconstruct it starting from the base block, i.e. 'Pipe (MA)' [8].

To simplify the modeling process, the overall heat transfer characteristics of the exchanger are determined based on a single channel (fig. 2) using a counterflow model approach and then extrapolated to the entire heat exchanger. The positive heat and moisture transfer direction is assumed from the ETA to the SUP side (control volume 3 to 1).

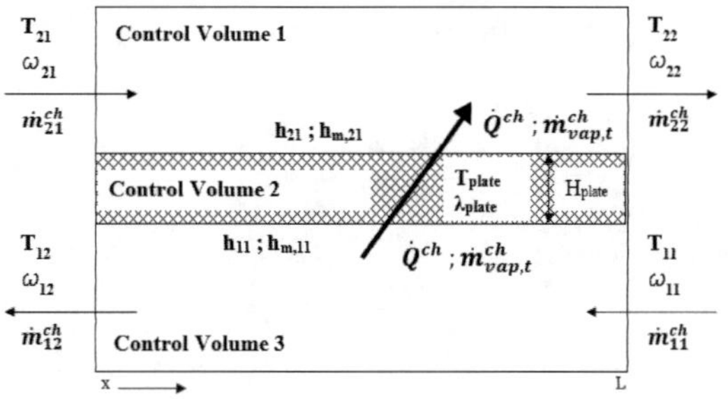

Figure 2. Counterflow model approach

3.2.1. Sensible Load

Regarding the sensible heat recovered in the exchanger, the effectiveness-NTU method has been implemented. The equations for the sensible heat recovery are derived from the available literature [2].

The thermal power exchanged between two channels ($\dot{Q}^{ch}$) is:

$$\dot{Q}^{ch} = \epsilon_s \dot{Q}^{ch}_{max} = \epsilon_s (\dot{m}^{ch} cp)_{min} (T_{11} - T_{21}) \qquad \epsilon_s = \frac{1 - e^{-NTU_s(1-C_r)}}{1 - C_r e^{-NTU_s(1-C_r)}} \tag{2}$$

Where ϵ_s is the sensible effectiveness for a counterflow exchanger, T_{11} is the inlet ETA flow temperature, T_{21} is the inlet SUP flow temperature, $\dot{Q}^{ch}_{max}$ is the maximum recoverable heat flow rate per channel and $\dot{m}^{ch}_{11}$ and $\dot{m}^{ch}_{21}$ are the inlet mass flow rate per channel, on the ETA side and SUP side respectively.

Given the very small size of the channels, for this type of ERV/HRV exchanger, the flow is laminar (Re = 90÷ 110), so the relationship derived by Shah et al. [2] is used to calculate the Nusselt number in rectangular channel heat exchangers:

$$Nu = 8.235[1 - 2.0421(\alpha^*) + 3.0853(\alpha^*)^2 - 2.4765(\alpha^*)^3 + 1.0578(\alpha^*)^4 - 0.1861(\alpha^*)^5] \quad (3)$$

Where α^* is the ratio between the two dimensions (a,b) of the channel cross-section. For a square channel $\alpha^* = \frac{a}{b} = 1$ and therefore Nu = 3.608.

Multiplying by the total number of channels per side ($n_{channel}$), it is possible to calculate the overall heat exchange in the ERV/HRV ($\dot{Q}$):

$$\dot{Q} = n_{channel}\dot{Q}^{ch} \quad (4)$$

To simulate the thermal inertia of the heat exchanger membrane and thus obtain a more accurate modeling of thermal exchange during transients, a thermal mass has also been introduced in the model.

3.2.2. Latent load For an ERV-type heat recovery system, it is necessary to introduce an amount of recovered moisture. The equations for moisture recovery are derived mainly from the work of Križo et al. [4] , Min et al. [5] and Zhang et al. [9].

In a similar fashion to equation (2) it is possible to define the amount of moisture transferred between two channels ($\dot{m}^{ch}_{vap,t}$) as:

$$\dot{m}^{ch}_{vap,t} = \epsilon_l(\dot{m}^{ch}_{dry})_{min}(\omega_{11} - \omega_{21}) \quad (5)$$

Where $\dot{m}^{ch}_{dry}$ is the mass flow rate of dry air per channel, and ω_{11} and ω_{21} are the humidity ratios of the two flows. So the latent effectiveness (ϵ_l) can be calculated as:

$$\epsilon_l = \frac{1 - e^{-NTU_l(1-R)}}{1 - Re^{-NTU_l(1-R)}} \qquad R = \frac{\dot{m}^{ch}_{min}}{\dot{m}^{ch}_{max}} \qquad NTU_l = \frac{U_l A^{ch}}{\dot{m}^{ch}_{min}} \quad (6)$$

The total moisture transfer resistance (U_l) is:

$$U_l = \left(\frac{1}{h_{m,11}} + \frac{H_{plate}}{P_m} + \frac{1}{h_{m,21}}\right)^{-1} \qquad P_m = \frac{C D_{wm}\omega_{max}e^{5294/T}}{10^6(1 - C + \frac{C}{RH})^2 RH^2} \quad (7)$$

Where H_{plate} is the thickness of the membrane, P_m is the membrane permeability, D_{wm} is the moisture diffusivity in membrane, ω_{max} represents the maximum moisture content of the water vapor in the membrane and C gives the shape of the sorption curve. For simplicity, these properties were assumed to be constant in the model, and the values at 20°C were used, allowing for a small error. Temperature T and relative humidity RH in equation (8) are chosen as the average values on the plate surface. The convective moisture transfer coefficient h_m can be obtained using the Chilton–Colburn analogy from the convective heat transfer coefficient, the Lewis number (Le is around 1.15 for air) [3], and from ρ and c_p of the air flow:

$$h_m = \frac{h}{\text{Le}^{2/3}\rho c_p} \quad (8)$$

Finally the total transferred moisture is:

$$\dot{m}_{vap,t} = n_{channel}\dot{m}^{ch}_{vap,t} \quad (9)$$

Table 1. Experimental vs Model data

$\dot{m}$	$\epsilon_{sensible}$			ϵ_{latent}		
[kg/s]	**Experimental**	**Model**	**Error [%]**	**Experimental**	**Model**	**Error [%]**
0.0155	0.968	0.97	0.21	0.943	0.933	-1.06
0.1002	0.814	0.836	2.70	0.666	0.68	2.10
0.1417	0.779	0.782	0.39	0.589	0.6	1.87

3.2.3. ERV/HRV Model Validation The experiments for validation were carried out with T_{ODA}=2°C, RH_{ODA}=82% and T_{ETA}=20°C, RH_{ETA}=61%:

Where experimental effectivenesses are calculated as:

$$\epsilon_s = \frac{(\dot{m}c_p)_{21}(T_{22} - T_{21})}{(\dot{m}c_p)_{min}(T_{11} - T_{21})} \qquad \epsilon_l = \frac{(\dot{m})_{21}(\omega_{22} - \omega_{21})}{(\dot{m})_{min}(\omega_{11} - \omega_{21})} \tag{10}$$

As shown in Table 1, since the error of the ERV/HRV model is contained within acceptable limits, it can be considered as validated.

3.3. Aeraulic model calibration

The aeraulic model is calibrated according to experimental results provided by a notified body, that tested the system in the past. Experimentally, a volumetric flow rate of 227 m^3h^{-1} was set for extraction and supply air, the pressure drops were fixed on the external ducts, and the rotation speeds of the three fans were measured.

Table 2. OEA Volume flow rate comparison (Model forecast vs Experimental data)

n Fan Model [rpm]	n Fan Experimental [rpm]	Volume flow OEA Model [m^3/h]	Volume flow OEA Experimental [m^3/h]	Percentage Error [%]
1930	1770	13.7	13	5.4
2060	1900	24.4	27	-9.6
2277	2117	46	48	-4.2
2477	2317	66	67	-1.5
2660	2500	85.3	85	0.3

We opted for a calibration approach based on the correction of the fan characteristic curves instead of the resistances; this appeared as the most convenient approach, given the high number of flow resistances involved and, on the other hand, the low number of fans. In this case, a correction factor of 160 rpm for the OEA fan allows for a faithful replication of the behavior of the real system. Similarly, a correction factor of 245 rpm was determined for the ETA fan, and 116 rpm were added to the SUP fan. These offsets were introduced to compensate uncertainties in the system:

- The correction on the OEA fan takes into account the uncertainties of the ODA, OEA, and EHA branch flow resistances, such as the ODA filter, the check valve, and T-junctions.
- In a similar fashion the correction on the ETA and SUP fans takes into account the uncertainties on the flow resistances of the ETA and SUP branches, and also accounts for the uncertainties on the fans operating curves, due to the presence of a scroll around the fans, designed to increase their performance.

It is possible to observe (Table 2) that the error decreases rapidly as the flow rate increases. This is mainly due to the fact that the dimensionless operating curve, at the small flow rates processed by the fan, deviates significantly from the actual operating fan curve. In this case, measurements were taken with deliberately high pressure drops on the external channels, to see the effect of these on the third fan; therefore the flow rate is in most cases, really small.

However, under nominal operating conditions, the system works with much higher flow rates. Therefore, the system can be considered to be calibrated with satisfactory accuracy, given the relatively low error on the higher flow rates.

4. Results and discussion

4.1. Effect of installation conditions

The focus of this paper is on the ventilation system and its performance; however a simple model of the thermodynamic cycle of the heat pump and of the refrigeration circuit is defined to complete the physical modeling of the whole system and perform sensitivity analyses. The description of the refrigeration model is omitted for brevity.

In this section, we analyze the results of the model by assessing different installation conditions:

Table 3. Installation conditions

	Installation		
	Ideal	Medium	Critical
ETA, SUP ducts lenght DN160 [m]	15	15	15
ODA, EHA ducts lenght DN200 [m]	1	0	0
ODA, EHA ducts lenght DN160 [m]	0	4	6
ETA, SUP number of bends 45°	2	4	6
ODA, EHA number of bends 45°	2	4	6
ETA, SUP Specific Flow Resistence [Pa s^2/kg^2]	10434	11868	13302
ODA, EHA Specific Flow Resistence [Pa s^2/kg^2]	721	5828	8462

The analyses were carried out in the winter season with $T_{ODA} = 2°C$, $RH_{ODA} = 50\%$, and indoor conditions of $T_{ETA} = 20°C$, $RH_{ETA} = 50\%$, $n_{OEA} = 2500$ rpm, $\dot{m}_{ETA} = \dot{m}_{SUP} = 0.025$ kg s^{-1} (~ 75 m^{3}h^{-1}).

The supply mass flow rate was purposely set as very low, precisely because the main advantage of this three-fan system lies in being able to provide high heat transfer rates even when the user requires reduced flow rates.

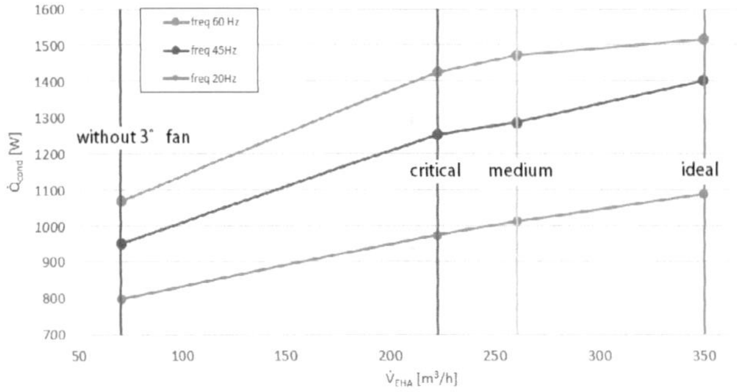

Figure 3. Condenser Heat transfer rate vs $\dot{V}_{EHA}$

For a compressor frequency of 45 Hz (fig. 3), it is observed that, for a constant supply flow rate, the presence of the OEA fan increases the heat transfer rate by 32% with respect to a traditional system in an ideal installation, owing to the increased flow rate at the evaporator. However, the heat transfer rate decreases as the external pressure drop on EHA and ODA channels increases, since the rotation velocity of the OEA fan is limited (due to noise) to 2500 rpm. Conversely, it has been ascertained that the impact of increasing external pressure losses

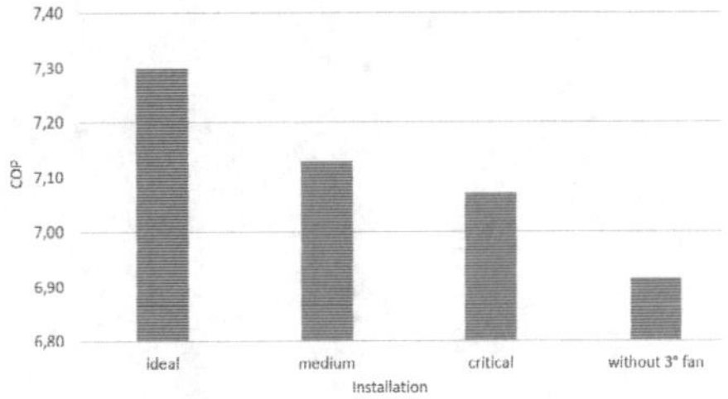

Figure 4. Variation of COP under different installations

on the ETA and SUP channels has an impact only on the ETA and SUP fans, which do not have limitations on their rotation velocity.

Finally, it can be observed from Figure 3 that the impact of reducing the available heat transfer rate to the condenser, as installation conditions vary, becomes increasingly severe at higher compressor frequencies.

An additional analysis that can be conducted is to study the influence of the OEA fan (and consequently, the installation) on the thermodynamic performance of the heat pump. In this case, the COP merely refers to the thermodynamic cycle.

The boundary conditions of the analysis under consideration are the same as in the previous case, apart from a supply flow rate of $\dot{m}_{ETA} = \dot{m}_{SUP} = 0.0752$ kg s^{-1} ($\sim$ 227 m^{3}h^{-1}). From Figure 4, a slight decrease in the COP is observed as installation conditions worsen.

To conclude the analysis, it is evident that the three-fan system presents significant advantages in terms of system performance. However, greater attention must be paid during the installation phase to ensure that these advantages are actually realized.

4.2. ODA Filter Fouling model (Maintenance)

The phenomenon of fouling within this work has been examined for the pre-filter located inside the ODA channel of the heat pump. This was done by evaluating the variation of system performance in terms of volumetric flow rate to the OEA fan, during the winter season (T_{ODA} = 2°C, RH_{ODA} = 50 %) for an ideal installation, with n_{OEA} = 2500 rpm. These analyses were carried out based on the following assumptions regarding fouling:

- The increase of pressure losses with time due to filter fouling is exponential [10].
- The maximum admissible value of pressure loss is three times the pressure loss of the clean filter [11].
- The considered period is 180 days from the installation of a new filter, after which fouling reaches the maximum admissible level.

A general pressure drop-fouling equation for the filter examined could then be defined:

$$\Delta p = R_{cleanfilter}(1 + e^{0.0294 \cdot d})\dot{m}^2 \tag{11}$$

Where d denotes the number of days past the installation, and $e^{0.0294 \cdot d}$ is the correction term accounting for filter fouling.

From the analysis of the data presented in Figure 5 (carried out with $\dot{V}_{SUP}$ = 227 m^{3}h^{-1}), it can be observed that, as time passes, the flow rate processed by the third fan in the heat pump (and consequently the heat transfer rate available at the evaporator) decreases by an amount which is not very high, but still significant if compared with the maximum expected

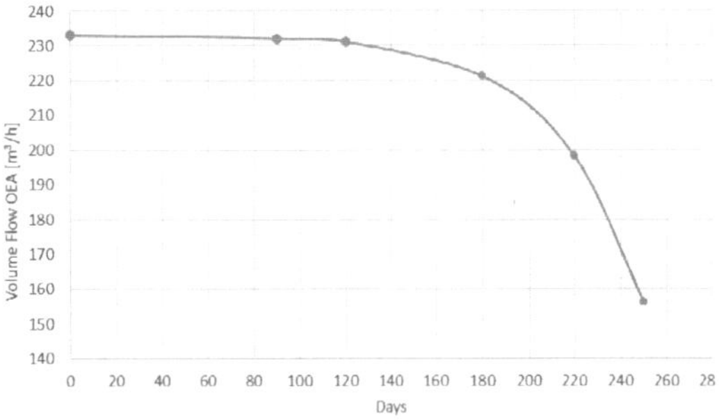

Figure 5. $\dot{V}_{OEA}$ fan degradation

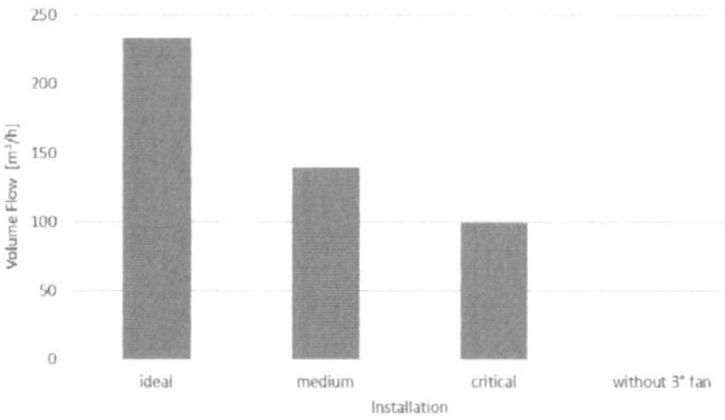

Figure 6. $\dot{V}_{OEA}$ vs Installation

performance. This behavior is due to the initial value of the filter pressure drop, which, being a large mesh filter, turns out to be quite small, even when triplicated at the 180th day.

After 180 days, the contribution of filter fouling begins to significantly degrade the system performance. For example, comparing Figure 6 and Figure 5, it can be observed that after 250 days the system behaves almost as if it were subjected to a medium installation, with all the consequences discussed in the previous section. It is therefore essential to respect the maximum 180-day interval between filter changes, as recommended by regulations, in order to always maintain acceptable system performance.

5. Conclusions

A comprehensive model of the air handling system of a ventilation machine coupled with a commercial heat pump has been successfully developed and experimentally calibrated. In particular, an ERV/HRV heat exchanger model was developed in Simscape and experimentally validated. A refrigeration circuit was then coupled to the system to show the effects of incorrect installation conditions, and thus the decay of the performance of the associated thermodynamic cycle.

6. References

[1] Dorer V and Breer D 1998 *Energy Build.* **27** 247–255
[2] Shah R K and Sekulic D P 2003 *Fundamentals of heat exchanger design* (John Wiley & Sons)
[3] Min J and Su M 2010 *Appl. Therm. Eng.* **30** 991–997
[4] Križo K, Kapjor A and Holubčík M 2022 *Energies* **15** 6021
[5] Min J, Hu T and Liu X 2010 *J. Membr. Sci.* **357** 185–191
[6] Manz H, Huber H, Schälin A, Weber A, Ferrazzini M and Studer M 2000 *Energy Build.* **31** 37–47
[7] Filis V, Kolarik J and Smith K M 2021 *Energy Build.* **234** 110689
[8] MathWorks URL `https://it.mathworks.com/help/simscape/physical-modeling.html`
[9] Zhang L Z, Liang C H and Pei L X 2008 *J. Membr. Sci.* **325** 672–682
[10] Eker O F, Camci F and Jennions I K 2016 *MSSP* **75** 395–412
[11] 2017 UNI EN ISO 16890-1:2017 Filtri d'aria per ventilazione generale Standard

Acknowledgement

We thank the Zehnder Group - Competence Center Campogalliano (MO) for the invaluable help throughout this research and for the resources made available to the purpose.

The role of SRF in achieving the EU landfill targets in the Umbria Region, Italy

Aron PAZZAGLIA[1], Beatrice CASTELLANI[2]*

[1] CIRIAF, University of Perugia, 06125 Via G.Duranti 67, Perugia, Italy
[2] CIRIAF, Department of Engineering, University of Perugia, 06125 Via G.Duranti 67, Perugia, Italy

* Corresponding author: beatrice.castellani@unipg.it

Abstract. In December 2015, the European Commission published the circular economy package "Closing the loop - An European action plan for the Circular Economy" with four quantitative objectives: long-term recycling targets for municipal waste (55% in 2025, 60% in 2030 and 65% in 2035) and landfill target of 10% by 2030. In Umbria (Italy) 43% of waste are landfilled in 2020, so the region is far from the European target. Hence, it is significant to analyse different strategies to improve the waste management in the region. In particular, the main issue is how to reduce waste landfill in the most efficient way. The SRF and waste-to-energy approaches are the most interesting to study as a substitute for landfilling. In the communication "The role of waste-to-energy in the circular economy", the European Commission clarifies the position covered by the different energy processes in the waste hierarchy and identifying the technologies with the highest potential in terms of efficient management of resources and environmental impact. Among the most efficient waste-to-energy technologies, the document mentions the gasification of recovered solid fuel (SRF) and co-incineration of the resulting synthesis gases in the combustion plant to replace fossil fuels in the production of electricity and heat and co-incineration in the cement production. The European Commission recognizes the important role of energy recovery in the transition to the circular economy, if this does not stop the improvement in recycling rates. In this paper, MSW management policy in Umbria (Italy) and possible solutions are discussed. The paper considers three scenarios in the management of MSW: i) direct combustion of residual waste in waste-to-energy plants; ii) combustion of SRF in waste-to-energy plants; iii) combustion of F-SRF in dedicated plants (cement plant). All three scenarios may be able to match the 10% landfill EU target.

Acronyms	Full form	Definition
MSW	Municipal solid waste	Waste generated by households, businesses, and institutions in urban areas
SRF	Solid recovered fuel	Solid fuel produced from from non-recyclable waste materials
F-SRF	Fuel – Solid recovered fuel	Category of SRF, in compliance with the provisions of Article 8, Paragraph 2 of Ministerial Decree 22/2013
MBT	Mechanical biological treatment plant	Facility that combines mechanical and biological processes to treat municipal solid waste.

 2685 (2024) 012046 doi:10.1088/1742-6596/2685/1/012046

1. Introduction

The Directive 2008/98/EC [1] establishes a hierarchy for waste management, with prevention as a priority and landfill disposal as the last choice. The European Commission's communication “The role of waste-to-energy in the circular economy” [2] recognizes the importance of these technologies in the transition to a circular economy, as long as they do not hinder recycling efforts. The most efficient waste-to-energy technologies include gasification of solid recovered fuel (SRF) and co-incineration in cement production.

The circular economy package "Closing the loop" [3] includes long-term recycling targets for municipal waste and a landfill target of 10% by 2030. In Umbria region (Italy), 43% of waste was landfilled in 2020 [4], so it is far from the European target and solutions are needed to lower this percentage. The study focuses on waste management in the Umbria Region and presents three alternative scenarios: i) direct combustion of residual waste in waste-to-energy plants; ii) combustion of SRF in waste-to-energy plants; iii) combustion of F-SRF in dedicated plants (cement plant). Finally, the paper compares the three scenarios in terms of their impact on landfill depletion, taking into account traditional fossil fuel and CO_2 savings.

2. Solid recovered fuel classification and production in Italy

2.1 Classification

Legislative Decree 205/2010 introduced, in Article 183, paragraph 1, letter cc) of Legislative Decree 152/2006, the definition of Solid Recovered Fuel (SRF): "the solid fuel produced from waste that meets the classification and specification characteristics identified of the UNI CEN/TS 15359 technical standards and subsequent amendments and additions; subject to the application of Article 184-ter, secondary solid fuel, is classified as special waste."

SRF is regulated at the European and national scales from two standards, UNI EN 15359:2011 [5] and BS EN ISO 21640:2021 [6] "Secondary Solid Fuels - Classification and Specifications," which sets out a classification system and scheme for defining the properties of SRF.

Fuel SRF (F-SRF) is a category of the various possible types of SRF, which has a classification and specification characteristics so that a declaration of conformity can be issued in compliance with the provisions of Article 8, Paragraph 2 of Ministerial Decree 22/2013 [7]. In this way, it is possible to distinguish SRF waste from SRF fuel. SRF fuel ends its waste status and becomes a product [1].

Hence, SRF produced from unsorted solid waste, depending on its characteristics, can be destined for waste-to-energy in dedicated plants or for co-firing in industrial plants, thus reducing the amount of material to be landfilled.

The classification of different types of SRF is based on the requirements of the technical standard UNI EN 15359 "Solid recovered fuels" (SRF), which considers 3 parameters:

- lower heating value (LHV). It is an index of energy value, expressed by the statistical measure of the average value, in MJ/kg;
- chlorine content (Cl). It is an index of the degree of aggressiveness on plants, expressed with the average value, in percentage units;
- mercury content (Hg) It is an index of the significance of environmental impact, expressed as the median or 80th percentile value, in mg/MJ.

For each of them, limit values are given to identify 5 classes, so this classification provides 125 different types of SRF depending on the combination of the three parameters and provides the user with immediate and clear information about the fuel. For the purposes of Ministerial Decree 22/2013, F-SRF must respect limits for LHV and Cl as defined by classes 1, 2, 3 and their combinations and, for the parameter Hg, as defined by classes 1 and 2 (table 1).

Table 1. SRF classification: green boxes show the characteristics needed to define F-SRF.

Parameter	Statistical measure	Unit of measurement	Classes				
			1	2	3	4	5
LHV	Mean	MJ/kg	≥ 25	≥ 20	≥ 15	≥ 10	≥ 3
Cl	Mean	% dry matter	≤ 0.2	≤ 0.6	≤ 1.0	≤ 1.5	≤ 3.0
Hg	Median	mg/MJ	≤ 0.02	≤ 0.03	≤ 0.08	≤ 0.15	≤ 0.50
	80th percentile	mg/MJ	≤ 0.04	≤ 0.06	≤ 0.16	≤ 0.30	≤ 1.00

Chemical and physical parameters are defined in Annex A, Part 1 of the UNI EN 15359 standard, expressed as the mean/median of the individual parameters. Only for F-SRF the Ministerial Decree 22/2013 defined thresholds (table 2).

Table 2. Chemical and physical thresholds. No limit values have been set for ash and moisture. Setting limit values for ash and moisture is left to specific agreements between manufacturer and user.

Parameter	Statistical measure	Unit of measurement	Thresholds
		Physical parameters	
Ash	Mean	% dry matter	-
Moisture	Mean	%	-
		Chemical parameters	
Antimony (Sb)	Median	mg/kg dry matter	50
Arsenic (As)	Median	mg/kg dry matter	5
Cadmium (Cd)	Median	mg/kg dry matter	4
Chromium (Cr)	Median	mg/kg dry matter	100
Cobalt (Co)	Median	mg/kg dry matter	18
Manganese (Mn)	Median	mg/kg dry matter	250
Nickel (Ni)	Median	mg/kg dry matter	30
Lead (Pb)	Median	mg/kg dry matter	240
Copper (Cu)	Median	mg/kg dry matter	500
Thallium (Tl)	Median	mg/kg dry matter	5
Vanadium (V)	Median	mg/kg dry matter	10
Σ metals [Sb,As,Cr,Cu,Co, Pb,Mn,Ni,V]	Median	mg/kg dry matter	-

Waste-to-energy plants can be considered as the proper destination of SRF, while F-SRF can be used in cement plants with a production capacity of more than 500 tons per day of clinker and/or thermal power plants with a thermal combustion capacity of more than 50 MW [7].

2.2 Production

The F-SRF production process comprises several stages that are operated in a mechanical biological treatment plant (MBT). Firstly, residual waste (unsorted waste from MSW) is sieved to divide fine fraction (organic waste) and coarse fraction (inorganic waste). The fine fraction undergoes a biological treatment (aerobic, anaerobic, or both) and then it is sent to landfill. Coarse fraction has the highest calorific value.

The material goes through an electromagnetic separator to remove any light ferrous fractions that can be recovered. Then, a parasitic current separator is used to extract any non-ferrous metallic fractions. An optical separator is then employed to reduce the presence of plastic fractions containing chlorine. Finally, the material is passed through a shredder-refiner to reduce its size before it is pressed obtaining F-SRF [8], [9]. SRF undergoes only sieving, electromagnetic separator, and parasitic current separator.

3. Data about Umbria Region, Italy

The total municipal solid waste (MSW) production in Umbria is ≈ 439 million kg in 2020, 66.2% of MSW was recycled. The residual waste represented 32.2% of MSW [4].

In 2020 in Umbria the residual waste was 141.5 million kg [10]. The management of residual waste involves MBT to reduce its organic content and weight, followed by landfill disposal. The first phase occurs at selection plants, which yield three fractions: metals, coarse fraction (inorganic dry material) sent to landfills, and fine fraction (organic wet material) biostabilized and then sent to landfills. In 2020, 99% (≈ 140 million kg) of the residual waste collected in Umbria underwent selection, resulting in 0.6% (≈ 849 miles kg) of metals sent for recovery, almost 61% (≈ 86 million kg) of coarse fraction disposed of in Umbrian landfills, almost 38% (≈53 million kg) of fine fraction biostabilized in Umbrian plants (with only 0.5% disposed of in biostabilization plants outside the region). Biostabilization produces biostabilized waste equal to the 24% (≈ 34 million kg) of starting residual waste. It is disposed in landfill. An amount of waste equal to the 37 % of the total production was disposed in landfill in 2020 (a total of 162.5 million kg) [4].

Considering this regional scenario, it is significant to study other pathways to treat residual waste in order to lower the amount of waste landfilled and reach European target of maximum 10% waste landfilled.

4. Materials and methods

4.1 Data

Data are collected in the biggest waste management company of the Region Umbria: when not available, literature data are used. Total amount of residual waste, coarse fraction and fine fraction waste, and LHV values are collected. Moreover LHV, chlorine, and mercury levels are shown in table 3 and taken from sample analyses to identify F-SRF code according to Ministerial Decree 22/2013.

4.2 Scenarios description

Three scenarios about residual waste management are described: i) direct combustion of residual waste in waste-to-energy plants; ii) combustion of SRF in waste-to-energy plants; iii) combustion of F-SRF in dedicated plants (cement plant). For these three pathways, energy production and CO_2 reduction are evaluated. A residual waste mass balance is analyzed. This study does not consider pollutants analysis, transportation and plant consumption cost.

5. Result and discussion

Three different samples of coarse fraction waste were characterized in order to measure their chlorine and mercury content and LHV value. All analysis were performed by ACCREDIA-accredited test laboratory n. 0181 L according to UNI CEI EN ISO/IEC 17025:2018 [11]. Table 3 shows the class code of coarse fraction waste analyzed in the considered mechanical treatment plant. The coarse fraction waste characterization defines a F-SRF with code 311. Residual waste LHV is 10 MJ/kg as received [12], [13].

Table 3. F-SRF identification by LHV, chloring, and mercury values following Ministerial Decree 22/2013.

LHV (MJ/kg) as received	17.9 ± 2.08	Class 3
Chlorine (% dry mass)	0.1795	Class 1
Mercury (median) as received	0.012	Class 1

5.1 First scenario

In the first scenario, residual waste is directly fired in a waste-to-energy plant. The outputs are 1415 million MJ, 88.5 million kg CO_{2eq} [12], 7 million kg ashes, 32 million kg of slags that are mostly recovered [14].

The disposed waste is 10.2 million kg. Figure 1 shows the flowchart of the process, simplified mass balance, energy and CO_2 output of the process.

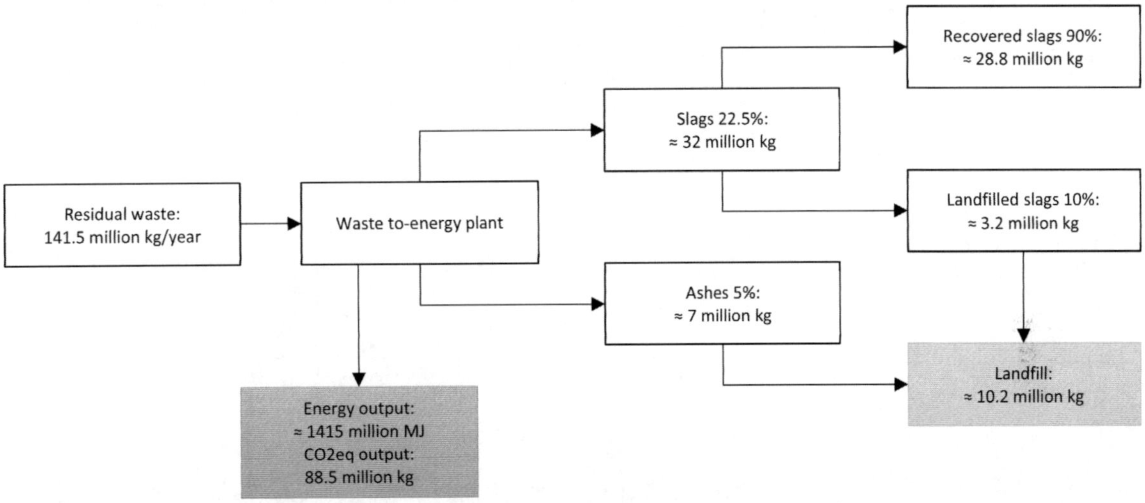

Figure 1. Residual waste directly fired in waste to energy plant

5.2 Second scenario

In the second scenario, residual waste undergoes a MBT. Fine fraction waste was biostabilized while coarse fraction is fired in waste-to-energy plant. Ashes and slags are produced as combustion residues. The disposed waste is equal to 40.2 million kg. Figure 2 shows the flowchart of the process, simplified mass balance, and energy and CO_2 output of the process.

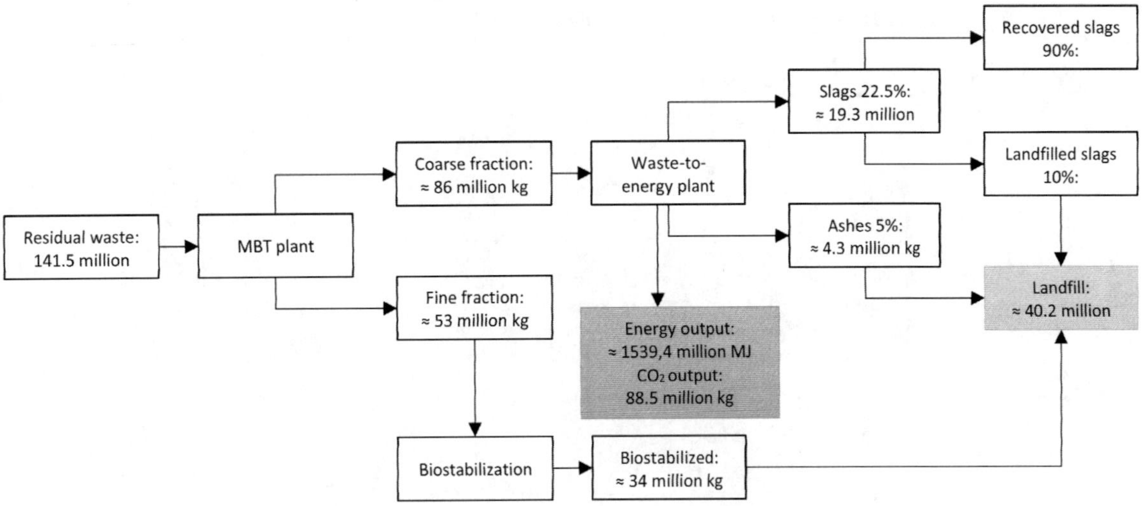

Figure 2. SRF fired in waste to energy plant

Journal of Physics: Conference Series **2685** (2024) 012046 doi:10.1088/1742-6596/2685/1/012046

5.3 Third scenario

In the third scenario, residual waste undergoes a MBT. Fine fraction materials were biostabilized while coarse fraction is burned in a cement plant. This process does not produce residues because they concur to produce the clinker structure. The disposed waste is 34 million kg. Figure 3 shows the flowchart of the process, simplified mass balance, and energy and CO_2 output of the process.

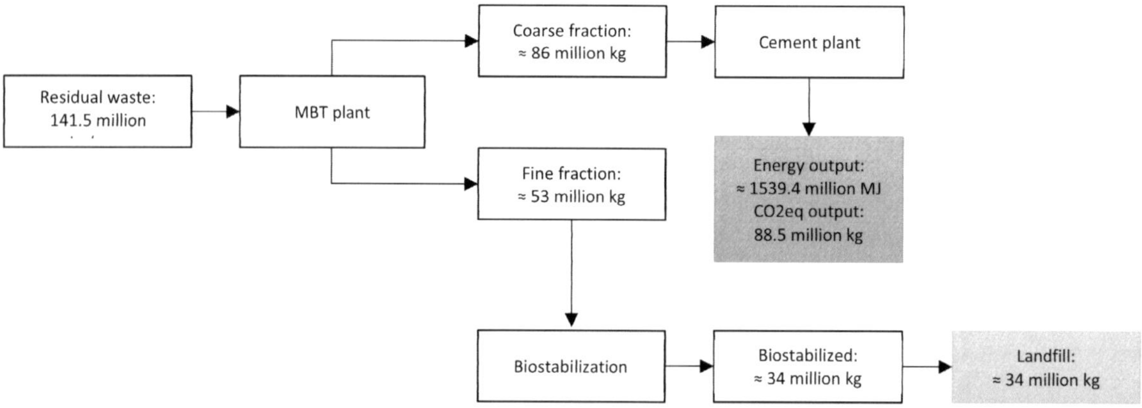

Figure 3. F-SRF fired in cement factory

Petcoke and CO_2 saving can be evaluated. By dividing the energy output for petcoke LHV (34 million MJ/kg [8]), the weight of petcoke saved in each scenario can be calculated. Considering fossil carbon (C_{fossil}) % in petcoke and SRF (87% and 28% mass respectively [8]) it is found the CO_2 output for each fuel multiplying C_{fossil} for factor conversion 3.67 [15]. Then subtracting CO_2 SRF output to CO_2 petcoke output, the CO_2 saved is found. In the first scenario the same value of SRF C_{fossil} is taken for residual waste. Figure 4 shows petcoke and CO_2 saved. Table 4 shows the calculation.

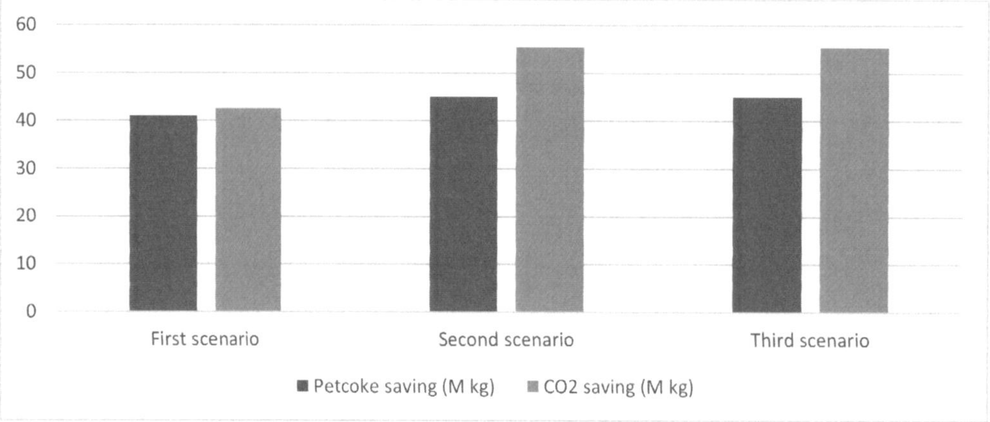

Figure 4. Petcoke and CO_2 saved in the three scenarios.

Table 4. Calculation to obtain petcoke and CO_2 saved.

	F-SRF	Residual waste	Petcoke scenario 1	Petcoke scenario 2	Petcoke scenario 3
LHV (MJ/kg)	17.9	10	34	34	34
Energy output (MJ)	1539.4	1415	-	-	-
Petcoke saved (M kg)	-	-	41	45	45
C_{fossil} (%)	28	17	87	87	87
C_{fossil} (M kg)	24.1	24.1	35.67	39.15	39.15
CO_2 output (M kg)	88.5	88.5	131	144	144
CO_2 saved (M kg)	-	-	42.5	55.5	55.5

Umbria landfill residual capacity is 808 thousand m^3 at the end of 2020 so it is possible to define landfill depletion in the three different scenarios. Considering 800 kg of waste are about 1 mc [16], table 5 shows Umbria landfill depletion in years, waste disposed (in kg and m^3), and % landfill on total waste considering only residual waste.

Table 5. Umbria landfill depletion.

	% landfill on total waste	Waste disposed (M kg)	Waste disposed (miles mc)	Years to landfill depletion (y)
Scenario 0*	37 %	162	202.5	≈ 4
Scenario 1	2.3 %	10.2	12.75	≈ 63
Scenario 2	9.2 %	40.2	50.25	≈ 16
Scenario 3	7.8 %	34	42.5	≈ 19

*this scenario is the current model in which biostabilized and coarse fraction were put in landfill.

6. Conclusion

The present paper proposes three scenarios to manage residual waste consisting in waste-to-energy solutions and petcoke thermal substitution in cement factory. A focus on SRF explains their production, legislation, and application in aforementioned waste managements. The first scenario shows the best result in terms of waste production but is the worst in fossil fuel and CO_2 reduction. All three scenarios are below the 10 % landfill EU target. This paper shows the potentiality of these three pathways in residual waste management that can be put into practice to produce energy and so reduce the emitted CO_2. Moreover, the diminishing of waste disposed in landfill is a crucial achieved goal. More studies are needed to make a more precise mass, energy and CO_2 balances. Finally, a life cycle analysis is needed to estimate environmental, economic, and social impact.

Acknowledgement

The authors would like to acknowledge GESENU SpA, Italy, for the support in data collection.

References

[1] European Parliament and the Council of the European Union, "DIRECTIVE 2008/98/EC OF THE EUROPEAN PARLIAMENT AND OF THE COUNCIL of 19 November 2008 on waste and repealing certain Directives," Nov. 2008.

[2] European Commission, "Communication from the Commission to the European Parliament, the Council, the European Economic and Social Commitee and the European Commitee of the Regions - The role of waste to energy in the circular economy," Jan. 2017.
[3] European Commission, "Communication from the Commission to the European Parliament, the Council, the European Economic and Social Commitee and the European Commitee of the Regions - Closing the loop - An EU action plan for the Circular Economy," Dec. 2015.
[4] ISPRA, "Rapporto rifiuti urbani 2021," 2021.
[5] UNI - Ente Italiano di Normazione, "UNI EN 15359:2011 Combustibili solidi secondari - Classificazione e specifiche," 2011.
[6] BSI - British Standards Institution, "BS EN ISO 21640:2021 - Solid recovered fuels — Specifications and classes," 2021.
[7] MINISTERO DELL'AMBIENTE E DELLA TUTELA DEL TERRITORIO E DEL MARE, "Decree 14 Febraury 2013, n. 22," Feb. 2013.
[8] M. Grosso, S. Dellavedova, L. Rigamonti, and S. Scotti, "Case study of an MBT plant producing SRF for cement kiln co-combustion, coupled with a bioreactor landfill for process residues," *Waste Management*, vol. 47, pp. 267–275, Jan. 2016, doi: 10.1016/j.wasman.2015.10.017.
[9] V. SERIFI, M. TARIC, and Z. DZUBUR, "ESTIMATION OF THE ENERGY POTENTIALS FOR THE REGION OF PRIZREN AND PROPOSAL SELECTION ON TECHNOLOGY, PROCESSES AND EQUIPMENT," *ANNALS OF THE ORADEA UNIVERSITY. Fascicle of Management and Technological Engineering.*, vol. XXIII (XIII), 2014/3, no. 3, 2014, doi: 10.15660/AUOFMTE.2014-3.3085.
[10] Arpa Umbria, "I rifiuti non differenziati (RND)," 2021.
[11] UNI - Ente Italiano di Normazione, "UNI CEI EN ISO/IEC 17025:2018 Requisiti generali per la competenza dei laboratori di prova e taratura," 2018.
[12] M. Horttanainen, N. Teirasvuo, V. Kapustina, M. Hupponen, and M. Luoranen, "The composition, heating value and renewable share of the energy content of mixed municipal solid waste in Finland," *Waste Management*, vol. 33, no. 12, pp. 2680–2686, Dec. 2013, doi: 10.1016/j.wasman.2013.08.017.
[13] Dieter Reimann, "CEWEP Energy Report III," 2010.
[14] Regione Umbria, "DGR 1135 del 2/11/2022 - Rapporto ambientale per la VAS Piano Regionale di Gestione Integrata dei Rifiuti della Regione Umbria," Nov. 2022.
[15] N. Permpool, S. Bonnet, and S. H. Gheewala, "Greenhouse gas emissions from land use change due to oil palm expansion in Thailand for biodiesel production," *J Clean Prod*, vol. 134, pp. 532–538, Oct. 2016, doi: 10.1016/j.jclepro.2015.05.048.
[16] K.-U. Heyer, K. Hupe, and R. Stegmann, "Methane emissions from MBT landfills," *Waste Management*, vol. 33, no. 9, pp. 1853–1860, Sep. 2013, doi: 10.1016/j.wasman.2013.05.012.

A new dry deposition model implemented in CALPUFF code to simulate contamination of radionuclides released into atmospheric environment post-nuclear accidental event

M Giardina[1],* P Buffa[1], A Cervone[2], C Lombardo[3], D Giaiotti[4], D Oshurok[5] and O Skrynyk[5]

[1] Department of Engineering, University of Palermo, Viale delle Scienze, Edificio 6, Palermo, 90128, Italy
[2] Department of Industrial Engineering, Alma Mater Studiorum University of Bologna, via dei Colli 16, 40136 Bologna, Italy
[3] Italian Agency for New Technologies, Energy and Sustainable Economic Development (ENEA), Via Martiri di Monte Sole, 4, Bologna, 40129, Italy
[4] ARPA FVG–CRMA, Regional Agency for Environmental Protection, Palmanova, Italy
[5] Ukrainian Hydrometeorological Institute, Kyiv, Ukraine

* Corresponding author: mariarosa.giardina@unipa.it

Abstract. Long-range transport and deposition analyses of 137Cs following a hypothetical incident at Gösgen nuclear power plant are studied by using CALMET-CALPUFF model system. Comparisons are performed with results obtained from a version modified of CALPUFF using a new dry deposition velocity model. This model is based on a combination of aerodynamic resistances and considers local features of the mutual influence of inertial impact and turbulent processes. The results show that the modified CALPUFF code seems to be an appropriate tool for performing impact assessments on long-range transport in complex terrain contexts or to support preparedness and response capabilities for nuclear and radiological accidents.

1. Introduction

As well known, CALMET/CALPUFF is an advanced non-steady state meteorological modelling system used to predict air quality and simulate meteorology [1, 2]. CALMET is a diagnostic meteorological model that reconstructs 3D wind and temperature fields starting from meteorological measurements, orography, land use data and CALPUFF assesses atmospheric pollution resulting from a wide number of species and a various range of emission sources. A new version of the CALPUFF tool allows for the assessment of particulate matter and the long-range transport of selected radionuclides and their decay products in several aerosol fractions.

In this paper, CALMET/CALPUFF system is used to study transport, dispersion, and deposition of radioactive particles released following a hypothetical accident at Gösgen nuclear power plant (in German Kernkraftwerk Gösgen, abbreviated in KKG), located in the canton of Solothurn, Switzerland. The analyses focus on the transport and dispersion of ^{137}Cs in Italian territory.

Meteorological data of European Centre for Medium-Range Weather Forecasts (ECMWF), surface stations measurements and data by upper-air stations, distributed throughout the domain of interest, are used.

In addition, comparisons are performed with results obtained from a modified CALPUFF version using a new dry deposition velocity model developed in [3].

Different physical processes interact to control the dry deposition of particles. Gravitational settling, turbulent diffusion, and surface collection are generally the most important, and their effects are usually accounted for by means of equivalent resistances (gravitational, aerodynamic, and surface collection resistances) that oppose to the deposition. The new model combines aerodynamic resistances that affect the particle flux in the quasi-laminar sub-layers and considers local features of the mutual influence of inertial impact processes and turbulent ones.

Results are compared to understand if the CALMET/CALPUFF tool has the capacity to provide data and recommendations in case of a post-nuclear accident in areas bordering with the Alps and beyond.

2. Dry deposition processes

The lifetime of particles in the atmosphere is determined by wet deposition, which removes particles by precipitation or in-cloud processing, and dry deposition, which directly uptakes particles to terrestrial surfaces.

Dry deposition remains a globally important first-order loss process that scales with concentration and is critical for estimating accurate spatial and temporal aerosol distributions in atmospheric models. The complexity of the fluid-dynamic processes that affect the deposition rate is governed by the following three main phenomena [4-6]:

- Atmospheric turbulence transport in the surface layer (SL). This process remains unaffected by the physical and chemical properties of the pollutant, relying solely on the degree of turbulence;
- Diffusion in the thin layer of air that overlooks the air-ground interface (i.e. quasi-laminar sublayer, QLS). It is molecular diffusion for gasses, Brownian motion for particles and gravity for heavier particles;
- Transfer to the ground (e.g. interception, impaction, and rebound) that is significantly influenced by surface characteristics such as urban context, grass, forest, etc.;
- Deposition process. For particles with intermediate diameters (dp) in the range of approximately 0.1÷1 μm, it strongly depends on atmospheric conditions, surface characteristics, and particle size. For dp above this range, deposition is primarily governed by the following interaction mechanisms;
  - Inertial impaction. It occurs when a particle has high inertia and is carried by the flow towards an obstacle. The particle becomes incapable of tracking the flow deviation, resulting in a collision with the obstacle;
  - Turbulent impaction. In this process, particles attain velocity high enough such that turbulent eddies can result in a transverse "free flight velocity". Thus, particles can reach the surface.

A diverse set of parameterizations of dry deposition is available and used in different models and monitoring networks. The main difference lies on the schematisation of the surface resistances in QLS. However, there remains significant controversy regarding which mechanisms dominate for various particle sizes, atmospheric conditions, and surface characteristics due to limitations in both measurements and models.

2.1. A new resistance model to calculate the dry deposition velocity

The phenomena described in previous section are typically described using models based on resistances approach, and the interested reader is referred to [4] for further investigations.

In this context, a new approach is suggested in [3], where the electrical analogy method has been adapted to incorporate processes like particle rebound or resuspension.

The resistances, that affect the particle flux in SL and QLS, are combined as described in figure 1:

- aerodynamic resistance, r_a, in SL (i.e. contribution to the deposition due to the atmospheric turbulence) is connected in series with the resistance r_{ql} that takes into account mechanisms of diffusion by Brownian motions and impaction phenomena in QLS;
- quasi-laminar sublayer resistance, r_{ql}, is evaluated by considering a parallel circuit. The contribution of Brownian diffusion motions and the impaction phenomena can be considered as follows:
  - r_{db} assess the contribution of the Brownian diffusion;
  - r_{ii} considers the single effect of the inertial impaction regime for large particles;

- r_{ii} and r_{ti} are linked in a series arrangement to account the local characteristics of the mutual influence of inertial impact processes (represented by r_{ii}) and turbulent impact phenomena (represented by r_{ti}).

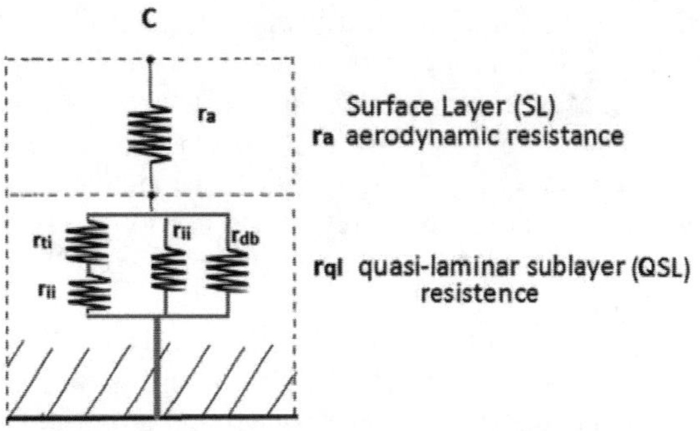

Figure 1. New resistances schematization for the particle deposition velocity calculation.

The total resistance r_t is evaluated using the following relationship:

$$r_t(z) = r_a + r_{ql} \tag{1}$$

where, r_{ql}, using the electric analogy, is evaluated as:

$$\frac{1}{r_{ql}} = \frac{1}{r_{bd}} + \frac{1}{r_{ii}} + \frac{1}{r_{ii} + r_{ti}} \tag{2}$$

The reader is referred to [3] for details on the calculation of resistances of eq. (2), not included here for the sake of brevity.

2.2. Dry deposition velocity in CALPUFF

CALPUFF is a multi-species non-steady-state puff dispersion model that simulates the effects of time and space varying meteorological conditions on pollutant transport, transformation, and removal [2].

It includes algorithms for near-field effects such as stack tip downwash, building downwash, transitional buoyant and momentum plume rise, rain cap effects, partial plume penetration, and coastal interactions effects as well as longer range effects such as pollutant removal due to wet scavenging and dry deposition.

- For dry deposition process, the technical algorithms are based on three options;
- Full treatment of space and time variations of deposition with a resistance model;
- User-specified diurnal cycles for each pollutant;
- No dry deposition.

The first option allows the computation of dry deposition of gases and particulate matter as a function of geophysical parameters, meteorological conditions, and pollutant species.

The resistance across the QLS is evaluated according to the following relationship:

$$r_{ql} = \frac{1}{u_*(Sc^{-2/3} + 10^{-3/St})} \tag{3}$$

where u_* is the friction velocity, St is the Stokes number and Sc is the Smith number [3, 4].

The second option allows user-specified, diurnally varying deposition velocities to be used for one pollutant instead of the resistance model. This option allows a "typical" time dependence of deposition to be incorporated but it does not include any spatial dependency. The last option bypasses the dry deposition model completely.

The CALPUFF deposition module has been modified to incorporate a fourth option that utilizes the resistance model described in [3].

DRY, DRYI, VDCOMP subroutines, that read data and process geophysical parameters and meteorological conditions, and a new VDPG subroutine, which allows the application of the new model, have been modified and incorporated in CALPUFF code.

3. Brief description of the Gösgen reactor

Gösgen KKG is a pressurized water nuclear reactor that generates about 15% of the Switzerland's electricity consumption, producing 8 TWh of electricity every year.

The main components of the reactor coolant system include the reactor itself, the pressuriser system, and three identical parallel circulation loops, each characterized by the presence of a steam generator (SG). The coolant enters the reactor at a temperature of 565 (K) and then moves downward through the annular gap, which is the space between the core barrel and the reactor pressure vessel. When the coolant reaches the semi-spherical base of the reactor vessel, it undergoes a 180-degree redirection, flowing through the reactor core and heating up to 598 K.

The thermal energy of the primary coolant in the SGs is transferred to the water on the secondary side of the SGs, resulting in the production of steam. This steam is then utilized in the turbine and subsequently passes through the condenser, where it condenses into liquid water. The heat from the condenser is released into the environment through a natural-draught cooling mechanism within the cooling tower, a reinforced concrete structure with a hyperbolic design, reaching a height of approximately 140 m.

4. CALMET-CALPUFF simulations

In this work, the computing domain for the CALMET simulation covers an area of 1.315.108 km^2 extended to a territory comprising the Switzerland and Italy countries, with cell resolution of 15x15 km.

The CALMET wind field module uses the following two-step approach:

- ***step 1***, involves refining the initial estimated wind field to account for the kinematic impacts of terrain, slope flows, and terrain blocking effects, resulting in an initial wind field;
- ***step 2***, an objective analysis procedure is employed to incorporate user-specified observational data from surface and upper air measurements into the wind field obtained in step 1, yielding the final wind field.

Meteorological conditions, with time step of 1 hour, are calculated by using meteorological data from 94 surface stations and measurements of 10 upper air sounding stations, for 5 days starting from the 2011-12-16 at 01:00 up to 2011-12-21 at 12:00.

Location of surface and upper-air stations, used in this study, are reported in figure 2. Surface stations are shown as white square points, upper-air stations as red circle points.

ERA5 reanalysis datasets, conducted by ECMWF, were acquired for the examined time frame and then processed with the FORCAL software [7] to enable their utilization in step 1 of the CALMET wind field module. FORCAL tool allows the reduction of grid cell sizes of the computational domain compared to the data initially supplied by ECMWF, enhancing the resolution of the meteorological information.

CALPUFF analyses have been performed by using CALMET results, and the input conditions reported in table 1 to schematize the emission source.

Radioactive emission has been hypothesized to occur at the exit of the cooling tower. The release is set to be of only a single isotope ^{137}Cs with a value of 10^{16} Bq over 1 hour (puff emission). This is a typical release for a PWR under severe accident conditions [8].

In the following, results of three different CALPUFF simulations, obtained by using the following options for the dry deposition process, are compared:

- CALPUFF default option for the dry deposition resistance eq. (3), that we call in the following section default option (DO);
- new option to use dry deposition resistance model of [3], referred as new option (NO);
- imposed deposition velocity of 0.005 m/s as suggested in [9], referred as imposed deposition velocity (IDV).

In [9] the constant for the deposition velocity of ^{137}Cs aerosol particles was taken from [10, 11] and it comes from the assessments made during the Chernobyl Nuclear Power Plant (CNPP) accident.

It is worth noting that the IDV approach does not account for phenomena such as eddy diffusion due to turbulent air movements or interception/impaction with the surface element structures. Additionally, near the surface, retention or rebound is influenced by a combination of surface and impact properties as described in section 2.

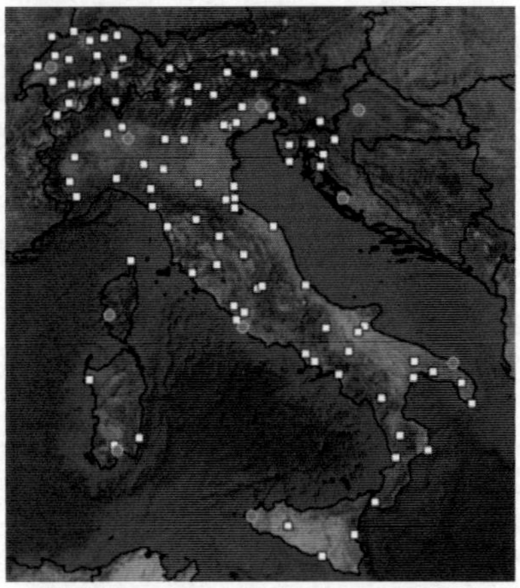

Figure 2. Surface stations (white square points) and upper-air stations (red circle points.

Table 1. Input data used for CALPUFF simulations.

Input data	
Source of ^{137}Cs	10^{16} Bq
Particle diameter of ^{137}Cs	1 μm
Time emission	2011/12/16 h=19:30
Duration of emission	1 h
Temperature of emission	423 K
Cooling tower height	140 m
Cooling tower diameter	70 m

4.1. Results

Figures 3 and 4 report results of wind field maps at 10 m above the ground.

For the day 2011-12-16 at 01:00 and 12:00, the northern part of Italy is characterized by low-intensity wind fields (figure 3), and cloudy weather. In the Tyrrhenian Sea, high-intensity wind blows from west toward south-east at the time 01:00. Over the next 24 hours, the wind direction changes, shifting from the west towards north-east (figure 3, 2011-12-16 at time 12:00).

This wind field behaviour takes places again the next days 2011-12-17 and 2011-12-18, and the low-intensity wind situation persists in the northern regions of Italy. However, the formation of wind fields characterized by vortical motions is predicted, initially affecting the central regions of Italy, and subsequently extending to the southern regions.

In the day 2011-12-21, a circular wind field forms at the center of the Tyrrhenian Sea at 00:00 and gradually moves towards the Calabria region by 12:00 (figure 4).

Note that, during the simulation period, no significant precipitation is observed in any region of Italy, indicating the prevalence of dry deposition processes.

Significant attention has been given to studying the evolution of atmospheric stability conditions during the simulated period. In CALMET, the Pasquill-Gifford-Turner (PGT) stability scheme is used to classify atmospheric conditions in the boundary layer. The classes range from unstable (Classes 1, 2, and 3) to neutral (Class 4) and stable (Classes 5 and 6). Unstable conditions typically occur during the day due to ground-level heating, leading to thermal turbulence in the boundary layer. Stable conditions, on the other hand, are associated with night-time cooling, resulting in reduced turbulence levels and the presence of temperature inversion at lower levels. Neutral conditions are often observed during high wind speeds or overcast sky conditions.

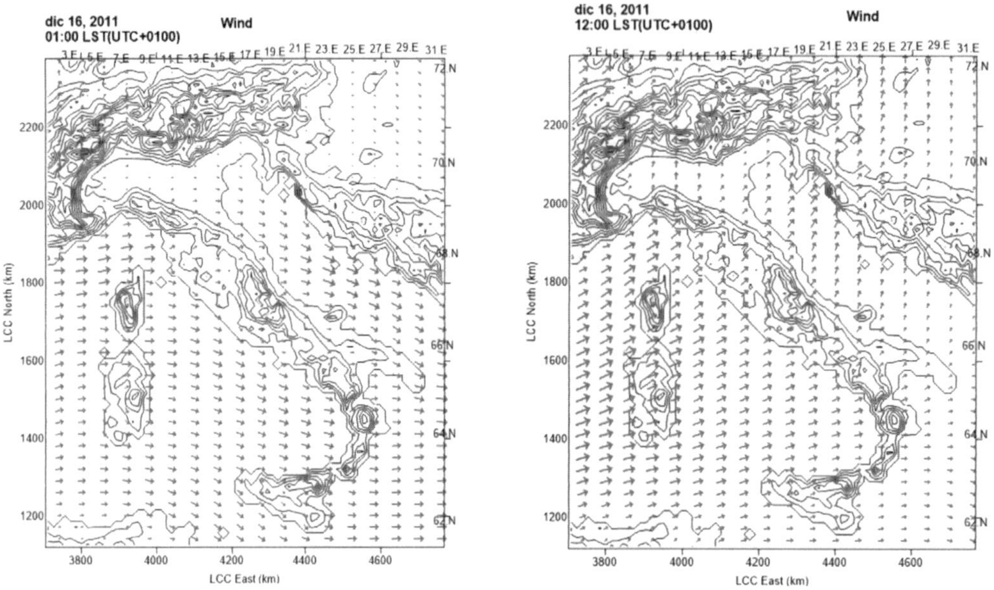

Figure 3. Wind field at 10 m on the 2011-12-16 at 01:00 and 12:00.

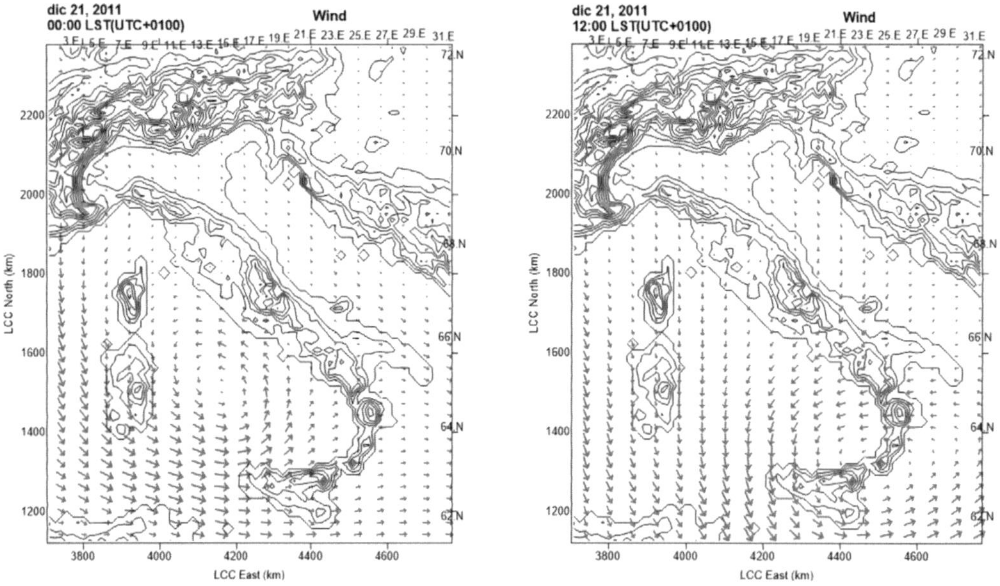

Figure 4. Wind field at 10 m on the 2011-12-21 at 00:00 and 12:00.

Analysing the temporal distribution of PTG, except for the first day of simulation that is characterized by class 5 (moderately unstable) and 4 (slightly unstable) conditions for almost the entire Italy, it can be observed that, classes 2 (slightly stable) and 3 (neutral condition) occur with high frequency in the northern regions. Class 4 is predominantly observed in Tyrrhenian Sea, Adriatic Sea and major islands.

Extended periods of atmospheric stability or low stability during winter, characterized by a lack of precipitation, do not favour the dispersion of pollutants in the air and promote their accumulation.

Figures 5 through 7 show the maps of cumulative dry deposition distribution of ^{137}Cs predicted by CALPUFF for the entire simulation period, applying DO (figure 5), NO (figure 6), and IDV (figure 7) options, respectively.

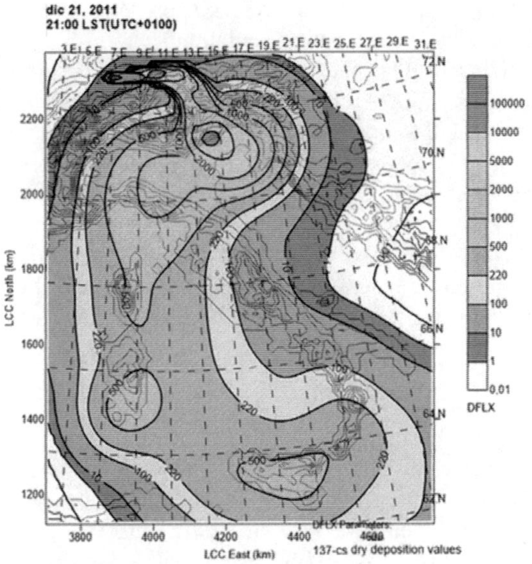

Figure 5. Cumulative dry deposition of ^{137}Cs obtained by using default resistance model of CALPUFF.

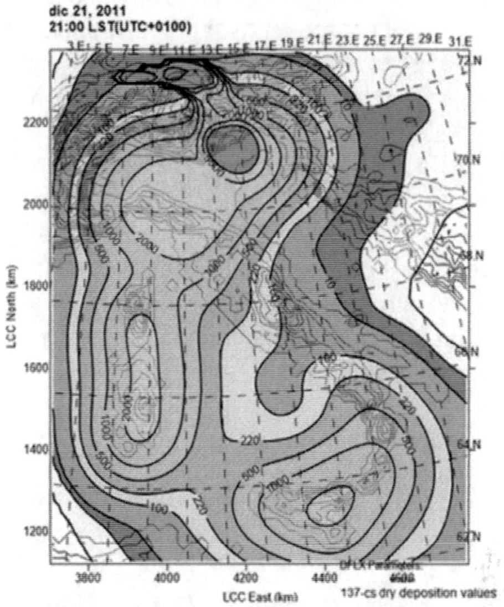

Figure 6. Cumulative dry deposition of ^{137}Cs obtained by using new resistance model.

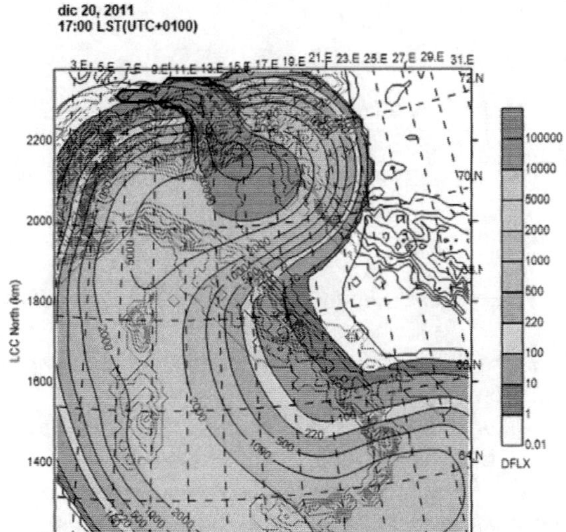

Figure 7. Cumulative dry deposition of ^{137}Cs obtained by using imposed deposition velocity.

All simulations highlight that the most critical condition takes place in the same regions of Italy with the following differences:

- The results obtained using DO option (figure 5) identify Lombardy, the eastern part of Liguria and Piemonte, Toscana, and western zone of Emilia-Romagna as the territories with the deposition in range 1.000÷10.000 Bq/m^2 and values higher than 5.000 Bq/m^2 in Veneto.
  Deposition of about 500 Bq/m^2 is predicted in the southern area of Sardinia and in the central zone of Sicily. Some territories of Switzerland, regions of Italy not listed before, and Mediterranean Sea are characterized by a pollutant deposition lower than 220 Bq/m^2;
- The results obtained by using NO option (figure 6) extend the areas with high pollution to Veneto, Friuli Venezia Giulia and northern area of Trentino Alto Adige. These areas experience deposition in the range 1.000÷10.000 Bq/m^2. ^{137}Cs deposition involves the regions of Sardinia and Sicily but with the difference that it is in the range of 1.000÷2.000 Bq/m^2. It is worth noting that there is an area to east of Gösgen KKG affected by a deposition exceeding 100.000 Bq/m^2.
- The analysis carried out by using IDV option (figure 7) highlights an extension of territories at east of the Gösgen KKG where the ^{137}Cs deposition exceeds 100.000 Bq/m^2. The comparison with the NO simulation shows that in this case the deposition is relatively uniform in the Tyrrhenian Sea with depositions in the range 1.000÷2.000 Bq/m^2.

In summary, the use of default resistance model, compared to the other two cases, predicts lower dry deposition and, consequently, underestimates health consequences for the population.

On the other hand, the use of the new model yields results that are more consistent with those obtained by setting the deposition velocity to 0.005 m/s. CALPUFF simulations with NO and IDV options results in an estimated deposition velocity that is at least one order of magnitude higher than the values calculated with DO option. However, NO and IDV options also yield significantly different deposition maps, particularly regarding the spatial coverage across the territory.

It is worth highlighting that, the result of CALPUFF simulations of Chernobyl nuclear accident reported in [9] with the default resistant model for the dry deposition velocity of ^{137}Cs aerosol particles have shown substantial underestimation in comparison with the measured data. Moreover, the Authors in [9] highlight that the magnitude of the deposition velocity is an order of magnitude smaller than in other previously published assessments performed for particle sizes ranging from approximately 0.1 up to1.0 µm [10, 11].

5. Conclusions

Research activities have concerned transport and dispersion analyses of ^{137}Cs radioactive particles released after a hypothetical severe accident in the Gösgen nuclear power plant by using CALMET/CALPUFF modelling system. Comparisons are performed by three different CALPUFF simulations. The most unfavourable conditions are predicted by CALPUFF results that use an imposed dry deposition velocity, while the least severe conditions are associated with the simulation that uses the default model of CALPUFF for dry deposition. The use of the new dry deposition model led to results falling into an intermediate condition. In future research activities, there are plans to further validate the model by simulating events such as the Chernobyl accident.

References

[1] Scire JS, Robe FR, Fernau ME, Yamartino RJ 2000 *A User's Guide for the CALMET Meteorological Model* Publisher: Earth Tech, Inc.

[2] Scire JS, Strimaitis DG, Yamartino RJ 2000 *A user's guide for the CALPUFF dispersion model* Publisher: Earth Tech, Inc.

[3] Giardina M and Buffa P 2018 *A new approach for modeling dry deposition velocity of particles*. Atmos. Environ., 180, 11-22.

[4] Seinfeld J and Pandis S 1998 *Atmospheric Chemistry and Physics, From Air Pollution to Climate Change,* John Wiley & Sons Inc.

[5] Giardina M et al. 2017 *Dry deposition models for radionuclides dispersed in air: a new approach for deposition velocity evaluation schema*, J. Phys.: Conf. Ser. 923 012057.

[6] Giardina M et al. 2019 *Atmospheric dry deposition processes of particles on urban and suburban surfaces: Modelling and validation works*, Atmos. Environ., Volume 214, 1 October.

[7] Martorana F, Giardina M, Buffa P, Beccali M 2021 *A new tool to process forecast meteorological data for atmospheric pollution dispersion simulations of accident scenarios: A Sicily-based case study, Journal of Sustainable Development of Energy, Water and Environment Systems*, 9(3), 1080377.

[8] Guglielmelli A and Rocchi F 2017 *Evaluation of the radiological impact on the Italian territory of a severe nuclear accident at Krško NPP by means of a statistical methodology*, 26th Int. Conf. Nuclear Energy for New Europe, NENE2017, Bled, Slovenia

[9] Giaiotti D, Oshurok D, Skrynyk O 2018 *The Chernobyl nuclear accident 137Cs cumulative depositions simulated by means of the CALMET/CALPUFF modelling system*, Atmos. Pollut. Res, 9(3), 502-512.

[10] Izrael YA et al. 1990 *Chernobyl: radioactive contamination of the environment*. Gidrometeoizdat, Leningrad.

[11] Talerko N 2005 *Mesoscale modeling of radioactive contamination formation in Ukraine caused by the Chernobyl accident* J. Environ. Radioact. 78(3), 311-329.

Acknowledgments

Acknowledgement is made for the use of ECMWF's computing and archive facilities in this research.

Control of CO_2 evaporation in an integrated photovoltaic module: experiments and modelling

M Azzolin*, E Zanetti, R Conte and D Del Col

Department of Industrial Engineering, University of Padova, Via Venezia 1, 35131 Padova, Italy

* Corresponding author e-mail: marco.azzolin@unipd.it

Abstract. The use of environmentally friendly refrigerants and the improvement of the system performance are two main topics in the research on heat pump technology. Given the recent limitations imposed on the usage of high global warming potential refrigerants, there is a growing demand in the market for the adoption of natural refrigerants, like CO_2. Nevertheless, CO_2 heat pumps operate under transcritical conditions, which can adversely affect its efficiency. Therefore, incorporating a solar heat source into a CO_2 heat pump can offer a solution to mitigate low system performance issues. In the present study, a photovoltaic module integrated with the evaporator of a CO_2 heat pump is experimentally studied. The PV-T evaporator exploits solar radiation to both generate electricity in the PV cells and ensure the evaporation of CO_2 in the collector's tubes. Simultaneously, the PV conversion efficiency is improved by the cooling effect of the evaporation. The present evaporator works in dry expansion mode, thus the refrigerant flow after the expansion device is sent to the solar collectors, where it evaporates before returning to the compressor. When using a PV-T evaporator, it is necessary to prevent superheating in the evaporator to keep a uniform and efficient cooling of the PV cells. The current heat pump design prevents the occurrence of superheated vapor at the PV-T evaporator's outlet. Beside the experimental activity a dynamic numerical model of the system has been developed and validated.

1. Introduction

The use of CO_2 as refrigerant in heat pump systems is growing due to the recent restrictions imposed by many countries and organizations such as the Regulation No 517 of the European Union of 2014 [1] and the Kigali Amendment of the United Nations of 2016 [2], that force a phasedown of hydrofluorocarbons to reduce the emissions of greenhouse gases. CO_2 is an environmentally friendly refrigerant with a low global warming potential ($GWP_{100yr}=1$) and zero ozone-depleting potential (ODP=0), it is non-toxic and non-flammable (ASHRAE class A1). However, due to its low critical point and higher operating pressure compared to synthetic refrigerants, the design of CO_2 systems should account for these issues [3]. An interesting solution to increase the efficiency of CO_2 heat pumps is to exploit solar radiation as a low-temperature thermal source [4]. In direct solar-assisted heat pumps (DX-SAHPs) the evaporator is a solar collector, which converts the incident radiation into heat, which is then exchanged with the refrigerant. In the literature, few experimental studies on direct solar-assisted heat pumps (DX-SAHPs) working with CO_2 as the refrigerant are available. Duarte et al. [5] analyzed the performance of a CO_2 heat pump with a sheet-and-tube solar thermal collector as evaporator. The surface temperature of the solar evaporator was experimentally measured with an infrared camera. The results showed that when the solar radiation is high, a high degree of superheat is observed at the collector's outlet and the performance of the heat pump is penalized by reduction in the evaporating area [6]. In a previous numerical work on the same heat pump system, Paulino et al. [7] suggested the use of an electronic expansion valve to control the refrigerant mass flow rate at the evaporator inlet, and so the value of the superheating at the collector's outlet.

Less works investigate CO_2 heat pumps with photovoltaic-thermal (PV-T) collector as direct expansion evaporator. Zanetti et al. [8] demonstrate that a solar assisted heat pump with a direct PV-T evaporator can reach higher performance in terms of COP compared to the air source. Paradis et al. [9] numerically analyzed the performance of a new geometry of PV/T solar collector composed by monocrystalline PV cells coupled on the back by an absorber plate with a stainless steel serpentine tube. The authors showed that in this geometry, used as evaporator in a CO_2 heat pump, the electrical efficiency can increase of about 2% compared to a standard solar PV.

The presence of superheated vapor in the solar evaporator is much more disadvantageous in a PV-T collector. In this case, both a decrease in the collector efficiency and photovoltaic efficiency is detected, because the refrigerant cannot effectively cool down the solar cells. In the literature, the problem of superheating control in PV-T evaporators of DX-SAHP is limitedly addressed. The present paper investigates the control of the evaporation pressure in a solar-assisted heat pump with CO_2 as refrigerant and PV-T evaporators. The layout of the refrigerant loop has been designed to avoid superheated vapor at the collectors' outlet and thus to operate at the optimal possible evaporation temperature. The heat pump has been studied both experimentally and numerically, through a dynamic model of the system components. After the validation, the dynamic model has been used to evaluate the transient performance and control of the heat pump.

2. Experimental set-up

2.1. *Heat pump prototype and sensors*

The prototype presented in this work is a solar-assisted heat pump working with CO_2 as the refrigerant. The unit is installed at the Department of Industrial Engineering, University of Padova, Italy (45° 24′ 23″ N, 11° 52 40″E). A picture of the prototype is reported in Figure 1a. The heat pump is equipped with three photovoltaic-thermal (PV-T) solar collectors which work as the evaporator of the system. The schematic of the refrigerant loop is reported in Figure 1b. The refrigerant operates a transcritical cycle, therefore after the inverter driven compressor (COMP), the CO_2 is sent to a brazed plate gas-cooler (GC), where hot water on the secondary loop is produced. The cooled refrigerant is then sent to an internal heat exchanger (INT) where its temperature is further lowered, and then it enters an electronic expansion valve (EEV). After the expansion device, the refrigerant is sent to the three PV-T evaporators. The vaporized CO_2 obtained at the outlet of the evaporator enters a receiver where the liquid and vapor phases are separated. The vapor phase is then extracted at the top of the receiver and sent back to the internal heat exchanger, where it is heated up before entering the compressor.

Each PV-T collector is composed by a sheet-and-tube heat exchanger coupled to a polycrystalline module (270 W nominal power). The absorber of the sheet-and-tube heat exchanger is an aluminum plate and it has been coupled to a fifteen tubes copper serpentine (pitch 8 mm). The internal diameter of the copper tubes is equal to 6 mm and the thickness is equal to 1 mm. Figure 1b) also reports the main sensors installed on the heat pump. Various temperature and pressure sensors have been positioned at the inlet/outlet of the above-mentioned components (on the refrigerant side). The temperature and mass flow rate of the water at the inlet/outlet of the gas-cooler are also measured. In addition, the air temperature and solar irradiance (positioned on the same plane of the solar evaporators) are recorded. Finally, the power consumption of the compressor is measured by means of a power analyzer.

2.2. *Data reduction*

Experimental tests have been performed both in steady-state and dynamic conditions. The collected data have been used to calculate the main heat pump performance indicators. On the water side, the heat flow rate at the gas-cooler is evaluated considering the water temperature increase $(\Delta T)_w$:

a) b)
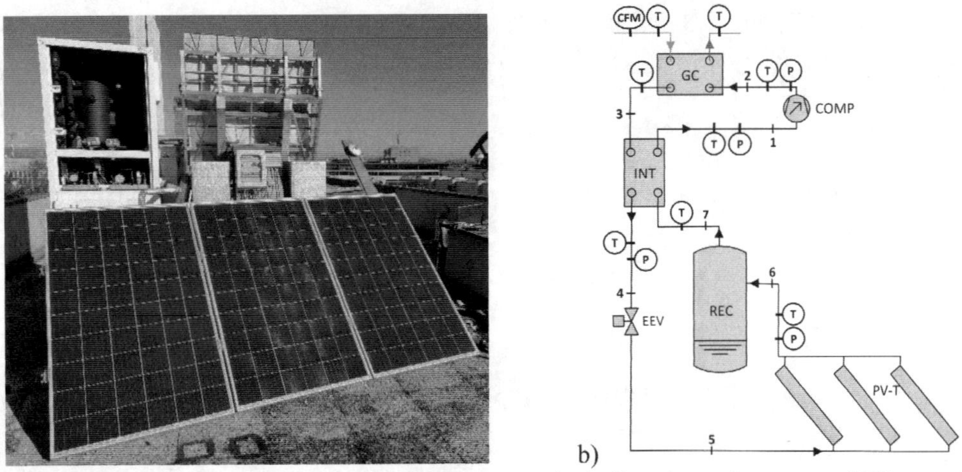

Figure 1. Schematic of the solar-assisted heat pump, including the main sensors ("T" temperature sensor, "P" pressure sensor, "CFM" Coriolis flow meter) installed on the refrigerant and water loops.

$$Q_{gc} = \dot{m}_w \, cp_w \, (\Delta T)_w \quad (1)$$

Where $\dot{m}_w$ and cp_w are the water mass flow rate and specific heat capacity.

On the refrigerant side, the values of the specific enthalpy h at the inlet/outlet of the main components where the refrigerant is found in single-phase conditions (h_1, h_2, h_3, h_4) are calculated with Refprop [10] using the measured values of pressure and temperature.

In addition, the evaporation temperature T_e can be determined with Refprop, by using the measured value of the low-pressure (p_6). The superheating at the outlet of the evaporator can be evaluated as:

$$SH = T_6 - T_e \quad (2)$$

Finally, the coefficient of performance of the heat pump COP is calculated with the ratio between the heating capacity Q_{gc} and compressor power consumption P:

$$COP = Q_{gc}/P \quad (3)$$

All the measured parameters are acquired with a sampling rate of 10 s through a data logger Agilent 34970-A and saved through LABView software. Each steady-state point is calculated as an average value considering a time interval equal to 5 minutes.

3. Experimental results

3.1. Low-pressure control

The control system of the present heat pump manages to fix the high-pressure level to the desired value by automatically varying the aperture of the expansion valve. Therefore, there is no active device in charge of the control of the low-pressure. However, from Figure 1b, it is possible to understand that, if the liquid level inside the low-pressure receiver is constant, the refrigerant at the outlet of the receiver is always found as saturated vapor (specific enthalpy equal to h_v), since it is extracted from the top:

$$h_7 = h_v \quad (4)$$

The hypothesis of constant level has been verified experimentally by checking the position of the liquid level inside the receiver through sight glasses. Furthermore, the value of the specific enthalpy in 7 (h_7) can be indirectly calculated by applying the energy balance to the refrigerant loop:

$$(h_2 - h_3) - (h_2 - h_1) = h_4 - h_7 \quad (5)$$

where the two terms on the left-hand side represent the specific energy fluxes at the gas-cooler and compressor respectively, whereas the term on the right-hand side represents the specific heat flux absorbed at the low-pressure side. From eq. (5):

$$h_7 = h_4 + h_3 - h_1 \tag{6}$$

The vapor quality x_7 can be calculated through Refprop, at the corresponding pressure. The calculated values of x_7 are reported in Figure 2 for all the experimental tests conducted under steady-state conditions at various compressor speeds and environmental conditions. It can be observed that as hypothesized in eq (4), the vapor at the outlet of the receiver is always between 0.94 and 1.

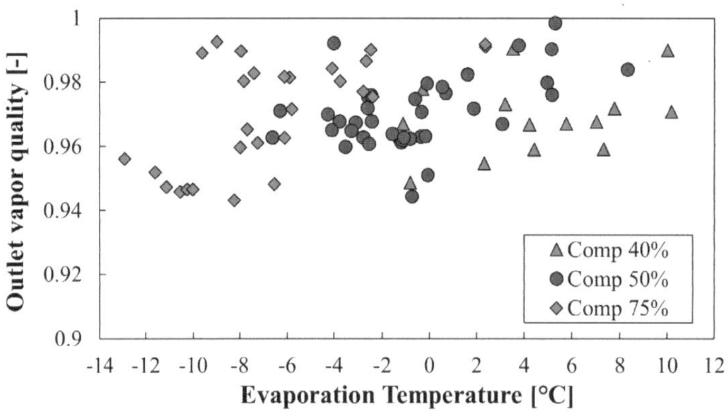

Figure 2. Vapor quality at the outlet of the receiver (state 7 in Figure 1), for the tests realized in steady-state conditions.

This result is of particular interest because, under the above hypothesis (null variation of the liquid level inside the receiver), it can also be concluded that, in steady-state conditions, the enthalpy at the outlet of the evaporator h_6 is equal to h_7 (thus, $x_6 = x_7 \approx 1$). This means that the PV-T evaporators are always forced to completely evaporate the CO_2 mass flow rate until saturated conditions: the low-pressure of the system is thus passively regulated to complete the evaporation process. This pressure control is very efficient, since it provides the highest evaporation temperature at each operative condition and allows the refrigerant to flow at constant temperature in the collectors, providing a uniform cooling of the cells. In fact, a higher value of evaporation temperature would cause an incomplete evaporation process due to the higher heat losses towards the environment, whereas a lower evaporation temperature would cause the presence of superheated vapor in the evaporator tubes. This finding (no presence of superheating at the outlet of the evaporators) has been confirmed also through the temperature measurements obtained with the thermocouples placed on the outlet tube of the PV-T collectors. Based on the data collected by these thermocouples, the difference between the measured temperatures and the saturation temperature was always within ± 1.0 K.

3.2. Heat pump performance

In this section, the experimental performance of the heat pump operating in steady-state conditions is presented. The efficiency of a heat pump is strictly related to the environmental conditions and in the case of a solar-assisted heat pump, solar radiation plays a significant role. Figure 3 reports the variation of the evaporation temperature for different values of solar irradiance. Experimental data have been recorded when the compressor was working at 50% of the maximum speed and the high pressure was equal to 80 bar. Furthermore, the temperature difference of the secondary water across the gas-cooler $(\Delta T)_w$ was maintained equal to 5 K and the inlet temperature equal to 30 °C. All the experimental points have been collected when the air temperature was between 7 °C and 7.5 °C, therefore the effect of air temperature is negligible when comparing the performance at different irradiance values. Figure 3 shows that solar irradiance has a positive effect on the evaporation temperature, which increases linearly with the radiation showing a growth rate equal to 0.8 K each 100 W m^{-2}.

Figure 4 reports the variation of the *COP* with the evaporation temperature obtained in steady-state conditions. The experimental points are referred to data collected when the high pressure was fixed at 80 bar and the water inlet temperature at the gas-cooler was equal to 30 °C, at different environmental conditions. It can be observed that the *COP* displays an increasing trend with the evaporation temperature. This effect is due to the lower pressure ratio, which decreases the compression work. The effect of the different compressor speeds is also reported: the coefficient of performance and evaporation temperature are lower at high compressor velocities. In fact, when increasing the compressor speed, the refrigerant mass flow rate increases and, since the energy flux incident on the panels is fixed, the evaporation of the CO_2 is guaranteed by decreasing the evaporation temperature.

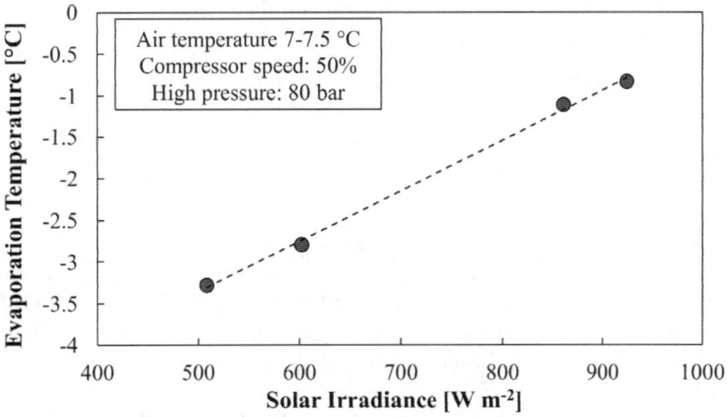

Figure 3. Evaporation temperature versus solar irradiance. Operating conditions: compressor speed equal to 50%, high pressure equal to 80 bar and air temperature between 7 °C and 7.5 °C.

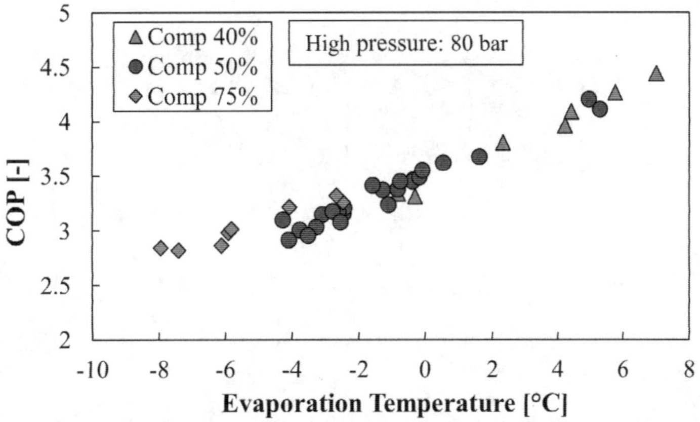

Figure 4. Experimental values of the coefficient of performance versus evaporation temperature obtained during steady-state conditions. The points refer to tests realized at fixed high pressure equal to 80 bar and water inlet temperature at the gas-cooler equal to 30 °C.

4. Numerical modelling

4.1. Model description

A dynamic model of the solar assisted heat pump has been developed in Matlab environment. The model is capable to solve the CO_2 transcritical cycle when the velocity of the compressor, the high-pressure, the inlet condition of the water and the environmental conditions are selected. For each time step, the model outputs are the main energy fluxes and the thermodynamic conditions of the refrigerant along the loop. More details about the model can be found in [11]. In this work, the equation used for the modelling of the PV-T evaporators and of the receiver are presented, since they represent the primary components for regulating the system.

The PV-T evaporator is discretized into the three main layers of the PV sandwich, namely the glass ("g"), the PV cells ("pv") and the absorber plate ("abs"). For each layer of thickness l and density ρ, the energy conservation equation can be written considering the energy fluxes between the various layers for an infinitesimal area $dA = dydz$ (see Figure 5).

$$l_g \rho_g c p_g \frac{\partial T_g}{\partial t} = G\alpha_g - q_{g-pv} - q_{loss,top} - (q_{cond,g,x} + q_{cond,g,z}) \tag{7}$$

$$l_{pv} \rho_{pv} c p_{pv} \frac{\partial T_{pv}}{\partial t} = G(\tau\alpha) + q_{g-pv} - q_{pv-abs} - p_{pv} - (q_{cond,pv,x} + q_{cond,pv,z}) \tag{8}$$

$$l_{abs} \rho_{abs} c p_{abs} \frac{\partial T_{abs}}{\partial t} = q_{pv-abs} - q_{loss,top} - q_e - (q_{cond,abs,x} + q_{cond,abs,z}) \tag{9}$$

In eq. (7) to (9), the last term represents heat flux due to conduction with the adjacent discretization. In eq. (8), p_{pv} is the specific electrical power generated by the PV cell. In eq. (9), q_e is the heat flux provided to the fluid flowing in the evaporator tubes:

$$q_e = K(T_{abs} - T_e) = \dot{m}_r \, dh_e \tag{10}$$

Where K is the global heat transfer coefficient between the plate and the fluid bulk (see [11]), $\dot{m}_r$ is the mass flow rate processed by the compressor and dh_e is the effective enthalpy drop of the refrigerant occurring in the tubes positioned under the considered discretization.

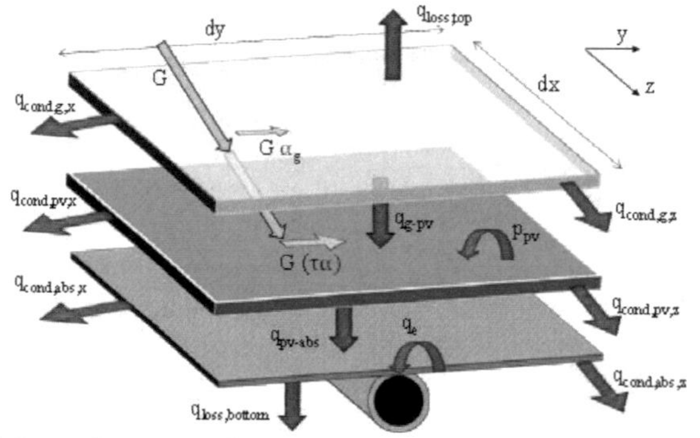

Figure 5. Schematic of the main energy fluxes considered in the modelling of the PV-T evaporator.

The receiver positioned at the outlet of the evaporator has been considered adiabatic. The mass and energy balance can be written:

$$\frac{dm_{rec}}{dt} = \frac{d}{dt}\left(\rho_v V_l + \rho_l (V_{rec} - V_v)\right) \tag{11}$$

$$\frac{dH_{rec}}{dt} = \frac{d}{dt}\left(\rho_v V_l h_l + \rho_l (V_{rec} - V_v)\right) = \dot{m}_r (h_v - h_6) \tag{12}$$

Where the pedix "l" refers to the liquid phase, m_{rec} is the refrigerant mass inside the receiver, V_{rec} is the receiver volume and H is the total enthalpy.

It can be noted that, as reported in Section 3.1, under steady-state conditions, eq (12) yields $h_6 = h_v$ (i.e., saturated vapor at the outlet of the evaporator). On the contrary, during transient conditions, the presence of a liquid/vapor mixture or superheated vapor can be found.

4.2. Model validation and results

The numerical model has been used to simulate the dynamic behavior of the solar assisted heat pump. Simulations have been run considering real operating conditions as reference. Figure 6 reports the main parameters recorded during an experimental test performed in dynamic conditions. The heat pump was working at 40% of the maximum compressor speed and high-pressure equal to 80 bar for several hours. Figure 6a shows the environmental conditions during 2 hours of that experimental test. It can be observed that the air temperature was between 14 °C and 17 °C whereas the solar irradiance was characterized by high variations due to the passage of clouds. The evaporation temperature at the beginning of the test was around 11 °C since before 12:50 there were clear sky conditions. At 13:15 the solar irradiance decreased from 1000 W m^{-2} to 300 W m^{-2} and the evaporation temperature reached 4°C. At 13:30 the sky was completely covered by clouds and the solar irradiance was below 200 W m^{-2} until 14:15. During this time interval, the evaporation temperature decreased, and the minimum value was equal to -1.9 °C. When the solar irradiance rose, the evaporation temperature increased accordingly. The response of the heat pump is not immediate, due to the thermal inertia of the PV-T evaporator. This is also confirmed by the numerical model, which is capable to reproduce the evolution of evaporation temperature with a mean deviation with respect to the experimental data equal to 0.63 K.

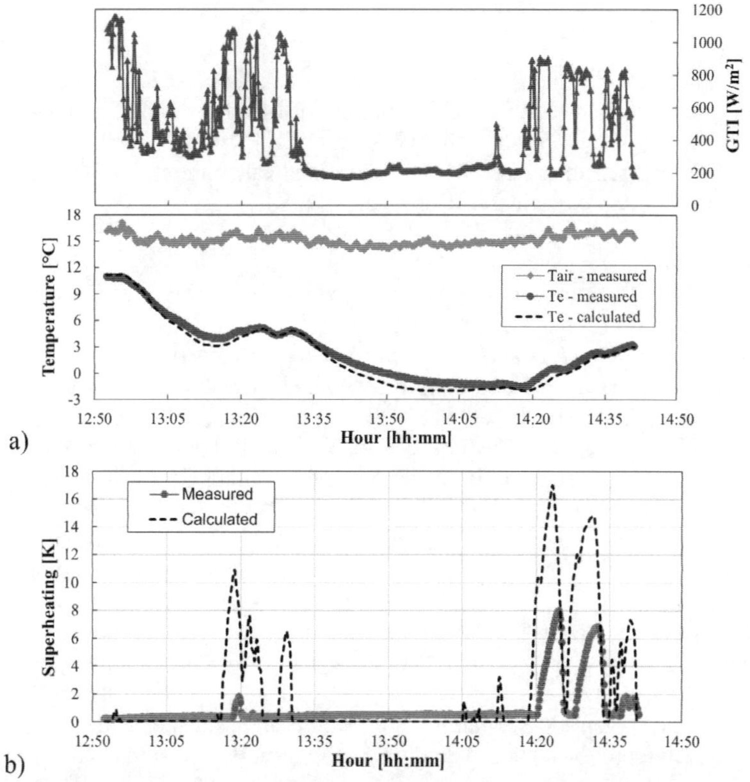

Figure 6. Comparison between the experimental and numerical results obtained in dynamic conditions during a partly cloudy day: (a) air temperature, solar irradiance and evaporation temperature versus time; (b) superheating at the evaporator's outlet.

Figure 6b displays the vapor superheating at the evaporator's outlet during the same test. It can be observed that the experimental superheating was below 1 K in the time intervals characterized by a decrease in the evaporation temperature (low irradiance). On the contrary, when the solar irradiance increased sharply, some superheating was detected. However, the superheating conditions at the outlet of the PV-T collectors were reached only for limited time intervals, until the regulation of the evaporation temperature. The numerical model is capable to reproduce when superheating occurs, but it overestimates its value, with a maximum deviation respect to the experimental data equal to 8 K.

5. Conclusions

In the present work, the performance and control of a dry expansion solar evaporator of a solar-assisted heat pump are presented. The heat pump prototype works with CO_2 as refrigerant and is equipped with three PV-T solar collectors that work as the evaporator. The system operates in dry expansion mode, meaning that the flow of CO_2 to the PV-T evaporators is controlled by the throttling valve. The experimental data show that, under steady-state conditions, the vapor quality at the evaporators' outlet was in a range between 0.94 and 1. The present configuration efficiently maximizes the evaporation temperature by fully exploiting all the available surface area for the evaporation of the CO_2 flow rate and providing a uniform cooling of the PV cells. The evaporation temperature has been found to increase linearly with the solar irradiance when all the other parameters are constant, and the increase rate, in the present operative conditions, is equal to 0.8 K per 100 W m^{-2}. The numerical model presented here can simulate the dynamic operation of the system. The model was found to predict with good accuracy the response of the heat pump during a partly cloudy day: it was further demonstrated that superheating at the outlet of the evaporator is found only in transient conditions when solar irradiance varies sharply.

Acknowledgements

Project funded under the National Recovery and Resilience Plan (NRRP), Mission 4 Component 2 Investment 1.3 - Call for tender No. 1561 of 11.10.2022 of Ministero dell'Università e della Ricerca (MUR); funded by the European Union – NextGenerationEU, Project code PE0000021, Concession Decree No. 1561 of 11.10.2022 adopted by MUR, according to attachment E of Decree No. 1561/2022, Project title "Network 4 Energy Sustainable Transition – NEST"

References

[1] EU No 517/2014 2014 European Regulation No 517/2014 on fluorinated greenhouse gases and repealing Regulation (EC) No 842/2006 *Off. J. Eur. Union*
[2] United Nations 2016 *Kigali Amendment to the Montreal Protocol*
[3] Rony R U, Yang H, Krishnan S and Song J Recent Advances in Transcritical CO 2 (R744) Heat Pump System: A Review
[4] Yang L W, Xu R J, Hua N, Xia Y, Zhou W Bin, Yang T, Belyayev Y and Wang H S 2021 Review of the advances in solar-assisted air source heat pumps for the domestic sector *Energy Convers. Manag.* **247** 114710
[5] Duarte W M, Rabelo S N, Paulino T F, Pabón J J G and Machado L 2021 Experimental performance analysis of a CO2 direct-expansion solar assisted heat pump water heater *Int. J. Refrig.* **125** 52–63
[6] Faria R N, Nunes R O, Koury R N N and Machado L 2016 Dynamic modeling study for a solar evaporator with expansion valve assembly of a transcritical CO2 heat pump *Int. J. Refrig.* **64** 203–13
[7] Paulino T de F, de Oliveira R N, Maia A A T, Palm B and Machado L 2019 Modeling and experimental analysis of the solar radiation in a CO2 direct-expansion solar-assisted heat pump *Appl. Therm. Eng.* **148** 160–72
[8] Zanetti E, Azzolin M, Girotto S and Del Col D 2023 Performance and control of a CO2 dual source solar assisted heat pump with a photovoltaic-thermal evaporator *Appl. Therm. Eng.* **218**
[9] Paradis P L, Rousse D R, Lamarche L and Nesreddine H 2018 A hybrid PV/T solar evaporator using CO2: Numerical heat transfer model and simulation results *Sol. Energy* **170** 1118–29
[10] Lemmon E W, Huber M L and Mclinden M O 2018 REFPROP 10.0 *NIST Stand. Ref. database*
[11] Zanetti E, Azzolin M, Conte R, Girotto S and Del Col D 2022 Experiments and dynamic modelling of dry expansion and flooded evaporators in a CO2 solar assisted heat pump *Appl. Therm. Eng.* **217**

Energy comparison analysis between direct and indirect dry saturated steam generation, thermally powered by EFPCs' solar fields

A Levrano[1], A V Anacreonte[1,2,*], E Gaudino[1,2], R Vitobello[2], S Sparano[2], R Russo[3], M Musto[1,3]

[1] Department of Industrial Engineering, University of Naples Federico II, P.le Tecchio, 80, Naples (Italy)
[2] TVP Solar SA, 10 rue du Pré-de-la-Fontaine ZIMEYSA, 1242 Satigny, GE, Switzerland
[3] Institute of Applied Sciences and Intelligent Systems, National Research Council of Italy, Via Pietro Castellino 111, Napoli, 80131, Italy

* Corresponding author: vitalianoalessandro.anacreonte@unina.it

Abstract. Steam is a key energy vector in the industrial sector and each application requires it at a specific pressure and temperature. In this paper the production of low pressure dry saturated steam for industrial use through high-vacuum flat plate solar collectors (HVFPCs) is discussed. This technology can produce steam from solar energy, hybridizing it with existing fossil powered steam generators to obtain significant energy savings and reduce CO_2 emissions. An energy comparison using the 0-D TRNSYS® software between numerical results of different plant configurations is made, which differ in the type of dry saturated steam production device. These devices are necessary as it is not possible to produce steam directly inside collectors. Two possible steam generation methods were analysed: direct steam production, using a Flash vessel, and indirect steam production, using a Kettle reboiler. Finally, each configuration was simulated by imposing a solar field ΔT of 10 °C and 20 °C. Dynamic results show that flash vessel configurations are generally the most efficient, with the same operating parameters, compared to the configurations with Kettle reboiler. Furthermore, configurations with certain ΔT, such as to determine lower operational solar field temperatures, lead to the best results due to the higher HVFPCs' efficiency.

1. Introduction

Dry saturated steam is widely used within global industry, the latter made up of various production sectors, each one requiring steam at a specific pressure level. Generally dry saturated steam at low and medium pressure is used in the construction sector, and in chemical, medical and food industries. This is generally produced on site using firetube steam generators, which allow to produce dry saturated steam up to pressures of 20 bar (upper limit of medium pressure dry saturated steam). Dry saturated steam at high pressure, up to 60 bar, is used, instead, in the metal and petrochemical industries, and is generally produced by direct flame generators, i.e., industrial water-tube boilers [1]. All these traditional steam production systems exploit the fossil fuel source (natural gas, coke, etc.). It is estimated that around 30% of the fossil fuel consumed in the world's industrial sector is burned to produce process steam [2]. This practice is the cause of significant CO_2 emissions into the environment and so far, has benefitted of limited decarbonization efforts. However, these could be partially reduced by resorting to innovative steam production devices that use renewable solar energy as a thermal input source, and which can well be included in the global industrial decarbonization discourse.

In recent years, an extremely efficient thermal production technology using solar energy is being developed. These are MT-Power ultra-high vacuum solar collectors, i.e., high efficiency solar collectors, operating in the temperature range of 80-180°C, manufactured by TVP Solar company. These collectors belong to the family of non-concentrating solar panels and are direct descendants of the much less efficient flat plate solar collectors [3]. The distinctive feature of these collectors is that they have an evacuated air gap between the absorber and the protective glass, in which an air-pressure of 10^{-4} mbar is reached [4]. This air evacuation is much deeper than in the traditional evacuated collectors [5] and allows convective heat exchanges between absorber and glass to be almost eliminated, resulting in a net reduction of the device's heat losses, and, consequently, in high heat absorption efficiencies even at relatively high operating temperatures [6].

Currently these collectors can only absorb thermal energy in the form of an increase in the absorbing fluid's sensible heat up to 180 °C, for which an absorption efficiency of no less than 50% is estimated, as some studies conducted in 2018 [7] and 2020 [8] have shown. This is since steam cannot be produced directly inside the collectors. A recent study [9] has analysed the possibility to directly generate steam, but the practical activities results have shown that under the effective working conditions it is not possible to reproduce turbulence conditions inside the copper pipes of the collectors such as to determine an efficient mixing between liquid and vapor phases produced, necessary to obtain an acceptable thermal absorption capacity, since the operating temperatures are too low to determine it. Therefore, because of the boundary conditions and the technological limits mentioned, the production of steam directly in the collectors would lead to a too low heat absorption efficiency.

In this paper alternative steam production systems which do not contemplate steam production directly inside collectors are analysed. The dry saturated steam production technique proposed involves the use of thermo-fluid dynamic devices which carry out the so-called sensible-to-latent transformation. Two possible steam generation methods were analysed: direct steam production, using a Flash vessel (FV), and indirect steam production, using a Kettle reboiler (KR). A comparison between numerical results of different plant configurations is made, which differ from each other in the type of dry saturated steam production device, and in the level of solar field operating temperature.

2. Case studies

A dynamic study is conducted on some existing TVP Solar plants, using the 0-D dynamic simulation software TRNSYS®. Within this software, the logical functioning of the solar field and steam production device is reproduced with appropriate "Types", and the behaviour of the entire plant is dynamically analysed as the boundary conditions, such as the hourly solar radiation and ambient temperature, and secondary climatic conditions such as wind, humidity, etc., vary.

The case studies analysed concern two TVP Solar plants: one installed at the operational company headquarters in Avellino, Italy, in which only a part of the available single side solar field is usable and equal to approximately 140 m^2, corresponding to 70 collectors arranged in 10 parallel strings of 7 panels each; the other located in Qurayyah, Saudi Arabia, having a double-side type solar field with a net receiving area of approximately 1000 m^2 and made up of 510 collectors.

In addition to evaluating the dynamic operational differences of the two devices for the two case studies, in this study, the plant production difference that occurs by modifying the design temperature difference between the fluid entering the solar field and the fluid leaving it is also evaluated. Two possible design ΔT are simulated for both configurations, equal to 10 °C and 20 °C.

3. Steam generation devices' main features

The analysed thermodynamic devices, i.e., Flash vessel (FV) and Kettle reboiler (KR), use the sensible heat of the fluid coming from the solar field as thermal input to produce steam. In the first case, steam is produced inside the flash vessel unit through a flash-type expansion [10] of the same solar field fluid (primary fluid), while in the second case, steam is produced thanks to the heat transferred by the primary fluid to a secondary evaporating fluid (which does not circulate in the solar field), using the Kettle reboiler. In both cases, the exclusive circulation of liquid fluid inside collectors is ensured.

However, both devices impose a certain reduction in the temperature necessary to produce steam with respect to the maximum temperature reachable in the solar field (currently 180 °C). For this reason, an upper limit of dry saturated steam production temperature of around 150 °C can be defined.

However, only low-pressure steam can be produced, as it is below 6 bar. This is the highest pressure achievable, and it is possible only under specific conditions of irradiance, i.e., in the Middle East and North Africa. Instead, under average Italian climatic conditions, an upper production temperature limit of 160 °C is assumed, which allows the production of dry saturated steam up to 2 bar and 130 °C.

3.1. *Flash vessel*

Flash vessel is a direct generation device, in which the primary flow entering-it is the same evaporating fluid coming out of it. This flow, which is necessarily hot pressurized osmotic water, undergoes a flash-type expansion thanks to a throttling valve, which imposes a designed pressure reducing. During the expansion, the amount of heat in the fluid is conserved (isenthalpic transformation), but the saturated liquid cannot hold this amount of heat, so it flashes. Therefore, saturated steam, composed of a mixture of steam and largely of water, is produced at pressure lower than the input one, and is sent to the flash separator, which allows to separate the dry saturated steam from the saturated condensate, thanks to the different densities of the two phases.

3.2. *Kettle reboiler*

This device is like a shell-and-tube heat exchanger, in which a primary fluid circulates in U-bundles and transfers its sensible heat to the secondary fluid. The generation of steam is made possible because the secondary fluid sits in the boiling device during a residence time sufficient to ensure the transfer of energy from the primary to the secondary fluid. The first difference with respect to the flash vessel is that two distinct fluids are required (heat transfer fluid & osmotic water to be evaporated). As is the case for the flash vessel, the steam generated is sent to the application of the customer plant. Once it has transferred to the user its latent heat (and eventually some of its sensible heat) it returns to the kettle reboiler in the form of pressurized condensate. In order to maximize the generation of steam by the kettle, the returning condensate is beforehand preheated in a heat exchanger so as to bring it as close as possible to the saturated temperature at the desired pressure. An important difference to point out is that kettle reboiler needs to impose a pinch point ΔT between primary and secondary fluid. This entails an increase in the maximum temperature required at the primary, with the same steam production temperature, compared to the configuration with flash vessel. In this paper a pinch point ΔT equal to 10 °C is considered.

3.3. *Dry saturated steam generation devices' installation issues*

Some industrial activities require the use of both low and medium pressure steam, so the thermal output from the boiler is often medium pressure dry saturated steam, while the low pressure dry saturated steam demand is satisfied using pressure reducers [11]. Therefore, steam supply lines connecting generator to utility/reducer are mostly designed to manage medium pressure steam, and, since these cannot be decommissioned, because of the immovability of traditional generators, necessary to carry out thermal integration in the hours of absence of solar availability, direct or stored, these would not be able to manage low pressure steam produced by collectors; so it is necessary to install new supply lines, parallel to the existing ones, the cost of which is to be considered in a technical-economic feasibility analysis.

4. **TRNSYS® implementation logic**

In the Figure 1 and Figure 2 the TRNSYS® plant's logic schemes respectively with flash vessel and kettle reboiler are shown. The solar field is highlighted in blue, while the application side of the flash vessel and kettle reboiler units is highlighted respectively in red and yellow. The main "Types" used in this logic schemes are shown in Table 1 e Table 2.

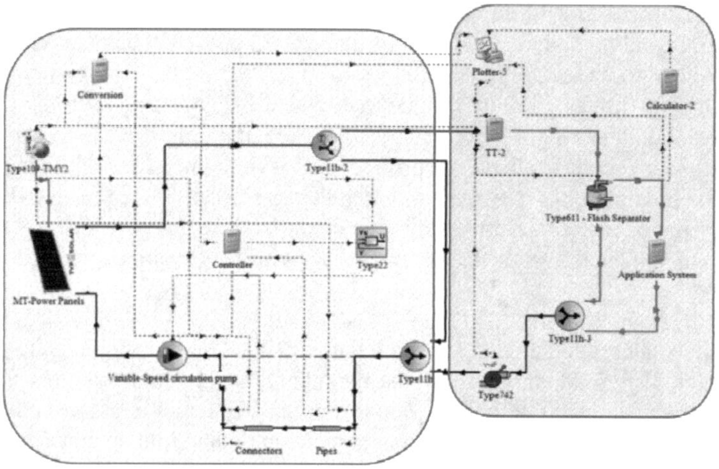

Figure 1. TRNSYS® template of plant configuration with flash vessel

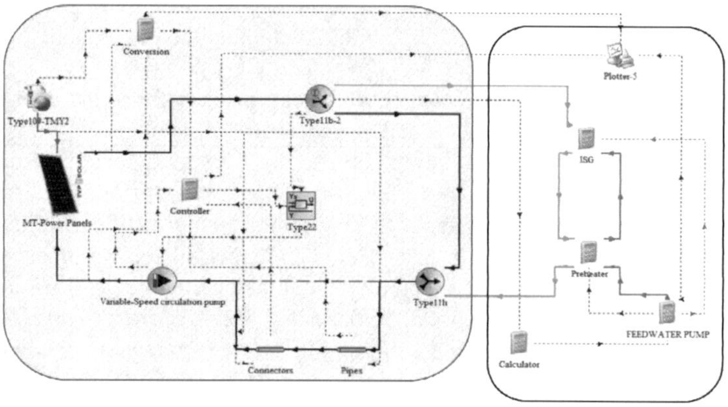

Figure 2. TRNSYS® template of plant configuration with Kettle reboiler

Table 1. Main TRNSYS® "Types" describing solar field, used in logic plant schemes implemented.

Component name	Type name	Unit group	Function
Weather	Type109	Solar field	Type containing Meteonorm® TMY files.
MT-power panel	Type432	Solar field	Type calculating the solar energy collected.
Circulating pump	Type110	Solar field	Type regulating solar field flow rate
Iterative controller	Type22	Solar field	Type generating a pump partialization value [0,1]
Piping	Type31	Solar field	Type calculating pipes thermal losses
Controller	Equa-Type	Solar field	Type calculating the electrical power consumed by the solar field pump
3-w-diverting-valve	Type11b	Solar field	Type diverting flow to application side or solar filed depending on its temperature
3-w-mixing-valve	Type11h	Solar field	Type mixing flows and directing it to the solar field

Table 2. Main TRNSYS® "Types" describing steam devices, used in logic plant schemes implemented.

Component name	Type name	Unit group	Function
Flash separator	Type611	Flash vessel	Type generating output dry saturated steam thermodynamic properties
Booster pump	Type742	Flash vessel	Type restoring the correct expansion pressure of the flash vessel returning flow
Application system	Equa-Type	Flash vessel	Type expressing the user's heat exchanger behaviour
ISG	Equa-Type	Kettle reboiler	Type expressing the kettle reboiler behaviour
Pre-heater	Equa-Type	Kettle reboiler	Type expressing the kettle reboiler pre-heater behaviour
Feedwater pump	Equa-Type	Kettle reboiler	Type expressing the kettle reboiler secondary flow feedwater pump behaviour

In Table 2, the kettle reboiler appears under the wording ISG (indirect steam generator) and modeled using the Equa-Type, as a customized type that simulates its behavior is not available in TRNSYS®. Equation (1) shows the logical calculation of the steam produced instant by instant by kettle reboiler:

$$SteamProduced = NetHeat * \left(\frac{1 - Thermal\ Losses\ ISG\%}{\Delta H_{ISG} + \Delta H_{ph}}\right) \quad (1)$$

where $NetHeat$ is the thermal energy available as input to the steam production device, $Thermal\ Losses\ ISG\%$ are the kettle reboiler percentage thermal losses towards the environment (imposed equal to 1% of the available thermal energy), and ΔH_{ISG} and ΔH_{ph} are the design enthalpy variations inside the kettle reboiler and preheater.

5. Results

The annual results of plant steam production, total production hours and electricity consumption due to the circulation and pressurization pumps, are presented. In the Figure 3 and Figure 4 the Avellino and Qurayyah TVP Solar plant simulation results, respectively, are reported. The comparison analysis between all the different plant configurations is unique for both case studies.

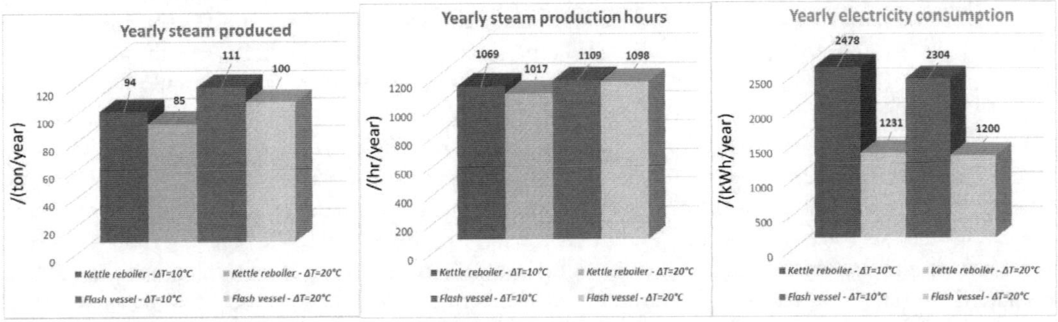

Figure 3. Avellino plant: yearly steam production, total annual working hours, and yearly electricity consumption by pumps

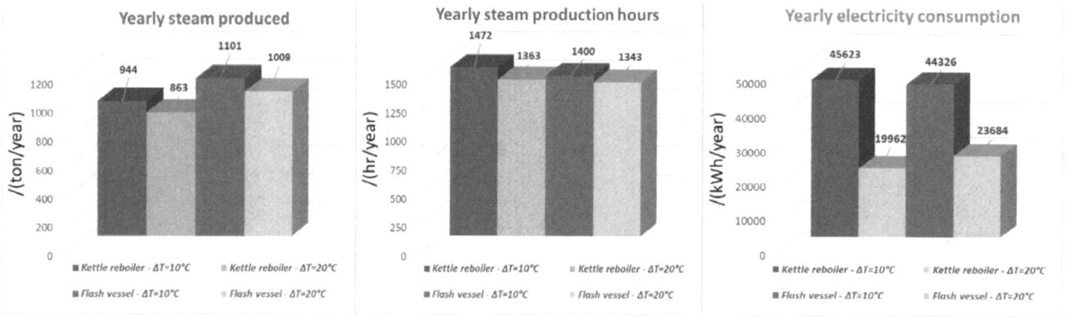

Figure 4. Qurayyah plant: yearly steam production, total annual working hours, and yearly electricity consumption by pumps

Plant steam production is a function of a net balance between the energy collected in the solar field and the thermal losses inside the pipes and devices. The configuration that determines a higher annual production of steam (111 ton/year for Avellino plant, and 1101 ton/year for Qurayyah plant) is the one with a flash vessel and primary flow ΔT equal to 10 °C (FV & ΔT=10 °C). That is since the flash vessel configuration allows for a lower maximum temperature at the primary side than the kettle reboiler one. Therefore, collectors tend to work with greater efficiency, determining greater productivity. This result is magnified by the fact that this configuration presents the lowest average solar field temperature, since, at the same solar field inlet fluid temperature, a ΔT of 10 °C is considered. Furthermore, with a lower average working temperature, less heat losses in the pipes are detected. Moreover, thermal dissipation from flash vessel tends to be lower because those devices are more compact than kettle reboiler because of the smaller residence time of the first device compared to the second. Everything said also explains why the plant configuration with kettle reboiler and ΔT equal to 20°C (KR & ΔT=20 °C) is the worst in terms of productivity.

Plants productive operating hours reach values of around 1000 h/year and 1400 h/year, respectively for Avellino and Qurayyah, plants, but results do not show objectively justifiable differences between the various simulated configurations.

Total electrical energy consumed by pumps is a parameter of little interest during plant design steps, but in this paper, it is fundamental since it is the only factor that can potentially generate CO_2 emissions in plant operational hours. As can be seen in Figure 3 and Figure 4, there is no net difference in electrical consumption between configurations with different steam production devices. On the other hand, evident differences are found in configurations with different ΔT. In particular, the configurations with lower ΔT are the most energy intensive. This result is related to the amount of circulating flowrate, which, considering same thermal energy collected in the solar field, in the configuration with $\Delta T = 10\ °C$ must necessarily be double that the one circulating in configurations with $\Delta T = 20\ °C$. Therefore, "ΔT=10 °C" configurations have double energy consumption compared to the "ΔT=20 °C" ones (around 2400 kWh/year vs 1200 kWh/year for Avellino plant, and around 45000 kWh/year vs 22000 kWh/year for Qurayyah plant). Anyway, taking the configuration "FV & ΔT=10 °C" as a reference for optimal production, all other configurations need to carry out thermal integrations to achieve the same steam productivity. Thermal integrations are generally made through a traditional medium-low pressure steam generator, i.e., a firetube boiler, for which a primary energy consumption of 650 kWh for each ton of steam produced can be assumed. Considering a firetube boiler having an overall conversion efficiency of 90% and burning natural gas, with an emission factor of 0.24 kgCO_2/kWh, Tables 3 and 4 show the CO_2 emission savings for the reference configuration, comparing it to the other analysed ones. On the other hand, the reference configuration has a higher electricity consumption, so extra emissions due to electricity purchased from the grid are also calculated. Electrical grid emission factors equal to 0.424 kgCO_2/kWh for Italy and 0.568 kgCO_2/kWh for Saudi Arabia are considered (Tables 5 and 6). As shown in Table 7 and Table 8, for both the Avellino and Qurayyah plants, the reference configuration always allows overall emission savings compared to all the other configurations analysed. Therefore, this configuration represents the best choice also from an environmental point of view. In the Tables 7 and 8 a balance between CO_2 emission savings due to the less thermal integration and extra CO_2 emissions due to the extra electricity-purchasing in the reference configuration, for both Avellino and Qurayyah plants, is shown.

Table 3. Comparison between CO_2 emissions due to thermal integration - Avellino plant

Steam production reference	FV & $\Delta T = 10\ °C$		
Comparison plant configurations	FV & ΔT=20 °C	KR & ΔT=10 °C	KR & ΔT=20 °C
Relative steam production /(ton/year)	+11	+17	+26
Relative thermal integration /(kWh/year)	-7150	-11000	-17000
Relative CO_2 emissions /(kgCO_2/year)	-1900	-2900	-4500

Table 4. Comparison between CO_2 emissions due to thermal integration - Qurayyah plant

Steam production reference	FV & ΔT=10 °C		
Comparison plant configurations	FV & ΔT=20 °C	KR & ΔT=10 °C	KR & ΔT=20 °C
Relative steam production /(ton/year)	+92	+157	+238
Relative thermal integration /(kWh/year)	-60000	-102000	-155000
Relative CO_2 emissions /(kgCO_2/year)	-16000	-27200	-41500

Table 5. Comparison between CO_2 emissions due to grid electricity usage - Avellino plant

Steam production reference	FV & ΔT=10 °C		
Other analysed plant configurations	FV & ΔT=20 °C	KR & ΔT=10 °C	KR & ΔT=20 °C
Relative electrical consumption /(kWh/year)	+1100	-170	+1070
Relative CO_2 emissions /(kgCO_2/year)	+460	-70	+450

Table 6. Comparison between CO_2 emissions due to grid electricity usage – Qurayyah plant

Steam production reference	FV & ΔT=10 °C		
Other analysed plant configurations	FV & ΔT=20 °C	KR & ΔT=10 °C	KR & ΔT=20 °C
Relative electrical consumption /(kWh/year)	+20500	-1300	+24300
Relative CO_2 emissions /(kgCO_2/year)	+11600	-740	+13800

Table 7. CO_2 emission balance for Avellino plant configurations

Steam production reference	FV & ΔT=10 °C		
Comparison plant configurations	FV & ΔT=20 °C	KR & ΔT=10 °C	KR & ΔT=20 °C
Thermal integration CO_2 /(kgCO_2/year)	-1900	-2900	-4500
Grid electricity CO_2 /(kgCO_2/year)	+460	-70	+450
Balance	*-1440*	*-2970*	*-4050*

Table 8. CO_2 emission balance for Qurayyah plant configurations

Steam production reference	FV & ΔT=10 °C		
Comparison plant configurations	FV & ΔT=20 °C	KR & ΔT=10 °C	KR & ΔT=20 °C
Thermal integration CO_2 /(kgCO_2/year)	-16000	-27200	-41500
Grid electricity CO_2 /(kgCO_2/year)	+11600	-740	+13800
Balance	*-4400*	*-27940*	*-27700*

6. Conclusions

This paper deals with the technical problem related to the production of dry saturated steam for industrial use by exploiting thermal energy produced by MT-Power technology high vacuum solar collectors. In detail, two possible dry saturated steam production technologies were investigated: direct production through a flash vessel, and indirect production through kettle reboiler. This study was carried out with dynamic analyses conducted using the TRNSYS® simulation software. Results showed that the optimal accessory device, to combining with an ultra-high vacuum collectors' solar field, to carry out the sensible-to-latent transformation is the flash vessel, as it allows the solar field to work at relatively low average temperatures and involves higher collectors' efficiency and less heat losses. Moreover, for the same reason, it results that is more convenient to keep the plant ΔT as low as possible. With same solar field ΔT, plant configuration with flash vessel has a productivity of +18%. While, with the same steam-device, configurations with a solar field ΔT equal to 10 °C have a productivity of +10%, but also a higher electricity consumption of about +100%. Comparing the two TVP Solar plants, the one placed in Avellino has a unitary steam productivity of about 0.7 $ton/m^2/year$, while the other in Qurayyah has unitary productivity of 1 $ton/m^2/year$.

Future studies will analyse the increase in plant productivity due to installation of a pressurized thermal storage tank, which accumulates the excess thermal energy produced during the day and then return it in the hours with lack of insolation. Moreover, the results obtained in this study will lead to the implementation in the Avellino facility of a flash vessel, which will be used as an experimental setup for the upcoming investigations.

References

[1] Hall S M 2018 *Rules of Thumb for Chemical Engineers – Sixth Edition* (Amsterdam: Elsevier).
[2] Einstein D, Worrell E and Khrushch M 2001 *Lawrence Berkeley National Laboratory*.
[3] Gao D, Zhong S, Ren X, Kwan T H and Pei G 2022 *Renew. Energy* **184**: 20.
[4] D'Alessandro C, De Maio D, Caldarelli A, Musto M, Di Giamberardino F, Monti M, Mundo T, Di Gennaro E, Russo R and Palmieri V G 2022 *Sol. Energy* **243**: 81.
[5] Benz N and Beikircher T 1999 *Sol. Energy* **65** I2.
[6] De Maio D, D'Alessandro C, Caldarelli A, De Luca D, Di Gennaro E, Casalino M, Iodice M, Gioffre M, Russo R, Musto M 2021 *Thin Solid Films* **735**: 138869.
[7] Moss R W, Henshall P, Arya F, Shire G S F, Hyde T and Eames P C 2018 *Appl. Energy* **216**: 588.
[8] Gao D, Gao G, Cao J, Zhong S, Ren X, Dabwan Y N, Hu M, Jiao D, Kwan T H and Pei G 2020 *Appl. Energy* **26**: 115129.
[9] Gao D, Li J, Ren X, Hu T and Pei G 2022 *Renew. Energy* **197**: 966.
[10] Mansour A and Müller N 2019 *Exp. Therm. Fluid Sci.* **107**: 146.
[11] Qian J, Wei L, Zhang M, Chen F, Chen L, Jiang W and Jin Z 2017 *Energy* **135**: 650.

Infrared measurements of fluid temperature in a polymeric Pulsating Heat Pipe

L Pagliarini[1,*], F Clemens[2], F Bozzoli[1], L Cattani[1], N Miche[2], M Bernagozzi[2], M Marengo [2,3], A A Alqahtani[4], V Bertola[4]

[1]Department of Engineering and Architecture, University of Parma, Parco Area delle Scienze 181/A, I-43124 Parma, Italy
[2]Advanced Engineering Centre, School of Architecture, Technology and Engineering, University of Brighton, Lewes Road BN24GJ, United Kingdom
[3]Department of Civil Engineering and Architecture, University of Pavia, Via Ferrata 3, Pavia, Italy
[4]Laboratory of Technical Physics, University of Liverpool, Brownlow Hill, Liverpool L69 3GH, United Kingdom

* Corresponding author: luca.pagliarini@unipr.it

Abstract. Pulsating heat pipes are two-phase passive heat transfer devices partially filled with a working fluid in saturation conditions. During operation, supplying heat to one end of the system (named evaporator) results in a local increase in temperature and pressure, which drives the fluid through a transport section (named adiabatic section) towards the cooled, opposite end (named condenser) for effective heat dissipation. The local thermo-fluid dynamic state of the working fluid is sometimes assessed by means of non-intrusive techniques, such as infrared thermography. In this case, the radiative properties of the systems in the infrared spectrum must be known a priori. Nevertheless, since pulsating heat pipes may be manufactured with different materials, wall thicknesses and channel geometries, the radiative properties of the walls and the confined flow are not always known or assessable by means of the available literature. Hence, the work proposes to design a straightforward calibration procedure for quantitative infrared fluid temperature measurements in a polymeric pulsating heat pipe charged with FC-72 and having unknown radiative properties. The emissivity and transmissivity of the walls and confined fluid are estimated with good accuracy. The results will allow repeatable and reliable fluid temperature measurements in future experimentations on the mentioned device.

1. Introduction

Effective and reliable heat transfer devices are required for a substantial reduction in energy consumption for electronics cooling. Two-phase systems are widely used in thermal management applications [1]. However, standard active two-phase devices, which rely on the employment of auxiliary equipment, are not efficient from an energetic standpoint [2]. On the contrary, passive devices are characterized by spontaneous heat dissipation, thus representing an attractive solution to the energy issue. In particular, Pulsating Heat Pipes (PHPs) are achieving resounding interest within the academic and industrial communities due to their high flexibility, high heat transfer efficiency and low manufacturing cost, when compared to other passive systems, e.g., Heat Pipes (HPs). In their most common and simplest layout, PHPs are constituted by a capillary evacuated tube bent in many turns and filled with a working fluid in saturation conditions, thus presenting an alternation of liquid slugs and vapor plugs. Three main sections are often identified: the evaporator zone, where heat is supplied, the condenser zone, where heat is dissipated, and the adiabatic section, where the device does not thermally interact with the external environment [3]. Heat dissipation is guaranteed by self-sustained oscillations of the working fluid, induced by temperature differences between the evaporator and the condenser. Such thermal instabilities result in pressure differences acting as pumping forces for a spontaneous fluid motion [4]. Nevertheless, the very complex non-equilibrium thermodynamics underlying the PHPs physics undermines a clear and full understanding of their working behaviour. In addition, the experimental approaches usually adopted for the study of PHPs lack of completeness and generality. This is due to the fact that the PHPs operation is generally assessed from global characterizations, based on measurements over time of both evaporator and condenser temperatures [5]. A local evaluation of the fluid thermodynamics state is instead rarely achieved ([6–8]) due to difficulties linked to sensors placement inside the fluid stream without increasing the set-ups complexity or perturbing the system [9]. On the other hand, devices having transparent inserts or windows might be investigated in terms of working fluid temperature by means of non-intrusive techniques, such as InfraRed (IR) thermography. Mameli et al. [10] carried out IR investigations on a sapphire insert (transmissivity provided by the manufacturer) in a multi-turn PHP having inner diameter equal to 3 mm. The device was filled with FC-72 (50% vol.) and tested under different gravity conditions. At both ends of the transparent channel, micro-thermocouples were inserted in the fluid stream to directly monitor the fluid temperature during operation. Being FC-72 (liquid phase) partially transparent to IR radiations, as suggested in [11], its temperature could be only assessed when it completely filled the tube. In fact, by comparing the fluid temperature signals recorded through micro-thermocouples and the ones recorded by the IR camera, an agreement between the two sets of data was present only when liquid passed through the section investigated by thermography. Due to the excellent sensitivity of the adopted IR camera, temperature gradients along liquid slugs could be also assessed with good accuracy, reflecting the local effects of both the evaporator and condenser sections on the fluid flow. Mangini et al. [12] performed IR acquisitions on the fluid flow through transparent inserts made of sapphire (thickness: 1 mm, inner diameter: 2 mm) in a multi-turn PHP partially charged with ethanol. Here, the transmissivity of the sapphire wall was provided by the manufacturer equal to 0.9. However, the fluid emissivity at different temperatures was unknown, preventing quantitative fluid temperature measurements. Hence, liquid ethanol was circulated at prescribed temperatures through a sapphire channel (same dimensions of the one mounted on the experimental set-up). The transparent branch was framed by an IR camera, and the outputted signals were compared with the ones obtained from the IR observation of a black body (emissivity = 0.997) at same temperatures. The two sets of IR signals were in good agreement, suggesting that ethanol (liquid phase) was perfectly opaque to infrared radiations for the considered thickness of 2 mm in between 15°C and 35°C. The vapour phase was instead found to be fully transparent. The calibration loop shown in [12] was similar to the one adopted in [11]. However, the estimation of IR properties by using a dedicated calibration loop is not always feasible due to possible difficulties in the manufacturing process or presence of complex layouts (e.g., micro or flexible geometries).

The present work proposes to design a straightforward technique to estimate radiative properties of a polymeric PHP partially filled with FC-72 for the sake of future experimentations, without substantial modifications on the layout or dedicated experimental apparatuses. The main challenge is represented by the fact that both the device walls and the confined fluid are partially transparent in the IR spectrum. Hence, the wall, fluid and background temperatures must be carefully monitored during calibration. The charged device is kept at prescribed temperature inside a thermal chamber. The background temperature is controlled by means of a thermostatic loop. IR acquisitions are carried out during steady state. By adopting a radiation model that considers emissions and transmissions of all the radiative elements in the system, emissivity and transmissivity of both working fluid and PHP walls (partially transparent in the IR spectrum) are estimated through a minimization approach. The

Journal of Physics: Conference Series **2685** (2024) 012050 doi:10.1088/1742-6596/2685/1/012050

proposed technique will be used to achieve quantitative fluid temperature measurements by means of thermography on the investigated device. The procedure could be replicated for IR analyses of other partially transparent devices, whose geometry and material do not allow feasible manufacturing of a calibration loop.

2. Experimental set-up

The studied PHP (6 turns in the evaporator section, 5 turns in the condenser section) is made of polypropylene, with 3 x 0.7 mm rectangular channels (Figure 1a). In particular, polypropylene is chosen for fabrication to guarantee high degree of foldability and consequent high flexibility for different kinds of application [13], as shown in Figure 1b.

a) b)

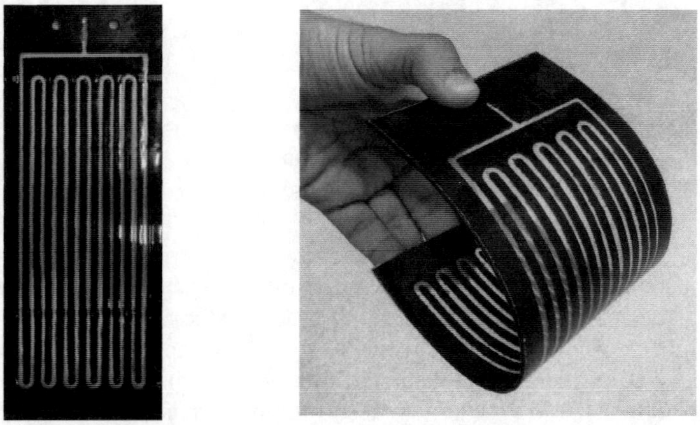

Figure 1. Studied device (a) and its potential foldability (b).

The serpentine channel is cut-out in a black sheet (0.7 mm thickness) using a commercial laser cutter (HPC® Laser LS1290 Pro). Then, the channel is sandwiched between two transparent sheets (0.4 mm thickness each), which were tightly bonded to the black sheet by selective transmission laser welding using a nanosecond pulsed fiber (SPI® Lasers G4 HS-L 20 W). A micro-metering valve, mounted on an aluminum clamp screwed on the PHP, is used for the filling/vacuuming procedure. For further details on the manufacturing processes, see [14].

A calibrated IR camera (TELOPS® FAST-IR M100K, pixel resolution: 1280x1024, spectral bandwidth: 3 - 4.9 μm) is used to perform acquisitions on the device. Before the tests, the PHP was filled with pure FC-72 (FR = 50% vol.), previously degassed by means of iteratively warming/cooling under vacuum conditions.

For successful fluid temperature measurements by means of thermography, the following radiative properties of the system needed to be estimated since practically unknown for the given application (i.e., for the PHP walls and fluid thicknesses):

- Emissivity and transmissivity of the polypropylene walls;
- Emissivity and transmissivity of FC-72 (liquid phase), when confined inside the device channels.

Specifically, to achieve a good characterization of the PHP radiative behaviour in the IR spectrum, the effects of the background and of the emissive elements constituting the PHP, such as the polypropylene walls and the working fluid, should be separately taken into consideration. Hence, the background emissivity and temperature must be known, together with the temperature of the emissive elements. To this aim, a copper plate, which surface was coated with a high-emissivity paint ($\varepsilon = 0.92$), and which temperature was controlled by a thermal bath, was placed behind the device (Figure 2a).

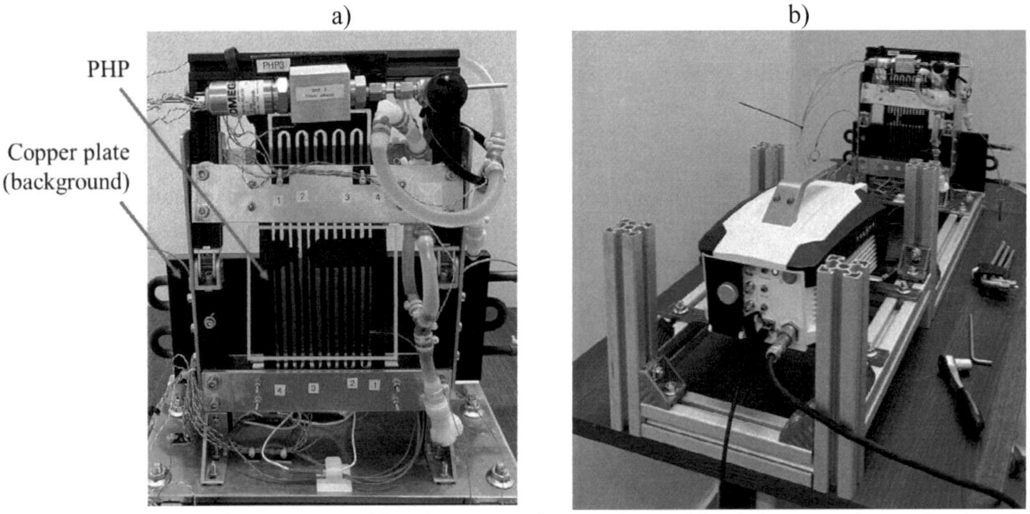

Figure 2. Experimental apparatus, with reference to the area framed by the camera (green box) and background position (a); test bench, located inside the thermostatic chamber (b).

The set-up was therefore introduced inside a thermostatic chamber (accuracy = ± 1°C), as shown in Figure 2a; this guaranteed that, at steady-state conditions, the temperature of all the emissive elements but the background was equal to the set-point temperature of the chamber.

The PHP was tested without triggering its activation by means of any heat load at the evaporator, and two IR pictures were taken when thermal equilibrium with the chamber environment was reached. Specifically, the portion framed by the IR camera during acquisitions is shown in Figure 2a (green box). In Table 1, the pairs of background and ambient temperature set-points imposed over a total of 7 different tests are listed.

Table 1. Temperature set-points imposed during the tests.

Test number	Background temperature set-point [°C]	Ambient temperature set-point [°C]
1	55	20
2	19	25
3	21	25
4	18	30
5	22	30
6	19	35
7	22	35

Two T-type thermocouples were placed on the background surface and on the PHP wall, respectively, to assess the accuracy of the given set-point temperatures.

3. Methods

3.1. Radiation model

The adopted radiation model is described by the energy flow diagram of Figure 4. Specifically, the IR radiation acquired by the IR camera sensors Q_{IR} is a combination of radiations emitted by the background Q_{BG}, the PHP walls Q_w' and Q_w'', and the working fluid Q_f, net of losses through the device elements (walls and fluid) due to their non-unitary transmissivity in the IR spectrum.

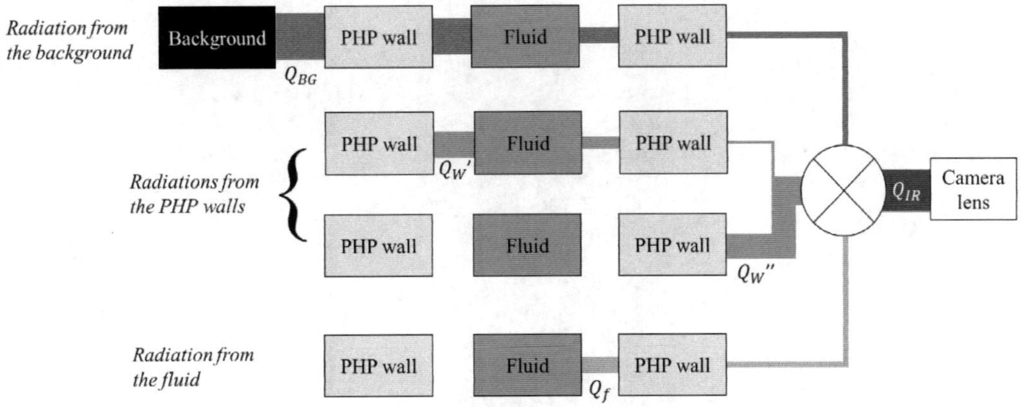

Figure 3: Radiation model.

Such a radiation model can be formalized by Equation (1), where the subscript *net* stands for net IR radiation reaching the camera lens.

$$Q_{IR} = Q_{BG,net} + (Q_{w,net}' + Q_w'') + Q_{f,net} \tag{1}$$

By furtherly considering the temperature and radiative properties of each emissive element, Equation (1) reads as:

$$Q_{IR} = \sigma\varepsilon_{BG}\tau_{PHP}\tau_f T_{BG}^4 + \sigma\left[\left(1+\tau_{PHP}^{1/2}\right)\tau_f\right]\varepsilon_w T_w^4 + \sigma\varepsilon_f\tau_{PHP}^{1/2}T_f^4 \tag{2}$$

where σ is the Stefan-Boltzmann constant, ε_{BG} is the emissivity of the background, τ_{PHP} is the transmissivity of both PHP walls ($\tau_{PHP}^{1/2}$ for the single wall), T_{BG} is the background temperature, τ_f is the transmissivity of the fluid, ε_w is the emissivity of the single PHP wall, T_w is the wall temperature, and ε_f is the fluid emissivity.

3.2. Radiative properties evaluation

The radiative properties of all the emissive elements were evaluated by considering areas containing both vapour and liquid phases. To discern the two phases, a coefficient *c* was introduced in (1):

$$Q_{IR} = \sigma\varepsilon_{BG}\tau_{PHP}\tau_f^c T_{BG}^4 + \sigma\left[\left(1+\tau_{PHP}^{1/2}\right)\tau_f^c\right]\varepsilon_w T_w^4 + c\sigma\varepsilon_f\tau_{PHP}^{1/2}T_f^4 \tag{3}$$

Such a coefficient was set to 0 when the vapour phase was observed, and to 1 when the liquid phase was instead investigated. Hence, τ_{PHP}, ε_w, τ_f and ε_f (liquid phase) were assessed through the resolution of the following minimization problem:

$$\min_{\tau_w,\varepsilon_w,\tau_f,\varepsilon_f \in R^+}\left[Q_{IR} - \sigma\varepsilon_{BG}\tau_{PHP}\tau_f^c T_{BG}^4 - \sigma\left[\left(1+\tau_{PHP}^{1/2}\right)\tau_f^c\right]\varepsilon_w T_w^4 - c\sigma\varepsilon_f\tau_{PHP}^{1/2}T_f^4\right] \quad ; \quad \tau_w, \varepsilon_w, \tau_f, \varepsilon_f < 1 \tag{4}$$

4. Results

As mentioned in Section 3.2, IR acquisitions were processed by discerning liquid phase from vapour phase. It has to be pointed out that, since the device was not fully filled with FC-72, the liquid phase tended to stratify due to gravity and vibrations at the bottom of the channels, as noticeable from the IR sample of Figure 4.

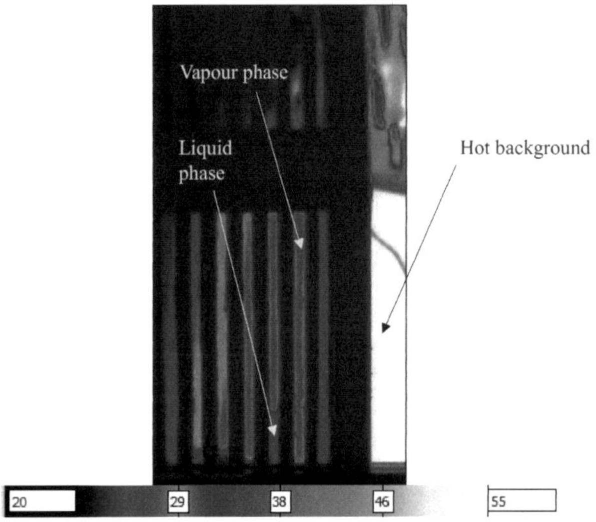

Figure 4. IR acquisition sample; system at 20°C, hot background at 55°C.

Q_{IR} referred to all pixels within each framed PHP channel was used as input for Equation (4). Specifically, the data collected during every test condition listed in Table 1 were considered at the same time in the minimization problem resolution. All the estimated radiative properties are shown in Figure 5, along with the 95% confidence intervals. In particular, the intervals have been estimated through a nonlinear regression, adopting the jacobian and residual outputs given by the combination of MATLAB® (v. R2022b) functions *nlinfit* and *nlparci*. To the Authors' knowledge, the literature does not provide radiative properties for the specific thicknesses and temperatures adopted in the present investigation, especially in the wavelength range of the adopted IR camera. Anyway, the estimated values are in good agreement with [15] for what concerns the FC-72 emissivity, and with [16] regarding the polypropylene wall transmissivity, confirming the goodness of the approach.

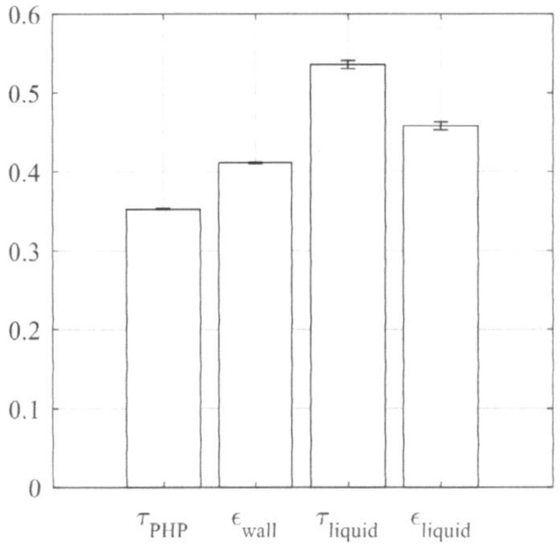

Figure 5: Evaluated radiative thermal properties of the PHP elements.

The estimation of the radiative properties of the system will be used, in future studies, for the fluid temperature evaluation during its operation in microgravity and standard-gravity conditions. It has to be stressed that, since during experimentations the wall temperature is not usually monitored in every part of the device, Equation (2) cannot be directly used to quantify T_f in the overall system. However, a simple yet effective assumption may be adopted, such as considering the wall temperature equal to the average between the fluid temperature and the environmental temperature T_{env}, $T_w = (T_{env} + T_f)/2$. Such assumption might be correct when focusing on the PHP adiabatic section (area of interest for the present investigation), according to [17], especially when the adiabatic zone is not thermally insulated for visualization purposes. T_f is therefore estimated by finding the real, positive root of the following fourth-order polynomial, coming from a rearrangement of Equation (2):

$$\left[\tau_f \varepsilon_w \left(1 + \tau_{PHP}^{1/2}\right) + \frac{\varepsilon_f \tau_{PHP}^{1/2}}{16}\right] T_f^4 + \left[\frac{T_{env} \tau_f \varepsilon_w}{4}\left(1 + \tau_{PHP}^{1/2}\right)\right] T_f^3 + \left[\frac{3}{8} T_{env}^2 \tau_f \varepsilon_w \left(1 + \tau_{PHP}^{1/2}\right)\right] T_f^2 + \left[\frac{T_{env}^3 \tau_f \varepsilon_w}{4}\left(1 + \tau_{PHP}^{1/2}\right)\right] T_f + \frac{T_{env}^4 \tau_f \varepsilon_w}{16}\left(1 + \tau_{PHP}^{1/2}\right) + \varepsilon_{BG} \tau_{PHP} \tau_f T_{BG}^4 - Q_{IR} = 0 \quad (5)$$

5. Conclusions

A polymeric pulsating heat pipe, partially filled with FC-72, was studied by means of thermography to estimate the radiative properties of the device walls and confined fluid (liquid phase). In fact, the literature does not generally provide references for the radiative properties of materials and fluids which are suitable for any investigated geometry and application. Hence, a calibration procedure, relying on a dedicated radiation model for the system, was proposed without fabricating complex calibration peripherals, such as calibration loops. To control the wall, fluid and background temperatures, a copper plate at prescribed temperature was placed behind the device, and the set-up was inserted into a thermostatic chamber. Different set-point temperatures were given to the chamber and the background, and infrared samples were taken once the system reached the steady state. By solving a minimization problem, the transmissivity and emissivity of the device walls and working fluid were found equal to 0.36, 0.42, 0.53 and 0.46, respectively. The estimated radiative properties will be adopted, in future experimentations, for fluid temperature measurements by means of thermography in the operating device.

6. References

[1] Mousa M H, Yang C M, Nawaz K, Miljkovic N, Review of heat transfer enhancement techniques in two-phase flows for highly efficient and sustainable cooling, Renewable and Sustainable Energy Reviews. 155 (2022) 111896. https://doi.org/https://doi.org/10.1016/j.rser.2021.111896.

[2] Sohel Murshed S M, Nieto de Castro C A, A critical review of traditional and emerging techniques and fluids for electronics cooling, Renewable and Sustainable Energy Reviews. 78 (2017) 821–833. https://doi.org/10.1016/J.RSER.2017.04.112.

[3] Marengo M, Nikolayev V S, Pulsating Heat Pipes: Experimental Analysis, Design and Applications, (2018) 1–62. https://doi.org/10.1142/9789813234406_0001.

[4] Spinato G, Borhani N, Thome J R, Understanding the self-sustained oscillating two-phase flow motion in a closed loop pulsating heat pipe, Energy. 90 (2015) 889–899. https://doi.org/https://doi.org/10.1016/j.energy.2015.07.119.

[5] Alhuyi Nazari M, Ahmadi M H, Ghasempour R, M.B. Shafii, Mahian O, Kalogirou S, Wongwises S, A review on pulsating heat pipes: From solar to cryogenic applications, Appl Energy. 222 (2018) 475–484. https://doi.org/https://doi.org/10.1016/j.apenergy.2018.04.020.

[6] Jun S, Kim S J, Experimental investigation on the thermodynamic state of vapor plugs in pulsating heat pipes, Int J Heat Mass Transf. 134 (2019) 321–328. https://doi.org/https://doi.org/10.1016/j.ijheatmasstransfer.2019.01.053.

[7] Sun Q, Qu J, Li X, Yuan J, Experimental investigation of thermo-hydrodynamic behavior in a closed loop oscillating heat pipe, Exp Therm Fluid Sci. 82 (2017) 450–458. https://doi.org/https://doi.org/10.1016/j.expthermflusci.2016.11.040.

[8] Mameli M, Marengo M, Khandekar S, Local heat transfer measurement and thermo-fluid characterization of a pulsating heat pipe, International Journal of Thermal Sciences. 75 (2014) 140–152. https://doi.org/10.1016/j.ijthermalsci.2013.07.025.
[9] Pagliarini L, Cattani L, Mameli M, Filippeschi S, Bozzoli F, Heat transfer delay method for the fluid velocity evaluation in a multi-turn pulsating heat pipe, International Journal of Thermofluids. 17 (2023) 100278. https://doi.org/https://doi.org/10.1016/j.ijft.2022.100278.
[10] Mameli M, Pietrasanta L, Fioriti D, Di Marco P, Filippeschi S, Large Diameter Pulsating Heat Pipe for Future Experiments on the International Space Station: Ground and Microgravity Thermal Response, in Joint 19th IHPC and 13th IHPS, Pisa, Italy, June 10-14, 2018.
[11] Catarsi A, Fioriti D, Mameli M, Filippeschi S, Di Marco P, Accuracy Analysis of Direct Infrared Temperature Measurements of Two-Phase Confined Flows, (2018) 8943–8953. https://doi.org/10.1615/ihtc16.tpm.024202.
[12] Mangini S, Marengo M, Araneo L, Mameli M, Fioriti D, Filippeschi S, Infrared analysis of the two phase flow in a single closed loop pulsating heat pipe, Exp Therm Fluid Sci. 97 (2018) 304–312. https://doi.org/https://doi.org/10.1016/j.expthermflusci.2018.04.018.
[13] Article R, Plane T G, Chamber V, A Review of Heat Pipe technology for Foldable Electronic Devices, (2019).
[14] Der O, Alqahtani A A, Marengo M, Bertola V, Characterization of polypropylene pulsating heat stripes: Effects of orientation, heat transfer fluid, and loop geometry, Appl Therm Eng. 184 (2021) 116304. https://doi.org/https://doi.org/10.1016/j.applthermaleng.2020.116304.
[15] Brutin D, Sobac B, Rigollet F, Le Niliot C, Infrared visualization of thermal motion inside a sessile drop deposited onto a heated surface, Exp Therm Fluid Sci. 35 (2011) 521–530. https://doi.org/https://doi.org/10.1016/j.expthermflusci.2010.12.004.
[16] Tsilingiris P T, Comparative evaluation of the infrared transmission of polymer films, Energy Convers Manag. 44 (2003) 2839–2856. https://doi.org/https://doi.org/10.1016/S0196-8904(03)00066-9.
[17] Mameli M, Catarsi A, Mangini D, Pietrasanta L, Michè N, Marengo M, Di Marco P, Filippeschi S, Start-up in microgravity and local thermodynamic states of a hybrid loop thermosyphon/pulsating heat pipe, Appl Therm Eng. 158 (2019) 113771. https://doi.org/10.1016/j.applthermaleng.2019.113771.

Acknowledgments
The Authors would like to acknowledge the European Space Agency (ESA) support through the grant 4000128640/19/NL/PG/pt, ESA MAP project TOPDESS.

Heat flux measurement approach for an enhanced thermometric method: preliminary tests

L Evangelisti[1,*], L Barbaro[1], E De Cristo[2], C Guattari[3], T D'Orazio[3], F Asdrubali[1] and R De Lieto Vollaro[1]

[1] Roma TRE University, Department of Industrial, Electronic and Mechanical Engineering, Via Vito Volterra 62, 00146 Rome, Italy
[2] Niccolò Cusano University, Department of Engineering, Via Don Carlo Gnocchi 3, 00166 Rome, Italy
[3] Roma TRE University, Department of Philosophy, Communication and Performing Arts, Via Ostiense 139, 00154 Rome, Italy

* Corresponding author: luca.evangelisti@uniroma3.it

Abstract. In situ tests are suitable to confirm the real thermal performance of building components, and several significant on-site measurement techniques have been studied in literature. However, among them the Thermometric (THM) method has been poorly examined by the scientific community, thus having opportunities for improvement, being a quite a new and non-standardized technique. The theory behind this technique is the Newton's law of cooling and the main issue is associated to the heat transfer coefficient for which there is no agreement about the value to use. Here, a simple experimental apparatus characterized by a vertical heated sample, suitably thermally insulated was realized. Sensors were installed and direct heat flux measurements through a heat flux plate were performed and compared with (i) the heat flows obtained through the THM method (test conducted using the internal heat transfer coefficient recommended by the ISO 6946) and (ii) the heat fluxes obtained through the proposal of an enhanced THM method based on dimensionless groups analysis, thus requiring data processing based on convective and radiative components.

1. Introduction

It is known that in situ tests are suitable to confirm the real thermal performance of building components [1]. Theoretical values can vary from the performance evaluated on site, under real thermal boundary conditions [2]. Consequently, experimental tests are necessary for understanding the behavior of building elements, by using measuring tools and methodological approaches appropriate for logging temperatures, heat fluxes and air velocities both for indoor and outdoor environments [3]. Then, a post-processing is required to achieve performance indicators, such as thermal resistance and/or thermal conductance of building components [4].

Different significant on-site measurement techniques have been studied in literature. In particular, the standardized Heat Flow Meter (HFM) method, the Quantitative Infrared Thermography approach (QIRT), the Simple Hot Box HFM method (SHB-HFM) and the so-called Thermometric method (THM) have been examined by several researchers. The research efforts related to these methods [5] are shown in figure 1 (obtained through Web of Science database).

The THM method covers only 4%, demonstrating that it has been poorly examined in the literature, thus having opportunities for development. The THM method is fairly new, and it is a non-standardized technique [6]. It needs a suitable total internal heat transfer coefficient for the heat flux calculation, according to the Newton's law of cooling. However, there is no agreement regarding the value to use for the heat transfer coefficient [7].

As a direct solution, considering horizontal heat flows, the heat transfer coefficient proposed by the standard ISO 6946 could be used. It is a constant value, equal to 7.69 W/m^{2}K, composed by a convective part of 2.5 W/m^{2}K and a radiative component of 5.19 W/m^{2}K [8]. Nonetheless, the issue is whether these values are representative for any indoor environmental conditions.

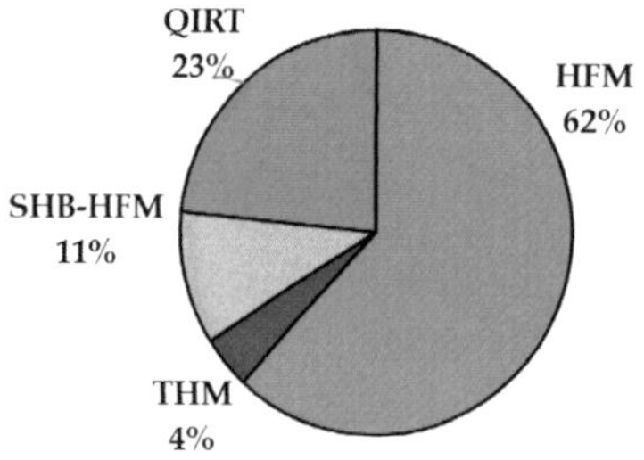

Figure 1. Literature analysis related to the in-situ measurement techniques.

The heat exchange between fluids and solids can be evaluated as a function of the surface thermal resistance, and its assessment requires the knowledge of different factors. Obtaining the correct value is challenging, involving appropriate experimental setup [9]. The identification of appropriate heat transfer coefficients is related to the issues of assessing thermal comfort and energy consumption of buildings.

Starting from this, here preliminary experimental tests were conducted as the first step of a wider research activity that aims to make a comparison between the standardized HFM method and the THM method. A simple low-cost experimental apparatus characterized by a vertical sample heated through a heating mat in the rear part, suitably thermally insulated to generate a one-dimensional heat flux was realized. Sensors were installed on the front free surface. Direct heat flux measurements through a heat flux plate were performed and compared with (i) the heat flows obtained through the THM method (test conducted using the internal heat transfer coefficient recommended by the ISO 6946) and (ii) the heat fluxes obtained through the proposal of an enhanced THM method based on dimensionless groups analysis, thus requiring a data processing based on convective and radiative components.

The aim of this work is to lay the groundwork for overcoming the disagreement related to the value of the total heat transfer coefficient in the THM method by proposing an alternative approach.

2. Materials and methods

2.1 The experimental setup

A simple low-cost experimental apparatus characterized by a vertical sample heated through a heating mat in the rear part, suitably thermally insulated was realized. In particular, 5 cm thickness XPS panels (characterized by a thermal conductivity of about 0.035 W/mK) were shaped and assembled to create an insulating structure where the sample (made of poplar wood, with an unknown thermal conductivity) can be put into. The whole assembly was enclosed using wood panels to guarantee a higher structural stability. The internal XPS insulating structure has a square shape (60 x 60 cm) and a thickness of 10 cm (overlapping of two 5 cm XPS layers). The central part has a thickness of 5 cm for housing the sample (whose dimensions are 30 x 30 cm). One side of the structure can be removed to enable the sample, the heating mat, and sensors installations, respectively. Figure 2 shows the inner XPS assembly, the external cladding, and the exploded view of the whole structure. The heating mat is characterized by a flexible polyester heating film manufactured by ALPER (dimensions equal to 28 x 28 cm, 16W, 230V, with a thickness of 0.1 cm).

The following sensors were applied and used within a methodological approach for evaluating direct and indirect heat flux measurement techniques: (i) surface temperature sensors were applied to measure the rear surface temperature of the wooden sample; (ii) a thermal imaging camera was used to evaluate the front surface temperature of the sample; (iii) a heat flux sensor was installed for direct heat flow measurements; (iv) a surface temperature probe and (v) a hot-wire anemometer were used to compute the heat flux through an indirect approach. Table 1 lists the main technical characteristics of the equipment. Figure 3 shows a schema of the overall experimental apparatus.

Figure 2. Internal XPS structure (a); external cladding (b); exploded view of the structure (c).

Table 1. Technical features of the equipment.

Sensor/measurement instrument	Manufacturer	Model	Measuring range	Resolution
Heat-flux sensor	Hukseflux	HFP01	-2000 to 2000 W/m^2	60 x 10^{-6} V/(W/m^2)
Surface temperature sensor	LSI	EST124	-60 to +80 °C	0.01 °C
Air temperature sensor	LSI	EST033	-50 to 70 °C	0.01 °C
Hot-wire anemometer	TESTO	0628 0152	0 to 5 m/s	0.01 m/s
Thermal imaging camera	Fluke	Ti480 PRO	-10 to 1000 °C	0.1 °C

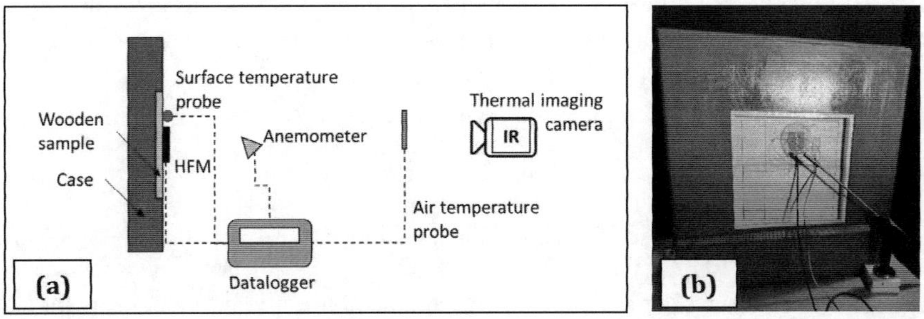

Figure 3. Schematic view of sensors installation (a); the actual experimental apparatus (b).

2.2 Methodology

The methodological approach is characterized by the following steps:

1) the heating mat was turned on until steady state conditions were reached, and 9 surface temperature sensors were preliminary applied on the rear surface of the wooden sample for measuring the surface temperature (3 equidistant sensors on 3 equidistant lines, within a 3x3 matrix characterized by a spacing

of 7 cm). Moreover, a thermal imaging camera was used to assess the front surface temperature of the sample for obtaining data about the temperature distribution of the front free surface. In order to obtain reliable results, the emissivity of the wooden sample was obtained by comparing surface temperature values measured through a surface temperature probe and the thermal imaging camera itself. This approach allowed us to identify an emissivity of the wooden sample equal to 0.84.

2) Due to the hypothetically inhomogeneous heating of the sample, a calibrated Comsol 3D model was created to evaluate possible heat flux deviations from one-dimensional conditions, thus identifying the best position for installing the heat flux sensor and the surface temperature probe. The calibration procedure was done by calculating the model efficiency (*EF*) index [10]:

$$EF = \frac{\sum_{i=1}^{N}(m_i - \overline{m})^2 - \sum_{i=1}^{N}(s_i - m_i)^2}{\sum_{i=1}^{N}(m_i - \overline{m})^2} \tag{1}$$

where m_i is the measured value at time t_i, s_i is the simulated value for each time t_i and N is the total number of samples. In addition, $\overline{m}$ is the mean value among the measured ones. *EF* ranges between 0 and 1. When it is equal to 1, the measured and simulated values are identical. Once the model was calibrated based on the temperatures measured by the thermal imaging camera (with an expected EF value greater than 0.9), Comsol was used to simulate the distribution of the heat flux on the sample. In this way, the best position for installing the sensors was identified.

3) Once the stationary conditions have been reached, the heat flux sensor was used for direct heat flow density measurements (q_{HFM}).

4) At the same time, data logged by the surface and air temperature probes and data acquired through the hot-wire anemometer was used to calculate the convective coefficient through the dimensionless groups analysis [9]. Three distances (5, 6 and 7 cm from the sample) were preliminary tested in order to comprehend the influence of the anemometer position. Distances of less than 5 cm have not been tested due to the protective structure of the anemometer. Grashof (*Gr*) and Reynolds (*Re*) numbers were estimated, and the Archimedes (*Ar*) number allowed to understand the specific convection conditions:

$$Ar = \frac{Gr}{Re^2} \tag{2}$$

It is known that if Ar is lower than 0.7 the convection is forced; if Ar is much greater than 10 the convection is natural. Lastly, if Ar is in the range of 0.7 to 10, the convection is mixed. In natural convection, the convective heat transfer coefficient (h_c) can be obtained through the Nusselt number (*Nu*), which is a function of the Rayleigh (*Ra*) number. Considering vertical surfaces, the equations below [11] can be applied:

$$Nu = 0.59 \cdot Ra^{\frac{1}{4}}, \ \ 10^4 < Ra < 10^9 \tag{3}$$

$$Nu = 0.10 \cdot Ra^{\frac{1}{3}}, \ \ 10^9 < Ra < 10^{13} \tag{4}$$

Nu is defined as a function of h_c and the geometrical characteristic length (equal to the vertical length of the wooden sample, of 0.30 m).

5) The radiative heat transfer coefficient (h_r) was computed according to the following equation:

$$h_r = 4\varepsilon\sigma T_m^3 \tag{5}$$

where ε is the emissivity of the sample, σ is the Stefan-Boltzmann constant and T_m is the average thermodynamic temperature of the surface and the surrounding surfaces (considered as the average between the wooden sample surface temperature and the air temperature measured far from the specimen) [8].

6) The heat flux density was calculated by applying the Newton's law of cooling (q_{THM}):

$$q_{THM} = h_{tot} \cdot (T_s - T_{air}) \tag{6}$$

where T_s and T_{air} are the surface temperature of the wooden sample and the air temperature, respectively, and h_{tot} is the total heat transfer coefficient, obtained as the sum of the convective and radiative component. The heat flux obtained from equation (6) was named q_{THM} because this formula must be applied when the THM method is used for wall thermal characterization.

7) Finally, q_{HFM} and q_{THM} were compared, also considering the widely used h_{tot} value equal to 7.69 W/m^{2}K, recommended by the standard ISO 6946.

A preliminary uncertainty analysis in terms of repeatability was carried out, considering a coverage factor of 2, which implies approximately 95% confidence.

3. Results and discussion

According to the methodological approach, stable operating conditions were reached during the experimental campaign. During the first step, 9 surface temperatures and thermographic images were acquired. In this phase, an inhomogeneous heating of the sample was observed, with higher temperatures in its upper part. Given the limitation of the heating mat's capacity to uniformly heat the sample, the thermal image was processed (adapting the temperature scale) to identify the best position for the installation of the heat flow sensor. In particular, this position was identified highlighting the sample area that had a temperature difference of less than 2°C (see figure 4a). Then, considering the thermal image with the original temperature scale (relative to the whole sample), two perpendicular virtual lines crossing the previously identified region (called L0 and L1, as shown in figure 4b) were created using the thermographic image processing software, thus obtaining the associated temperatures. Starting from the obtained temperature data, the wooden sample was modelled through Comsol aiming at simulating the heat fluxes crossing the sample. The 3D model was calibrated using as input the rear temperatures and exploiting the front temperatures measured by the thermal imaging camera on L0 and L1. The obtained values for the EF index were respectively 0.945 referring to the vertical line (L0) and 0.988 referring to the horizontal one (L1), thus highlighting a good agreement between experimental measurements and simulation.

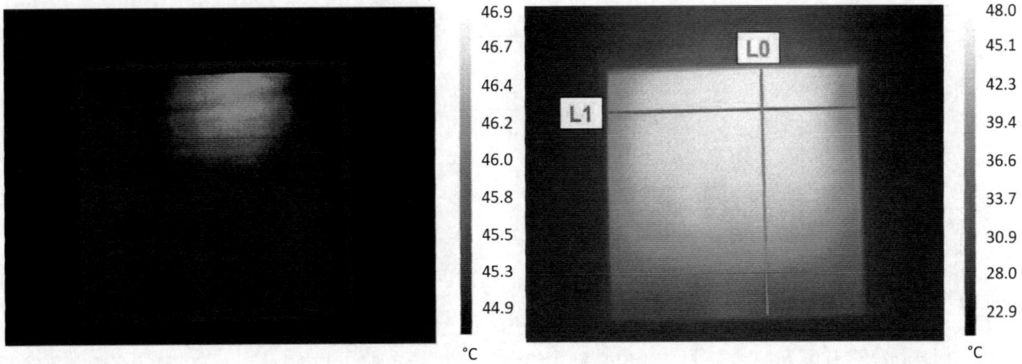

Figure 4. Area where the ΔT is at most 2°C (a); virtual lines for the calibration procedure (b).

Consequently, the heat fluxes along L0 and L1 were simulated using Comsol, and the *x*, *y* and *z* heat flux components (see figure 5a) were evaluated to identify the best position for installing the sensors. In particular, the condition applied to evaluate the heat flux distortion was that the *x* and *z* components of the heat flux had to be less than 5% of the *y* component. This condition was satisfied within the green area shown in figure 5b.

Data derived from the heat flux sensor (q_{HFM}), the surface (T_s) and air temperature (T_{air}) probes and data acquired through the hot-wire anemometer (u), were processed to define the convection mode. The obtained values are reported in Table 2. For an average film temperature of 30.94 °C, it was considered a thermal conductivity equal to 0.02641 W/mK, a kinematic viscosity equal to $1.61 \cdot 10^{-5}$ m²/s, a thermal expansion coefficient of $3.29 \cdot 10^{-3}$ 1/K and a Prandtl number (*Pr*) equal to 0.714. The dimensionless parameters approach allowed to obtain the results listed in Table 3, where only average values are listed.

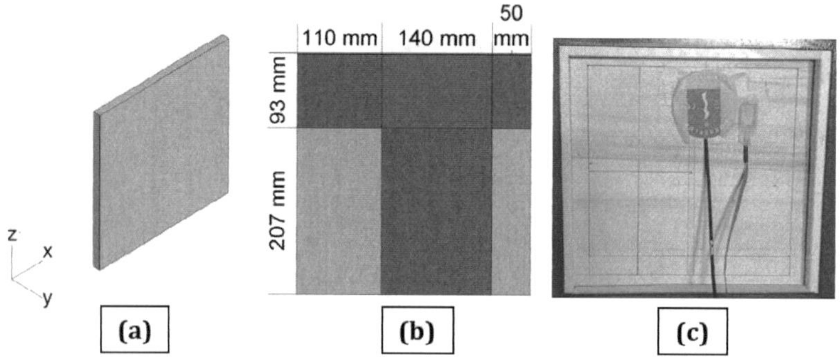

Figure 5. Sample and coordinate system (a); green area, characterized by one-dimensional heat flux according to the Comsol model (b); heat flux and surface temperature sensors (c).

Table 2. Operating conditions of experimental campaign.

Anemometer distance [cm]	q_{HFM} [W/m²]	T_s [°C]	u [m/s]	T_{air} [°C]
5	180.34±0.37	40.27±0.01	0.08±0.01	21.57±0.02
6	182.33±0.34	40.39±0.01	0.08±0.01	21.50±0.01
7	183.78±0.23	40.43±0.01	0.08±0.01	21.45±0.01

Table 3. Dimensionless parameters average values.

Anemometer distance [cm]	*Gr*	*Re*	*Ar*	*Ra*	*Nu*
5	$6.27 \cdot 10^7$	$1.40 \cdot 10^3$	34.24	$4.47 \cdot 10^7$	48.25
6	$6.33 \cdot 10^7$	$1.42 \cdot 10^3$	34.05	$4.52 \cdot 10^7$	48.37
7	$6.36 \cdot 10^7$	$1.43 \cdot 10^3$	33.90	$4.54 \cdot 10^7$	48.42

The convection is purely natural for each tested distance of the anemometer because the *Ar* is constantly much greater than 10. Starting from this, the convective heat transfer coefficients (h_c) were found through the Nusselt number (*Nu*), considering vertical surfaces. From the calculation of the *Ra* number, equation (3) was applied. The radiative heat transfer coefficient (h_r) was computed according to equation (5) and the heat flux density (q_{THM}) was calculated by applying the Newton's law of cooling by means of equation (6). The results listed in Table 4 allow to observe that the distance of the anemometer has a negligible influence. As stated before, distances less than 5 cm have not been assessed due to the protective structure of the anemometer (see figure 3b).

It is worthy to observe that the obtained average total heat transfer coefficient of about 9.61 W/m²K is quite different from that suggested by the standard. It is related to the specific temperature differences between sample and air which in this case differ widely from the values that can occur when a real wall is analyzed. However, this comparison highlights the need for an experimental approach to identify suitable convective and radiative heat transfer coefficients, useful for obtaining appropriate values to be used in the thermal characterization of envelope components when the THM method is to be applied.

All this brings us back to the main problem of the THM method, associated with the heat transfer coefficient for which there is no agreement on the value to use. In this case, using the value recommended by the standard would imply percentage differences equal to about -20%. For this reason, q_{THM} values were compared both with the data logged by the heat flux sensor (q_{HFM}) and the heat flux computed applying the h_{tot} value equal to 7.69 W/m^{2}K, recommended by the ISO 6946 (named q_{ISO}). The results are shown in Table 5.

The obtained results allowed to calculate extremely low percentage differences between the THM and HFM method. For a distance of 5 cm of the anemometer, the percentage differences ranged between about -2% and +2%. For the distances of 6 cm and 7 cm similar values were found, with percentage variations ranging between about -4% and +1%.

Higher percentage differences were found by comparing the use of the h_{tot} suggested by the standard and the HFM method. In this case, variations ranging from about -23% to -18% were obtained.

Table 4. Convective and radiative heat transfer coefficients, and heat flux for the THM approach.

Anemometer distance [cm]	h_c [W/m^2 K]	h_r [W/m^2 K]	h_{tot} [W/m^2 K]	q_{THM} [W/m^2]
5	4.25±0.01	5.356±0.002	9.60±0.01	179.64±0.56
6	4.26±0.01	5.357±0.001	9.62±0.01	181.67±0.25
7	4.26±0.01	5.357±0.001	9.62±0.01	182.59±0.19

Table 5. Heat fluxes obtained through the direct and indirect approach.

Anemometer distance [cm]	q_{HFM} [W/m^2]	q_{THM} [W/m^2]	q_{ISO} [W/m^2]
5	180.34±0.37	179.64±0.56	143.85±0.34
6	182.33±0.34	181.67±0.25	145.30±0.16
7	183.78±0.23	182.59±0.19	145.95±0.12

4. Conclusions

The aim of this work was to lay the grounds for overcoming the disagreement regarding the value of the total heat transfer coefficient when the THM method is applied. Here, an alternative method based on the dimensionless groups' evaluation was proposed. A simple experimental setup was built, and direct and indirect heat flux measurements were conducted. The measurements were made by applying a heat flux sensor and the results were compared with the heat fluxes obtained through the THM method. The tests were performed both using the conventional total heat transfer coefficient suggested by the ISO 6946 standard, and through the proposal of an advanced THM method based on the evaluation of the convective and radiative heat transfer coefficients.

The results showed that the THM method can provide satisfying results in terms of indirect heat flux measurements, showing extremely low percentage differences if compared with the HFM method (variations ranging from -4% to +2%). Consequently, these preliminary results allowed to affirm that the THM method has chances for improvement, needing further investigations and aiming at providing new measurement systems able to evaluate heat fluxes through an indirect approach. A measurement system able of autonomously calculating the heat transfer coefficient according to the specific thermo-fluid dynamic conditions would allow users to be freed from potentially inappropriate choices. The results highlighted that using the total heat transfer coefficient value suggested by the ISO 6946 lead to lower heat flux values, with percentage variations of about -20% on average.

Future developments will be related to the optimization of the experimental apparatus thus obtaining a more homogeneous distribution of temperatures on the sample, and the application of this methodological approach to real case studies. On the other hand, due to the thermal and fluid dynamic phenomena, the position of the hot wire anemometer will be examined more in depth.

5. References

[1] Flores Larsen S, Hongn M, Castro N and González S 2019 Comparison of four in-situ methods for the determination of walls thermal resistance in free-running buildings with alternating heat flux in different seasons *Constr. Build. Mater.* **224**, pp 455-473

[2] Desogus G, Mura S and Ricciu R 2011 Comparing different approaches to in situ measurement of building components thermal resistance . *Energy Build.* **43** pp 2613–20

[3] Nardi I and Lucchi E 2023 In Situ Thermal Transmittance Assessment of the Building Envelope: Practical Advice and Outlooks for Standard and Innovative Procedures. *Energies* **16** 3319

[4] Evangelisti L, Guattari C, De Lieto Vollaro R and Asdrubali F 2020 A methodological approach for heat-flow meter data post-processing under different climatic conditions and wall orientations. *Energy Build.* **223** 110216

[5] Bienvenido-Huertas D, Moyano J, Marí D and Fresco-Contreras R 2018 Review of in situ methods for assessing the thermal transmittance of walls. Renew. *Sustain. Energy Rev.* **102** pp 356–371

[6] Teni M, Krstić H and Kosiński P 2019 Review and comparison of current experimental approaches for in-situ measurements of building walls thermal transmittance. *Energy Build.* **203** 109417

[7] Evangelisti L, Scorza A, De Lieto Vollaro R and Sciuto S A 2022 Comparison between Heat Flow Meter (HFM) and Thermometric (THM) Method for Building Wall Thermal Characterization: Latest Advances and Critical Review *Sustainability* **14** 693

[8] *ISO 6946:2017* 2017 Building Components and Building Elements—Thermal Resistance and Thermal Transmittance—Calculation Methods. International Organization for Standardization (ISO): Geneva Switzerland

[9] Evangelisti L, Guattari C and De Rubeis T 2021 Preliminary analysis of the influence of environmental boundary conditions on convective heat transfer coefficients, *J. Phys. Conf. Ser.* **1868** 012024.

[10] Loague K and Green R E 1991 Statistical and graphical methods for evaluating solute transport models: overview and application, *J. Contam. Hydrol.* **7** pp 51–73.

[11] Bergman T L, Lavine A S, Incropera F P and Dewitt D P 2017 *Fundamentals of Heat and Mass Transfer* John Wiley & Sons.

Permeability measurement of a 3D-printed AlSi10Mg porous medium: comparisons between first results with different experimental and numerical techniques

L Vitali[1], G Brambati[2], R Caruana[1], S Foletti[2], M Guilizzoni[1] and A Niro[1]*

[1] Department of Energy, Politecnico di Milano, Via Lambruschini 4, 20156 Milan, Italy
[2] Department of Mechanical Engineering, Politecnico di Milano, Via La Masa 1, 20156 Milan, Italy
* Corresponding author: alfonso.niro@polimi.it

Abstract. In this work, the permeability of a 3D-printed, AlSi10Mg porous medium, with porosity $\varepsilon = 0.3$ and an effective pore radius of 48 μm, developed to operate as wick in a sinter-like heat pipe, has been investigated by means of two different experimental approaches, and of two different numerical methods. The two experimental methods are the capillary rise tests, from which permeability was estimated by fitting the theoretical capillary rise curve to the experimental data, and the direct measurement of the the mass flow rate across the porous sample at an imposed pressure difference. The numerical simulations were performed too using two different approaches and software tools, namely, Lattice-Boltzmann with Palabos, and Finite-Volumes with OpenFOAM. In both cases, the simulation domain was reconstructed from a micro-computer aided tomographic scan of a porous medium sample. Preliminary simulations were run on a simple configuration, both to check simulation setup and validate results, and mesh independence was assessed. Then, pressure-driven and velocity-driven simulations of an incompressible fluid flow across the domain were performed, from which the permeability was estimated using Darcy and Darcy-Forchheimer equations. The results show that the methods, while not in complete agreement, provide a useful estimate. The numerical methods also complement the information given by the experimental techniques by highlighting the flow paths, and allow to analyze scenarios of increased and decreased porosity.

1. Introduction

Metallic porous media are of interest in a wide range of engineering fields, as chemical [1], biomedical [2], and heat transfer [3,4]. Many fabrication techniques are used for this type of media, including sol-gel, polymer porosifier method, foaming method, and additive manufacturing (AM) [5]. In particular, additive manufacturing, also known as 3D-printing, have become increasingly used in recent years, as it gives the possibility to manufacture embedded components in structures.

A common application of porous media in heat transfer field is using them as wick for heat pipes [6], as their structure promotes the capillary rise of the liquid, thus allowing the device operation also against gravity. Heat pipe performances are strongly influenced by the wick material, which must be compatible with the working fluid [7] and the structure, and it must adequately satisfy the need of a trade-off between the requirements of high capillary pressure and high permeability [8]. The first one can be estimated based on the pore radii and the contact angle. The latter, quantifying the medium's resistance to fluid flow, can be measured or estimated in several ways.

One experimental way to measure permeability of a porous media is to perform a capillary rise test. This method has been used by both Zhang et al. [9] and Deng et al. [10] to characterize the performance of sintered copper wicks, for cylindrical heat pipes, and sintered nickel and copper wicks for loop heat pipes, respectively. Furthermore, Darcy's law [11] can be used to estimate the permeability value of a porous medium by directly measuring the mass flow rate across the sample, and consequently the flow velocity, at imposed pressure difference [12].

Another possibility for estimating permeability of porous media is performing numerical simulations. This has been done by Eshghinejadfard et al. [13], who used the Lattice Boltzmann method with either single or multiple relaxation time to calculate the permeability of three-dimensional hydrodynamics porous media at low Reynold numbers, as well as by Narváez et al. [14], who used the Lattice Boltzmann method as well, but focusing on the choice of the simulation set-up parameters, and on their influence on the final results. Moreover, Jeon and Byon [15] used CFD to estimate the permeability of a dual-height micro-post wick while considering the meniscus shape. The domain for the CFD simulations can be reconstructed from micro-computer aided tomography scans of real samples, as for example in the work by Otaru and Samuel [16], in which the reader can also find additional references to significant papers about this topic.

This work aims at estimating the permeability of a 3D-printed aluminum porous medium, developed to be used as wick in a sintered-like heat pipe, using both experimental and numerical methods. More specifically, two experimental techniques, namely the capillary rise test and the direct measurement of mass flow rate at different pressure differences, and two numerical methods, namely Lattice Boltzmann and Finite Volume, have been investigated and the results are compared and discussed.

2. Experimental setups and results

2.1. Samples

The additive-manufactured-samples have been made by Beam-IT S.p.A in different printing sessions. In order to produce a sinter-like porous structure, also referred to as "wick" in this paper, several preliminary trials were necessary to find a combination of printing parameters which allows to form a mechanically-resistant aggregate of the AlSi10Mg powder with a significant porosity. Such parameters are summarized in table 1. The measurements presented below use a 10-cm-high half-pipe sample, with a 4 mm wick thickness, and printing direction orthogonal to the sample's vertical axis for the capillary rise test, and a 2-cm-long cylinder with a 1 cm diameter wick, externally filleted to be attached to a 1/2" pipe fitting, for the direct mass flow rate measurement. The μCT section used for the numerical models, described in Section 3.1, has been retrieved from a 15-cm-long heat pipe sample, with a 1.6 cm internal diameter and a wick thickness of 2.4 mm. All these samples have been printed using the same parameters, however the geometrical differences between samples resulted in different porosity values, which are 0.30 and 0.26 for the capillary rise and the permeability samples, respectively. The tomographed section's porosity is dependent on the choice of a threshold parameter, which has been chosen to set a porosity value of 0.36. Figure 1 shows the heat pipe section and the samples used for the permeability tests, as visual reference.

Table 1. Porous medium printing parameters

Print parameter	Unit	Value	Print parameter	Unit	Value
Layer thickness	μm	60	Speed	mm/s	2350
Hatch Distance	μm	140	Energy	J/mm^3	10.13
Power	W	200	Focal / spot size	μm	160

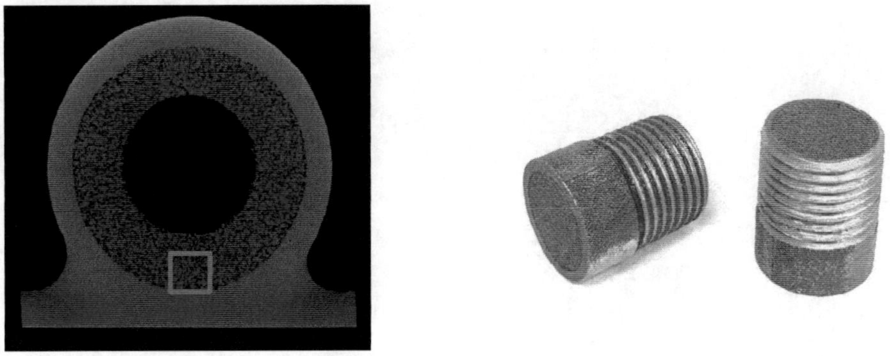

(a) (b)

Figure 1. Tomographed heat pipe section with highlight of the wick portion used for computational analyses (a) and samples for the direct measurement permeability test (b).

2.2. Capillary rise test

The capillary maximum rise and the capillary rate-of-rise tests can be used to characterize the capillary performance parameter of a wick structure. At any time, capillary pressure ΔP_c is supposed in equilibrium with hydrostatic head ΔP_h, and friction pressure drop ΔP_f,

$$\Delta P_c = \Delta P_h + \Delta P_f \tag{1}$$

Assuming:

$$\Delta P_c = \frac{2\sigma \cos\theta}{r_p} \qquad \Delta P_h = \rho g h \qquad \Delta P_f = \frac{1}{K}\mu\varepsilon h \frac{dh}{dt} \tag{2}$$

where g is the gravity acceleration, h the liquid height, ε the porosity, σ the surface tension, ρ the mass density, μ the dynamic viscosity, θ the contact angle, r_p the average pore radius, and K the permeability. Substituting expressions 2 into equation 1 and, finally, solving the latter for the rise rate, the following equation is obtained:

$$\frac{dh}{dt} = \frac{2\sigma}{\mu\varepsilon}\frac{K\cos\theta}{r_p}\frac{1}{h} - \frac{\rho g K}{\mu\varepsilon} \tag{3}$$

As it can be seen from equation 3, the relation between time and the height reached by the liquid is a function of both the thermophysical and capillary properties of the used liquid, and the porous medium's properties, i.e. the pore radius r_p, the permeability K and the porosity ε. The porosity is measured by weighing the sample after being dried (4h at 50 °C in a ventilated oven), and after being saturated with ethanol, as described by Jafari et al. [17], hence the two unknowns are r_p and K. Assuming that all the properties are constant, numerical fitting of the experimental capillary rise trend *h(t)* can be performed to determine both. Different techniques can be used, as explained in detail by Elkholy et al. [18], and the outcome may vary depending on the method of choice. In this work the equation was numerically integrated using a forward finite difference approach, and the Matlab® function *fminsearch* – implementing a Nelder-Mead algorithm – was used to perform the fitting and to obtain the values of r_p and K that give the best agreement between the theoretical and experimental curves. This method has been found to provide more consistent results with respect to the other test methods for our experiments than the ones described in [18].

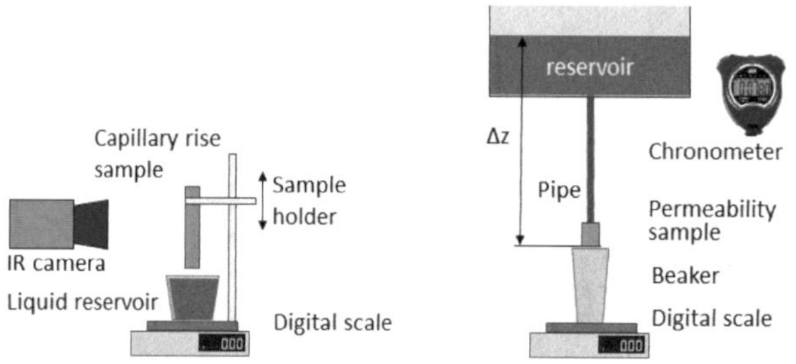

Figure 2. Capillary rise test set-up scheme (left) and permeability set-up scheme (right)

The capillary rise tests have been carried out with ethanol using simultaneously two different methods, for cross-validation. The first one consists of recording the capillary rise with a FLIR T450sc IR camera, for increased contrast between the wet and the dry surface, and considering the mass rise equal to the product of wetted height, sample cross-section, porosity, and fluid density. The second method consists of recording the mass decrease of ethanol contained in a liquid reservoir where, as showed in figure 2 (left), the tip of the sample is dipping. To achieve meaningful measurements, corrections have to be carried out to consider evaporation of ethanol both from the reservoir, as calibrated before the trial, and from the sample, considered proportional to the wetted height. The corrections are then applied, resulting in the curves shown in figure 3 (left). The comparison between the two complementary measurement techniques, i.e. balance and thermography, is also shown in figure 3 (left), where the saturation mass, i.e. the maximum mass of liquid that the porous material is able to absorb from dry conditions, is also evidenced. The main uncertainties associated with the two methods are due to the retrieval of the wet-dry interface for the IR camera data, and the corrections due to evaporation for the balance method. In this paper, solely the latter is used for the retrieval of the results. Figure 3 (right) shows the experimental mass rise curve and the finite difference solution of equation 3 using r_p and K resulting from the fitting procedure. Such values may vary with the total experiment time, i.e. the resulting values are 50, 43 and 40 μm for r_p and 3.7, 2.3 and 2.0 μm^2 for K, considering 500, 1000 and 1500 s of the rising curve.

2.3. *Flow rate based measurement*

As permeability measures the "resistance" of a fluid to flow through a material, usually it is measured by recording the mass flow rate m of a fluid with known properties while imposing a pressure difference ΔP across a sample long L.

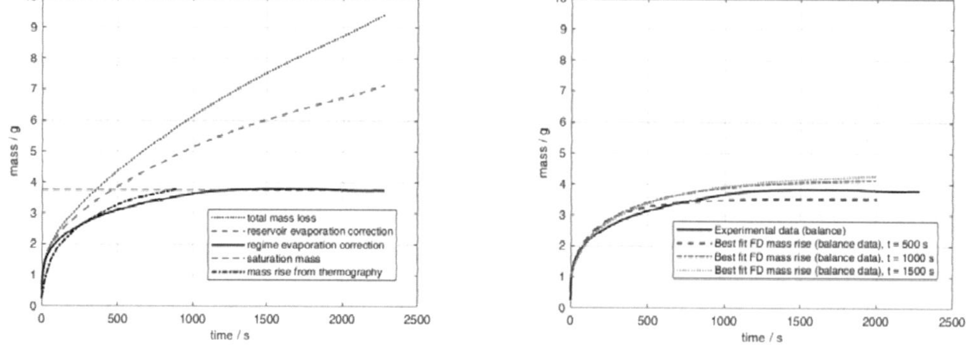

Figure 3. Measurement of the mass absorbed by the specimen before and after corrections for evaporation and comparison with thermographic data (left), comparison with a finite difference (FD) solution of equation 3 with r_p and K obtained from curve fitting

In general, a Darcyan permeability coefficient K_1 [m^2] and a Forchheimer term K_2 [m], which accounts for the pressure losses, can be introduced according to the following equation:

$$\frac{\Delta P}{L} = \frac{1}{K_1}\mu w + \frac{1}{K_2}\rho w^2 \tag{6}$$

However, when inertial effects are negligible, i.e. the flow velocity w is less than 0.1 m/s, as always verified in this work, only the first term of equation 6 is considered, and hence $K = K_1$.

The measurement of K is carried out by simply putting a reservoir of acetone, large enough so that its free surface does not significantly lower during the trial, at known heights Δz, namely 1.1, 1.5 and 1.9 m respectively, and by measuring the resulting steady-state mass flow rate across the sample. Equation 6 is then applied to retrieve K. A scheme of the experimental setup used is depicted in figure 2 (right) showing the sample and the reservoir connected by a 2-m-long pipe with an internal diameter of 1.6 mm. A corrective term ΔP_{loss} to take into account the friction loss across the pipe can be easily calculated with Poiseuille's equation for friction factor, as flow is laminar across the whole experimental range. The results of these experiments are reported in table 2, which shows that the permeability has a high degree of independency from ΔP, and the average value for the tested sample is 3.10 μm^2.

Table 2. Results of the direct permeability measurement experiment

H [m]	1.1	1.5	1.9
ΔP [Pa]	8294	11550	14728
$\Delta P - \Delta P_{loss}$ [Pa]	6904	9611	12262
m_{vol} [ml/s]	0.545	0.761	0.968
K [μm^2]	3.10	3.10	3.09

3. Numerical setups and results

3.1. Computational domain

The domain for the final simulations was created on the basis of a micro-computer-aided tomographic (μCT) scan of the same porous material used for the experimental test. The porous sample was scanned with a resolution of 10 μm and reconstructed. The reconstruction was segmented into solid and air regions according to a threshold which was selected to have a final porosity consistent with some independent porosity measurements. The selected porosity for the numerical simulations is 0.36, even though a sensitivity analysis is carried out with the Lattice-Boltzmann method as described in paragraph 3.3. From such segmentation an iso-surface, representing the solid boundaries, was extracted and exported as an STL file. To facilitate the setup of the boundary conditions and the post-processing to calculate the permeability, two thin regions were added at the beginning and at the end of the porous domain extracted from the μCT reconstruction. The bounding box of the domain has dimensions 1 mm × 1 mm × 4 mm and it is shown along with CFD results in figure 5.

3.2. Finite Volume simulations

The numerical simulations using the Finite Volume method were performed using the CFD toolbox OpenFOAM® [19]. For the present simulations, the steady-state, incompressible, simpleFoam solver was selected and preliminarily validated by simulating a single duct and an array of circular mini-channels, and comparing the pressure-drop results to the estimates of the analytical models. The mean absolute error (MAE) on the developed flow region along the single duct was 0.33 %, with a mean absolute deviation (MAD) equal to 0.13%. On the developed flow regions of the array of parallel mini-channels the MAE was 0.69% with a MAD of 0.46%. Mesh independence was assessed by using different meshes with and without mesh refinement near the walls for the porous medium. Meshes up to 4.5 million cells were tested, and the final used mesh consisted of 1.3 million cells. For all simulations, the fluid was acetone, whose values of density and viscosity were taken at 50 °C and 0.1 MPa, namely, 756 kg/m^3 and 2.451×10^{-4} Pa·s, respectively. Simulations were performed for both velocity-driven case, with an inlet velocity of 0.1–5 mm/s, and pressure-driven case, for a 1 Pa pressure difference between inlet and outlet. Both types of simulations easily converged, with very stable values for the quantities of interest and residuals lower than 10^{-6}, in less than 1000 iterations. At all the investigated velocities, flow is laminar, so no turbulence models were used. For the wall of both porous medium and channels, a classic no-slip boundary condition was set. Steady-state simulations were performed using the SIMPLE algorithm.

Focusing the attention on the simulations on the porous medium, figure 5 shows an example of the results that were obtained, in terms of pressure distribution in the domain and velocity field and streamlines in a slice of the same.

Figure 5. Examples of the finite volumes simulation results on the porous medium: pressure distribution (left) and flow streamlines (right) in the domain.

From the pressure differences, obtained in the velocity-driven simulations, or from the superficial velocity, obtained in the pressure-driven simulations, permeability was calculated with both the Darcy and the Darcy-Forchheimer approaches, i.e. by equation 6. As velocities are very low, as expected, the inertial term has a negligible influence and hence the two approaches returned practically identical values. With the Darcy approach the mean value of the calculated permeability is 11.8 μm^2, with a standard deviation of 0.06 m^2 when varying the superficial velocity; with the Darcy-Forchheimer approach – performing a quadratic fitting with zero intercept of the pressure drop as a function of the superficial velocity – the value is 11.9 μm^2, with a percentage difference between these two values of 1.4%.

3.3. Lattice-Boltzmann simulations

In this study, the open-source code Palabos (Parallel Lattice Boltzmann Solver) [21] was used to solve Lattice-Boltzmann equations. At the end of the simulation, the value of permeability can be calculated in accordance with Darcy's law, which is defined for a single-phase 1D fluid flow as follows:

$$K_{LB} = \nu_{LB} \frac{< u_{LB} >}{\Delta P_{LB}/L_{LB}} \tag{7}$$

where < u > is an averaged velocity of the fluid and ν the kinematic viscosity.. The subscript LB in equation 9 indicates that the quantities are expressed in lattice units, since Palabos operates entirely in not-dimensional units. Therefore, it is necessary to determine conversion coefficients to retrieve the physical meaning of the results. The spatial unit conversion coefficient Δx is readily available as it corresponds to the voxel size used in the CT-scans, whereas the temporal unit coefficient Δt can be determined by combining the physical value of the fluid's kinematic viscosity with Δx. To impose a pressure difference ΔP as a boundary condition, the physical pressure difference must first be divided by the fluid density to obtain a value expressed solely in terms of length and time. Subsequently, Δx and

Δt are used to calculate the pressure difference in lattice units. In regard to permeability, the physical value can be obtained by multiplying the simulation result by $(\Delta x)^2$, whereas the physical velocities multiplying the simulation results by the Δx-to-Δt ratio. Pressure-independency and validation tests have been again conducted on capillary bundles composed by straight tubes, before applying the method to the wick and a slight overprediction of the permeability, circa 6%, has been observed with respect to analytical pressure-drop solutions.

Multiple simulations were performed with varying pressure differences in order to verify the pressure-independence of the permeability value obtained. Throughout the simulation, the average energy standard deviation was monitored, and convergence was considered achieved when its value fell below the threshold of 10^{-8}. The computed permeability for the analysed porous sample was 16.3 μm². Figure 6 (left) shows a visualization of the velocity field. Additional porous samples were also extracted from the CT-scan of the wick to investigate the influence of varying porosity on permeability. The results of this study are depicted in figure 6 (right).

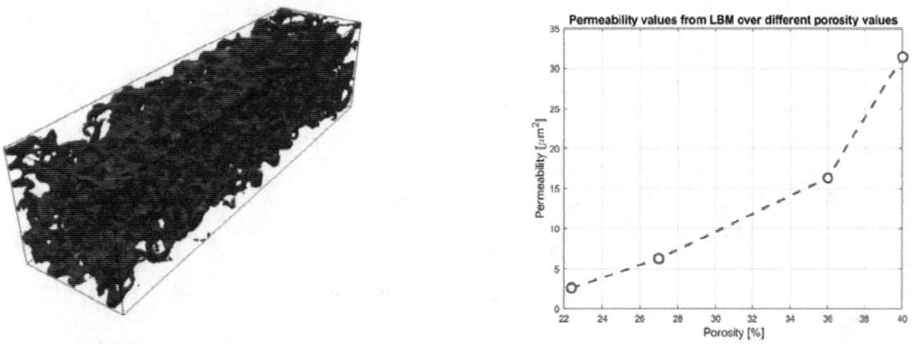

Figure 6. Fluid field visualization from LBM simulations (left) and permeability values for samples with different porosity (right)

4. Discussion and conclusions

Two experimental and two numerical methods have been applied to estimate the permeability of additively-manufactured porous media, in the context of evaluating a wick structure for heat pipes. The considered samples, and one virtual reconstructions by μCT, are characterized by different porosities due to the different geometry of the samples, which in turn causes different resulting structures, even for the same printing parameters. Furthermore, the selection of a threshold value for the computerized geometry may significantly vary the porosity, hence the permeability of the sample. A recap of the results is presented in table 3.

Table 3. Permeability results for different measurement / estimation methods

Porosity	K [μm²]	Method
0.22	2.7	Numerical, lattice-Boltzmann
0.26	3.1	Experimental, mass flow rate measurement
0.27	6.0	Numerical, lattice-Boltzmann
0.29	2.0-3.7	Experimental, capillary rise
0.36	11.8	Numerical, finite volumes
0.36	16.3	Numerical, lattice-Boltzmann
0.40	32.0	Numerical, lattice-Boltzmann

As a general note, the estimation of K by measuring the mass flow rate shows good internal consistency with variations of imposed pressures, while the capillary rise test, due to the high evaporation rates of ethanol with respect to the measured data, needs either a controlled environment or a precise model of such phenomenon to retrieve qualitative data, but at the moment provides a reasonable estimate for the purpose of evaluating the thermal models of heat pipes. The experimental data confirm the range of permeabilities measured by Gotoh et al. [22]. Finally, it should be pointed out that water cannot be employed in this test due to the low wettability of the aluminium porous structure.

Concerning the numerical methods, the lattice-Boltzmann method provides a value of K 40% higher than the finite volumes simulation, a difference which is higher than the 6% variation with respect to analytical solutions in simple test cases showed by the method's validation. The finite volume method performs well against validation cases, and should be the method of choice for any preliminary test based on design only, especially on porous media consisting of lattice structures, with production outcomes similar to the design. It should be pointed out that the used computational domain, while having its smaller side 20 times larger than the average estimated pore size, could not be statistically representative of the entire sample. For random porous media as the sinter-like structure presented in this paper, consistency issues between specimens, print direction and sample shape may arise, especially when confronting measurement techniques with different samples, however, the estimates obtained with the presented techniques are often accurate enough to evaluate the range of operating limits of the components in which they are used.

References

[1] Heydari A and Peyvandi K. 2019 *J. of Pet. Sci. and Eng.* **181**, 106235.
[2] Khanafer K and Vafai K. 2006 *Heat and Mass Trans.*, **42** 10, 939-53.
[3] Wang C Y and Cheng P 1997 *Adv. in heat trans.* **30** 93-196
[4] Badruddin I A, Khan, T Y and Baig M A A 2020 *Materials Today: Proceedings*, **24,** 1318-21.
[5] Zhao B, Gain A K, Ding W, Zhang L, Li X, and Fu, Y 2018 *Int. J of Adv. Manufacturing Tech.* **95**, 2641-2659.
[6] Iverson B D, Davis T W, Garimella S V, North M T, and Kang S S 2007 *J. of Thermophy. and Heat Trans.* **21**(2), 392-404.
[7] Caruana R, Gallazzi L, Iazurlo R, Marcovati M and Guilizzoni M 2022 *Fluids*, **7**(3), 109.
[8] Mwaba M G, Huang X, and Gu J 2006 *Int. J. of en. res.* **30** 7, 489-499.
[9] Zhang J, Lian L X, Liu Y and Wang R Q 2021 *Int. J. of Heat and Mass Trans.* **164** 120536.
[10] Deng D, Liang D, Tang Y, Peng J, Han X and Pan M. 2013 *Exp. Therm. and Fluid Sci.* **50** 1-9.
[11] Masoodi R, Pillai K M and Varanasi P P 2007 *AIChE Journal*, **53** 11, 2769-82.
[12] Deng D, Tang Y, Huang G, Lu L and Yuan D 2013 *Int. J. of Heat and Mass Trans.* **56** 1-2, 283-293.
[13] Eshghinejadfard A, Daróczy L, Janiga G and Thévenin D 2016 *Int. J. of Heat and Fluid Flow* **62** 93-103.
[14] Narváez A, Zauner T, Raischel F, Hilfer R and Harting J 2010 *J. of Stat. Mech.: Theory and Exp.* **2010** 11, P11026.
[15] Jeon S and Byon C 2017 *Int. J. of Heat and Mass Trans.* **115** 879-85.
[16] Otaru A J and Samuel M B 2021 *Mat. Res. Express* **8** 046516, 1-15.
[17] Jafari D, Wits W W and Geurts B J 2018 *App. Therm. Eng.* **143** 403–14
[18] Elkholy A, Durfee J, Mooney J P, Robinson A J and Kempers R 2023 *Meas. Sci. Technol.* **34** 045301
[19] OpenFOAM and The OpenFOAM Foundation, https://openfoam.org. Last access: April 2023.
[20] Guilizzoni M, Santini M, Lorenzi M, Knisel V and Fest-Santini S 2014 *J. of Phys: Conf. Series* **547** 012028
[21] Latt J, Malaspinas O, Kontaxakis D, Parmigiani A, Lagrava A, Brogi F, Ben Belgacem M, Thorimbert T, Leclaire S, Li S, Marson F, Lemus J, Kotsalos C, Conradin R, Coreixas C, Petkantchin R, Raynaud F, Beny J and Chopard B 2021 Palabos: Parallel Lattice Boltzmann Solver. In *Computers & Mathematics with Applications* **81** 334-350
[22] Gotoh R, Furst B I, Roberts S N, Cappucci S, Daimaru T, Sunada E T 2022. *Prog. Addit. Manuf.* **7**(5), 943–55.

An overview of the Transient Hot Wire and preliminary studies for its implementation

R Ricciu[1,*], P M Congedo[2], C Baglivo[2], R Baccoli[1], P L M Buonomo, C C Mastino[1], F Floris[3], F Cambuli[3], and A G Ostrogorsky[4]

[1] Dipartimento di Ingegneria Civile Ambientale e Architettura, Università di Cagliari - Cagliari IT
[2] Dipartimento di Ingegneria dell'Innovazione, Università del Salento Lecce IT
[3] Dipartimento di Ingegneria Meccanica, Chimica e dei Materiali, Università di Cagliari - Cagliari IT
[4] Faculty Mechanical, Materials, and Aerospace Engineering Armour College of Engineering, Chicago Illinois USA

* Corresponding author: e-mail: ricciu@unica.it

Abstract. The opportunity to design a new portable system for measuring thermal conductivity in fluids arose from the need to measure the phase state (liquid and/or vapour). As is well known, thermal conductivity varies in relation to the phase state of the fluid. The Transient Hot Wire Method was used to measure thermal conductivity. First, the work includes innovations in the analytical model of heat transfer, progressing over time to contemporary numerical models used in standards. The applications in the bibliography span a wide range, including gases, liquids, and the latest advances in the field of nanofluids. Secondly, the study delves into the transient model of heat exchange, focusing on conduction. The authors used standardized correction equations for the prototype, ensuring accuracy and relevance. Finally, the work includes the design and construction of the transducer. This process begins with a Monte Carlo simulation, predicting the expected results. This simulation helps to define measurement system components based on their performance characteristics and in-depth analysis.

1. Introduction

The practical need for increasingly accurate results has driven the development of knowledge regarding the physical phenomenon related to heat transport. In this work, the authors present a brief history of the evolution over time of research and the reference regulations, starting from1936 when Pfriem, H. [1] addressed both analytical theory and the instrumental practice of measuring thermal conductivity in fluids (gases), contributing to some of the equations used today.
In 1949, E. F.M. Van der Held and F. G. Van Drunen, in [2], reported the hot wire model that is still used today. They wrote the energy balances for the diffusion of heat for the wire and the equation that relates the temperature of the wire to that of the surrounding fluid (to some extent) at a distance r from the wire. In addition to the basic theory, they incorporated considerations on the influence of the thickness of the hot wire, the influence of the diameter of the tube filled with the liquid and the uncertainties due to direct measures such as time and electric current.

In 1981, S.S. Mohammadi et al. [3] proposed a mathematical model with ramp forced corrections for the absolute determination of fluid thermal conductivity using a transient Hot-Wire Technique. The work reports the energy balance equations in the wire and the equation to derive the temperature in the fluid as a function of time and distance from the Hot Wire. Analytical solutions for three heat generation functions (Dirac, step and ramp) are presented, and the exponential integral is presented to define the Truncation error for obtaining the long-time solution. The authors also presented the analysis of approximations to the ideal line-source model.
In 1986, P. G. Knibbe [4] studied the apparatus of thermal conductivity measuring instruments based on the transient line source technique. A theoretical analysis show that the error caused by the end-effect can be kept very small, so that it is not always necessary to correct for this in a practical manner. In the analysis a distinction is made between the conductive and the convective end-effects. The conductive end-effect, which is dominant in the first part of the measurement, is almost constant during this period. The convective end-effect causes an error that increases roughly with the time squared. In 1995, M.L.V Ramires et al. [5] proposed new standard reference data for the thermal conductivity of water based on ITS 90, covering the normal liquid range along the saturation line. These recommendations are based on previously selected as well as on new highly accurate data obtained with transient hot wire instruments, where the wires were coated in the case of water.
In 1995, Xin-Gang Liang [6] discussed the Transient Hot Wire (THW) method, establishing new criteria for error analysis. The analysis includes the influence of dimensions on the accuracy of thermal conductivity measurement and explores the radial and axial effects of the phenomenon.
In 2008, A. G. Ostrogorsky et al. [7] showed how semiconductor melts are low Prandtl number fluids, characterized by high thermal conductivity yielding broad boundary layers and low viscosity that cannot stabilize the buoyancy-driven convection. They measured thermal conductivity by the HWT method during the production of Te-doped InSb crystals, both on the ground and in Microgravity conditions at the International Space Station.
Today, the reference book for the analytical treatment on THW is still that of H.S. Carslaw and J.C. Jaeger [8]. In particular, the authors summarize the analytical theory known in 1959 on heat transmission by conduction. Chapters X and XIII specifically address the linear heat source and the application to theoretical measurement of thermal conductivity.
This paper lays the theoretical foundation for the design of a portable (not inside a cell) and inexpensive instrument for measuring thermal conductivity in fluids. The aim is to achieve a measurement uncertainty compatible with that of benchtop measuring instruments by applying the principles of heat and mass transfer.

2. The principle of the non-stationary method.

2.1. Conceptual and mathematical introduction

The usual methods for measuring the thermal conductivity of a material are based on solving the Fourier differential equation using the stationary method, specifically in cases where the temperature depends only on one coordinate.

$$dq = -\lambda \frac{dT}{dx} dSdt \tag{1}$$

If the material is in a fluid state, this approach introduces challenges related to convection currents. However, mitigating convection currents by employing an extremely thin layer of the fluid makes the measurement of temperature and layer thickness exceptionally difficult.
Referring to Figure 1, imagining the localized heating of a line of a fluid for a very short duration, results in the radial transfer of heat from the axis to the periphery. For example, an electric current can be passed through a thin linear electrical wire drowned in the homogeneous material whose thermal conductivity is to be measured. The consistent generation of heat in the wire induces a cylindrical temperature field

in the material. The temperature rise in the material (indicated by M in Figure 1) depends on the thermal properties of the material. The feasibility of measuring thermal conductivity based on this principle was initially proposed by researchers such as H. S. Carslaw and J. C. Jaeger [8], as well as E. F. M. Van Der Held, J. Hardebol, and J. Kalshoven [2].

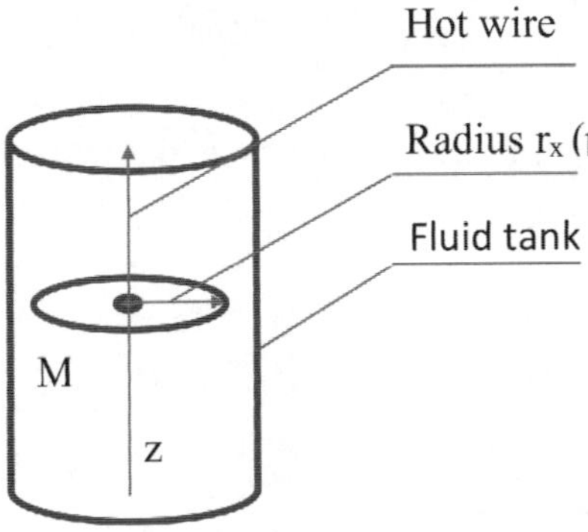

Figure 1. Geometrical scheme of the THW method.

where: M is the material (i.e. water) into the fluid tank, z is the normal to the xy-plane of heat propagation. z is the geometric center of symmetry of the hot wire; r is the radius of the heat propagation in the material (i.e. water).
The approach is based on observing the temperature rise at a given distance "r" from an electrically heated cylindrical wire, which generates a constant heat flow in the liquid. The assumptions are that the wire is infinitely long and negligibly thin. Under these conditions, the Fourier equation and associated boundary conditions apply:

Fourier in radial coordinate system:
$$\frac{\partial T(r,t)}{\partial t} = a\left(\frac{\partial T}{\partial r^2} + \frac{\partial T}{r\partial r}\right) \quad (2)$$

where: a = thermal diffusivity of the liquid (m^2/s); r = distance from the Hot Wire (m); t is the time (s) T is the temperature (K).

Fourier in cartesian coordinate system:
$$\frac{1}{k}\frac{\partial T}{\partial t} = \left(\frac{\partial^2 T}{\partial x^2} + \frac{\partial^2 T}{\partial y^2} + \frac{\partial^2 T}{\partial z^2}\right) \quad (2')$$

where: k = 1/a
If we have the continuous line source and we suppose heat to be liberated at the rate $\rho c_p \phi(t)$ per unit length of a line parallel to the z-axis and through the point (x', y') in the plane xy in figure 1.
where: ρ is the density (kg/m^3); c_p is the specific heat (W/kgK); $\phi(t)$ is W/s
If supply of heat starts at t = 0 when the solid is at zero °C, the temperature at time t is:

The solution of Equation 1 is given by

$$T(r,t) = \frac{1}{4\pi k}\int_0^t \emptyset(t) \cdot e^{\frac{-r^2}{4k(t-t')}} \frac{dt'}{(t-t')} \quad (2'')$$

where: $r^2 = (x-x')^2+(y-y')^2$
If $\phi(t) = q$, constant, this becomes

$$T(r,t) = \frac{1}{4\pi k}\int_{\frac{r^2}{4kt}}^{\infty} \frac{e^{-u}}{u}\, du \quad (2''')$$

$$T(r,t) = -\frac{q_l}{4\pi\lambda} \cdot E_i\left(-\frac{r^2}{4kt}\right) \tag{2''''}$$

$$-E_i(-x) = \int_x^{\infty} \frac{e^{-u}}{u} du \tag{2'''''}$$

where: E_i = Exponential integral; q_l = linear heat production of the hot wire per unit time and unit length (W/m)
For small values of x:

$$E_i(-x) = -\gamma - \ln x + \frac{x}{1.1!} - \frac{x^2}{2.2!} + \cdots \tag{1}$$

where: γ = Euler's constant: 0.5772

$$\int_0^{\infty} (z)^{-1}(e)^{-z}dz = -\gamma + \ln\frac{1}{x} + \frac{x}{1.1!} - \frac{x^2}{2.2!} + \cdots \tag{2'}$$

Thus, for large values of t:

$$T(r,t) = \frac{q_l}{4\pi \cdot k} \cdot \left[\ln\left(\frac{4 \cdot k \cdot t}{r^2}\right) - \gamma\right] \tag{3''}$$

If $\rho\ c_p$ = constant

$$T(r,t) = \frac{q_l}{4\pi \cdot \lambda} \cdot \left[\ln\left(\frac{4 \cdot k \cdot t}{r^2}\right) - \gamma\right] = -\frac{q_l}{4\pi \cdot \lambda} \cdot \left[\ln\left(\frac{r^2}{4kt}\right) + \gamma\right] \tag{4'''}$$

by taking the difference in temperature at two times we obtain:

$$T(r,t_1) - T(r,t_2) = \frac{q_l}{4\pi \cdot \lambda} \cdot ln\left(\frac{t_2}{t_2}\right) \tag{4}$$

And obtaining the thermal conductivity from Equation 5:

$$\lambda = \frac{q_l}{4\pi \cdot} \cdot \frac{ln(t_2) - ln(t_1)}{T(r,t_1) - T(r,t_2)} \tag{5}$$

This relation is the base of the non-stationary method of measuring the thermal conductivity of fluids.

2.2. Error causes

Most authors [9 - 14] have cited factors such as the Knudsen effect, the limited cell size, the compressibility and pressure of the fluid, or the loss of heat due to radiation, among other causes of error. According to Healy, Groot and Kestin [15], in the set of experiments conducted by the authors, the "other causes" did not exert a significant influence. For the calculation, three causes of error were considered to calibrate the measurement equipment using the thermal conductivity of distilled water as a benchmark. In particular:

1) *Finite Radius of the Wire:* The finite radius, the finite conductivity, and the heat capacity per unit volume of the platinum wire influence the results. The standard solution for the heat transmission equation assumes a null radius of the wire; thus, in the real case, a small correction must be applied. The correction is not applied directly to the final value of the thermal conductivity but to the value of the final temperature that is reached by the hot wire [16]. The correction value can be calculated by (6):

$$\delta T_{\propto} = r^2 \frac{\left[(\rho c_p)_w - \rho c_p\right]}{2\lambda t} \Delta T - \frac{q}{4\pi\lambda} \cdot \frac{r^2}{4\lambda t} \cdot \left[2 - \frac{\lambda}{\lambda_w}\right] \tag{6}$$

where: α = radius of the Hot Wire; ρ is the density of fluid (kg/m^3); c_p is the specific heat of the fluid (J/kgK); λ = thermal conductivity of the wire; w is referred to the water

2) *Finite length of the hot wire (end effect):* This error is generated by a nonuniform distribution of temperature in the wire, peaking at the center and decreasing at both edges. Correcting this error is challenging because it is not possible to analytically calculate the temperature distribution as a function of time and space. Therefore, a worst-case scenario correction is employed. Knibbe [4] proposes both a conductive and a convective end-effect correction. In this case, the error correction is calculated directly for the value of the thermal conductivity via:

$$E = \frac{r}{2L} \sqrt{\left(\frac{\lambda_1}{\lambda}\right) \ln\left(\frac{4at}{r^2\gamma}\right)} \tag{7}$$

where: L is length of the wire; λ_1 is thermal conductivity of the wire; λ is thermal conductivity of fluid

3) *Natural convection*: In the proposed case, the change in the density of the fluid caused by the difference in temperature becomes a complex problem when measuring liquids with diverse kinematic properties. A graphical and numerical method is suggested to appropriately consider the lapse of time for the analysis of thermal conductivity. The analytical model in Equation 4 hypothesized that the values of temperatures are a function only of the thermal properties of the examined fluid and the heat flux generated by the hot wire. By appropriately choosing the start and end points of the curve (T, t), representing the beginning and the apex of the curve obtainable through the concentric cylinder method, it is proposed to obtain the two instants:

$$t > t_{min} = r^2 c_p \rho / 2\lambda \tag{8}$$

where r is the radius of the cylinder (hot wire), λ is the thermal conductivity, ρ is the density, c_p is the specific heat.

3. Apparatus design

3.1. The model to the propagation of uncertainty: Monte Carlo simulation

To define a priori the value of the final uncertainty for thermal conductivity, an uncertainty propagation model was chosen that could reduce the errors of the symbolic calculation as much as possible. The software Matlab was used to implemented manually a Monte Carlo according to ISO/IEC 98-1:2009 and "GUM" [17] using the method of generating 106 samples for each variable. The input quantities of the Monte Carlo simulation are characterized by accuracy and type of distribution summarized in Table 1:

Table 1. Input data of the Monte Carlo simulation

	L (m)	r (m)	I (A)	V (V)	t (s)
U[a]	$1*10^{-5}$	$5*10^{-7}$	$1*10^{-3}$	$380*10^{-6}$	$1*10^{-9}$
D[b]	gaussian	gaussian	rectangular	rectangular	rectangular

[a] Uncertainty.
[b] Distribution.

To the evaluation of Standard Deviation and kurtosis, proceeded according to the block diagram in Figura 2:

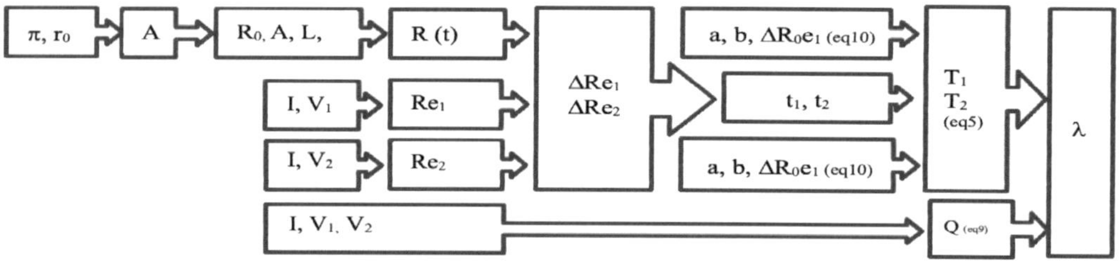

Figure 2. Block diagram to evaluate the uncertainty of the thermal conductivity.

where: r = radius of the hot wire (m); A = wire cross section surface (m^2); t_1, t_2 = times in which the reference quantities are measured (s); Re_1 = electrical resistance of the wire when powered with current I and voltage V_1 (W); Re_2 = electrical resistance of the wire when powered with current I and voltage V_2 (W); $Q_{t1,t2}$ = power average between t_1 and t_2 (W/m).

$$Q(t) = V(t) \cdot I \tag{9}$$

$$R(t) = R_0[1 + aT(t) + bT(t)^2] \tag{10}$$

The equipment was selected (in Table 2) based on their uncertainty characteristics. The uncertainty values of the equipment were used in the uncertainty propagation simulation with Monte Carlo [9] to obtain the design value of the thermal conductivity λ which is about $15*10^{-3}$ (W/mK).

Table 2. Reference of the equipment.

	Accuracy	*Field*
I1[a]	±1(V)	0 - 380 (mV)
I2[b]	10^{-3} [A]	0 – 10 [A]
I3[c]	$5*10^{-7}$ (m)	$25*10^{-3}$ (m)
I4[d]	10^{-5} (m)	$18*10^{-2}$ (m)

[a] DAQ NI USB 6221.
[b] Agilent multimeter model n. 34410.
[c] Micrometre screw gauge.
[d] Calliper gauge Agilent.

The Kurtosis value with Pearson equation of λ is equal to 3.02. The normal distribution is equal to 3.00.

Table 3. Output data of the Monte Carlo simulation.

	A (m^2)	Re (W)	T (°C)	Q (W/m)	k (W/mK)
S. D[a]	$12*10^{-11}$	$55*10^{-5}$	8,28 (1%)	$89*10^{-4}$	$15*10^{-3}$
K[b]	3,01	2,12	3,01	1,98	3,02

[a] Standard Deviation.
[b] Kurtosis.

3.2. Probe designer
The sensor probe (transducer) is designed with the sensor (depicted as a thin line in Figure 3) and the support lying in a plane containing the axis from Figure 1. The sensor wire, constructed from platinum wire (99.99%) with a 38 mm radius, has a minimum length of 33 mm to minimize errors caused by heat conduction to the supports. To reduce the influence of the thermal boundary layer from the supports, the sensors are positioned at least 10 mm away from the walls of the experimental fluid tank (refer to Appendix A). Ensuring maximum verticality during mounting minimizes convection effects on the transducer.
In the described tests, the hot wire is centrally placed between the chamber's center and walls, preventing it from encountering the thermal boundary layer at the wall. The probe's support wire, made of 1.0 mm radius copper wire, is electrically isolated from each other. Copper is chosen for its low temperature coefficient of resistance, crucial for immersion in a bath at a precisely controlled temperature of 0°C, using water in the solid-liquid transition. Throughout the test, the support wires maintain the initial test fluid temperature due to their substantial thermal inertia. Copper's minimal change in resistance over the temperature range prevents measurement errors in length (l) due to resistance variation in the support wires. Bath temperature is measured with a thermometer, accurate to 0.1 K, validated against thermometers calibrated at DIN EN 10204.

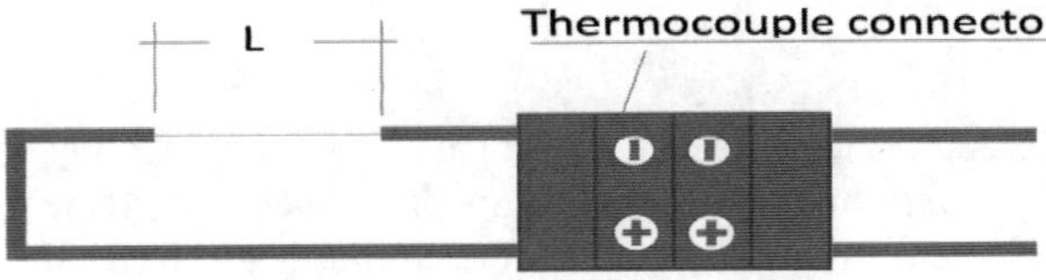

Figure 3. Scheme of the transducer

3.3. Conceptual scheme of connection and Instrumentation

The Hot Wire unit will be including the references equipment and a power supply, DC TENMA 72-10480, with 0-30 [V] DC and 0-3 [A]. The electrical circuit will be performed on power in constant current at 1.0 A.
The conceptual scheme of connection of instrumentation in Figure 4: the DC power supply is connected in series with the ampere-meter (AMP) and with the / Hot wire. The data acquisition system (DAQ) will be connected in parallel to measure the instantaneous variation of the electrical potential (V) at the ends of the platinum hot wire and the data storage (PC).
Next, the current generator (DC) is activated to provide constant current [A]. The values of the voltage at the ends of the transducer (measured by the DAQ board) depend on the temperature (resistance) of the platinum wire (HW).
During the transient state, the voltage acquisition system will be records the changes of the electrical potential at the ends of the platinum wire. The values of the electrical potential at the ends of the

transducer will be recorded the temperature calculate according to the empirical relationship between the resistivity and the temperature, as expressed in Equation 10.
The values presented are the mean value of 30 measurements on 10 experimental procedures, the confidence interval is 95%.

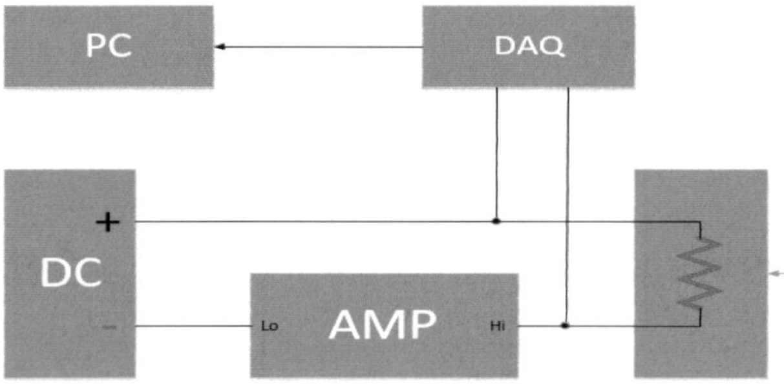

Figure 4. conceptual scheme of connection of instrumentation

4. Conclusions

In this paper, the authors designed a portable transducer based on the thermophysical principles of transient hot wire for measuring thermal conductivity in fluids. They presented the analytical model of heat transfer from the hot wire to the homogeneous fluid around the wire. The causes of systematic errors that most affect the phenomenon of heat and mass transfer in wire/fluid interaction were reported. Through the Monte Carlo simulation proposed in GUM [17], the measuring equipment and DAQ were chosen based on the design value of the uncertainty of thermal conductivity. The next step involves the fabrication and testing of the portable hot-wire transducer.

Acknowledgements

The research activity presented in this paper was financially supported by the Italian Ministry of Research (MIUR) and by Fondazione di Sardegna in the project: Strategies and Technologies for Scientific Education and Dissemination, and the authors would like to thank the CIPNES Authority of Olbia and in particular Dr. Aldo Carta and Eng. Giovanni Maurelli for the financial and technical support to the institutional research project: analysis and programmatic scenarios for the innovation and eco-sustainability of energy infrastructures to support the local production system.

Appendix

From equation 2'''':

$$\frac{T(r,t)}{\frac{q_l}{4\pi\lambda}} = -E_i\left(-\frac{r^2}{4kt}\right) \quad \text{(A.1)}$$

For small values of x, the exponential integral in equation 2'''' can be simplified by truncating its serial development to the first two terms when the r^2 / 4kt argument is much less than 1 and is approximated as,
For x < 0,5

$$Ei(-x) = 0.5772 + lnx - x + \frac{x^2}{4}(\text{Error } 2.132*10^{-5} \text{ at } x{=}0.1) \quad \text{(A.2)}$$

For x < 0.1

The exponential integral $Ei(-x) = 0.5772 + lnx$(Error 5.3% at x=0.1) (A.3)
Table A.1 respectively.
In Figure A.1 the Exponential Integral exact (full lines) and two-term approximation (dashed line with squares, equation A.4) and four term approximation (dashed line triangles, equation A.2).

$$\frac{T(r,t)}{\frac{Q}{4\pi k}} = -\left[-0,5772 + \ln\left(\frac{r^2}{4a_t t}\right) - \frac{r^2}{4a_t t} + \frac{1}{4}\left(\frac{r^2}{4a_t t}\right)^2\right] \quad (A.4)$$

Note that r < 0,5, correspond to Ei > 0,56, i.e. the high temperature region closes to the continuous line source, which is of interested. One is usually not interested in the low temperature region, away from the source (large r and low t),

$$\frac{r^2}{4a_t t} > 0,5$$

Therefore, equation A.4 can be used in most practical situations.
For values of r>0.5, the exponential integral is given as chart and a table A.1.

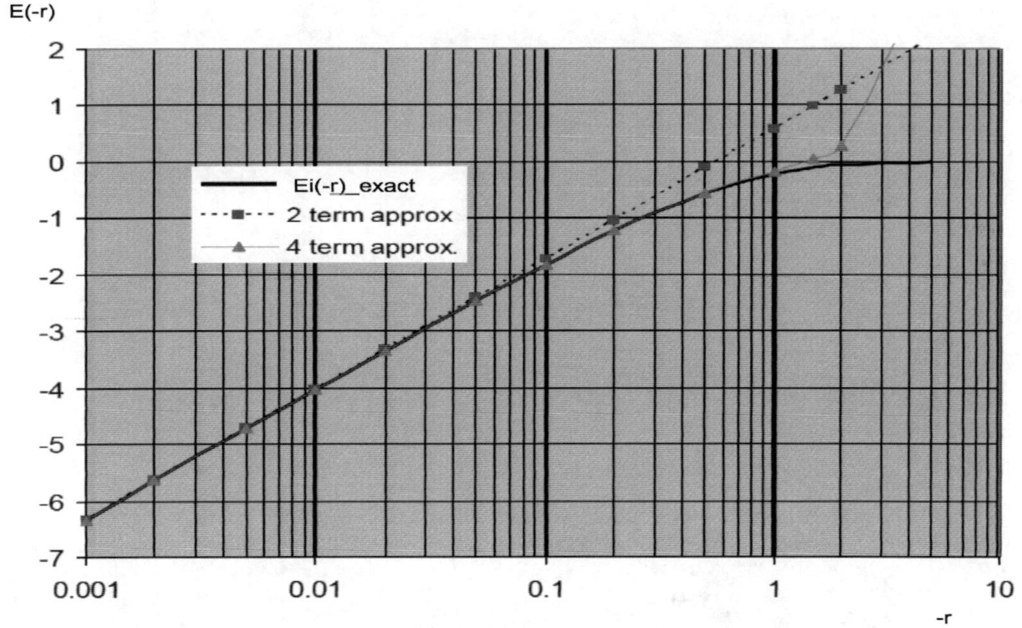

Figure A.1. Exponential Integral approximation.

From the equation 9 at t=0.5 s, the four-term approximation, equation A.4 becomes inaccurate for>0.7 mm.
The authors propose figure A.2 in a pure heat conduction case.

Table A.1. Ei approximation function of -r

-r (mm)	-Ei(-r)	_(0.3)	_(0.4)	Error_(0.3)	Error_(0.4)
0.01	4.037930	4.037945	4.027970	-3.86559E-06	0.002466460
0.02	3.354708	3.354723	3.334823	-4.53763E-06	0.005927425
0.05	2.467898	2.467907	2.418532	-3.55953E-06	0.020003341
0.1	1.822924	1.822885	1.725385	2.13201E-05	0.053506821
0.2	1.222651	1.222238	1.032238	0.000337489	0.155737576
0.5	0.559774	0.553447	0.115947	0.011301738	0.792867721
0.8	0.310597	0.285944		0.079373145	
1	0.219384	0.172800		0.212339771	
1.2	0.158408	0.080478		0.491956080	
1.5	0.100020				
2	0.048901				
3	0.013048				
4	0.003779				
5	0.001148				
8	0.000115				
10	0.000004				

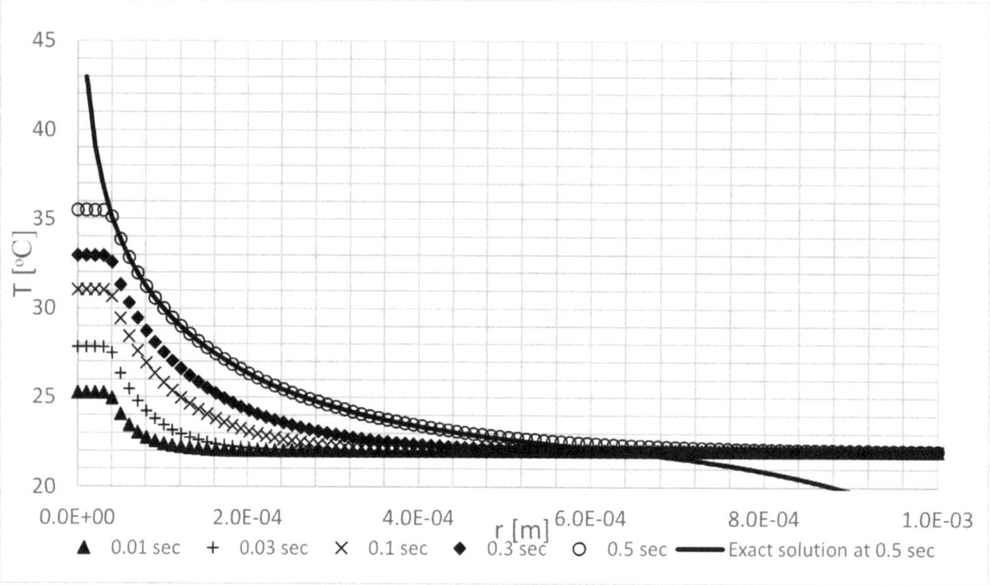

Figure A.2. Radial temperature distribution for the case of pure heat conduction

5. References

[1] H. Pfriem, 1936 Zur Messung veränderlicher Temperaturen von Gasen und Flüssigkeiten *Ingenieur-Archiv, vol. 7*, pp. 85 - 92

[2] E. F. M. v. d. Held and F. G. V. Drunen 1949 A method of measuring the thermal conductivity of liquids *Physica* vol. XV, no. 10, pp. 865 - 881

[3] S. S. Mohammadi, M. S. Graboski and E. D. Sloan 1981 A Mathematical Model of a ramp forced hot-wire thermal conductivity instrument *International Journal of Heat and Mass Transfer,* vol. 24, no. 4, pp. 671 - 783

[4] P. G. Knibbe 1986 The end-effect error in the determination of thermal conductivity using a hot-wire apparatus *International Journal of Heat and Mass Transfer,* vol. 29, no. 3, pp. 463 - 473

[5] M. L. V. Ramires, C. A. N. d. Castro, Y. Nagasaka, A. Nagashima, M. J. Assael and W. A. Wakeham 1995 Standard reference data for thermal conductivity of water *Journal of physical and chemical reference data,* vol. 24, pp. 1377 - 1381

[6] X.-G. Liang 1995 The boundary induced error on the measurement of thermal conductivity by transient hot wire method *Meas. Sci. Technol.,* vol. 6, pp. 467 - 471

[7] A. G. Ostrogorsky, C. Marin, A. Churilov, M. Volz, W. A. Bonner and T. Duffar 2008 Reproducible Te-doped InSb experiments in Microgravity Science Glovebox at the International Space Station, *Journal of Crystal Growth,* vol. 310, no. 2, pp. 364 - 371

[8] H. S. Carslaw and J. C. Jaeger 1959 Conduction of Heat in Solids *London, UK: second ed., Oxford University Press*

[9] J. W. Haarmant 1971 A Contribution to the Theory of the Transient Hot-Wire Method *Physica,* vol. 52, pp. 605 - 619

[10] J. J. Healy, J. J. D. Groot and J. Kestin 1976 The Theory of the Transient Hot-Wire Method for Measuring Thermal Conductivity," *Physica,* vol. 52, pp. 392 - 408

[11] AAVV 1999 *Standard Test Method for Thermal Conductivity of Refractories by Hot Wire (Platinum Resistance Thermometer Technique)* ASTM Designation C1113-99 (American Society for Testing and Materials,, Philadelphia, PA, USA

[12] X. Zhang, W. Hendro, M. Fujii, T. Tomimura and N. Imaishi 2000 Measurements of the Thermal Conductivity and Thermal Diffusivity of Polymer Melts with the Short-Hot-Wire Method,, in *Fourteenth Symposium on Thermophysical Properties*, Boulder, Colorado, U.S.A

[13] S. Y. Kwon and S. Lee 2016 Reliability of Thermal Conductivity Measurement of Liquids by Using Transient Hot-Wire, Photon-Correlation Spectroscopy and the Laser Flash Method," *Journal of the Korean Physical Society,* vol. 68 , no. 10, pp. 1145 - 1155

[14] AAVV 2018 *Standard test method for thermal conductivity of liquids* American Society for Testing Materials ASTM D2717, , Philadelphia, PA, USA

[15] J. J. Healy, J. J. D. Groot and J. Kestin 1976 The Theory of the Transient Hot-Wire Method for Measuring Thermal Conductivity," *Physica,* vol. 52, pp. 392 - 408

[16] A. I. J ohns, A. C. Scott, J. T. R. Watson, D. Ferguson and A. A. Clifford 1988 Measurement of the thermal conductivity of gases by the transient hot-wire method *Philosophical transactions of the Royal society of London, A. mathematical and physical sciences volume 325 pages 295-356 number 1585*

[17] AAVV, ISO/IEC GUIDE 98-1:2009 *Uncertainty of measurement — Part 1: Introduction to the expression of uncertainty in measurement* ISO/IEC GUIDE, Geneva, Switzerland

An overview on Finite Different Method (FDM) in multilayer thermal diffusion problem

R Ricciu[1,*], P M Congedo[2,*], R Baccoli[1], C C Mastino[1], C Baglivo[2], P L M Buonomo[1] and G Popolano[3]

[1] Dipartimento di Ingegneria Civile Ambientale e Architettura, Università di Cagliari - Cagliari IT

[2] Dipartimento di Ingegneria dell'Innovazione, Università del Salento, Lecce IT

[3] Ricercatore indipentente e consulente tecnico aziendale Cagliari IT

* Corresponding author: ricciu@unica.it

Abstract. Numerous mathematical models have been developed for thermal diffusion through single-layer materials, while further developments are needed for multi-layer materials. Diffusion processes through a multilayer material are of interest in industrial, applied thermodynamics, physics, electrical and civil engineering applications. The proposed scheme is easily applicable in various fields, demonstrating that Finite Different Methods (FDMs) are flexible, simple to implement and help to represent multilayer materials even without associating them with other numerical methods. The numerical method studied is a FDM in a civil engineering application. Through the study of a multilayer wall, its criticalities are highlighted, and a possible implementation is proposed.

1. Introduction

Diffusion is an occurrence that draws interest from a wide range of fields such as heat, mass, and electric charge transport [1-4]. It is possible to use the Finite Different Method (FDM) alone or with mixed methods [1], analytical [5], or semi-analytical solutions [6]. Diffusion processes through a multilayered material are of interest for a wide range of applications, including industrial annealing steel coils [7-9], semiconductors [10,11], geological profiles [12,13], or medical applications [14,15]. A dynamic thermal model of a building wall must include a method for obtaining solutions to the diffusion equation for heat flow in solid multilayers. When a finite difference method is used, it is necessary to represent the continuous materials by a network of nodes and associated thermal capacities. The walls of buildings usually contain materials of widely differing properties in close conjunction. In 1985, Waters and Wright [16] studied the criteria for the distribution of nodes in multilayer walls in finite difference thermal modeling and suggested a simple criterion for the distribution. They suggested a given number of nodes throughout a model and discussed the distribution of nodes with respect to the internal boundaries of a multilayer wall. They studied the discontinuity created by an internal boundary using Taylor expansions, and this approximation in the time derivative was applied at a point on the boundary. This method is used to calculate the spatial changes in temperature. The results of the work are interesting in terms of the distribution of nodes and provide evidence of the difference in temperature obtained from the analytical model in boundary layers and the temperature obtained from a finite difference model. In

1999, Antonopoulos and Koronaki [17] proposed a procedure for analyzing the effective thermal capacitance, time constant, and thermal delay of buildings into components corresponding to discrete sections of an envelope to develop different parts of an architectonic composition and even the layers of exterior multilayer walls. The developed procedure was based on a finite-difference solution. The focus of their work [17] was to estimate the thermal parameters of building components corresponding to different parts of exterior walls and develop a procedure for the effective thermal parameters of an entire house. In 2000, Iliev [18] notably used an averaging of diffusivities across an interface, and during a lecture [19] to the Scientific School of Applied Environmental Thermodynamics in 2008, Orioli A. proposed an efficient distribution of nodes and provided evidence of the difference from the temperature obtained from an analytical model in layers and the temperature obtained from a finite difference model. In [19], Orioli proposed using the finite difference method – in an explicit form - in a multilayer wall. In particular, about heat conduction, Orioli proposed a solution in the case of internal physical boundary layers, suggesting that the dynamic parameters of the wall are weighted linearly with the distance between the left and right nodes; this is applicable when the layers have similar thermal properties but not when the thermal parameters are different, as often occurs in a wall, like better explained with more details in the following sections. Among the many disseminated books and papers regarding the numerical solutions of equations, Carlslaw in 1950 [20], Lapidus in 1999 [21], Canale in 2015 [22], and Tabares-Velasco and Griffith in 2012 [23] proposed a finite element method, as a numerical solution, for many science and engineering problems; however, the application of conduction phenomena in multilayer walls does not include the calculation of the temperature in the plane between two inner wall layers. At present, it is easy to believe that commercial software implements the best technical and scientific solutions. The most popular commercial building energy simulation software programs in research applications, such as CoDyBa [24], ESP-r [25–27], and TRNSYS [28–30], propose similar approaches to calculate the thermal resistivity (mK/W) and heat capacity (J/K) of walls. In recent years, new numerical models have been developed because of the study of early PCM (phase change material) models added to the layer walls [31,32], and the initially modified empirical models were obtained using an equivalent heat transfer coefficient [30] to fully implement finite difference models [29,33] and control volume models [27]. These models are used as prediction models [33]. TRNSYS and EnergyPlus are software programs that implement a routine FDM but do not easily validate the model, particularly with a multilayer wall. Kuznik et al [34] presented good results with these methods. In particular, Tabares-Velasco et al [35] tested EnergyPlus version 8 with a PCM application in a multilayer wall and noted improvements over version 6, which showed at most a half degree of difference between the experimental and simulation results. In addition, the difference between the simulated and measured heat flux was approximately 1 W/m^2 from approximately 4 W/m^2.To evaluate the thermal conductivity at the interface between nodes (layers), EnergyPlus uses a linear interpolation between nodal points [36] and the Conductance Finite Difference (CondFD) algorithm outputs the heat flux at each node and the heat capacitance of each half-node. During the CondFD solution iterations, the heat capacitance of each half node is stored similarly to the method discussed in [18]. In this work, an application of the FDM is proposed to model the diffusion of one-dimensional heat in a multilayer wall and is verified through an experimental test of a civil engineering application.

2. Theory

With an evolution of temperature over time and space in a wall, it is possible to apply the finite difference method (FDM) at the heat equation written as a continuous version of energy balance equation (1) and in the case of a single layer with a homogeneous and uniform material:

$$\rho c_p \frac{\partial T}{\partial t} = \lambda \frac{\partial^2 T}{\partial x^2} \qquad (1)$$

Where:

- $\rho c_p \frac{\partial T}{\partial t}$ is the heat stored
- $\lambda \frac{\partial^2 T}{\partial x^2}$ is the variation of the temperature gradient in the x direction (temperature Laplacian operator with a vector in the x direction).

In terms of finite difference elements, it is possible to have a discrete version in equation (2):

$$\rho c_p \frac{T_i^{n+1} - T_i^n}{\Delta t} = \lambda \frac{T_{i+1}^n - 2T_i^n + T_{i-1}^n}{\Delta x^2} \tag{2}$$

where “n” refers to the evolution over time Δt and “i” refers to the evolution over space Δx.
In explicit finite difference schemes, the temperature at time n+1 depends explicitly on the temperature at time n. The explicit finite difference discretization of equation (3) is:

$$T_i^{n+1} = T_i^n + \frac{\lambda}{\rho c_p} \Delta t \frac{T_{i+1}^n - 2T_i^n + T_{i-1}^n}{\Delta x^2} \tag{3}$$

the internal diffusivity coefficient k [m2/s] is equal to:

$$k = \frac{\lambda}{\rho C_p} \tag{4}$$

Where:

- ρ is the density [kg/m^3]
- Cp is the specific heat = heat capacity [J/kgK]

It is thus possible to calculate the interface temperature of each individual layer.
Implementing equation (3) in time and space x, it is possible to discern a trend of temperature, for example, in a homogeneous and isotropic wall or in general condition in a solid introduced in the first part of this paragraph.

3. Stationary model

When considering a multilayer wall, some authors [8-11] use EnergyPlus with a linear interpolation between nodal points equidistant from the interface plane between the two layers of different and a constant Δx distance for the whole thickness of the wall. Some authors [8 -11] modify equation (8) by introducing average values for the following elements:

$$k_M = \frac{\lambda_M}{\rho_M C_{pM}} = \frac{1}{2}\left(\frac{\lambda_1}{\rho_1 C_{p1}} + \frac{\lambda_2}{\rho_2 C_{p2}}\right) \tag{5}$$

where k_M is the average value of the internal diffusivity coefficient (m^2/s) by two layers, 1 and 2 in Figure 1, or:

$$\rho_M C_{pM} = \frac{1}{2}\left(\rho_1 C_{p1} + \rho_2 C_{p2}\right) \tag{6}$$

$$\lambda_M = \frac{1}{2}(\lambda_1 + \lambda_2) \quad (7)$$

If the Δx distance in the whole thickness of the wall is not constant, equation (7) becomes equation (9):

$$\rho_M C_{pM} = \left(\frac{\rho_1 C_{p1} \Delta x_1 + \rho_2 C_{p2} \Delta x_2}{\Delta x_1 + \Delta x_2}\right) \quad (8)$$

$$\lambda_M = \left(\frac{\lambda_1 \Delta x_1 + \lambda_2 \Delta x_2}{\Delta x_1 + \Delta x_2}\right) \quad (9)$$

$$T_{M-SM} = \left(\frac{T_{cold}\frac{\lambda_2}{\Delta x_2} + T_{Hot}\frac{\lambda_1}{\Delta x_1}}{\frac{\lambda_1}{\Delta x_1} + \frac{\lambda_2}{\Delta x_2}}\right) \quad (10)$$

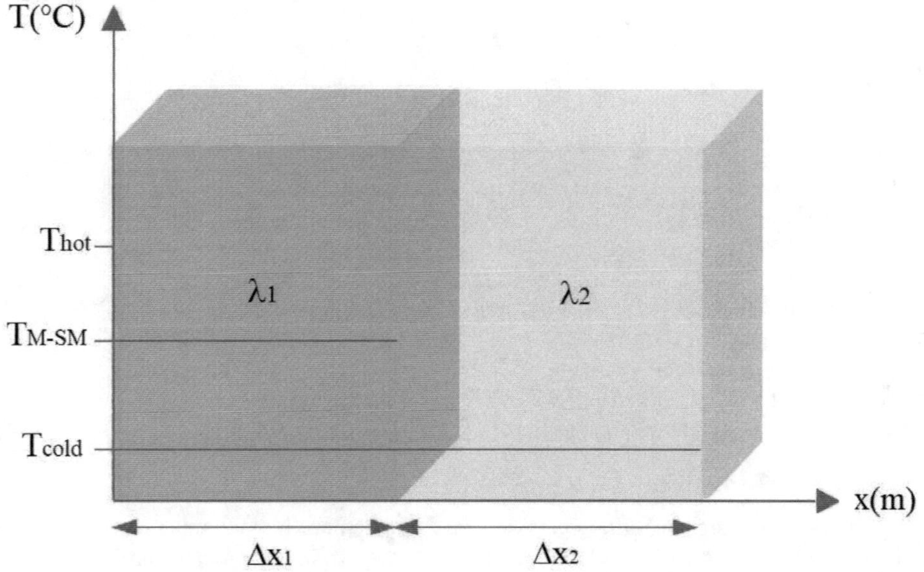

Figure 1. Scheme and equation to apply the stationary model.

4. FDM and stationary methods comparison

By compiling a MATLAB script, the FDM was applied to equation (3). Given an initial thermal state, the variation in time t and in space x of the temperatures inside a wall consisting of two layers with very different thermal properties between them was calculated. The two layers, having a different thermal conductivity, are defined according to a classification of building materials: one is an insulating layer, polyurethane, and the other is a good heat conductor, brick.

Table 1 shows the characteristic parameters of the finite difference model, the initial thermal state, and the thermophysical properties of the two materials and the temperature value for the node in the middle of the two layers for the wall after the results have thermally stabilized.

The specific heat flux (W/m^2) can be calculated by applying the Fourier law, the values of the specific heat fluxes on the two faces of the interface plane should be equal. Table 1 shows that the specific thermal fluxes calculated with the method of [19] does not have the same values.

Table 1. Data and results of stationary and dynamic thermal model.

$\Delta\tau^a$ (s) = 60	Δx^a (m) = 0.01	$T_{Hot}{}^a$ (°C) = 30	$T_{cold}{}^a$ (°C) = 20	Tw^a (°C) = 10
$x_1{}^b$ (m) = 0.1	$\lambda_1{}^b$ (W/mK) = 0.04	$Cp_1{}^b$ (J/kgK) = 1400	$\rho_1{}^b$ (kg/m^3) = 100	
$x_2{}^c$ (m) = 0.1	$\lambda_2{}^c$ (W/mK) = 0.55	$Cp_2{}^c$ (J/kgK) = 1000	$\rho_2{}^c$ (kg/m^3) = 1600	
$T_{Hot}{}^d$ (°C)	$T_{M\text{-}WA}{}^d$ (°C) (Fourier)	$T_{M\text{-}SM}$ (°C) ref.[19] ^{d}	$T_{Cold}{}^d$ (°C)	
(30.0)	(24.5)	(20.7)	(20.0)	

	layer 1 (W/m^2)	layer 2 (W/m^2)
$\mathbf{q_x}{}^e$	0.037	0.037
$\mathbf{q_x}{}^f$	0.02	0.25

[a] Data of finite differences model parameters
[b] Data layer 1 (polyurethane)
[c] Data layer 2 (brick)
[d] Data at regime Temperatures
[e] Results by Fourier Law
[f] Results by weighted average ref.[19]

Were:

- $\mathbf{\Delta\tau}$ is the discrete time (s)
- **q** is the specific heat flux (W/m^2)
- $\mathbf{\Delta x}$ is the discrete distance (m)
- $\mathbf{T_{M\text{-}WA}}$ is the Weighted Average Temperature (K)
- $\mathbf{T_{M\text{-}SM}}$ is the Stationary Method Temperature (K)

At the established temperatures in Table 1, it is evident the difference in the temperature values between the standard FDM application in the middle node between the two layers and the stationary model.

5. Discussion of the results and conclusions

The authors propose the calculation of the dynamic temperature for a multilayer wall according to a unidimensional heat flux model through a numerical method. The authors were interested in determining the values of the temperature in the transient time until the temperature stabilizes over time (steady state) in the interface between the physical layers. The proposed numerical method uses the FDM without a mixed method, i.e., FDM with a semi-analytical model.

The temperature simulated by the FDM in the literature (in particular in [5]) in the interface between physical layers of a wall provides results at a steady state, with a difference of 5% compared to the measured value in the laboratory in the same multilayer wall.

The referenced model in [19] is compared with the stationary model (Fourier) the difference at about 25°C of the middle temperature from both surfaces temperature are of about 4°C.

This difference could be critical when studying a point of humidity condensation between the layers or if linked to the internal comfort temperature which, as we know, is limited to a few degrees.

To the future, the authors intend to find a mathematical/numerical physical model of the heat equation that best models the temperature variation over time and space in the specific case of layers with specific heat values and thermal conductivity and density very different that will be a little computational capacity.

6. References

[1] Baccoli R, Kumar A, Concas A, Gatto G, Pintus N, Medda A, Rodriguez G, Thermal diffusivity from Fourier's inverse problem supervised by an optimization model: Theoretical analysis and experimental validation, Case Studies in Thermal Engineering, Volume 40,2022, 102533, ISSN 2214-157X, https://doi.org/10.1016/j.csite.2022.102533.

[2] De Monte F, An analytic approach to the unsteady heat conduction processes in one-dimensional composite media, International Journal of Heat and Mass Transfer, Volume 45, Issue 6, 2002, Pages 1333-1343, ISSN 0017-9310, https://doi.org/10.1016/S0017-9310(01)00226-5.

[3] Becker S M, One-Dimensional Transient Heat Conduction in Composite Living Perfuse Tissue, ASME. J. Heat Transfer. July 2013; 135(7): 071002. https://doi.org/10.1115/1.4024063.

[4] Furmański P, Łapka P, Evaluation of a human skin surface temperature for the protective clothing – Skin system based on the protective clothing–skin imitating material results, International Journal of Heat and Mass Transfer, Volume 114, 2017, Pages 1331-1340, ISSN 0017-9310, https://doi.org/10.1016/j.ijheatmasstransfer.2017.06.033.

[5] Simões N, Tadeu A and Godinho L, Transient heat transfer across a multi-layer wall, International Journal for Housing Science and Its Applications, 27(4), pp. 287-300, 2003.

[6] Carr E J, March N G, Semi-analytical solution of multilayer diffusion problems with time-varying boundary conditions and general interface conditions, Applied Mathematics and Computation, Volume 333, 2018, Pages 286-303, ISSN 0096-3003, https://doi.org/10.1016/j.amc.2018.03.095.

[7] Barry S, Sweatman W, Modelling heat transfer in steel coils, ANZIAM Journal (E) 50 (2009) C668–C681.

[8] Mc Guinness M, Sweatman W, Boawan D, Barry S, Annealing steel coils, in: T. Marchant, M. Edwards, G. Mercer (Eds.), Proceedings of the 2008 MISG, 2009.

[9] Yuen W Y D, Transient temperature distribution in a multilayer medium subject to radiative surface cooling, Applied Mathematical Modelling, Volume 18, Issue 2, 1994, Pages 93-100, ISSN 0307-904X, https://doi.org/10.1016/0307-904X(94)90164-3.

[10] Muñoz Aguirre N, González de la Cruz G, Gurevich Y G, Logvinov G N and Kasyanchuk M N (2000), Heat Diffusion in Two-Layer Structures: Photoacoustic Experiments. phys. stat. sol. (b), 220: 781-787. https://doi.org/10.1002/1521-3951(200007)220:1<781::AID-PSSB781>3.0.CO;2-D

[11] Gurevich Y G, Lashkevich I, Gonzalez de la Cruz G, Effective thermal parameters of layered films: An application to pulsed photothermal techniques, International Journal of Heat and Mass Transfer, Volume 52, Issues 19–20, 2009, Pages 4302-4307, ISSN 0017-9310, https://doi.org/10.1016/j.ijheatmasstransfer.2009.03.068.

[12] Grossel P, Depasse F, Alternating heat diffusion in thermophysical depth profiles: multilayer and continuous descriptions, Journal of Physics D: Applied Physics 31 (1998) 216–223, https://doi.org/10.1088/0022-3727/31/2/007.

[13] Nowamooz H, Nikoosokhan S, Lin J, Chazallon C, Finite difference modeling of heat distribution in multilayer soils with time-spatial hydrothermal properties, Renewable Energy, Volume 76, 2015, Pages 7-15, ISSN 0960-1481, https://doi.org/10.1016/j.renene.2014.11.008.

[14] Mulholland G P, Cobble M H, Diffusion through composite media, International Journal of Heat and Mass Transfer, Volume 15, Issue 1, 1972, Pages 147-160, ISSN 0017-9310, https://doi.org/10.1016/0017-9310(72)90172-X.

[15] Simpson M J, McInerney S, Carr E J et al. Quantifying the efficacy of first aid treatments for burn injuries using mathematical modelling and in vivo porcine experiments. Sci Rep 7, 10925 (2017). https://doi.org/10.1038/s41598-017-11390-y.

[16] Waters J R, Wright A J, Criteria for the distribution of nodes in multilayer walls in finite-difference thermal modelling, Building and Environment, Volume 20, Issue 3, 1985, Pages 151-162, ISSN 0360-1323, https://doi.org/10.1016/0360-1323(85)90010-1.
[17] Antonopoulos K A, Koronaki E P, Thermal parameter components of building envelope, Applied Thermal Engineering, Volume 20, Issue 13, 2000, Pages 1193-1211, ISSN 1359-4311, https://doi.org/10.1016/S1359-4311(99)00090-3.
[18] Iliev O P A finite-difference scheme of second-order accuracy for elliptic equations with discontinuous coefficients. Diff Equat 36, 928–930 (2000). https://doi.org/10.1007/BF02754419.
[19] Orioli A, https://dokumen.tips/documents/1-scuola-estiva-di-fisica-tecnica-termofisica-il-ritardo-del-fattore-di-decremento.html?page=15.
[20] Carlslaw H, Jaeger J, “Conduction of heat in solids” Clarendon Press, Oxford.
[21] Lapidus L, Pinder G F, Numerical solution of partial differential equations in science and engineering 1999 A Wiley - Interscience Publication John Wiley & Sons, inc. ISBN 0-471-09866-3.
[22] Canale R P, Chapra S C, Numerical Methods for Engineers, Seventh edition, 2015 McGraw-Hill Education ISBN 978–0–07–339792–4.
[23] Tabares-Velasco P C, Griffith B, Diagnostic test cases for verifying surface heat transfer algorithms and boundary conditions in building energy simulation programs, Journal of Building Performance Simulation, 5:5, 329-346, (2012) https://doi.org/10.1080/19401493.2011.595501.
[24] Kuznik F, Virgone J, Experimental investigation of wallboard containing phase change material: Data for validation of numerical modeling, Energy and Buildings, Volume 41, Issue 5, 2009, Pages 561-570, ISSN 0378-7788, https://doi.org/10.1016/j.enbuild.2008.11.022.
[25] Heim D, Clarke J A, Numerical modelling and thermal simulation of PCM–gypsum composites with ESP-r, Energy and Buildings, Volume 36, Issue 8, 2004, Pages 795-805, ISSN 0378-7788, https://doi.org/10.1016/j.enbuild.2004.01.004.
[26] Schossig P, Henning H M, Gschwander S, Haussmann T, Micro-encapsulated phase-change materials integrated into construction materials, Solar Energy Materials and Solar Cells, Volume 89, Issues 2–3, 2005, Pages 297-306, ISSN 0927-0248, https://doi.org/10.1016/j.solmat.2005.01.017.
[27] Heim D, Isothermal storage of solar energy in building construction, Renewable Energy, Volume 35, Issue 4, 2010, Pages 788-796, ISSN 0960-1481, https://doi.org/10.1016/j.renene.2009.09.005.
[28] Stovall T K and Tomlinson J J (1995). "What are the potential benefits of including latent storage in common wallboard?" Journal of Solar Energy Engineering, Nov 1995, 117(4): 318-325 (8 pages), https://doi.org/10.1115/1.2847868
[29] Koschenz M, Lehmann B, Development of a thermally activated ceiling panel with PCM for application in lightweight and retrofitted buildings, Energy and Buildings, Volume 36, Issue 6, 2004, Pages 567-578, ISSN 0378-7788, https://doi.org/10.1016/j.enbuild.2004.01.029.
[30] Ibáñez M, Lázaro A, Zalba B, Cabeza L F, An approach to the simulation of PCMs in building applications using TRNSYS, Applied Thermal Engineering, Volume 25, Issues 11–12, 2005, Pages 1796-1807, ISSN 1359-4311, https://doi.org/10.1016/j.applthermaleng.2004.11.001.
[31] Tomlinson J J, Heberle D P, Analysis of wallboard containing a phase change material, 25th Intersociety Energy Conversion Engineering Conference, Reno, NV, 12-17 Aug. 1990, https://doi.org/10.1109/IECEC.1990.716494.
[32] Stovall T K and Tomlinson J J (1995). "What are the potential benefits of including latent storage in common wallboard?" Journal of Solar Energy Engineering, Transactions of the ASME 117(4): 318-325.
[33] Curtis O P, Advanced zone simulation in EnergyPlus: incorporation of variable properties and phase change material (PCM) capability, Proceedings of building simulation 2007, p. 1341-

1345.

[34] Kuznik F, Virgone J, Roux J J, Energetic efficiency of room wall containing PCM wallboard: A full-scale experimental investigation, Energy and Buildings, Volume 40, Issue 2, 2008, Pages 148-156, ISSN 0378-7788, https://doi.org/10.1016/j.enbuild.2007.01.022.

[35] Tabares-Velasco P C, Christensen C, Bianchi M and Booten C, Verification and Validation of EnergyPlus Conduction Finite Difference and Phase Change Material Models for Opaque Wall Assemblies, 2012, http://www.osti.gov/bridge

[36] https://energyplus.net/sites/all/modules/custom/nrel_custom/pdfs/pdfs_v8.8.0/EngineeringReference.pdf (p.69)

Acknowledgments
The authors would like to thank the CIPNES Authority of Olbia and in particular Dr. Aldo Carta and Eng. Giovanni Maurelli for the financial and technical support to the institutional research project: analysis and programmatic scenarios for the innovation and eco-sustainability of energy infrastructures to support the local production system.

BEPU analysis of an Upgraded ICE facility test by TRACE/DAKOTA coupling

A. Bersano[1], G. Agnello[2], P. A. Di Maio[2], G. Grippo[1] and Fulvio Mascari[1]

[1] FSN-SICNUC, ENEA C.R. Bologna, Via Martiri di Monte Sole 4, 40129, Bologna, Italy
[2] Dipartimento di Ingegneria, Università degli studi di Palermo, Viale delle Scienze, Edificio 6, 90128, Palermo, Italy
[*] e-mail: gianmarco.grippo@enea.it

Abstract. The Ingress of Coolant Event (ICE) in the plasma chamber is one of the safety issues in fusion nuclear plants. The best estimate thermal-hydraulic system codes adopted to perform deterministic safety analysis should be validated against the phenomena typical of accidental transients in fusion installations. TRACE (TRAC/RELAP Advanced Computational Engine), best estimate thermal-hydraulic system code developed by USNRC, has been adopted to simulate an ICE. The calculated results have been compared to the experimental data obtained in one test performed in the upgraded Integrated ICE facility at JAERI. In this updated configuration the pressure suppression system is connected to the top of the plasma chamber instead of the bottom of the vacuum vessel. The facility nodalization has been developed in the SNAP environment/architecture. To qualify the code and the nodalization, an accuracy evaluation has been performed both from a qualitative and quantitative point of view. Then, considering the presence of some uncertainties in the input-deck development, an uncertainty analysis has been carried out. The probabilistic method to propagate the input uncertainties has been selected and the analysis has been carried out with the DAKOTA toolkit coupled with TRACE code in SNAP. In the uncertainty analysis, some relevant statistical parameters have been considered to characterize the dispersion of the results and the correlation between the uncertain input parameters selected and the PC pressure chosen as figure of merit.

Keywords: Ingress of Coolant Event; TRACE; Code accuracy; Uncertainty Analysis; DAKOTA

Acronyms: ICE (*Ingress of Coolant Event)*; PC (*Plasma Chamber*); JAERI (*Japan Atomic Energy Research Institute*); TRACE (*TRAC/RELAP Advanced Computational Engine*); USNRC (*United State Nuclear Regulatory Commission*); SP (*Suppression Tank*); DT (*Drain Tank*); VV (*Vacuum Vessel*); SNAP (*Symbolic Nuclear Analysis Package*); FFTBM (*Fast Fourier Transform Based Method*); UA (*Uncertainty Analysis*); DAKOTA (*Design Analysis Kit for Optimization and Terascale Application*); SD (*Simulated Divertor*); MV (*Magnetic Valve*); RP (*Relief Pipe*); SoT (*Start of Transient*); PhW (*Phenomenological Window*); RTP (*Relevant Thermohydraulic Phenomena*); RTA (*Relevant Thermohydraulic Aspects*); AA (*Average Amplitude*); WF (*Weighted Frequency*); FoM (*Figure Of Merit*); PDF (*Probability Density Function*).

Introduction

The Ingress of Coolant Event (ICE) in fusion reactors may occur due to the break of the cooling tubes, installed in the plasma-facing components, causing the loss of vacuum in the Plasma Chamber (PC). The Integrated ICE facility was built at JAERI (Japan Atomic Energy Research Institute – Naka Laboratories) to study the thermal-hydraulic behavior of this accident [1-4]. The best estimate thermal-hydraulic system code TRACE (TRAC/RELAP Advanced Computational Engine) [5], developed by US Nuclear Regulatory Commission (USNRC), was applied in previous activities to simulate a test conducted in the Integrated ICE facility [3]. In this activity, the TRACE code has been applied to the Upgraded ICE configuration, having the Suppression Tank (ST) connected to the top of the PC; moreover, a Drain Tank (DT) was added below the Vacuum Vessel (VV). The preliminary results were presented in [4]; this paper describes the final results. The nodalization of the Upgraded ICE facility has been developed using the Symbolic Nuclear Analysis Package (SNAP) [6]. Then the qualitative and quantitative accuracy of the code results has been evaluated [7]. For the quantitative accuracy evaluation, the Fast Fourier Transform Based Method (FFTBM) [8] has been applied. Finally, an Uncertainty Analysis (UA) has been developed applying the probabilistic method to propagate input uncertainty [3,9]. The UA has been carried out using the DAKOTA (Design Analysis Kit for Optimization and Terascale Application) toolkit [10] coupled with TRACE in the SNAP environment/architecture.

1. ICE upgraded facility

The Integrated ICE facility is a scaled-down experimental facility (volumetric scaling factor 1:1600 with respect to ITER reactor [1,2]) designed to study the thermal-hydraulic behavior during an ICE. Other parameters have been scaled accordingly to the volumetric scaling factor [11]. The main components are the PC, the VV, the Simulated Divertor (SD), the ST and the DT (Figure 1). The break of the cooling tubes is simulated in the facility with up to three injection lines connected on the side of the PC. The water for the injection is provided by a pressurized boiler with electrical heaters. On the drain line it is installed a Magnetic Valve (MV) [2]. The pressure suppression system is composed by the ST connected to the top of the PC by three Relief Pipes (RP) and activated by other three MV.

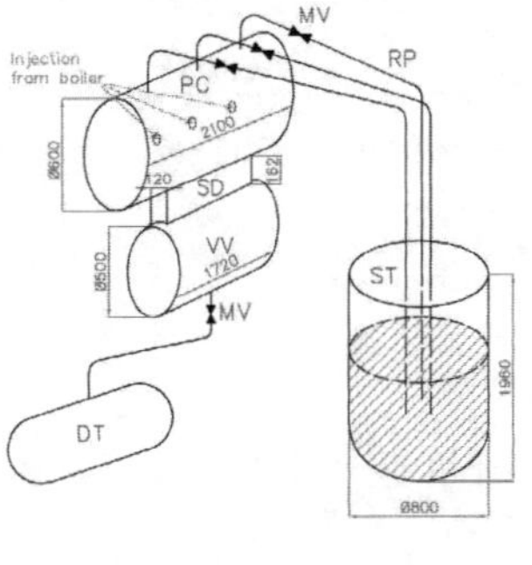

Figure 1. Schematic view of the ICE upgraded facility.

2. Description of the selected test

For the present analysis, test P1 has been selected since it has a relatively high injected mass flow rate and consequent PC peak pressure. The injection is performed adopting all three injection lines for 45 s. The water injection temperature and pressure are 150°C and 2 MPa respectively [1,2]; the total injected mass flow rate is around 5.5 kg/s [2]. The electrical heaters in the PC, SD and VV walls are activated to compensate for the heat losses [1]. At the Start Of Transient (SOT), the water injection starts, and the PC pressure rises. The PC over pressurization is limited by the activation of the pressure suppression system at the opening of the MV on the RP. To study and characterize the TRACE code accuracy both qualitatively and quantitatively, the transient has been subdivided by the authors into three main Phenomenological Windows (PhWs), mainly considering the PC pressure behavior (Table I), identifying the Relevant Thermohydraulic Phenomena (RTP) and Relevant Thermohydraulic Aspects (RTA).

3. TRACE nodalization of ICE upgraded facility

In this analysis TRACE V5 patch 4 has been adopted and SNAP was used to develop the nodalization (Figure 2). The TRACE/SNAP environment/architecture is presented in [12,13]. The PC, SD, VV and DT, where multidimensional phenomena could occur, have been simulated adopting the TRACE 3D

Vessel component. The PC is subdivided into 6 axial levels, 4 radial rings and 12 azimuthal sectors; the VV is subdivided into 6 axial levels, 2 radial rings and 6 azimuthal sectors. The SD is simulated with a Vessel component in Cartesian coordinates having 4 axial levels, 2 volumes in the x-direction and 1 volume in the y-direction. Each axial level corresponds to one slit of the SD. Three Fill components are connected on the side of the PC to simulate the nozzles for water injection from the boiler. The DT is simulated by a horizontal Vessel with 1 axial levels, 1 radial ring and 6 azimuthal sectors. The VV is connected to the DT by a Valve and a Pipe component. The three RP are modeled with two Pipe components each and the MV with a Valve component controlled by a trip (i.e. the pressure set-point in the PC). An opening delay of the MV of 2.5 s has been assumed as in [2]. Finally, the ST is simulated by a Pipe component having a single volume and the RP are connected to the ST through crossflow junctions. Heat Structures have been included to consider the solid structures of the facility: PC, SD, VV, DT, RP and ST walls, flanges and insulating material. To simulate the heat losses, the ambient temperature and a heat transfer coefficient have been imposed on the outer surface of the heat structures.

Table I. Identified PhWs, RTP and RTA.

PhW	Time [s]	RTP	RTA
1	0 - 45	- Water flashing in the PC and VV - Water condensation in the SD [1] - Pressure suppression in the ST - Condensate discharge in the DT	- Opening of the MV - Maximum PC pressure 0.373 MPa (at 2 s)
2	45 - 172.5	- Water condensation in the SD [1] - Pressure suppression in the ST - Condensate discharge in the DT	- Minimum PC pressure 0.0455 MPa (at 172.5 s)
3	172.5 - 300		- ST and DT pressure increment

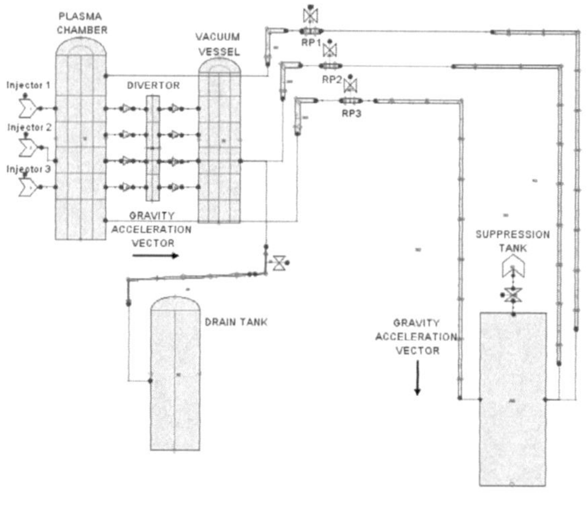

Figure 2. ICE upgraded nodalization developed through SNAP.

4. Methodology

The code calculation results should be assessed through a process to evaluate their accuracy. Accuracy is usually considered as the discrepancy between the experimental data and the calculated results. In particular, the accuracy evaluation is an element of the code independent qualification process [7] that is carried out by the code users. The accuracy evaluation is done in two stages: firstly, it is evaluated the qualitative accuracy and then the quantitative one.

4.1. Accuracy Evaluation

The qualitative accuracy evaluation is a judgment of the code results subjectively performed by the code-user, taking into account the capability of the code to predict the involved phenomena through the behavior of selected parameters. Initially, the PhWs of the test are defined and the RTPs in each PhW are identified with the characterizing RTA. Then, each RTP is subjectively judged by a visual comparison of the experimental and calculated results. In this analysis, the qualitative accuracy evaluation is performed with four subjective judgment marks (Excellent (+), Reasonable (o), Minimal (NA), Unqualified (-)), which are given both to the experimental data and the calculated results (Table II) [14]. The quantitative accuracy evaluation provides a numerical indication of the performance of a calculation. In this analysis, it has been performed through the FFTBM [7]. The results are provided by two parameters: the Average Amplitude (AA) and the Weighted Frequency (WF) [15,16]. The accuracy evaluation is mostly based on the AA (the lower is the AA, the more accurate is the result), while the WF is an additional qualitative information that may be considered for the accuracy evaluation [7]. The total AA and the total WF are calculated using proper weighting factors (in this case set equal to one for all parameters [1]). The JSI FFTBM Add-In 2007 developed at Jožef Stefan Institute (Slovenia) [15-17] has been adopted as tool to perform the FFTBM. The AA reference threshold values to evaluate the accuracy are set as in [18].

Table II. Qualitative accuracy evaluation judgement marks [14]

Data	+	o	NA	-
Experimental	Phenomenon occurred in the test and it is directly measured	Phenomenon occurred in the test and it is indirectly measured	Phenomenon occurred during the test but there is no instrumentation to detect (lack of instrumentation)	Phenomenon not occurred in the test
Calculated	Phenomenon is clearly predicted by the code (**Excellent**)	Phenomenon is partially predicted (i.e. the answer of the code is **reasonable** but closure code relations are not appropriate, etc.)	Models are not appropriate to predict (i.e. nodalization strategy, etc.) (**Minimal**)	Phenomenon is not predicted by the code (**Unqualified**)

4.2. Uncertainty Analysis

4.2.1. DAKOTA tool in the SNAP environment/architecture

DAKOTA [**Error! Reference source not found.**0] is developed by Sandia National Laboratories to execute parametric and uncertainty analyses in an automatic and fast way. The purpose of this toolkit is to bridge computer codes and analysis methods for parametric evaluation, uncertainty qualification and system optimization [3]. DAKOTA is also available as a plug-in [19] for SNAP. The workflow of the TRACE/DAKOTA coupling in the SNAP environment/architecture is presented in [3,20]. Starting from the TRACE reference input-deck, DAKOTA samples the selected input uncertain parameters, creating a set TRACE input-decks. Based on Wilks [21,22] and as reported in [23], the minimum number of code runs depends on the probability content, confidence level and number of Figure Of Merits (FOMs) [24]. Through DAKOTA it can be performed the statistical analysis of the FOM and characterized, e.g. through Pearson and Spearman coefficients [3,20,25], the relationship between the uncertain input parameters and the FOMs.

4.2.2. Hypothesis adopted in the UA

In the present application, the uncertain input parameters have been taken as in [3], which considered previous references openly available on similar applications. The PC pressure has been selected as FOM because it is the most important safety parameter. Considering one FOM, with a probability content and a confidence level of 95%, a total of 93 calculations were necessary, based on Wilks, for the two-sided tolerance interval [9].

5. Results

5.1. Reference calculation results description

Before the SOT, a steady state calculation has been performed to obtain the initial conditions of the test. The reference calculation results are shown against experimental data in Figure 3 to Figure 4. The PhWs are highlighted by dashed lines. In PhW1 a fast pressurization occurs in the PC (Figure 3). Then, after the opening of the MV the pressure suppression in the ST begins; therefore, in the remaining part of the PhW two counteracting phenomena are present: the PC pressurization and the pressure suppression in the ST. The variation of the weight of the two phenomena along the PhW determines the pressure evolution, with an initial peak followed by a gradual reduction. The code predicts the pressure peak in the PC (maximum value 0.386 MPa) and the subsequent reduction, even if with a higher rate after the peak. Considering the PC temperature (Figure 4) at the SOT there is a quick reduction (with a minimum value of 385 K) due to the income of colder water from the boiler, followed by a slight increment and a subsequent gradual reduction. The temperature behavior is predicted by the code with a higher temperature increase after the initial drop; however, the temperature value at the end of PhW1 is correctly predicted. In the second PhW the water injection is over; therefore, the occurring phenomena are the pressure suppression in the ST and the condensate discharge in the DT. They both contribute to the reduction of the pressure in the PC, which are predicted by the code, even if with a lower rate at the beginning of PhW2 (Figure 3). The PC temperatures gradually reduce in this PhW (Figure 4). The behavior in the PC is predicted by the code, even if with a slight overestimation, following the pressure one. In PhW3, the PC and VV pressure and temperature are almost constant (Figure 3 and Figure 4 respectively). The qualitative and quantitative behavior is correctly predicted by the code.

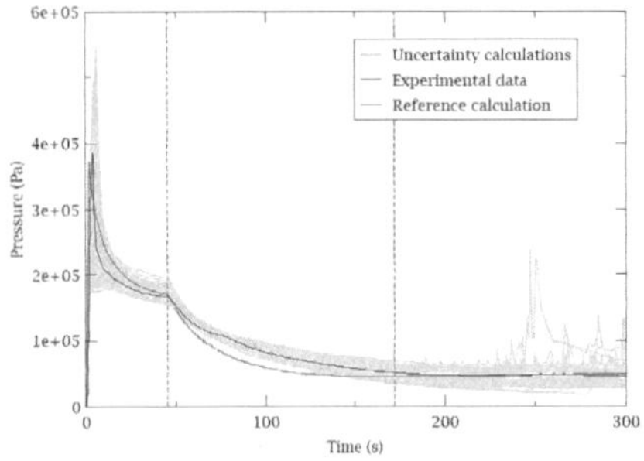

Figure 3. PC pressure (experimental data vs reference and uncertainty calculations results).

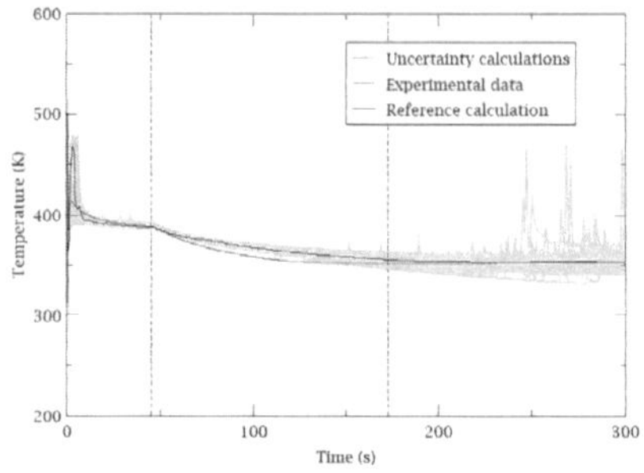

Figure 4. PC temperature (experimental data vs reference and uncertainty calculations results).

5.2. Reference calculation accuracy evaluation

Table III shows the qualitative accuracy evaluation results of the reference calculation. The code is able to qualitatively predict all the phenomena identified in the various PhWs. The results of the FFTBM are presented in Table IV. For the quantitative accuracy evaluation, the total AA is below 0.3 in all PhWs (Table IV); therefore, the calculation can be classified as very good [18]. In addition, all the parameters have an AA below 0.3 except for the PC pressure in PhW1 and PhW3 and the VV pressure in PhW1, which are however below 0.5. In general, considering both the AA of each parameter and the total AA in all the PhWs, the quantitative accuracy results can be classified as "very good".

5.3. Uncertainty Analysis

5.3.1. Dispersion of the results

In PhW1 the PC pressure peak presents a results dispersion band from around 0.17 MPa to 0.55 MPa (Figure 3). Then the results dispersion reduces along the PhW, with a final width of around 0.04 MPa.

Considering the PC temperature (Figure 4), the drop at the SOT is present in all the calculations and the subsequent increase presents a results dispersion from around 385 K to 480 K. In the second PhW the PC pressure result dispersion width is relatively narrow (Figure 3), and all the calculations slightly overestimates the experimental data in the central part of the PhW. Similar considerations can be drawn for the PC temperature (Figure 4). In the final PhW, the PC pressure dispersion band width slightly increases (reaching a final width of around 0.06 MPa), with the experimental data comprised among the calculated results (Figure 3). Some slight pressure peaks are present in the second half of the PhW. Similar considerations can be drawn for the PC temperature with a final band width of around 30 K (Figure 4). A scalar statistical analysis has been performed on the maximum value of the FOM, which is relevant for the system safety. The mean and median values are 0.381 MPa and 0.395 MPa respectively, close to the experimental value (0.373 MPa); the standard deviation is 0.108 MPa. Figure 5 shows the PDF of the FOM at its maximum.

Table III. Qualitative accuracy evaluation results for the reference calculation.

Phenomenon	Experiment		TRACE
	Phenomena	Measurement	Phenomena
Water flashing in the PC and VV	+	PC and VV pressure; PC and VV temperature	+
Water condensation in the SD	o/NA*	Only visual observation, not experimentally quantified	+
Pressure suppression in the ST	+	PC, VV and ST pressure	+
Condensate discharge in the DT	+	DT pressure	+

* The phenomenon occurs in the facility; it is visually observed, but it is not directly or indirectly measured.

Table IV. FFTBM results for the reference calculation.

Variables	PhW1		PhW2		PhW3	
	AA	WF	AA	WF	AA	WF
PC pressure	0.42	0.15	0.16	0.03	0.24	0.10
VV pressure	0.46	0.15	0.17	0.03	0.33	0.10
ST pressure	0.09	0.13	0.19	0.07	0.13	0.04
DT pressure	0.22	0.08	0.18	0.03	0.13	0.05
PC temperature	0.16	0.17	0.04	0.04	0.02	0.09
VV temperature	0.21	0.14	0.07	0.06	0.02	0.09
Total	0.26	0.14	0.13	0.05	0.15	0.08

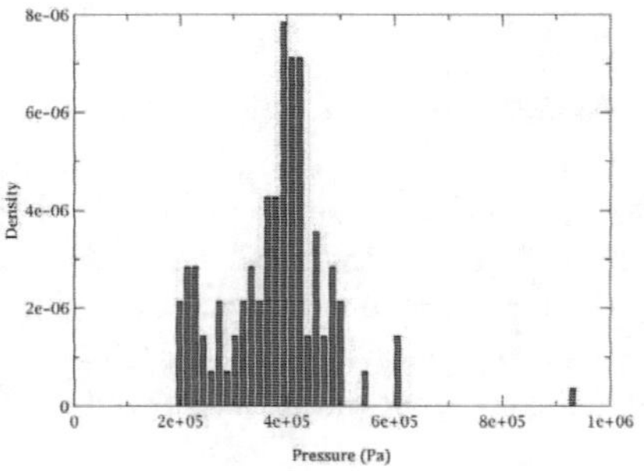

Figure 5. *Probability density function of the FOM maximum.*

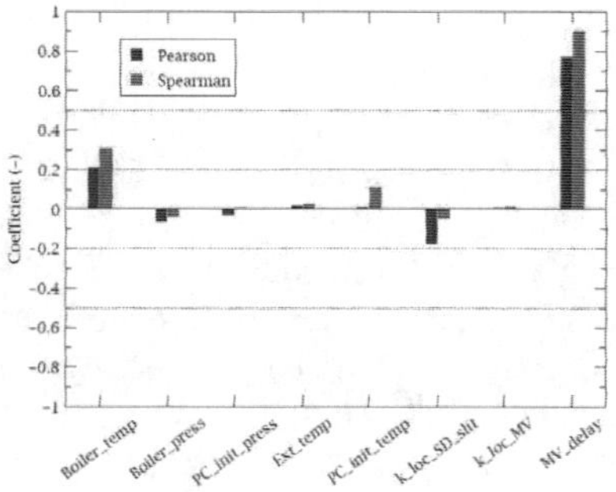

Figure 6. *Pearson and Spearman correlation coefficients for the maximum PC pressure.*

5.3.2. Response correlation
Correlation coefficients have been computed to characterize the relationship between the uncertain input parameter and the PC pressure, selected as FOM. In particular, it has been considered the Pearson coefficient, which is an indication of the linear relationship between an input and an output, and the Spearman coefficient, which is an indication of the monotonic relationship between an input and an output. If the coefficient is greater than 0.5 (or lower than -0.5) there is a significant correlation; if the coefficient is between 0.2 and 0.5 (or -0.2 and -0.5) there is a moderate correlation, otherwise the correlation is low [3,25]. The Pearson and Spearman correlation coefficients have been evaluated at the maximum value of the FOM (Figure 6). The boiler temperature shows a coefficient close to the threshold between a low and a moderate correlation. The MV opening delay presents the highest coefficient (0.77 and 0.90 for Pearson and Spearman respectively) and has a significant correlation with the PC maximum pressure.

Conclusions
The ICE is one of the possible accidents in nuclear fusion installations. In [3] it has been analyzed the prediction of the TRACE code against the Integrated ICE facility data, with the ST connected at the bottom of the VV. As a follow up activity, the present paper analyzes the capability of the TRACE code to predict the main thermal-hydraulic phenomena of the Integrated ICE facility in the upgraded configuration. The code resulted to be able to predict all the involved thermal-hydraulic phenomena. The quantitative accuracy evaluation through FFTBM confirmed the agreement between the experimental and calculated data, with a maximum total AA of 0.26 in PhW1. The behavior of the DT, specific of the upgraded ICE configuration, has been correctly predicted by TRACE, as the PC pressure evolution and its maximum. After the qualification of the nodalization and the code, an UA has been performed adopting the probabilistic method to propagate input uncertainties. The UA has been carried out by coupling the DAKOTA toolkit with TRACE in the SNAP environment/architecture. Nine input uncertain parameters have been selected and the PC pressure has been chosen as FOM. The purpose of this UA is not to be exhaustive in terms of input uncertain parameters, but to provide some insights to characterize the results dispersion against the available experimental data and the correlation between the selected uncertain input parameters and the FOM. The experimental data considered resulted to be mostly comprised among the UA calculations. Pearson's and Spearman's correlation coefficients show similar results and the parameter with the highest correlation with the FOM is the MV delay (Pearson and Spearman coefficients respectively 0.77 and 0.90).

Acknowledgements
The authors gratefully acknowledge USNRC for their valuable comments and suggestions during the preparation of the manuscript.

References
1. G. Caruso et al., "Numerical study on Ingress of Coolant Event experiments with CONSEN code", *Fusion Engineering and Design*, **100**, pp. 443-452 (2015)
2. S. Paci et al., "Analysis of the Ingress of Coolant Event tests performed in the upgraded ICE facility aimed at the ECART code validation", *Fusion Engineering and Design*, **152** (2020)
3. A. Bersano et al., "Ingress of Coolant Event simulation with TRACE code with accuracy evaluation and coupled DAKOTA Uncertainty Analysis", *Fusion Engineering and Design*, **159** (2020).
4. A. Bersano et al., TRACE qualification for Ingress of Coolant Event and DAKOTA uncertainty analysis, *The 19th International Topical Meeting on Nuclear Reactor Thermal Hydraulics (NURETH-19),* Brussels, Belgium, March 6 - 11 (2022)
5. U. S. Nuclear Regulatory Commission, "TRACE V5.840 Theory Manual, Field Equations, Solution Methods and Physical Models" (2013)
6. Applied Programming Technology, Inc., "Symbolic Nuclear Analysis Package (SNAP) User's Manual" (2021)

7. F. Mascari et al., "Scaling issues for the experimental characterization of reactor coolant system in Integral Test Facilities and role of system code as extrapolation tool", *Proceedings of 16th International Topical Meeting on Nuclear Reactor Thermal Hydraulics (NURETH-16)*, **6**, pp. 4921-4934 (2015)
8. W. Ambrosini et al., "Evaluation of accuracy of thermalhydraulic code calculation", *Energia Nucleare*, **7**, pp. 5-16 (1990)
9. H. Glaeser, "GRS Method for Uncertainty and Sensitivity Evaluation of Code Results and Applications", *Science and Technology of Nuclear Installations* (2008)
10. B.M. Adams et al., Dakota, "A Multilevel Parallel Object-Oriented Framework for Design Optimization, Parameter Estimation, Uncertainty Quantification, and Sensitivity Analysis: Version 6.7 User's Manual", SAND2014-4633 (2017)
11. K. Takase et al., "Experimental verification of effectiveness of integrated pressure suppression systems in fusion reactors during in-vessel loss of coolant events", *Nuclear Fusion*, **41**, pp. 1873-1883 (2001)
12. F. Mascari et al., "Sensitivity analysis of the MASLWR helical coil steam generator using TRACE", *Nuclear Engineering and Design*, **241**, pp. 1137–1144 (2011)
13. J. Staudenmeier, "TRACE reactor system analysis code, MIT Presentation, Safety Margins and Systems Analysis Branch", Office of Nuclear Regulatory Research, U.S. Nuclear Regulatory Commission (2004)
14. F. Mascari et al., "Analysis of the OSU-MASLWR 001 and 002 Tests by Using the TRACE Code", NUREG/IA-0466 (2016)
15. A. Prošek et al., "Quantitative assessment with improved fast Fourier transform based method by signal mirroring", *Nuclear Engineering and Design*, **238**, pp. 2668-2677 (2008)
16. A. Prošek et al., "Use of FFTBM by signal mirroring for sensitivity study", *Annals of Nuclear Energy*, **76**, pp. 253-262 (2015)
17. A. Prošek, "JSI FFTBM Add-In 2007 User's Manual", IJS-DP-9752 (2007)
18. F. D'Auria et al., RELAP/MOD3.2 Post Test Analysis and Accuracy Quantification of SPES Test SP-SB-04, NUREG/IA-0155 (1999)
19. Applied programming Technology Inc., "Uncertainty analyses User manual, Symbolic Nuclear Analyses Package (SNAP)" (2012)
20. A. Bersano et al., "Evaluation of a Double-Ended Guillotine LBLOCA Transient in a Generic Three-Loops PWR-900 with TRACE Code Coupled with DAKOTA Uncertainty Analysis", *ATW - International Journal for Nuclear Power*, **11/12**, pp. 526-532 (2019)
21. S.S. Wilks, "Determination of sample sizes for setting tolerance limits", *Annals of Mathematical Statistic*, **12**(1), pp. 91-96 (1941)
22. S.S. Wilks, "Statistical prediction with special reference to the problem of tolerance limits", *Annals of Mathematical Statistic*, **13**(4), pp. 400-409 (1942)
23. A. Guba, M. Makai, L. Pál, "Statistical aspects of best estimate method-I", *Reliability Engineering and System Safety*, **80**, pp. 217-232 (2003)
24. S.M. Bajorek et al., "Uncertainty Methods Framework Development for the TRACE Thermal-Hydraulics Code by the U.S.NRC", *OECD/CSNI Workshop on Best Estimate Methods and Uncertainty Evaluations*, NEA/CSNI/R(2013)8/PART2 (2013)
25. K.A. Gamble et al., Uncertainty Quantification and Sensitivity Analysis Applications to Fuel Performance Modeling, SAND2016-4597C (2016)

Retro-reflective plaster coatings after outdoor aging and soiling: an in-lab optical performance characterization

A Di Giuseppe[1,*], A Pazzaglia[1], B Castellani[1,2], A Nicolini[1,2], F Rossi[1,2] and F Cotana[1,2]

[1] CIRIAF, Interuniversity Research Centre on Pollution and Environment "M. Felli", Via G. Duranti 67, 06125, Perugia, Italy

[2] University of Perugia, Department of Engineering, Via G. Duranti 93, 06125, Perugia, Italy

* Corresponding author: alessia.digiuseppe@crbnet.it

Abstract. The potential benefits of highly reflective and retro-reflective (RR) materials on urban context have been performed by several studies through their optical characterization. During their lifetime, outdoor aging and soiling can affect the RR optical behaviour. To this aim, this study investigates the performance of RR plaster coatings after outdoor aging and soiling from May to October 2022 in Perugia, Italy. The same RR plaster coatings in pristine conditions were already characterized in lab in terms of optical performance. In particular, RR samples obtained by using glass beads with different diameters (i.e., 40 ÷ 70 μm and 70 ÷ 110 μm) distributed in three superficial density ranges (i.e., 0.30 ÷ 0.40 kg/m^2, 0.20 ÷ 0.30 kg/m^2 and 0.10 ÷ 0.20 kg/m^2) have been investigated in this study through spectrophotometric and angular reflectivity analyses. In all cases, the post-aging RR samples exhibit lower global reflectance values with respect to the same RR samples in pristine condition. Concerning the angular reflection distribution analysis, a stronger relative RR component was found for the post-aging RR sample with an average diameter of 70 ÷ 110 μm and superficial density distribution equal to 0.10 ÷ 0.20 kg/m^2 for each investigated incident direction.

1. Introduction

Researchers have given considerable focus to Urban Heat Islands (UHI) due to their noteworthy impact on urban thermal surroundings and human well-being. Urban population growing and consequent urbanization have led to an increase of city temperature and a worsening of UHI phenomenon [1].

In particular, a numerous cluster of buildings with enclosed spaces emerged due to urbanization, showcasing a variety of typical forms such as the canyon, semi-closure, and courtyard designs [2]. The arrangement of buildings has a substantial influence on the urban climate caused by multiple reflections of solar radiation in comparison to more open structures or land [3]. The consequence of this building configurations is that the urban surfaces absorb a huge amount of solar radiation, increasing urban temperature [4].

In order to solve this issue, numerous researches have been conducted towards investigating and creating effective mitigation approaches to mitigate the UHI effects, i.e. urban overheating [5–7]. Creative solutions, including the application of high reflective and diffusive materials on roofs have been extensively studied. Additionally, measures aimed at enhancing the earth surface reflectivity have been proposed for building implementation. These innovative strategies have demonstrated a notable ability to maintain significantly lower surface temperatures than traditional configurations due to their high albedo characteristics [8-9].

Another relevant technology is the use of retro-reflective (RR) materials, which have been proposed due to their ability to reflect sunlight primarily in the incidence direction [10].

Several studies have demonstrated the cooling benefits of RR materials, particularly in urban canyons where they are effective in reducing the energy absorbed by building surfaces [11-12].

However, the performance of RR materials may be diminished over time due to soiling and aging, which may lead to a material degradation. Many studies have tested the effect of aging on RR materials. In Paolini et al. [13], roofing membranes were exposed to natural aging for two years, resulting in an average reflectance decrease of 0.18. Yuan et al. [14] conducted extended outdoor tests on various glass bead RR samples spanning 368 days, and they found no substantial alterations in solar reflectance and angular reflection patterns post-outdoor aging. Meanwhile, Morini et al. [15] employed accelerated aging tests in a laboratory setting, following ASTM G154 standards, to evaluate the long-term viability of RR tiles and paints designed for building applications. The findings indicated that the RR properties remained consistent following the aging trials, with minimal alterations detected in directional reflectivity and overall reflectance.

The objective of this paper is to investigate the impact of natural aging and soiling exposure on the optical characteristics of RR coatings, which have been fabricated using glass beads of different diameters distributed according to three ranges of superficial density. In a previous work, the Authors [16] have characterized the optical performance of these same samples in their original state. In order to determine the optical performance of the aged samples, spectrophotometric and angular reflectivity measurements were conducted in this study. Finally, a comparative analysis of the optical properties between the post-aging and pre-aging samples was carried out.

2. Materials and Methods

The initial part of this section entails a description of the samples under investigation, emphasizing their optical attributes. Subsequently, the section outlines the methodology implemented to assess the optical efficacy of the samples.

2.1 Samples

The analysed samples were manufactured and partially investigated in a previous work [16]. Six RR samples were produced by painting a 10 x 10 cm mortar square with white high-reflective paint and embedding RR glass beads with different diameters and superficial densities, as shown in Table 1. The two different diameter ranges are: $\phi 1 = 40 \div 70$ μm; $\phi 2 = 70 \div 110$ μm. The three ranges of superficial density are: $\rho s1 = 0.30 \div 0.40$ kg/m^2, $\rho s2 = 0.20 \div 0.30$ kg/m^2, and $\rho s3 = 0.10 \div 0.20$ kg/m^2.

Table 1. Characteristics of the RR samples that were tested, including data of glass bead diameters, the range of superficial densities, and the calculated weight of the embedded glass beads.

Name of the sample	Bead diameter, ϕn [μm]	Range of superficial density, ρsj [kg/m^2]	Weight of embedded glass beads [g]
$RR_{\phi1,\ \rho s1}$	$\phi 1 = 40 \div 70$	$\rho s1 = 0.30 \div 0.40$	2.97
$RR_{\phi1,\ \rho s2}$	$\phi 2 = 70 \div 110$	$\rho s1 = 0.30 \div 0.40$	3.35
$RR_{\phi1,\ \rho s3}$	$\phi 1 = 40 \div 70$	$\rho s2 = 0.20 \div 0.30$	2.26
$RR_{\phi2,\ \rho s1}$	$\phi 2 = 70 \div 110$	$\rho s2 = 0.20 \div 0.30$	2.02
$RR_{\phi2,\ \rho s2}$	$\phi 1 = 40 \div 70$	$\rho s3 = 0.10 \div 0.20$	1.53
$RR_{\phi2,\ \rho s3}$	$\phi 2 = 70 \div 110$	$\rho s3 = 0.10 \div 0.20$	1.17

2.2. Methodology
From 24th May 2022, the rooftop of CIRIAF building at the University of Perugia (located at 43° 7' 9.449'' N, 12° 21' 27.451'' E) was used to test the samples to various aging climate drivers such as relative humidity (RH), air temperature (T), UV radiation, rain precipitation, and dust soiling (Figure 1). The aging process ended on 26th October 2022. After that, the samples were analysed by spectrophotometric and angular reflectivity measurements. The obtained results were compared with those of the pre-aging samples.

Figure 1. Samples during the outdoor aging and soiling in summer.

2.3. Spectrophotometric and angular reflectivity analysis
A PerkinElmer UV/Vis/NIR Spectrophotometer LAMBDA™ 1050+ [17], equipped with a 150 mm InGaAs Integrating Sphere, was employed for spectrophotometric analysis. To ensure device accuracy, calibration was performed using a Spectralon™ [18] certified reflectance standard from Labsphere. Three measurements were conducted for each sample to determine the mean and standard deviation of their spectral reflectance. The wavelength range analyzed spanned from 300 to 2500 nm, with a data interval of 5 nm. Data collection was performed using UV WinLab Software, and reflectance measurements were executed in accordance with the ASTM E 903–20 method. Additionally, ASTM G 173–03 was employed to obtain terrestrial solar irradiance values, enabling the calculation of Solar Reflectance (SR) for each sample.

To assess the angular distribution of reflected radiation by each sample for different incident radiation angles, a directional reflectivity analysis was carried out. However, the conventional testing equipment available in the laboratory or on-site was not suitable for investigating the angular reflectivity characteristics of the samples. Consequently, the Authors implemented a specialized experimental setup previously introduced in their prior research [19-20]. This setup comprises a semi-circular structure, graduated at 10° intervals, and a LED light source. Irradiance measurements at each angular position along the semi-circular structure, i.e., every 10°, were obtained using a Delta-OHM HD 9221 photo radiometer equipped with an LP9221/RAD probe [21]. Two consecutive irradiance measurements were taken exclusively at the position where the light source was fixed, i.e., one on the right and one on the left of the light source. As such, the average irradiance value was considered representative of the reflected radiation towards the incident direction.

3. Results and discussion
In this paragraph, results of the investigation of the samples' optical characterization are presented and discussed. The obtained data are analysed and then compared with data of the same samples in pristine conditions already discussed in a previous work by the Authors [16].

3.1 Spectrophotometric characterization

As previously outlined, the spectral reflectance of the samples was determined by conducting three spectrophotometric measurements for each sample, allowing the calculation of the average spectral reflectance.

Figure 2 shows a comparison of the average spectral reflectance of all RR samples in pre- and post-aging conditions. It can be observed that in the visible (VIS) range, all RR post-aging samples exhibit lower reflectance compared to the same RR samples in pristine condition.

Moving to the near-infrared (NIR) region, a lower reflectance difference is observed for all RR post-aging samples compared to the same RR pre-aging samples.

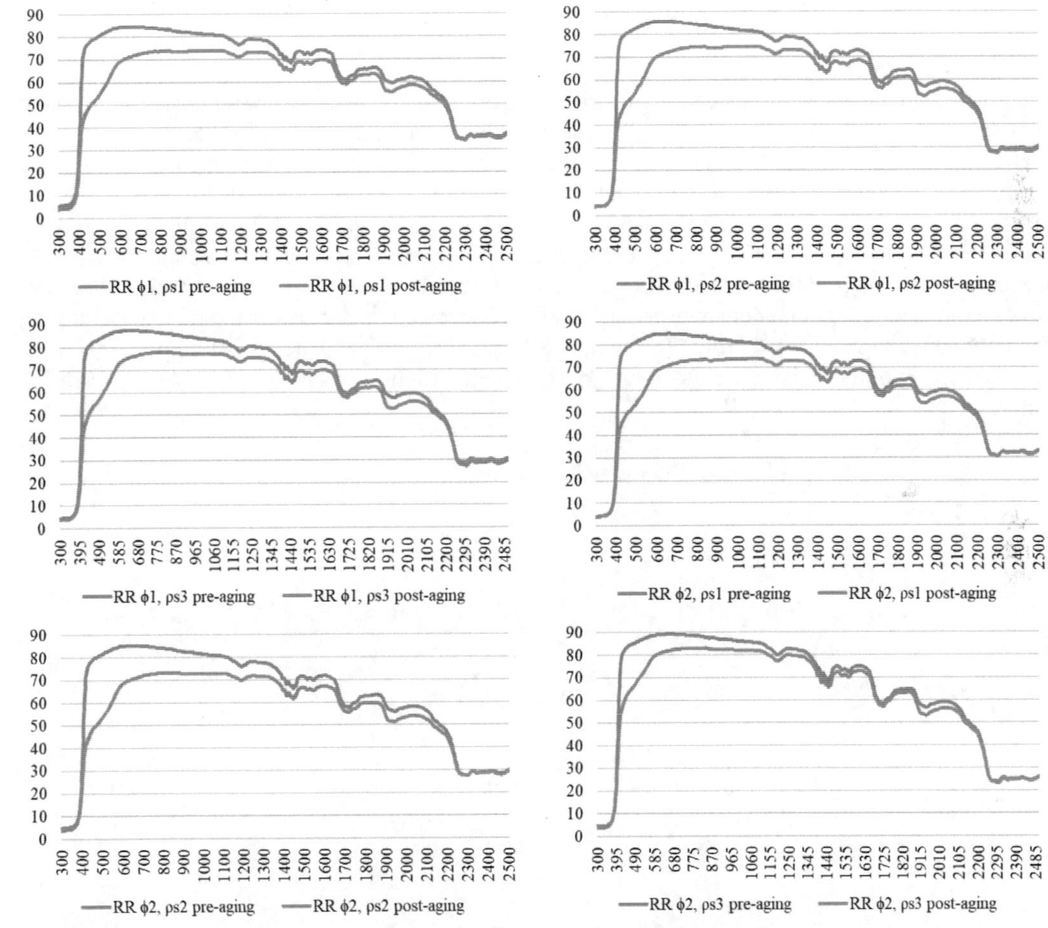

Figure 2. Comparison between the pre- and post-aging RR samples in terms of average spectral reflectance over three measurements.

Thus, SR [%] of each sample was calculated by the UV WinLab Software. SR values are shown in Table 2. The Δ values (%) are calculated as the difference between the RR samples in pristine condition and the same RR samples after aging in terms of SR values.

The aging process seems to negatively affect the total SR in all RR samples. The maximum Δ value is 13.8% for RR sample characterized by glass bead diameter equal to 70 ÷ 110 m and superficial density equal to 0.2 ÷ 0.3 kg/m^2 ($RR_{\phi 2, \rho s2}$).

Journal of Physics: Conference Series **2685** (2024) 012056 doi:10.1088/1742-6596/2685/1/012056

Table 2. Total SR [%] of pre- and post-aging RR samples.

Samples	Solar reflectance (%) Pre-aging[a]	Solar reflectance (%) Post-aging	Δ (%)
$RR_{\phi1, \rho s1}$	75.9	63.3	12.6
$RR_{\phi1, \rho s2}$	76.6	63.2	13.4
$RR_{\phi1, \rho s3}$	78.1	66.2	11.9
$RR_{\phi2, \rho s1}$	75.8	62.6	13.2
$RR_{\phi2, \rho s2}$	76.0	62.2	13.8
$RR_{\phi2, \rho s3}$	79.7	71.8	7.9

[a] Results from a previous work by the Authors [16].

3.2 Angular reflectivity characterization

Within this section, the angular reflectivity distribution for all samples under investigation is presented. It delved into the radiation reflected (W/m^2) by each sample across incident angles ranging from 0° to -70°, at 10° intervals, along the designed semi-circular gonio-reflectometer array. In Figure 3, a visual comparison of the angular reflectivity distribution for both pre- and post-aging RR samples at incident angles of 0° and -50° is shown. The direction of the incoming radiation striking the test sample is indicated by the black arrow. The presentation centers on the relative angular reflection distribution, denoting the percentage ratio between the radiation reflected by each sample in a specific direction, relative to the total reflection contributions across all directions by the same sample for each examined incident radiation angle. Consequently, the comparison of reflection percentages is feasible exclusively within the context of the same sample's angular reflection distribution and is not applicable for different samples.

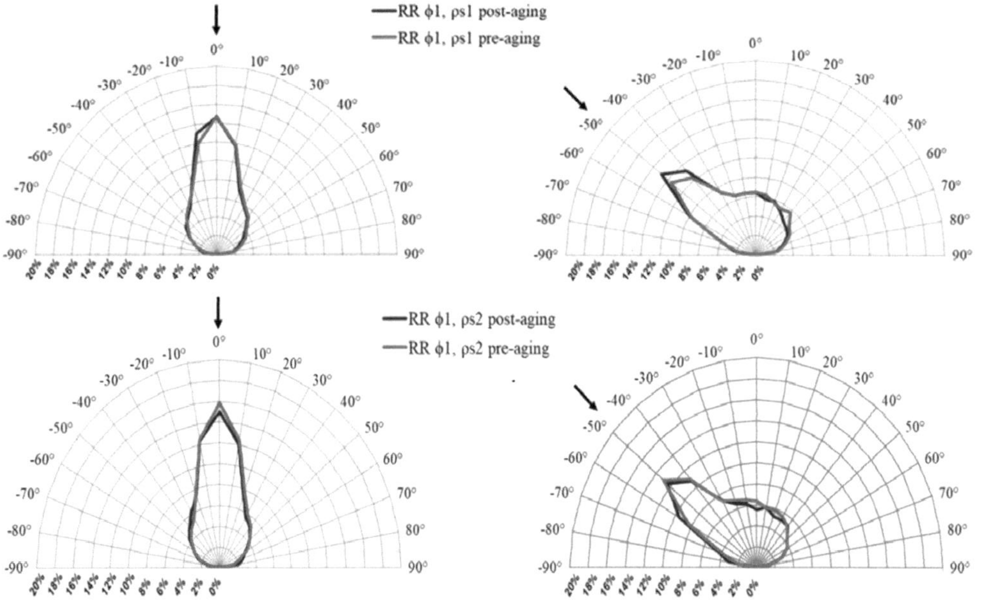

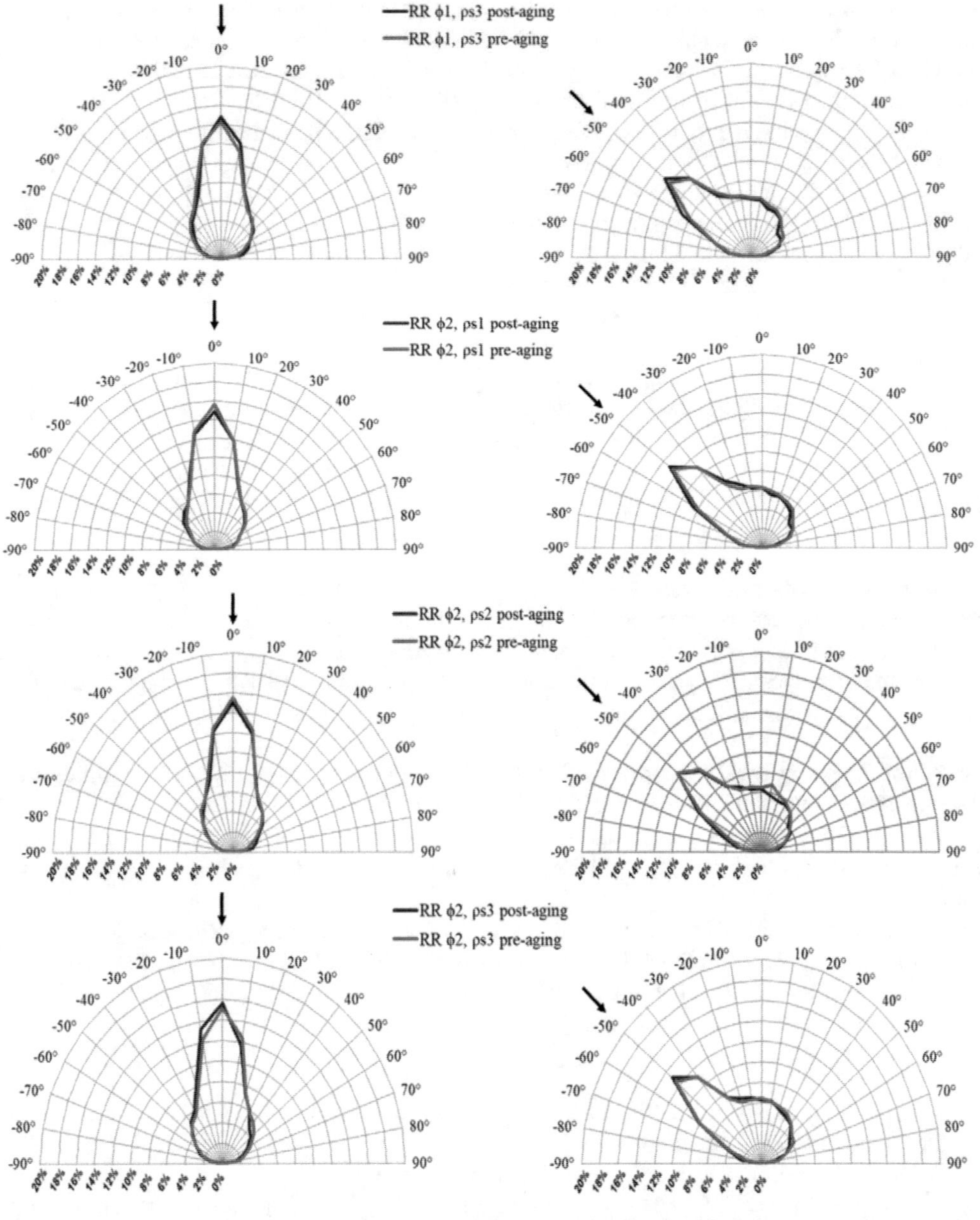

Figure 3. Comparison in terms of angular reflection distribution for all pre- and post-aging RR samples for incident radiation angles of 0° and -50°.

All RR samples in pristine conditions exhibit a good RR capability, as described in [16]. $RR_{\phi1,\ \rho s2}$ exhibited the highest RR capability among the RR samples in pristine conditions for each incident radiation angles investigated.

Considering the pre-aging RR samples characterized by glass bead diameter of 40 ÷ 70 μm, $RR_{\phi1,\ \rho s2}$ provided the highest RR capability with a RR percentage equal to 15.9% and 12.9% for 0° and -50° incident radiation angles, respectively.

Otherwise, among the RR samples in pristine conditions with glass bead diameter of 70 ÷ 110 μm, RR with a superficial density equal to 0.30 ÷ 0.40 kg/m^2 (ρs1) exhibited the highest RR component values, equal to 15.6% and 12.7% for 0° and -50° incident radiation angles, respectively.

The RR behaviour was confirmed for all post-aging RR samples. The $RR_{\phi1,\,\rho s1}$ pre- and post-aging exhibited the same RR behaviour at 0°C with a RR percentage of 14.7%, whilst the $RR_{\phi1,\,\rho s1}$ post-aging showed the highest RR capability for lower angle of incidence (i.e. -50°) with a RR percentage of 12.9%.

Concerning the $RR_{\phi1,\,\rho s2}$ post-aging, it exhibited the lowest RR performance at 0° and -50° with respect to the same RR sample in pristine condition.

The $RR_{\phi1,\,\rho s3}$ post-aging exhibited the best RR performance both at 0° and -50° angles of incidence, equal to 14.8% and 12.7%, respectively, with respect to the pre-aging sample.

Among the RR characterized by glass bead diameter of 70 ÷ 110 μm (ϕ2) in post-aging conditions, it is clear that: $RR_{\phi2,\,\rho s1}$ post-aging showed the highest RR capability only for lower angle of incidence (i.e. -50°) with a RR percentage of 13.1% compared to the same pre-aging RR sample; $RR_{\phi2,\,\rho s2}$ post-aging as well as $RR_{\phi1,\,\rho s2}$ exhibited the lowest RR capability for each incident radiation angle compared to the same RR sample in pristine condition; $RR_{\phi2,\,\rho s3}$ post-aging as well as $RR_{\phi1,\,\rho s3}$ exhibited the highest RR capability at 0° and -50° with a RR percentage equal to 15.6% and 13.1%, respectively.

Thus, among the post-aging RR samples, $RR_{\phi2,\,\rho s3}$ performed the highest RR capability for each incident radiation angles investigated.

4. Conclusions and further developments

The present work aims at investigating the effect of aging processes on RR materials' optical performances. RR samples were developed using a highly reflective paint and glass beads. Two RR glass beads diameter ranges (i.e., ϕ_1=40÷70 μm; ϕ_2=70÷110 μm) and three RR glass bead superficial densities (i.e., ρs1 = 0.30 ÷ 0.40 kg/m^{2,} ρs2 = 0.20 ÷ 0.30 kg/m^2, and ρs3 = 0.10 ÷ 0.20 kg/m^2) were investigated.

The effect of the aging processes on the RR optical behaviour was assessed comparing the results with those obtained for the same RR materials in pristine conditions. The main findings are:

- Concerning spectrophotometric analysis, all post-aging RR samples exhibited lower spectral reflectance values both in VIS and NIR ranges compared to the same RR samples in pristine conditions;
- Concerning the angular reflectivity analysis, among the post-aging RR samples, $RR_{\phi2,\,\rho s3}$ performed the highest RR capability, with a RR percentage of 15.6% at 0° and 13.1% at -50°.

Future developments may involve the application of a protective layer to investigate its effect on the optical performances of RR materials after aging processes.

5. References

[1] M. Santamouris, "Analyzing the heat island magnitude and characteristics in one hundred Asian and Australian cities and regions," Science of The Total Environment, vol. 512–513, pp. 582–598, Apr. 2015, doi: 10.1016/j.scitotenv.2015.01.060.

[2] Y. Gao, R. Yao, B. Li, E. Turkbeyler, Q. Luo, and A. Short, "Field studies on the effect of built forms on urban wind environments," Renew Energy, vol. 46, pp. 148–154, Oct. 2012, doi: 10.1016/j.renene.2012.03.005.

[3] L. Da-long, Z. Hui-hui, and L. Jia-ping, "Rule of long-wave radiation in enclosed building space," Energy Build, vol. 182, pp. 311–321, Jan. 2019, doi: 10.1016/j.enbuild.2018.09.044.

[4] E. Jamei, P. Rajagopalan, M. Seyedmahmoudian, and Y. Jamei, "Review on the impact of urban geometry and pedestrian level greening on outdoor thermal comfort," Renewable and Sustainable Energy Reviews, vol. 54, pp. 1002–1017, Feb. 2016, doi: 10.1016/j.rser.2015.10.104.

[5] H. Akbari et al., "LOCAL CLIMATE CHANGE AND URBAN HEAT ISLAND MITIGATION TECHNIQUES – THE STATE OF THE ART," JOURNAL OF CIVIL ENGINEERING AND MANAGEMENT, vol. 22, no. 1, pp. 1–16, Dec. 2015, doi: 10.3846/13923730.2015.1111934.

[6] E. Enríquez, V. Fuertes, M. J. Cabrera, J. Seores, D. Muñoz, and J. F. Fernández, "New strategy to mitigate urban heat island effect: Energy saving by combining high albedo and low thermal diffusivity in glass ceramic materials," Solar Energy, vol. 149, pp. 114–124, Jun. 2017, doi: 10.1016/j.solener.2017.04.011.
[7] U. Kitrattana, P. Phadungbut, S. Watechagit, and W. Jonglertjunya, "Dependence of cool roof paint colors on household cooling load and analysis of roof paint deterioration," 2022, p. 020042. doi: 10.1063/5.0118116.
[8] A. Synnefa, A. Dandou, M. Santamouris, M. Tombrou, and N. Soulakellis, "On the Use of Cool Materials as a Heat Island Mitigation Strategy," J Appl Meteorol Climatol, vol. 47, no. 11, pp. 2846–2856, Nov. 2008, doi: 10.1175/2008JAMC1830.1.
[9] J. Yuan, K. Emura, and C. Farnham, "Is urban albedo or urban green covering more effective for urban microclimate improvement?: A simulation for Osaka," Sustain Cities Soc, vol. 32, pp. 78–86, Jul. 2017, doi: 10.1016/j.scs.2017.03.021.
[10] S. Liu, J. Wang, and X. Meng, "Spectral properties of retro-reflective materials from experimental measurements," Case Studies in Thermal Engineering, vol. 39, p. 102434, Nov. 2022, doi: 10.1016/j.csite.2022.102434.
[11] F. Rossi et al., "Experimental evaluation of urban heat island mitigation potential of retro-reflective pavement in urban canyons," Energy Build, vol. 126, pp. 340–352, Aug. 2016, doi: 10.1016/j.enbuild.2016.05.036.
[12] F. Rossi et al., "Beneficial effects of retroreflective materials in urban canyons: results from seasonal monitoring campaign," J Phys Conf Ser, vol. 655, p. 012012, Nov. 2015, doi: 10.1088/1742-6596/655/1/012012.
[13] R. Paolini, M. Zinzi, T. Poli, E. Carnielo, and A. G. Mainini, "Effect of ageing on solar spectral reflectance of roofing membranes: Natural exposure in Roma and Milano and the impact on the energy needs of commercial buildings," Energy Build, vol. 84, pp. 333–343, Dec. 2014, doi: 10.1016/j.enbuild.2014.08.008.
[14] J. Yuan, K. Emura, and C. Farnham, "A study on the durability of a glass bead retro-reflective material applied to building facades," Prog Org Coat, vol. 120, pp. 36–48, Jul. 2018, doi: 10.1016/j.porgcoat.2018.03.009.
[15] E. Morini, B. Castellani, A. Nicolini, F. Rossi, and U. Berardi, "Effects of aging on retro-reflective materials for building applications," Energy Build, vol. 179, pp. 121–132, Nov. 2018, doi: 10.1016/j.enbuild.2018.09.013.
[16] M. Cardinali, A. Di Giuseppe, B. Castellani, M. Filipponi, A. Nicolini, and F. Rossi, "An investigation towards the optimum design of retro-reflective materials as building envelopes for the enhancement of optical performance," Constr Build Mater, vol. 358, p. 129466, Dec. 2022, doi: 10.1016/j.conbuildmat.2022.129466.
[17] PerkinElmer, "https://www.perkinelmer.com/it/product/lambda-1050-2d-base-inst-no-sw-l6020055 ," PerkinElmer UV/Vis/NIR Spectrophotometer LAMBDATM 1050+.
[18] Labsphere, "https://www.labsphere.com/ ," SpectralonTM Labsphere.
[19] A. Di Giuseppe et al., "The Effect of the Substrate on the Optic Performance of Retro-Reflective Coatings: An In-Lab Investigation," Energies (Basel), vol. 14, no. 10, p. 2921, May 2021, doi: 10.3390/en14102921.
[20] A. M. Gambelli, M. Cardinali, M. Filipponi, B. Castellani, A. Nicolini, and F. Rossi, "A normalization procedure to compare retro-reflective and traditional diffusive materials in terms of UHI mitigation potential," 2019, p. 020085. doi: 10.1063/1.5138818.
[21] Delta OHM, "https://docs.rs-online.com/0053/0900766b80115ee0.pdf," Delta OHM HD9221 photo-radiometer with an LP9221/RAD probe.

Acknowledgments
The authors Alessia Di Giuseppe's and Aron Pazzaglia's acknowledgments are due to the PhD school in Energy and Sustainable Development from the University of Perugia.

Design and thermal test of high-vacuum insulator for heat delivery pipes

F. Capolupo[1], C. D'Alessandro[2], P. Strazzullo[1,3*] R. Russo[3] and M. Musto[1,3]

[1] Industrial Engineering Department, University of Napoli "Federico II", Piazzale Vincenzo Tecchio, 80, Napoli, 80125, Italy
[2] TVPSOLAR SA, 10 rue du Prè-de-la-Fontaine, 1242 Satigny, Geneva
[3] Institute of Applied Sciences and Intelligent Systems, National Research Council of Italy, Via Pietro Castellino 111, Napoli, 80131, Italy

[*] e-mail: paolo.strazzullo@unina.it

Abstract. Thermal piping insulation of implants is crucial for heat delivery, production, collection, or storage at high temperature values. It is currently obtained by enveloping low thermal conductivity materials such as rockwool, fiberglass, polyurethane, polystyrene, and aerogel. However, better performances can be reached by adopting vacuum technology. In this case, conductive losses are annihilated, and the radiative heat transfer mechanism represents the only loss mechanism. Here, we compare a high vacuum-based novel solution and the traditional insulation for heat delivery applications. We propose a high vacuum-based solution consisting of an evacuated gap that surrounds the hot pipe coated by a thin aluminium foil. Experimental results using this novel solution show a fivefold reduction of the thermal radiation losses compared to the traditional solutions when in the temperature range between 100 °C and 250 °C.

1. Introduction

Thermal insulation is crucial in various applications and systems where a temperature sensibly above or below ambient is required, together with low energy consumption. In the building sector, proper wall insulation can represent a satisfying trade-off to find a compromise between people's comfort and energetic efficiency [1, 2]. Indeed, sanitary hot water requires well-insulated pipes to avoid thermal losses along the network [3] and storage of cryogenic or hot fluids demands high-performance insulators to limit the heat exchange over time [4].

Depending on the application and temperature ranges, different technological solutions have been investigated by changing materials and/or geometries [5]. Usually, to quantitatively define the performance of an insulator, two parameters are considered: thermal conductivity λ (W/m·K) and thermal transmittance U (W/m^2·K). The first parameter derives from the linear Fourier's equation of conduction as an intrinsic parameter of the insulating material, whereas the second quantifies the performance of the insulator based on its geometry, materials, and total exchanged heat. The most used materials for insulators are fiberglass, polyurethane, polystyrene, rockwool [6], aerogel [7], with the last reaching the lowest value of 17 (mW/m·K). Better performances, with an equivalent thermal conductivity of 4 (mW/m·K), can be achieved by adopting composite insulation and vacuum technology, as demonstrated by Vacuum Insulation Panel (VIP) [8, 9, 10],where thermal radiation represents the main thermal exchange mechanism, minimizing the role of conduction of residual gases.

Between 100 °C and 250 °C, rockwool is the most used insulation material for heat delivery piping, as it achieves thermal conductivity values of 0.040 (W/m·K). However, the thermal insulation efficiency of rockwool is influenced by environmental conditions: in the presence of 5% moisture level, the thermal conductivity of the material can reach a maximum value of 0.065 (W/m·K) [11].

Journal of Physics: Conference Series **2685** (2024) 012057 doi:10.1088/1742-6596/2685/1/012057

Different technologies have been applied to get rid of the dependence on atmospheric conditions. Among these, the best results have been obtained with pipes wrapped in VIP, which have reached lower thermal conductivity than traditional rockwool [12,13].

In this work, we present a comparison between a high vacuum-based novel solution and traditional insulators for heat delivery applications. The adopted solution involves isolating the hot pipe with high-vacuum technologies and a low-emissivity material coating to minimize radiative heat transfer. The work also proposes a new measurement technique for axial losses compensation inspired by the guarded plate method [14, 15, 16]. Experimental results show that this novel solution reduces the thermal losses by a factor of 5 compared to the traditional solutions, ensuring a primary energy saving [17] in the temperature between 100 °C and 250 °C and considering the same geometry and boundary conditions.

2. Design of the vacuum insulator

The designed prototype consists of two coaxial pipes (figure 1), both made of stainless steel AISI 304. The choice of using steel as the preferred material is based on several factors. First, compared to aluminium and copper pipes, steel has a lower thermal conductivity without compromising the insulation performance. Additionally, steel has excellent mechanical strength properties. The inner pipe has a diameter of 54 mm and a length of 1.7 m, while the outer pipe is 1.4 m long with a diameter of 100 mm. At this stage, we assumed the thickness of pipes equal to zero.

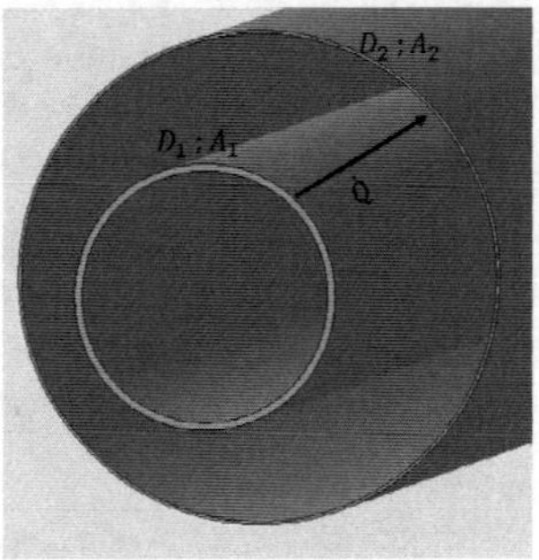

(1)

Figure 1. Scheme of the simulated prototype (radial section).

Figure 2 shows the gas thermal conductivity trend as a function of the pressure. At decreasing pressure, the gas thermal conductivity decreases until reaching a threshold that depends on the characteristic dimensions δ of the vacuum container. For the designed prototype, we generate a vacuum gap between the inner and outer pipes.

The characteristic dimension δ, in this case, is equal to 23 mm. For pressure lower than 10^{-4} mbar, gas thermal conductivity is less than 1% compared to the value at atmospheric pressure, thus the conduction is negligible, and radiation is the predominant heat exchange mechanism [18], resulting in a vacuum thermal insulation condition.

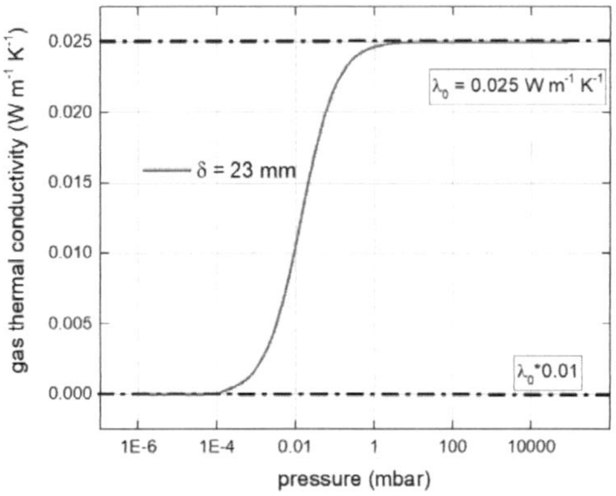

Figure 2. Gas thermal conductivity vs pressure trend depending on vacuum vessel size.

Considering a cylindrical geometry and the one-dimension hypothesis of the temperature field, we can express the radiative thermal power $\dot{Q}$ in equation (1) [19]:

$$\dot{Q} = \sigma A_1 \varepsilon_{eq}(T_1^4 - T_2^4) \quad (1)$$

where A_l is the lateral area of the inner pipe (m^2), T_1 and T_2 are the temperatures (K) of the internal and external surfaces, σ is the Stefan-Boltzmann constant equal to $5.67 \cdot 10^{-8}$ ($W/m^2 \cdot K^4$), and ε_{eq} is the equivalent thermal emissivity between the internal and external surface, expressed as reported in equation (2). The factor ε_{eq} depends on both internal and external thermal emissivity and a shape factor, i.e., the diameter ratio D_l/D_2 that represents the view factor.

$$\varepsilon_{eq} = \frac{1}{\frac{1}{\varepsilon_1}+\frac{D_1}{D_2}\left(\frac{1}{\varepsilon_2}-1\right)} \quad (2)$$

The combination of equations 1 and 2 marks the relevance of low thermal emissivity to reduce the thermal exchange in high vacuum conditions.

The convective heat exchange between the outer pipe surface and the environment, as defined by equation (3), completes the analytical model:

$$\dot{Q} = h_{\text{limi}} A_2 (T_2 - T_{env}) \quad (3)$$

where A_2 is the lateral area (m^2), h_{limi} is the liminar coefficient which considers both convective and radiative heat exchange, and T_{env} is the environmental temperature of the laboratory, equal to 24 °C.

The thermal insulation performance of the adopted solution is evaluated in terms of equivalent thermal conductivity (W/m·K) expressed in equation (4):

$$\lambda_{eq} = \frac{\dot{Q}}{2\pi L(T_1 - T_2)} \ln\left(\frac{D_2}{D_1}\right) \quad (4)$$

where $\dot{Q}$ (W) is the total thermal power exchanged between internal and external pipe, L (m) is the axial length, and T_l (K) and T_2 (K) are the temperatures of the inner and the outer surface, respectively.

Figure 3 shows theoretical curves of the equivalent thermal conductivity as a function of inner temperature, obtained for different power supply $\dot{Q}$.

We choose a Vacuum Insulation Panel (VIP) wrapped around a stainless-steel pipe, with a thermal conductivity $\lambda = 0.012$ (W/m·K) as reference: it represents the lowest value found in the literature for heat delivery systems in the temperature range between 50 °C and 100 °C [12]. The literature suggests the use of conventional thermal insulating materials above a temperature of 100 °C; among these, rockwool is one of the most used materials with a thermal conductivity of 0.040 (W/m·K). All the other curves are referred to the λ_{eq} trend of the designed prototype for different internal emissivity ε_1 values: 0.15 for stainless steel [20], 0.045 for aluminium coating [21], and 0.02 for copper coating [22]. The simulations show no advantage at high vacuum thermal insulation between two stainless steel surfaces, as the emissivity of stainless steel is too high. Therefore, the use of a thin sheet coating on the inner pipe is found to be necessary to reduce emissivity while enhancing the performance of the system, as highlighted in equation (2). We have chosen aluminium as a coating to test the prototype experimentally, as it is more economically advantageous than copper.

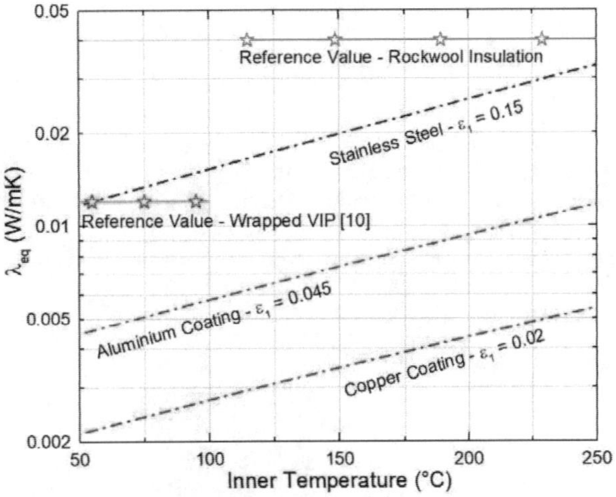

Figure 3. Theoretical curves of equivalent thermal conductivity vs inner temperature.

3. Experimental setup

Figure 4 shows the setup used for the experimental analysis. In particular, the internal pipe section, where thermal-insulation performance is measured, is hidden within the vacuum gap.

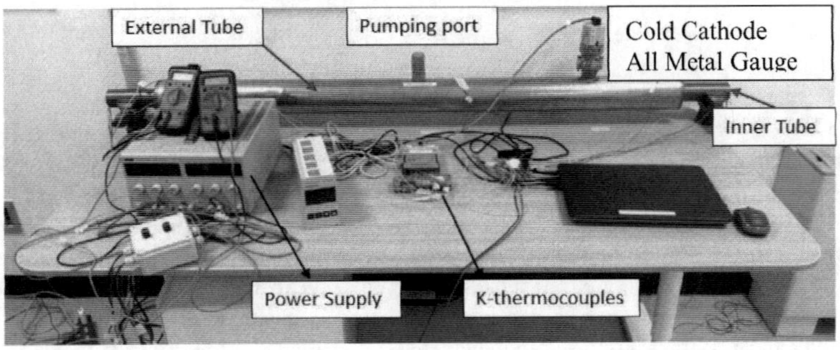

Figure 4. Experimental setup.

The prototype design has provided significant improvements in thermal insulation performance achieved by using an internal emissivity value equivalent to that of aluminium. The adopted solution involves a 15 μm thickness aluminium foil around the inner pipe, as shown in Figure 5, attempting to achieve complete adhesion to the surface and simulate the behavior of a thick coating.

Figure 5. Tested sample.

The vacuum gap between the two pipes is created by removing the interstitial gas using a pumping system connected to a copper pipe called "pumping port" (see Figure 4). A Cold Cathode All Metal Gauge is installed on the outer pipe monitoring in real time the pressure during thermal tests, ensuring that it is always below the threshold that defines the high vacuum regime, thereby retaining the advantages of minimal gas thermal conductivity (Figure 2).

To ensure the vacuum seal of the prototype, two flat washers are welded to the ends of the two pipes, as shown in Figure 6, defining a length of the sample to be tested for vacuum thermal insulation of 1.4 m.

Figure 6. Flat welded washer.

Once the prototype was built up, it was necessary to verify the experimental measuring conditions to evaluate λ_{eq} in accordance with equation (3). The predictive model assumes the one-dimensionality of the temperature field in the radial direction for pipes of infinite length.

Preliminary tests have shown that conduction losses at the ends of the pipe have led to measurement errors in the evaluation of the equivalent conductivity λ_{eq}. Moreover, the errors increase with decreasing value of equivalent thermal conductivity because the two contributions, radiative and conductive, cannot be distinguished. Therefore, to avoid conductive losses, it has been necessary to apply a temperature compensation technique: the lateral elements of the test specimen are supplied with additional energy to unify the temperature field in the axial direction, with a temperature difference of less than 1 °C.

In particular, a schematic view of the custom-made heating cartridge designed by the authors to realize the compensation is shown in Figure 7. The object is divided into three sections: the central heater (1 m long), whose length represents the effective tested one, the side heaters (15 cm long), and two Teflon spacers that provide thermal isolation between them. Power is supplied by a DC generator connected to a series of electrical resistors.

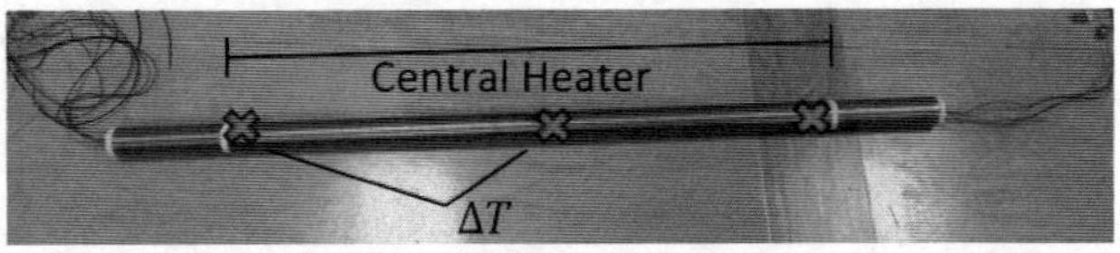

Figure 7. The heater cartridge.

The power of the three heating elements is controlled individually. Figure 8, on the left, shows the electrical resistance that crosses each element inside a copper pipe that provides uniform heat transfer thanks to its high thermal conductivity. Each section of the cartridge is made of a steel housing that serves as a temperature gauge for the inner pipe. The thermocouples are placed in grooves as close as possible to the inner surface of the hot pipe, as shown in Figure 8 on the right. Three different temperatures are monitored along the axis by means of three type K thermocouples: one placed in the center used as reference value for calculating λ_{eq}, and two placed at the ends of the central zone, which allow to assess the effects of compensation. Similarly, thermocouples are placed on the outer surface of the vacuum chamber pipe.

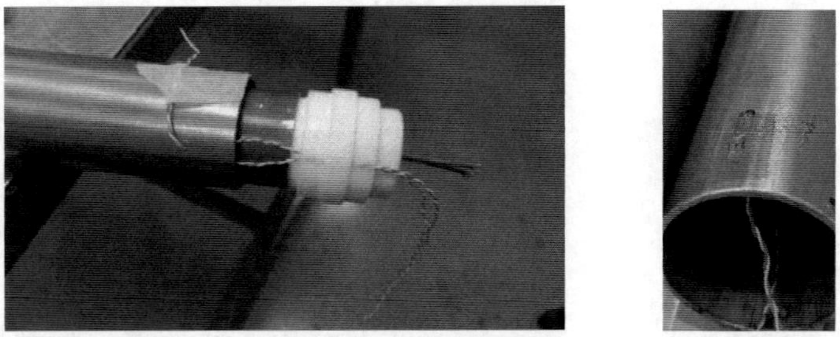

Figure 8. On the left, the copper pipe inserted into a stainless steel one, and the Teflon spacer. On the right, thermocouples groove on stainless steel casing.

4. Results and discussion

In this work, an experimental analysis was performed to characterize the insulation performance of piping for thermal collectors operating at mid temperatures. The high vacuum technology combined with a low emissivity layer was studied to find an alternative solution for the insulation of piping in the medium temperature range with lower thermal conductivity than the solutions known in the literature.

Tests were conducted by heating the prototype with electrical power for three different temperature values: 100 °C, 160 °C, and 220 °C. The experimental points, shown in Figure 9, were recorded in steady-state conditions. We considered only measurements with an axial temperature gradient ΔT lower than 1 °C, resulting in good compensation.

The system operates in high vacuum with a maximum pressure of $3.87 \cdot 10^{-6}$ mbar at 220 °C; in this condition, the gas thermal conductivity can be neglected, and radiation is the only heat transfer mechanism during all thermal tests.

Figure 9 shows a comparison of the thermal test results for three temperature conditions using three different solutions:

- analytic model based on the hypothesis of full contact (dashed line) between Al layer and pipe;
- an analytic model based on the hypothesis of no contact (dotted line) between Al layer and pipe, where a net internal emissivity ε_{IN} =0.020 between the inner pipe and the aluminium foil is assumed and evaluated as in equation (2) with a view factor equal to 1;
- a reference solution based on VIP until 100 °C (λ_{eq} = 0.012 (W/m·K)) and on rockwool (λ_{eq} = 0.04 (W/m·K)) for higher temperatures.

Experimental equivalent conductivity values (red squares) are in good agreement with the model based on the hypothesis of no contact between Al layer and pipe (see figure 9).

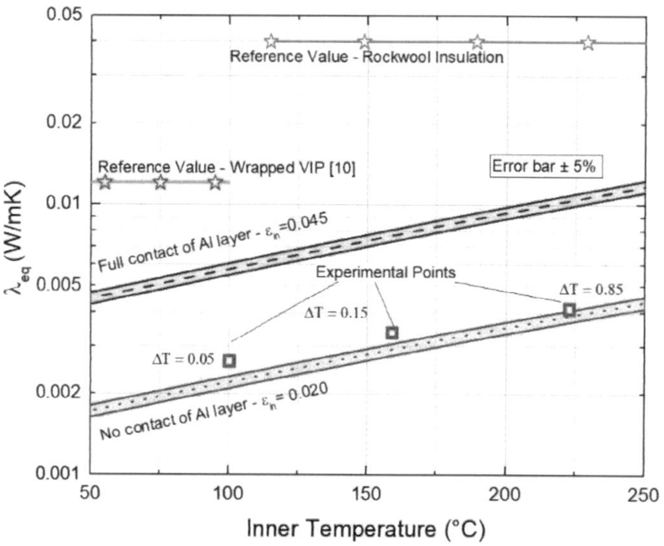

Figure 9. Experimental results carry out an equivalent thermal conductivity about three times lower than VIP reference (between 50 °C and 100 °C) and about two times lower than theoretical value for full contact hypothesis.

5. Conclusions

In the temperature range 100 °C - 250 °C, commercially available solutions to contain thermal losses along heat pipelines are rockwool, fiberglass, and polyurethane with respectively 0.04 W/mK, 0.052 W/mK, 0.02 W/mK as thermal conductivity. In this study, the authors conducted thermal tests to evaluate the thermal insulation (i.e. thermal conductivity) of a high vacuum enclosure (pressure < 10^{-6} mbar) with an aluminium layer covering the stainless steel pipe under test. This solution takes advantage of the high reflectivity of the aluminium layer, which contains the emitted radiation from the pipe, acting as a radiation barrier. Experimental results showed a fivefold reduction of the thermal radiation losses compared to the traditional solutions when in the temperature range between 100 °C and 250 °C. In the following are summarized the results obtained:

- The measured equivalent thermal conductivity of the aluminium coating solution is lower than that of common commercial solution rockwool and developing technologies described in the literature for the temperatures between 100 °C and 250 °C temperature range, reducing the equivalent thermal conductivity from 0.04 (W/m·K) to 0.003 (W/m·K) at 150 °C.
- The reduction of the equivalent thermal conductivity obtained by using the proposed technology allows primary energy savings due to the lower required heat input. For the same thickness of insulation of 23 mm, the thermal losses were reduced from 56 W to 4 W per unit of pipe length with the use of the proposed technology.
- A robust high-vacuum tight enclosure solves the issue of degradation of thermal insulation performance due to atmospheric conditions, returning a stable lifetime performance. In a few years, thermal losses with rockwool insulation could become 50% higher than nominal.
- The use of vacuum insulation allows to save space that might be critical in some applications. This is because thermal radiation depends utterly on the thermal emittance of surfaces, whereas traditional insulating materials use their low thermal conductivity, such as rockwool, where the thicker, the better.

Further investigation is needed to optimize the geometry of the tested prototype and find a solution that also reduces its size. The analytical curves suggest that copper coating can improve thermal insulation under high vacuum conditions. In the future, the economic compromise between low thermal conductivity and cost-saving should be evaluated to find the proper coating material.

6. References

[1] Simona, Paraschiv Lizica, Paraschiv Spiru, and Ion V. Ion. "Increasing the energy efficiency of buildings by thermal insulation." *Energy Procedia* 128 (2017): 393-399.
[2] Baldinelli, Giorgio, et al. "Thermal conductivity measurement of insulating innovative building materials by hot plate and heat flow meter devices: A Round Robin Test." *International Journal of Thermal Sciences* 139 (2019): 25-35.
[3] Lund, Rasmus, and Soma Mohammadi. "Choice of insulation standard for pipe networks in 4th generation district heating systems." *Applied Thermal Engineering* 98 (2016): 256-264.
[4] Monkam, Linus Kameni, et al. "Feasibility analysis of a new thermal insulation concept of cryogenic fuel tanks for hydrogen fuel cell powered commercial aircraft." *International Journal of Hydrogen Energy* 47.73 (2022): 31395-31408.
[5] Șova, Daniela, Mariana Domnica Stanciu, and Sergiu Valeriu Georgescu. "Design of thermal insulation materials with different geometries of channels." *Polymers* 13.13 (2021): 2217.
[6] Abu-Jdayil, Basim, et al. "Traditional, state-of-the-art and renewable thermal building insulation materials: An overview." *Construction and Building Materials* 214 (2019): 709-735.
[7] Sen, Sourav, et al. "Recent developments in biomass derived cellulose aerogel materials for thermal insulation application: A review." *Cellulose* 29.9 (2022): 4805-4833.
[8] Chen, Zhou, et al. "Preparation and characterization of vacuum insulation panels with super-stratified glass fiber core material." *Energy* 93 (2015): 945-954.
[9] Zach, Jiří, et al. "Development of vacuum insulating panels (VIP) with non-traditional core materials." *Energy and Buildings* 199 (2019): 12-19.
[10] Kan, Ankang, et al. "Thermal performance assessment of cold chain chamber with vacuum insulation panel envelope layer." *Cleaner Engineering and Technology* 4 (2021): 100157.
[11] Hayrullin, Aidar, Aigul Haibullina, and Alex Sinyavin. "Insulation thermal conductivity heating networks during transportation thermal energy under dry and moisturizing condition: a comparative study of the guarded hot plate and guarded hot pipe method." *Transportation Research Procedia* 63 (2022): 1074-1080.
[12] Berge, Axel & Adl-Zarrabi, Bijan. (2012). Using High Performance Insulation in District Heating Pipes.
[13] Berge, A., and B. Adl-Zarrabi. "Long term performance of vacuum insulation panels in hybrid insulation district heating pipes." *Energy Procedia* 116 (2017): 334-342.
[14] Thermal conductivity measurement of insulating materials with a three layers device.
[15] Mirković-Marjanović, Milica, et al. "Verification and validation of an advanced guardedhot plate for determination of thermal conductivity." *Thermal Science* 26.6 Part A (2022): 4735-4747.
[16] Huang, Jianhua. "Sweating guarded hot plate test method." *Polymer testing* 25.5 (2006): 709-716.
[17] Kayfeci, Muhammet. "Determination of energy saving and optimum insulation thicknesses of the heating piping systems for different insulation materials." *Energy and buildings* 69 (2014): 278-284.
[18] Dushman, Saul, and Sanborn C. Brown. "Scientific foundations of vacuum technique." *American Journal of Physics* 30.8 (1962): 612-612.
[19] Siegel, Robert. *Thermal radiation heat transfer*. CRC press, 2001.
[20] Roger, C. R., S. H. Yen, and K. G. Ramanathan. "Temperature variation of total hemispherical emissivity of stainless steel AISI 304." *JOSA* 69.10 (1979): 1384-1390.
[21] Russo, Roberto, et al. "The absorptance of selective solar absorber working in high vacuum." (2018): 56-4.
[22] De Maio, Davide, et al. "Multilayers for efficient thermal energy conversion in high vacuum flat solar thermal panels." *Thin Solid Films* 735 (2021): 138869.

Fluid dynamic parameters of naturally derived hydroxyapatite scaffolds for *in vitro* studies of bone cells

E Salerno[1,*], A d'Adamo[2], G Corda[2], C Ongaro[2], B Zardin[2], G Orlandi[3], A Ruffini[4], J Bertacchini[3] and D Angeli[1,5]

[1] DISMI – Department of Sciences and Methods for Engineering, University of Modena and Reggio Emilia, Via Amendola 2, Reggio Emilia 42122, Italy
[2] DIEF - Department of Engineering "Enzo Ferrari", University of Modena and Reggio Emilia, Via Vivarelli 10, Modena 41125, Italy
[3] Department of Surgery, Medicine, Dentistry and Morphological Sciences with Interest in Transplant, Oncology and Regenerative Medicine, University of Modena and Reggio Emilia, Modena 41125, Italy
[4] Institute of Science and Technology for Ceramics (ISTEC), National Research Council (CNR), Via Granarolo 64, Faenza 48018, Italy
[5] INTERMECH MO.RE. Interdepartment Research Center, University of Modena and Reggio Emilia, Via Amendola 2, Reggio Emilia 42122, Italy

* Corresponding author: elisabetta.salerno@unimore.it

Abstract. Hydroxyapatite scaffolds obtained from the biomorphic transformation of wood are characterized from a fluid dynamic point of view. Such material of recent introduction offers great advantages for the *in vitro* study of bone cells, mostly in virtue of its peculiar porous structure. Determining the flow resistance and morphological parameters of these scaffolds is an essential step towards their practical use in bioreactors and microfluidic devices. To this aim, a series of tests involving a draining fluid are performed on a set of disc-shaped scaffolds, followed by the microscopy analysis of the pores visible on the sample faces. Contrarily to what expected, a temperature dependence is observed for the flow resistance, even after normalizing it by the fluid properties. The interpretation of the experimental results is assisted by numerical outcomes from Computational Fluid Dynamics modelling, which underline some limitations in the application of classical laws to the present problem. While the complex and variable internal structure of the scaffolds prevents the systematic use of simplified formulae, a correlation is found between the flow resistance and the pore geometry, which can facilitate the characterization of further samples.

1. Introduction

Understanding the mechanisms governing bone transformation and healing is at the basis of bone regenerative medicine. To this aim, it is of great importance to try to re-create *in vitro* the same environment found inside living organisms, including the flow conditions. In fact, it is well ascertained that bone cells, i.e. osteoblasts, osteoclasts and osteocytes, regulate the process of bone growth and resorption in response to external mechanical stimuli, which are sensed in the form of fluid shear stresses (FSS) acting on their surface [1]. These are in the order of the Pascal or lower, depending on the cell type and on the physiological conditions [2]. Fluidic systems thus need to be designed to be able to generate the required levels of FSS.

One of the key elements to be integrated in culturing systems is the so-called scaffold, that is, an extracellular matrix providing the attachment sites for the cells. Three-dimensional scaffolds are considered superior to planar scaffolds, because they allow for cell expansion and re-arrangement in all directions, but their fluid dynamic characterization is not always straightforward. In the current investigation, hydroxyapatite scaffolds resulting from the biomorphic transformation of wood are adopted as the substrate for bone cells to be cultured under perfusion. This naturally derived material, termed as B-HA, has been recently developed [3] and presents a 3D structure with interconnected, hierarchically organized pores that closely resembles the microstructure of real bone. Its inherent flow resistance is an essential parameter that must be determined in order to calculate and control the flow rate and the shear stresses the cells inside the pores are exposed to. However, as the result of the specific fabrication process, the internal geometry of B-HA is anisotropic, not predictable and specific of the individual scaffolds. Furthermore, it is hard to be directly inspected, since B-HA is prone to shatter upon sectioning with common laboratory instrumentation.

Since those factors prevent the application of simple formulae, a series of experiments is carried out here to accurately derive the fluid dynamic parameters of a set of B-HA samples. The flow resistance is reconstructed by exposing the scaffolds to a draining volume of fluid and the scaffold morphology is analysed via microscope observations. The collected data on pore number and size are used in combination with the values relative to flow resistance to extract useful correlations in order to simplify the characterization of B-HA. The experimental campaign is supported by a Computational Fluid Dynamics (CFD) study, aimed at investigating in detail the flow development inside the pores. The numerical outcomes obtained on one of the available samples are compared with the experimental results and used to explain the observed trends.

2. Experimental analysis

2.1. Experimental methods

The naturally-derived B-HA scaffolds adopted in the present study are in the form of round pills nearly 10-11 mm in diameter and 4-5 mm in height. To obtain accurate values of the flow resistance, perfusion tests are performed on a set of 12 scaffolds. As depicted in Figure 1a, each sample is wedged in the orifice of a hourglass-type transparent container and exposed to a draining volume of bi-distilled water (nearly 500 ml). An imaging system is used to capture the temporal variation of the liquid level in time lapse mode. The acquired frames are analysed by means of an image processing software (ImageJ, NIH, USA) to extract the instantaneous liquid height. More details on the experimental apparatus and data elaboration can be found in [4]. The scaffold flow resistance, R_f, is then computed for each frame from the following formula, after integration between the starting time and the considered instant:

$$R_f = \frac{\Delta p}{Q_v} = -\frac{4\,\rho\, g\, h(t)}{\pi\, D^2\, \partial h/\partial t} \tag{1}$$

In the above equation, Δp is the pressure difference over the scaffold dictated by the hydrostatic head, being ρ the fluid density and h the hydraulic head with respect to the scaffold. The volumetric flow rate, Q_v, is expressed in terms of the descending liquid level and the diameter of the liquid free surface, D, which is a function of h too (the portion of interest of the container has a conical geometry). The symbols used, together with the relative definition and units, are reported in Table 1. An average value of R_f is then extracted for the whole sequence with the associated standard deviation as the uncertainty. The experiment is repeated 4-5 times both at room temperature and inside an oven set at 37 °C, i.e. the temperature typically imposed during *in vitro* studies of cell cultures. In this way, any possible temperature dependence of the scaffold flow resistance is taken into account. The actual temperature of the liquid is measured with a thermocouple immediately before and at the end of each test.

Journal of Physics: Conference Series **2685** (2024) 012058 doi:10.1088/1742-6596/2685/1/012058

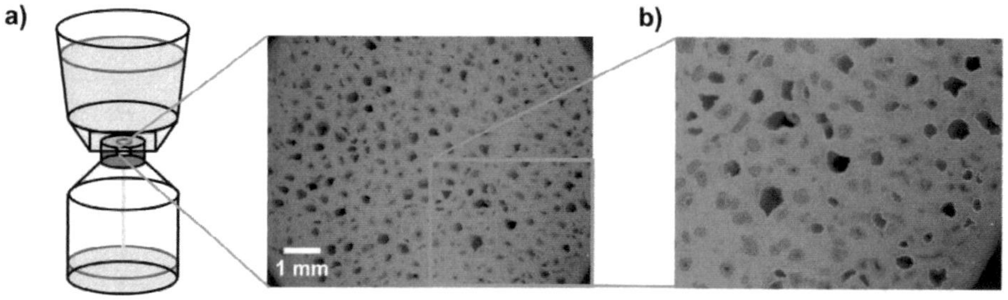

Figure 1. a) Schematic of the test device and microscope image of one of the scaffolds (upper face) inserted into the device orifice. b) Pores identified on the scaffold face (detailed view) via repeated processing procedure.

Information on the pore number and size is collected by inspecting the top and bottom surfaces of the samples with a stereomicroscope (Nikon SMZ800, Plan 1X objective). It is observed that fluid immersion induces swelling in the B-HA solid matrix, causing a general contraction of the pores which affects the scaffold flow resistance. Hence, the morphological analysis of each sample is postponed until the completion of the relative permeation tests in order to get more representative geometrical data. Microscope images are processed with ImageJ: image segmentation is performed first via filtering, rolling-ball background subtraction and semi-automatic thresholding to separate the features of interest (i.e. the pores) from the background. The obtained binary image is then scanned with the automated particle detection algorithm included in the software, which counts and measures the cross-sectional area of the detected pores. Due to the presence of pores significantly different in size, the overall procedure is repeated twice or three-times on each image by adjusting threshold as well as the particle size range in the detection algorithm. In this way, both large and tiny pores, as small as 47 μm^2 in area, are detected while reducing area over-/under-estimation and duplicated counts. In Figure 1b, the output of the processing technique applied to one of the images is displayed as an example. Eventually, pore cross-sections are approximated as circles, and the equivalent diameters are computed. Results are grouped in 11 classes of increasing width to have a proper representation of diameter distribution.

Table 1. List of symbols.

Symbol	Definition	Units	Symbol	Definition	Units
A	Scaffold cross-sectional area	m^2	L	Scaffold height	m
A_{mean}	Mean pore area	m^2	N	Number of pores	-
A_{sc}	Maximum imaged scaffold area	m^2	Q_v	Volumetric flow rate	m^3/s
D	Diameter of the descending liquid level	m	R_f	Scaffold flow resistance	$Pa{\cdot}s/m^3$
Da_L	Darcy number	-	R_f/μ	Normalized scaffold flow resistance	$1/m^3$
g	Acceleration of gravity	m/s^2	Re	Reynolds number	-
h	Hydraulic head	m	t	Time	s
k	Scaffold permeability	m^2	T	Temperature	°C
	Other subscripts/superscripts			Greek letters	
Symbol	Definition		Symbol	Definition	Units
eq	equivalent		Δp	Pressure difference over the scaffold	Pa
i	relative to i-th pore		μ	Fluid dynamic viscosity	$Pa{\cdot}s$
ov	relative to tests inside oven		ρ	Fluid density	kg/m^3
RT	relative to tests at room temperature				

2.2. Experimental results

The thermo-physical properties of culture media commonly adopted for *in vitro* cell growth are slightly different from those of water [5]. As a consequence, the measured scaffold flow resistances should be conveniently made independent of the test liquid properties. By supposing laminar flow conditions (as confirmed below), the Darcy's law can be considered valid on first approximation for the present problem, reading as:

$$\Delta p = \frac{L \mu}{A k} Q_v \tag{2}$$

From a comparison with equation (1), it can be seen that the fraction on the right hand of equation (2) must be equal to R_f, as given by the dynamic viscosity of the liquid (μ) and by the height (L), cross-sectional area (A) and permeability of the scaffold (k). Then, flow resistance information related solely to scaffold properties can be obtained if scaling R_f by the water viscosity at the corresponding test temperature. The R_f/μ so derived for the investigated samples distribute over one order of magnitude, ranging from $(1.51 \pm 0.06) \cdot 10^{11}$ to $(1.254 \pm 0.007) \cdot 10^{12}$ m^{-3}, relatively to room temperature conditions (21-23 °C). Such a wide interval of data is symptomatic of a great variability of B-HA internal structure, directly connected to its natural origin.

Contrarily to what expected, the R_f/μ values do not remain fixed when moving to higher temperatures (35-38 °C), though the variation of water viscosity could be considered the main responsible for any temperature dependence. A general increase is observed, with the shifted range going from $(1.83 \pm 0.02) \cdot 10^{11}$ to $(1.53 \pm 0.04) \cdot 10^{12}$ m^{-3}. This cannot be attributed to a swelling of the scaffold due to hygroscopic effects, since the two test conditions are not always applied in the same order. In Figure 2a, the relative variation of R_f/μ with respect to the room temperature value is reported for each sample, normalized by the specific temperature difference. A total mean increment of $(1.08 \pm 0.13) \cdot 10^{-2}$ °C^{-1} is estimated when passing from an average liquid temperature of 22 °C to 37 °C.

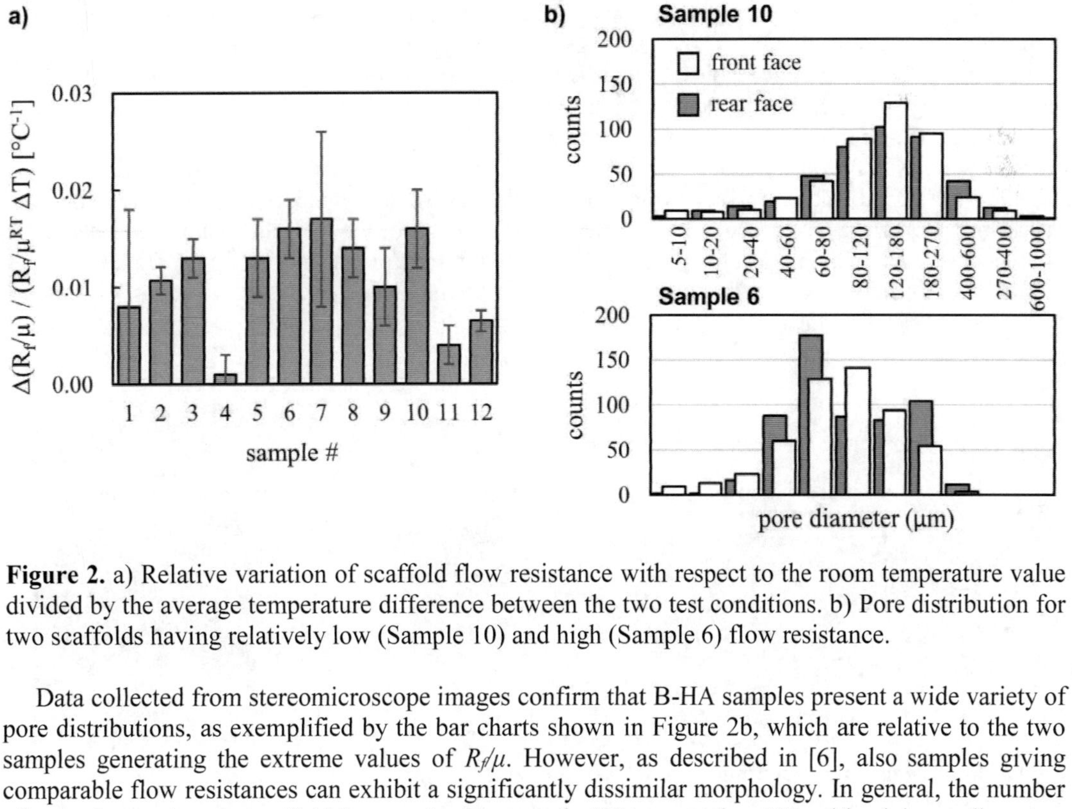

Figure 2. a) Relative variation of scaffold flow resistance with respect to the room temperature value divided by the average temperature difference between the two test conditions. b) Pore distribution for two scaffolds having relatively low (Sample 10) and high (Sample 6) flow resistance.

Data collected from stereomicroscope images confirm that B-HA samples present a wide variety of pore distributions, as exemplified by the bar charts shown in Figure 2b, which are relative to the two samples generating the extreme values of R_f/μ. However, as described in [6], also samples giving comparable flow resistances can exhibit a significantly dissimilar morphology. In general, the number of pores detected on the scaffold faces varies from nearly 400 to more than 900, with minimum diameters of 8 μm observable in all the cases. The maximum pore diameter, instead, differs from sample to sample, ranging from nearly 260 to 910 μm. Interestingly, the pore distribution on one face of the scaffolds roughly replicates the opposite one, suggesting that pores extend preferentially along the vertical direction. This allows for the samples to be approximated to some extent as a network of cylindrical channels.

3. Numerical analysis

3.1. Numerical model

In order to support the interpretation of the experimental results and to inspect in detail the flow development inside the pores, numerical simulations of the perfusion tests are performed. A 3D model is created of one of the samples, having intermediate fluid dynamic and morphological characteristics within the set of the measured scaffolds. The pore diameter distributions extracted from the front and rear face of the scaffold are averaged, and the resulting distribution (Figure 3a) is used to populate the model with cylindrical pores of length equal to the scaffold height (4.3 mm). The measured scaffold diameter (10.40 mm) is preserved too. The model is built by imposing circumferential symmetry, so that a 30° slice, like the one sketched in Figure 3b, hosts the same relative pore distribution than the full sample. A 1 mm thick fluid plenum is added above and below the scaffold to provide numerical stability and to reproduce the spontaneous flow subdivision into channels of different size, according to the individual channel resistance.

The computational tool SIMCENTER STAR-CCM+ (SIEMENS SISW, v. 2021.3), based on the Finite Volume method, is used to solve the continuity and Navier-Stokes equations numerically. The experimental configurations are replicated by specifying thermo-physical properties of water at two different temperatures, namely 22 and 37 °C, which correspond to the average liquid temperature in flow resistance tests at room temperature and in an oven, respectively, on the selected sample. For simplicity, the liquid level is kept fixed, associated to a mean pressure difference of Δp = 947 Pa over the scaffold. This implies a Reynolds number inside the pores not greater than Re = 80, as computed from the mean experimental data. Hence, laminar flow conditions are applied while adopting a steady-state solver and second-order schemes for discretization.

By taking advantage of the model axial symmetry and the preferential direction of the flow, only a 30° slice is meshed and solved, allowing for a reduction of the computational requirements. An unstructured quadrilateral surface mesh is generated separately for the scaffold/channel region and the two plenums, with a minimum cell thickness of 1 μm at the walls of the smallest pores. Both grid types are axially extruded with a uniform spacing of 0.1 and 0.05 mm, respectively, for a total of 1.1 M cells. In Figure 3c, a detail of the mesh in the channel region can be observed.

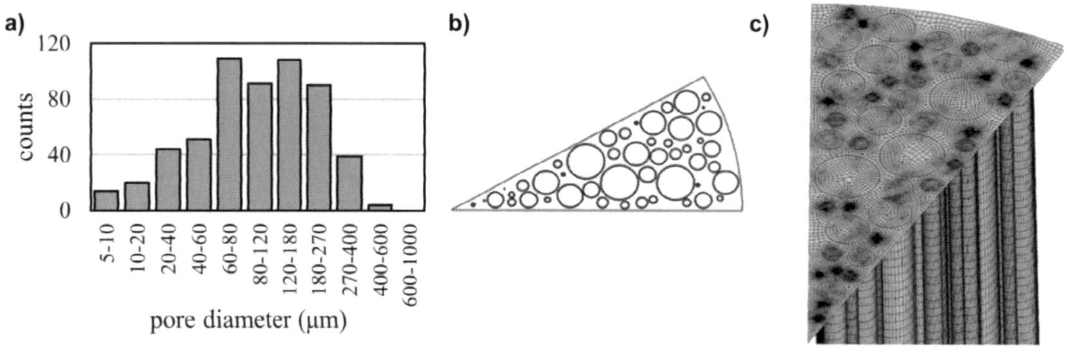

Figure 3. a) Average pore distribution of the sample selected for the numerical analysis. b) Arrangement of the pores in a 30° slice of the model (top view). c) Detail of the grid generated in the scaffold region.

Stagnation and static pressure are imposed at the inlet of the upper plenum and at the outlet of the lower plenum, respectively, whereas a no-slip condition is specified at the liquid/solid boundaries. The reader is referred to [6] for further details on the numerical model, including the assessment of the mesh quality.

3.2. Numerical results
Normalized flow resistances are extracted from the simulation outputs, by dividing the imposed pressure difference by the obtained volumetric flow rate and by the temperature-dependent water viscosity. Numerical and experimental results for the selected sample are compared in Figure 4a. It can be noted that CFD predictions tend to overestimate R_f/μ by 20-30%, possibly because the boundary condition specified at the liquid/solid interface does not adequately account for the absorption mechanisms that are responsible for the swelling of the scaffolds. Dedicated investigations are needed in this sense for future improvements of the numerical model. Nevertheless, the increase of R_f/μ with temperature is still evident.

To better elucidate this point, contours of the velocity magnitude plotted at the two temperatures are depicted in Figures 4b and 4c. Higher flow velocities are associated with the highest temperature, following a reduction in water viscosity. More noticeably, a non-negligible entry region, extending over 1/4 of the channel length, can be seen at the channels inlet already in the case at the lowest temperature, and the temperature increment induces its further expansion. The absence of a fully-developed flow all along the pores poses some limitations to the applicability of the Darcy's law to the present problem, and justifies the preservation of the temperature dependence in the R_f/μ values.

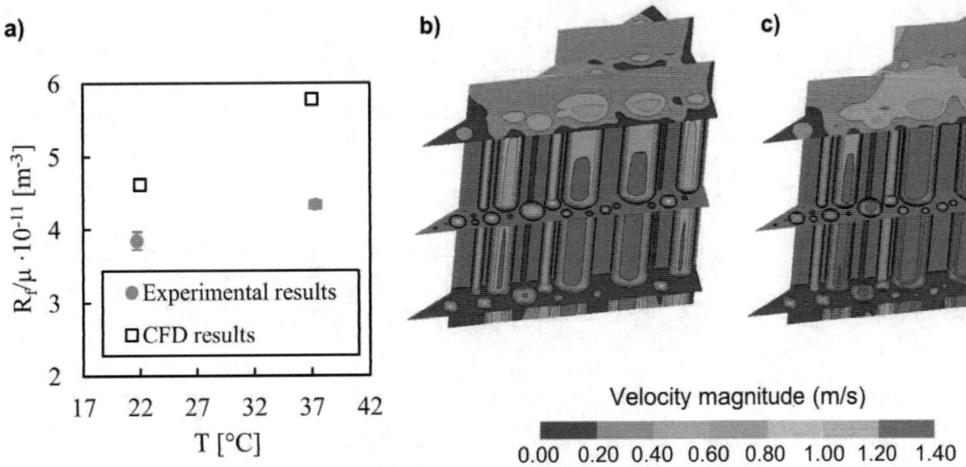

Figure 4. a) Comparison of the experimental and numerical results for R_f/μ for the selected sample. b) and c) Contours of the velocity magnitude for two different temperatures imposed to the water properties: b) $T = 22$°C and c) $T = 37$ °C.

4. Discussion
Under the approximation of parallel cylindrical pores having the same length, the geometrical data acquired from the microscope images could be used to estimate an equivalent normalized flow resistance ($R_{f\,eq}/\mu$), according to the following formula:

$$\frac{R_{f\,eq}}{\mu} = \left(\sum_i \frac{1}{R_{f\,i}/\mu}\right)^{-1} = \left(\sum_i \frac{\pi\, D_i^4}{128\, L}\right)^{-1} \tag{3}$$

R_{fi} is the resistance of the i-th pore having diameter D_i, and is calculated from the Hagen-Poiseuille's equation. For each sample, equation (3) is applied separately to the pores detected on the front and rear face, and results are averaged. However, the vast majority of resistances obtained in this way seem to underestimate the actual values, showing discrepancies up to 87% relatively to the measured ones, as demonstrated by the data appended in Table 2, where measured data referred to tests in oven are reported. One of the causes of such differences is identified in the limited field of view of the stereomicroscope in use, which does not allow for the detection of the most peripheral pores even at the lowest magnification. Though pores near the specimen borders are less abundant, as shown in [4], and do not significantly affect average outputs, their contribution in determining the equivalent flow resistance is lost. To fix this point, a microscope mounting a lens with a slightly lower magnification would be needed. On the other hand, less resolved images would impair the detectability of the smallest pores, and would aggravate some errors already introduced by the image processing technique. These include, for example, the interpretation of adjacent pores as a single, merged structure, which implies a reduction of the value calculated with equation (3). Another explanation for the observed discrepancies comes from the complex internal structure of B-HA, which in some cases can invalidate the geometrical simplifications made.

In light of the above considerations, perfusion tests stand out as the best method for an accurate determination of B-HA flow resistance. However, this appears as an unfeasible solution when multiple scaffolds need to be analysed. Introducing appropriate correlations between the fluid dynamic and statistical morphological properties of the scaffold would represent a helpful alternative. For this purpose, flow resistances are further normalized in order to get non-dimensional quantities. In particular, Darcy number (Da_L) is computed referred to the scaffold height, as given by equation (4):

$$Da_L = \frac{1}{A\,L\,R_f/\mu} \tag{4}$$

Table 2. Comparison of the normalized flow resistances for the set of scaffolds: values measured from the tests in oven ($R_{f\,ov}/\mu$) against values obtained from equation (3) ($R_{f\,eq}/\mu$).

Sample	$R_{f\,ov}/\mu$ [$\cdot 10^{11}$ m^{-3}]	$R_{f\,eq}/\mu$ [$\cdot 10^{11}$ m^{-3}]	$\frac{(R_{f\,eq}/\mu - R_{f\,ov}/\mu)}{R_{f\,ov}/\mu}$ [-]	Sample	$R_{f\,ov}/\mu$ [$\cdot 10^{11}$ m^{-3}]	$R_{f\,eq}/\mu$ [$\cdot 10^{11}$ m^{-3}]	$\frac{(R_{f\,eq}/\mu - R_{f\,ov}/\mu)}{R_{f\,ov}/\mu}$ [-]
1	4.34 ± 0.06	2.07	-52%	**7**	7.3 ± 0.6	3.10	-57%
2	3.23 ± 0.03	0.43	-87%	**8**	2.60 ± 0.03	1.12	-57%
3	6.79 ± 0.11	1.58	-77%	**9**	7.3 ± 0.2	6.54	-10%
4	11.63 ± 0.07	2.91	-75%	**10**	1.83 ± 0.02	1.14	-38%
5	4.52 ± 0.14	3.36	-26%	**11**	9.9 ± 0.3	11.4	15%
6	15.3 ± 0.4	6.85	-55%	**12**	3.39 ± 0.03	5.64	66%

Since the operating temperature in typical scaffold applications is T = 37 °C and temperature dependence has been observed for R_f/μ, resistance values in equation (4) are inserted pertaining to tests carried on in oven only.

On the other hand, different non-dimensional statistical parameters describing pore geometry are considered, including: (i) the mean pore diameter scaled by the scaffold diameter, (ii) the mean pore area (A_{mean}) scaled by the captured scaffold area (A_{sc}), and (iii) surface porosity (NA_{mean}/A_{sc}, being N the number of pores). For their determination, averaged data from the front and rear face are used for each sample. It is pointed out that the scaffold area entering the surface porosity and the A_{mean}/A_{sc} ratio is only the visible portion of the scaffold cross-sectional area in the microscope images. Moreover, surface porosity and scaled pore diameter/area are parameters related to the scaffold surface. Their involvement as the independent variable in correlations is reasonable only for samples which have an approximately regular internal structure. For this reason, samples exhibiting $R_{f\,eq}/\mu$ largely different from the measured values (> 70%) are excluded in the search of correlations.

Among the explored possibilities, the best correlation is found for Da_L as a quadratic function of A_{mean}/A_{sc}, which is reported in Figure 5. The achieved relation returns flow resistance values that, in the vast majority of cases, are more reliable than those resulting from equation (3), but still have some discrepancies with respect to the measured ones. Hence, while using this function would allow for a sensible reduction of experimental and analysis efforts in the extraction of B-HA fluid dynamic properties, its validity needs to be evaluated by considering the scaffold final application.

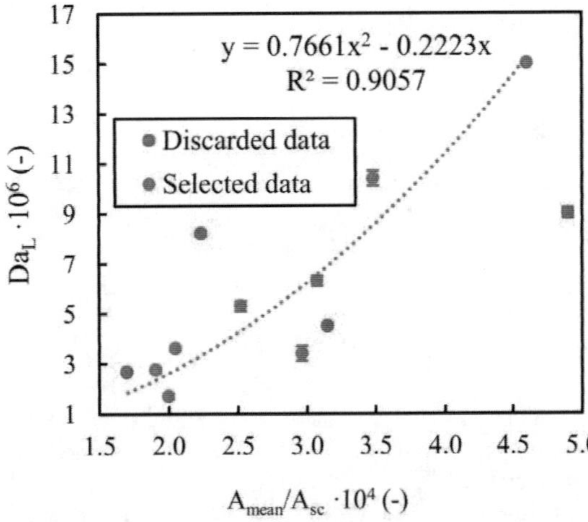

Figure 5. Plot of the Darcy number as a function of the normalized pore mean area. The reported correlation is found from green data ("selected data") only.

5. Conclusions

The flow resistance and pore geometrical parameters of naturally derived hydroxyapatite scaffolds are determined as functional parameters to control the stimuli over bone cells cultured under dynamic conditions. The wide scattering of the extracted data directly reflect the variable internal structure of this type of scaffold. For modelling purposes, this can be approximated to a network of cylindrical pores, simplifying the investigation of flow development by CFD analysis. The inspection of velocity contours from numerical simulations reveals the presence a non-negligible entry region inside the pores, which undermines the validity of the classical Darcy's law when decoupling flow resistance from the fluid properties. On the other hand, for predictive purposes, the simplification of the internal pore geometry can lead to quite unreliable results. A correlation between the fluid dynamic properties of the scaffolds at their typical operating temperature and pore statistical parameters is proposed as a better alternative. The correlation found still suffers from some limitations, consequent to the assumptions made, but can be reasonably employed to streamline the preparatory characterization of the scaffolds to be used in *in vitro* studies.

6. Reference

[1] Stewart S *et al.* 2019 *Bone Joint Res.* **9** 1
[2] Wittkowske C *et al.* 2016 *Front. Bioeng. Biotechnol.* **4**:87
[3] Tampieri A *et al.* 2017 Patent WO/2017/021894. https://patentimages.storage.-googleapis.com/94/d9/35/da5164da603a00/ AU2016304431B2.pdf
[4] Salerno E *et al.* 2022 *Regen. Biomater.* **9** rbac034
[5] Poon C 2022 *Mech. Behav. Biomed. Mater.* **126** 105024
[6] d'Adamo A *et al.* 2023 *Regen. Biomater.* **10** rbad002

The impact of the distribution of the condensation energy adsorbed in the water pool on the DCC efficiency at sub-atmospheric pressure

D Aquaro[1], L Berti[1], R Lo Frano[1*], S A Cancemi[1], F d'Errico[1], B Sarkar[2]

[1] Department of Civil and Industrial Engineering, University of Pisa, Largo Lucio Lazzarino, 56122 Pisa ITALY
[2] ITER Organization, Saint Paul Lez Durance - France

*corresponding author: rosa.lo.frano@unipi.it

Abstract. The safety of a fusion reactor like ITER relies on the reliability of the pressure suppression system (for keeping pressure below 0.15 MPa during accidental scenarios), and, in turn, on the efficiency of Direct Contact Condensation (DCC).
In this study, the role played by the distribution of the condensation energy in the water is analysed considering the experimental results of temperature and pressure recorded during steam condensation tests in a closed volume at sub-atmospheric conditions.
The investigation of the mechanisms involved in the condensation at the water-steam interface is fundamental to verify that the subcooled water mass fully participates into the steam condensation for all the foreseen condensation regimes. Particularly, the paper analyses the distribution of the condensation energy in a water pool for different test conditions and with reference to the large-scale facility under operation at the DICI-University of Pisa.
The subcooled water volume of the condensation tank in this system was assumed subdivided in 104 annular discrete volumes corresponding to the location of temperature and pressure sensors (testing monitoring points). Moreover, this discretization was used to implement a model to visualize the experimental data and in particular the accumulation of the condensation energy in the different part of the water pool and of the vacuum space.
The experimental results showed that most of the tests had a participating fraction of water mass greater than 50%. Nevertheless, the non-uniformity and stratification of the temperature in the water pool requires a greater water mass for condensing a given steam mass. The analyses of steady state pure steam condensation tests determined that it needs to increase 1.39 times the water mass respect to the case of full participation of the water at the steam condensation.

1. Introduction

The International Thermonuclear Experimental Reactor (ITER), which is under construction in Saint Paul Lez Durance (France), manages Loss of Coolant Accidents (LOCA) in the Vacuum Vessel (VV) by means of a Pressure Suppression System (PSS). This safety system is made of four tanks partially filled with water working at sub-atmospheric pressure. The steam flows inside the tanks through multi-holed tubes (spargers) one for each tank [1]. Conceptually the PSS is similar to system used in the fission reactors (particularly in the Boiling Water Reactor), although it differs substantially from this in terms of the steam condensation that occurs in a closed volume starting from a sub-atmospheric pressure.

These steam condensation conditions have not been tested previously and very few experimental studies can be found in the scientific literature. An extensive experimental test campaign, funded by

ITER Organization, was carried out at the B. Guerrini laboratory of the University of Pisa on two scale facilities [2]-[3]. In the Small-Scale Facility (SSF) (1/22 geometrical scale and 1/100 scale considering the steam mass flow rate) tests were performed using spargers having 1 up to 24 holes and 5 g/s as maximum steam mass flow rate per hole. The Large-Scale Facility (LSF) (full scale from geometrical point of view and 1/10 reduced scale for the steam mass flow rate) allowed to test a 100 holes sparger (sparger A) and a 1000 holes full-scale sparger (sparger B) [4]-[5]. Figure 1 shows the two spargers assembled inside the Experimental Testing Tank (ETT). The ETT (92 m^3 of volume, partially filled with 55 m^3 of water) is the main component of the LSF in which the steam condenses by direct contact with water. It is fully instrumented by 60 temperature sensors and 18 pressure sensors, and 6 video cameras. The spargers are instrumented by one temperature and one pressure sensors in the internal volume. Four strain gauges and one accelerometer are mounted on the sparger external surface.

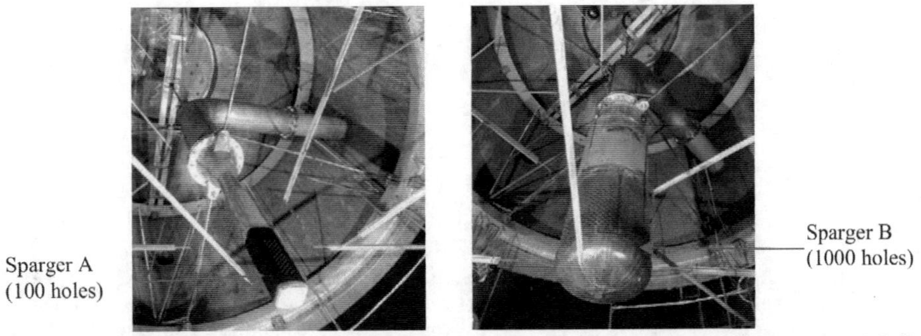

Figure 1. Spargers tested inside the Large-Scale Facility: sparger A (right figure), and sparger B (left figure).

The qualification of the pressure suppression system consists in the demonstration of the ability to condense the steam mass, foreseen in the accidental scenario, without to overcome the pressure limit of 0.15 MPa. In this context, it is necessary to verify that the water mass fully participates to the steam condensation for all the possible expected condensation regimes. Obviously, the temperature stratification contrasts with this requirement.

This paper illustrates the distribution of the condensation energy in the water pool for different LSF test conditions and for the two spargers' configurations. The ETT has been subdivided in discrete volume correspondent to the location of the temperature and pressure sensors in the horizontal and vertical direction. The water pool and the vacuum space have been discretized in ring volumes for implementing a model of accumulation of the steam condensation.

2. Steam condensation regimes in a closed volume at sub-atmospheric pressure

The steam condensation tests, performed in the LSF employing 100 holes and 1000 holes spargers, experimentally analysed the following conditions:

- Steam mass flow rate, Q: up to 0.5 kg/s
- Water temperature, T_w: 15-100°C
- Pressure in front of the sparger holes (downstream pressure), P: 0.25-1.2 bars

Moreover, tests were performed injecting steam flux mixed with a non-condensable gas flux (up to a 20% fraction in mass). Tests were performed at constant (steady state tests) or variable (transient tests) steam mass flow rate.

The map of steam condensation regimes (CR) at atmospheric pressure is defined in the plane T_w – Q_h being T_w, the water temperature and Q_h, the steam mass flow rate per area of one hole [6]-[7]. In the steam direct contact condensation in a closed volume starting from a sub-atmospheric pressure, the parameter which defines the condensation regimes is, instead, the ratio Q_h/P (Kg/(sm^2kPa)) between the steam mass flow rate per hole (Q_h) and the downstream pressure (P) [5]. In particular, the following steam condensation regimes can be identified (Figure 2):

-unstable CR for $Q_h/P < 1$ kg/(s m^2 kPa) called at increasing water temperature Chugging (C), Transient Chugging (TC), and Bubbling Condensation Oscillation (BCO)
- oscillatory CR for $1 < Q_h/P < 2$ kg/(s m^2 kPa) called Condensation Oscillation (CO) or Interfacial Oscillation Condensation (IOC)
-stable CR for $Q_h/P > 2$ kg/(s m ^{2}kPa), called Stable Condensation (SC).

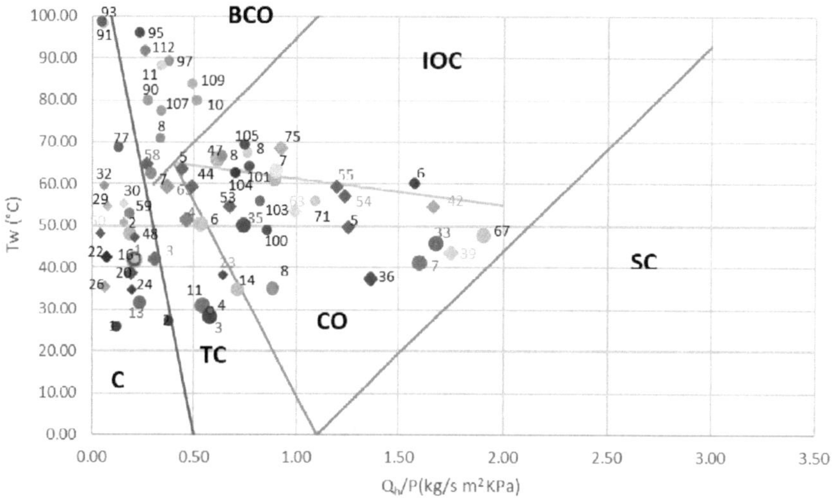

Figure 2. Map of the CRs for steady state pure steam condensation tests with sparger A: the inclined straight lines and the numerical points represent the CRs boundaries and the test conditions, respectively.

3. Experimental distribution of condensation energy in the water pool of the ETT

The ETT volume has been subdivided in the radial and axial direction by means of discrete volumes, to be consistent and considering the position of the temperature and pressure sensors. Radially the volumes have annular shape, the internal and external radius correspond to the average distance between the sensors. Along the height, the ETT volume is subdivided in fourteen layers (ten in the water pool and four in the vacuum space).

Generally, four temperature sensors are located along the radius except in the holed area of the sparger where six sensors have been installed. Moreover, in this zone, the axial layers (6 and 7) have been subdivided in further two sublayers to locate a greater number of sensors. This same volume discretization was used for implementing the numerical model to study the accumulation of the condensation energy in the different parts of the water pool and vacuum space. The water pool was discretized by 104 ring volumes, while the vacuum space by 55 ring volumes. The energy released in the ETT, $E_V(t_J)$, until an instant t_J, is determined by the summation of the steam mass flow rate $Q(t_j)$ per the time span and per its enthalpy, $H_{022}(t_j)$, calculated at the temperature and pressure of the steam in the inlet nozzle ($T_{022}(t_j)$ and $P_{022}(t_j)$):

$$E_v(t_J) = \sum_1^J Q(t_j)\,(t_j - t_{j-1})\,H_{022}(t_j) \quad (1)$$

For each water volume, V_{wi}, the correspondent mass $M_{wi}(t_J)$ and thermal capacity $C_{Wi}(t_J)$ are calculated by:

$$M_{wi}(t_J) = V_{wi}\rho_i(T_{wi}(t_J)) \qquad\qquad C_{wi}(t_J) = M_{wi}(t_J)c_{pi}(T_{wi}(t_J))$$

being $\rho_i(T_{wi}(t_J))$ and $c_{pi}(T_{wi}(t_J))$ the density and the specific heat of the generic i-th water volume, respectively. They are calculated at the instantaneous value of temperature, $T_{wi}(t_J)$, of the generic i-th water volume. The accumulated condensation energy $C_{AEi}(t_J)$ in the V_{wi} is given as:

$$C_{AEi}(t_J) = \sum_1^J C_{wi}(t_j)(T_{wi}(t_j) - T_{wi}(t_{j-1})) \tag{2}$$

where $T_{wi}(t_{J-1})$ is the temperature of i-th volume at the previous step j-1. At each instant, the percentage of the injected steam energy accumulated as condensation energy in each volume, is calculated by:

$$F_{ASEi}(t_J) = 100\frac{C_{AEi}(t_J)}{E_v(t_J)} \tag{3}$$

In the vacuum space, where is globally accumulated an energy smaller than 0.5% of the total one, the calculation starts with the determination of the initial masses of steam and non-condensable gas (air). From the initial average temperature of vacuum space, $T_{vs}(t_o)$, we derived the initial saturation pressure $p_{satvs}(t_o)$ and the initial steam mass $M_{svs}(t_o)$:

$$M_{svs}(t_o) = \frac{18V_{vs}(t_o)p_{satvs}(t_o)}{(8.31(T_{vs}(t_o)+273.15)} \tag{4}$$

where $V_{vs}(t_o)$ is the initial value of the vacuum space volume. The initial mass of air, $M_{avs}(t_o)$ contained in vacuum space is obtained known the initial average pressure of vacuum space $p_{vs}(t_o)$ as:

$$M_{avs}(t_o) = \frac{1000V_{vs}(t_o)(p_{vs}(t_o)-p_{satvs}(t_o))}{287(T_{vs}(t_o)+273.15)} \tag{5}$$

At a generic instant t_J, the steam mass $M_{svs}(t_J)$ and the air mass $M_{avs}(t_J)$ are always calculated from the same equations 4) and 5) substituting the instantaneous values of $T_{vs}(t_J)$, $p_{satvs}(t_J)$, $p_{vs}(t_J)$ and $V_{vs}(t_J)$.
The instantaneous values of the vacuum space volume, $V_{vs}(t_J)$ is given as: $V_{vs}(t_J)=V_{tot}- V_w(t_J)$, being V_{tot} the total volume of ETT, $V_w(t_J)$ the instantaneous value of the water volume which changes for the variation of the water density, $\rho_w(T_{wave}(t_J))$ and for the condensation of steam mass $M_s(t_J)$ according to the following formulas:

$$V_w(t_J) = V_w(t_o) + \frac{M_w(t_o)+M_s(t_J)}{\rho_w\left(T_{wave}(t_J)\right)} \tag{6}$$

$T_{wave}(t_J)$ is the average temperature of the water at the instant t_J. In each volume of the vacuum space, V_{vsi}, the masses of air, $M_{avsi}(t_J)$, and of steam, $M_{svsi}(t_J)$, are assigned proportionally to the relative volumes:

$$M_{avsi}(t_J) = \frac{V_{vsi}}{V_{vs}(t_J)}M_{avs}(t_J)\; M_{svsi}(t_J) = \frac{V_{vsi}}{V_{vs}(t_J)}M_{svs}(t_J) \tag{7}$$

Defining the thermal capacity of the steam and air in each i-th volume of the vacuum space at a generic instant t:

$$C_{svsi}(t_J) = M_{svsi}(t_J)c_{psi}(t_J) \qquad C_{avsi}(t_J) = M_{avsi}(t_J)c_{pai}(t_J)$$

the energy accumulated in each volume of the vacuum space, V_{vsi}, is given by:

$$C_{AEvsi}(t_J) = \sum_1^J(C_{svsi}(t_j) + C_{avsi}(t_j))(T_{vsi}(t_j) - T_{vsi}(t_{j-1})) \tag{8}$$

where c_{psi} and c_{pai} the specific heat of the steam and air calculated at the temperature $T_{vsi}(t_J)$, respectively. The percentage of the injected steam energy that is accumulated as condensation energy in each volume can be obtained from the equation 3). The elaboration of the huge amount of data was performed by implementing in-house programs in Fortran 90: they allowed also to visualize the results by contour bands using the post-processor of a FEM code (MSC-Mentat [8]).

4. Experimental Results and Discussion

Figure 3 provides the results obtained with the sparger B in a condensation test defined by the parameter Q_h/P= 0.171 Kg/(s m^2 kPa). The condensation regime is Chugging (Figure 3-a) that is characterized by an intermittent steam flux.

A large temperature stratification occurred in this test (see Figure 3-b). At the final instant of the test, the water in front of the sparger holes has a temperature of about 59 °C while in the lower zone the temperature ranges between 16-17°C. Figure 3-c) shows the percentage of energy accumulated in the discretized volumes at the final instant, t_f. In this test, the water participating to the steam condensation is a fraction of the whole water pool.

To determine the actual participating water mass, F_{Mw}, all the volumes in which the adsorbed energy is smaller than $10^{-5}E_v(t_f)$ were discarded (considered not participating). This limit, assuming during the test an injection of 3.5 t steam mass, corresponds at a temperature increase of 0.005°C and 0.1°C in the greatest or smallest volume, respectively. Based on that, the participating water mass in the test N.152, shown in Figure 3, was the 57.9 % of the total mass in the ETT.

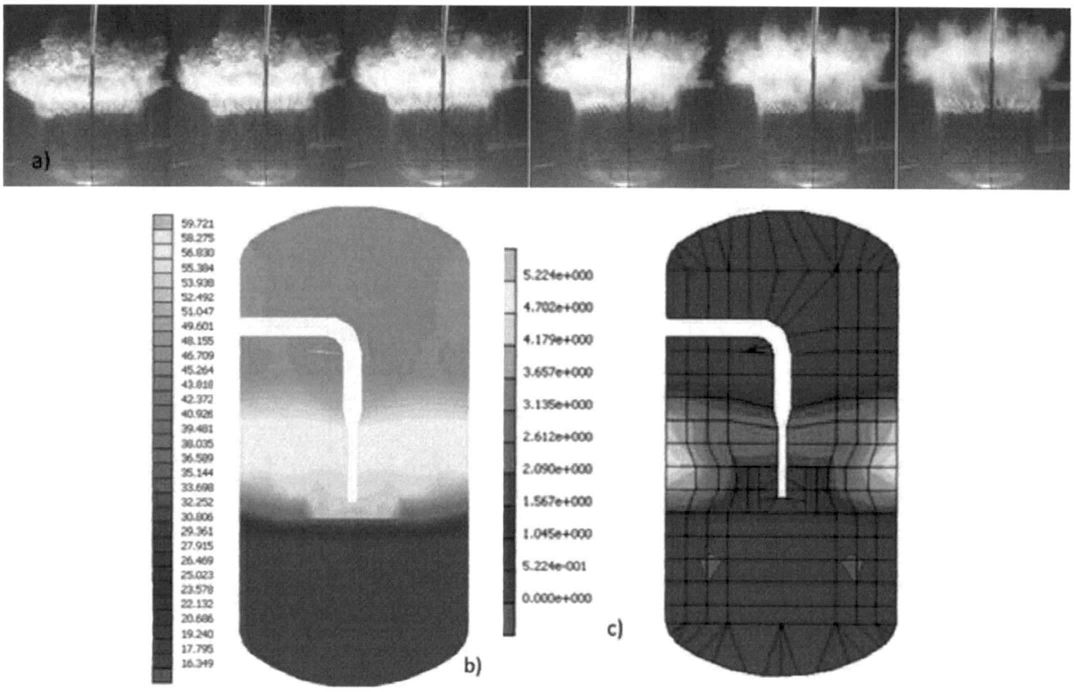

Figure 3. Condensation test of pure steam with the sparger B (test n.152) for Q_h/P=0.171 kg/(sm^{2}kPa): a) steam jet; b) temperature distribution (in °C) in the ETT at t_f; c) percentage of condensation energy accumulated in the discretized water volumes.

Figure 4 provides results of test N.36 of pure steam condensation carried out with the sparger A. The test conditions correspond to the parameter Q_h/P= 1.54 Kg/(s m^2 kPa). Therefore, the condensation regime is the Condensation Oscillation (CO). Figure 4-b) shows the distribution of the temperature in the water pool and in the vacuum space at the final instant of the transient. The temperature is almost uniform. The maximum temperature difference is 5.2°C (T_{min}=36.5 °, T_{max}=41.7°C). In this test condition, a large part of the water participates to the steam condensation as shown in Figure 4-c). On the base of previous assumption, the participating water mass at the steam condensation is the 86 % of the total mass in the ETT. The elaboration of all experimental data permitted to determine the distribution of the participating water fraction, F_{Mw}, versus Q_h/P for both the spargers and for all the test conditions (condensation of pure steam and of the mixture of steam and non-condensable gas in transient and steady state operation).

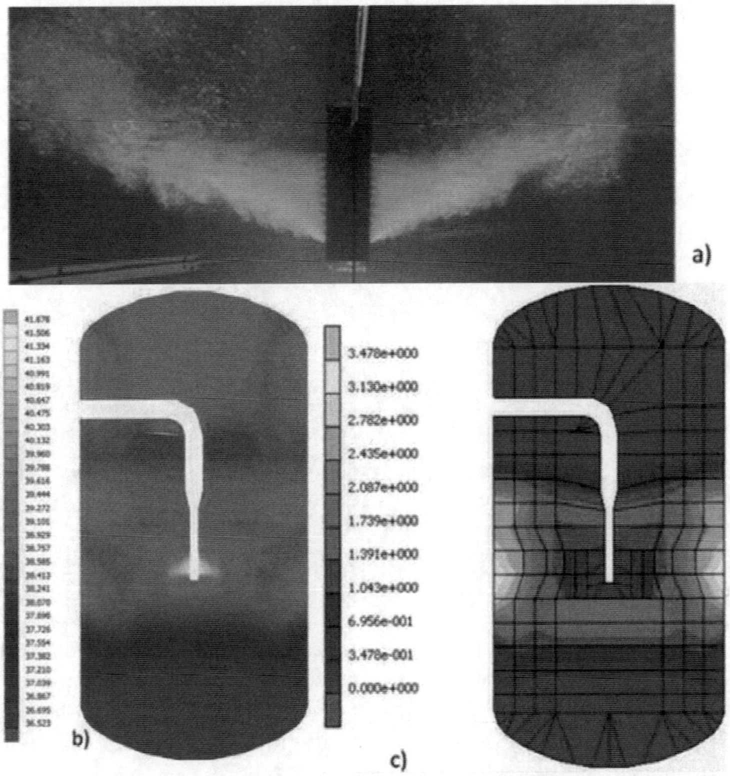

Figure 4. Condensation test of pure steam with the sparger A (test n.36) for Q_h/P= 1.54 kg/(sm^{2}kPa): a) steam jet; b) temperature distribution (in °C) in the ETT at t_f; c) percentage of condensation energy accumulated in the discretized water volumes.

Figure 5 (a) shows the participating water fraction, F_{Mw}, versus Q_h/P, for the tests of pure steam condensation with the sparger A. The orange line represents the minimum participating water fraction at different condensation regimes. This diagram can be considered a quantitative description of the performances of the ETT and of the sparger A in the direct contact condensation of pure steam. For the different CRs, the participating water fraction, F_{Mw} has the following values:

- F_{Mw} ranges between 0.25-1 per Q_h/P <0.2. CRs correspond to smallest values of steam mass flow rate or highest values of downstream pressure (pure steam tests after non condensable gases injection). F_{Mw}=1 is obtained for very low water temperatures. In these conditions, large turbulence is produced by degassing effects. In fact, the water temperature increase reduces the air solubility and air bubbles move vertically in the water towards the vacuum space. F_{Mw} ranges between 0.25 and 0.8 for most of the tests, the minimum values correspond to higher temperatures.
- F_{Mw} ranges between 0.5 and 1 per 0.2< Q_h/P <1.6. Also, for these tests F_{Mw}=1 is obtained at low temperatures and due to degassing effects. Discarding these tests, F_{Mw} ranges between 0.5 and 0.8.
- F_{Mwp}=1 per Q_h/P >1.6. All the water mass participates to the steam condensation. CRs are CO or SC.

Journal of Physics: Conference Series **2685** (2024) 012059 doi:10.1088/1742-6596/2685/1/012059

The determination of the experimental value of the participating water mass to the steam condensation permitted to calculate global parameters to assess the efficiency of the steam condensation, in particular the average temperatures of the actual and participating water mass and the correspondent average increment of the temperature for a given injected steam energy. Figure 5 (b) shows, for the sparger A and steady state tests with pure steam, the experimental data, and the derived correlation between the participating water average temperature, T_{wp}, and the average temperature of the water pool, T_{wave}. A linear best fit of experimental data is given by: $T_{wave} = 0.8877\ T_{wp}$. The average temperature of the water pool, T_{ave}, is 11.23% and smaller than T_{pw} (error bands -30%-+20%).

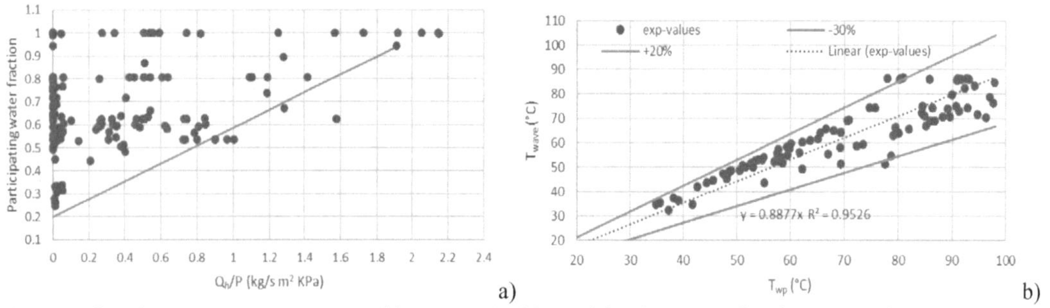

a) b)

Figure 5. Steady state pure steam tests with sparger A: (a) Participating water fraction versus Q_h/P, (b) average temperature (T_{wave}) versus the participating water average temperature T_{wp}.

Figure 6 (a, b) show the water temperature increment versus the steam energy per unit of the actual or participating water mass, respectively. The experimental linear best fits are given by the following formulas:

$$\Delta T\ (°C) = \frac{0.3321 E_v(kJ)}{M_{wa}(Kg)} \qquad\qquad \Delta T\ (°C) = \frac{0.2385 E_v(kJ)}{M_{wp}(Kg)}$$

The coefficient of linear best fit can be interpreted as the inverse of an average effective value of the water specific heat. The average value of the water specific heat in the temperature range 40-100°C is 4.188 kJ/kg°C and the inverse value is 0.2388 Kg °C/kJ. This value is almost equal to the coefficient of the correlation relative to the participating water mass. In the other case (normalizing the steam energy at the actual water mass) the coefficient is 1.39 times greater.

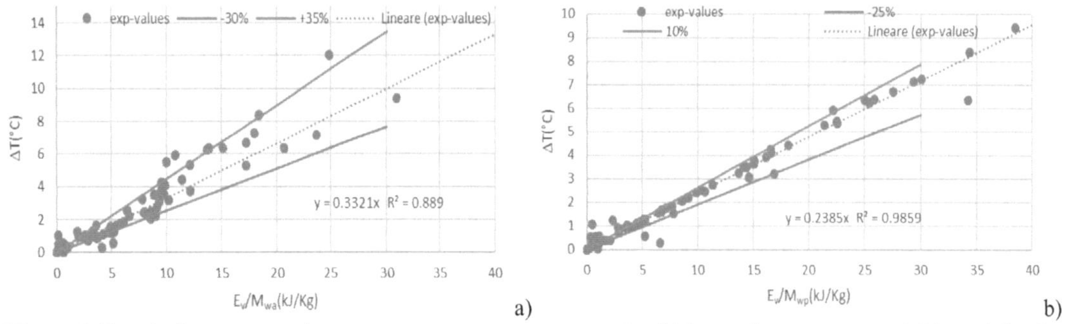

a) b)

Figure 6. Trend of temperature increment vs. steam energy per unit of (a) actual water mass and (b) participating water mass.

This means that fixed the maximum pressure in the vacuum space and, therefore, the maximum temperature of the water, the non-uniformity and stratification of the temperature in the pool required a water mass 1.39 times greater than that that occurring in the case of full participation of the water into the steam condensation.

5. Conclusions

The paper described the implemented model for the analysis of the experimental results obtained in steam direct contact condensation in a closed volume starting from sub-atmospheric pressure.
The measurements of temperature and pressure in the water pool and in the vacuum space permitted to determine the fraction of the water mass in which the condensation energy is accumulated, considering the non-uniformity and stratification of the temperature.
Considering the experimental results obtained in pure steam steady state condensation tests, the provided elaboration model permits to assess the performances of the ETT and the sparger A.
In fact, assuming that the final values of the vacuum space pressure and water temperature in ETT are 1.5 bar and 110°C, respectively, a pool of 60 m^3 of water at 20°C condenses with a temperature increment of ΔT=90°C about 5800 kg of steam. This value is greater than that foreseen in the Large LOCA which could occur in the Vacuum Vessel of ITER.

6. References

[1] Giambartolomei G et Alii, 2020 *Journal of Physics: Conference Series* Volume 1599, doi:10.1088/1742-6596/1599/1/012024
[2] Mazed D et Alii, 2018 *Nucl. Eng. Des.*, **335**(January), pp. 241–254.
[3] Pesetti A et Alii, 2021, *Fusion Eng. Des.*, **171**
[4] Pesetti A et Alii 2021, *Fusion Eng. Des.*, **168**
[5] Aquaro D et Alii 2021, *Nuclear Fusion* Vol. 61, https://doi.org/10.1088/1741-4326/ac2bbe
[6] Sik Kim Y et Alii 1997 *Journal of the Korean Nuclear Society* Vol. 29, Number 1, pp. 45-57
[7] Song C H et Alii, 2012- *Journal Heat Transfer* Vol.134
[8] MSC Marc Mentat 2016 MSC Software Corporation - www.mscsoftware.com

Preliminary thermal-hydraulic deterministic safety analysis of an in-vessel LOCA for the DTT facility

G. Grippo[1*], A. Bersano[1] and F. Mascari[1]

[1] FSN-SICNUC, ENEA C. R. Bologna, Via Martiri di Monte Sole 4, 40129, Bologna, Italy

* e-mail: gianmarco.grippo@enea.it

Abstract. The DTT (Divertor Tokamak Test) facility is a new experimental tokamak under development by an Italian consortium in cooperation with several high-standard European laboratories. The ENEA division FSN-SICNUC is involved in the project for carrying out deterministic safety analysis of postulated accidents. In the first year of activity, it has been analyzed an In-Vessel LOCA (IVLOCA) scenario. The IVLOCA scenario selected is caused by a break in the divertor's cassette cooling tubes with the consequent release of coolant inside the Vacuum Vessel (VV). The best estimate thermal-hydraulic system code TRACE, developed by USNRC, was selected to conduct the preliminary thermal-hydraulic analysis of this accident. DTT is currently under design, so not all the data are frozen. Therefore, based on some engineering assumptions and scaling considerations from similar facilities, the nodalization of DTT VV was developed using the three-dimensional TRACE component "VESSEL". Then, a sensitivity analysis was carried out to simulate the break and the consequent water injection in the VV. This was done to compare the system behavior and test different nodalization approaches. Subsequently, the data of the DTT divertor cooling system were used to run additional simulations. The results allow to compare different nodalization approaches and to have a preliminary estimate of the pressure peak and temperature behavior in the VV for an IVLOCA. Finally, a first uncertainty analysis was carried out using the DAKOTA toolkit, coupled with TRACE code in SNAP. Two uncertain input parameters were selected: the rupture area of a cooling divertor tube and the temperature of the divertor coolant. The uncertainty analysis allows having a wider spectrum of system behavior and a preliminary insight on the dispersion of the VV pressure, selected as figure of merit. This paper aims to show the results of this preliminary analysis, characterizing the phenomena that occurred during the selected transient.

1. Introduction

The DTT (Divertor Tokamak Test) facility is a new Italian 6T, 5.5 MA superconducting tokamak [1]. The aim of the DTT project is to test different divertor configurations and possible solutions for future design. The performance extrapolation of different divertor configurations, based on the physical knowledge obtained thanks to ITER experience, is sufficiently reliable for the implementation of a solution on DEMO. Considering that some plant data have not yet been fully defined, this paper presents a preliminary deterministic safety analysis of a postulated accident in a nuclear fusion reactor such as DTT. Among the possible accidents, the first scenario that has been considered is the In-vessel LOCA [2]. Initially, the postulated initiating events (PIE), which can cause this scenario, have been evaluated: LFV1 (Large First Wall rupture) [3] and LDV1 (In-Vessel LOCA caused by a divertor cassette large rupture) [3]. In this work, it has been selected the LDV1 case; in particular, it has been performed the simulation of a break of the divertor primary cooling circuit, assuming a rupture of the tube apparatus inside the divertor component itself [4].

Therefore, the purpose of this study is to conduct a preliminary thermal-hydraulic analysis of in-vessel LOCA accident, performed with the best estimate thermal-hydraulic system code TRAC/RELAP Advanced Computational Engine (TRACE) developed by USNRC [5,6]. The adopted nodalizations were developed through the Symbolic Nuclear Analysis Package (SNAP) [14,15].

2. Input development

The development of the DTT model for the TRACE code is based on the available geometric and thermohydraulic design data. Considering that the plant is currently under design, some data, necessary to develop the model and qualitatively evaluate the results, are not already present in the current collection of data, summarized in the Safety Data List (SADL) [7]. Therefore, in order to develop this preliminary analysis and test the TRACE model, some data have been extrapolated from the existing experimental Integrated ICE facility [4,8], where the in-vessel LOCA has been investigated. The Integrated ICE facility, built at the Japan Atomic Energy Research Institute (JAERI), is used for the characterization of two-phase flow during an Ingress of Coolant Event (ICE) in fusion reactors; in particular, it simulates a plasma facing component tube rupture. The facility is geometrically scaled 1:1600 from the ITER-FEAT design with prototypical fluid conditions [4]. ICE has a pressure suppression system (Suppression Tank, ST) connected with the VV through a Relief Pipe (RP). The suppression system is activated by a magnetic valve (MV), installed on the RP [8].The Integrated ICE facility is equipped with three injection nozzles, from a boiler to the Plasma Chamber (PC), to simulate the rupture of the coolant piping and the consequent water injection in the PC. The chosen ICE experiment, considered to test the DTT model, is case 4, in which 2 out of 3 nozzles are in function: the main thermal-hydraulic parameters of the test are summarized in Table 1 [8].

Table 1. ICE case 4 main thermohydraulic parameters [8]

Parameter	Unit	Value
Initial wall temperature (PC, VV)	*K*	503.15
Initial wall pressure (PC, VV)	*bar*	0.01
Water temperature injected from the boiler	*K*	423.15
Boiler pressure	*bar*	20
Injection time	*s*	45

2.1. Geometrical and thermohydraulic data

Since some in-vessel components' geometrical data are currently under design, the most important parameters on which the TRACE nodalization is based are the plasma side VV volume and, consequently, single torus geometrical quantities [1,7,9,10].

Regarding the thermal-hydraulic data of DTT, the considered ones are reported in **Table 2** [11].

Table 2. DTT thermohydraulic parameters [7,9]

Quantity	Unit	Value
Vacuum Vessel nominal pressure	*Pa*	5.0E-06
Divertor cooling fluid temperature	°*C*	30 - 130
Divertor cooling water volume	m^3	30

2.2. Vacuum Vessel modelling

In order to develop the VV model, it was used the 3D TRACE component "VESSEL" [5,6]. The nodalization has been created (Figure 1) to fit the first radius of the grid with the internal one of the VV structure, while the last radius of the grid with the most external one of the VV. Seven radial rings have been created in total. In relation to the axial 3D nodalization, ten axial levels have been used, distributed with higher density on the top and on the bottom of the VV to find a compromise between the geometric detail and the calculation time. In addition, the coolant injection to simulate the rupture of the divertor cassette occurs at the bottom of the PC [4]. Regarding the azimuthal coordinate, four azimuthal sectors have been defined. The flow area fractions and the volume fractions have been settled to define the internal geometry of the VV and replicate the characteristic "D shape"; in particular, the volume fraction has been settled to zero where the cell was not comprehensive of a VV volume portion as well as the flow area fraction for the boundary [5,6]. About the geometrical data of the PC and the plasma facing components, it has been done a volume scaling from the Integrated ICE facility data [8]. To simulate the PC, the TRACE model flow areas have been settled to define the VV internal region representative of the DTT PC; in addition, the cell R4A4 (radial mesh 4, axial mesh 4) has a non-zero lower axial flow area (Figure 2), to simulate the PC area in which is assumed the divertor. In fact, the nodalization has been done to recreate the position in which is located the divertor and, consequently, where the rupture of the coolant piping takes place.

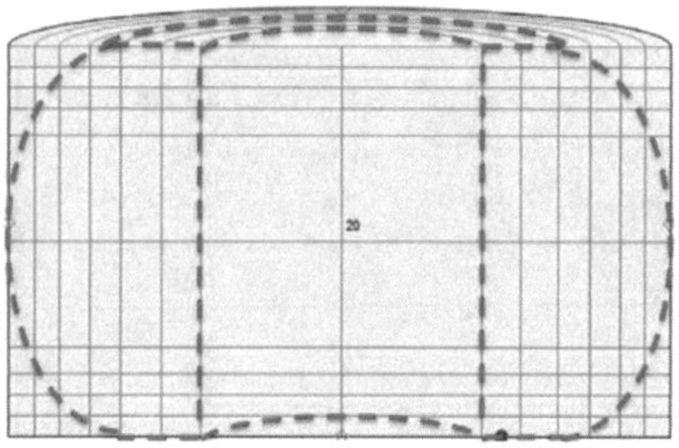

Figure 1. Scheme of VV TRACE nodalization developed through SNAP

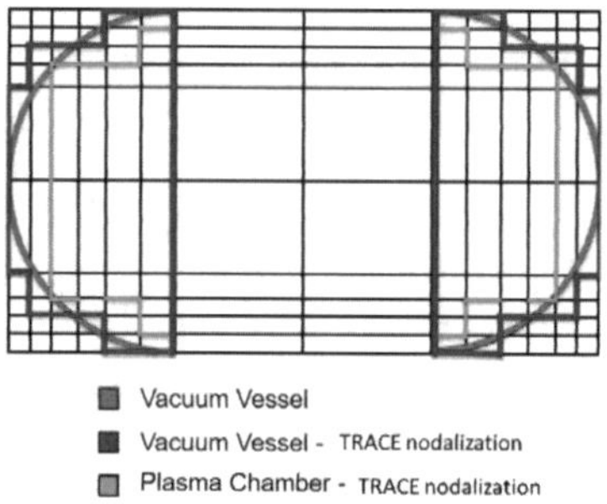

Figure 2. PC and VV discretization in the TRACE 3D vessel nodalization

2.3. System injection – Sensitivity analysis

After the development of the TRACE VV and PC nodalization (Figure 1, Figure 2), it was needed to properly simulate the water inlet due to the rupture of the divertor cooling tubes. To simulate the release of coolant inside the VV, several injection models were created, initially using the Integrated ICE thermal-hydraulic data (Table 1) to evaluate the code results.

Fill model – Mass flow ramp

The adoption of the TRACE FILL injection [5,6] allows the definition of the mass flow rate and the thermal-hydraulic boundary conditions. For this model, the ICE experimental injection curve [8] is settled by the FILL component, properly scaled from the ICE data.

BREAK model

The TRACE BREAK component [5,6] is normally used to settle the boundary thermal-hydraulic conditions specifying the inlet/outlet fluid temperature and pressure. The BREAK is connected to a valve to reproduce the same ICE injection time (45 s) [8]. The difference between the current model and the previous one is that the boundary condition implemented is the BREAK constant pressure instead of the injection mass flow rate (Table 1). The valve that provides the BREAK connection to the VV has a hydraulic diameter that represents the rupture of the coolant pipe, properly scaled from the Integrated ICE data considering the presence of two active nozzles in the test (ICE case 4 [8]). The obtained value has been applied to all the models derived from the Integrated ICE facility with an injection valve.

BREAK model – Pressure ramp

The BREAK component allows also the setting of a time-dependent pressure trend. Therefore, it has been created a model in which is implemented the ICE injection pressure curve [8].

Accumulator model

The water volume inside the divertor cooling circuit is a very important data for the simulation of an IVLOCA (Table 2). Therefore, it has been created also a model with a volume containing the actual amount of cooling water. It does not act as a FILL and BREAK components but simulates the water injection inside the VV without specific predefined behaviours. To fulfil the requirements, it has been used the TRACE PIPE component [5,6] modelled considering the divertor cooling water volume data (Table 2). Initially it has been created a pipe full of liquid water with the ICE boiler thermal-hydraulic conditions. The moment in which the connecting valve opens, both VV and the pipe (Injection Tank, IT) are subject to a strong pressure difference. During the water discharge from the IT to the VV, the pressure equalizes causing the stop of the flow through the valve. To simulate the discharge of a larger water mass in the VV, it has been adopted, for the IT, the accumulator configuration: 75% of the total volume occupied by the coolant and 25% of the total volume filled with non-condensable gases [12]. In thus way during the water discharge the non-condensable gasses expand permitting the fluid flow.

2.3.1. *Results of the injection sensitivity analysis*

One of the aims of this work is to verify if DTT plant needs a pression suppression system. Since the presented results are referred to preliminary analysis, the target of this study is to make some qualitative and phenomenological consideration on the DTT behaviors in transient conditions and all the data are presented in adimensional form. When the design data of the DTT will be consolidate quantitative evaluation on the pressure peak will be done. Figure 3 to Figure 5 show the results of the sensitivity analysis on the different injection options considered. The pressurized water flashes at the entrance of the VV at lower pressure and enters in contact with the hotter walls of the VV, increasing the VV pressure (Figure 3). Figure 4 shows the pressure peak occurring at the end of the injection phase, the accumulator model has the lower peak. The entrance of pressurized cold water in the VV brings to a sudden drop of the temperatures in the VV atmosphere (Figure 5), that then rises again, following the saturation value. After the initial pressure peak at the end of the injection the pressure rises. Considering the results of this analysis on the various injection options, for the DTT simulation it has been decided to use the accumulator model (Figure 6). This option has been selected since the accumulator model provides the more conservative results, with the final highest PC pressure. In addition, this option simulates the water injection inside the VV without specific pre-imposed code component behaviours (e.g. fixed pressure or mass flow rate).

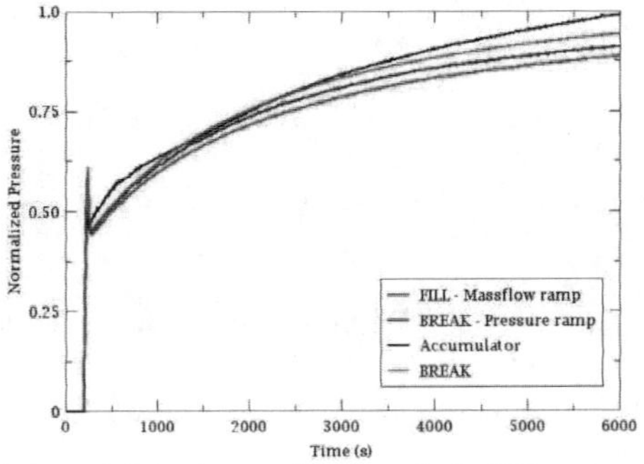

Figure 3. Injection sensitivity analysis, VV pressure

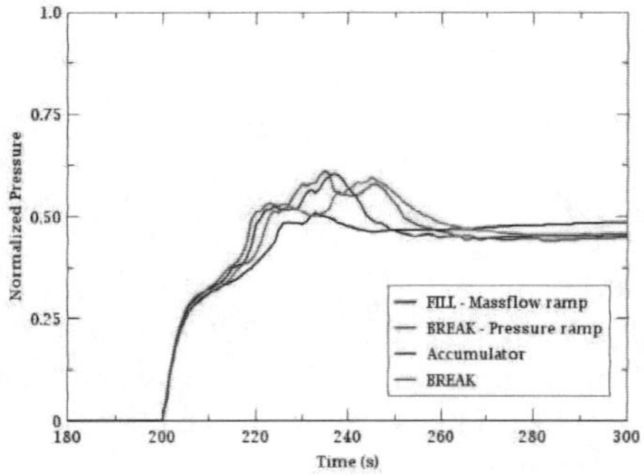

Figure 4. Injection sensitivity analysis, initial VV pressure peak

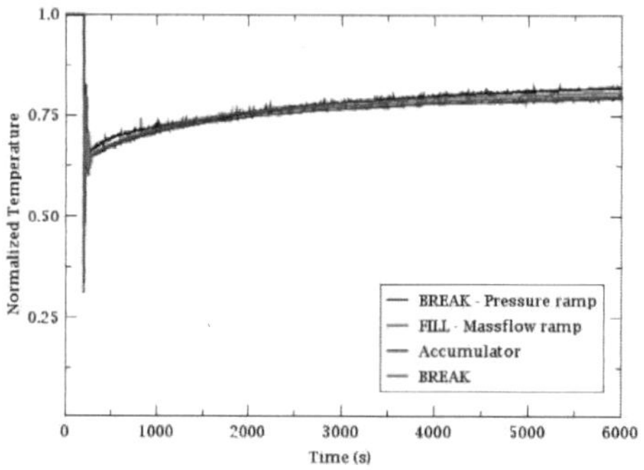

Figure 5. Injection sensitivity analysis, VV temperature

2.4. *DTT model*

To simulate the DTT conditions, only the thermal-hydraulic data have been changed since the IT has been modelled considering the water volume of the DTT divertor cooling system (Table 2). The pressure present in the Safety Data List [7] could be implemented since TRACE code does not allow systems simulation with pressure below $1e^{-3}\ Pa$ [6,13]. Therefore, as first approximation, it has been settled the initial pressure condition of the ICE facility (Table 1). For the cooling system thermal-hydraulic data, it has been considered the highest temperature case (Table 2): 403.15 K. Regarding the injection valve, it remains opened after the start of transient and the pressure difference between the VV and the IT regulates the mass flow and the amount of injected coolant. The valve flow area has been set to simulate the complete rupture of a cooling divertor pipe.

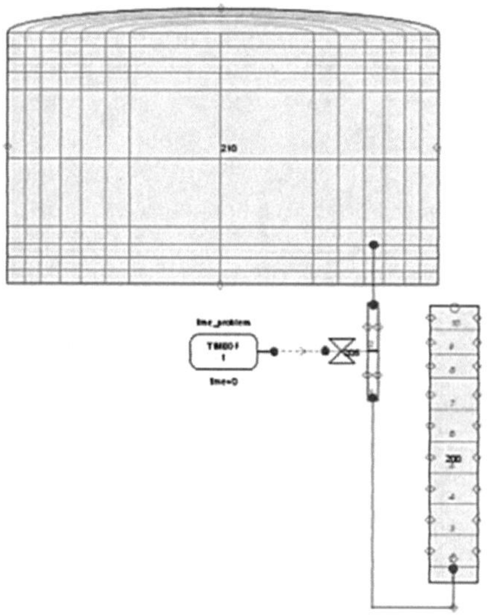

Figure 6. DTT nodalization developed through SNAP

3. Results

The simulations performed with ICE conditions have been used to test different options for the coolant liquid entrance in the VV and observe the behaviour predicted by the TRACE code. Considering the obtained results, it has been selected for the DTT model the PIPE component (Figure 6) modelled as an accumulator (IT).

Figure 7 shows the VV pressure behaviour along the simulation. At the beginning of the transient, when the valve opens, the VV pressure rises rapidly up to 500 s; then, there is a slower increase of the pressure up to the final maximum value. Figure 8 shows the pressure equalization between the VV and the IT after the opening of the connection valve. The IT pressure sharply decreases at the break opening and the equalization is reached at around 400 s; then, the two pressure trends are the same with a slight increase in the remaining part of the transient (Figure 8).

Regarding the VV temperature, at the beginning of the test, when the valve opens, the inlet of the coolant from the IT to the hotter VV environment causes a sudden decrease of the VV (Figure 9). Then the temperature slightly increases following the saturation value.

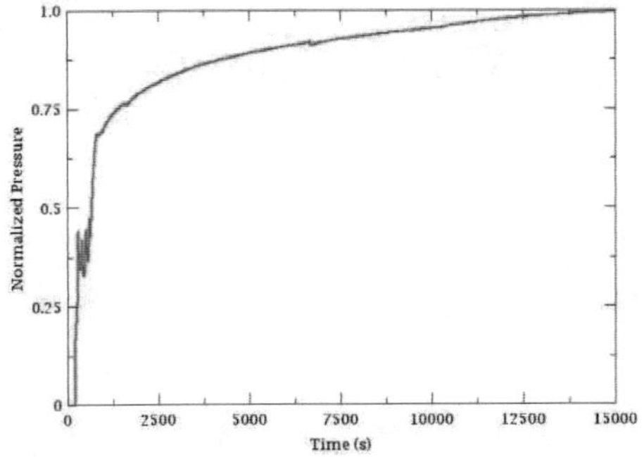

Figure 7. DTT nodalization results, VV pressure

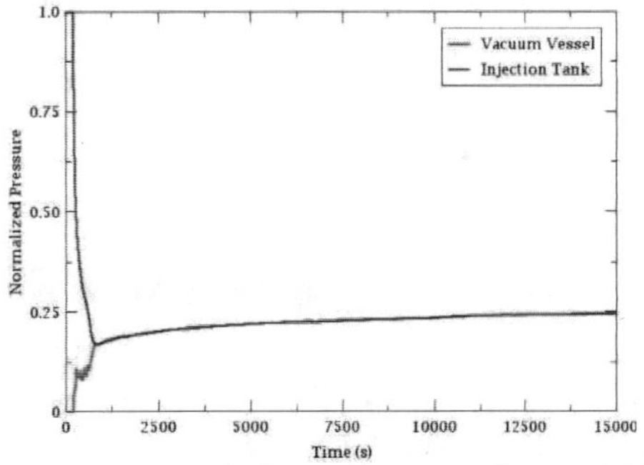

Figure 8. DTT nodalization results, accumulator model VV and IT pressure equalization

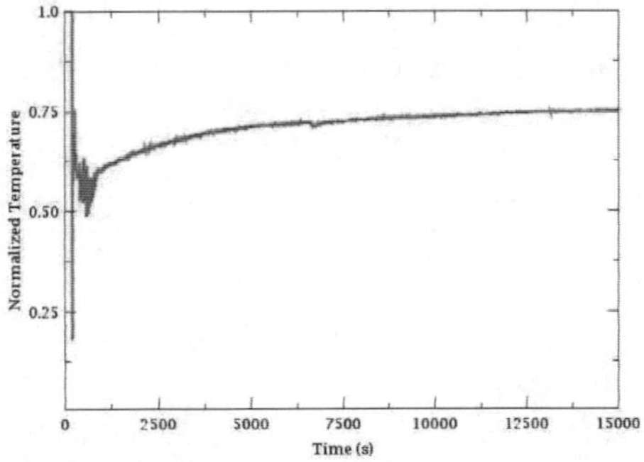

Figure 9. DTT nodalization results, VV temperature

3.1. *Uncertainty analysis*

Since some data of the DTT facility are still under design, an Uncertainty Analysis (UA) has been performed applying the probabilistic propagation of input uncertainties method. The UA has been carried out with the Design Analysis Kit for Optimization and Terascale Application (DAKOTA) toolkit, coupled with TRACE code in SNAP [14,15]. More information on DAKOTA can be found in references [16,17]. In the first instance, two Figures Of Merit (FOM) have been selected: the VV pressure and the VV temperature. Subsequently, two uncertain input parameters have been chosen: the divertor cooling pipe rupture area and water temperature.

Table 3. Uncertainty analysis input parameters

Parameter	Min value	Max Value	Distribution
Pipe break flow area [$\mathbf{m^2}$]	4.05366e-4	2.02683e-3	Uniform
DIV Cooling temperature [K]	303.15	403.15	Uniform

Table 3 shows the assumed variation range of both uncertain input parameters with a uniform distribution. Regarding the pipe rupture area, the maximum value corresponds to the complete break of the pipe (100% pipe flow area), while the minimum is equal to 20% pipe's total flow area. Considering the described range, it represents a large spectrum that covers different possible rupture in the system. Related to the divertor coolant temperature, it is presented in the Safety Data List as a range [7], therefore the UA has been performed using this range. In order to perform the UA, the probabilistic propagation of input uncertainties method has been used; considering 2 FOM, a probability content and a confidence level of 95%, 93 calculations have been required for the one-sided tolerance interval [17].

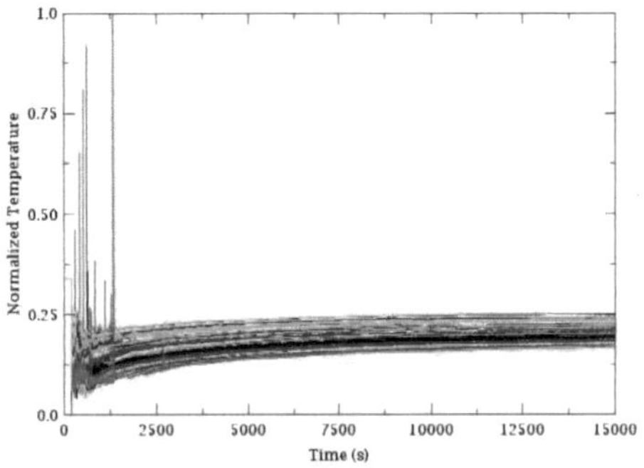

Figure 10. Uncertainty analysis results, VV temperature

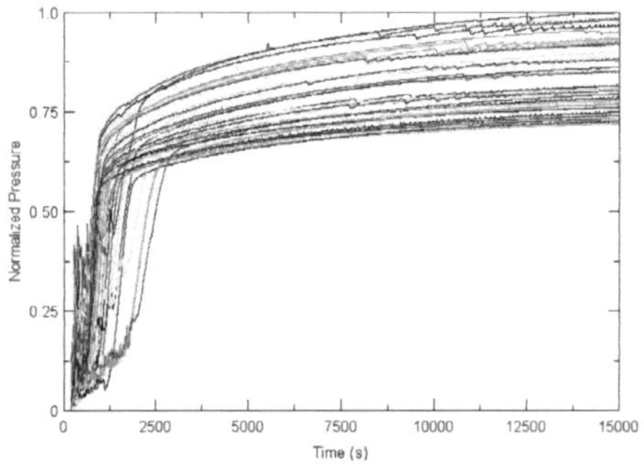

Figure 11. Uncertainty analysis results, VV pressure

To characterize the correlation between the uncertain input parameters and the FOMs, in the present work it has been considered the Pearson coefficient and the Spearman coefficient. The Pearson correlation coefficient, calculated between an input and an output variable, represents the linear correlation between the two variables (Table 4), while the Spearman one represents the monotonic correlation [8,16].

Table 4. Correlation coefficients results

	VV Temperature		*VV Pressure*	
	Pearson	*Spearman*	*Pearson*	*Spearman*
Pipe break flow area	-0.31	-0.39	0.67	0.96
DIV Cooling temperature	-0.02	0.07	0.01	-0.02

At the beginning of simulation, the VV pressure (Figure 11) rapidly increases, with a results dispersion width reaching around 0.75 (normalized pressure); then, at the end of the simulation, there is a results dispersion between around 0.75 and 1.0. Observing the correlation coefficients for the VV pressure FOM (Table 4), the pipe break flow area presents a positive correlation, while the divertor cooling water temperature is very low. In particular, the rupture area presents a significant correlation coefficient (i.e. >0.5) while the cooling temperature has a low correlation (<0.2) according to the threshold values used in [8]. Considering the VV temperature (Figure 10), some simulations show initially high peak values; then, the results dispersion width is almost constant with final values which range between around 0.125 and 0.25 (normalized temperature). The correlation coefficients (Table 4) are both negative for the pipe break flow area (moderate correlation according to the threshold values used in [8]) while the correlation is low for the divertor cooling temperature.

4. Conclusions

In previous activities of the authors, the thermal-hydraulic best estimate system code TRACE has been validated for fusion applications against the data developed in the Integrated ICE facility. Starting from that base, this work presents a preliminary analysis of an accident in the DTT facility caused by a break in the divertor cooling circuit, with consequently release of coolant inside the VV. The analysis was carried out with the TRACE code, adopting a 3D modeling of the VV and PC. The obtained results provide, in first approximation, the trend of pressure and temperature in VV during the accident. Then an uncertainty analysis was performed with the probabilistic propagation of input uncertainty method varying two input parameters, which were uncertain in the reference calculation. However, as mentioned above, these results are to be considered preliminary because some data of the facility are still under design. In future studies, the model will be updated considering: the divertor cooling circuit with updated geometrical dimensions following the finalized design, and several systems concerning the facility (e.g First Wall cooling circuit,...) exploring different scenarios and transients; finally, the input deck assumption will be substituted with all the final thermal hydraulic data in order to reduce the uncertainty on the results (specifically to the VV pressure peak at the beginning of the transient).

5. References

[1] E. R. Martone *et al.*, «DTT Divertor Tokamak Test facility – Interim Design Report».
[2] R. Caporali *et al.*, «Parametric analysis of an in-vessel LOCA for the ITER plant», in *Proceedings of 16th International Symposium on Fusion Engineering*, Champaign, IL, USA1995
[3] T. Pinna *et al.*, «Identification of accident sequences for the DEMO plant», *Fusion Eng. Des.*, 2017.
[4] K. Takase *et al.*, «Experimental verification of effectiveness of integrated pressure suppression systems in fusion reactors during in-vessel loss of coolant events», *Nucl. Fusion*, 2001.
[5] Division of Safety Analysis Office of Nuclear Regulatory Research U. S. Nuclear Regulatory Commission, «TRACE V5.0 PATCH 6 USER'S MANUAL Volume 1: Input Specification».
[6] Division of Safety Analysis Office of Nuclear Regulatory Research U. S. Nuclear Regulatory Commission, «TRACE V5.0 PATCH 6 USER'S MANUAL Volume 2: Modeling Guidelines».
[7] Samuele Meschini *et al.*, «Safety Data List».
[8] A. Bersano *et al.*, «Ingress of Coolant Event simulation with TRACE code with accuracy evaluation and coupled DAKOTA Uncertainty Analysis», *Fusion Eng. Des.*, 2020.
[9] DTT, «Plant Integration Document v3.2». 2022.
[10] M. Dalla Palma, «DTT Vacuum Vessel». Industrial Opportunity Days, Napoli 2022.
[11] E. Martelli *et al.*, «Design status of the Vacuum Vessel of DTT facility», *Fusion Eng. Des.*, 2021.
[12] I. A. E. Agency, «Passive Safety Systems and Natural Circulation in Water Cooled Nuclear Power Plants», International Atomic Energy Agency, Text, 2009.
[13] Division of Safety Analysis Office of Nuclear Regulatory Research U. S. Nuclear Regulatory Commission, «TRACE V5.0 PATCH 6 THEORY MANUAL: Field Equations, Solution Methods, and Physical Models».
[14] Applied Programming Technology, Inc, «Symbolic Nuclear Analysis Package (SNAP): User's Manual».
[15] K. Jones *et al.*, «TRACE Plug-in Users Manual».
[16] S. S. Wilks, «Statistical Prediction with Special Reference to the Problem of Tolerance Limits», *Ann. Math. Stat.*, 1942.
[17] A. Guba *et al.*, «Statistical aspects of best estimate method-I», *Reliab. Eng. Syst. Saf.*, vol. 80, fasc. 3, pp. 217–232, 2003.

Journal of Physics: Conference Series **2685** (2024) 012061 doi:10.1088/1742-6596/2685/1/012061

Two Phase Bubble Columns: the Determinants of the Flow Regime Transitions

N Varallo[1*], G Besagni[1], R Mereu[1] and F Inzoli[1]

[1] Politecnico di Milano, Department of Energy, Via Lambruschini 4, Milan 20156, Italy

* Corresponding author: nicolo.varallo@polimi.it

Abstract. The fluid dynamics in large-diameter bubble columns can be described by an analytical relation between two global flow parameters, the drift flux and the gas holdup. This relation, named bubble column operating curve, builds on five flow regime transitions. In order to determine the variables influencing the flow regime transitions, a statistical approach was derived by coupling: (1) the ordinary least squares method (OLS) to determine the relationship between the variables, (2) the variance inflation factor (VIF) to check for multicollinearity issues, and (3) the least absolute shrinkage and selection operator (LASSO), to select suitable variables. It was found that the geometrical characteristics of the sparger strongly influence the flow regime transitions, and uniform aeration is essential for all the regimes to exist. Increasing the superficial liquid velocity in the counter-current mode destabilises the mono-dispersed and poly-dispersed homogeneous flow regimes. As for the aspect ratio, an increase in the column aspect ratio slightly destabilises the existing flow regimes. The statistical method identifies viscosity as the only significative variable concerning the liquid phase properties.

1. Introduction

Bubble columns are gas-liquid reactors widely used in many industrial applications. They are of considerable interest in chemical processes involving reactions like oxidation, chlorination, alkylation, polymerization, and hydrogenation, as well as in the production of synthetic fuels and biochemical processes such as fermentation and biological wastewater treatment. Bubble columns offer several advantages, such as excellent heat and mass transfer between the phases, high durability, and low operating and maintenance costs due to the absence of moving parts [1].

In its simplest configuration, a bubble column consists of a vertical vessel without internals where a gas distributor disperses the gas phase into bubbles or coalescence-induced structures entering the column. The liquid flow rate can be fed co-currently or counter-currently to the rising bubbles, or it may be zero (batch operating mode). Despite the simple column layout, bubble column hydrodynamics is very complex due to the interactions between the continuous and the dispersed phases. The coupling between the phases physically manifests in the flow regimes.

Besagni (2021) [2] proposed a novel physical-based theory for the description of the fluid dynamics in large-diameter bubble columns (bubble column diameter greater than 0.15m at ambient operating conditions). The theory states that the fluid dynamics in large-diameter systems can be described by an analytical relation (bubble column operating curve) between two global flow parameters: the drift flux and the gas holdup. The drift flux, J, is defined as the volumetric flux of either component relative to a surface moving at volumetric average velocity expressed as follows.

$$J = U_G(1 - \varepsilon_G) \pm U_L\varepsilon_G \tag{1}$$

In Equation (1), ε_G is the global gas holdup, U_G is the superficial gas velocity, and U_L is the superficial liquid velocity. The sign on the right-hand side of Equation (1) depends on the bubble column operating mode: co-current mode (+) or counter-current mode (-); in the batch mode $U_L = 0$.

The bubble column operating curve (Figure 1) builds on five flow regime transitions. Consequently, six flow regimes can be encountered in large-diameter bubble columns. The six flow regimes emerging upon an increase in the gas flow rate at fixed system design parameters, operating mode, and phases properties are: (1) mono-dispersed homogeneous flow regime, (2) poly-dispersed homogeneous flow regime, (3) transition flow regime without coalescence-induced structures, (4) transition flow regime with coalescence-induced structures, (5) pseudo-heterogeneous flow regime, and (6) pure-heterogeneous flow regime (Figure 1).

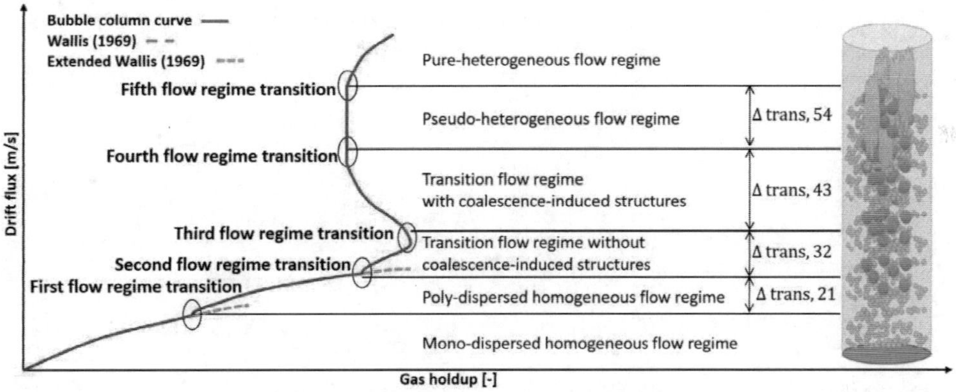

Figure 1. Flow regime and flow regime transitions in a large diameter bubble column [2]. The symbol Δ indicates the difference in drift flux and gas holdup between a flow regime transition and the previous one.

Changing the system design and/or the phase properties and/or the operating mode induces a change in the boundaries between the flow regimes without influencing the flow regime properties themselves [2]. Consequently, the bubble column fluid dynamics can be described and predicted a-priori via correctly modelling the flow regime transitions.

This paper is organised as follows. Section 2 describes the variables used to conduct the statistical analysis. Section 3 presents the statistical approach. The results obtained are presented and discussed in Section 4. Finally, conclusions are drawn, and future studies are proposed.

2. Dependent Variables and Predictors

As for the dependent variables, the flow regime theory [2] was applied to determine the flow regime transitions coordinates ($\varepsilon_{G,\mathrm{trans}}$ and $U_{G,\mathrm{trans}}$) of different large-diameter bubble columns operating in batch and counter-current mode with different gas distributors, aspect ratio values, and liquid phase properties (Table 1).

From a practical perspective the flow regimes transitions can be identified as follows.

- **First flow regime transition.** The transition can be detected by comparing the experimental drift flux values (J_E) with the theoretical drift flux curve (J_T) proposed by Wallis (1969) [3], written in terms of bubble swarm velocity (v_b):

$$J_T = v_b(1 - \varepsilon_G) \tag{2}$$

The evaluation of the bubble swarm velocity is a matter of discussion; in this study, the approach proposed by Krishna et al. (1999) [4] is considered:

$$v_b = U_t(1 - \varepsilon_G)^n \tag{3}$$

Where n is a fluid-dependent variable ($n \approx 2$ for water-based solutions) and U_t is the terminal velocity of an isolated bubble obtained fitting the experimental data. In the mono-dispersed homogeneous flow regime J_E equals J_T and the transition to the poly-dispersed homogeneous flow regime occurs when:

$$J_E \neq J_T \tag{4}$$

- **Second flow regime transition**. The flow regime theory states that the poly-dispersed homogeneous flow regime builds upon the mono-dispersed one. Consequently, the approach previously described can be applied to determine the second flow regime transition after a correct change in the variables (the first flow regime transition must be considered as the new origin):

$$\varepsilon_G' = \varepsilon_G - \varepsilon_{G,\text{trans},1} \tag{5}$$

$$U_G' = U_G - U_{G,\text{trans},1} \tag{6}$$

Therefore, the bubble column drift flux referred to the new coordinates is:

$$J_E' = U_G'(1 - \varepsilon_G') \pm U_L \varepsilon_G' \tag{7}$$

Considering a new theoretical drift flux curve (J_T'), the transition point can be detected when:

$$J_E' \neq J_T' \tag{8}$$

- **Third flow regime transition.** The third flow regime transition coincides with the maximum J_E with respect to ε_G:

$$\varepsilon_{G,\text{trans},3} = \max\left(\varepsilon_G(J_E)\right) \tag{9}$$

- **Fourth flow regime transition**. When the transition flow regime with coalescence-induced structures is established, an increase in the gas flow rate (in terms of drift flux) is followed by a decrease in the gas holdup. On the contrary, the pseudo heterogeneous flow regime is characterized by a constant value of gas holdup with respect to an increase in the gas flow rate. Consequently, the fourth flow regime transition can be determined based on the constant value of ε_G at high drift flux values:

$$\frac{\partial \varepsilon_G(J_E)}{\partial J_E} = 0 \tag{10}$$

- **Fifth flow regime transition**. The fifth flow regime transition is observed when the gas holdup begins to increase again due to an increment in the drift flux:

$$\varepsilon_G(J_E) \neq \varepsilon_{G,\text{trans},4} \tag{11}$$

Given the flow regime transitions coordinates, only the first flow regime transition was considered in "*absolute terms*" (i.e. $\varepsilon_{G,\text{trans},1}$ and $U_{G,\text{trans},1}$) since the others and the corresponding dependent variables were defined relative to the previous transition (i.e. $\Delta\varepsilon_{G,\text{trans},ij} = \varepsilon_{G,\text{trans},i} - \varepsilon_{G,\text{trans},j}$ and $\Delta U_{G,\text{trans,ij}} = U_{G,\text{trans},i} - U_{G,\text{trans},j}$, with $j = i - 1$ and $i = 2 \ldots 5$). In this way, if a flow regime does not exist, $\Delta\varepsilon_{G,\text{trans}}$ and $\Delta U_{G,\text{trans}}$ equal zero; they are different from zero otherwise. $\Delta\varepsilon_{G,\text{trans},54}$ is useless to describe the transition to the pure heterogeneous flow regime [2] and so only $\Delta U_{G,\text{trans},54}$ was considered.

Concerning the predictors (i.e., independent variables), they were divided into three different macro-categories: (1) geometrical characteristics of the column, (2) operating conditions, and (3) liquid phase properties (Table 2).

Table 1. Bubble column configurations analysed.

Ref.	Liquid phase	D_H [m]	AR [-]	Sparger	d_0 [mm]	Operating conditions
[6]	Water	0.2	1	Perforated plate	1	Open tube; batch
[7]	Water	0.158	0.105	Perforated plate	2	Open tube; batch
[8]	Water	0.156	0.125	Perforated plate	0.5	Open tube; batch
[9]	Water	0.29 – 0.4	0.34 → 4.14	Perforated plate	0.5	Open tube; batch
[10]	Water	0.15 – 0.4	10.67	Perforated plate	0.5	Open tube; batch
[11]	Water	0.29	0.176	Perforated plate	0.5	Open tube; batch
[12]	Water	0.15 – 0.38	2.857	Perforated plate	0.5	Open tube; batch
[13]	$CaCl_2$ solution	0.15	2.857	Perforated plate	0.5	Open tube; batch
[2]	Water	0.24	1 → 15	Perforated plate	0.5 - 1	Open tube; batch
[2]	Water	0.24	1 → 15	Needle	0.5	Open tube; batch
[2]	Water	0.24	1 → 15	Spider	2 → 4	Open tube; batch
[2]	Water	0.24	12.5	Pipe	3	Open tube – annular gap; Batch – counter-current
[2]	NaCl solution	0.24	5 → 10	Spider	2 → 4	Open tube; batch
[2]	EtOH solution	0.24	5 → 12.5	Spider	2 → 4	Open tube; batch
[2]	MEG solution	0.23	5 → 12.5	Spider	2 → 4	Open tube; batch

Table 2. Independent variables.

Column geometrical characteristics	Operating conditions	Liquid phase properties
Column diameter (D_H)	Superficial liquid velocity (U_L)	Viscosity (μ)
Sparger type	Presence of internals	Density (ρ)
Sparger hole diameter (d_0)		Surface tension (σ)
Free area[a] (A_f)		
Aspect ratio (AR)		

[a] The free area is the ratio between the total area of the sparger openings and the cross-sectional area of the column.

3. Statistical Methods

Following the approach proposed by Varallo et al. (2023) [14], the statistical method consists of OLS to determine the relations between dependent and independent variables, VIF to check for multicollinearity issues, and LASSO to select the significative predictors in describing the flow regime transitions. Subsequently, a segmentation of bubble columns with similar flow regime transitions coordinates is proposed, by using a CART approach.

The statistical approach consists of five steps (Figure 2a). In the first step, a dependent variable is selected out of $\varepsilon_{\text{trans},1}$, $U_{G,\text{trans},1}$, $\Delta\varepsilon_{G,\text{trans,ij}}$ and $\Delta U_{G,\text{trans,ij}}$. In the second step, a class of predictors is selected out of the three groups listed in Section 2. In the third step, the regression procedure is applied to the selected dependent variable and the selected class of predictors ("*partial regression models"*). Subsequently, the above-described steps (#1 to #3) are iterated until all the dependent variables have been considered and all the "*partial regression models"* have been obtained. In the fourth step the predictors found significant after the regression procedure are coupled, and the regression analysis is performed again to obtain the "*aggregate regression models*". Steps 2-3 are iterated until all the classes of predictors have been evaluated with respect to the dependent variables and all the "*aggregate regression models*" have been obtained. The OLS-VIF-LASSO procedure applied at step#3 and step#4 is shown in Figure 2b.

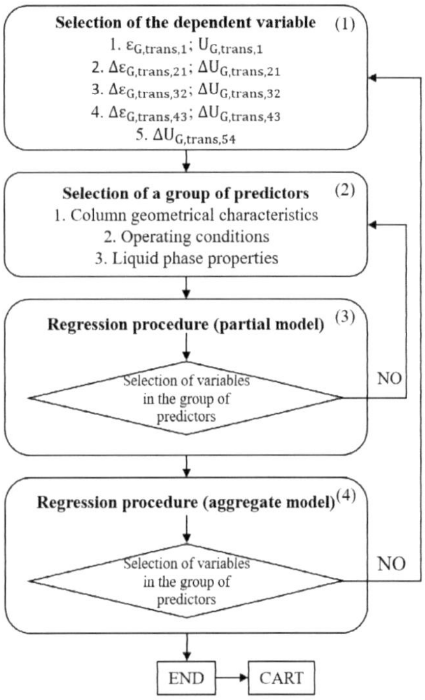

Figure 2a. Procedure overview.

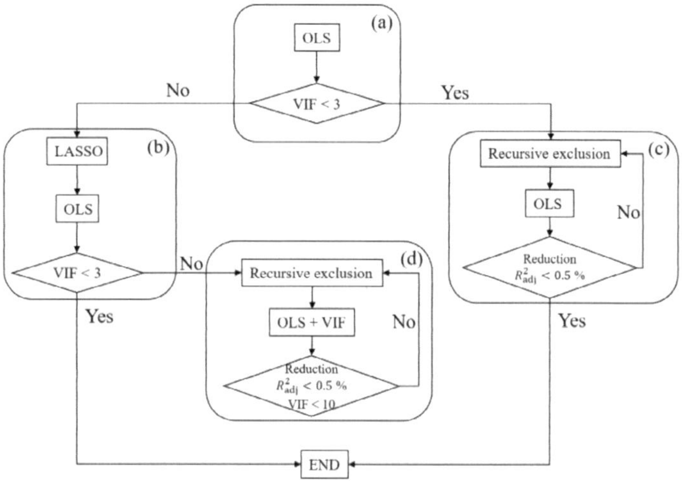

Figure 2b. OLS-VIF-LASSO procedure overview.

4. Results and Discussion

Table 3 and Table 4 display the adjust coefficient of determination (R^2_{adj}) for the partial and aggregate regression models. The geometrical characteristics of the column mainly influence the flow regime transitions, expect the second flow regime transition, where the influence of the operating conditions is more significant than the column geometrical characteristics. The liquid phase properties primarily influence $U_{G,\text{trans},1}$ ($R^2_{\text{adj}} = 14.28\ \%$) rather than $\varepsilon_{G,\text{trans},1}$ ($R^2_{\text{adj}} = 2.94\ \%$). In contrast, it does not explain the variance of the second transition ($R^2_{\text{adj}} = 0.62\ \%$ and $R^2_{\text{adj}} = 0.63\ \%$ for $\Delta\varepsilon_{G,\text{trans},21}$ and $\Delta U_{G,\text{trans},21}$, respectively). The partial regression models referred to the operating conditions and liquid phase properties for the third, fourth and fifth flow regime transitions show a negative or zero value of R^2_{adj}. The reason is that these transitions are present only when the column operates with the perforated plate and needle spargers, which provide uniform aeration. In the experimental data analysed to create the dataset, the perforated plate sparger was tested only in batch mode with water as the liquid phase (Table 1). Therefore, the influence of the operating conditions and liquid phase properties on the third and fourth flow regime transitions cannot be analysed.

Table 3. Coefficient of determinations: gas holdup.

R^2_{adj}	$\varepsilon_{G,trans,1}$	$\Delta\varepsilon_{G,trans,21}$	$\Delta\varepsilon_{G,trans,32}$	$\Delta\varepsilon_{G,trans,43}$
Geometrical characteristics	56.75 %	29.64 %	34.49 %	52.41 %
Operating conditions	8.34 %	22.50 %	-0.06 %	0.00 %
Liquid phase properties	2.94 %	0.62 %	-3.00 %	-2.72 %
Aggregated models	60.63 %	34.87 %	34.49 %	52.41 %

Table 4. Coefficient of determinations: superficial gas velocity.

R^2_{adj}	$U_{G,trans,1}$	$\Delta U_{G,trans,21}$	$\Delta U_{G,trans,32}$	$\Delta U_{G,trans,43}$	$\Delta U_{G,trans,54}$
Column geometrical characteristics	54.99 %	17.20 %	39.94 %	51.06 %	58.70 %
Operating conditions	26.68 %	40.13 %	-0.57 %	0.00 %	1.20 %
Liquid phase properties	14.28 %	0.63 %	-2.60 %	-2.67 %	-0.83 %
Aggregated models	64.65 %	43.41 %	39.94 %	51.06 %	58.70 %

The regression model results (Table 5 and Table 6 present an example) indicate that $\varepsilon_{G,trans,1}$ and $U_{G,trans,1}$ decrease as the distributor's holes diameter and aspect ratio increase, leading to a destabilization of the mono-dispersed homogeneous flow regime. A counter-current motion of the liquid causes the same effect, decreasing $U_{G,trans,1}$ concerning the batch mode.

Regarding the second flow regime transition, the variables of interest are the sparger holes diameter, the aspect ratio, the superficial liquid velocity, and the free area. An increase in the sparger holes diameter decreases $\Delta\varepsilon_{G,trans,21}$ and increases $\Delta U_{G,trans,21}$. The column aspect ratio influences only $\Delta\varepsilon_{G,trans,21}$, and an increase in its value lead to a decrease in $\Delta\varepsilon_{G,trans,21}$. An increase in the free area causes a reduction in $\Delta U_{G,trans,21}$, but it does not influence $\Delta\varepsilon_{G,trans,21}$. Finally, the poly-dispersed homogeneous flow regime is destabilized by an increase in the superficial liquid velocity in the counter-current mode.

Regarding the other flow regime transitions, the only variables of interest are the geometrical characteristics of the column. In particular, a dominant role is played by the type of gas distributor. The uniform aeration provided by the needle and perforate plate spargers is necessary to ensure that the third, fourth, and fifth flow regime transitions exist.

The graphical representation of the results, provided by the regression trees (Figure 3 and Figure 4 show an example), can be immediately used to identify the boundaries between the flow regimes.

Table 5. Output of the regression procedure: aggregate regression model for $\Delta\varepsilon_{G,trans,21}$.

	Coefficient	**Standard error**	**t-value**	**Pr (> t)**	**VIF**
Intercepts	0.123	0.012	10.015	<2E-16	-
Sparger holes diameter	-0.008	0.006	-1.376	0.171	2
Aspect ratio	-0.005	0.001	-3.615	4.4E-04	1.82
Superficial liquid velocity	0.751	0.228	3.287	0.001	1.33

Table 6. Output of the regression procedure: aggregate regression model for $\Delta U_{G,trans,21}$.

	Coefficient	**Standard error**	**t-value**	**Pr (> t)**	**VIF**
Intercepts	0.220	1.12E-01	1.973	0.051	-
Sparger hole diameter	0.002	1.48E-03	1.274	0.205	1.67
Free area	0.02	0.550	2.146	0.038	1.82
Superficial liquid velocity	0.469	0.006	7.606	7E-12	1.34

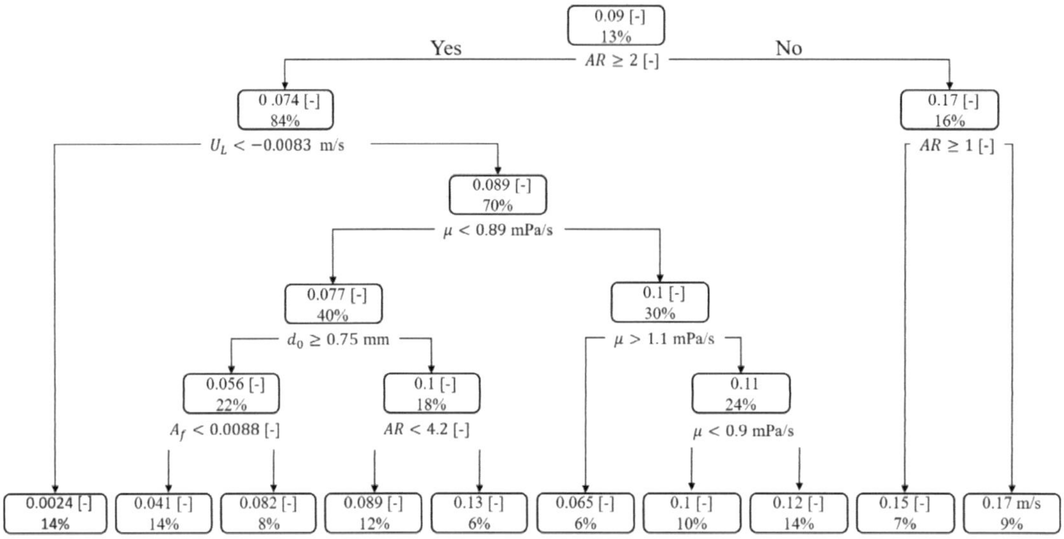

Figure 3. Regression tree: $\Delta\varepsilon_{G,\mathrm{trans},21}$

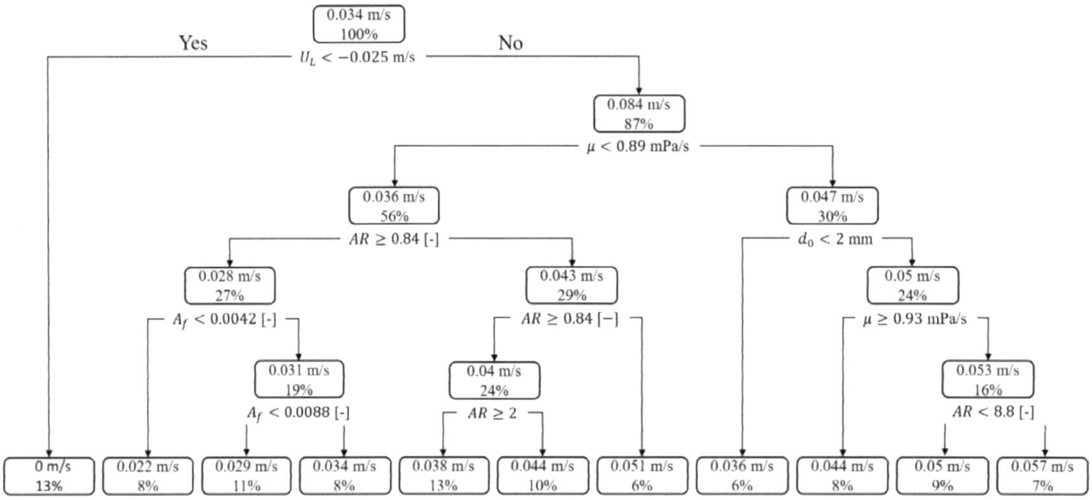

Figure 4. Regression tree: $\Delta U_{G,\mathrm{trans},21}$

5. Conclusions

The statistical method defined in this study has proven to be an interesting and powerful tool for describing bubble columns fluid dynamics. The graphical representation of the results, provided by the regression trees, can be immediately used to identify the boundary between the flow regimes and their fluid dynamics properties. Once the design characteristics of the column and the working fluids are defined, the regime transitions can be identified, and an approximate bubble column characteristic curve can be drawn (Figure 5).

Future studies should extend the applicability of the approach proposed by expanding the dataset considered in this study. In particular, the dataset should be expanded as follows.

- Perforated plate and needle spargers should be considered with liquid phases different from water to include the effects of the liquid properties on the third, fourth and fifth flow regime transitions.
- Columns operating in co-current mode should be added for better understanding the influence of the surface liquid velocity on flow regime transitions.
- Columns operating at different temperatures and pressures should be considered since they are of interest for industrial applications.

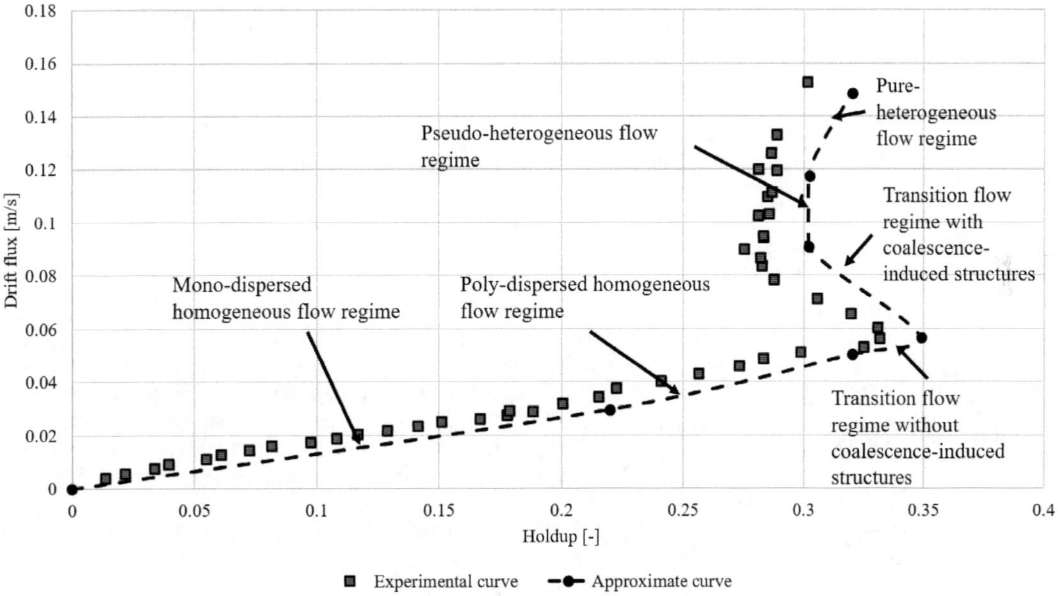

Figure 5. Example of statistically derived bubble column characteristics curve.

6. References

[1] Shaikh A and Al-Dahhan M H 2007 Int. J. *Chem. React. Eng.* **5**(1)
[2] Besagni G *Int. J. Multiph. Flow* **135** 103510
[3] Wallis G.B 1969 One-Dimensional Two-Phase Flow (New York: McGraw-Hill)
[4] Krishna R, Ellenberger J and Maretto C 1999 Int. Commun. Heat Mass 26 467-75
[5] Krishna R, Urseanu M I and Dreher A J 2000 *Chem. Eng. Process* **39**(4) pp 371–378
[6] Deckwer W D, Burckhart R and Zoll G 1974 *Chem. Eng. Sci.* **29**(11) pp 2177–2188
[7] Wilkinson P M, Spek A P and van Dierendonck L L 1992 *AIChE* **38**(4) pp 544–554
[8] Letzel H M, Schouten J C, Krishna R and Van Den Bleek C M 1997 *Chem. Eng. Sci.* **52**(24) pp 4447–4459
[9] Ruzicka M, Drahoš J and Thomas N H 2001 *Chem. Eng. Sci.* **56**(21-22) pp 6117–6124
[10] Vandu C O, Koop K and Krishna R 2004 *Chem. Eng. Technol.* **27**(11) pp 1195–1199
[11] Chilekar V P, Singh C, Van Der Schaaf J, Kuster B F M and Schouten J 2007 *AIChE* **53**(7) pp 1687–1702
[12] Ruzicka M C, Vecer M M, Orvalho S and Drahos J 2008 *Chem. Eng. Sci.* **63**(4) pp 951–967
[13] Yang N, Chen J, Ge W and Li J 2010 *Chem. Eng. Sci.* **65**(1) pp 517–526
[14] Varallo N, Besagni G and Mereu R 2023 *Chem. Eng. Trans.* **100**, 361-366.
[15] Besagni G and Borgarello M 2018 *Energy* **165** pp 369–386

Convective condensation of R449a inside a smooth tube

A Lucchini[1,*], I M Carraretto[1], L P M Colombo[1], D Mazzeo[1], P G Pittoni[2] and G Lipori[3]

[1] Politecnico di Milano, Via Lambruschini 4, 20156 Milan, Italy
[2] Iowa State University, 2529 Union Drive, Ames, IA 50011, U.S.A.
[3] Lu-Ve S.p.A., via Caduti della Liberazione 53, 21040 Uboldo (VA), Italy

* Corresponding author: andrea.lucchini@polimi.it

Abstract. The HVAC sector has started the phase-out of refrigerants characterized by high values of global warming potential and atmospheric lifetime. Drop-in replacement requires that the new, environmentally safe fluids also show comparable heat transfer performances. This work addresses R449a, a low GWP zeotropic mixture (components: R32, R125, R1234yf, R134a, mass fractions: 24.3%, 24.7%, 25.3%, 25.7%, respectively), suitable to replace both R404A and R507A. Experiments were carried out on condensation in horizontal smooth tubes (outer diameter: 9.52 mm, thickness: 0.3 mm). The range of operating conditions meets the standard for HVAC devices (operating pressure: 14.46 bar, bubble temperature: 30°C, temperature glide: approximately 5 K refrigerant mass flux ranging from 136 to 202 kg m^{-2} s^{-1}, quality change -0.8 and -0.2, mean quality ranging from 0.2 to 0.8). The test section is the inner pipe in a tube-in-tube counter-flow heat exchanger, where the refrigerant is cooled by a demineralized water stream in the annulus. Both the pressure drop and the heat transfer coefficient were measured across a length of 1.3 m and 1.1 m, respectively.

Keywords: *Convective condensation, R449a, heat transfer coefficient, pressure drop, smooth tube*

1. Introduction

Refrigerants are widely used in many industrial and domestic applications, such as refrigeration, air conditioning and power generation. Since the 1987 Montreal Protocol, the environmental issue has been one of the most important, leading to the development of new refrigerants. The HVAC sector is undergoing the replacement of hydrofluorocarbons (HFCs) with lower GWP, environmentally friendly refrigerants. The hydrofluoroolefins (HFOs) and the mixtures of hydrofluorocarbons and hydrofluoroolefins (HFC+HFO) are among the candidates. The mixtures allow for the mitigation of the flammability concerns of some pure HFOs, meeting the constraints on the GWP, and keeping similar performances to their HFC predecessors. On the other hand, many mixtures show zeotropic behavior. That makes the description of the phase change more complex compared to pure fluids. Condensation starts at the dew point temperature and ends at the bubble point temperature. Furthermore, coupling between heat and mass transfer takes place. Close to the vapour-liquid interface, concentration gradients develop in the vapor phase due to the preferential condensation of the less volatile component. The

Table 1. R449A composition.

Component		R32	R125	R1234yf	R134a
Atmospheric lifetime	[days]	1790	10600	11	5100
Mass fraction	[$kg_i \cdot kg_{tot}^{-1}$]	24.3%	24.7%	25.3%	25.7%
Molar mass	[$kg \cdot kmol^{-1}$]	86.5	120.0	114.0	102.0
GWP	[-]	675	3500	4	1430

Table 2. R449A properties.

Critical pressure	[bar]	44.47
Critical temperature	[°C]	81.5
Temperature glide	[°C]	4.47
Molar mass	[$kg \cdot kmol^{-1}$]	87.2
GWP.	[-]	1397

Table 3. Smooth tube geometrical features

Inner diameter	[mm]	D_I	8.92
Outer Diameter	[mm]	D_O	9.52
Wet perimeter	[mm]	P	28.0
Cross section area	[mm^2]	A	62.5
Hydraulic diameter	[mm]	D_H	8.92

Table 4. Uncertainties affecting the heat transfer coefficient

Mass flow rate (water/refrigerant)	[-]	± 0.15% of the reading
Temperature	[K]	± 0.1
Operating pressure	[bar]	± 0.025
Pressure drop	[kPa]	± 0.037
Evaporator power	[kW]	± 0.09

increased concentration of the more volatile components reduces the interface temperature, resulting in a lower driving temperature difference and degrading the heat transfer.

To properly design HVAC devices, a lot of efforts have been made to test the heat transfer features of the new refrigerants and to develop suitable correlations to predict the pressure gradient and the heat transfer coefficient during boiling and condensation. Muller-Steinhagen and Heck [1] and Nozu et al. [2] developed correlations to predict the former quantity for smooth tubes. Cavallini et al. [3] [4] [5] proposed a method to predict both the quantities for pure fluids flowing into horizontal smooth or micro-fin tubes. Similarly, Kedzierski and Goncalves [6], working on four different refrigerants (R134a, the near-azeotropic refrigerant mixture R410A and its pure components R125 and R32), provided a model to compute both the quantities during convective condensation inside micro-fins tube. On the other hand, Kumar et al. [7] experimentally evaluated heat transfer during condensation of pure R134a inside a micro-fin tube for different tube inclinations and developed a new correlation. The authors found that, as the condensation proceeds, for all the possible inclination angles a decrease in the heat transfer coefficient is depicted: this is explainable by looking at the augmentation in the average thickness of condensate film which offers higher thermal resistance to the heat flux. A further correlation to compute the heat transfer coefficient was developed by Deng et al. [8].

As many correlations were developed working on pure fluids, they are unable to account for the mass transfer resistance at the liquid-vapor interface and the corresponding reduction in heat transfer. To extend their feasibility to zeotropic mixture Bell and Ghaly [9] proposed a correction.

2. Experimental setup

The tests focused on R449a (a zeotropic mixture made of 4 pure refrigerants, composition in Table 1, properties in Table 2), which is one of the possible drop-in replacements for the refrigerants R404A (GWP 3922) and R507A (GWP 3985). A concise description of the experimental setup (Figure 1), made up of three circuits, follows (an extended version is available in Colombo et al. [10]).

In the refrigerant loop (red line in Figure 1), a shell-and-tube condenser feeds the pump with liquid refrigerant. To prevent cavitation, after the condenser, a plate heat exchanger (sub-cooler) chills the liquid refrigerant. The mass flow rate is fixed by a gear pump with an inverter drive while its value is measured by a Coriolis flow meter (range: 0÷400 $kg \cdot h^{-1}$, uncertainty: ±0.15% of the reading). The thermodynamic state of the refrigerant is monitored by a thermocouple (K type, uncertainty: ±0.1 K) and a pressure transducer (range: -1÷30 bar, uncertainty: ±1% of full scale) as it enters the electric evaporator (power 9 kW). A software tunes its thermal power to get a two-phase flow in suitable conditions. To reach thermal equilibrium between liquid and vapor, a coiled tube (length: 12m, coil diameter: 0.7m, coil axis perpendicular to gravity) is installed at the evaporator outlet. Then, for the development of the flow regime, a straight adiabatic calming section (length: 4.7 m. inner diameter 8.92 mm) is adopted. Subsequently, the refrigerant enters the test section, (tube-in-tube counter flow heat exchanger, distance between the pressure taps: L=1.3 m, heat transfer length: l=1.11 m, thermal insulation: rubber foam, thickness: 100 mm). The refrigerant flows in the inner tube (geometrical features in Table 3) while demineralized water flows in the annulus. The refrigerant inlet pressure is measured by an absolute pressure transducer (range: 0÷16 bar, uncertainty: ±0.25% of full scale) while the pressure drop is read by a differential pressure transducer (range: -1.0÷1.0 bar, uncertainty ±0.1% of full scale). The refrigerant inlet and outlet temperatures are provided by two thermocouples (K type, uncertainty: ±0.1 K) while the wall temperature is computed as the average value of three thermocouples glued in grooves (length: 50mm, depth: 0.15mm, width: 0.4mm) machined on the outside of the inner tube (top, side, bottom positions). A group is at the inlet and another is at the outlet. The reference junctions of all the thermocouples installed in the test section are placed in Dewar flasks filled with melting ice. Then the refrigerant returns to the condenser.

The water circuit (green line in Figure 1) exchanges thermal power with the refrigerant in the test section to induce phase change. The water mass flow rate is measured by a Coriolis flowmeter (range: 0÷400 $kg \cdot h^{-1}$, uncertainty of ±0.15% of reading). A thermally insulated tank supplies the water flow, which is controlled using a bypass or needle valve. Subsequently, the water is cooled in a plate heat exchanger (glycol flushed on the other side for cooling) then a PID-driven electric heater (6 kW), sets the inlet temperature in the test section, which is measured, like the outlet temperature, using a group of 3 thermocouples connected in series (K type, uncertainty: ±0.1 K). The power supply causes a prescribed temperature change in the water flow.

In the glycol loop (blue line in Figure 1), a tank stores a mixture of water and glycol (30% volume concentration, freezing temperature -17°C, temperature -10°C) produced by a commercial chiller (cooling capacity of 21 kW). Two separate circuits, one for water and the other for the refrigerant, supply the heat exchangers. The former loop cools the demineralized water. The latter controls the refrigerant pressure at the test section inlet and prevents cavitation in the refrigerant pump. To achieve the former goal, a manual needle valve sets the mass flow rate and a PID-driven electric heater (3 kW) tunes the condenser inlet temperature to match the refrigerant saturation pressure at the test section inlet. A bypass drains the glycol to cool the liquid refrigerant leaving the condenser.

3. Data sampling

During a test, 181 samples were collected for each quantity (sampling frequency: 1 Hz) and their mean values are the output. The acquisition procedure, to check the repeatability, prescribes, for each experiment, to repeat 10 tests, and their mean values are the output. The operating conditions are univocally defined by four parameters:

1. the refrigerant pressure at the test section inlet (p_{rTi});
2. the refrigerant mass flux (G);
3. the quality variation in the test section (Δx);
4. the mean quality in the test section (x_m).

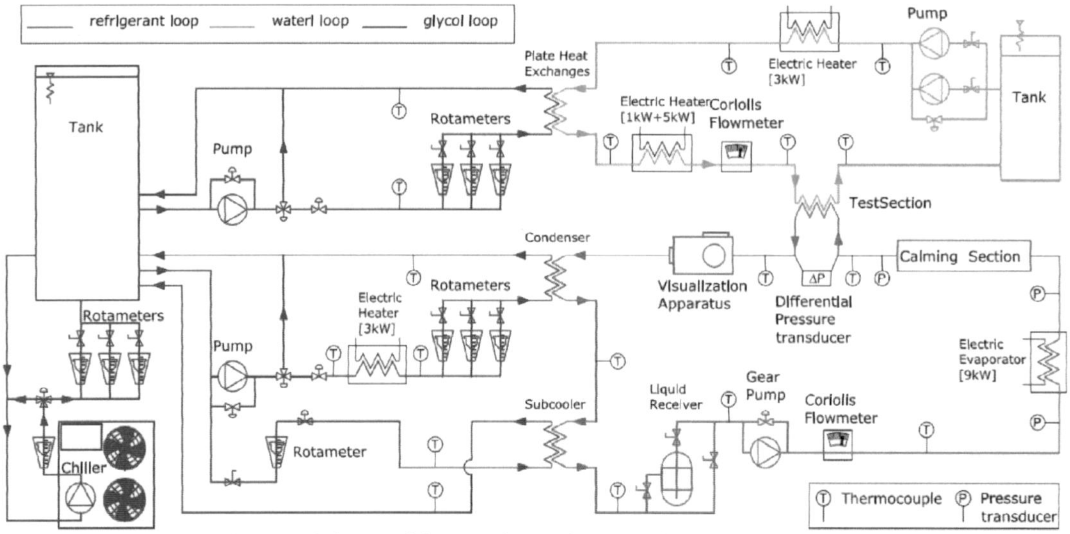

Figure 1. Scheme of the experimental setup and three circuit loops.

4. Data processing

Once an experiment is completed the readings of the instruments are processed assuming:

- steady-state operating conditions (the data are processed only if the oscillations of the quantities are smaller than ±3% and the temperatures oscillate within ±0.2 K);
- negligible thermal dispersions (single phase tests proved that, in the test section, the refrigerant side thermal power and the waterside thermal power differ at most of 5%);
- negligible thermal resistance of the copper tube (a posteriori calculation shows that it is two orders of magnitude smaller than the refrigerant thermal resistance);
- negligible fouling effect (the refrigerant circulation pump is gear-type magnetically driven and does not require lubricant).

The post-processing provides three outputs: the operating conditions (p_{rTi}, G, Δx, x_m), the total pressure drop per unit length (Z) and the heat transfer coefficient (h). Their uncertainties are evaluated with the uncertainty propagation algorithm described by Moffat [11]. For a quantity y which depends on n independent variables x_j, the absolute uncertainty U_y is:

$$U_y = \sqrt{\sum_{j=1}^{n} \left(U_{x_j} \frac{\partial y}{\partial x_j} \right)^2} \quad (1)$$

The refrigerant pressure at the test section inlet p_{rti} is the reading of the absolute pressure transducer while the temperature at the same location T_{rti} is provided by a thermocouple.

Mass flux G is the ratio of the refrigerant mass flow rate m_r and the duct cross-sectional area (A_c):

$$G = \frac{m_r}{A_c} \quad (2)$$

The refrigerant quality change in the test section is computed from the refrigerant conditions, defined by pressure and enthalpy, at the inlet and outlet sections.

- The outlet pressure p_{rto} is determined by subtracting the pressure drop ($\Delta p>0$) measured by the differential pressure transducer from the inlet pressure:

$$p_{rTo} = p_{rTi} - \Delta p \quad (3)$$

- The test section inlet enthalpy i_{rti} is computed by the energy balance at the evaporator. The enthalpy of the subcooled liquid at the evaporator inlet i_{rei} is the output of NIST software RefProp 10 on the base of the temperature T_{rei} and pressure p_{rei} at the evaporator inlet:

$$i_{rEi} = i(T_{rEi}, p_{rEi}) \quad (4)$$

The thermal power Q_E, provided by the evaporator, is measured by a net-analyzer and the evaporator outlet enthalpy, which matches with the enthalpy at the test section inlet, is:

$$i_{rTi} = i_{rEi} + \frac{Q_E}{m_r} \quad (5)$$

- The test section outlet enthalpy i_{rto} is determined by the energy balance at the test section:

$$i_{rTo} = i_{rTi} + \frac{m_a c_{pa}(T_{aTi} - T_{aTo})}{m_r} \quad (6)$$

The inlet quality x_{Ti} and the outlet quality x_{To} are computed employing RefProp 10 providing, as inputs, the pressure and the enthalpy at the corresponding position:

$$x_{Ti} = x(p_{rTi}, i_{rTi}) \quad (7)$$

$$x_{To} = x(p_{rTo}, i_{rTo}) \quad (8)$$

The difference between the last two quantities is the quality change Δx, while their average is the mean quality x_m

$$\Delta x = x_{Ti} - x_{To} \quad (9)$$

$$x_m = \frac{x_{To} + x_{Ti}}{2} \quad (10)$$

A differential pressure transducer provides the total pressure drop Δp (accelerative component and frictional component) on the test section while the distance between the pressure taps (L=1.3 m) was measured when the test section was built. The pressure gradient Z (uncertainty smaller than 4%) is:

$$Z = \frac{\Delta p}{L} \tag{11}$$

The heat transfer coefficient h, referred to the tube's inner surface, is computed using the logarithmic mean temperature difference ΔT_{ml}, which calculation relays on the refrigerant and wall temperatures. The wall temperatures are computed as the averages of the readings of the thermocouple glued on the outside of the copper tube separating the refrigerant and the water:

$$T_w = \frac{T_b + T_s + T_t}{3} \tag{12}$$

Accordingly, the logarithmic mean temperature difference is:

$$\Delta T_{lm} = \frac{(T_{wo} - T_{rTo}) - (T_{wi} - T_{rTi})}{\ln \frac{T_{wo} - T_{rTo}}{T_{wi} - T_{rTi}}} \tag{13}$$

Such quantity allows computing the refrigerant heat transfer coefficient:

$$h = \frac{Q_T}{\pi D_I l \Delta T_{lm}} \tag{14}$$

As the thermocouple providing T_{rTo} is close to the test section outlet, the distance between them might be too short to guarantee thermal equilibrium between liquid and vapor, which could make the reading not reliable. A viable option, to overcome that issue, is represented by computing the temperature difference between the inlet and outlet, assuming thermodynamic equilibrium between the phases at both locations:

$$t_{rT} = T(p_{rTo}, i_{rTo}) - T(p_{rTi}, i_{rTi}) \tag{15}$$

Then, such difference is added to the refrigerant inlet temperature:

$$T_{rToc} = T_{rTi} + t_{rT} \tag{16}$$

The corrected refrigerant outlet temperature T_{rToc} replaces the measured refrigerant outlet temperature T_{rTo} in equation (14) to compute the corrected logarithmic mean temperature difference ΔT_{lmc} and then, the corrected heat transfer coefficient h_c using equation (15). Accordingly to the data reported in Table 4, the uncertainty related to the heat transfer coefficient, regardless of the procedure adopted, is not larger than 8%.

5. Results

The tests can be grouped into two categories: complete condensation and partial condensation, both performed by setting the test section inlet pressure at pressure p_{rTi}=14.46 bar (bubble temperature 30°C).

- Complete condensation aims to simulate real operating conditions, different mass fluxes were tested, while the quality change and the mean quality were fixed ($G \in [136;202]$ kg·m^{-2}·s^{-1}, $q \in [27.6;70.2]$ kW m^{-2}, Δx=-0.8, x_m=0.5, red circles in Figure 2).
- Partial condensation aims to highlight the effect of the mean quality, while the mass flux and the quality change were fixed (G = 202 kg·m^{-2}·s^{-1}, q=17.5 kW m^{-2}, Δx=-0.2, $x_m \in [0.2;0.8]$, blue circles in Figure 2).

As reported in the open literature [12] on a technical point of view, with respect to heat transfer, the best operating conditions are for mass fluxes larger than 150 kg·m^{-2}·s^{-1}, for which a consistent portion of the perimeter is wetted by the liquid (almost uniform liquid thickness).

During complete condensation (Figure 3), the total pressure gradient Z versus the mass flux G is a nearly quadratic function (power law fitting provides: n=2.19, cn=4.32·10-3 s-1).

Figure 4 shows, for complete condensation tests, the heat transfer coefficients h and h_c and their percentage difference Δh% versus the mass flux G, while Figure 5 displays, for the partial condensation tests, the same quantities versus the mean quality x_m. In both cases, the percentage difference between

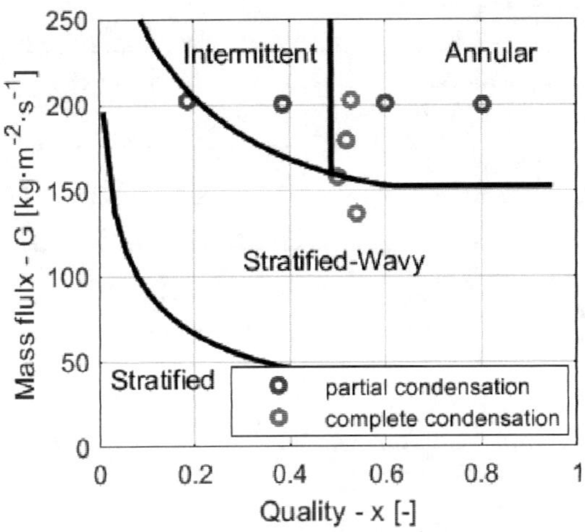

Figure 2. Operating conditions tested reported on the Kattan flow pattern map [13].

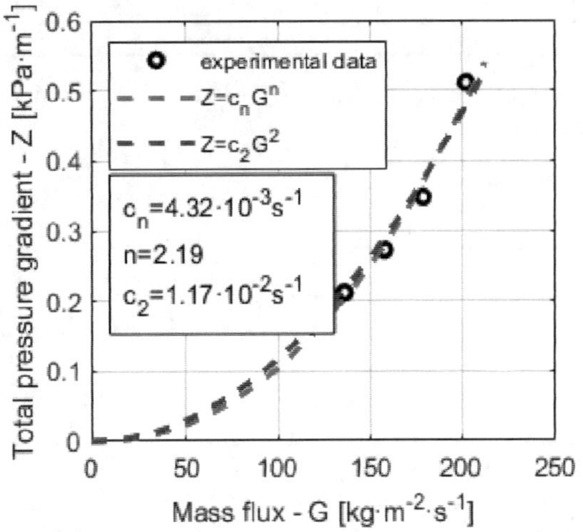

Figure 3. Total pressure gradient Z vs mass flux G during complete condensation (x_m=0.5, Δx=0.8).

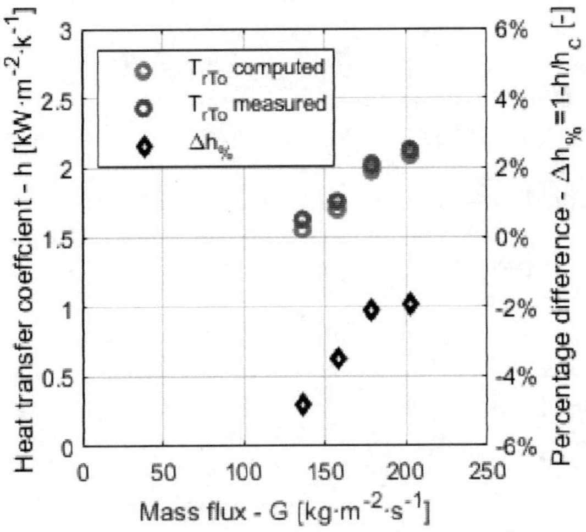

Figure 4. Heat transfer coefficient during h vs mass flux G during complete condensation (x_m=0.5, Δx=0.8).

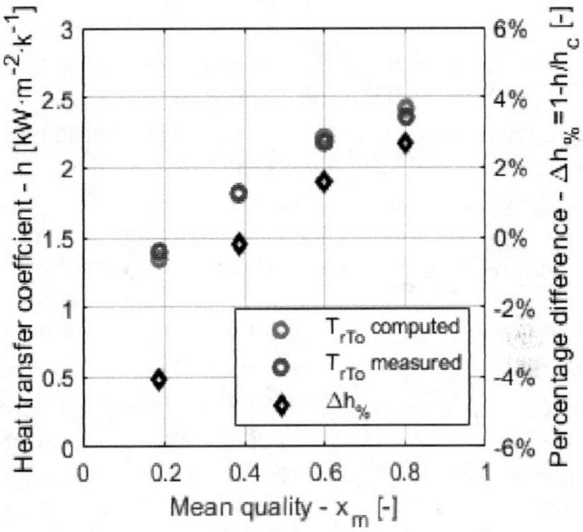

Figure 5. Heat transfer coefficient h vs mean quality x_m during partial condensation (G=202 kg·m^{-2}·s^{-1}, Δx=0.2).

the heat transfer coefficients is always within ±5%, which means that it is acceptable to assume thermal equilibrium at the test section outlet.

To analyze the heat transfer performances it is useful to highlight some remarks.

A. The density of the liquid phase (ρ_l=1070kg·m^{-3}) is almost twenty times larger than the density of the vapor phase (ρ_v=58.9kg·m^{-3}).
B. The thermal conductivity of the liquid phase (k_l=7.72·10^{-2}W·m^{-1}·K^{-1}) is almost five times larger than the thermal conductivity of the vapor phase (k_v=1.55·10^{-2}W·m^{-1}·K^{-1}).
C. Because of remark A, during the experiments, the vapor occupied the largest part of the cross-sectional area (e.g. case x=0.1, the volume quality is 67%).
D. In the stratified-wavy flow, the liquid lies in the lower part of the tube because the gravity effect overcomes the shear stress effect. The heat transfer performances of the flow are mainly related to the vapor thermal conductivity.
E. In the intermittent flow the effects of gravity and shear stress are comparable and, periodically, the liquid is capable of adjoining a significant portion of the cross-sectional

Table 5. parameters used to benchmark the prediction capability of the correlations

Meaning	equation	n°	meaning	equation	n°
Percentage error of the j-th operating conditions	$E_{j\%} = \frac{g_{Cj}-g_{ej}}{g_{ej}}$	(17)	Mean absolute percentage error	$E_{A\%} = \sum_{j=1}^{n} \frac{\lvert E_{j\%} \rvert}{n}$	(19)
Mean percentage deviation	$E_{\%} = \sum_{j=1}^{n} \frac{E_{j\%}}{n}$	(18)	Standard deviation of the mean percentage error	$s = \sqrt{\sum_{j=1}^{n} \frac{\left(E_{j\%} - E_{\%}\right)^2}{n}}$	(20)

Table 6. Pressure drop: correlations performances

Correlation	$E_{\%}$	$E_{A\%}$	$s_{\%}$
Cavallini et al.	-3.0%	19.0%	21.3%
Kedzierski and Goncalves	-32.4%	32.6%	25.5%
Muller-Steinhagen and Heck	-23.6%	23.6%	16.7%
Nozu et al.	11.7%	33.7%	39.9%

Table 7. Heat transfer coefficient: correlations performances

Correlation	$E_{\%}$	$E_{A\%}$	$s_{\%}$
Cavallini et al.	13.9%	13.9%	7.5%
Deng et al.	10.5%	10.5%	4.3%
Kedzierski and Goncalves	-6.9%	7.9%	7.6%
Kumar et al.	-6.2%	7.5%	6.6%

perimeter. The heat transfer performances of the flow are influenced by the thermal conductivities of both liquid and vapor.

F. In the annular flow the shear stress effect overcomes the gravity effect and the liquid arranges, in an almost uniform layer, all around the cross-sectional perimeter. The heat transfer performances of the flow are mainly related to the liquid thermal conductivity.

G. As long as the quality change Δx and the mean quality x_m are the same, because of remarks B, E and F, it could be expected that the heat transfer performance of intermittent flow is better than the stratified-wavy flow and worse than the annular flow.

Complete condensation tests (Figure 4) highlight a slope change of the heat transfer coefficient trend in the range G=[150; 200] $kg \cdot m^{-2} \cdot s^{-1}$. That could be related, as the mass flux increases (Figure 3), to the transition from the stratified-wavy flow to the intermittent flow and annular flow because of remark G. The experiment concerning partial condensation (Figure 5) further supports the explanation. As the mean quality increases the flow regime shifts from stratified-wavy to intermittent flow and, in the end, to annular flow. In agreement with remark G, the heat transfer coefficient grows. The slope reduction shifting from x_m=0.6 to x_m=0.8 could be explained because annular flow onset in both operating conditions and the heat transfer coefficient increase is related only to the larger vapor velocity and thinner liquid annulus, as mentioned by Kumar et al. [7].

The correlations' capability to predict the experimental data is quantified using the parameters reported in Table 5. The results concerning the total pressure gradient Z are summarized in Table 6. Only two of the four selected correlations can provide predictions of the experimental data within ±30%, the best performing is the Cavallini et al. [5] correlation (mean absolute percentage error +19%) even though the Muller-Steinhagen and Heck [1] provides similar absolute deviation but shows a tendency towards underestimation. On the contrary, as shown in Table 7, all the selected correlations, with the Bell and Ghaly [9] correction, provide good performances, the absolute percentage error, for all of them, is smaller than 15% and for the Kumar et al. [7] correlation, which is the best performing (mean absolute percentage error 7.5%).

6. Conclusions

The manuscript reports the investigation of the heat transfer performance (total pressure gradient and heat transfer coefficient) of the refrigerant R449A during convective condensation inside a horizontal smooth tube. Complete condensation showed that a nearly quadratic power-law interpolation curve fits well with the total pressure gradient as a function of the mass flux. Moreover, both the complete and partial condensation experiments highlight the effect of the flow regime transition on the heat transfer coefficient. The correlation benchmark pointed out that the predictions of the pressure gradient showed a limited agreement with the data, the Cavallini et al [5] correlation, which is the best performing, has a 19% mean absolute percentage error. On the contrary, the correlations for the heat transfer coefficient, with the Bell and Ghaly [9] correction, prove a good predictive capability, mean absolute percentage error lower than 15%.

7. Nomenclature

Latin symbols

symbol	meaning	units	symbol	meaning	units
A	cross sectional area	[m^2]	n	number of elements	[-]
c_p	specific heat capacity	[$J \cdot kg^{-1} \cdot K^{-1}$]	p	refrigerant pressure	[Pa]
D	diameter	[m]	P	wet perimeter	[m]
$E_{\%}$	mean percentage error	[-]	Q	thermal power exchanged	[W]
G	refrigerant mass flux	[$kg \cdot m^{-2} \cdot s^{-1}$]	s	standard deviation	[-]
g	generic quantity	[-]	T	temperature	[K]
h	heat transfer coefficient	[$W \cdot m^{-2} \cdot K^{-1}$]	t	temperature difference	[-]
i	specific enthalpy	[$J \cdot kg^{-1}$]	U	absolute uncertainty	[-]
L	pressure taps distance	[m]	x	refrigerant quality	[-]
l	heat transfer length	[m]	y	independent variable	[-]
m	mass flow rate	[$kg \cdot s^{-1}$]	z	pressure gradient	[$Pa \cdot m^{-1}$]

Greek symbols

symbol	meaning	units	symbol	meaning	units
Δp	pressure drop	[Pa]	Δx	test section quality change	[-]

Subscripts

symbol	meaning	symbol	meaning
a	water	j	j-th element
A	absolute	lm	log mean
b	bottom	m	mean
C	correlation	o	outlet
c	corrected	O	outer
E	evaporator	r	refrigerant
e	experimental	s	side
H	hydraulic	t	top
i	inlet	T	test section
I	inner	w	wall

8. References

[1] H. Muller-Steinhagen and K. Heck. A simple friction pressure drop correlation for two-phase flow in pipes. Chemical Engineering and Processing, 20:297–308, 1986.

[2] H. Nakata S. Nozu, H. Katayama and H. Honda. Condensation of a refrigerant cfc11 in horizontal microfin tubes (proposal of a correlation equation for frictional pressure gradient). Experimental Thermal and Fluid Science, 18:82–96, 1998.

[3] L. Doretti G.A. Longo A. Cavallini, D. Del Col and L. Rossetto. Pressure drop during condensation and vaporisation of refrigerants inside enanched tubes. Heat and Technology, 15(1):3–10, 1997.

[4] L. Doretti M. Matkovic L. Rossetto C. Zilio G. Censi A. Cavallini, D. Del Col. Condensation in horizontal smooth tubes: a new heat transfer model for heat exchanger design. Heat Transfer Eng., 27:31–38, 2006.

[5] S. Mancin L. Rossetto A. Cavallini, D. Del Col. Condensation of pure and near- azeotropic refrigerants in microfin tubes: A new computational procedure. International Journal of Refrigeration, 32:162–174, 2009.

[6] M.A. Kedzierski and J.M. Goncalves. Horizontal convective condensation of alternative refrigerants within a micro-fin tube. Journal of Enhanced Heat Transfer, 6:161–178, 1999.

[7] R. Kumar M.A. Akhavan-Behabadi and S.G. Mohseni. Condensation heat transfer of r- 134a inside a microfin tube with different tube inclinations. International Journal of Heat and Mass Transfer, 50:4864– 4871, 2007.

[8] M. Fernandino H. Deng, M. Rossato and D. Del Col. A new simplified model for condensation heat transfer of zeotropic mixtures inside horizontal tubes. Applied Thermal Engineering, 153:779–790, 2019.

[9] M. Ghaly K. Bell. An approximate generalized design method for multicomponent/partial condensers. AIChE symposium series, 69:72–79, 1973.

[10] L. P. M. Colombo, A. Lucchini, T. Nhan Phan, L. Molinaroli, and A. Niro, "Design and assessment of an experimental facility for the characterization of flow boiling of azeotropic refrigerants in horizontal tubes," J. Phys. Conf. Ser., vol. 1224, no. 1, 2019

[11] Moffat, R.J., "Describing the uncertainties in experimental results,"Exp.Therm. and Fluid Sci., 1(1), pp 3-17, (1988).

[12] Hewitt, G. F. and Barbosa, J., "Heat exchanger design handbook", 98. Begell house New York, 2008.

[13] Kattan, N., Thome, J.R., Favrat, D.,"Flow boiling in horizontal tubes: part 1-development of a diabatic two-phase flow pattern map,"J. Heat Transfer, 120(1), pp. 140–147, (1998).

Acknowledgements

This study was carried out within the NEST - Network 4 Energy Sustainable Transition (D.D. 1243 02/08/2022, PE00000021) and received funding under the National Recovery and Resilience Plan (NRRP), Mission 4 Component 2 Investment 1.3, funded from the European Union - NextGenerationEU. This manuscript reflects only the authors' views and opinions, neither the European Union nor the European Commission can be considered responsible for them.

Experimental data for flow boiling of R450A in a horizontal tube

R Mastrullo[1], A W Mauro[1,*], A F Passarelli[1], I Viscardi[1] and L Viscito[1]

[1] Department of Industrial Engineering, Federico II University of Naples, P.le Tecchio 80, 80125 Naples, Italy

* Corresponding author: wmauro@unina.it

Abstract. According to the new European policies aimed at the replacement of highly-pollutant greenhouse gases refrigerants, the scientific community has focused on new synthetic environmentally friendly substances to be employed in vapor compression cycles for the refrigeration and the air conditioning fields.
On this regard, R450A is a new blend made up of R134a (42%) and R1234ze (58%), having a GWP equal to 604, and therefore represents an attractive solution as pure R134a substitute. In this paper, new flow boiling heat transfer and pressure drop data of R450A collected at the refrigeration lab of Federico II University of Naples are presented. The data refer to a horizontal stainless-steel tube having an internal diameter of 6.0 mm. The effects of mass flux (from 150 to 400 kg $m^{-2}s^{-1}$), heat flux (from 10 to 20 kW m^{-2}) and saturation temperature (from 30 to 50 °C) are presented and discussed, together with the assessment of the most quoted two-phase heat transfer and pressure drop prediction methods.

1. Introduction

Refrigeration and air conditioning systems are an essential part of modern life, providing cool air in homes, offices, vehicles, and many other applications. However, the use of halogenated fluids in vapor compressionsystems has provided a considerable contribution to the release of greenhouse gases in the atmosphere [1] [2] and therefore tothe undesired rise of the Earth temperature. In fact, most of the synthetic refrigerants present a high Global Warming Potential (GWP), and thus act as pollutant gases in case of accidental leakages [3] or during the plant dismission procedure. So far, the hydro-fluoro-carbon (HFC) R134a is the most common fluid in medium-temperature applications of both developed and developing countries. Its GWP is about 1480 [4] and is already banned in several fields according to the new European Regulations.Among possible substitutes, R450A is a blend of two refrigerants (mass fraction 42%/58% of R134a/R1234ze(E)), with a considerably lower GWP of 604 [4], non-toxic and non-flammable. It can be considered as a direct drop-in replacement of R134a [5] in existing commercial and industrial medium and high temperature refrigeration equipment that uses positive displacement compressors and direct expansion systems. It can also be suitable for heat pumps, vending machines, drink dispensers, centrifugal air and water chillers, and for replacing R134a in medium-temperature circuits of two-stage hybrid cascade systems with CO_2.

The rising interest towards this blend is testified by numerous scientific publications related to the performance comparison of several systems employing R450A as possible R134a substitute. Mota-

Babiloni et al. [6,7] tested R450A in a vapor compression plant equipped with a variable compressor and R134a as original working fluid. They found that R450A could lead to higher Coefficient of Performance (COP) values, lower discharge temperatures and lower pressure drop for the internal heat exchanger. Similar results were obtained by Makhnatch et al. [8] in a small refrigeration unit, reporting only a slight reduction in the cooling capacity when employing R450A. The same research group [9] experimentally investigated the use of R134a and alternative fluids in operating conditions typical of countries with high ambient temperatures, finding a lower performance of R450A with respect to both R134aand R513A. For different applications, Gatarìc and Lorbek [10] tested R450A in household heat pump tumble dryers and Molinaroli et al. [11] in a water-to-water heat pump, with similar outcomes related to the energy consumption and efficiency (not much different from R134a) and heating capacity (slightly reduced because of the lower condensation pressure with respect to R134a). Several numerical approachesincluding Artificial Intelligence have also been followed to investigate the possible use of R450A in refrigeration systems (Belman-Flores et al., 2017 and Zendehboudi et al., 2019) [12] [13] and in waste heat-solar driven ejector heat pumps for simultaneous heating and cooling (Al-Sayyab et al., 2021) [14], all with encouragingand positive outcomes.

Besides system analyses, the literature review presents fewer examples related to the two-phase thermal and hydraulic characteristics of R450A, and most of them are related to condensation heat transfer. Jacob et al. [15] conducted experiments on R450A condensing in a 4.7 mm tube, using a novel distributedtemperature sensor. More recently, Azzolin et al. [16] experimented flow condensation of R450A in smalltube channels having internal diameters of 0.96 mm and 3.38 mm, providing also flow visualization results. Diani et al. [17] tested the new blend in a 7 mm outer diameter microfin tube during flow condensation, highlighting the significant effect of mass velocity, especially for in the low vapor quality region. A comparisonwith R134a was proposed by Morrow and Derby [18], that experimented condensation heat transfer in a set of 7 parallel minitubes with a diameter of 0.95 mm. They found that the heat transfer coefficients of R450A could be up to 25% lower than the ones obtained for the reference fluid. To the best of our knowledge,the only work in literature concerning flow boiling heat transfer of R450A is that of Kedzierski and Kang [19] in a microfin tube. The authors found that both two-phase heat transfer coefficient and pressure drop were15% lower than those obtained for R134a in the same conditions and proposed a correlation to improve thestatistics of their assessment.

Since no studies are aimed at the investigation of flow boiling characteristics of R450A in smooth conventional channels, more data on these configurations could be useful to assess the existing correlation and to correctly design evaporators in vapor compression systems. For this reason, the present study explores thermal and hydraulic characteristics of R450A in a horizontal stainless-steel tube having an internal diameter of 6.0 mm. The experiments are aimed to evaluate the heat transfer coefficient and the frictional pressure gradient in different operating conditions in terms of mass flux, imposed heat flux, saturation temperature and vapor quality, finally proposing an assessment with the most quoted prediction methods.

2. Experimental apparatus

2.1. *Refrigerant loop and test tube*

Figure 1 shows a schematic view of the experimental plan. The refrigerant in sub-cooled condition flows thorough a magnetic gear pump and the flow rate is measured by a Coriolis flow meter before the preheating section. The latter is made up of a copper tube where the heatis provided by four fiberglass heating tapes.The applied heat at the preheater section is modulated through a solid-state relay to obtain a specific local vapor quality at the inlet of the test section. The following plate heat exchanger and tube-in-tube heat exchanger work respectively as condenser and sub-cooler before the pump suction head that closes the loop. Several throttling valves are positioned on the liquid and vapor lines, respectively, and a bypass loop is placed at the pump outlet to have multiple mass flow rate control options. The coolant is demineralized water, with its temperature set and remotely controlled by a thermostatic bath.

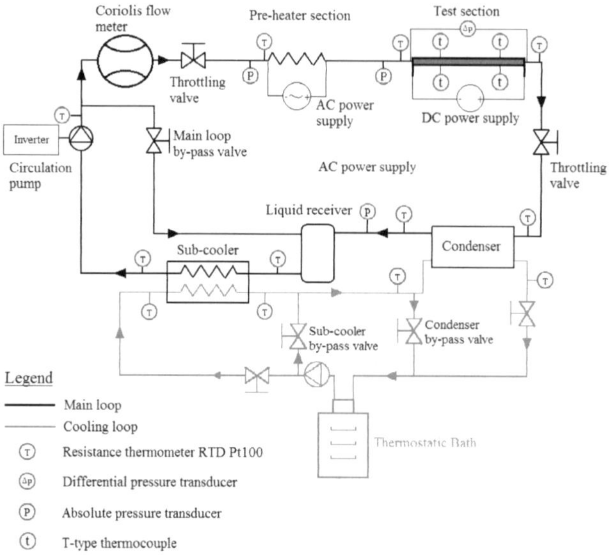

Figure 1. Experimental apparatus.

The test section employed for the present experimental campaign is a smooth, horizontal, circular stainless steel (AISI SS316) tube with an internal diameter of 6.0±0.05 mm and an outer diameter of 8.0±0.05 mm. The heat flux is obtained through Joule effect provided by a modulable DC power supply unit with 193.7±0.79 mm representing the total measured heated length (see points A and E in Figure 2). Two pressure taps are placed at 237.5±0.91 mm from point A and are useful for the inlet pressure and pressure drop measurements. For the evaluation of the local heat transfer coefficient, point C (see Figure 2) is placed at 146.7 ±0.64 mm from the inlet section. More details are available in some authors' previous works [20, 21, 22] (Lillo et al., 2018, Arcasi et al., 2021, 2022).

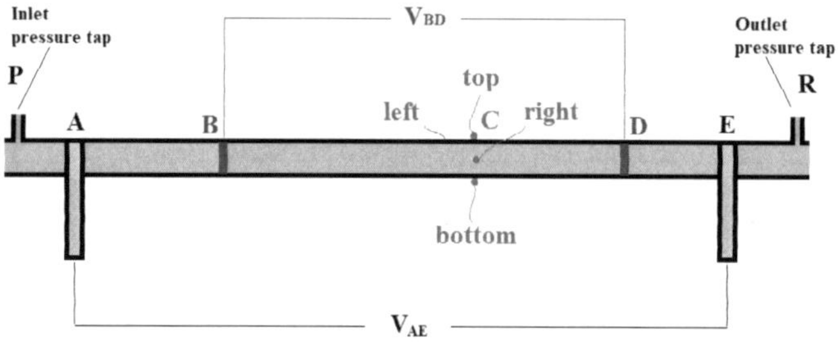

Figure 2. Test tube.

2.2. Measurement instrumentation

Both the main and secondary loops are equipped with several resistance thermometers (RTDs) having an overall uncertainty of ±0.180 °C, able to measure the working and cooling fluid temperatures. Four T-type thermocouples are instead placed at the measurement point C and evaluate the outer wall temperature forthe calculation of the two-phase heat transfer coefficient. Following a dedicated in-situ calibration that includes two high-accuracy resistance thermometer (one placed on the tube and the other placed in the environment for the compensation of the cold junction temperature) and adiabatic tests, the overalluncertainty that also includes the residual errors of the calibration curve is estimated as ±0.1 °C.

The test section inlet pressure is obtained through an absolute pressure transducer having a measuring rangeof 0-35 bar and an overall uncertainty of ±0.5% of the read value, whereas the pressure drop is estimated with a differential pressure transducer, carrying an instrumental uncertainty of ±0.06 kPa.

A Coriolis flow meter provides the value of the mass flow rate with a maximum uncertainty of ±1% of the measurement. For the preheating section, the AC voltage (100 mV-500 V) and current (1 mA-16 A) are separately measured, and the total uncertainty of the estimated ohmic electrical power is ±1.0% of the reading, as provided by the manufacturer. Finally, the DC ohmic power for the test section is obtained through separate measurements of voltage (transducer within 0-5 V and an uncertainty of ±0.03% of the reading) and current (directly measured by the DC power unit within 0-300 A and an uncertainty of ±1.0%).

3. Method

3.1. Data reduction process

All the calculations have been implemented in MATLAB [23] environment, and the thermodynamic and transport properties, when not differently detailed, are obtained through the software Refprop 10.1 [24]. The local heat transfer coefficient is calculated according to the Newton law, Eq. (1):

$$h = \frac{q}{T_{wall} - T_{sat}} \tag{1}$$

The inner T_{wall} is obtained from the outer wall temperature T_{th} by considering the analytical solution to the heat transfer phenomenon with the hypothesis of 1-D heat flux in the radial direction, uniform heat generation inside the tube, as shown in Equation (2):

$$T_{wall} = T_{th} + \frac{V_{BD} \cdot I}{4\pi\lambda_{tube}\overline{BD}} \cdot \left[\left(\frac{D}{d}\right)^2 \cdot \left(1 - ln\left(\left(\frac{D}{d}\right)^2\right)\right) - 1\right] \cdot \left[\left(\frac{D}{d}\right)^2 - 1\right]^{-1} \tag{2}$$

T_{sat} is instead the local saturation temperature in the measurement point, obtained as a function of the local pressure and enthalpy, by interpolating the saturation curve experimentally obtained with adiabatic experiments, in which the temperature is estimated with a resistance thermometer. The local heat flux is instead the measurement of an electric quantity, as shown in Equation (3):

$$q = \frac{V_{BD} \cdot I}{\pi d \overline{BD}} \tag{3}$$

To keep into account the non-negligible glide of the tested mixture [25] (Mauro et al., 2020), the local vapor quality at the measurement point is obtained as a function of both saturation temperature and local specific enthalpy. This latter value is the result of an energy balance over the preheating and the test section up to the measurement point, as shown in Equation (4).

$$i_c = i_{in,preh} + \frac{\dot{Q}_{preh}}{\dot{m}} + \frac{\dot{Q}_{TS} \cdot \left(\frac{\overline{AC}}{\overline{AE}}\right)}{\dot{m}} \tag{4}$$

The frictional pressure drop has been calculated as the measured pressure gradient divided by the distance between the two pressure ports (see points P and R in Figure 2). In fact, all pressure drop values presented in this work are obtained at adiabatic conditions and with a horizontal configuration; therefore geodetical and momentum contributions are null.

3.2. Uncertainty analysis and validation

The law of propagation of error [26] was used to estimate the combined uncertainty of all measured and derived parameters, by also using a coverage factor of 2 to ensure a confidence level higher than 95%. The average uncertainty related to the heat transfer coefficient and pressure drop is ± 16.87% and ± 14.57%, respectively.

Liquid single-phase experiments by using refrigerant R134a were carried-out to evaluate the correct insulation and the proper functioning of the measurement instrumentation. The electrical power in the preheater and test section was compared to the heat absorbed by the sub-cooled refrigerant, with good results. Also, the liquid single-phase heat transfer coefficients were evaluated and compared to benchmark correlations. Further information on the validation procedure can be obtained from previous works with the same test facility [27, 28] (Lillo et al., 2018b, Mastrullo et al., 2017).

4. Results

R450A is the working fluid employed for the convective boiling experiments. Mass fluxes of 150 and 400 kg $m^{-2}s^{-1}$, heat fluxes from 10 to 20 kW m^{-2} and bubble saturation temperatures from 30 to 50 °C have been investigated and the effect of their variation on the local heat transfer coefficients and pressure drop values(obtained at adiabatic conditions) is discussed in the following diagrams.

4.1. Heat transfer coefficient

The effect of the operating conditions on the R450A flow boiling heat transfer coefficients is shown in Figure 3a-c.

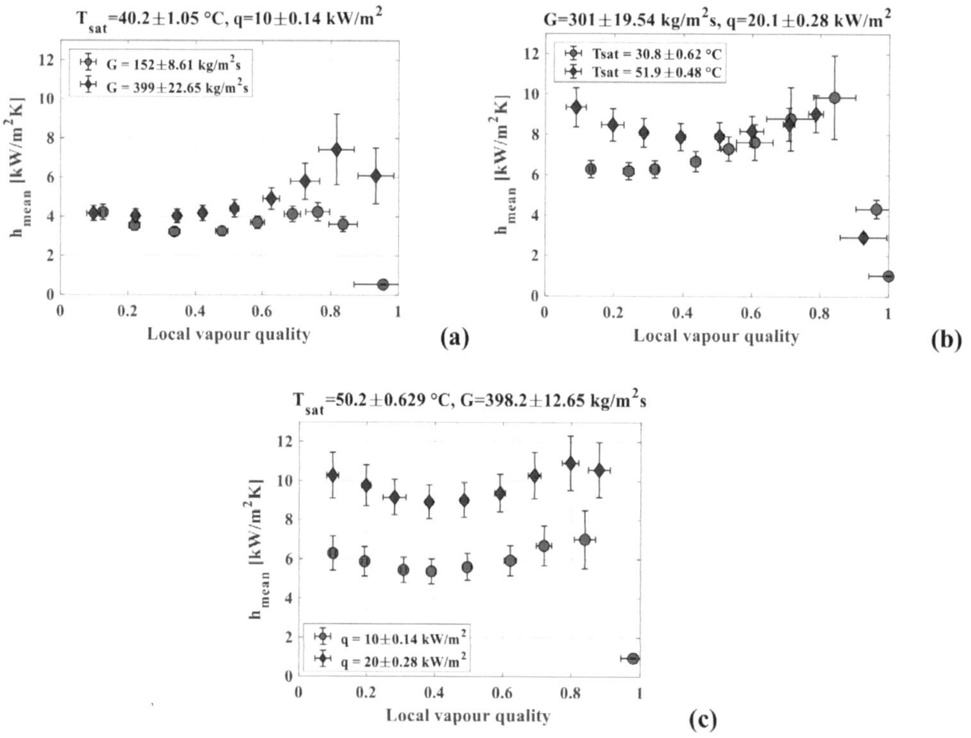

Figure 3. Effect of the operating conditions on the R450A flow boiling heat transfer coefficient. (a) Effect of mass velocity; (b) Effect of bubble saturation temperature; (c) Effect of the heat flux.

As regards the influence of the mass flux (see Figure 3a), the increasing trend with vapor quality is found to be representative of the heat transfer coefficient behaviour only for high mass velocities. In case of G equalto 150 kg $m^{-2}s^{-1}$, instead, the trend is somehow decreasing and may imply a possible early dry-out phenomenon or a stratified flow regime. The local peripheral heat transfer coefficient values validate this assumption, since it was observed a sudden drop of the top heat transfer coefficient already at low vapor qualities. The effect of the saturation temperature is shown in Figure 3b. The increase of the system pressureleads to lower heat transfer coefficients for high vapor qualities, in which the convective contribution plays a major role. In fact, the increase of the reduced pressure lowers the difference betweenliquid and vapor phase densities and therefore decreases the shear stress at the interface with a consequent penalization on the convective heat transfer contribution. At lower vapor qualities, instead, the heat transfer coefficient increases with the saturation temperature, implying a significant nucleate boiling contribution, that is enhanced with reducing surface tension for higher system pressures. Finally, the effect of the imposed heat flux is shownin Figure 3c, with a significant increase of the boiling heat transfer coefficient (up to 200%) whenpassing from 10 to 20 kW m^{-2}. This implies that the nucleate boiling contribution is preponderant for such conditions and the increase of the heat flux enhances the nucleation sites activity, with more bubble per unit time at the wall.

The experimental data have been compared to some well-known correlations to test their accuracy within the present range of operating conditions. The best results have been found with the prediction methods of Li and Wu (2010) [29] and Del Col (2010) [30], as shown in Figure 4a-b, respectively. The first provides a Mean Absolute Percentage Error (MAPE) of 26.50% and a Mean Relative Percentage Error of 0.17%.The latter provides a MAPE of 20.36% and a MRPE of -9.89%. The prediction method of Wojtan et al. [19] shows better fitting of the experimental data for high values of vapour quality, while for low values of vapour quality the method does not work properly. Other methods were also tested with even larger errors, thus being excluded from this graphical comparison.

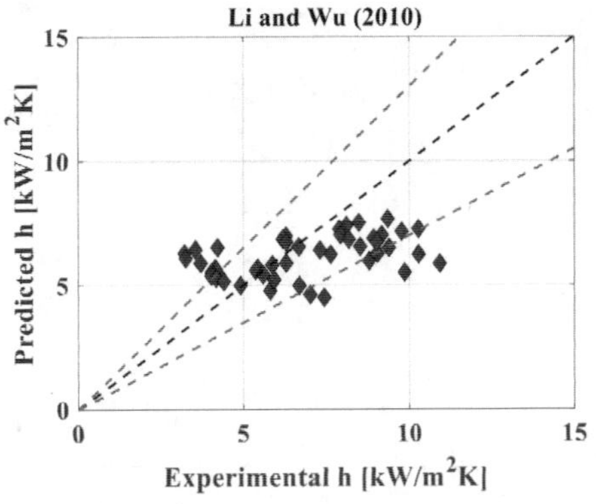

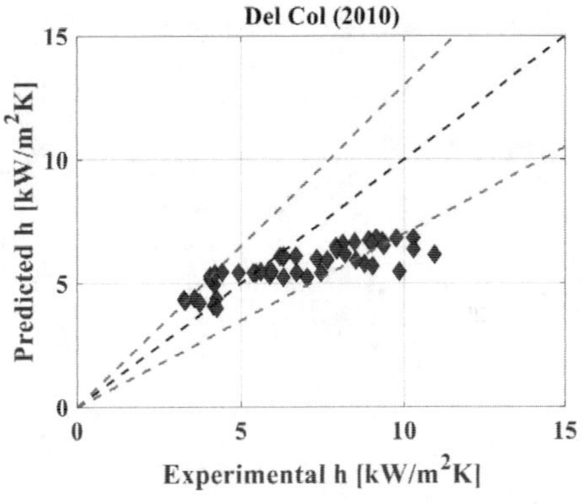

Figure 4. Predicted vs experimental heat transfer coefficient. (a) Li and Wu (2010) correlation; (b) Del Col (2010) correlation.

4.2. Pressure drop

All pressure drop experiments have been obtained within the same range of parameters used for the heat transfer coefficient tests, except for the adiabatic conditions maintained for such analysis. The effect of the mass flux and of the bubble saturation temperature on the frictional pressure gradient is provided in Figure 5a-b. Larger pressure drop values are obtained, as expected, for higher values of the mass velocity. In fact, its increase leads to higher flow velocities and therefore higher friction. On the contrary, the flow reduces itsspeed with the increase of the saturation temperature, due to a progressively higher vapor density when approaching the critical point.

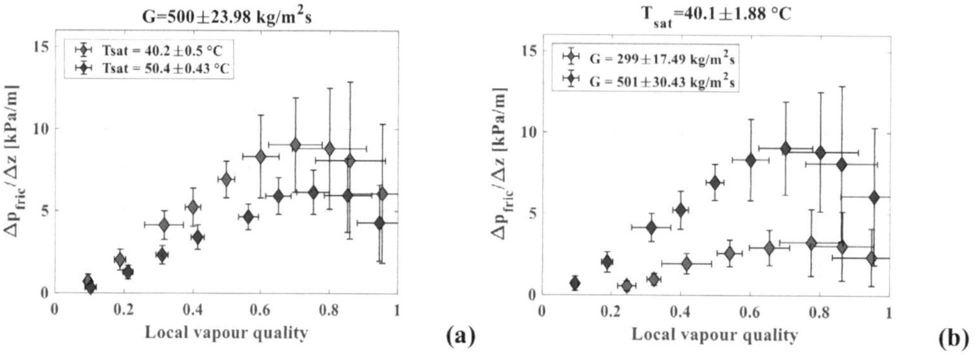

Figure 5. Effect of the operating conditions on the R450A two-phase pressure drop. (a) Effect of mass velocity; (b) Effect of bubble saturation temperature.

The assessment of the two-phase frictional pressure gradient prediction methods is graphically shown in Figure 6a-b for the correlations of Cicchitti (1960) [31] and Müller-Steinhagen and Heck (1986) [32], respectively. The best accuracy was obtained with the homogeneous flux model of Cicchitti [31] having a MAPE of 33.15% and a MRPE of 8.88%. Similarly, the Müller-Steinhagen and Heck correlation (1986) [32] provides a MAPE of 29.92% and a MRPE of 21.69%.

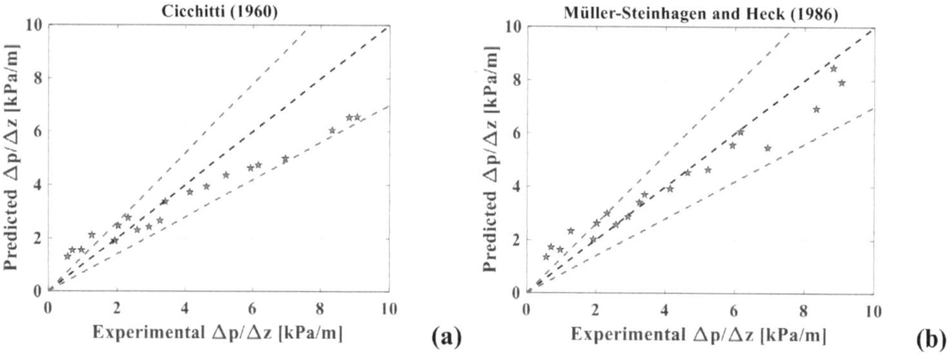

Figure 6. Predicted vs experimental pressure drops. (a) Cicchitti's model (1960); (b) Müller-Steinhagen and Heck (1986) correlation.

5. Conclusions

Two-phase flow boiling heat transfer coefficient and pressure drop data of refrigerant R450A have been collected in this paper. The effect of the operating conditions imposed has been analyzed and the assessmentof some correlations has been performed. The main outcomes are listed as follows:

- The vapor quality has a positive effect on the boiling heat transfer coefficient up to the dry-out condition, especially with high values of the mass velocity. For low mass fluxes, instead (150 $kgm^{-2}s^{-1}$), an early dry-out can be detected right after the onset of boiling. The saturation temperature has a negative effect on the heat transfer coefficient, due to a reduction of the flow velocity, whereas the heat flux significantly enhances the heat transfer performance, thus showing a non-negligible nucleate boiling contribution.
- The correlation of Del Col et al. [30] best fits the heat transfer coefficient experimental data, with a MAPE of 26.50% and a MRPE of -0.17%.
- The frictional pressure gradients at adiabatic conditions are increased with increasing mass flux and decreasing saturation temperature. The model of Müller-Steinhagen and Heck works

best with the present database with a MAPE of 29.92% and a MRPE of 21.69%.

6. References

[1] Abas N, Kalair A R, Khan N, Haider A, Saleem Z, Saleem M S 2018 *Renew. Sustain. Energy Rev.* **90** 557–569
[2] Choi S, Oh J, Hwang Y, Lee H 2017 *Appl. Therm. En.* **120** 88–98
[3] Pelella F, Viscito L, Mauro A W 2022 *Energy Conversion and Management* **260** 115646
[4] Metz B, Davidson O, Bosch P, Dave R, Meyer L 2007 *Fourth Assessment Report of the Intergovernmental Panel on Climate Change*
[5] Khan M, Wen J, Shakoori M.A, Tao, W 2022 *J. of Molecular Liquids* **353** 118795
[6] Mota-Babiloni A, Navarro Esbrì, J, Barragàn-Cervera A, Moles F, Peris B 2015 *Int. J. Refrigeration* **51** 52-58
[7] Mota-Babiloni A, Navarro Esbrì J, Barragàn-Cervera A, Moles F, Peris, B 2015 *Energy* **90** 1636-1644
[8] Makhnatch P, Mota-Babiloni A, Kodabandeh R, 2017 *Int. J. Refrigeration* **84** 26-35
[9] Makhnatch P, Mota-Babiloni A, Lopez-Belchi A, Kodabandeh R *Energy* **166** 223-235
[10] Gataric P, Lorbek L 2021 *Int. J. Refrig.* **128** 22–33
[11] Molinaroli L., Lucchini A, Colombo L P M 2021 *Int. J. Refrig.* **135** 139–147
[12] Belman-Flores J M , Mota-Babiloni A, Ledesma S, Makhnatch P 2017 *Applied Thermal Engineering* **127** 996-1004
[13] Zendehboudi A, Mota-Babiloni A, Makhnatch P, Saidur R, Sait S M 2019 *Int. J. Refrigeration* **100** 141-155
[14] Al-Sayyab A K S, Mota-Babiloni A, Navarro-Esbrì J 2021 *Energy Conversion and Management* **228** 113703
[15] Jacob T A, Matty E P, Fronk B M 2019 *Int. J. Refrigeration* **103** 274-286
[16] Azzolin M, Berto A, Bortolin S, Del Col D 2022 *Int J. Refrigeration* **137** 153-165
[17] Diani A, Liu Y, Wen J, Rossetto L 2022 *Int. J. Heat Mass Transfer* **196** 123260
[18] Morrow J A, Derby M M 2022 *Int. J. Heat Mass Transfer* **192** 122894
[19] Wojtan L, Ursenbacher T, Thome J R 2005 *Int. J. Refrigeration* **48** 2970-2985
[20] Lillo G, Mastrullo R, Mauro A W, Viscito L 2018 *Int. J. Heat MassTransfer* **126** 1236-1252
[21] Arcasi A, Mastrullo R, Mauro A W, Viscito L, 2021 *Int. J. Heat Mass Transfer* **164** 120604
[22] Arcasi A, Mauro A W, Napoli G, Viscito L 2022 *Int. J. Heat Mass Transfer* **188** 122599
[23] MATLAB 2019a Release *Mathworks*
[24] Lemmon E W, Mc Linden M O, Huber M L 2009 *REFPROP NIST* Database **23**
[25] Mauro A W, Napoli G, Pelella F, Viscito L 2020 *Int. J. Refrigeration* **119** 195-205
[26] Moffat J R 1985 *J. Fluids Eng.* **107** 173-178
[27] Lillo G, Mastrullo R, Mauro A W, Viscito L 2018 *Energy Procedia* **148**,1034- 1041.
[28] Mastrullo R, Mauro A W, Viscito L 2017 *Journal of Physics: Conference Series* **923** 012015
[29] Li W, Wu Z 2010 *Int. J. Heat Mass Transfer* **53** 1967-1976
[30] Del Col D 2010 *Exp. Therm. Fluid Sci.* **34** 234-245
[31] Cicchitti A, Lombardi C, Silvestri M, Soldaini G, Zavalluilli R 1960 *Energia Nucleare* **7** 407-425
[32] Müller-Steinhagen H, Heck K, 1986 *Chem. Eng. Proceedings* **20** 297–308

Pool boiling performances comparison of FC-72 and Novec 649 in the presence of a DC electric field

A I Garivalis[1,*], **P Fanani**[1] **and P Di Marco**[1]

[1] DESTEC, University of Pisa, Largo Lucio Lazzarino 1, Pisa 56122, Italy

* Corresponding author: alekos.garivalis@unipi.it

Abstract. Electronics cooling by direct immersion of the components in a dielectric liquid to obtain pool boiling is a valuable solution to dissipate the high heat fluxes of latest and increasingly challenging requirements. In the last decades, perfluorocarbons such as FC-72 were intensively studied and used. However, such fluids will be replaced by other refrigerants with lower Global Warming Potential. The Novec 649 is a suitable alternative to FC-72 as most of the thermophysical properties are similar. Enhanced boiling of FC-72 due to electric fields has been a subject of studies in the past years, for ground and space applications. The aim of this work is to compare the boiling performances of the two fluids in the presence of a DC electric field. Boiling curves were obtained at ambient pressure with fully degassed fluids and at different subcooling levels. Results showed that the performances of the two fluids are comparable in the absence of the electric field, but there are some differences when the electric field is on: electric field enhances the heat transfer coefficient of Novec 649 at low heat fluxes and the Critical Heat Flux at high subcooling. Boiling patterns and bubbles dynamics appear quite different between the two fluids in the presence of the electric field. These observations are explained with the fact that while FC-72 behaves as an electric insulator, while Novec 649 is a leaky-dielectric in the examined conditions. Thus, Novec 649 is a valid replacement of FC-72 and further investigations are needed to better quantify its boiling performances under the action of an electric field.

1. Introduction

Boiling heat transfer is recognized as one of the most effective ways to remove heat. In the last decades, the cooling needs of microelectronics and power electronics increased considerably. An alternative to traditional air and indirect liquid cooling consists of the direct immersion cooling, where the electronic components dissipate substantial amounts of heat by boiling of a surface in direct contact of a dielectric liquid. The refrigerant fluid should have good dielectric and thermophysical properties, chemical compatibility with the materials in contact with, chemical stability, low flammability, low toxicity, and low environmental impact [1].

Concerning the last point, refrigerants must avoid ozone depletion and greenhouse effect. In the last four decades, the perfluorohexane (C_6F_{14} that can be found on the market as FC-72 Fluorinert^{TM}) has been studied and used for cooling of power electronics. However, in atmosphere FC-72 is a greenhouse gas having GWP of 9000 over a one-hundred-year lifetime [2]. For this reason, it will be replaced by other coolants, such as the Novec^{TM} 649 Engineered Fluid, commercial name of the $CF_3CF_2C(O)CF(CF_3)$. Thermophysical properties of the two fluids are comparable (see Table 1), but the Novec 649 has global warming potential of one over a one-hundred-year lifetime. Thus, it is interesting to compare the boiling performances of the two fluids in view of a replacement.

The very first pool boiling experiments performed with Novec 649 are due to Forrest et al. [1] in 2009: they used a nickel wire and compared with FC-72, founding that the performances are comparable. Novec 649 showed slightly higher heat transfer coefficient but moderately lower Critical Heat Fux (CHF). More recently, other authors compared the two fluids using enhanced surfaces such as copper surfaces with deep minichannels [3], microporous copper surfaces and hydrophilic and hydrophobic nanostructured surfaces [4]. The results are similar to the ones by [1], i.e., similar performances between the two fluids were observed.

Table 1. Physical and thermophysical properties for Novec 649 and FC-72.

	Novec 649	**FC-72**
Boiling point [°C]	49.2	56.6
Freezing point [°C]	-108	-90
Critical temperature [°C]	169	178
Critical pressure [bar]	18.7	18.3
Surface tension [mN/m]	10.8	11.0
Thermal conductivity [W/(mK)]	0.059	0.056
Liquid specific heat [J/(kgK)]	1103	1050
Liquid density [kg/m^3]	1602	1680
Kinematic viscosity [cSt]	0.39	0.38
Latent heat [kJ/kg]	88.0	93.1
Vapor pressure [kPa]	40.4	30.9
Dielectric constant [-]	**1.8**	**1.7**
Dielectric strength [kV at 2.54 mm]	>40	38
Resistivity [Ohm·cm]	**10^{10}**	**10^{13}**
Ozone Depletion Potential (ODP)	0	0
Global Warming Potential (GWP)	1	9000

In addition to the enhanced boiling surfaces, the dielectric properties of dielectric refrigerants allow to impose an external electric field to enhance boiling performances. Indeed, the presence of two phases and the liquid-vapor interface provide an electric force on bubbles, which can improve the heat transfer coefficient and delay the Critical Heat Flux [5]. These effects are particularly efficient in microgravity conditions, where buoyancy force is negligible. The use of an external electric field can significantly improve the boiling performances in space applications, restoring or exceeding the ground ones [6]. When a fluid is in the presence of an electric field, it is subjected to a volumetric electric force $\mathbf{f}_e$ [7]:

$$\mathbf{f}_e = \rho_E \mathbf{E} - \frac{\varepsilon_0 E^2}{2} \nabla \varepsilon_R + \frac{\varepsilon_0}{2} \nabla \left(\rho \left(\frac{\partial \varepsilon_R}{\partial \rho} \right)_T E^2 \right) \tag{1}$$

The first term of the right side of equation (1) is named electrophoresis and depends on the electric field $\mathbf{E}$ and the free charge density ρ_E; it is present when charge build-up occurs, and it is usually neglected in dielectric fluids. The second term is the dielectrophoresis, that is proportional to the electric field squared and the gradient of electric permittivity ε_R (ε_0 is the vacuum permittivity). The third term is called electrostriction (ρ is the mass density), and it is an irrotational term, thus it does not induce fluid motion. According to [8], the relaxation time $\tau = \varepsilon_0 \varepsilon_R / \sigma$ is the ratio between the electric permittivity and the electric conductivity of a fluid σ, and represents the characteristic time that charges need to move from the bulk to the external boundaries - such as the liquid-vapor interfaces – excluding the electric field from the bulk. The value of τ must be compared with a characteristic time t_c of the system, such as the period of electric field frequency or the interface oscillation. If $\tau \gg t_c$ the charges have no time to move towards the boundaries and thus the fluid behaves as a perfect insulator: electric field penetrates in the bulk and the dielectrophoretic and electrostriction terms of equation (1) dominate (electrophoretic term is negligible). If $\tau \ll t_c$ the charges reach the system boundaries within the characteristic time of the system, excluding the electric field from the bulk and making the fluid a conductor. In this case, the electrophoretic term can overcome the other terms. There is another condition, when $\tau \sim t_c$ and the electric field is present in the bulk but there is also free charges build-up; the fluid in these conditions is named leaky-dielectric [9]. In general, the electrophoretic term (due to the presence of free charges) is larger than the other two, when present. For insulating fluids, high electric permittivity gradients and electric field intensities influence the fluid dynamic, as dielectrophoretic and electrostriction terms are involved.

Electric fields have been studied in the past to enhance boiling performances of FC-72. The aim of this work is to experimentally investigate the enhancement capability of the electric field with Novec 649 also compare it with the FC-72.

2. Experimental methods

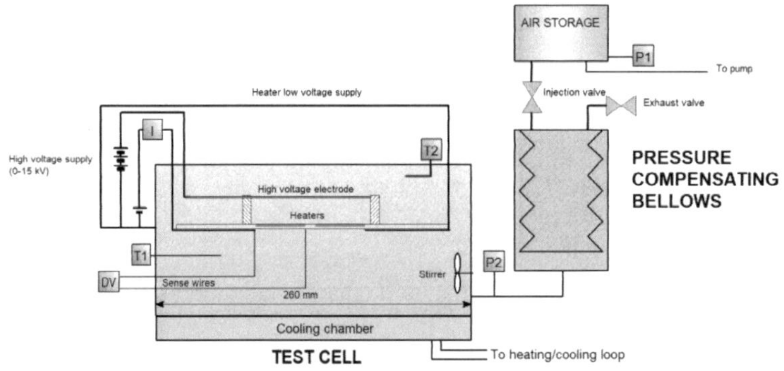

Figure 1. Schematic of the boiling support, the test cell and the and the pressurization systems.

Experiments were conducted using a dedicated apparatus, capable to obtain boiling curves controlling pressure, subcooling level and Non-Condensable Gases (NCG) concentration; in this framework, ambient pressure and complete degassed fluids were used, while different subcooling levels were set. The experimental facility is made of five subsystems: the boiling section (including the heater and the electrode to produce the electric field), the test cell, the pressure control, the temperature control and the power supplies and data acquisition systems. Part of the apparatus is shown in figure 1.

The boiling section is a 3D printed polycarbonate (PC) support fixed with screws to the test cell top; thermophysical properties of the PC allow to reach temperatures above 120 °C and ensure proper thermal insulation (k = 0.2 W/m/K). The square 10 × 10 × 0.5 mm^3 P-doped N-type silicon heaters that produce boiling are glued to the support by means of thermal and electrical insulating glue; the doped silicon used is electrically conductive allowing Joule heating by means of a DC power supply (ELC ALR3206D) connected with two copper wires soldered by ultrasonic welding technology. Current and voltage across the heater were measured to calculate the average heat flux provided to the heater during boiling; the good insulation properties of the support and the glue ensured that only a negligible amount of power was dissipated by conduction. The maximum temperature measurement error is calculated as 0.6 K. The complete uncertainty analysis is described by Liu et al. [10]. Temperature signal was also connected to a safety system that cat off the power when an abrupt increase of temperature occurred, indicating that the CHF value was reached. Above the heater, a metal grid made of 7 stainless-steel rods of 1 mm diameter and 5 mm spaced was placed at 6 mm from the grounded surface located at the heater level; this grid configuration produces a nearly uniform electrostatic field in the boiling region allowing the vapor to escape. A high DC of 15 kV was applied producing an electric field with average value of 2.5×10^6 V/m.

The test cell is a 2.3 L aluminium box that contains the boiling section and is filled with liquid. Two lateral windows allow to observe the boiling process using a high-speed camera (XIMEA xiQ-MQ022MG-CM, up to 450 fps) and a LED as backlight illumination. The cell is externally insulated with foam. The temperature of the liquid inside the cell is monitored by two temperature sensors (AD590) placed in different locations. During the experiments, the temperature of the liquid was kept within ±0.5 °C of the setpoint.

The pressure control system consists of a compensating bellows connected to the test cell. The liquid fills one side of the bellows chamber, while the other side contains air. Air can be injected from a high-pressure storage (2 bar) or ejected in the ambient operating two electro-valves. With this subsystem the pressure can be adjusted compensating liquid volume variations when temperature changes.

The temperature control system is a water closed loop operated by a pump. Water flows in a heating/cooling chamber below the test cell and in a heat exchanger on whose sides electric heater pads (200 W) and 8 Peltier cells are arranged. A PID control system switches the heaters to heat and the Peltier cells to cool depending on the test cell liquid temperature. A stirrer inside the test cell ensures the homogenization (not during the experiments).

Finally, a National Instruments (NI) cDAQ-9174 data acquisition device was used to collect the signals of the sensors. Images from the camera were collected to observe the boiling patterns.

3. Results

3.1. Tests at saturation

In figure 2 the performances of the two fluids are showed. It can be noticed that the CHF of FC-72 is slightly higher than the Novec 649, while the heat transfer coefficient is better for Novec. Similar trends were obtained by Forrest et al. [1] with the same fluids boiling on a Nickel wire. Even if almost the same superheat is required, Novec 649 has a lower saturation temperature, resulting in a lower wall temperature. The theoretical values of CHF calculated with Zuber-Kutateladze correlation [11] are reported. Measured values without electric field are predicted within 6% error, which is the same order of magnitude of the uncertainty. At saturation, the effect of the electric field on the FC-72 is to increase the CHF of about 8%, while for Novec 649 remains the same. However, electric field has a strong effect on the heat transfer coefficient of Novec 649 for low heat fluxes. Such effect is not present for FC-72. Finally, the VDI correlation [12] is used to predict the slope of the heat transfer coefficient (using 0.82 as exponent). It is in good agreement for both fluids in the absence of the electric field and for the FC-72 with the electric field, while electric field induces an enhancement effect for Novec 649.

3.2. Effect of subcooling

In figure 3 the boiling curves at different subcooling levels (saturation, 5 K and 25 K) are compared for the two fluids. For the FC-72 the CHF always increases with subcooling; moreover, the wall superheat (that in turn determines the heat transfer coefficient) does not change significantly with subcooling level or electric field. On the other hand, for the Novec 649 the presence of the electric field produces a decrease of the wall superheat especially at low heat fluxes and high subcooling, indicating that the combination of subcooling and electric field enhances the heat transfer coefficient. Concerning the CHF, while close to saturation there is no significant enhancement, a clear increase is observed with 25 K of subcooling.

3.3. Critical Heat flux

The CHF data are collected in figure 4. Generally, the CHF of FC-72 is higher than Novec 649 in the absence of electric field. The effect of the electric field for FC-72 is a delay of the boiling crisis, for all the subcooling levels considered. On the other hand, small or negligible enhancement of CHF is observed with Novec 649 at low subcooling levels. However, increasing the subcooling, the enhancement capability of the electric field increases, and the critical heat fluxes of both fluids is comparable.

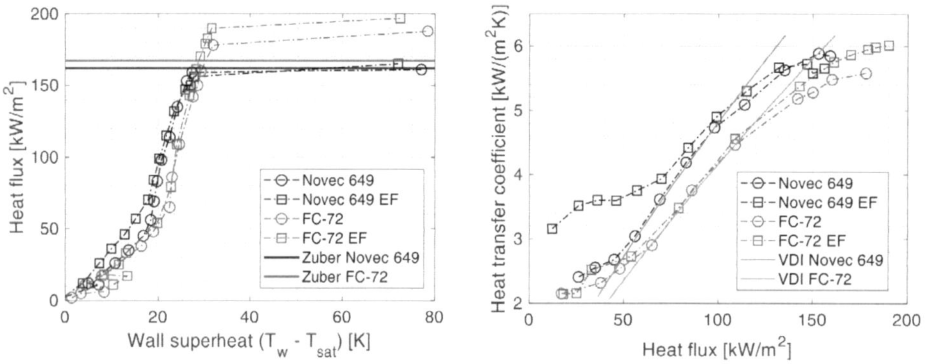

Figure 2. Boiling curves and heat transfer coefficients comparison at saturation with and without electric field. The VDI correlation is also showed [3].

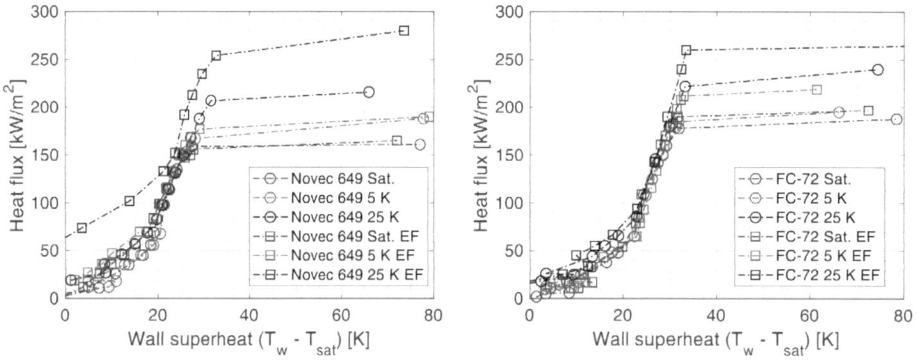

Figure 3. Boiling curves of Novec 649 and FC-72 with and without electric field and at different subcooling levels.

4. Discussion

Despite FC-72 and Novec 649 present similar thermophysical properties and the boiling performances are comparable, there are some important differences in terms of electric field effects that were presented. Such aspects are associated to differences in the boiling patterns that were observed during the tests with the high-speed camera.

In figure 5 the boiling patterns of the two fluids in the same conditions are compared. FC-72 shows no significant differences between the two cases (figures 5c and 5d). Bubbles are elongated vertically due to the electric stresses (like the bubble at the left corner of figure 5d) but the large amount of vapor makes them to coalesce into the columns. Being the boiling pattern similar, also similar heat transfer coefficients were observed (figure 2).

The boiling pattern of FC-72 is very similar to the one of Novec 649 in the absence of electric field (figures 5a and 5c). However, a great difference appears when the electric field is applied: bubbles are vigorously pushed towards the heater and spreads laterally (figure 5b). In the authors opinion, this lateral agitation and spread of vapor outside the heater boundaries is the reason why the heat transfer coefficient is enhanced, especially at lower heat fluxes, when the amount of vapor is less. The shape of single bubbles is less elongated and more similar to a bullet (see figure 5a on the left). Lower bubbles density allows the electric field to distribute and penetrate maximizing the electric stresses at the interface [8,11] and producing forces that repel each other [13]. On the other hand, when there is a large amount of vapor at high heat fluxes, bubbles coalesce in columns cancelling the previous benefits. Electric forces tend to push the bubbles to the heater, until the vapor reaches the proximity of the grid where there is a high gradient of the electric field, that induces vapor to escape towards the bulk liquid [14]. However, Novec 649 vapor is subjected to higher electric stresses that obstacles the vapor escaping at low subcoolings, reducing the CHF enhancement (figure 4).

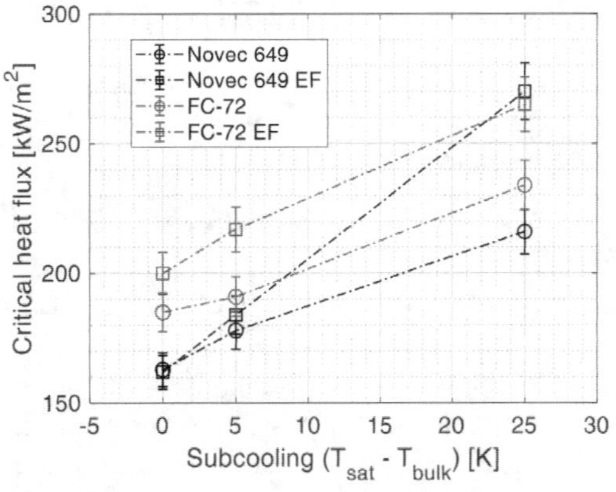

Figure 4. Critical Heat Flux for both fluids as a function of subcooling level in the absence and presence of the electric field.

Another proof of that is provided in figure 6. Increasing the subcooling, the amount of vapor below the grid is reduced. Vapor bubbles are always pushed towards the heater and laterally, but when the subcooling is higher (figures 6a and 6b) and bubbles are small and far from the grid, a reduction of the bubbles size is observed in the presence of the electric field.

High subcooling means higher temperature gradients, that in turn produce electroconvection. This fact, together with the improved penetration of the electric field (no vapor that shields it) can explain the CHF and heat transfer coefficient enhancement of Novec 649 at 25 K subcooling.

In summary, Novec 649 and FC-72 behave differently in the presence of an electric field. According to Table 1, the electric permittivity of the two fluids is similar. However, the electrical resistivity of Novec is three orders of magnitude lower than FC-72. This seems to be the most important difference together with the GWP. Electric resistivity and dielectric permittivity influence the relaxation time of the electric charges, and the resulting electric behaviour of the fluid: the relaxation time for the two fluids is $\tau = 155$ s for FC-72 and about $\tau = 0.2$ s for Novec 649. It is possible to estimate from the high-speed videos the bubbles detachment, which occurs between 0.1 and 1 seconds; this can be considered as the range of the characteristic time of the system t_c. Thus, while FC-72 can be considered an insulating fluid (verified by [14]), the Novec 649 behaves as a leaky-dielectric. In a leaky-dielectric, free charges accumulate at the interfaces modifying the field; free charges usually dominate the electric force, and their effect is stronger than dielectrophoresis and electrostriction forces.

5. Conclusions

A comparison of the pool boiling performances at ambient pressure and degassed conditions of the two dielectric fluids FC-72 and Novec 649 was performed. Results showed that the fluids are mostly comparable in the absence of the electric field, confirming previous works available in literature. Zuber-Kutateladze and VDI correlations are verified for Critical Heat Flux and heat transfer coefficient respectively. Increasing the subcooling level produces an increase of the CHF.

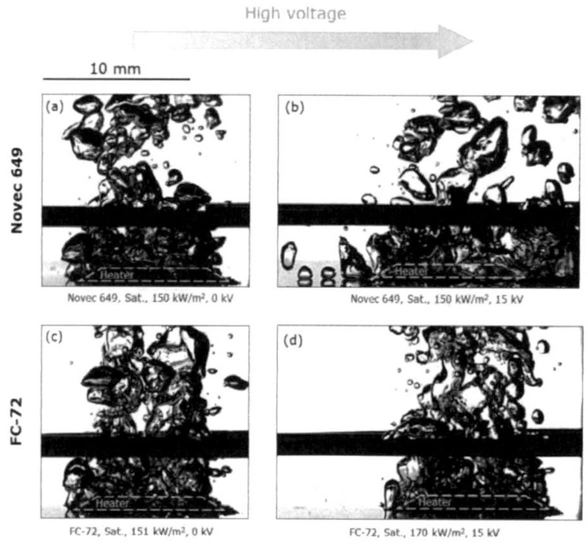

Figure 5. Comparison of the boiling patterns in saturation conditions. (a) Novec 649 in the absence of electric field. (b) Novec 649 in the presence of the electric field. (c) FC-72 in the absence of electric field. (d) FC-72 in the presence of the electric field.

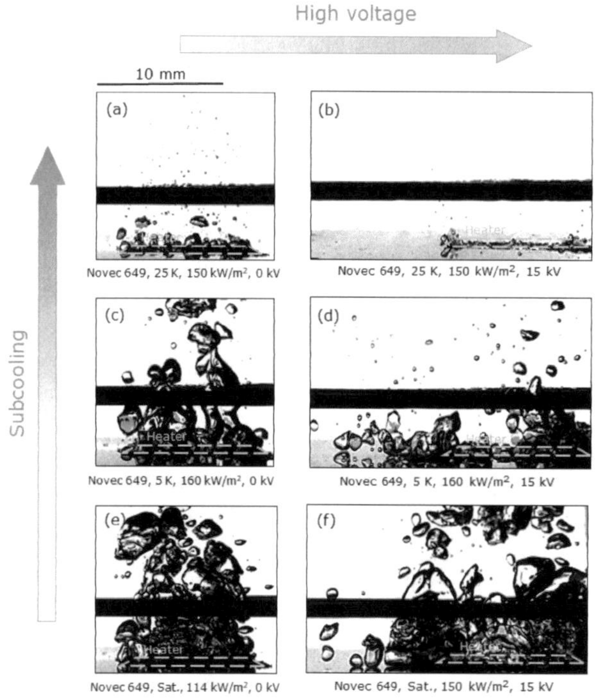

Figure 6. Comparison of the boiling patterns of Novec 649. (a) 25 K subcooling in the absence of electric field. (b) 25 K subcooling in the presence of the electric field. (c) 5 K subcooling in the absence of electric field. (d) 5 K subcooling in the presence of the electric field. € Saturation conditions in the absence of electric field. (f) Saturation conditions in the presence of the electric field.

Indeed, a difference is noted when the electric field is applied by means of a grid electrode: bubbles motion and agitation is quite more intense for Novec 649 with respect to FC-72, leading to a reduction of the wall superheat in the first part of the nucleate boiling curve (the isolated bubbles regime). Noteworthy, an accurate measure of electric conductivity for Novec 649 is currently not available in literature. However, according to the manufacturer data and the experimental observations, we concluded that the Novec 649 behaves as a leaky-dielectric.

In conclusion, based on experimental evidence the Novec 649 is a suitable candidate to replace FC-72 in boiling applications and showed good potentialities for the enhancement through an external electric field – especially if combined with high subcooling levels. A better measure of the electrical conductivity, a detailed analysis of the electric forces and of the influence of the electrode shape will be performed.

6. References

[1] Forrest E, Hu L W, Buongiorno J and McKrell T 2009 Pool Boiling Performance of Novec 649 Engineered Fluid *ECI International Conference on Boiling Heat Transfer, Florianopolis-SC-Brazil, 3-7 May 2009*

[2] Anon 2006 *EPA-430-R-06-90, Uses and emissions of liquid PFC heat transfer fluids from the electronics sector.*

[3] Pastuszko R, Kaniowski R, Dadas N and Bedla-Pawlusek M 2021 Pool boiling enhancement and a method of bubble diameter determination on surfaces with deep minichannels *Int. J. Heat Mass Transf.* **179**

[4] Može M, Vajc V, Zupančič M and Golobič I 2021 Hydrophilic and hydrophobic nanostructured copper surfaces for efficient pool boiling heat transfer with water, water/butanol mixtures and novec 649 *Nanomaterials* **11**

[5] Di Marco P and Grassi W 1993 Saturated Pool Boiling Enhancement by Means of an electric Field *Enhanc. Heat Transf.* **1** 99–114

[6] Garivalis A I, Manfredini G, Saccone G, Di Marco P, Kossolapov A and Bucci M 2021 Critical heat flux enhancement in microgravity conditions coupling microstructured surfaces and electrostatic field *npj Microgravity* **7** 37

[7] Landau L D and Lifshitz E M 1984 *Electrodynamics of Continuous Media* (New York: Pergamon)

[8] M. C. Zaghdoudi, M. Lallemand 2001 Nucleate Pool Boiling Under Dc Electric Field *Exp. Heat Transf.* **14** 157–80

[9] Saville D A 1997 ELECTROHYDRODYNAMICS:The Taylor-Melcher Leaky Dielectric Model *Annu. Rev. Fluid Mech.* **29** 27–64

[10] Liu B, Garivalis A I, Cao Z, Zhang Y, Wei J and Marco P Di 2022 Effects of electric field on pool boiling heat transfer over microstructured surfaces under different liquid subcoolings *Int. J. Heat Mass Transf.* **183** 122154

[11] Liang G and Mudawar I 2018 Pool boiling critical heat flux (CHF) – Part 1: Review of mechanisms, models, and correlations *Int. J. Heat Mass Transf.* **117** 1352–67

[12] Gorenflo D and Kenning D 2010 H2 pool boiling *VDI Heat Atlas*

[13] Siedel S, Cioulachtjian S, Robinson A J and Bonjour J 2016 Lateral coalescence of bubbles in the presence of a DC electric field *Int. Commun. Heat Mass Transf.* **76** 127–32

[14] Garivalis A I and Di Marco P 2022 Isolated bubbles growing and detaching within an electric field in microgravity *Appl. Therm. Eng.* **212** 118538

Acknowledgments

This work is supported by the Joint ESA-CMSA Project (TGMTYY00-RW-05-1.00), ESA (European Space Agency) under the MAP-MANBO project (AO-2004-111), Second batch of scientific experiment proposals aboard China Space Station (TGMTYY00-JY-53-1.00), and the National Natural Science Foundation of China (No.51636006).

Gravity induced shape effects on the time-dependent evaporation of pendant drops

S Tonini*, G E Cossali

Department of Engineering and Applied Sciences, University of Bergamo, via Marconi 6, 24044 Dalmine (BG), Italy

* Corresponding author: simona.tonini@unibg.it

Abstract. The paper presents a method to model the time-dependent evaporation of pendant drops taking into account the effect of drop deformation induced by gravity. The model is based on the solution to the time-dependent drop mass and energy conservation equations, where the mass and energy fluxes through the gas mixture are numerically evaluated for a range of Bond numbers and contact angles. The evaporation characteristics of pendant and sessile drops on hydrophobic and hydrophilic substrates are compared in terms of evaporation times and evaporative cooling, for both constant contact angle and constant contact radius modes.

1. Introduction

The interest of the scientific community on the evaporation of a drop in contact with a solid surface is witnessed by the huge literature available on this topic. This is mainly due to its importance in many applications (like, printing technology [1], cooling systems [2], surface patterning [3], biomedical processes, etc.; see [4] for a recent review), but also to the complexity of the phenomena occurring in this process. Pendant drops, which are drops hanging from a surface, are often used for experimental studies on drop evaporation [5, 6], and they are important for surface tension measurements of air-liquid surfaces [7, 8], in heat transfer processes [9] or in large volume protein single-crystals growth [10]. Differently from sessile drops, pendant drops become unstable when their size increases, falling from the substrate, a fact that limits the size of the drops. The shape of a drop in contact with a solid surface has been deeply studied since the pioneering work of [11] and a copious literature on this subject is available (see for example [12] and the cited literature therein).

The effect of gravity on the processes involving pendant drops has been often related to the natural convection caused by buoyancy in the gas surrounding the drop [13]. Gravity also influences the flow inside the drop, leading to fluid structures with interesting practical implications (see for example [14, 15]). The first study on the specific effect of gravity induced deformation on the evaporation of a pendant drop is possibly that of [16], who found experimentally an increment in the evaporation rate up to about 30%. In an experimental study presented in [17], it was shown that a pendant drop on super-hydrophobic substrates evaporates faster than the sessile one of equal volume, suggesting that this may be partially ascribed to the different drop shapes, caused by the opposite effect of gravity, considered the non-negligible value of the Bond number ($Bo = 0.2$).

Pendant and sessile drop evaporation can be considered a typical example of a diffusion-controlled phenomenon. Quasi-steadiness is often assumed, considering that the drop life-time is orders of magnitude larger than the transport time-scale. In the absence of external convection, and assuming constant gas mixture thermophysical properties, the analytical problem reduces to the solution to a Laplace equation [18, 19]. For drops with a spherical-cap shape (i.e. with negligible effect of gravity on their shape), the analytical solution given by [20] in terms of Mehler-Fock transform is almost universally used (see [21] for some different types of solution). Even when thermophysical properties are temperature dependent, the phenomenon can be modelled through the solution of a Laplace problem (see [22]). The analytical-numerical model reported in the next sections considers the evolution of the shape and temperature of evaporating pendant drops, and quantifies the effect of the actual drop shape and of the evaporative cooling for different evaporation modes and substrate wettability.

2. Mathematical modelling

In this paper, a mixed analytical-numerical approach will be used to model the diffusion-limited processes taking place during the evaporation of a pendant drop deformed by the effect of gravity.

2.1. *Modelling the energy and mass fluxes through the gas phase*

In the following, the gas phase is assumed to be a binary mixture of the vapourising species and the ambient gas, behaving as a single component ideal gas. Under quasi-steady conditions, the species conservation equations can be written as.

$$\nabla_j N_j^{(p)} = 0 \tag{1}$$

where p=0,1 for the ambient gas and the vapour, respectively, and the molar fluxes are $N_j^{(p)} = N_j^T y^{(p)} - Mm^{(p)} c D_{10} \nabla_j y^{(p)}$, where $N_j^T = N_j^{(0)} + N_j^{(1)}$ is the total molar flux, c the molar density, D_{10} the binary diffusion coefficient, $Mm^{(p)}$ the molar mass of the p-component and $y^{(p)}$ is the molar fraction of the p-component. The steady-state energy conservation equation is

$$\rho U_j \nabla_j \left(c_p^{(1)} T \right) = \nabla_j \left(k_{mix} \nabla_j T \right) \tag{2}$$

where k_{mix} is the gaseous mixture conductivity, $c_p^{(1)}$ is the vapour specific heat capacity, ρ is the mass density of the mixture and U_j is the mass average velocity. In equation (2) minor terms (like viscous dissipation), assumed to be irrelevant for the present analysis, are neglected (refer to [24] p. 465, or [25] p. 589 for a more complete form). The effect of external convection is neglected, but the Stefan flow is fully considered (U_j is the Stefan flow velocity). Assuming negligible absorption of the gas in the liquid drop and uniform Dirichlet boundary conditions on the drop surface, an analytical solution to the PDE system (1, 2) can be found, also when the gas mixture thermophysical properties (ρ, $c_p^{(1)}$ and k_{mix}) depend on temperature (see [22] and Appendix), and the solution is given in terms of an auxiliary function, Φ, solution of the following Laplace problem

$$\nabla^2 \Phi = 0; \quad \Phi(\text{drop surface}) = 1; \quad \Phi(\text{infinity}) = 0 \tag{3}$$

In analytical modelling, pendant and sessile drops are often assumed to have the shape of a spherical cap, since the drop size is usually smaller than the capillary length, $L_c = (\sigma/\rho g)^{1/2}$. However, it was shown [26] that the evaporation rate of a sessile drop may be perceptibly influenced by the gravity induced drop deformation even when the drop size is smaller than L_c. The analytical solution of the problem (3) for spherical caps was given in [20] (see also [21] for other types of solutions), but when the drop is deformed by gravity, analytical solutions are not available and numerical tools are needed (see next subsection). The shape of sessile and pendant drops in the gravitational field is fully defined by its volume and the substrate wettability [26], that is by the Bond number $Bo = R_{eq}^2 / L_c^2$ (where $R_{eq} = (3V/4\pi)^{1/3}$ and V is the drop volume,) and the contact angle, θ_c. Since the evaporation and the heat rates can be found from the integral $\int_A \nabla_n \Phi dA$ (see Appendix), both rates depend on Bo and θ_c and, to find such dependence, a parametric analysis is performed, using the numerical methods described in the following subsections.

2.2. *Energy transfer through the solid substrate*

The heat transfer from the solid substrate can be modelled assuming quasi-steadiness, and assuming that the substrate is much thicker than the drop, the problem is equivalent to the heat conduction from a semi-infinite solid to a disc of radius equal to the drop basis radius, R_c, laying on the solid surface. This problem can be solved using the conductive shape factor method obtaining

$$\dot{Q}_c = k_w \left(T_w - T_d \right) S \tag{4}$$

where S is the conductive shape factor. For the given configuration (a disk on a semi-infinite body) the shape factor is equal to $S=4R_c$ [27]. This simple method can be considered acceptable when the drop temperature is almost constant, and this happens just after the first period of temperature variation, which is usually very short compared to the whole evaporation period.

2.3. Numerical methods for the solution of the Laplace problem

The shape of a pendant drop in a gravitational field is found solving numerically the classical Bashford-Adams equation [11] by a Runge-Kutta method with variable time steps. A double iteration cycle was necessary to yield the equilibrium drop shape for given Bo and θ_c, and the relative accuracy on the drop volume was better than 2×10^{-4}, while the average accuracy on the contact angle was better than 0.4deg. The obtained drop shape was compared to the numerical solution of the integral method proposed in [28] (see [26] for details), and relative differences of the drop profile coordinates were less than 10^{-4}. The drop profiles where then used to set the numerical domain to solve the problem (3) using the commercial code COMSOL Multiphysics®. Figure 1 shows some samples of the pendant drop shape on two substrates. The upper limit of the Bond number for the two contact angles is due to the stability constrains for pendant drops (see section 3).

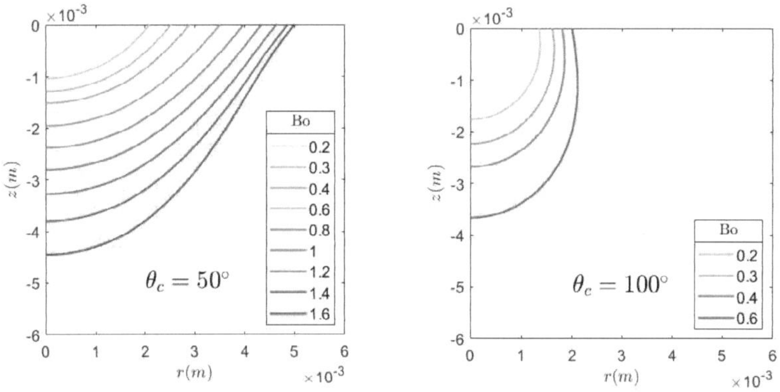

Figure 1. Pendant drop shapes on hydrophilic (θ_c = 50°) and hydrophobic (θ_c = 100°) substrates.

A finite element discretisation approach is used on 2D axis-symmetric grids. The computational domain is a half-sphere with a radius $R_\infty=30R_{eq}$ (see figure 2 for a schematic representation); this choice was suggested by previous works [29, 30]. The Dirichlet condition at infinity is substituted by an equivalent condition on the outer boundary, $R= R_\infty$, through an iterative procedure that accounts for the finiteness of the domain size (more details can be found in [26, 29, 30]). The mesh was unstructured, made by triangular tessellation, with grid refinements around the triple line. An analysis of grid independence was performed (more details can be found in [29, 31]) and in the present study grids made of about 10^5 cells were used. The evaporation and the sensible heat rates are evaluated by the integral $\int_A \nabla_n \Phi dA$ on the drop surface (see Appendix), and also on a spherical surface containing the drop, as an additional check. To quantify the effect of drop deformation on evaporation, a parameter G

$$G(\theta_c, Bo) = \frac{\dot{m}_{ev}(\theta_c, Bo)}{\dot{m}_{ev}(\theta_c, 0)} \quad (5)$$

is used, where $\dot{m}_{ev}(\theta_c, Bo)$ is the evaporation rate from a deformed drop with a given Bond number and $\dot{m}_{ev}(\theta_c, 0)$ is the corresponding value for an equivalent volume spherical cap, which is given by the Picknett-Bexon correlation [23]. Figure 2 shows the parameter G as a function of Bo for pendant drops on twelve substrates, with contact angle ranging from 20° up to 110°.

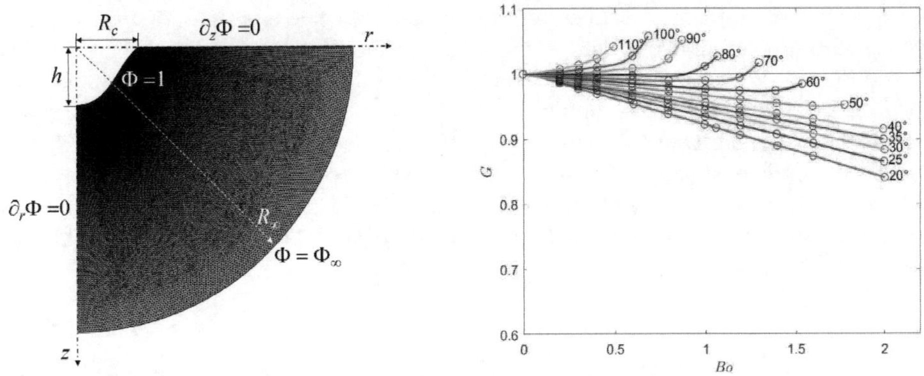

Figure 2. Schematic of the numerical domain and sample of the numerical mesh (size not to scale) (left). Non-dimensional evaporation rate G (equation 5) versus the Bond number for pendant drops on twelve hydrophilic and hydrophobic substrates (right).

For a fixed Bond number, the parameter G increases with the contact angle. For a fixed contact angle θ_c the parameter G reaches a minimum value at a specific Bond number, depending on the substrate wettability. On hydrophobic substrates, pendant drops deformed by gravity show higher evaporation rate than the corresponding iso-volumic spherical caps. For hydrophilic substrates, the minimum value of G is obtained with deformed pendant drops. The smaller the contact angle, the larger the value of Bo corresponding to the minimum value of G. Within the range of the selected operating conditions, neglecting the effect of deformation can lead to an underestimation of the evaporation rate from pendant drops on hydrophobic substrates up to about 7%, while for hydrophilic substrates the overestimation can reach about 16%. Validation of the numerical method was done on drop geometries where analytical solutions to the problem (3) exist [20], and the agreement on the values of the evaporation rates was of the order of 10^{-3} for hydrophilic substrates and 10^{-4} for hydrophobic substrates. This numerical tool was then used to build up a database for the quantity $\int_A \nabla_n \Phi dA$ that will be used to solve the mass and energy balance on the pendant drop, as described in the next subsection.

2.4. The mass and energy balance on the drop

The mass and energy conservation equations for the pendant drop, neglecting the non-uniformity of the inner temperature distribution, are

$$\frac{dM}{dt} = -\dot{m}_{ev}\,; \quad M\frac{d\left(c_L T_d\right)}{dt} = -\dot{m}_{ev} h'_{LV} + \dot{Q}_s + \dot{Q}_c \tag{6}$$

where M is the drop mass, c_L is the liquid specific heat, h'_{LV} is the modified latent heat of vaporisation [32, 33], $\dot{m}_{ev}$ is the evaporation rate, $\dot{Q}_s$ is the sensible heat rate exchanged through the drop/gas interface and $\dot{Q}_c$ is that exchanged with the substrate. $\dot{Q}_s$ and $\dot{m}_{ev}$ are calculated from the quantity $\int_A \nabla_n \Phi dA$ (see Appendix), which can be found from the above-mentioned database by a bivariate interpolation, given the instantaneous value of Bo and θ_c. An implicit Euler method was used to numerically solve equations (6).

3. Results and discussion

Time-dependent evaporation of water pendant and sessile drops is analysed, to point out the effect of drop deformation on the evaporation characteristics. The two main evaporation modes of sessile and pendant drops are the so-called constant contact radius (CCR) and constant contact angle (CCA). The CCR mode, often observed for hydrophilic substrates, is characterised by the pinning of the triple line and the constancy of the drop base radius. The CCA is often observed for hydrophobic substrates, the contact angle remains constant while the drop shrinks and the triple line slides on the substrate surface. However, when a drop evaporates on real substrates, both modes may appear and also a combination of the two is observed. With the aim of evidencing the sole effect of the drop shape on the evaporation, simplified scenarios will be used, assuming that the drop evaporates in one of the two modes during the whole observation period. One hydrophobic (θ_c=100°) and one hydrophilic (θ_c=50°) substrates are considered, and pendant, sessile and spherical-cap shaped drops are assumed to evaporate under both modes. Figure 3 shows the shape and the stability map (see [12] for a clear explanation) of sessile and pendant drops, respectively, and the paths followed by drops on hydrophobic and hydrophilic substrates are reported for both evaporation modes (thick black lines).

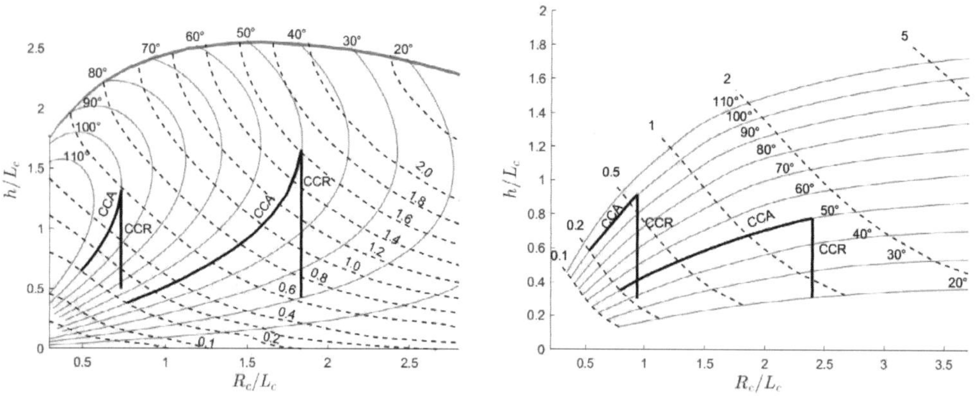

Figure 3. Stability map for pendant drops (left) and shape map for sessile drops (right). The top grey curve in the graph on the left is the stability limit for pendant drops. The dashed lines are iso-*Bo*, the continuous lines are iso-θ_c, the thick black curves show the paths of CCA and CCR evaporating drops. The meaning of the geometrical parameters h and R_c is explained in figure 2.

The initial Bond number is 0.6 for the hydrophobic substrate and 1.6 for the hydrophilic substrate. When the CCA evaporation mode is set, the evaporation process is recorded until the Bond number reduces to 0.2, while with the CCR evaporation mode the process is analysed until the contact angle reduces to 20° for the hydrophilic substrate ($\theta_{c,i}$=50°) and until the Bond number reduces to 0.25 for the hydrophobic substrate ($\theta_{c,i}$=100°). The evaporative cooling effect is considered, setting the wall thermal conductivity equal to 1.05 W/mK (glass substrate) and it is further analysed later in this section. The initial drop temperature is fixed equal to 20°C in equilibrium with the wall and the surrounding gas. Figure 4 shows, for the CCA evaporation mode and $\theta_{c,i}$=50°, the transient profiles of the non-dimensional equivalent drop radius squared versus the non-dimensional time, $\tau = tD_{10} / R_{eq,0}^2$, where D_{10} is the mass diffusion coefficient of the vaporizing species in air, calculated at the initial conditions, for sessile, pendant and a spherical cap drops, with the same initial volume. The evaporation process follows the well-known D^2-law for all the drop shapes (with the exception of the short initial evaporative cooling period), with the sessile drop vaporizing faster. Figure 4 reports the corresponding profiles of the drop temperature, which evidences the initial cooling (about 0.5 degrees) followed by an almost constant temperature period. The time to reach *Bo*=0.2 is 3.3% and about 7% greater for the spherical and the pendant drops, respectively. For the CCA evaporation mode and $\theta_{c,i}$=100° the differences in the time needed to reach the Bond number equal to 0.2 differ by less than 0.9%. Figure 5a shows the transient profiles of the heat rates as function of τ for the three drop shapes vaporizing in CCR mode on a hydrophilic substrate. When the drop contact angle reduces to 20°, the final Bond number is equal to 0.874, 0.673 and 0.8 for the sessile, the pendant and the spherical drops, respectively. For the CCR mode too, the pendant drop vaporizes slower, always experiencing smaller evaporation and heat rates. Differently from the CCA cases, the evaporation and heat rates of the spherical-cap drop are initially lower than those of the sessile drop; at about τ =5000, the spherical-cap drop starts to show higher evaporation and heat rates. The time to reach Bo=0.9 is about 12% longer for the pendant drop compared to the sessile drop, and about 0.7% shorter for the spherical drop. The drop temperature varies in a range of less than 0.5 degrees for all the three drop shapes (not shown here). On a hydrophobic substrate ($\theta_{c,i}$=100°), the effect of drop shape is rather limited, as for the corresponding CCA case. When the Bond number reduces to 0.25, θ_c is equal to 40.7°, 67.9° and 47.8° for the sessile, the pendant and the spherical drops, respectively. The lifetime to evaporate up to this point is 1.5% and 0.2% higher for the pendant and the spherical drops, respectively, compared to the sessile one.

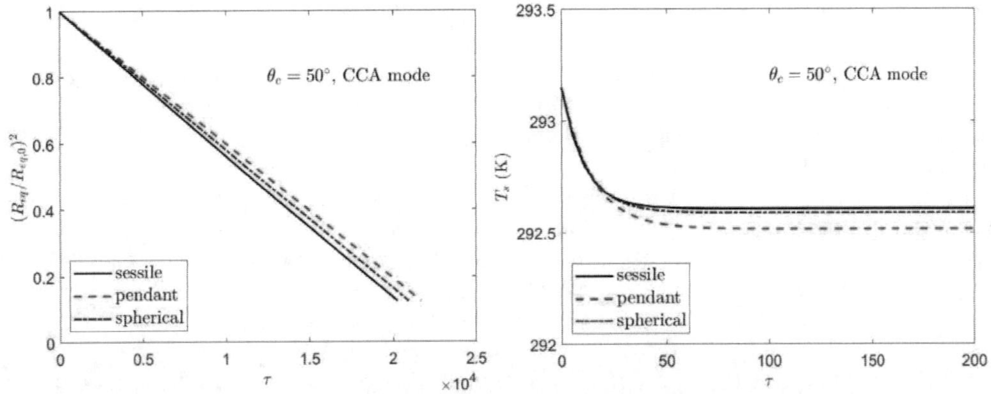

Figure 4. Non-dimensional drop radius squared (left) and drop temperature (right) versus the non-dimensional time, for a sessile, a pendant and a spherical cap drop, on a hydrophilic substrate.

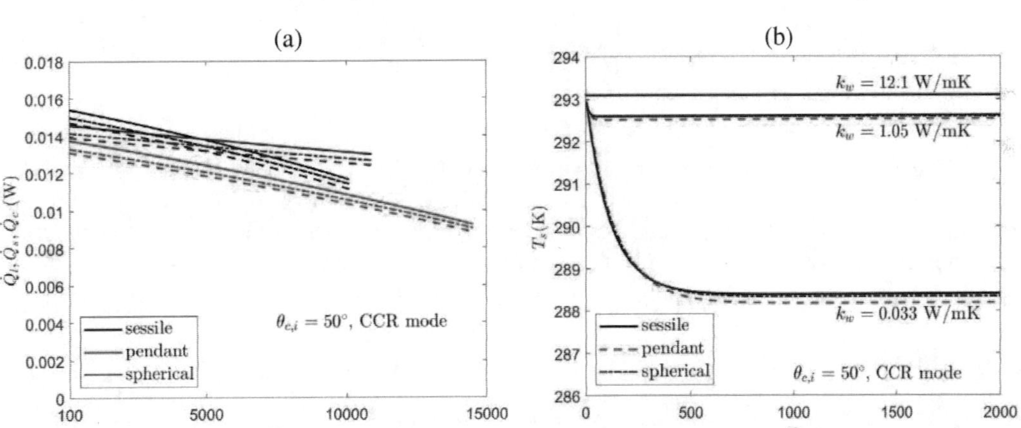

Figure 5. (a) Sensible heat rate exchanged with the gas, Q_s (dash line), heat rate from the substrate, Q_c (dash-point line) and latent heat rate, $Q_l = m_{ev}h'_{LV}$ (solid line), versus the non-dimensional time. (b) Drop temperature versus the non-dimensional time on a substrate with three thermal conductivities.

The effect of wall thermal conductivity on the cooling and evaporation characteristics for the CCR mode and θ_c=50° is briefly analysed, for three values of the wall thermal conductivity, namely 0.033 W/mK (polyurethane), 1.05 W/mK (glass) and 12.1 W/mK (stainless steel). Figure 5b shows the profiles of drop temperature as function of the non-dimensional time within the initial evaporative cooling period. The cooling effect is important for the polyurethane substrate, with drop temperature reducing of about 4.4°C. On more conductive walls the cooling effect is less evident and the evaporation becomes faster. The effect of drop shape is similar for all the scenarios with the three selected thermal conductivities. The larger the thermal conductivity, the smaller the time when the spherical-cap shaped drop starts vaporizing faster than the sessile drop, while the pendant drop always experiences the slowest vaporization.

Conclusions

The paper reports a theoretical investigation of the evaporation of sessile and pendant water drops with the specific aim to show the effect of drop deformation due to gravity on the evaporation characteristics. Drops on hydrophobic (θ_c=100°) and hydrophilic (θ_c =50°) substrates were studied. A comparison between the evaporation characteristics of pendant and sessile drops, which are differently deformed by gravity, and a spherical non-deformed drop, deposed on the same substrates, was performed. Two evaporation modes (constant contact angle, CCA, and constant contact radius, CCR) were considered. For all the selected cases, the pendant drops always show the slowest vaporisation. Differences are larger for hydrophilic substrates than for the hydrophobic ones, mainly due to the fact that, for stability issues, the initial drop sizes (and therefore the drop deformation) are smaller. The maximum difference in the evaporation time between the pendant and the sessile drops was found to be about 12% in case of evaporation on hydrophilic substrate in CCR mode.

The effect of substrate conductivity of drop evaporative cooling was briefly analysed, by comparing the effect of three very different substrates (insulating material, glass and stainless steel). As expected, the larger effect was found for the case of insulating material with a cooling of about 4.4°C, while with the more conductive material (stainless steel) the effect was much lower (less than 0.1°C).

Appendix

The main features of the model proposed in [22], which allows to consider the dependence of the gas mixture thermo-physical properties on temperature are briefly summarised here (refer to [22, 19] for more details). The gas mixture properties were assumed to depend on the absolute temperature through power laws as

$$D_{10} = \left(D_{10,ref}\hat{T}_{ref}^{-m}\right)\hat{T}^{m};\ c = \left(c_{ref}\hat{T}_{ref}\right)\hat{T}^{-1};\ k_{mix} = k_{ref}\frac{a_0\hat{T}^{q_0} + a_1\hat{T}^{q_1}}{a_0\hat{T}_{ref}^{q_0} + a_1\hat{T}_{ref}^{q_1}};\ c_p^{(1)} = \left(c_{p,ref}^{(1)}\hat{T}_{ref}^{-b}\right)\hat{T}^{b} \qquad \text{(A1)}$$

where $\hat{T} = T/T_\infty$, T_∞ is the free stream temperature and the pedex *ref* refers to an arbitrary reference condition. The mixture thermal conductivity is obtained by the Wassilieva [34] relation, and the exponents m, q_p and b depend on the species involved. Under uniform Dirichlet boundary conditions on the drop surface and at infinity, the solution to the conservation equations for the vapour concentration and the temperature fields can be written in a closed form when the harmonic auxiliary function Φ (see equations 3) is known. Defining the function $\hat{H} = \ln\left(1 - y^{(1)}\right)\left(a_0\hat{T}_{ref}^{q_0} + a_1\hat{T}_{ref}^{q_1}\right)\hat{T}_{ref}^{1-m-b}Le_M^{-1}$, where $y^{(1)}$ is the vapour molar fraction and $Le_M^{-1} = (b+1)Mm^{(1)}c_{p,ref}^{(1)}D_{10,ref}c_{ref}/k_{mix,ref}$, then

$$\hat{H} = -\frac{1}{K_0}\left[\frac{a_0}{h_0}\hat{T}^{2+q_0-m}W_{h_0}\left(\frac{\hat{T}^{1+b}}{K_0}\right) + \frac{a_1}{h_1}\hat{T}^{2+q_1-m}W_{h_1}\left(\frac{\hat{T}^{1+b}}{K_0}\right)\right] + \hat{H}_0$$
$$\Phi = -\frac{\Phi_1}{K_0}\left[\frac{a_0}{1+q_0}\hat{T}^{1+q_0}W_{g_0}\left(\frac{\hat{T}^{1+b}}{K_0}\right) + \frac{a_0}{1+q_1}\hat{T}^{1+q_1}W_{g_1}\left(\frac{\hat{T}^{1+b}}{K_0}\right)\right] + \Phi_0 \qquad \text{(A2)}$$

where K_0, H_0, Φ_0 and Φ_1 are constants to be obtained from the boundary conditions, $h_p = (2+q_p-m)/(1+b)$ and $g_p = (1+q_p)/(1+b)$. The function $W_s(x)$ is related to the hypergeometric function by $W_s(x) = {}_2F_1(s,1,1+s,x)$. The sensible heat flux and the vapour mass flux on the drop surface can then be calculated in terms of the gradient of Φ as

$$n_j^{(1)} = f_n^{(1)}\nabla_j\Phi = -\frac{k_{mix,ref}\hat{T}_{ref}^{b}}{c_{p,ref}^{(1)}\left(a_0\hat{T}_{ref}^{q_0} + a_1\hat{T}_{ref}^{q_1}\right)\Phi_1}\nabla_j\Phi;\ q_j = f_q\nabla_j\Phi = -T_\infty\frac{k_{mix,ref}\left(K_0 - \hat{T}_s^{1+b}\right)}{\left(a_0\hat{T}_{ref}^{q_0} + a_1\hat{T}_{ref}^{q_1}\right)\Phi_1}\nabla_j\Phi \qquad \text{(A3)}$$

The heat and evaporation rates are then found by integrating the normal component of the fluxes over the drop surface

$$m_{ev} = f_n^{(1)} \int_A \nabla_j \Phi dA; \quad \dot{Q}_s = f_q \int_A \nabla_j \Phi dA \tag{A4}$$

References

[1] Park J, and Moon J 2006 *Langmuir*, **22**(8) 3506-3513
[2] Kim J 2007 *Int. J. Heat Fluid Flow* **28**(4) 753-767
[3] Feng W, Ueda E, and Levkin P A 2018 *Adv. Mater.* **30**(20) 1706111
[4] Wilson S K and D'Ambrosio H M 2023 *Annual Review of Fluid Mechanics* **55**, 481-509
[5] Radhakrishnan S, Srivathsan N, Anand T N C, Bakshi S 2019 *International Journal of Thermal Sciences* **140** 368-376
[6] Erbil H Y 2012 *Advances in colloid and interface science* **170**(1-2) 67-86
[7] Morita A, Carastan D and Demarquette N 2002 *Colloid Polym. Sci.* **280** (9) 857-864
[8] Berry J D, Neeson M J, Dagastine R R, Chan D Y C and Tabor R F 2015 *J. Colloid Interface Sci.* **454** 226-237
[9] Sikarwar B S, Muralidhar K and Khandekar S 2013 *Interfacial Phenomena and Heat Transfer*, **1** (4) 339-356
[10] Sandu I, Iordache I, Fleaca C T, Dumitrache F and Niculescu A M 2014 *Journal of Crystallization Process and Technology* **4** 206-211
[11] Bashforth F and Adams J C 1883 *An attempt to test the theories of capillary action* (Cambridge Press)
[12] Kumar A, Gunjan M R, Jakhara K, Thakurc A and Raja R 2020 *Colloids and Surfaces A* **597** 124619
[13] Somasundaram S, Anand T N C and Bakshi S 2015 *Phys. Fluids* **27**, 112105
[14] Devlin N R, Loehr K and Harris M T 2016 *AIChE Journal* **62**(3) 947-955
[15] Li Y, Diddens C, Lv P, Wijshoff H, Versluis M and Lohse D 2019 *Physical Review Letters* **122**, 114501
[16] Hsu N T, Sato K and Sage B H 1954 *Industrial and Engineering Chemistry* **46**(5) 870-876
[17] Starinskaya E, Miskiv N, Terekhov V, Safonov A, Li Y, Lei M-K and Starinskiy S 2023 *Water* **15** 273
[18] Sazhin SS 2022 *Droplets and Sprays: Simple Models of Complex Processes* (Springer Nature)
[19] Cossali G E, Tonini S 2021 *Drop Heating and Evaporation: Analytical Solutions in Curvilinear Coordinate Systems* (Springer)
[20] Lebedev N N 1972 *Special Functions & Their Applications* (Dover Publications)
[21] Cossali G E and Tonini S 2023 *Applied Mathematical Modelling* **114** 61-77
[22] Cossali G E and Tonini S 2019 *Int. J. Heat Mass Transfer* **138** 1166-1177
[23] Picknett R G and Bexon R 1977 *J. Colloid and Interface Science* **61**(2) 336-350
[24] Slattery J C 1981 *Momentum, energy and mass transfer in continua* (R. Krieger Publ.)
[25] Bird R B, Stewart W E and Lightfoot E N 2002 *Transport Phenomena* (Wiley, New York)
[26] Tonini S and Cossali G E 2023 *Phys. Fluids* **35** 032113
[27] Çengel Y A and Ghajar A J 2015 *Heat and Mass Transfer: Fundamentals and Application* (5th Edition McGraw-Hill, NY)
[28] Robertson W M and Lehman G W 1968 *J. Appl, Phys.* **39**(4), 1994-1996
[29] Tonini S and Cossali G E 2022 *Phys. Fluids* **34**, 073313
[30] Tonini S and Cossali G E 2022 *Phys. Rev. E* **105**, 054803
[31] Cossali G E and Tonini S 2023 *Applied Mathematical Modelling* **114** 61–77
[32] Tonini S and Cossali G E 2012, *Int. Journal of Thermal Sciences* **57** 45-53
[33] Crowe C T 1976 *Lawrence Livermore Lab Rep* (CaliforniaUniv., Livermore, USA)
[34] Wassiljewa A 1904 *Physikalische Zeitschrift* **5** (22) 737-742

Journal of Physics: Conference Series **2685** (2024) 012066 doi:10.1088/1742-6596/2685/1/012066

Multi-scale analysis of respiratory droplets transport within the breathing cloud

L Campanelli[1,*], M Cavazzuti[1] and P Tartarini[2]

[1] Dipartimento di Scienze e Metodi dell'Ingegneria, Università degli Studi di Modena e Reggio Emilia, via Giovanni Amendola 2, Reggio Emilia, 42122, Italia

[2] Dipartimento di Ingegneria "Enzo Ferrari", Università degli Studi di Modena e Reggio Emilia, via Pietro Vivarelli 10, Modena, 41125, Italia

[*] Corresponding author: ludovico.campanelli@unimore.it

Abstract. Infectious diseases are transmitted primarily through pathogen laden droplets commonly exhaled during respiratory events such as breathing, coughing, or sneezing. The transport and evaporation of droplets are governed by the fundamental laws of fluid mechanics and convection-diffusion. From these, analytical models can be created helping in better understanding pathogens transmission from a mechanical perspective. The droplet transport within the humid air breath cloud and the local ventilation are crucial for accurately predicting the droplet fate. Different levels of complexity are possible for modelling the breath cloud, from simple 1D models to complex unsteady 3D simulations including discrete phase models for the droplets simulation. The former are too simple to capture the fluid dynamics of intermittent jets in detail, while the latter are too computationally expensive. The current work presents a novel multi-scale approach where an analytical model of the droplet transport and evaporation is coupled to unsteady CFD simulations of warm humid puffs of exhaled air. The proposed model has the advantage of the accuracy of an analytical model and the computational cost of a relatively standard unsteady CFD simulation, and can be used to predict the trajectory of the exhaled droplets for a variety of respiratory events.

1. Introduction

The outbreak of SARS-CoV-2 has resulted in significant changes to our daily lifes, including the use of face masks, regular hand and surface disinfections, and the need to maintain social distance. World Health Organization (WHO) recommended a social distance of at least 1 meter, but many researchers have shown how in most cases this distance is not necessarily enough to prevent infection. Respiratory droplets can range from approximately a millimetre down to a fraction of micrometre in size and can either deposit on surfaces or remain airborne, giving rise to different transmission routes, either by contact with an infected surface, or by droplet inhalation on behalf of a susceptible host. An important role is played by the ambient conditions in terms of relative velocity between the droplet and the surrounding air, and water vapour partial pressure. This is particularly true for smaller droplets that evaporate rapidly and whose inertia is negligible so that they can be carried arbitrarily far by the local air flow, dangerously contributing to the chance of airborne transmission, and easily breaking the one metre social distancing rule. In this framework, it is apparent how the droplet transport within the humid air breathing cloud and/or within the ventilation air flow in a confined environment becomes essential for the correct prediction of the droplet fate. While for the latter case results cannot be generalised easily as much depends on the room and on the specific ventilation system layouts, for the former case general models of the exhaled air cloud as a sequence of jets and puffs are possible for different reference respiratory events and could be of great aid in predicting the risks associated to short-distance direct inhalation airborne transmission route.

In the last few years several studies investigated the effects of indoor ventilation on the droplets trajectory. Ji *et al.* [1] showed how mixing ventilation - due to the augmented turbulent heat and mass transfer - induce a faster droplet evaporation with respect to displacement ventilation. Motamedi *et al.* [2] performed CFD simulations to evaluate the risk of infection with different types of indoor ventilation. The study shows how the risk, depends on both the position of the infected individual and the type of ventilation, whereas no ventilation strategy can be clearly deemed better than others. In all the cases considered, it has been demonstrated that social distancing alone is not enough to contain indoor virus spread. Other works focused on the breath cloud. Xie *et al.* [3] used an empirical model of non-isothermal jet to model the breath cloud and estimate the falling distance for droplets of different size in a quiescent environment. It was shown that droplets can be carried even more than six metres away in case of sneezing. Bourouiba *et al.* [4] after a series of experiments on human coughs and sneezes, highlight how respiratory events should not be modelled as steady jets but rather as sequence of turbulent jets and puffs. An analytical model based on momentum conservation is proposed. Other important aspects that influence the disease transmission in indoor enviroments are the ambient thermohygrometric conditions: Bahramian *et al.* [5] show how a higher indoor temperature results in an increase of aerosolized droplet lifetime, and hence an increased airborne transmission. Busco *et al.* [6] validated a series of experimental results on sneezes with a CFD model accounting for the angular head motion and the dynamic pressure response during the sneeze. Their model is compared to simpler sneeze models having costant head angle and ejection velocity. Their results show a much larger exhaled droplets spatial spread.

In the literature a rather net distinction is found between analytical and CFD models. The latter have superior breath cloud modelling capabilities, while the former, being characterised by a continuous approach, provide more accurate trajectory and evaporation predictions. This calls for a multi-scale approach that could exploit the advantages of the two methods.

In a previous work, the authors developed an analytical model of droplet transport and evaporation [7], also addressing the role of its non-volatile components, such as salts and enzymes, on the droplet evaporation. An elaborated, although simplified, numerical model of the buoyant breath cloud as a sequence of jets and puffs was included and provided the local velocity field and thermo-hygrometric boundary conditions required by the analytical model.

In this work, a multi-scale approach is proposed and preliminarily tested where the aforementioned breath cloud model is replaced by unsteady CFD simulations of intermittent warm humid air jets. In this way, the intrinsic limits of the 1D breath cloud model can be overcome, thus making the model prediction more accurate, yet without the need of computationally expensive Discrete Phase Model (DPM) CFD simulations. With the model proposed, the trajectory of exhaled droplets can be predicted for any kind of respiratory event and ambient conditions. As such, the model could be used to assess the risks associated to different pathogen transmission routes under different real life scenarios. Preliminary results for the case of periodic mouth breathing in a confined environment representing a room at a given temperature and relative humidity are shown and discussed.

2. Analytical model

A brief overview of the mathematical model, that describes the behaviour of the exhaled droplets, is now given. In the following, it is assumed that the droplets have spherical shape and uniform thermophysical properties.

2.1. Transport

The droplet trajectory is computed from the balance of the forces acting on the droplet itself as it moves through the air. These are its weight (including buoyancy) and drag, while other forces are assumed negligible [8]. From Newton's second law the droplet acceleration can be written as

$$\begin{cases} \dot{u}_{\mathrm{r,x}} = -\dfrac{u_{\mathrm{r,x}}}{\tau_{\mathrm{d}}} \\ \dot{u}_{\mathrm{r,z}} = \dfrac{u_{\mathrm{r,t}}}{\tau_{\mathrm{d,t}}} - \dfrac{u_{\mathrm{r,z}}}{\tau_{\mathrm{d}}} \end{cases} \tag{1}$$

where where x and z are horizontal and vertical directions respectively, $\mathbf{u}_{\mathrm{r}}=\mathbf{u}_{\mathrm{p}}-\mathbf{u}_{\mathrm{m}}$ is the relative velocity between the droplet and the air stream, τ_{d} the characteristic time associated to drag calculated as the ratio between the vertical momentum of the droplet and the vertical component of the drag force, and the subscript 't' denotes the terminal condition. By terminal the zero acceleration condition the particle will attain in the limit is meant.

If a spherical droplet with initial relative velocity $\mathbf{u}_{\mathrm{r,0}}$ falling in a moist air stream having velocity $\mathbf{u}_{\mathrm{m}}$ is considered, by integrating equation (1)

$$\begin{cases} u_{\mathrm{p,x}}(t) = u_{\mathrm{m,x}} + u_{\mathrm{r,x,0}}\mathrm{e}^{-t/\tau_{\mathrm{d}}} \\ u_{\mathrm{p,z}}(t) = u_{\mathrm{m,z}} + u_{\mathrm{r,z,0}}\mathrm{e}^{-t/\tau_{\mathrm{d}}} + u_{\mathrm{r,t}}\dfrac{\tau_{\mathrm{d}}}{\tau_{\mathrm{d,t}}}\left(1-\mathrm{e}^{-t/\tau_{\mathrm{d}}}\right) \end{cases} \tag{2}$$

For the case of Stokes' flow, *i.e.* when the Reynolds number is low, drag force is a linear function of the relative velocity, as a consequence τ_{d} is constant and equal to $\tau_{\mathrm{d,t}}$. In case of non Stokes' flow, instead, τ_{d} changes during the droplet flight and tends to $\tau_{\mathrm{d,t}}$.

By further integrating equation (2) with initial position $\mathbf{s}_{\mathrm{p,0}}$, the droplet trajectory results

$$\begin{cases} s_{\mathrm{p,x}}(t) = s_{\mathrm{p,x,0}} + u_{\mathrm{m,x}}t + u_{\mathrm{r,x,0}}\tau_{\mathrm{d}}\left(1-\mathrm{e}^{-t/\tau_{\mathrm{d}}}\right) \\ s_{\mathrm{p,z}}(t) = s_{\mathrm{p,z,0}} + \left(u_{\mathrm{m,z}} + u_{\mathrm{r,t}}\dfrac{\tau_{\mathrm{d}}}{\tau_{\mathrm{d,t}}}\right)t + \left(u_{\mathrm{r,z,0}} - u_{\mathrm{r,t}}\dfrac{\tau_{\mathrm{d}}}{\tau_{\mathrm{d,t}}}\right)\tau_{\mathrm{d}}\left(1-\mathrm{e}^{-t/\tau_{\mathrm{d}}}\right) . \end{cases} \tag{3}$$

From equation (2) it can be noted that as time passes, the influence of the initial condition on the droplet velocity vanishes. In addition, in case of small terminal velocity the particle is entirely transported by the local air flow, and therefore can travel over arbitrarily long distances.

2.2. Evaporation

Considering convection-diffusion equation in stationary form without source or sink terms and neglecting cross-diffusion of air in water, the mass transfer from the droplet surface to the surrounding moist air due to evaporation can be written in terms of the droplet diameter reduction rate in time as

$$\frac{\mathrm{d}D_{\mathrm{p}}}{\mathrm{d}t} = -\frac{2D_{\mathrm{v,s}}\mathrm{Sh}}{\rho_{\mathrm{p}}D_{\mathrm{p}}R_{\mathrm{v}}T_{\mathrm{s}}}\frac{p}{p_{\mathrm{a,s}}}\left(p_{\mathrm{v,s}} - p_{\mathrm{v,\infty}}\right) \tag{4}$$

where the subscripts 's' and '∞' indicate that the related quantity is to be evaluated by the droplet surface or far from it in the surrounding air, and the subscripts 'a' and 'v' denote dry air and water vapour respectively. D_{p} is the droplet diameter, Sh the Sherwood number (that can be evaluated using Ranz-Marshall empirical correlation), p the pressure, ρ_{p} the droplet density, R_{v} the water vapour ideal gas constant, and T the temperature. Finally, D_{v} is the mass diffusivity of water in air that can be evaluated with the correlation proposed in [9]

$$D_{\mathrm{v}} = 21.2 \cdot 10^{-6}\left[1 + 0.0071\left(T_{\mathrm{s}} - 273.15\right)\right] . \tag{5}$$

By introducing the evaporation rate K defined as

$$K = -\frac{\mathrm{d}D_\mathrm{p}^2}{\mathrm{d}t} = \frac{4D_\mathrm{v,s}\mathrm{Sh}}{\rho_\mathrm{p}R_\mathrm{v}T_\mathrm{s}}\frac{p}{p_\mathrm{a,s}}\left(p_\mathrm{v,s} - p_\mathrm{v,\infty}\right) , \tag{6}$$

separating the variables, and integrating equation (6) in time from an initial droplet diameter $D_\mathrm{p,0}$, is it possible to write the droplet diameter as a function of time

$$D_\mathrm{p}(t) = \sqrt{D_\mathrm{p,0}^2 - Kt} . \tag{7}$$

In general K is not constant during the droplet evaporation, yet it can be shown that for small diameters K tends to a constant value, function of the particle density and the ambient thermohygrometric conditions alone.

2.3. Energy

When a droplet is released in the air, its evaporation starts immediately. At the beginning the droplet is gradually cooled (or warmed) from its initial temperature to the ambient wet bulb temperature. At a later stage, the droplets keeps shrinking at constant temperature. In case the droplet contains a non-volatile fraction, the evaporation slows down as it progresses according to Raoult's law, and the droplet temperature gradually grows up to when thermal equilibrium with the ambient is met and evaporation terminates.

In terms of thermal energy balance, at any time the sensible thermal power stored by the particle is balanced by evaporative cooling and convective heating

$$\dot{Q}_\mathrm{sn} = \dot{Q}_\mathrm{ev} + \dot{Q}_\mathrm{cv} . \tag{8}$$

Expanding these terms and recasting equation (8) in terms of the droplet temperature rate of change

$$\frac{\mathrm{d}T_\mathrm{p}}{\mathrm{d}t} = \frac{6\lambda_\mathrm{s}\mathrm{Nu}\left(T_\infty - T_\mathrm{s}\right)}{c_\mathrm{p,p}\rho_\mathrm{p}D_\mathrm{p}^2} - \frac{3h_\mathrm{lt}K}{2c_\mathrm{p,p}D_\mathrm{p}^2} \tag{9}$$

is found, where Nu is the Nusselt number, λ the thermal conductivity of the moist air, c_p the droplet specific heat, and h_lt the water latent heat of vaporisation. To be noted that in case the droplet contains some non-volatile compounds its density and specific heat will change over time and should be computed as weighted average between the corresponding values of the solid and the liquid fractions.

The fact that the droplet temperature variation is small and limited in time allows the dependency of K upon the droplet temperature to be neglected, where T_p=T_s. equation (9) can thus be integrated assuming an initial droplet temperature $T_\mathrm{s,0}$ to give

$$T_\mathrm{p}(t) = T_\mathrm{s,0}\mathrm{e}^{-t/\tau_\mathrm{ev}} + T_\mathrm{s,t}(1 - \mathrm{e}^{-t/\tau_\mathrm{ev}}) \tag{10}$$

where τ_ev is the characteristic time associated to evaporation and $T_\mathrm{s,t}$ is the droplet terminal temperature, *i.e.* the ambient wet bulb temperature when the droplet is evaporating. To be noted how the solutions in equation (2) and in equation (10) are of the same form.

3. Methods

From the analytical model outlined above, it is apparent how the local air velocity and thermohygrometric conditions in terms of temperature and relative humidity play a fundamental role on the droplet transport, evaporation, and thermal equilibrium. These quantities, for instance, are found in the definitions of the characteristic times and terminal quantities in

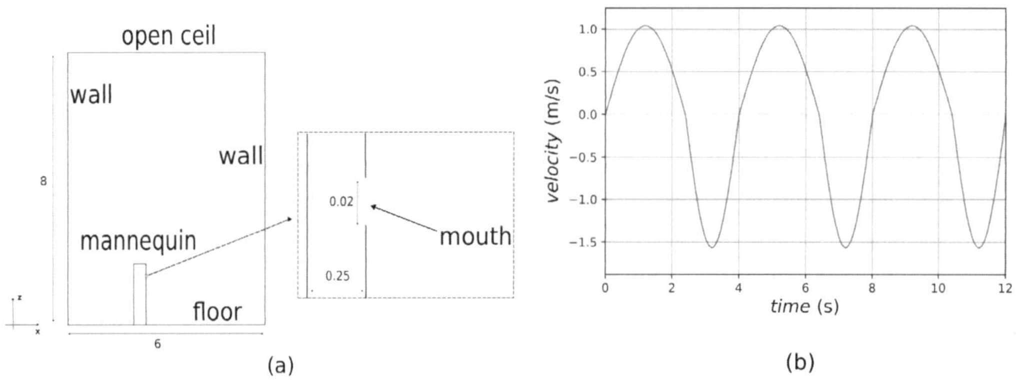

Figure 1. CFD model geometry and boundary patches (a), and normal velocity component at the mouth (b).

equation (2) and in equation (10), and largely affect the partial pressures in equation (6). If the local air properties were constant and uniform things would be easier, but this is not the case for respiratory droplets. These in fact are exhaled within the breath cloud, a warm and humid buoyant air intermittent puff that gradually mixes with the colder ambient air, so that the ambient conditions seen by the droplet change continuously, at least as long as the droplet remains trapped inside the breath cloud.

A mean is needed to retrieve the required information on the ambient air and feed it to the analytical model. In the following a preliminary coupling between the analytical model and an unsteady CFD simulation is introduced and discussed.

3.1. Case study

The case study for the CFD analysis consists of a simplified 2D open room geometry, 6 m long and 4 m high, where a breathing mannequin is placed. The breathing process is modelled as a time-periodic inlet/outlet boundary having period of 4 s and composed of a 2.4 s long exhalation followed by a 1.6 s inhalation [10]. The flow is normal to the mouth surface, a circle of 2 cm in diameter, and its velocity varies sinusoidally whereas the period of the positive and negative halves of the sinusoid is not the same as mentioned (see figure 1b). Maximum positive velocity during exhalation is 1.042 m/s while minimum negative velocity during inhalation is -1.563 m/s, so that a tidal volume of 0.5 l per breath is obtained.

3.2. CFD model

The CFD model is built in OpenFOAM. A multiphase solver is adopted to account for moisture in the air, together with *k-ω SST* turbulence model. For the preliminary results presented in this work a 2D structured mesh was chosen. To minimise the impact of the boundaries on the results, the outlet patch (named *open ceil* in figure 1a) is located sufficiently far from the area of interest, at a height of 8 m from the *floor*. Slip condition is imposed at all solid boundaries but the *floor* and the *mannequin*. At the *mouth* inlet the velocity function in figure 1 is imposed, inlet temperature and relative humidity being 307.15 K and 95% respectively. A fixed temperature is imposed on the *mannequin* while the remaining solid patches are assumed adiabatic. The *mannequin* surface temperature was estimated using the formula $T_{\text{clo}} \approx T_{\text{skin}} + q''_{\text{skin}}/(0.155 I_{\text{clo}})$ where the insulation coefficient (I_{clo}) for coverall clothes and the body heat flux for an individual standing at rest (q''_{skin})

Table 1. Mesh sensitivity test.

Elements [$\times 10^3$]	**Size** [mm]	***d*** [mm]	***T*** [°C]	***RH*** [%]
70	6.67–71.7	1003.8	22.55	65.36
282	3.34–35.9	1412.0	23.38	68.45
637	2.22–23.9	1411.7	23.16	68.09

are provided by [11]. Assuming I_{clo}=0.72 clo, T_{skin}=36 °C, q''_{skin}=70 W/m^2 the clothes surface temperature is $T_{\text{clo}}\approx$28 °C. Ambient initial conditions are still air at 293.15 K and 50% relative humidity. Since the breath cloud and the body surface are warmer than the surroundings, both cloud buoyancy and thermal plume by the body occur.

The total simulated time is 50 s, and data are saved every 0.1 s. Table 1 displays the results of the mesh sensitivity test. The first column reports the mesh elements number, while the second gives the minimum and maximum diagonal cell size in the area of interest, where it must be considered that the mesh is finer close to the mouth and gets coarser with the distance. The remaining columns show the mean values, for a coupled simulation of a 10 µm droplet, of the distance travelled by the droplet (d), and the local temperature (T) and relative humidity (RH) encountered at three instants of time taken at the end of the first three exhalations (*i.e.* at 2.4, 6.4, and 10.4 s). The intermediate mesh is chosen as a reasonable trade-off between accuracy and computational time.

3.3. Coupling

The CFD simulation is run once before the analytical model. When all the results are saved, the analytical model is launched to simulate a series of droplets of different size. The boundary conditions for the analytical model is limited to the droplet initial position, velocity, temperature, size, and solid fraction properties, whereas all the information related to the ambient and the breath cloud is taken from CFD.

At each time step the analytical model provides the simulated time and the droplet coordinates to the coupling script. Since the current time and position will not match one of the CFD saves and mesh cell centres, interpolation will be needed both in space and time: the CFD results of the two closest saves in time are read and queried using the probe filter of the VTK module [12]. This filter locates the closest cell centres to the query point and takes care of the interpolation returning the desired ambient air values. This information is then linearly interpolated by the coupling script before being fed back to the analytical model where all the remaining simulation parameters are updated. These include the droplet position, velocity, diameter and temperature with their rate of change, chemical composition, characteristic times, and forces, together with the thermophysical properties of both droplet and ambient air.

Figure 2 shows the logical flow of the multi-scale method proposed. Of course, the accuracy of the method is proportional to the number of CFD saves, and the size of the grid.

4. Results

Figure 3 compares the results of the previous analytical model developed by the authors with those of the current multi-scale approach in terms of droplet trajectory and evaporation progress. Droplets of three different sizes are simulated: namely, 10 µm, 50 µm, and 100 µm, so that the latter will be mostly governed by gravity, the first by drag, and the remaining, to some extent, by

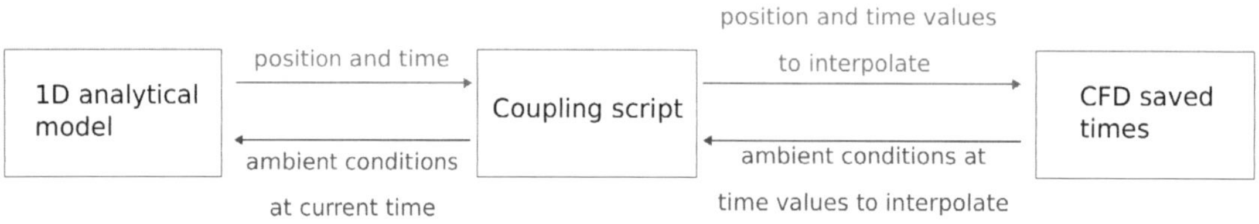

Figure 2. Logical flow of the multi-scale method.

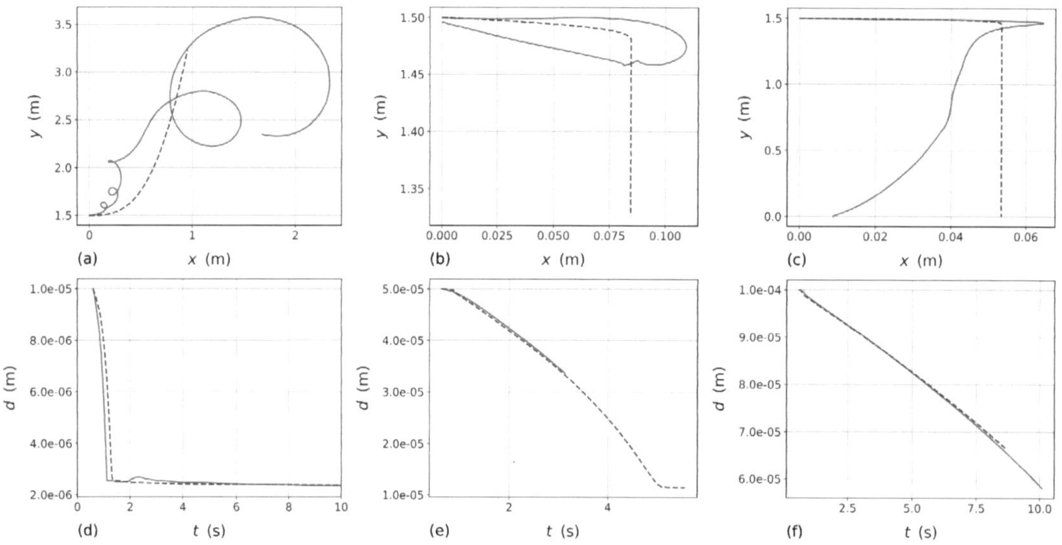

Figure 3. Droplets trajectory (top) and evaporation progress (bottom): multi-scale approach (solid red line) *vs* analytical model (dashed blue line). From left to right the images refer to the 10 μm, the 50 μm, and the 100 μm droplet, respectively.

both. Simulations either stop when the droplet reaches the ground, or exits from the domain, or remains airborne (*i.e.* the thermal equilibrium is reached out of the breath cloud area of influence while the droplet relative velocity becomes sufficiently small), or the end time is reached.

Simulation times can be rather different depending on the droplet size. On average, the analytical model requires ≈1 minute of CPU time per simulated droplet on an ordinary workstation equipped with Intel Core i7-2600 CPU at 3.4 GHz, while the multi-scale model takes ≈5 minutes in addition to the time required by the preliminary CFD run. The particle trajectories in an equivalent DPM CFD analysis are highly dependent on the mesh size and the time step, thus implying much larger computational costs to achieve a comparable accuracy with respect to the multi-scale model.

While the agreement is remarkable for what concerns the droplet evaporation, the trajectories computed by the two methods are rather different. The more realistic trajectory of the multi-scale approach is apparent for the droplet in figure 3a which due to its small size is largely carried around by the vortices generated in the breath cloud; vortices that cannot be modelled analytically. Even though, both methods predict that the droplet is brought upward by the buoyant breath cloud, the path and the distance travelled are much different, and so the risks of virus transmission. Similarly, the droplet in figure 3b remains lingering in front of the exhaler before being inspired again. This particular behaviour cannot be caught by the analytical approach where inspiration is not modelled. Last, the droplet in figure 3c is caught by entrainment vortices suddenly reversing its course after crossing the jet shear layer, it is then attracted towards the no-slip mannequin surface due to the Coandă effect, reaching the ground before complete evaporation occurs. Also in this case, the difference in the predicted trajectory is due to physical phenomena that a simple analytical model cannot handle.

5. Conclusions

Exhaled droplets pose a potential risk for infectious diseases spreading. Larger droplets, with a diameter of 100 μm or more follow a parabolic trajectory governed by gravity, while smaller droplets are mainly transported by drag by the local airflow. Several works in the literature have investigated the role of the breath cloud on the droplet transport, either developing analytical models or performing complex DPM CFD analysis.

The weak point of analytical models is clearly the difficulty in implementing a simple but reliable breath cloud model. In this work, a multi-scale approach has been proposed to overcome this difficulty by means of a one-way coupling between an analytical model previously developed by the authors, and a CFD simulation of an unsteady breath cloud in a large room carried out with OpenFOAM; the local air velocity and thermohygrometric conditions needed by the analytical model are thus provided by the CFD analysis. This approach combines the advantages of CFD modelling capabilities in capturing the jet dynamics more in detail while dropping the need for DPM, and that of continuous analytical model accuracy in predicting the droplet transport and evaporation.

Preliminary results are presented where the multi-scale approach is compared to the purely analytical model for the case of three droplets of different diameters. While the two approaches agree very well on the prediction of the droplet evaporation, things are rather different on transport prediction, where the role of turbulence within the breath cloud or in the jet shear layer may have a dramatic impact over the droplet fate.

To the author's knowledge, such a multi-scale approach has never been investigated in the literature, and in this preliminary analysis it has been proven to be an interesting feasible alternative to complex and computationally intensive DPM CFD analyses, or simpler analytical methods with limited modelling capabilities.

In future developments, the authors plan to implement the analytical model coupling to a 3D CFD model featuring a more realistic mannequin geometry. Investigation on possible means to further speed up the CFD query procedure will also be carried out.

6. References

[1] Ji Y,Qian H,Ye J and Zheng X 2018 *J Aerosol Sci* **115** 164–72

[2] Motamedi H, Shirzadi M, Tominaga Y and Mirzaei P A 2022 *Sustain Cities Soc* **76** 103397

[3] Xie X, Li Y, Chwang AT, Ho P L and Seto W H 2007 *Indoor Air* **17** 211–25

[4] Bourouiba L, Dehandschoewercker E and Bush J W M 2014 *J Fluid Mech* **745** 537–63

[5] Bahramian A, Mohammadi M and Ahmadi G 2023 *Sci Total Environ* **858** 159444

[6] Busco G, Yang S R, Seo J and Hassan Y A 2020 *Phys Fluids* **32** 073309

[7] Cavazzuti M and Tartarini P 2023 Transport and evaporation of exhaled respiratory droplets: an analytical model *Phys Fluids* **35** 103327

[8] Elghobashi S and Truesdell G C 1992 *J Fluid Mech* **242** 655–700

[9] Holterman H 2003 *Kinetic and Evaporation of Water Drops in Air* (Wageningen: IMAG Report)

[10] Abkarian M, Mendez S, Xue N, Yang F and Stone H A 2020 *P Nat Acad Sci* **117** 25237–45

[11] ASHRAE Standard 2013 *ANSI/ASHRAE Standard 55-2013: Thermal Environmental Conditions for Human Occupancy* (ASHRAE)

[12] Schroeder W, Martin K M and Lorensen W E 2006 *The Visualization Toolkit, an Object-Oriented Approach to 3D Graphics* (New York: Kitware Inc.)

Biochar powders coating to improve evaporative cooling in Maisotsenko-cycle systems

N Morselli[1,*], F D Fracasso[1], M Cossu[1], F Ottani[1], M Puglia[1], S Pedrazzi[1], G Allesina[1], A Muscio[1] and P Tartarini[1,2]

[1] Department of Engineering "Enzo Ferrari", University of Modena and Reggio Emilia, Via Vivarelli 10/1 – 41125 Modena, Italy

[2] INTERMECH—Inter-Departmental Center, University of Modena and Reggio Emilia, Via Vivarelli 2 – 41125 Modena, Italy.

[*] Corresponding author: nicolo.morselli@unimore.it

Abstract. This work presents an experimental study on the performance of biochar powder coatings on aluminum surfaces for use in indirect evaporative coolers based on the Maisotsenko cycle. The performance of the biochar coated samples was compared to cellulose-coated aluminum samples and uncoated ones. Results showed that biochar coatings improved the performance of uncoated aluminum, with the 150 μm particle size coating offering performance comparable to cellulose. However, wetting times were longer, which has implications for spraying strategies.

1. Introduction

A rediscovered alternative to vapor compression systems lays on indirect evaporative cooler based on the Maisotsenko cycle. These systems exploit the latent heat of water evaporation to cool an air flow, thus eliminating the compression work and the need to use refrigerant fluids with positive impact on the environment.

Despite reaching coefficients of performance 5-6 times higher than traditional cooling systems [1], evaporative coolers are often penalized by continuous pump operation for the constant spray of water in the secondary channel in order to obtain a homogeneous humidification of the heat transfer surface [2]. Considerable efforts have been made over the years to improve the water retention and water diffusion of the materials used aiming at reducing the energy consumption for water pumping and further improve the COP.

Several researches focused on the study of porous materials for the wet duct of the heat exchanger, such as: cellulose fiber coating for metal foils, cotton wool, honeycomb structures, coconut fiber, polymer sheets, stainless steel with sintered nickel layer, etc [3, 4].

This work aims to study the water retention and water diffusion capability of biochar powder deposited on aluminum surfaces.

Biochar is a byproduct of the thermochemical process of gasification of wood biomass and is characterized by high porosity and surface area, as well as high resistance over time.

Biochar has a particle size that depends on the type of biomass used in the gasification process, ranging from tens of millimetres to a few microns. In this work it was test the performance of biochar coatings with a particle size of less than 250 μm.

Aluminum supports measuring 50x50 mm and with a thickness of 100 μm were specifically made for the deposition of biochar powders. This was achieved by utilizing an epoxy resin layer as an adhesive.

A prototype of evaporative heat exchanger was used to test the specimens. This heat exchanger had the ability to connect to a thermal imaging camera (IR camera), which allowed for the observation of temperature distribution on the lower surface of the specimen.

The analyses of temperature and relative humidity at the inlet and outlet led to the measurement of the temperature drop of the air stream generated by the tested specimen, as well as the water evaporation rate offered by each specimen. Through IR images were gathered data on the homogenization of the water distribution on the coated surface and on the stabilization time of the surface temperature, an index of the diffusion speed of the water on the specimen. Performance of biochar-coated specimens were compared to non-coated and cellulose-coated aluminium specimens.

The next section describes the test methods and the test rig used.

2. Materials and methods

2.1 Maisotsenko's cycle

In the Maisotsenko cycle, ambient air is cooled in a heat exchanger. In the simplest configuration of the cycle, a counter flow heat exchanger can be used (box in Fig. 1).

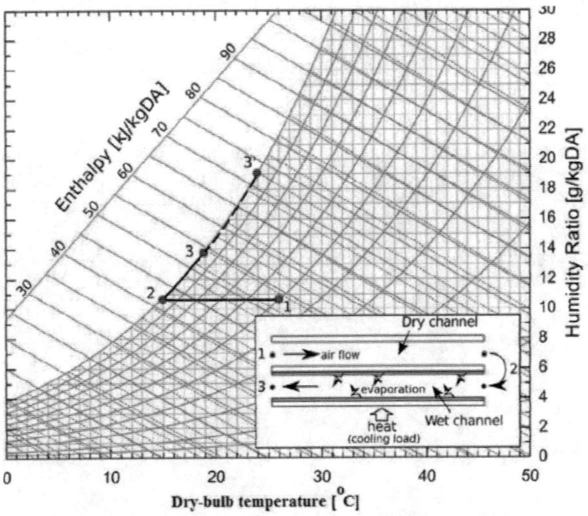

Figure 1. Psychrometric chart and principle of the Maisotsenko cycle: 1) ambient air; 2) released cooled non-humidified air; 3) wet-bulb temperature at ambient conditions; 3') released moist air. (Adapted from Dizaji et al, 2018 [13])

The Maisotsenko cycle involves cooling ambient air in a heat exchanger, typically by using a counter flow heat exchanger with a "dry" duct and a "wet" duct separated by a wall.

Air is introduced into the dry duct, where it is cooled by transferring heat through the separating wall. Some of the cooled air is released as non-humidified air, while the rest is diverted into the wet duct, where it is saturated with liquid water that is continuously vaporized. The air is warmed in the wet duct by heat transferred from the cooled dry air through the separating wall, and it is released into the ambient environment at close to saturation. The Maisotsenko cycle can theoretically cool the non-humidified air to the dew point temperature, which is lower than the temperature achieved by direct and indirect evaporative cooling systems. This process is illustrated on a psychrometric chart, where 1-2 represents the cooling process in the dry duct, 2 is the release temperature, and 3 is the dew point temperature. The chart also shows the progressive saturation process in the wet duct from 2 to 3', where the moist air is released.

For this cycle, it therefore becomes important to have a heat transfer surface which, on the one hand, guarantees homogeneous water distribution and a high evaporation rate, and on the other, does not add excessive thermal resistance in order to keep the heat flow between the dry and wet ducts high.

2.2 Tested surfaces

The surfaces were prepared using 100 μm thick aluminum foil which was subsequently coated with biochar of different particle sizes.

The biochar used in these tests comes from a downdraft gasification reactor and has an elemental composition shown in Table 1 along with the BET area (porosity indicator).

Table 1. Biochar chemical-physical characterization

C	H	N	O	S	ASH	BET
42.25 $\%_{w/w}$	0.80 $\%_{w/w}$	0.17 $\%_{w/w}$	56.78 $\%_{w/w}$	0.00 $\%_{w/w}$	33.2 $\%_{w/w}$	420 m^2/g

The as received biochar was sieved according to the ISO 3310-1 reference standard and through a RETSCH model AS200 vibrating sieve [5], in order to obtain three different particle sizes: < 75 μm, (75÷150) μm, (150÷250) μm hereinafter referred to as **75 μm**, **150 μm** and **250 μm**. The biochar powder was glued to a 50x50x0.1 mm aluminum support by spreading a layer of low viscosity two-component epoxy resin (175-225 cps at 25°C). SEM images of the three surfaces obtained are shown in Figure 2.

(a) (b) (c)

Figure 2. SEM images of the 75 μm (a), 150 μm (b) and 250 μm (c) biochar coated specimens.

2.3 Experimental setup

The test rig (Figure 3) consists of a rectangular measuring channel delimited at the bottom by the surface to be tested (1) and at the top by a TPU structure (2) which allows the insertion of the micrometric needle (3) used to drip water at controlled flowrate. Each 50x50x0.1 mm surface to be tested is glued to a 3D printed support in order to interface correctly with the test rig (Figure 4). Air with controlled temperature and humidity enters the inlet duct (4) and exits through the outlet duct (5).

The measurement of temperature and humidity is performed in points (6) and (7) through two SHT31 [6] sensors. The lower part of the test rig consists of a closing flange (8) and a housing (9) to insert the IR camera optics which will be directed towards the lower surface of the specimen, painted with matt black. The thermal emissivity of the black surface was measured through the TIR 100-2 emissometer [7] while the IR camera is a FLIR model 640T [8].

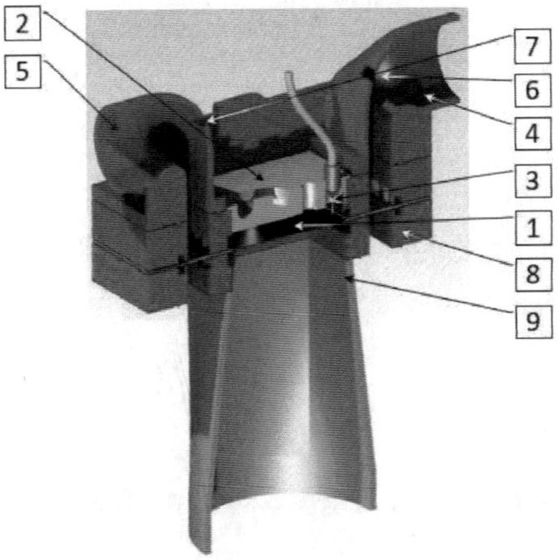

Figure 3. 3D render of the test rig.

(a) (b)

Figure 4. Front (a) and back (b) of the test specimen support.

2.4 Test methodology

The test methodology adopted aims to reproduce the conditions that occur inside an indirect evaporative heat exchanger. Dry bulb temperature (T_{db}) and inlet air humidity (RH_{in}) were chosen as 40°C and 26.5%RH, respectively. The temperature is controlled through a PTC resistor while the humidity is regulated with an ultrasonic humidifier. The inlet dry air flow ($\dot{m}_{a,dry}$) was chosen equal to 1.44 kg/h in order to have an average air speed in the rectangular duct equal to 1.2 m/s and within the velocity range found in literature that lays between 1 and 2 m/s [9, 10]. Once the test rig reaches steady conditions, the dripper is activated. The inlet and outlet T_{db} and RH are measured through the SHT31 sensors connected to an Arduino UNO board [11], while the temperature of the lower surface of the test specimen (averaged on the entire area of the painted aluminum) is measured through the IR camera software.

Each specimen is subjected to a preheating phase, after which the injection of water is activated at a flow rate ($\dot{m}_{w,in}$) of 3.7 g/h. The water settles on the biochar-coated aluminum specimen and begins to expand while it evaporates. This is followed by an evaporative cooling phase of the air leaving the test channel which continues until it stabilizes at a minimum value. The same happens for the back surface of the specimen, whose average temperature is recorded through the IR camera.

The IR images are used to both assess the uniformity of the water distribution and to measure the surface temperature of the specimen that can be used as comparison between different coatings.

Through the measured data it is possible to calculate the water evaporation rate $m_{w,\dot{e}vap}$ as reported in Eq. 1 [12].

$$m_{w,\dot{e}vap} = m_{a,\dot{d}ry} \cdot (x_{out} - x_{in}) \qquad [\frac{g_{vap}}{h}] \tag{1}$$

Where x_i is the absolute humidity of the moist air calculated according to Eq. 2

$$x_i = 0.622 \cdot \frac{\phi_i \cdot p_{v,sat}}{p_{amb} - \phi_i \cdot p_{v,sat}} \qquad \left[\frac{g_{vap}}{kg_{as}}\right] \tag{2}$$

Where *i* generically represent the *inlet (in)* or *outlet (out)* conditions, ϕ_i is the relative humidity measured through the SHT31 sensors, $p_{v,sat}$ is the water vapor saturation pressure calculated according to the correlation in Eq. 3, calculated at the given air temperature T_i and p_{amb} is the atmospheric pressure considered constant and equal to 101325 Pa.

$$p_{v,sat} = 6.11 \cdot 10^{\frac{7.5 \cdot T_i}{237.2 + T_i}} \cdot \phi_i \qquad [Pa] \tag{3}$$

In the following section the results obtained by comparing the different specimens tested are discussed.

3. Results

The plot in Figure 5 shows the time trends of the $\Delta T_{air} = T_{out} - T_{in}$ generated by the evaporation of water on the specimen being tested.

The two benchmarks are represented by the uncoated aluminum specimen (*Al*) and the cellulose coated specimen (*Cellulose*). It can be observed that the *Al* specimen generates the lowest ΔT_{air} while the *Cellulose* specimen presents the most uniform trend among all the samples analysed, reaching a maximum ΔT_{air} equal to 1.4 °C.

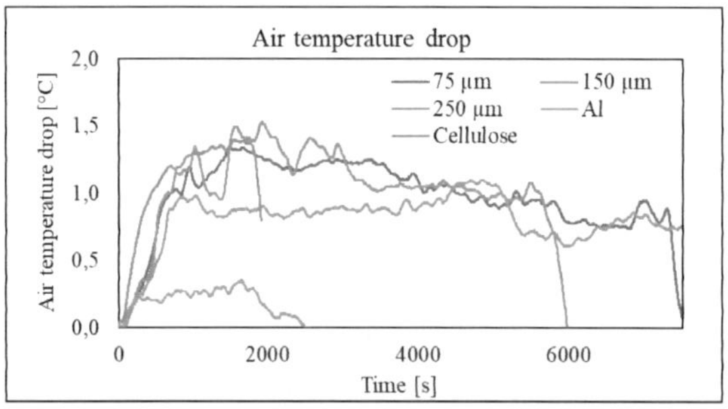

Figure 5. Air temperature drop generated by the different tested specimens.

Although the ΔT_{air} values obtained are comparable to *Cellulose*, less constant trends are observed for biochar-coated samples. The 150 µm coating seems to perform better offering a maximum deltaT equal to 1.5 °C and behaving in a similar way to the 75 µm. The 250 µm instead seems to behave differently in the initial stages, offering a maximum deltaT equal to 1.2 °C. What seems to unite all the biochar coated samples is the slower deltaT rise compared to that obtained with the cellulose sample. This is mainly due to the worse diffusion of water as discussed below.

Figure 6 shows the trends of the water evaporation rate generated by the different samples. The differences in relative humidity that have generated these trends are in the order of 10% and the uncertainty on relative humidity measurement equal to 2% does not lead to strong quantitative assessments. However, the average improvement of the evaporation rate given by the biochar coating compared to the untreated aluminum can be observed. In comparison with *Cellulose*, the biochar offers similar evaporation rates, however generating more discontinuous trends due, as will be seen later, to the partial distribution of water on the surface.

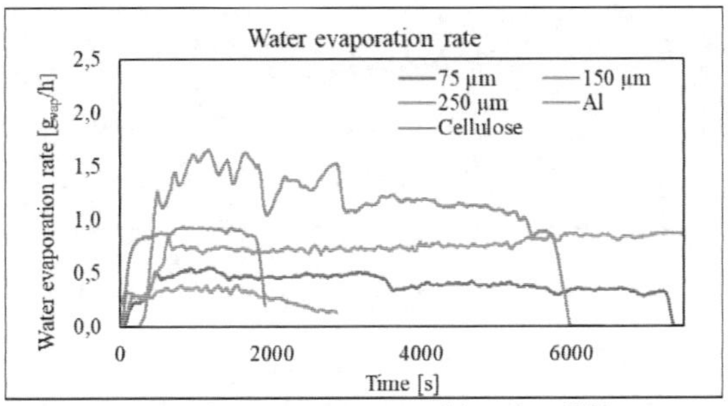

Figure 6. Water evaporation rate generated by the different tested specimens.

The thermographic analyzes returned the temperatures of the lower surface of each sample. Figure 7a shows the temperature of the 75 µm sample at the end of the preheating phase. The sample is heated homogeneously. Figure 7b shows the same sample at the end of the evaporation test: the uneven distribution of water can be seen which tends to create different patterns depending on how the drops descend along the sample. On the other hand, Figure 7c shows the cellulose-coated sample at the end of the same test: the temperature distribution, and consequently the water distribution, is more uniform, which validates the consistent data observed in the previous tests.

(a) (b) (c)

Figure 7. IR images of the 75 µm specimen at the end of the preheating phase (a), 75 µm (b) and Cellulose (c) at the end of the evaporation test *.

*The temperature scale is $(30-40)°C$ for pictures (a) and (b), while $(20-30)°C$ for (c).

The time required for the surface temperature to stabilize was determined through IR camera recordings. Figure 8 shows the temperature trends of the surfaces by averaging the values over the recording area (45x45 mm due to the presence of the PLA support). For each specimen, the graph was truncated at the point where a stable temperature was reached within a 1% tolerance.

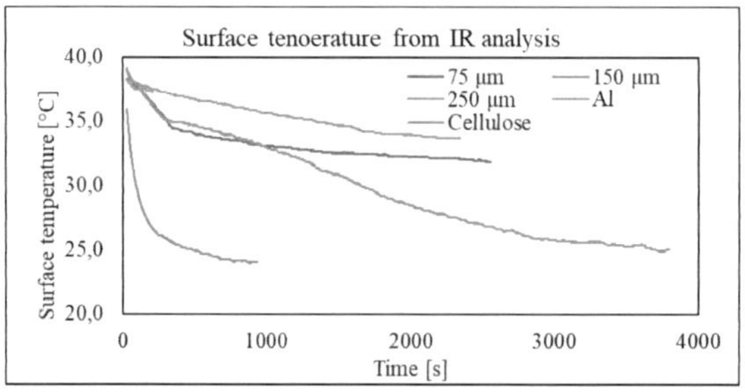

Figure 8. Stabilization time of the surface temperature of the specimens.

Table 2 shows the values of stabilization time and minimum surface temperature for each specimen. The 150 µm specimen reaches a minimum temperature of 24.9 °C, slightly higher than the cellulose specimen, but taking four times longer due to the lower wettability of the biochar-coated surface. The 75 µm specimen has an initial trend similar to the 150 µm specimen, but the minimum temperature is higher at 31.9 °C. The 250 µm specimen reaches a surface temperature of only 33.7 °C. It can be observed that the 150 µm sample achieved a uniform wetting, unlike the 75 µm and 250 µm specimens. Figure 9 shows the IR image of the 150 µm specimen after reaching the stabilization time.

Table 2. IR analysis results

Specimen	Temperature stabilization time [s]	Minmum average surface temperature [°C]
75 µm	2557	33,7
150 µm	3800	24,9
250 µm	2344	33,7
Al	218	36,5
Cellulose	938	24,1

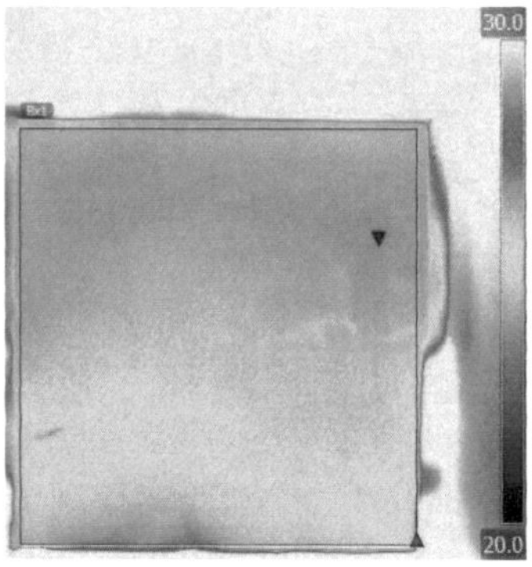

Figure 9. Surface temperature of the specimen 150 µm at the stabilization time.

4. Conclusions

This study experimentally evaluated the performance of aluminum surfaces coated with biochar at different grain sizes, for application in indirect evaporative coolers with particular reference to the Maisotsenko cycle. The biochar coatings always improved the performance of the uncoated aluminum, and in the case of the 150 µm grain size, the performance was comparable to the benchmark identified in the aluminum specimen coated with cellulose, a material commonly used in evaporative coolers. However, the wetting times were much longer, which has direct implications for spraying strategies.

From the IR analysis, it was observed that the high performance of the 150 μm specimen was enabled by its high wettability, which allowed for a homogeneous diffusion of water and an increased wetted surface area for heat transfer. In the 75 μm and 250 μm specimens, however, the wetting of the sample was only partial.

5. References

[1] Zhu G, Chen W, Lu S 2019. *Chemical Engineering and Processing - Process Intensification*, 145
[2] Ez Abadi A M, Sadi M, Farzaneh-Gord M, Ahmadi M H, Kumar R, & Chau K 2020. *Engineering Applications of Computational Fluid Mechanics*, 14(1), 1-12.
[3] Wani C, Ghodke S, Shrivastava C 2012 *International Journal of Advance Research in Science*, 1(3), 233-237.
[4] Shi W, Min Y, Ma X, Chen Y, Yang H. 2022 *Journal of Building Engineering*, 48, 103898.
[5] RETSCH AS200 datasheet; 2023. Accessed on 2023.04.12
[6] Sensirion SHT31-ARP-B datasheet; 2023Accessed on 2023.04.12
[7] INGLAS GmbH. 2023. TIR 100-2 [Datasheet]. Accessed on 2023.04.12
[8] FLIR. 2023. T640 [Datasheet]. Accessed on 2023.04.12
[9] Riangvilaikul B, Kumar S. 2010. *Energy and Buildings*, 42(5), 637-644.
[10] Duan Z, Zhan C, Zhao X, Dong X. 2016. *Building and Environment*, 104, 47-58.
[11] Arduino. 2023. UNO [Datasheet]. Accessed on 2023.04.12
[12] Çengel YA, Boles MA. 2006. *McGraw-Hill.*
[13] Sadighi Dizaji H, Hu EJ, Chen L, Pourhedayat S. 2018. *Applied Energy*, 228, 2176–2194.

Acknowledgments
N.M. acknowledges the financial support of the FAR Dipartimentale 2023 program of the Dept. of Engineering "Enzo Ferrari" - UniMORE.

A test bench for the assessment of flow meters accuracy for fuel consumption measurement in highly dynamic drive cycle tests

M. Martino[1], L. Postrioti[1*], A. Malengo[2]

[1] Department of Engineering - University of Perugia
[2] INRIM, Torino

* Corresponding author: lucio.postrioti@unipg.it

Abstract. The transport sector is one of the most significant contributors in terms of NOx, particulate matter and net CO_2 emission. In order to mitigate this environmental impact, severe homologation limits have been introduced by the different regional Authorities. Both pollutant emissions and CO_2 production of vehicles are measured during standardized laboratory test cycles and road tests, both characterized by relatively abrupt powertrain load and fuel rate transients. This dynamic approach results in severe issues in terms of fuel flow measurement capabilities as flow meters are typically calibrated in steady fuel flow conditions. In the present activity, an experimental test bench was developed with the aim of offering a flexible and affordable test system to evaluate the performances of different flow meters in dynamic conditions. The test bench consists of a complete fuel injection system managed to reproduce an assigned cycle; by this approach the highly-dynamic engine-like fuel flow rate is replicated, offering the opportunity to compare the accuracy of the flow meter under test with both a master Coriolis-type flow meter and the nominal fuel flow time-profile. This bench allows the execution of flow meter accuracy tests in controlled operating conditions with reduced efforts with respect to conventional vehicle test benches and with a high reproducibility level.

1. Introduction

The environmental impact in terms of pollutants and CO_2 emission is the major concern of thermal powertrain-based vehicles used in the transport sector. As a matter of fact, with the current composition of the vehicles fleet in Europe, the on-road and off-road transport sector is the largest contributor to NOx emissions (39% in 2018) and one of the most significant in terms of PM 2.5, CO and HC [1]. In parallel, significant efforts are being made to reduce the CO_2 impact of the transport sector, that can be evaluated to be around 14.5% of the total emission in Europe [2], with the final goal of zero-equivalent carbon emissions set by the European Commission by 2035 [3]. In this complex scenario, two different paths are being developed by the Legislator, acting on different time-scales.

In the short-medium term, the automotive industry is sustaining significant efforts to reduce its environmental impact in order to comply with the upcoming Euro 7 standard [4] by July 2025 in terms of pollutant emissions, implementing increasingly sophisticated after treatment systems (e.g. SCR-on-filter, gasoline particulate filter, burner and electric heaters to speed-up the aftertreatment system warm-up), on-board emission measurement capability and innovative engine management technologies. In the same time, the CO_2 emissions reduction request is supported by pushing the on-going implementation of the Corporate Average Fuel Economy limits in terms of CO_2/km [3] for vehicles adopting thermal or hybrid powertrains (with the development of innovative technologies, *e.g* ultra-lean combustion systems) and by spreading the electrification of the fleets. In parallel and on a longer time-scale, the adoption of the European Climate Law introducing the concept of overall climate neutrality by 2050, imposed the challenging target for the automotive industry of zero-equivalent carbon emissions by 2035 in Europe. This target is pushing the research on cheaper energy storage systems to enable an effective diffusion of pure electric vehicles and, in the same time, is forcing the development of carbon-neutral fuels like bio-fuels, e-fuels and the research to obtain economically competitive, green hydrogen.

In the above described frame, characterized by an unparalleled fast (and in some aspects tumultuous) adoption of emission targets, a more and more significant role is being played by the test cycle standards during which both pollutants and CO_2 emissions are measured. To control the vehicles emission, the Legislator has developed laboratory and road test cycles that need to be carried out during the vehicle homologation procedure. Up to 2017, in Europe, the final emissions have been measured through the NEDC (New European Driving Cycle), composed by a sequence of constant vehicle speed phases separated by slow transients, resulting in moderate powertrain load conditions with respect to on-road actual operation. The observed relevant gap between the type-approval and the real-world operation in terms of both emission and fuel consumption, has led to the introduction of the WLTP (Worldwide harmonized Light vehicles Test Procedure) and more recently of the RDE (Real Drive Emissions). Unlike the NEDC, these last tests are more representatives of the vehicle actual on-road operation, investigating a wider range of engine operation points.

The adoption of these new test procedures resulted in a significantly higher level of measured emissions during the vehicles homologation test cycles, that made more challenging the vehicles compliance to test limits and induced the development of advanced emission control technologies by the automotive industry. In the same time the new, highly dynamic test procedures requested an important upgrade of the measuring instruments (*e.g.* torque meters, air and fuel flow meters, gas analyzers) to be appropriate for the wider range of powertrain operating conditions and, mostly, to improve the measurement accuracy in the transients.

As a matter of facts, all the new standardized procedures are characterized by a relatively high dynamicity in terms of powertrain load, in order to reproduce the actual, real-world vehicle dynamic. This approach results in particular in severe issues in terms of fuel flow measurement capabilities, being marginal during the homologation tests the steady fuel flow conditions in which flow meters are normally calibrated. In the present activity, an experimental test bench was developed with the aim of offering a flexible and affordable test system to dynamically evaluate the performances of different flow meters in dynamic conditions. The developed test bench is based on a so-called "wet-system" in which a complete fuel injection system (composed of low- and high-pressure pumps, fuel pipes, fuel rail and injectors) is managed reproducing its actual operation during an assigned test cycle in terms of injection frequency, rail pressure and injected fuel quantity. The transient operation fuel flow rates are in this way accurately reproduced, offering the opportunity to compare the accuracy of the flow meter under test with both a master Coriolis-type flow meter and the nominal fuel flow time-profile. The flow meter accuracy tests can be carried out with time and cost efforts significantly lower than it would be required on the complete vehicle. Further, in the developed bench the reproducibility level of the calibration is inherently higher being the bench operation fully controllable in terms of operating conditions sequence and in terms of test fluid and injection hardware temperature.

2. Experimental Set-Up

The proposed test bench is intended for supporting the development of both hardware and actuation strategy of the main components of a fuel injection system feeding a thermal powertrain (*i.e.* high-pressure pump, injectors, rail, connecting pipes) and the appropriate calibration of sensors and instruments used for the injection system management (*e.g.* temperature and pressure transmitters, fuel flow meters). With these goals, the bench can mimic the actual operation of a complete injection system in static and dynamic operating conditions, reproducing with a high-fidelity level a prescribed sequence of operating conditions in terms of injection pressure, injection frequency and injection strategy (phase and duration with respect to a virtual engine crankshaft). During the bench operation, different measurements can be carried out to completely characterize the actual injection system operation and/or the bench itself can be used as driving force to produce and un-steady, intermittent fuel flow to be used for sensors/instruments development and calibration. The control system of the proposed hydraulic dynamic bench has been developed at the University of Perugia [5], [6], [7], [8] and its flexible architecture allows the operation with gasoline port injection (PFI), gasoline direct injection (GDI) and Diesel injection systems.

2.1. Dynamic Hydraulic Bench Global Structure

In Figure 1 (a) the dynamic hydraulic bench is represented in a Diesel configuration, while in (b) the overall bench layout is sketched. As depicted in Figure 1 (b), the bench fuel circuit starts from a fuel tank, eventually featuring a fuel heating/cooling auxiliary circuit to control the fluid temperature. From the tank, a low-pressure DC pump supplies the fuel to the high-pressure pump (or directly the fuel rail in PFI systems), which is driven by an AC motor controlled by an inverter. A high resolution, wide frequency range torque meter (Kistler 4503) is used to measure the input torque to the pump. The rail pressure level is controlled by a metering valve at the high-pressure pump inlet (Pressure Control Valve, PCV) for GDI systems or by the combined action of a metering valve and a discharge valve installed on the rail for Diesel systems (typically named SCV and DRV respectively).

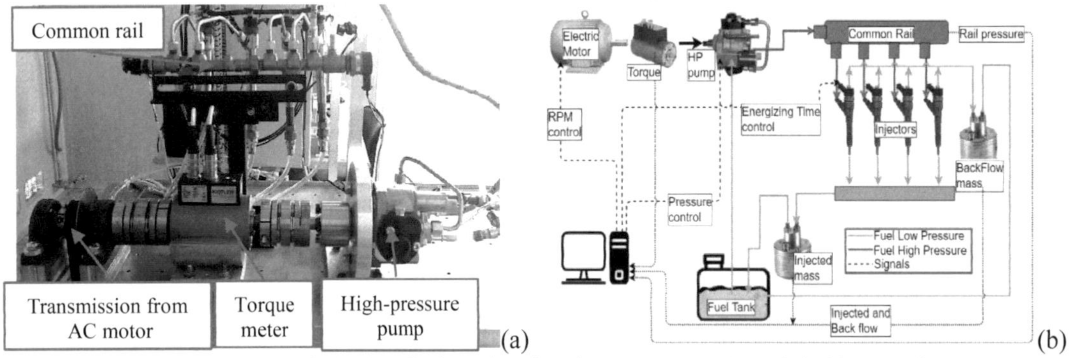

Figure 1. Dynamic hydraulic bench: Diesel arrangement (a); global layout (b).

The injected fuel is collected in a low-pressure rail and metered by a fast response, Coriolis-type mass flow meter used as reference or master flow rate instrument. An eventual flow meter under calibration is connected directly at the master flow meter exit, so to minimize any delay among the two instruments signals. Only for Diesel injection systems, a second mass flow meter is typically required to evaluate the so-called back-leak flow, a secondary flow produced by the injectors during their normal actuation cycle. The back-leak clearly represents a significant energetic loss as the corresponding fuel was compressed at the rail level by the high-pressure pump. Depending on the hardware under test, the back-flow rate can be comparable to the injection flow and should hence minimized. Both the injection flow and eventually the back-leak flow are returned to the tank completing the fuel circuit.

2.2. Dynamic Hydraulic Bench Control Functions and Software Structure

The main logic functions of the bench control system (mentioned in Figure 1 (b)) are:

- RPM control: the control system defines the command to the AC motor inverter to manage the high-pressure pump rotation speed so to define the injection frequency. An incremental encoder installed on the pump shaft is used to generate the general trigger of the system and to determine the actual pump shaft phase, so to actuate all the injectors and the metering valve commands.
- Pressure Control: for GDI and for second generation common rail Diesel systems, the pump shaft position is used to actuate the metering valve (PCV or SCV); all the control valve architectures can be managed by the bench rail pressure control system according a PID regulation strategy

operated in the Real Time layer of the control software (see below). Given the relevance of the back-leak pressure level for Diesel common rail injectors, a PID control strategy is also used to actively control this parameter in the eventual back-leak circuit, using an ancillary PFI injector as discharge valve.

- Energizing Time Control: this Real-Time software function uses the pump shaft position to determine the timing in which the pulse-train or sequence must be initiated for each injector following the proscribed engine sequence. Each injector injection sequence is composed on a number of injection events per each cycle (up to 8 events per cylinder are frequently used for automotive Diesel engines), separated by appropriate dwell times.
- Fast Signals Acquisition: a separate software module is used to operate at high sampling rate (typically 100 kHz) the acquisition of a number of wide frequency spectrum physical quantities, among which the instantaneous torque, the rail pressure in several positions along the circuit, the injectors current and voltage shape and eventually the injector needle lift. Signals provided by additional instruments (such as an injection analyzer measuring the shot-to-shot resolved, instantaneous flow rate produced by one or more injectors) can complete the analysis.

The control software was developed in LabVIEW environment, using a three-layer structure to implement the above-mentioned functions; each of the software layers runs on a different hardware depending on the required process and timing accuracy. In the following Figure 2 the three-layer structure of the control system is briefly described:

PC/Windows: the user interface is implemented in this top-level layer, so to allow the test plan definition, the on-line signals monitoring and data archive management. In this environment, the Fast Signal Acquisition task is also accomplished using a PCIe-6537 multifunction acquisition card.

Remote Controller/Real-Time LabView: all time critical functions, among which Pressure Control and ET Control, are implemented in this layer. The Real-Time controller allows to obtain the system robustness, reliability and timing accuracy required by critical tasks such as the PID regulation for rail pressure. The Real-Time controller typically receives macro inputs from the user interface and exchanges data with the low-level FPGA layer to generate hardware commands and to acquire data from the field.

FPGA: in this low-level layer, using specific input/output modules, basic tasks such as TTL pulse train generation to actuate injectors and valves, encoder position decoding and analog input are accomplished. All TTL pulse trains are supplied to external programmable drivers to generate the required current time profiles to actuate the injectors and the control valves.

2.3. Dynamic Hydraulic Bench Master Flow Meter: Calibration

The master flow meter adopted on the Dynamic Hydraulic Bench is a Coriolis-type mass flow meter by Siemens (MassFlow 2100 DI 1.5) operating in a range 0.1 to 30 kg/h with a 0.1% nominal measuring accuracy. The master flow meter was calibrated at INRIM laboratories by using a new test bench, recently developed in the framework of the EMPIR Project 20IND13 SAFEST [9]. Although it is currently in the testing phase and not yet running with fluids having similar characteristics to fuels, such as the calibration oil ISO 4113, it was possible to perform calibrations with water at ambient temperature of about 21 °C.

The calibration was performed at constant flow rate in the range from 1 kg/h to 30 kg/h by using the gravimetric method. Two calibration methods were used, the standing start and stop method, and the dynamic weighing method developed at INRIM [10]. The weighing system consists of a balance of 10 kg, with resolution of 1 mg, with a weigh-tank of 3 L which are enclosed inside an evaporation trap. The flow generation was performed by a pressure vessel maintained at constant pressure with an air pressure regulator, and a control valves located upstream and downstream of the meter under calibration.

The comparison between the static and dynamic weighing method was performed at the flow rates of 10 kg/h and 20 kg/h, the results have been consistent considering the uncertainties, but with a systematic difference of about 0.04 %. This difference is probably due to the difference in calibration methods, or the fact that the tests were not carried out with degassed water, further studies will be carried out when it will be possible to use the oil as a calibration fluid. With the aim of reducing the uncertainty contribution due to the repeatability of the measurements the minimum amount of water for each run was at least of 1 kg with a duration of the run of at least 5 min. Each calibration point, at constant flow rate, q, was repeated five times, and the percentage error E and standard deviation $s(E)$ were calculated, the results are shown in table 1 and Figure 3.

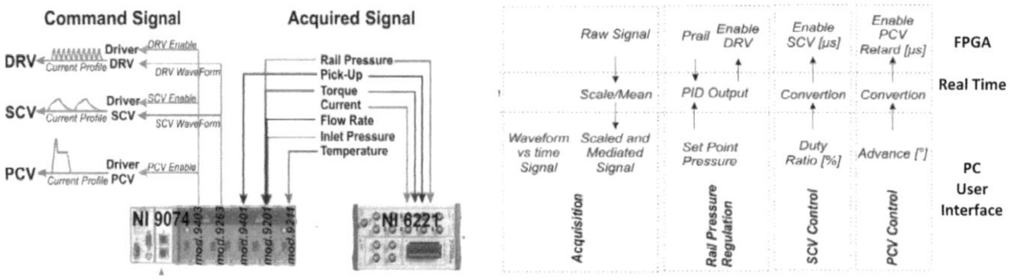

Figure 2. Dynamic hydraulic bench – control system structure.

A number of calibration repetitions were performed on different days, from which the contribution of uncertainty u_{inst} due to the stability was assessed. The uncertainties of the reference standard u_{ref} , shown in table 1, include the uncertainty contributions due to the weighing, the evaporation and the standing start and stop method, this later evaluated at 0.015%. For the evaluation of the standard uncertainty of the error $u(E)$, the uncertainty contribution due to the repeatability has been estimated as $s(E)$/sqrt(5), where 5 is the number of repetitions.

Table 1. Calibration results and associated uncertainties.

Water flow rate [q / kg h^{-1}]	E [%]	$s(E)$ [%]	u_{inst} [%]	u_{ref} [%]	$u(E)$ [%]	$U(E)$ [%]
0.97	0.057	0.010	0.015	0.023	0.028	0.06
2.07	-0.080	0.010	0.015	0.017	0.023	0.05
4.90	-0.007	0.010	0.015	0.016	0.022	0.05
9.50	0.061	0.010	0.010	0.015	0.019	0.04
14.90	-0.039	0.010	0.010	0.015	0.019	0.04
19.93	-0.165	0.010	0.010	0.015	0.019	0.04
25.05	-0.196	0.010	0.010	0.015	0.019	0.04
29.36	-0.186	0.010	0.010	0.015	0.019	0.04

The results have shown good behaviour of the meter, even at low flow rates. The errors are always between 0.2 %, with a very good repeatability, which are more than adequate for the requirements of the hydraulic dynamic bench at the University of Perugia. It is planned to repeat the calibration of the flow meter as soon as possible using as fluid a calibration oil and in addition evaluate the measurement performance of the flow meter under dynamic loads

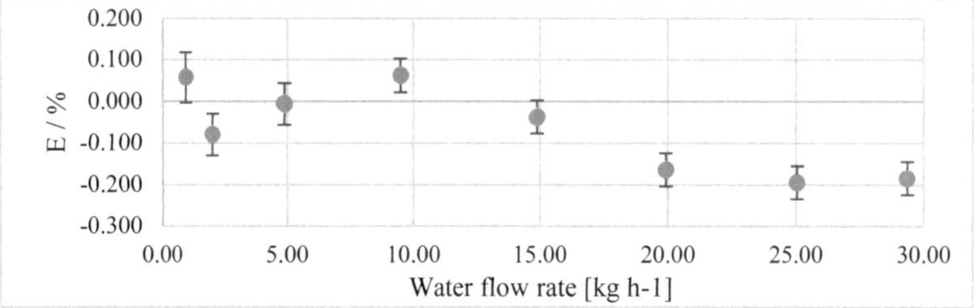

Figure 3. Relative errors with expanded uncertainty bars, as a function of the water flow rate.

3. Results and Comments

As an example of the Dynamic Hydraulic Bench capabilities, in the following some acquisitions are reported relevant to the simulation of an WLTP homologation test carried out for a 1.3 liters displacement, 4-cylinder passenger car in the power range 55 - 65 kW/l. In all the following plots, the following nomenclature is adopted:

- Nominal quantities: data recorded during the WLTP test executed in the homologation cell with the complete vehicle are named as "nominal" (injection pump speed, injection pressure level, injection pulse train as produced by the engine ECU) and are used as target quantities to be replicated on the Dynamic Hydraulic Bench.
- Experimental quantities: data recorded during the WLTP test on the Dynamic Hydraulic Bench with the injection system alone are named as experimental.

The injection system installed on the Dynamic Hydraulic Bench replicates the components installed on the tested vehicles, namely: Injection Pump: Denso HP3 (2 pistons); Rail: Bosch 0445214224; Injectors: Bosch CRI 2.20

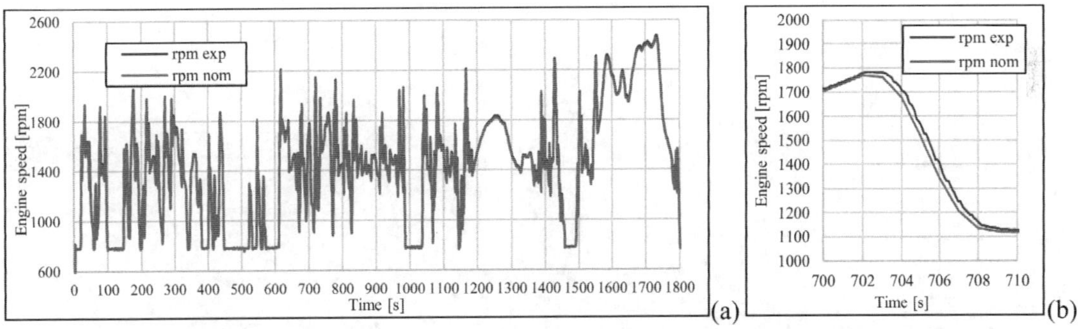

Figure 4. WLTP test, Nominal *vs* Experimental engine speed (a); zoom on (b)

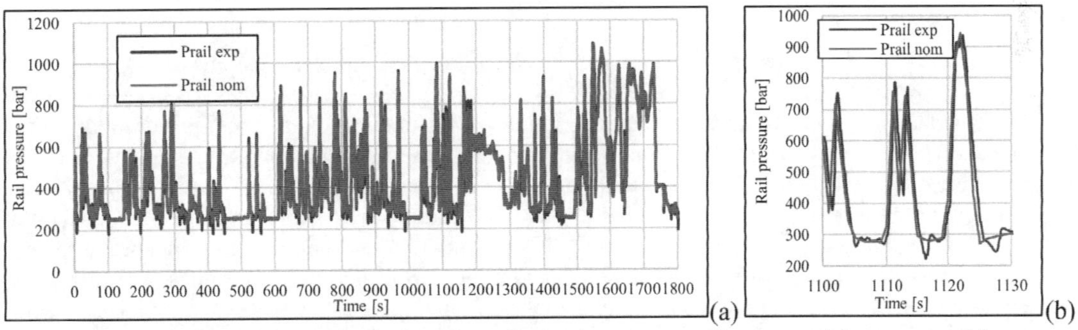

Figure 5. WLTP test, Nominal *vs* Experimental common rail pressure (a); zoom on (b).

In Figure 4 the comparison among the equivalent engine speed (doubled pump speed) obtained during the entire WLTP test for the vehicle and on the dynamic hydraulic bench is reported. As can be seen, the hydraulic bench has the capability of reproducing the engine behavior with an appropriate accuracy. In detail (see Figure 4 (b) for a typical speed transient), the experimental line (blue) seems to be in delay (about 0.35 s) with respect to the nominal signal (red). This delay corresponds to the RPM control system reaction and actuation times, as the red line is used directly as target for PID control system. Conversely, the same quantity is a logic output for the complete vehicle system.

In Figure 5 the comparison among the rail pressure time history measured on the vehicle by the engine ECU during the WLTP test and the corresponding injection pressure obtained on the dynamic hydraulic bench is reported. As for the pump speed control system, the comparison between the target (red) and the obtained (blue) P_{rail} time-profiles evidences globally a good efficacy of the pressure regulation system, based in this test only on the SCV actuation (DRV closed during the WLTP test). As for the pump speed, the rail pressure control system seems to follow the nominal trace with an acceptable delay of about 0.3 s. At the same time, some undershooting can be appreciated during the fast pressure decrease transients, possibly to be ascribed to the decision of excluding the DRV from the regulation strategy.

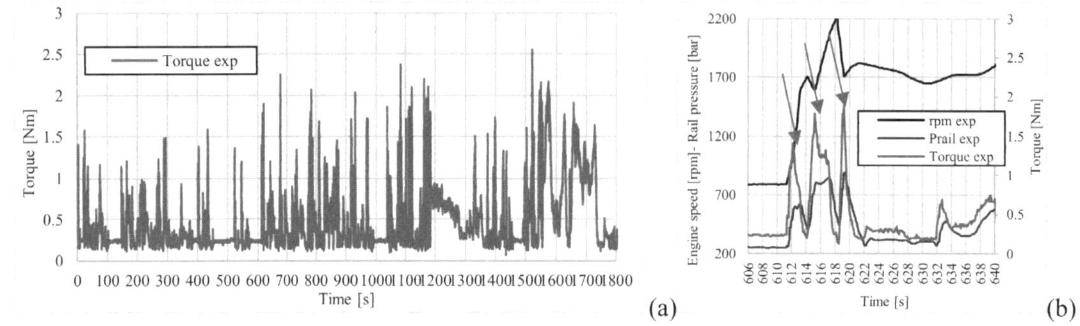

Figure 6. WLTP test, input torque to the high-pressure pump (a) torque, RPM and rail pressure (b).

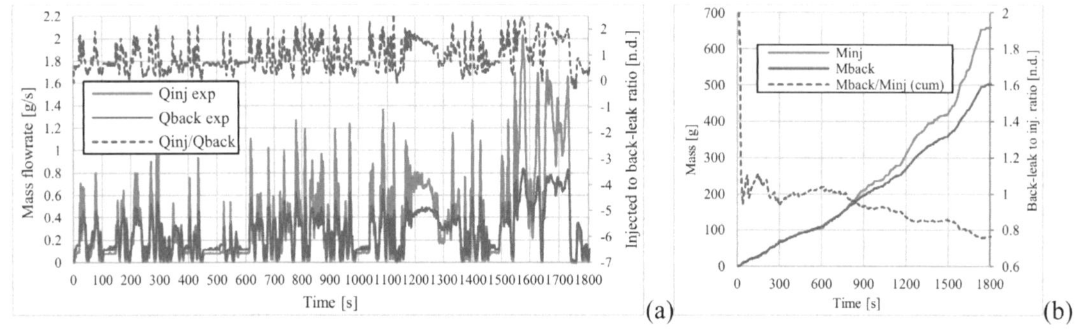

Figure 7. WLTP test, dynamic (a) and cumulated (b) injected and back-leak mass flow rate.

The instantaneous torque input to the pump (Figure 6) measured on the Dynamic Hydraulic Bench during the WLTP test was acquired by a Kistler 4503 meter, set to its wider operating range (0-50 Nm). The corresponding measurement on the complete vehicle is clearly not feasible. As can be observed in Figure 6, only for short periods during the test the actual torque request was higher than 2 Nm, with a peak of 2.56 Nm and a cycle average value of 0.47 Nm. The peculiar analysis of torque, pressure and rpm time history plot can reveal interesting details about the analyzed system dynamics, well-captured by the present experimental setup. As an example, in the plot in Figure 6 (b) the correct timing agreement among the acquired quantities is reported: the 3-step rail pressure increase at 611, 613 and 617 s (about 350 bar globally) is well captured in terms of pump torque. It is interesting to note that while for the first of these three steps also the pump shaft speed is being increased, during the last 2 pressure steps the pump speed is decreasing. Notwithstanding, the torque signal reaches the highest peaks in the test, suggesting the rail pressure to be the primary mechanical load in the considered system while the pump inertia plays a secondary role. As above mentioned, the simultaneous measurement of both the injected and the back-leak flows from the four injectors was carried out. As reported in Figure 7 (a), with the analyzed injection system and actuation strategy, the maximum instantaneous back leak flow rate was 0.82 g/s while the maximum injected flow rate was 2.01 g/s. Both the flow traces evidenced an appreciable capability of the used Coriolis-type flow meters and circuit configuration to follow the cycle dynamics. The instantaneous injected to back-leak ratio ranges from 0 up to 2.2 for the maximum load conditions

The cumulative flow injected and backflow mass measurement (Figure 7 (b)) evidenced how the two flows are distributed over the WLTC cycle. During the first 600 s the back-leak mass is slightly bigger than the injected one due to the low load and the long idle periods. The globally injected mass was 659 g, the global back-leak mass was 504 g i.e. 76% of the injected mass, confirming the relevance of the energetic loss associated with the back-leak flow required to operate common rail injectors.

4. Conclusions

In the global trend of adopting more and more severe legislations to limit both pollutant emissions and CO_2 production in the transport sector, the introduction in recent years of highly dynamic homologation test cycles for the automotive sector arose significant technical issues for the fuel flow measurement. While commercial fuel flow meters are normally calibrated in steady flow conditions, current test procedures include significant powertrain load dynamics, resulting in considerable fuel flow transients that could cause significant measurement errors. Consequently, meters accuracy should be verified in operating conditions similar to those faced in the vehicle operation. In the present paper, the capability of a dynamic hydraulic bench of replicating the operation of a complete injection system in realistic operating conditions is discussed. The bench, composed of low- and high-pressure pumps, rail and injectors, can be managed replicating a highly dynamic test cycle by imposing the high-pressure pump speed, rail pressure time-history and injection sequence. The reported results relevant to the WLTP homologation cycle of a B-segment vehicle equipped with a 4-cylinder Diesel engine evidence the bench capability of reproducing the injection system dynamic behavior, enabling the comparison of different type of fuel flow meters with both a master Coriolis-type flow meter and with the nominal fuel flow time-profile.

5. References

[1] https://ec.europa.eu/commission/presscorner/detail/en/ip_22_6495.

[2] https://climate.ec.europa.eu/eu-action/transport-emissions/road-transport-reducing-co2-emissions-vehicles/co2-emission-performance-standards-cars-and-vans_it

[3] https://www.europarl.europa.eu/news/en/press-room/20230210IPR74715/fit-for-55-zero-co2-emissions-for-new-cars-and-vans-in-2035

[4] COM_2022_586_

[5] Postrioti, L. et al. An experimental and numerical analysis of pressure pulsation effects of a Gasoline Direct Injection system. (2016) Fuel, 173, pp. 8-28. DOI: 10.1016/j.fuel.2016.01.012

[6] Piano, A. et al. Experimental and Numerical Assessment of Multi-Event Injection Strategies in a Solenoid Common-Rail Injector, (2017) SAE International Journal of Engines, 10 (4), pp. 2129-2140. DOI: 10.4271/2017-24-0012

[7] Postrioti, L. et al. Momentum flux measurement on single-hole GDI injector under flash-boiling condition, (2015) SAE Technical Paper, Volume 24, Issue 2480. DOI: 10.4271/2015-24-2480.

[8] Cavicchi, A. et al. Experimental Analysis of Fuel and Injector Body Temperature Effect on the Hydraulic Behavior of Latest Generation Common Rail Injection Systems, (2018) SAE Technical Papers, 2018-April. DOI: 10.4271/2018-01-0282

[9] Buker, O. et al, Characterization of a Coriolis flow meter for fuel consumption measurements in realistic drive cycle tests, (2023) Flow Measurement and Instrumentation, Volume 93. DOI 10.1016/j.flowmeasinst.2023.102424.

[10] Saba, F. et al. A dynamic gravimetric standard for liquid flow measurements Metrologia (2021), 58 015007

Acknowledgments

This work was supported within the Joint Research Project ‘‘Sustainable advanced flow meter calibration for the transport sector” (20IND13 SAFEST) and has received funding from the European Metrology Program for Innovation and Research (EMPIR) co-financed by the Participating States and from the European Union’s Horizon 2020 research and innovation program.

Experimental analysis of flow boiling heat transfer in multi-microchannel evaporators

F Riccardi[1], G Zummo[2,*], L Saraceno[2], L Gugliermetti[2], G Caruso[1]

[1] Dep. of Astronautical Electrical and Energy Engineering (DIAEE)
NERG - Nuclear Engineering Research Group, Corso Vittorio Emanuele II, 244, 00186 Rome (Italy)

[2] ENEA C.R. Casaccia
Laboratory of Processes and Systems Engineering for Energy Decarbonisation, Via Anguillarese, 301, 00123 Rome (Italy)

* Corrisponding author: giuseppe.zummo@enea.it

Abstract. This paper provides an experimental analysis of flow boiling heat transfer in four finned evaporators with different aspect ratio and heated length have been tested. R245fa was the refrigerant chosen as working fluid for its good thermal characteristics. The imposed heat flux at the footprint covered a range from 30 to 330 kW/m^2, while the fluid mass fluxes within the channel varied in a range between 17 and 225 kg/m^2 s. The hydraulic diameters of the investigated evaporators were 1.15 mm and 2.09 mm, with aspect ratio (channel width/channel height) respectively of 0.3 and 0.72. Thanks to the measuring apparatus, heat transfer data were collected and then processed to calculate the local heat transfer coefficients, using the fin array heat transfer model. These local values of the heat transfer coefficient were then compared with those calculated from existing correlations for mini-micro channels flow boiling and the Mean Absolute Percent Error was finally assessed.

1. Introduction

In recent years, given the trend toward miniaturization of components and simultaneous increase in their performance, the use of mini-microchannel heat sinks has been of great interest to cope with the increased heat fluxes and to ensure an adequate thermal control of the components where they are installed.

There are many applications of this technology in several industrial fields such as the cooling of electronic power converters for electric vehicles, but also next-generation microprocessors such as Data Centres' CPUs and GPUs, whose cooling process is to date an extremely energy-intensive activity.

IGBT (Insulated-Gate Bipolar Transistor) modules for electric vehicles represent an example of high-heat flux electrical component and they are widely used inside electrical vehicles inverters and charging stations. They generate heat fluxes in the range from 6.5 to 50 MW/m^2(based on the die size of the chip [1]). These devices should be maintained at a temperature lower than 100-115°C to avoid early failures, while , in the ICT (Information and Communication Technology) field the threshold temperature of the processors is set around 80-85°C. In 2007 Yang et al. [2] found that the CPU's processing speed is severely affected by the chip temperature decreasing by 10-15% only due to local temperature surges, which may exceed by 30°C the average chip temperature. Using organic refrigerants with saturation temperature at operative pressure lower than the threshold value imposed for the correct operation of the devices, it is possible to combine the advantages of micro-channel heat sinks with those of the two-phase heat transfer, such as the temperature uniformity above the cooled surfaces and the enhanced heat transfer coefficient due to the evaporation process, reducing the required pumping power, with respect to an equivalent single-phase cooling system. The major drawback of this technology is the high degree of complexity. Furthermore, in this kind of heat sinks the heat flux decreases steadily with increasing distance from the base and the top side of channel is an unheated surface. Since the two-phase heat transfer coefficient is heat-flux dependent, this unevenness in the heating condition strongly affects the heat transfer mechanism. A very comprehensive review of the various transition criteria and of the main issues concerning microchannel flow-boiling was conducted by Mahmoud and Karayiannis in 2017 [1].

In 2022 Criscuolo et al. [3] studied the flow boiling of R1234yf, R1234ze(E), R134a finding that the boiling development in high aspect ratio micro-channels can be divided in three regimes. In the first one the heat transfer coefficient is only determined by the heat flux, regardless of the mass flux. In the second one both the heat flux and the mass flux have an influence on the heat transfer coefficient, due to the superimposition of the nucleation and forced convection heat transfer mechanisms. The last regime is the one in which occurs the dry-out of the channel and the heat transfer coefficient rapidly drops. In 2016 Gugliermetti et al. [4] studied the saturated flow boiling of FC-72 in a 1 mm tube, testing several correlations for the heat transfer coefficient. The best result was obtained with the Li and Wu correlation for micro-channels.

2. Experimental test rig and measurement uncertainty

2.1 Experimental plant and fluid loops

The experimental facility is provided by the Casaccia research centre of ENEA (Agenzia Nazionale per le Nuove Tecnologie, l'Energia e lo Sviluppo Economico Sostenibile). The whole experimental system mainly consists of two fluid circuits: a primary circuit filled with the coolant and a secondary circuit for the cooling water. There is also a third loop, with a fan coil cooler to dissipate excess heat. A simplified diagram of the experimental setup is showed in Figure 1.

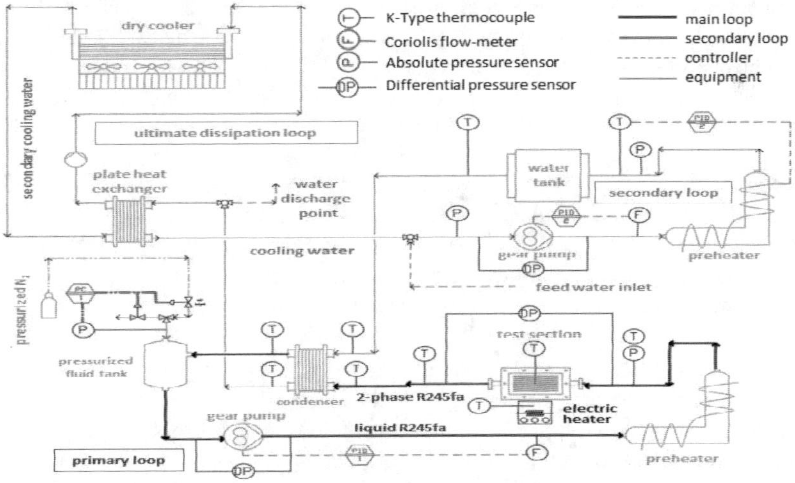

Figure 1. Piping and Instrumentation Diagram of the experimental plant

Liquid refrigerant is circulated by means of a gear- pump. The tank is placed in the upper part of the loop, to guarantee an adequate suction head avoiding pump cavitation. The pressure inside the tank is regulated by a solenoid valve that injects and expels N_2. Before entering the multichannel-evaporator (with a sub cooling varying between 5 and 20 °C), the liquid flow rate is measured by a Coriolis flowmeter. Two K-type thermocouples are placed at the evaporator inlet and outlet sleeves to measure the fluid bulk temperatures. An absolute pressure transducer is placed at the inlet of the evaporator, while the pressure drops across the test section are measured with a differential pressure sensor. The evaporator footprint temperatures are measured by means of calibrated K-type thermocouples inserted in the metallic base. The input heat-rate is provided by three cartridge heaters (400W) each, placed inside an aluminium cube, to simulate the surface of the chip to be cooled. The temperature of this surface is measured by a thermocouple inserted in a hole on the top of the aluminium block. Once evaporated, the fluid is condensed in a plate heat exchanger, cooled by the water coming from the secondary loop. Then, the condensed fluid finally returns to the storage tank.

2.2 Test section
The test section consists of the aluminium multichannel evaporator covered with a polycarbonate cover, the heating block, the instrumented inlet and outlet fluid sleeves, and the thermocouples to measure the wall temperatures. The whole heated part is insulated with low conductivity bricks made of vermiculite (~ 0.10 W/m°C), to reduce thermal losses. Figure 2 shows the evaporator with all its main dimensions

Main Dimensions	*E50*	*E50-2*	*A100*	*A100-2*
Channel's width	0.75	1.8	1.8	0.75
Channel's height	2.5	2.5	2.5	2.5
Hydraulic diameter	1.15	2.09	2.09	1.15
Channel's length	50	50	100	100
Fin thickness	0.3	0.5	0.5	0.3

Figure 2. Evaporator and main dimensions in *mm*.

3. Method
The real-time acquisition of all measured variables is achieved by means of a program developed in LabVIEW, within which are also set safety thresholds for maximum operating pressure, minimum flow rate, and maximum temperatures. Beyond these threshold values the power supply is cut off by the system. The same program also allows to control the flow rate, pressure, and power during the tests.

The measured quantities are acquired at a frequency of 1000 Hz and the program automatically calculates the average value recording one line containing these stored averages each second. Acquisition occurs only when surface temperatures have stabilized, and their standard deviation is low.

4. Data reduction
The collected data are then analysed using Python scripts and Excel spreadsheets, while the thermodynamic quantities are calculated using REFPROP 10 software, developed by NIST [5].

Below it is reported a summary of the calculations performed.

To get rid of the channel dimension and to consider the flow velocity within it, the mass flux has been calculated as the ratio between the mass flow rate $\dot{m}$ and the transversal flow area A_{flow}.

By heating the evaporators in vacuum conditions and waiting for steady temperatures, the heat losses towards the external environment have been calculated and the net heat rate is then calculated as the difference between the measured input power and the calculated heat losses.

To consider the actual position of the inserted thermocouples inside the channel, the dimensionless axial position ($\tilde{z}_{ad_i}$) has been calculated as the ratio between the axial coordinate of the relative thermocouple z_i and the channel length L_{ch}. Thermodynamic equilibrium quality is obtained from its definition, knowing the mixture enthalpy by the energy balance.

$$x_i = \frac{h_{in} - h_f}{h_{fg}} + \tilde{z}_{ad_i} \cdot \frac{\dot{q}_{net}}{h_{fg} \cdot \dot{m}} \tag{1}$$

From the assumption of uniform power distribution over the heater surface, it follows that the bulk temperature of the fluid linearly increases along the channel until saturation is reached. The quality has been used as reference value to calculate the real bulk temperature T_{∞_i} .

$$T_{\infty_i} = T_{sat}\ (x_i \geq 0) \qquad ; \qquad T_{\infty_i} = \tilde{z}_{ad_i} \cdot \frac{\dot{q}_{net}}{\dot{m}\, c_p} + T_{in}\ (x_i < 0) \tag{2}$$

The Fourier law of conductive heat transfer has been applied to get the temperature of the channel's base in contact with the fluid starting from the measured temperature of the bottom surface.

The ΔT used to calculate the heat transfer coefficient in boiling condition, has been calculated as the difference between the temperature of channel's base and that of the fluid bulk. Due to the non –uniformity of the heat flux in vertical walls, the heat flux at the channel's footprint (q''_{fp}) is taken as reference and is calculated as the ratio between the net input power and the footprint area (A_{fp}) of the multichannel array. The fin model has been applied to consider the heat transferred by the vertical walls, and the fin efficiency has been calculated imposing adiabatic boundary conditions at the top (equation 3).

$$\eta_{fin} = \frac{\tanh\,(m \cdot H_{ch})}{m \cdot H_{ch}} \qquad \text{where:} \quad m = \sqrt{\frac{2 \cdot htc}{k_{material} \cdot s_{fin}}} \tag{3}$$

The heat transfer coefficient is also involved in the expression of the fin efficiency and for this reason an iterative procedure has been performed for its calculation, solving the thermal balance of the fin array.

The heated surface has been divided into the finned part and the remaining base surface which is exposed to the fluid. The formula can be rearranged into a more suitable form for iterative computation of the heat transfer coefficient htc, whose value has been calculated by solving the following non-linear equation 4:

$$\frac{\dot{q}_{net}}{\left(A_{exposed\ base\ ch} + \eta_{fin}(htc) \cdot A_{finned}\right) \cdot \Delta T_{wall\ i}} - htc = 0 \tag{4}$$

The first term in the previous equation represents the heat transfer coefficient calculated from Newton's law, imposing the heat flux corrected with the fin efficiency.

5. Uncertainties

The average uncertainty over measured quantities was calculated using the root sum of squares formula. The statistic component was assumed to follow a Gaussian distribution with coverage factor of 95% being twice the standard deviation. The propagation of uncertainty on calculated quantities, such as the heat flux has been estimated using equation 5. The average values of the uncertainty were reported in Table 1.

$$u_y = \sqrt{\left[(dy/dx_1) \cdot u_{x_1}\right]^2 + \cdots + \left[(dy/dx_n) \cdot u_{x_n}\right]^2} \tag{5}$$

Table 1. Average uncertainty on measured and computed quantities

Quantities	*E50*	*E50-2*	*A100*	*A100-2*	*Units*
Twall	± 0.13	± 0.12	± 0.562	± 0.575	°C
Psat	± 0.075	± 0.075	± 0.05	± 0.05	bar
Mass flow rate	± 0.079	± 0.0783	± 0.13	± 0.2	Kg/h
Net heat flux at the footprint	± 0.07%	± 0.06%	± 1.5%	± 1.5%	W/m^2

6. Experimental Results

The evaporators E50 and E50-2 were tested at different working pressures of 2 and 3 (T_{sat} equal to 34 and 46°C, respectively) and at three different mass fluxes of 64, 94 and 122 kg/m^{2}s. Results are graphically shown in the curves below (Figure 3), which show how in the evaporator with smaller hydraulic diameter, the heat transfer coefficient (at the centreline) reaches a maximum value, then tends to decrease with the quality. The peak value of the *htc* increases with the pressure and slightly with the mass flux. This is due to the reduction of the specific volume of the vapour phase with increasing pressure, and thus, to the lower void fraction inside the channel (at equal quality). This leads to a lesser suppression of the nucleation mechanism which, instead, is slightly promoted by the reduction of latent heat. The decreasing arc of the curves is related to the nucleate boiling dominant stage of boiling, as reported also in the study carried out by Cheng et al. [6]. Moreover, it can be noticed that in the evaporator with wider channels the *htc* reaches values higher than in the other cases and it monotonically increases with the quality with a gradually decreasing slope. In the subcooled region, the heat transfer coefficient is almost not affected by the mass flux, because the fluid flow at the channel inlet is in laminar condition, with Nu number which tends to be constant (when fully developed). This is consistent with the three stages of boiling highlighted in the study of Criscuolo et al [3] and confirmed experimentally in the present study.

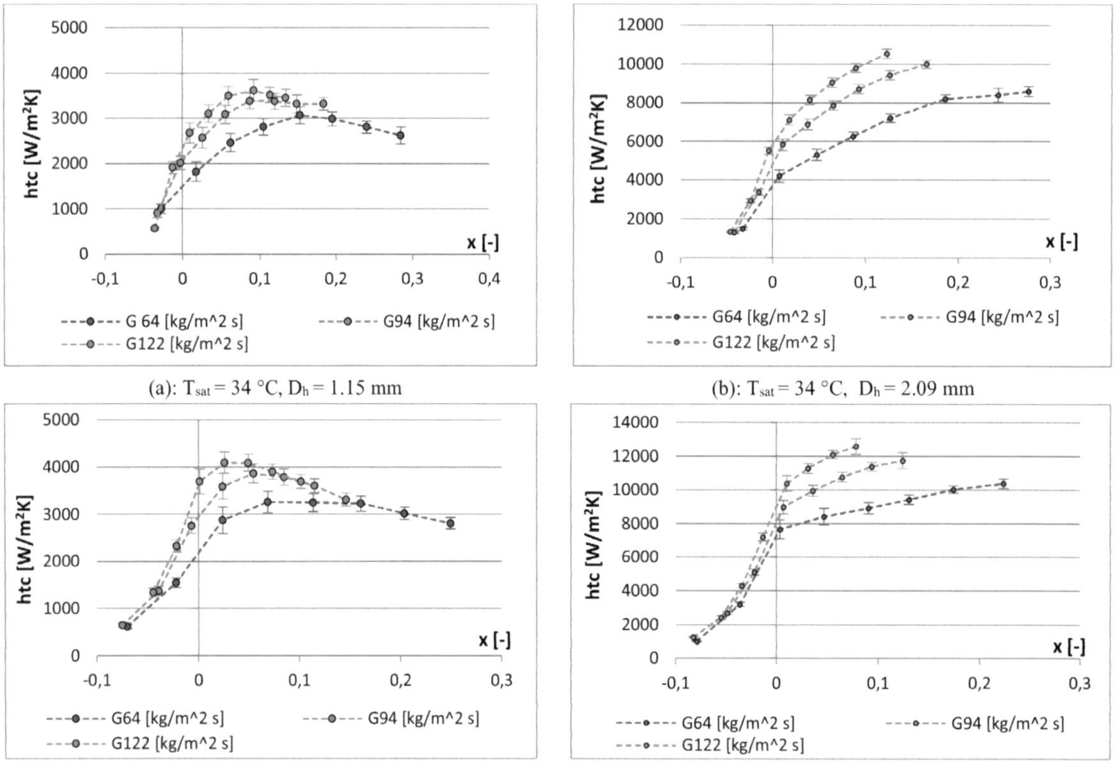

(a): T_{sat} = 34 °C, D_h = 1.15 mm

(b): T_{sat} = 34 °C, D_h = 2.09 mm

(c): T_{sat} = 46 °C, D_h = 1.15 mm

(d): T_{sat} = 46 °C, D_h = 2.09 mm

Figure 3. Experimental heat transfer coefficient as function of the local quality. L_{ch}=50 mm; $\tilde{z}_{ad_i}$= 0.5

The evaporators A100 and A100-2, with larger heated lengths have been tested with a working pressure of 1.8 bar (T_{sat} = 30.3 °C) and with different mass fluxes (32, 64, 97 and 128 kg/m^{2}s). In this case the *htc* has been measured near the channel outlet, at the dimensionless axial position of 0.87 and for this reason there are no experimental points in the subcooled region. At high heat fluxes, the heat transfer coefficient rapidly drops due to the incipience of the dry-out phenomenon, especially for the evaporator with a smaller channel diameter, as shown in Figure 4.

Journal of Physics: Conference Series **2685** (2024) 012069 doi:10.1088/1742-6596/2685/1/012069

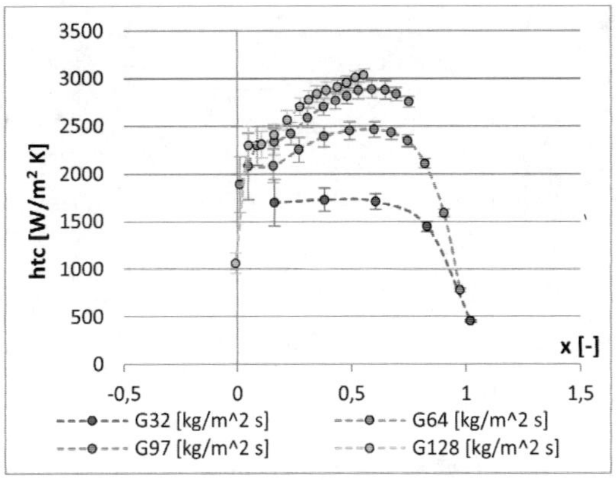

(c): T_{sat} = 30.3 °C, D_h = 1.15 mm

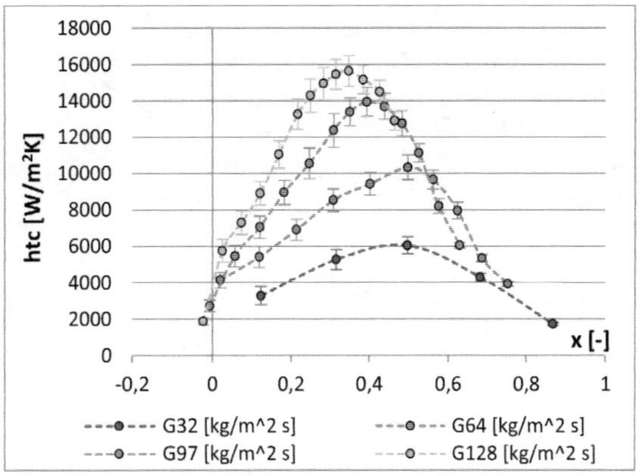

(d): T_{sat} = 30.3 °C, D_h = 2.09 mm

Figure 4. Experimental heat transfer coefficient as function of local quality. L_{ch}=100 mm; $\tilde{x}_{d_i}$= 0.87.

7. Comparison with available correlations

The experimental htc has been also compared to the one calculated from available correlations developed for mini and micro-channels to verify their applicability to the measured data. The Mean Absolute Percent Error (MAPE), reported in Table 2, has been used to find out the correlation which gives better results in the tested range of heat and mass fluxes. The parity-plot diagrams have been developed and the most representative are shown in Figure 5. The heat transfer coefficient obtained from correlations developed for rectangular channels and small tubes with the whole perimeter heated has been multiplied by a factor $Nu_3\ /Nu_4$ as done by Lee and Mudawar in [11].Where Nu_3 and Nu_4 are the laminar fully developed Nusselt numbers for a rectangular channel with respectively three and four heated walls. This coefficient is useful to estimate the effect of having only three heated walls during the htc calculations.

The tested correlations have been developed for small circular ducts or rectangular channels, with uniform heat flux.

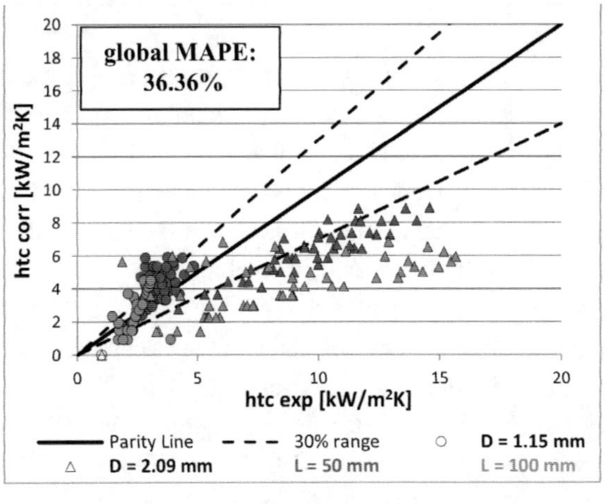

(a): Cooper

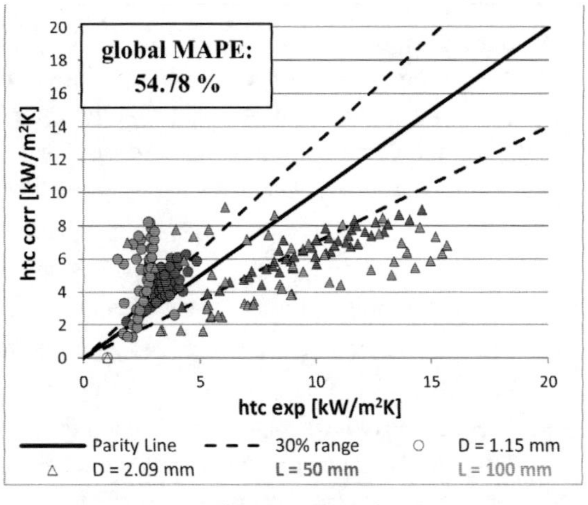

(b): Bertsch et al.

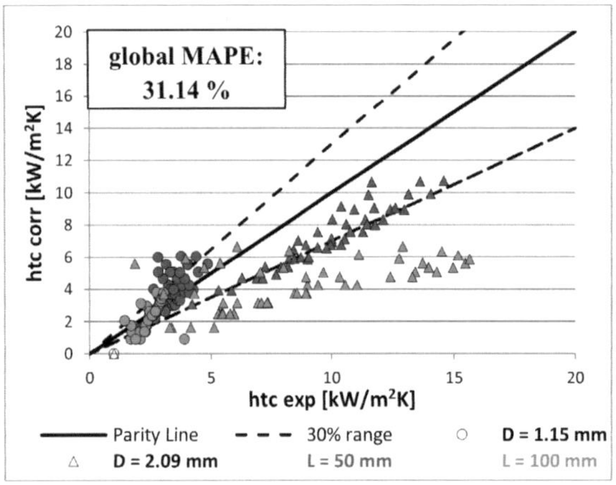

(c): Tran et al.

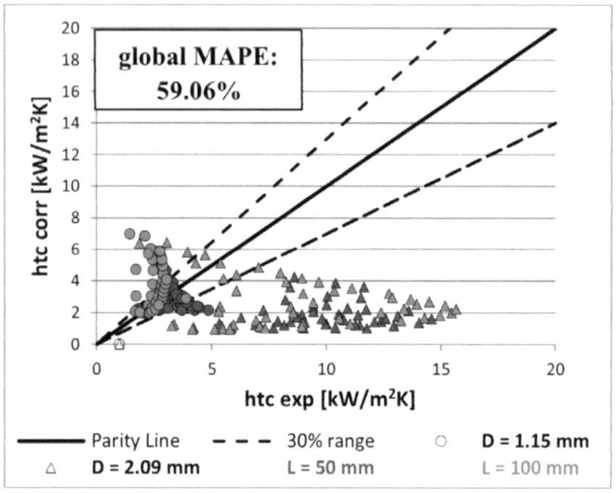

(d): Warrier et al.

Figure 5. Parity plot diagrams for some of the selected correlations

Table 2. Calculated Mean Absolute Percent Error for the selected correlations.

Tested correlation	*E50* *L/D = 43.48*	*E50-2* *L/D = 23.92*	*A100* *L/D = 47.85*	*A100-2* *L/D = 86.96*
Cooper [7]	26.26 %	35.22 %	54.65 %	29.33 %
Bertsch et al. [8]	32.48 %	33.55 %	52.20 %	100.91 %
Li and Wu [9]	72.80 %	18.93 %	36.67 %	87.61 %
Lazarek and Black [10]	175.94 %	21.14 %	42.54 %	153.18 %
Lee and Mudawar [11]	390.88 %	59.70 %	77.35 %	110.59 %
Mahmoud and Karayiannis 2 [12]	44.53 %	31.19 %	57.36 %	192.41 %
Tran et al. [13]	23.96 %	27.32 %	53.15 %	20.12 %
Warrier et al. [14]	31.26 %	80.50 %	72.19 %	52.32 %

8. Conclusions

This paper provides an experimental evaluation of the heat transfer coefficient in multichannel evaporators with three-sides-heated channels, using the fin heat transfer model to consider the unevenness of the heat flux over the vertical walls of the channels. The results showed the following main aspects:

- Smaller channels have lower “fluid-to-wall” heat transfer coefficients, decreasing with the quality;
- Three stages of boiling can be identified: a nucleation-dominated one; an intermediate one; a film forced-convective one. The htc profiles are similar to those studied by Cheng et al. in [6];
- In all the cases heat transfer coefficient increases with the pressure and with the mass flux;
- The Tran correlation and the Cooper’s one give the best results, especially for smaller channels, this confirms the strong influence of the nucleate boiling contribution, which is proportional to $q''^{0.6\div0.7}$.
- The Li and Wu correlation gives good results only for 2.09 mm diameter, giving the lowest MAPE (of 18.9%) for the evaporator with a diameter of 2.09 mm and a channel’s length of 50 mm. This is probably due to aspect ratio (closer to 1) which is more representative of the dataset used for the correlation development.

Flow instability and unequal flow distribution in lateral channels represents the main criticality in predicting the overall thermal resistance of the evaporator. The temperature distribution in vertical channel walls should be better investigated to develop more accurate models, especially for high aspect ratio micro-channels with adiabatic top side, where the wall heat flux is progressively reduced as the distance from the base increases and the heat transfer coefficient on vertical walls can strongly differ from that at the channel's base.

9. References

[1] Karayiannis T, & Mahmoud M M (2017). Flow boiling in microchannels: Fundamentals and applications. *Applied Thermal Engineering*, 115, 1372-1397

[2] Yang B, Wang P, & Bar-Cohen A (2007). Mini-contact enhanced thermoelectric cooling of hotspots in high power devices. *IEEE Trans.Compon. Packag. Technol.* , 3(30), 432-438

[3] Criscuolo G, Markussen W B, Meyer K E, & Kaern M R (2022). High heat flux flow boiling of R1234yf, R1234ze(E) and R134a in high aspect ratio microchannels. *International Journal of Heat and Mass Transfer.*

[4] Gugliermetti L, Caruso G, Saraceno L., Zummo G, & Celata G. (2016). Saturated flow boiling of FC-72 in 1mm diameter tube. *International Communications in Heat and Mass Transfer*(75), 115-123.

[5] NIST (s.d.). Reference Fluid Thermodynamic and Transport Properties Database. (REFPROP Version 10) Tratto da https://www.nist.gov/srd/refprop

[6] Cheng L, & Xia G (2017). Fundamental issues, mechanisms and models of flow boiling heat transfer in microscale channels. *International Journal of Heat and Mass Transfer*(108), 97-127.

[7] Cooper M (1984). Heat Flow Rates in Saturated Nucleate Pool Boiling - A Wide -Ranging Examination Using Reduced Properties. *Advances in Heat Transfer*, 16, 157-239.

[8] Bertsch S, Groll E, & Garimella, S (2009). A composite heat transfer correlation for saturated flow boiling in small channels. *International Journal of Heat Mass Transfer.*

[9] Li W, Wu Z (2010). A general correlation for evaporative heat transfer in micro/mini-channels. *International Journal of Heat & Mass Transfer*(53), 1778-1787.

[10] Lazarek G M, Black S H (1982). Evaporative heat transfer pressure drop and critical heat flux in a small vertical tube with R-113. *International Journal of Thermal Science*, 7(25), 945-960.

[11] Lee J, & Mudawar I (2005). Two-phase flow in high heat flux microchannel heat sink for refrigeration cooling applications: Part II-heat transfer characteristics. *International Journal of Heat & Mass Transfer*(48), 941-955.

[12] Mahmoud M, Karayiannis T G (2013). Heat transfer correlation for flow boiling in small to micro tubes. *International Journal of Heat & Mass Transfer*(66), 553-574.

[13] Tran T N, Wambsganss M, & D M France (1996). Small circular and rectangular channel boiling with two refrigerants. *International Journal of Multiphase Flow*, 3(22), 485-498.

[14] Warrier G R, Dhir V K, & Momomda L (2002). Heat transfer and pressure drop in narrow rectangular channels. *Experimental Thermal and Fluid Science*(24), 53-64.

The DEPLOY! Project: Development of a Deployable Pulsating Heat Pipe experiment on a parabolic flight

A. Billi[1,*], R. Perna[2], S. Picchi[2], V. Rosellini[5], E. Saltmarsh[3], M. Bocelli[4], N. Ricci[2], N. Miché[3], F. Bozzoli[4], M. Mameli[2], S.Filippeschi[2]

[1] Department of industrial engineering, University of Pisa, Pisa, Italy
[2] Department of Energy, Systems Land and Construction Engineering, University of Pisa, Pisa, Italy
[3] School of architecture, technology and engineering, University of Brighton, Brighton, UK
[4] Department of Engineering and Architecture, University of Parma, Parma, Italy
[5] Department of information engineering, University of Pisa, Pisa, Italy

[*] a.billi@studenti.unipi.it

Abstract. DEPLOY! Project aims at analysing the behaviour of Deployable Pulsating Heat Pipe (PHP) shaped as a helicoidal torsional spring in the adiabatic section on board a Parabolic Flight platform. A PHP is a passive thermal control device where the heat is efficiently transported by means of the self-sustained oscillatory fluid motion driven by the phase change phenomena. The microgravity environment allows to eliminate the buoyancy force contribution in the liquid phase momentum. Consequently, it is possible to isolate the contribution of the pressure drop caused by the 3D arrangement and infer on their effect on the PHP performance. As a result, a proper design based on the previous considerations would increase the flexibility of the PHP for use in space applications without significant reductions in efficiency. The goal of DEPLOY! is to demonstrate the functionality of a Deployable Pulsating Heat Pipe in various unfolding configurations by analysing its thermal-hydraulic response throughout a Parabolic Flight. The presented Deployable PHP is composed of an aluminium tube (inner/outer diameters 1.6mm/2.6 mm) and filled with HFE-7000. It is heated at the evaporator using a flat heater and cooled at the condenser with a water-cooled cold plate. T-type thermocouples are used to measure the wall temperature in several locations, while two pressure transducers (one located in the evaporator section and another in the condenser section of the same pipe) measure the local fluid pressure. Additionally, an IR Camera will be used to observe a section of the pipe for further analysis of the flow frequency. The device operation will be tested on ground and 0-g at different heat loads (24W, 34W), in multiple static positions corresponding to different opening angles (0°, 45°, 90°, 135°, 180°) and during its dynamic opening from 0° to 180°, thanks to a remotely controllable motion system.

Keywords: Deployable Pulsating Heat Pipe, Parabolic flight, Microgravity, infrared thermographic acquisitions, Inverse heat conduction problem

1 Introduction

In space missions, the persistent threat of irreparable damage presents an ongoing challenge. Addressing risks in the harsh deep space environment involves two primary strategies: minimizing the likelihood of risk occurrence and implementing effective risk mitigation measures. Retractable radiators play a pivotal role in reducing risks associated with micrometeoroid impacts, not only alleviating space constraints at launch but also minimizing the vulnerable surface area when inactive [1]. Passively deployable radiative panels, incorporating shape memory alloys for unfolding dynamics control, are currently under investigation as an innovative technological solution [2]. The deployable Pulsating Heat Pipe (PHP) represents a noteworthy advancement in thermal management after over two decades of research [3]. This experiment focuses on exploring the thermal response of deployable

PHP in microgravity conditions. Limited scientific literature addresses various flexible PHP devices, such as the flexible flat plate PHP, examined in different configurations, including a polycarbonate copper multi-layer structure [4] and a polypropylene semi-flat double sheet [5]. The latter, tested during the ESA 77th parabolic flight campaign, demonstrated promise. Another flexible PHP, developed in Japan with stainless steel flexible micro pipes, yielded valuable outcomes on microgravity platforms like the REXUS ballistic rocket and parabolic flight plane [6] [7]. The device under study, tested in three unfolding configurations (0°, 90°, and 180°) and three heating power levels (14W, 24W, and 34W) during the 79th ESA Parabolic flight campaign, showed encouraging, albeit inconclusive, results. Notably, the PHP configuration had minimal impact on working conditions when the evaporator temperature exceeded 50 °C. The current experiment aims to gather crucial information on device operation, particularly under more demanding conditions with higher heating power inputs. Key novelties and upgrades from previous experiences will be explored.

i) investigation of intermediate unfolding angle configurations (45 and 135°) to produce clearer indication on the effect of the device configuration on its working conditions;
ii) implementation of a remotely controlled actuation subsystem to analyze the performances of the device during its unfolding dynamic (from closed/0°, to open/180°) in microgravity conditions;
iii) inclusion of an IR Camera which will record the temperature variation on the outside of portions of multiple channels close to the evaporator section. Such observation outcomes will be useful for future analyses.

The test campaign will take place in November 2023 and will comprehend static and dynamic conditions testing.

2 Experiments

The experimental setup, shown in **Figure 1**, comprises a containment box and a control rack, designed to satisfy the platform interfaces requirements and safety guidelines. The containment box houses the Deployable PHP (i.e., the test cell) supported by a structural framework. Additionally, the containment box accommodates components of the actuation and thermal subsystems, as well as an IR Camera. The control rack serves as the central control centre for the experiment. It incorporates power supplies, the cooling, data acquisition and actuation management subsystems. These elements provide the necessary control and measurement capabilities to tune the experiment and collect relevant data.

2.1 Test cell

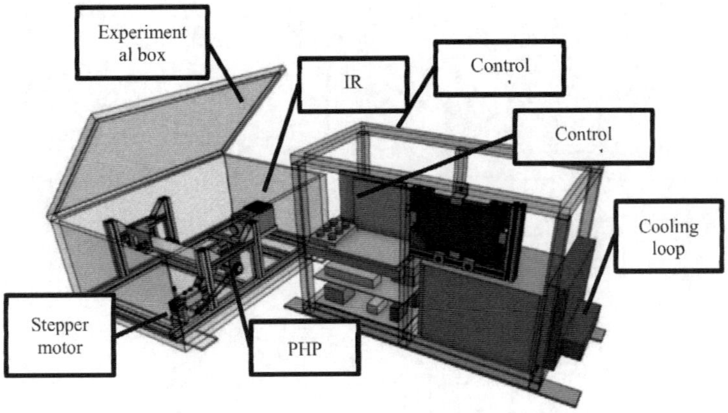

Figure 1 Experimental set up: on the right the control rack, on the left the experimental box.

The Deployable Pulsating Heat Pipe under study, depicted in **Figure 2**, features a capillary tube (1.6/2.6 mm I.D./O.D.) made of aluminum 6063. It is bent into a closed-loop serpentine with 11 U-turns in the evaporation section, resulting in 22 channels. The adiabatic, evaporator, and condenser sections are 715mm, 10mm, and 75mm long, respectively. Shaped as a torsional spring in the adiabatic section, the foldable PHP has a coil diameter (d) of 65mm, N = 3.5 coils per connection, totaling Lcoil = 715mm

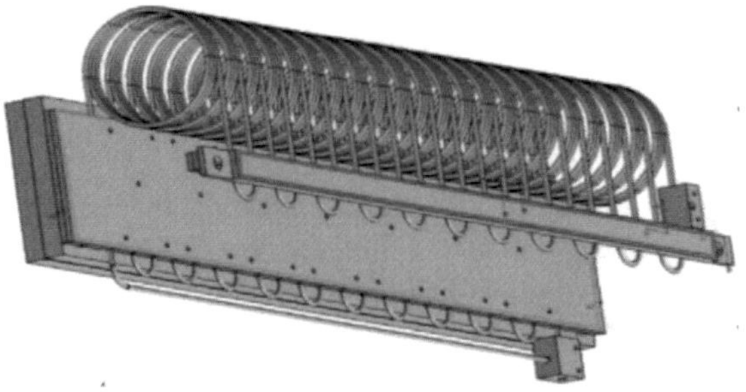

Figure 2 Deployable pulsating heat pipe (folded configuration). The heater is in red, the cold plate is in blue.

(90% of the channel length). The PHP is partially filled with HFE-7000 at a volumetric filling ratio of 70%. The evaporator section, as shown in **Figure 2**, is situated between two aluminum heat spreaders fastened to the tubes with screws. A flat, 500mm long heater (Joule Heater Hotspring®/Maxi - type WRP) connected to the spreaders provides heat input. A cold plate extracts heat from the condenser section and maintains a constant temperature, connected to a Liquid Loop System (LLS). The actuation subsystem (explained in section 2.2.2) deploys the condenser section while the evaporator remains fixed to the support structure.

2.2 Experimental setup

The experimental apparatus comprises multiple subsystems situated in two primary support structures. . As shown in **Figure 1**, the experimental setup consists of a control rack on the right and an experimental box on the left. The control rack houses the cooling loop (blue), automation and heating subsystem power supplies (small shelves on the left), a control panel with laptops (light blue), a monitoring screen (black), and control switches. The experimental box includes the Deployable PHP supported by an aluminium structure (light grey), the stepper motor with torque transmission, and an IR camera (blue).

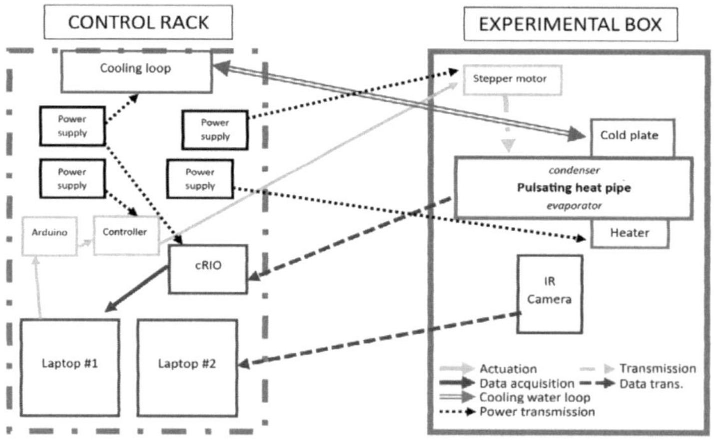

Figure 3 Experiment assembly - block description.

The main novelties of this experiment reside in the data acquisition subsystem composition and the actuation subsystem, that are described in more detail in the next sections. The block scheme in **Figure**

3 helps visualizing the connections between within and between subsystems. In red and blue the PHP and the thermal subsystem, the actuation subsystem in yellow, the data acquisition subsystem in purple.

2.2.1 Data acquisition subsystem

The system is composed by two main data paths. The first is the one connected to the IR Camera, the second one is related to the sensors connected to the PHP.

The IR Camera (Flir 7000) is in front of the evaporator section of the PHP, at 254 mm of distance. The device captures images at a 25Hz frequency, focusing on five pipes in direct contact with the evaporator section (60 mm height). Subsequently, thermal imaging data undergoes processing to extract a rectangular pixel section (approximately 20 pixels wide and 40 pixels high), representing the central image of a selected pipe. The acquired data is then filtered through the Altair acquisition program. Boundary conditions recorded during the campaign are integrated into Altair. A thermocouple (TC) strategically placed on the PHP in the observed area serves as a reference for IR Camera readings. Further processing occurs in a MATLAB script, transforming records into n-dimensional matrices, with 'n' corresponding to a single frame. Finally, the n-dimensional matrix undergoes processing using the Inverse Heat Conduction Problem (IHCP) algorithm. This algorithm estimates heat flow exchanged by the device and provides a reading on the pulsation frequency of the filling fluid, starting from a known temperature [8]. During the test campaigns, real-time sensor data are displayed and monitored to ensure the safe and optimal performance of the experimental setup. All collected data are stored for further analysis in multiple external storages. Pressure and temperature readings are plotted, and the results obtained from ground testing and parabolic flight campaign are compared to evaluate the influence of altered gravity on the PHP working conditions.

Temperature and pressure readings are performed during both the ground tests and the three parabolic flight whole duration. A total of 15 T-type thermocouples (±1 °C) are used to measure temperature. These thermocouples are positioned on the surfaces of 5 selected pipes: 5 at the evaporator section, 5 at the condenser section, and 5 at the adiabatic section. Additionally, 2 pressure transducers (Keller® PAA-33X, ± 0.5 kPa) are placed at the evaporator and condenser extremities of the same loop. To collect the data, a cRIO data acquisition system is utilized. All the collected data is transmitted to a computer equipped with data visualization software, allowing for real-time display and storage of the data. Following the completion of the parabolic flight campaign, the collected data is analysed and compared with the results obtained during the ground tests. Both the pressure transducers and the thermocouples readings are collected with a 1000Hz frequency.

2.2.2 Actuation subsystem

This subsystem is composed by a Nema 34 stepper motor (5.9 Nm torque), a transmission, a controller, and their power supplies. The motor changes the PHP configuration by opening it and closing it by 45° steps from a closed configuration (0°) to an open configuration (180°). In **Figure 4** the PHP is depicted in all the tested configurations. The stepper motor is connected to the PHP condenser section support shaft with a pulley + belt system with a gear ratio of 1 to 4. The motor is controlled by the driver board, which redirects power coming from a PSH 6066-A power supply and controls from an Arduino MEGA to the motor. The Arduino MEGA is powered and controlled by the data acquisition laptop, but its primary input source is a switch control panel located on the control rack next to said laptop. Safety measures are implemented to prevent damage to the experiment caused by accidental extra rotations of the condenser section via limit switches located at the end of the rotation span. The condenser weight is of 5.5 Kg and the maximum torque required to hold its position is of approximately 15N*m, well below the top torque transmission from the actuation of over 20 N*m. To avoid complications, no rotation is planned to happen during the hyper gravity phase of the parabolic flight, and the rotation speed is kept at lowest possible value of $\pi/10$ rad/s at the PHP shaft.

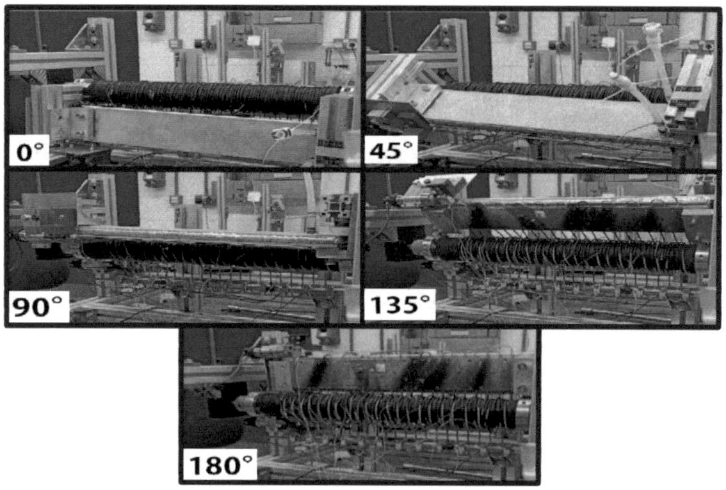

Figure 4 PHP unfolding positions. from top left to bottom: 0°-closed, 45°, 90°, 135°. 180°-open

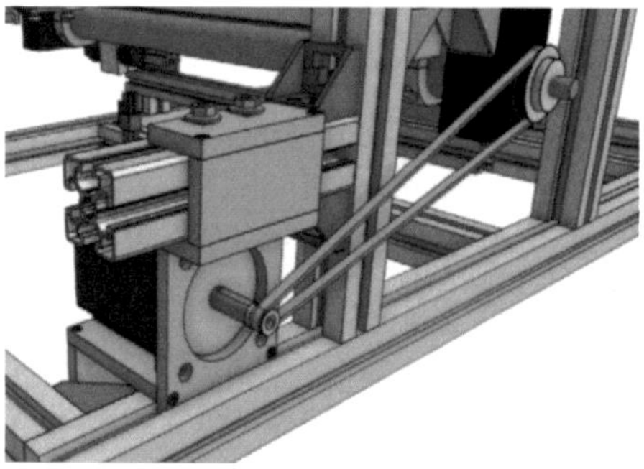

Figure 5 Actuation subsystem particular: from the left stepper motor, motor support, transmission (pulleys and timing belt), PHP shaft.

2.3 Experimental procedures

The experiment consists of two different test campaigns: a parabolic flight test and a ground test. The first is the main focus of the investigation, allowing an observation of the working conditions of the device in altered gravity, while the ladder is performed before it to set up a baseline for the tested values and establish the most functional test matrix to be designed for the PFC.

Ground tests serve as a baseline for subsequent Parabolic Flight Campaign tests. The objective is to establish a consistent comparison framework and all activities performed during the flight will be simulated.

The Parabolic Flight Campaign (PFC) is scheduled to take place in November 2023 aboard the Airbus A310 Zero-G aircraft of Novespace. The campaign is designed to be conducted over a span of three flight days, with each day consisting of 30 microgravity intervals lasting for 20 seconds each, interspersed with hyper-gravity intervals of 20 seconds at 1.8g. During a Zero-G flight, the parabolic maneuver is divided into three stages. Firstly, there is the pull-up stage, where the pilot lifts the nose of the A310 Zero G airbus to an angle of 45 degrees. This action subjects the experimental apparatus to hyper gravity, a pull that is 1.8 times than Earth's gravity. The pull-up, or "nose-up," lasts for approximately 20 seconds. Next is the parabola stage. As the aircraft ascends, the pilots reduce engine speed, and the aircraft follows a ballistic trajectory. Weightlessness begins when the aircraft enters a parabola, during which it experiences free fall for about 20 seconds. Finally, there is the pull-out stage. The nose of the plane is tilted downward by 45 degrees, and the pilots gradually level off the aircraft by increasing engine speed. Like the pull-up stage, the experimental apparatus is subjected to hyper gravity, 1.8 times stronger than Earth's gravity. After 20 seconds, the A310 Zero G returns to a horizontal trajectory. This maneuver is performed 30 times for each flight, in groups of 5 parabolas each. After performing 5 maneuvers, the aircraft enters a phase of 5 to 10 minutes of steady horizontal flight before proceeding to the next stage.

The PFC test sessions follow a predefined procedure to ensure compliance with safety and technical requirements in the environment. All experiments, including the assembly for this experiment on the plane, are powered up after take-off and must be shut down before the landing maneuver begins. The first set of 5 parabolic flight maneuvers begins 20 to 30 minutes after take-off, which aligns with the expected preparation time for the experiment to become fully operational before the injection of the first maneuver. Ground tests are conducted to verify this expectation. If the system takes longer than the allocated interval, the first 1 or 2 parabolic maneuvers of the first set may not be used for data collection, but 3 to 4 parabolic maneuvers will still be available for testing. Additionally, the parabolic flight consists of six groups of maneuvers, and the experiment is planned to be tested in five different configurations, reserving one group of parabolic maneuvers for a backup attempt.

On the first and second day, the tests resemble the static tests conducted during ground. The system is powered up, and the PHP is in its initial position (closed at 0°). When operational conditions are achieved, the test session begins, and data from the first set of 5 parabolic maneuvers are collected. At the end of this batch, the actuation subsystem adjusts the PHP angle from 0° to 45°. It is expected that the system will be minimally affected by this change, and the PHP should reach new steady-state conditions in a few seconds. The second set of parabolic maneuvers is used to test the device in the second configuration. This procedure repeats until the end of the fifth batch. The last parabolic session can serve as a backup in case any of the previous measurements were not optimally recorded. At the end of the final batch, the PHP is returned to its closed starting position, and the system is shut down before landing. This process is repeated similarly on the first and second flight days. The difference between the two flight tests lies in the thermal conditions to which the PHP is subjected. On the first day, the evaporator input power is 34W, resulting in a heater temperature of 110°C, while the cold plate temperature remains at 23°C throughout the test. The thermal conditions for the second day are expected to be different, with specific values determined after analysing data from the ground test campaign.

The third test session, on the final day of PFC, focuses on analysing the behaviour of the PHP in a dynamic situation. The pre-test procedures are the same as on previous days, with the system expected to reach nominal conditions within the same timeframe. The dynamic action is described in **Figure 6**. At the start of the first viable microgravity interval, the automation subsystem unfolds the device by rotating the condenser section 180° around the coil axis, transitioning the PHP from its initial position (open at 0°) to its open position (180°). This deployment occurs between 2 and 4 seconds after the injection and lasts for 10 seconds. Once the final position is reached, the PHP remains still until the end of the hyper gravity interval (approximately 30 seconds). During the subsequent steady horizontal flight interval, the condenser is returned to its starting position by the actuation subsystem. It may take several minutes for the PHP to return to its initial nominal conditions, making the following parabola unsuitable for data collection. This test is expected to be repeated during the first, third, and fifth parabolas of each batch. This test matrix allows the collection of three relevant samples in six different thermal configurations. For each batch, thermal conditions such as heating input and condenser temperature can be adjusted, creating a diverse set of data points. The initial thermal conditions match those of the first flight day, with subsequent conditions determined based on ground test results.

Journal of Physics: Conference Series **2685** (2024) 012070 doi:10.1088/1742-6596/2685/1/012070

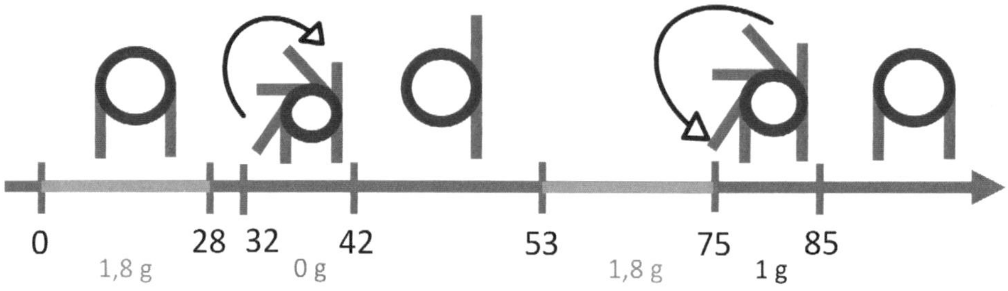

Figure 6 PFC day 3: dynamic test configuration.

3 Data analysis and expected results

The device's thermal resistance, without a working fluid, is determined to be 3 K/W based on prior measurements [6]. This resistance arises from heat conduction through the tubes and fluid flow resistance during operation. With a working fluid, the effective resistance is influenced by lower resistance from internal motions. Computation of thermal resistance involves mean temperatures at the evaporator and condenser and power input. The combined standard uncertainty considers diverse parameters, ensuring a comprehensive evaluation. The pressure drop model assumes incompressible fluid, no oscillation, full power transfer to the PHP, and 1D heat transfer. Slug motion is described by the momentum equation, considering gravity, hydrostatic pressure, and shear forces. The simplified equation includes contributions from gravity and shear forces, defining the pressure drop.

$$\Delta P = \rho \vec{g} \cos(\theta) + l(1-\alpha)\left(\frac{f_s}{2d_i}\rho_l \vec{v_l}\right)$$

To determine the fluid velocity, a wavelet analysis will be conducted [5] employing a complex Morlet wavelet. The rationale behind this selection is the resemblance between the wavelet shape and the signal profile [6]; a strong correlation between the two shapes leads to improved quality outcomes. Using the Morlet wavelet as the mother wavelet, the transformation equation, with scale factor s and time shift factor u, takes the following form.

$$WT(u,s) = \int_{-\infty}^{+\infty} f(t)\frac{1}{\sqrt{s}}\pi^{\frac{1}{4}} \exp\left[i\omega_0\left(\frac{t-u}{s}\right)\right] \exp\left[-\frac{(t-u)^2}{2s^2}\right] dt$$

Once the dominant frequency is determined, it is possible to assume that this frequency leads to the mean velocity of the individual liquid slug. It is then straightforward to calculate $\vec{v}$ as $\vec{v} = l \cdot f_z$.

Thermal imaging data is employed to extract a specific rectangular pixel section, measuring 5 pixels in width and 3 pixels in height, representing the central image of a chosen pipe. The records are filtered using the Altair acquisition programme, integrating campaign-recorded boundary conditions and referencing a thermocouple connected to a PHP point observed in the IR camera window. Processed through a MATLAB script, records transform into n-dimensional matrices (n for a single frame). This n-dimensional matrix feeds an Inverse Heat Conduction algorithm, estimating device flow and outputting the filling fluid's pulsation frequency [8].

4 Conclusions

The DEPLOY! Project aims at the technological advancement of the deployable PHP field. The device at study is an innovative foldable pulsating heat pipe which has been tested once before in a parabolic flight in 2022. This experiment is presented with novelties such as an actuation subsystem which allows to change the configuration of the device during the tests and an IR camera to analyse the behaviour of the working fluid starting from its surface temperature. This experiment contributes to enhance the knowledge on the deployable PHP, by investigating its working conditions in new configurations and at different temperatures. The dynamic deployment test, performed in simulated microgravity, is a novelty of this project; such test will shed light on an important working life condition that the device will face. The results of this test are important for a first estimate of the influence of the peculiar dynamic of this device on its working conditions and will offer insights on future testing configurations worth investigating for a complete understanding of this correlation. The experiment will be tested during a parabolic flight campaign in November 2023. The results are expected to enhance the knowledge on this device and generate further testing opportunities for the future.

Acknowledgements
The present work has been carried out as part of the ESA Academy Experiments Programme, supported by the European Space Agency (ESA) through the ESA Education Office, the European Low Gravity Research Association (ELGRA) and the UK Space Agency. The test campaign will take place in Bordeaux (FR) in the Novespace company facilities. The test cell is developed in and based on components of the laboratories of University of Pisa Energy, Systems, Territory and Construction Engineering Department (DESTeC), University of Brighton Advanced Engineering Centre, and University of Parma Engineering and Architecture Department.

References

[1] NASA, “State-of-the-Art of Small Spacecraft Technology - Thermal Control,” 2022.

[2] A. Bacciotti, F. Bucchi, F. Frendo, M. Mameli, R. Perna, and S. Filippeschi, “On the use of shape memory alloys for deployable passive heat radiators in space satellites,” *IOP Conf. Ser. Mater. Sci. Eng.*, vol. 1038, no. 1, p. 012061, 2021, doi: 10.1088/1757-899x/1038/1/012061.

[3] M. Mameli, G. Besagni, P. K. Bansal, and C. N. Markides, “Innovations in pulsating heat pipes: From origins to future perspectives,” *Appl. Therm. Eng.*, vol. 203, no. October 2021, p. 117921, 2022, doi: 10.1016/j.applthermaleng.2021.117921.

[4] C. Jung, J. Lim, and S. J. Kim, “Fabrication and evaluation of a high-performance flexible pulsating heat pipe hermetically sealed with metal,” *Int. J. Heat Mass Transf.*, vol. 149, 2020, doi: 10.1016/j.ijheatmasstransfer.2019.119180.

[5] O. Der, A. A. Alqahtani, M. Marengo, and V. Bertola, “Characterization of polypropylene pulsating heat stripes: Effects of orientation, heat transfer fluid, and loop geometry,” *Appl. Therm. Eng.*, vol. 184, no. November 2020, p. 116304, 2021, doi: 10.1016/j.applthermaleng.2020.116304.

[6] R. Perna *et al.*, “Experimental investigation of a coil shaped deployable pulsating heat pipe,” pp. 1–8, 2023.

[7] S. Filippeschi *et al.*, “Thermal performance analysis of a Pulsating Heat Pipe with a long adiabatic section tested with different working fluids,” *THERMINIC 2022 - 28th Int. Work. Therm. Investig. ICs Syst. Proc.*, vol. 72, pp. 22–26, 2022, doi: 10.1109/THERMINIC57263.2022.9950642.

[8] L. Pagliarini *et al.*, “Thermal characterization of a multi-turn pulsating heat pipe in microgravity conditions: Statistical approach to the local wall-to-fluid heat flux,” *Int. J. Heat Mass Transf.*, vol. 169, p. 120930, 2021, doi: 10.1016/j.ijheatmasstransfer.2021.120930.

Estimation of Internal Heat Flux on Pulsating Heat Pipes using Kalman Filter: Numerical and Experimental Results

B H M Margotto[1], M J Colaço[1], C E P Kopperschmidt[1], F Bozzoli[2,3], L Pagliarini[2], L Cattani[2,*], W B da Silva[4], M Hamada[5] and H Nagano[5]

[1] Department of Mechanical Engineering, Federal University of Rio de Janeiro, Rio de Janeiro, RJ, 21941-972, Brazil
[2] Department of Engineering and Architecture, University of Parma, Parco Area delle Scienze 181/A, Parma, Italy
[3] SITEIA.PARMA Interdepartmental Centre, University of Parma, Parco Area delle Scienze 181/A, Parma, Italy
[4] Department of Chemical Engineering, Federal University of Espírito Santo, Alegre, ES, 29500-000, Brazil
[5] Department of Mechanical Systems Engineerin, Nagoya University, Furo-cho, Chikusa-ku, Nagoya, Aichi, 464-8603, Japan

[*] e-mail: luca.cattani1@unipr.it

Abstract. Pulsating Heat Pipes (PHPs) are two-phase passive thermal devices characterized by the presence of significant fluid oscillations inside the tubes that permit to fast and efficiently transfer heat from a hot region to a cold one. They are present in many engineering applications, e.g., electronics, sustainable energies, aerospace. They are very attractive due to their high heat removal capability, flexibility, and low manufacturing cost. Although they have been widely studied their working principles are still not fully understood. To better comprehend their thermal behaviour recent works presented different approaches to assess the internal heat flux in PHPs. However, all the adopted approaches are based on whole-domain techniques that require a higher computational cost compared with sequential estimations. Therefore, to allow heat flux estimation with lower computational effort, in this work a procedure based on the Kalman filter has been adopted. The heat flux has been estimated by solving the inverse heat conduction problem in the PHP's wall adopting as input data the temperature measurement on the external surface of the pipe acquired by an infrared camera. Firstly, the procedure has been validated adopting synthetic data. Then, experimental data referring to actual operating conditions have been employed to estimate the internal heat flux. The Kalman filter has been adopted as technique to solve the classical instability intrinsically present in inverse problems. It is relatively simple to implement and requires only previous time information, requiring low computational costs. Moreover, the Kalman filter allows the real time estimation of the heat flux: assessing the heat flux on PHPs during operation could be a useful instrument to provide information about on-time working conditions and thus avoid any possible malfunctioning.

Keywords: *Pulsating Heat Pipes, Inverse Heat Conduction Problem, Kalman filter, Two-phase flow devices.*

1. Introduction

Pulsating Heat Pipes (PHPs) are devices that transfer heat from a hot region to a cold one by fluid oscillations in a two-phase passive flow. Due to their efficient and high heat transfer rate, geometrical flexibility, and low manufacturing cost, these devices are promising for many engineering areas, such as sustainable energy, electronics, fuel cells and battery, and aerospace applications [1]. Usually, PHPs thermal performance are evaluated by computing an overall device thermal resistance [2]. However, such approach evaluates only globally the thermal performance and only in pseudo-steady state condition. Many studies try to obtain deeper evaluation and comprehension of the thermal behavior along the tubes to characterize the fluid working conditions [3]. In this sense, recent works showed strong relation of the fluid motion inside the tube and the heat flux exchanged between the fluid and the pipe wall. In these works, the internal heat flux was non-intrusively assessed considering the temperature measurements on the external pipe wall and solving the inverse heat conduction problem (IHCP) in the PHP's wall. Cattani *et al.* [4] used the Tikhonov Regularization to estimate the heat flux on a PHP portion made of Sapphire and filled with ethanol, providing stunning results related to the fluid flow pattern inside the tube. Pagliarini *et al.* [5] proposed the use of Gaussian filter to evaluate the heat flux exchanged between the fluid and the pipe wall on a multi-turn PHP of aluminum under microgravity conditions. The work considered a transient one-dimensional heat conduction model and showed remarkable results regarding the fluid motion inside the tubes. Recently, Colaço *et al.* [6] employed the Gradient Conjugated Method with Adjoint Operator to solve the IHCP with the same experimental setup from [4] and [5], and it showed good agreement with the heat flux obtained when compared to the previous works. Recently, some works (see [7, 8]) provided a deep insight of PHP mechanism using the Gaussian filter to estimate the heat flux in PHPs. These works presented a whole-domain approach to solve the inverse problem, nevertheless it is attractive to obtain the thermal performance of such devices during operation. Such information could predict the device malfunction and failure, and it is crucial to avoid undesired system crashing due to overheating. For online control purposes, the sequential estimation with Bayesian filters is a suitable choice that correlates information of the desired states and measurements by Bayes' theorem [9]. Recursive filtering provides the estimation of the state sequentially, and its advantage is that it does not require storing all the complete dataset from the past. Furthermore, the use of Kalman Filter (KF) provides the optimal estimation when the problem under question is linear-Gaussian [10], that it is the case of the current study. Tuan *et al.* [11] first used the KF to estimate the heat flux with the Recursive Least-Square Estimator (RLSE). Daouas *et al.* [12] proposed the augmented vector technique to reduce the two-steps calculation from the previous approach that considers the temperature and heat flux on the target estimated state vector. Due to its simplicity and low computational effort, such approach gained attention and will be used in this work to assess the internal heat flux in PHPs. Therefore, the current work aims at evaluating the capability of the internal heat flux estimation in PHPs with KF for online monitoring purpose. The suggested technique for the internal heat flux estimation solves an IHCP considering available the temperature measurements on the external surface of the pipe wall. The obtained results showed good accordance when compared with the synthetic data and with the experimental data from previous references. The proposed technique is promising to real time performance monitoring since it is non-intrusive and needs lower computational effort compared to the whole-domain techniques since it performs the estimation sequentially.

2. Forward Problem

The physical problem is represented by a hollow cylinder (Figure1) indicating one of the pipes that constitute the PHP. The domain is a solid wall, defined by Ω, with a length of L and inner and outer radii r_{int} and r_{ext}, respectively. The inner surface, Γ_{int}, is submitted to a heat flux $q(z,t)$, that represents the heat exchanged between the pipe wall and the working fluid. The external surface, Γ_{ext}, exchanges heat by natural convection with the environment with a convective heat transfer coefficient of h_∞ and environmental temperature of T_∞. The other surfaces Γ_0 and Γ_L are considered adiabatic. The boundary conditions are set according to the PHP operation conditions and were experimentally verified in [4].

The study presumes the thermal properties constant and neglect the temperature variation over the angular direction to model the physical problem, which has been verified experimentally from [4].

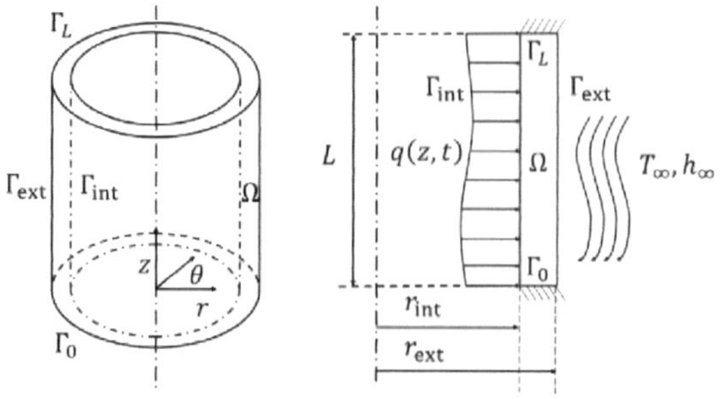

Figure 1. PHP geometry (left) and the two-dimensional axisymmetric PHP wall domain (right).

Therefore, a 2D heat conduction model in radial coordinate rules the physical description as follows:

$$\frac{1}{\alpha}\frac{\partial T}{\partial t} = \frac{1}{r}\frac{\partial T}{\partial r} + \frac{\partial^2 T}{\partial r^2} + \frac{\partial^2 T}{\partial z^2} \quad \text{in} \quad \Omega \tag{1}$$

$$-\lambda \left.\frac{\partial T}{\partial r}\right|_{r=r_{int}} = q(z,t) \quad \text{on} \quad \Gamma_{int} \tag{2}$$

$$-\lambda \left.\frac{\partial T}{\partial r}\right|_{r=r_{ext}} = h_\infty (T - T_\infty) \quad \text{on} \quad \Gamma_{ext} \tag{3}$$

$$\left.\frac{\partial T}{\partial r}\right|_{z=L} = 0 \quad \text{on} \quad \Gamma_0 \tag{4}$$

$$\left.\frac{\partial T}{\partial r}\right|_{z=L} = 0 \quad \text{on} \quad \Gamma_L \tag{5}$$

$$T(r,z,t=0) = T_\infty \quad \text{in} \quad \Omega \quad \text{for} \quad t = 0 \tag{6}$$

This work adopts the Finite Difference Method to solve the two-dimensional model. We consider the explicit scheme for time discretization, and we restrict the discretization according to the equation (7) to guarantee the stability of the solution [13].

$$\frac{\alpha \delta t}{\delta r^2} + \frac{\alpha \delta t}{\delta z^2} < \frac{1}{2} \tag{7}$$

3. Inverse Problem

The purpose of solving the inverse problem is to estimate the internal heat flux in PHP devices with temperature measurement available on the external surface. The inverse problem under question is also known as nonstationary inverse problem [10] since the desired quantities in this work vary over the time. Its solution in this work is the state estimation approach that assess the states quantities correlating them with measurements. Firstly, it is defined the discrete-time stochastic *evolution* and *observation* models as:

$$\mathbf{x}_{k+1} = \mathbf{f}_k(\mathbf{x}_k, \mathbf{u}_k, \mathbf{v}_k) \tag{8}$$

$$\mathbf{z}_{k+1} = \mathbf{h}_{k+1}(\mathbf{x}_{k+1}, \mathbf{w}_{k+1}) \tag{9}$$

where the subscript k and k+1 represent the kth and $(k+1)^{th}$ instant, respectively, $\mathbf{f}(\cdot)$ and $\mathbf{h}(\cdot)$ are the evolution and observation functions, possibly nonlinear, which describe the evolution of the states and return the measurable states, respectively. The state vector $\mathbf{x}$ contains all the information about the states to be estimated, and in this work, it contains the temperature and heat flux nodes, while $\mathbf{u}$ is known as input vector. The vector $\mathbf{z}$ is the measurement vector, and $\mathbf{v}$ and $\mathbf{w}$ are mutually independent and represent the process and measurement noises, respectively. When the problem is linear and it considers the Gaussian error distribution, equation (8) and (9) provide the following system:

$$\mathbf{x}_{k+1} = \mathbf{F}\mathbf{x}_k + \mathbf{u} + \mathbf{w}_k \tag{10}$$

$$\mathbf{z}_{k+1} = \mathbf{H}\boldsymbol{x}_{k+1} + \mathbf{v}_{k+1} \tag{11}$$

where **F** is the evolution matrix; both **F** and **u** are constructed with the finite difference method coefficients. **H** is the observation matrix and returns the measurable states. The evolution and observation models do not vary over time and the subscripts k and k+1 are neglected for matrices **F** and **H** and input vector **u**. The sequential estimation with Kalman Filter consists in two basic steps: *prediction* and *update*. The first step consists of evolving the temperature and the heat flux in the nodes from instant k to k+1 with the evolution model, represented in equation (10). When the measurement for instant k+1 is available, represented by equation (11), the predicted vector is corrected. Then, with the corrected state, a new prediction of the state can be performed for the next instant and consequently be updated. Figure 2 presents the flowchart of the described sequential estimation.

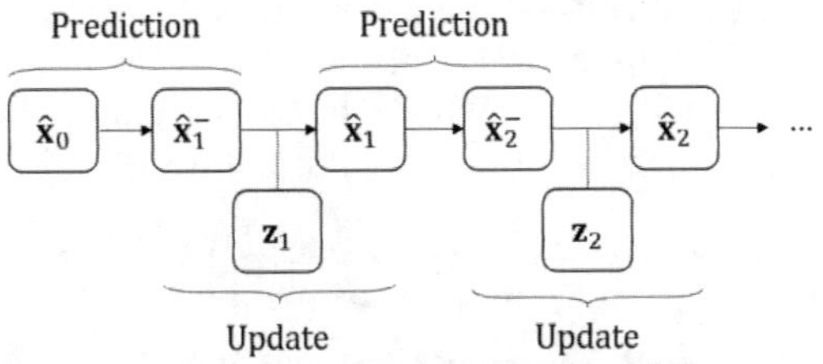

Figure 2. Flowchart of the sequential estimation with the Kalman Filter (Adapted from [10]).

Considering the described calculation procedure, the prediction and update steps comprehend therefore the equations below [9]:

Prediction:

$$\hat{\mathbf{x}}_{k+1}^{-} = \mathbf{F}\hat{\mathbf{x}}_k + \mathbf{u}_k \tag{12}$$

$$\mathbf{P}_{k+1}^{-} = \mathbf{F}\mathbf{P}_k\mathbf{F}^{\mathrm{T}} + \mathbf{Q}_k \tag{13}$$

Update:

$$\mathbf{K}_{k+1} = \mathbf{P}_{k+1}^{-}\mathbf{H}^{\mathrm{T}}(\mathbf{H}\mathbf{P}_{k+1}^{-}\mathbf{H}^{\mathrm{T}} + \mathbf{R}_{k+1})^{-1} \tag{14}$$

$$\hat{\mathbf{x}}_{k+1} = \hat{\mathbf{x}}_{k+1}^{-} + \mathbf{K}_{k+1}(\mathbf{z}_{k+1} - \mathbf{H}\hat{\mathbf{x}}_{k+1}^{-}) \tag{15}$$

$$\mathbf{P}_{k+1} = (\mathbf{I} - \mathbf{K}_{k+1}\mathbf{H})\mathbf{P}_{k+1}^{-} \tag{16}$$

where **P** represents the covariance matrix of the state vector, $\mathbf{Q}_k$ and $\mathbf{R}_{k+1}$ are the covariance matrices related to $\mathbf{w}_k$ and $\mathbf{v}_{k+1}$, respectively. The matrix **I** is the identity matrix and $\mathbf{K}_{k+1}$ is known as *Kalman Gain matrix*, which correct the predict states by correlating them with the measurement. The superscript "T" denotes the transpose of the matrix while "-" means the predicted state or covariance matrix. To estimate the temperature and the heat flux simultaneously, the evolution model considers the *augmented state vector* approach [12], where the estimation for both temperature and heat fluxes adopt them as a state. In this sense, the state vector is constructed by:

$$\mathbf{x}_{\mathrm{aug}} = [\mathbf{T} \quad \boldsymbol{q}]^{\mathrm{T}} \tag{17}$$

where **T** comprises all the temperature nodes originated from the finite difference method discretization and ***q*** contains all the heat flux nodes on the inner surface. Since the evolution model of the heat flux is unknown, it is common to describe unknown functions by a black-box random model [12]. The equation for such approach follows as:

$$q_{k+1} = q_k + \sigma_{\mathrm{rw}}\,\boldsymbol{\omega} \tag{18}$$

where $\boldsymbol{\omega}$ is a Gaussian random number and the parameter σ_{rw} is the standard deviation of the heat flux which controls the heat flux search window. High values of σ_{rw} admit higher range of the heat flux, but on the other hand it generates higher estimation instabilities. Therefore, this parameter should be analysed for each studied case.

4. Results and Discussions

To study the estimation accuracy of the here proposed procedure, the working conditions were considered equal to the ones of the experimental investigation performed in Ref. [4]. The tube is made of Sapphire with density ρ of 3970 kg/m^3, specific heat c_p of 419 J/kg·K and thermal conductivity λ of 27 W/m·K. The tube length L is 40mm and inner and outer radius are $r_{\text{int}} = 1$ mm and $r_{\text{ext}} = 2$ mm respectively. The environmental temperature T_∞ is 21°C and the convective heat transfer coefficient h_∞ is considered to be 5 W/m^2·K. Temperature was acquired with a frequency of 50Hz, which means 0.02 s between two measurements, while on the spatial point of view values were taken every 1 mm.

4.1. Synthetic data

Firstly, the procedure was validated considering synthetic data. They comes from the solution of the forward problem with COMSOL® Multiphysics software. The different method for forward and inverse problem alleviates the inverse crime [15]. To emulate the heat exchanged between the fluid and the wall, the heat flux function varies over time and space, and it is defined as [4]:

$$q(z,t) = A\cos(p\pi t)\left(1 + \frac{z}{8}L\right) \tag{19}$$

where p rules the heat flux frequency variation over time and A is the heat flux amplitude. Then, a random Gaussian noise sequence was added to exact data. Therefore, the definition of the synthetic temperature follows the equation below:

$$\mathbf{T}_{\text{meas}} = \boldsymbol{T}_{\text{exact}} + \sigma_{\text{meas}}\boldsymbol{\xi} \tag{20}$$

where $\boldsymbol{\xi}$ is a random Gaussian sequence and the value of the standard deviation σ_{meas} is 0.1K. The equation below evaluates the estimation error between the estimated and the exact heat flux:

$$\mathbf{E}_{\text{ave}} = \frac{\|\boldsymbol{q}_{\text{estimated}} - \boldsymbol{q}_{\text{exact}}\|_2}{\|\boldsymbol{q}_{\text{exact}}\|_2} \tag{21}$$

Figures 3 and 4 present the exact and estimated heat flux distributions for the two cases of A equal to 2000 and 5000 W/m^2 respectively. In both the cases the frequency parameter p was considered equal to of 1.2. Figures 3a and 4a represent the numerical distribution of q (Eq. 19) that have been used as input data in the solution of the forward problem. Solving forward problem with COMSOL® Multiphysics software it was possible to obtain distributions of temperature that after being spoiled with random noise they have been used as input data for the inverse problem to obtain the distributions of q showed in Figs. 3b and 4b. It is possible to notices that the KF presented a small delay when compared to the exact solution, which is intrinsic to the filter [10]. In addition, it can be seen that the estimation was less accurate near to the boundaries. Table 1 presents the estimation errors for the Kalman Filter and Tikhonov Regularization for the two considered cases.

Table 1. Average Estimation Error for Kalman Filter and Tikhonov Regularization [4].

p [Hz]	A [W/m^2]	Method	E_{ave} [%]
1.2	2000	KF	32.24
		Tikhonov	19.4
	5000	KF	29.94
		Tikhonov	12.8

The parameter for the evolution model of the heat flux from Eq. (18) σ_{rw} was 20 W/m^2 for A equal to 2000 W/m^2 and 50 W/m^2 for A equal to 5000 W/m^2. Even if the estimation error shows better estimation with the Tikhonov Regularization technique than with KF, the Tikhonov Regularization considers all the time window of the experiment, and consequently it should provide better results than the sequential estimation with KF. Even so, the KF was able to recover the heat flux with good accuracy and the computation effort was lower. The computational code for the inverse problem with KF was implemented on FORTRAN language with LAPACK package [15]. The CPU used for the simulations was an Apple Silicon M1 chip. The computational time, considering the smoothing procedure for the temperature measurement of each instant and the KF calculation, was about 0.12s.

Moreover, in both techniques the estimation error reduced when the amplitude parameter A increases. This can be explained since the difference of temperature between two measurement instants is higher when the value of A increases, and consequently the measurement errors influence less during the estimation.

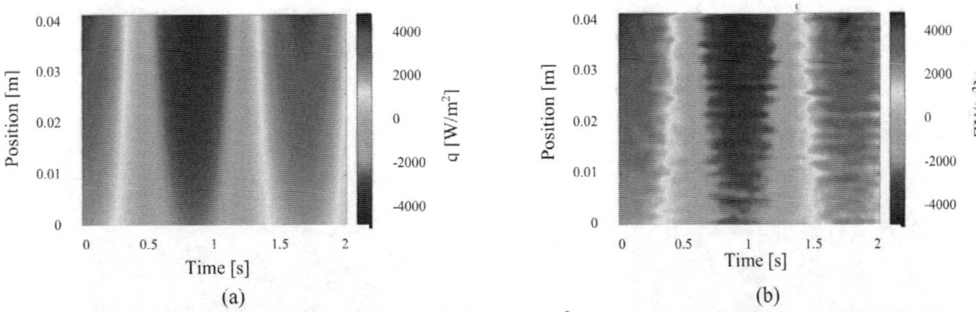

Figure 3. Results with synthetic data for A=2000 W/m^2; (a) Exact and (b) Estimated heat fluxes.

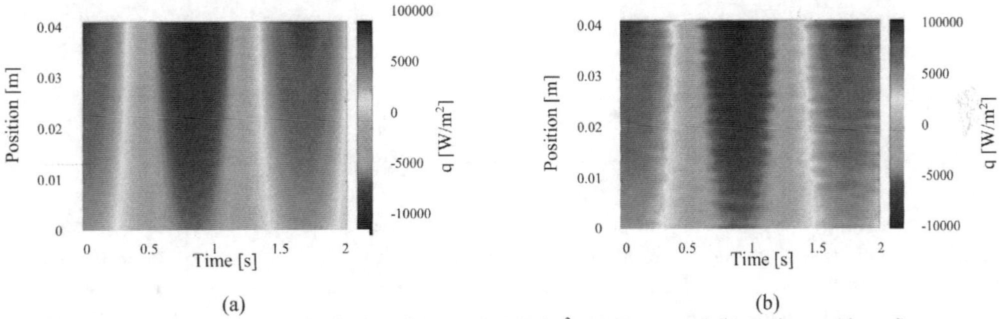

Figure 4. Results with synthetic data for A=5000 W/m^2; (a) Exact and (b) Estimated heat fluxes.

4.2. *Experimental data*

The experiment was performed according to reference [4]. The experiment consisted in a single loop PHP where the evaporator and condenser were made of copper with inner and outer diameters of 2 mm and 4 mm, respectively, connected by two sapphire tubes with axial length of 110 mm. The temperature measurements were done over a 40 mm portion of the Sapphire tube painted with high-emissivity black paint (ε=0.95) by using a highspeed and high resolution MWIR camera (AIM® TEC-MMG from ESA/ESTEC) that allowed to acquire infrared images at 50 fps. The experiment used pure ethanol for the working fluid and the tube was partially filled (filling ratio = 60%). The experimental tests were made varying the supply heating power to the evaporator to observe typical flow patterns that occurs in PHPs. The smooth temperature improved the quality for the accuracy estimation with a Cubic Smoothing Spline with Cross Validation following the Discrepancy Principle [16]. In the present work, the heat flux estimation considers the Oscillation and Startup flow results. Figure 5 and 6 present the results with Tikhonov Regularization and KF. The results with the KF show good accordance when compared to the Tikhonov Regularization results. As observed in synthetic results, the KF presented a little delayed response to the heat flux, approximately 0.15s on the peaks and valleys. For practical purpose, this delayed response with KF estimation could be considered and does not compromise the accuracy of the method.

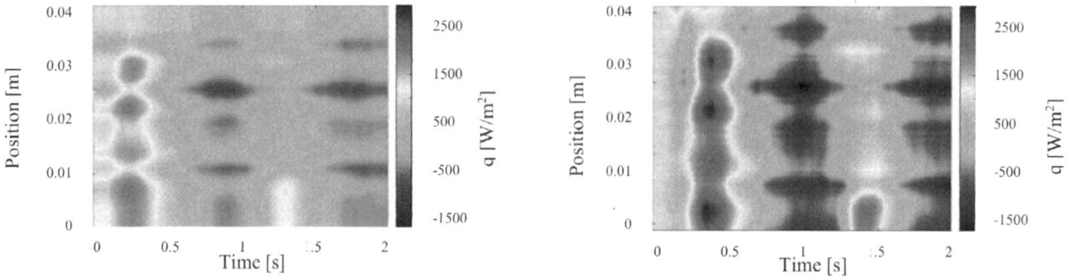

(a) (b)

Figure 5. Results with experimental data (a) Estimated heat flux with Tikhonov and (b) Estimated heat flux with Kalman filter for the Oscillating flow.

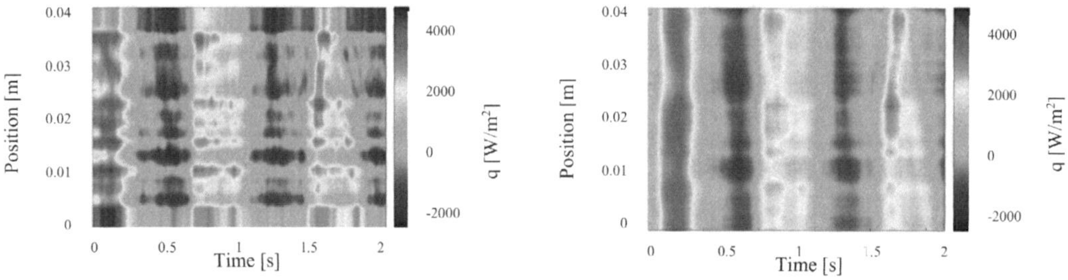

(a) (b)

Figure 6. Results with experimental data (a) Estimtaed heat flux with Tikhonov and (b) Estimated heat flux with Kalman filter for the Startup flow.

5. Conclusions

The present work evaluated the use of Kalman Filter to estimate the internal heat flux on Pulsating Heat Pipes by solving an inverse heat conduction problem. A high-speed and high-resolution infrared camera acquired the temperature along the adiabatic section. A two-dimensional heat conduction model was used for the forward problem. First the study evaluated the heat flux estimation by considering synthetic data and, consequently, experimental data. The sequential estimation is very sensitive to errors and to improve the accuracy of the results the estimation procedure adopted the smoothed temperature with Cubic Smoothing Spline with Cross Validation following the Discrepancy Principle. The results showed good agreement with the exact solution when considering the synthetic data. The estimation errors were higher with the Kalman Filter when compared to the Tikhonov Regularization method due to the different time domain window considered, but on the other hand, lower computational cost was required due to the non-necessity of storing all time domain. Then, the results with Kalman Filter considering experimental data were in good correspondence with the Tikhonov Regularization, whose results were a reference to evaluate the accuracy. The proposed methodology based on Kalman Filter encourages its application to heat flux estimation during Pulsating Heat Pipes operation to control the correct working of such devices and avoid undesired malfunctioning.

Acknowledgments

This paper was partially funded by the following Brazilian agencies: CNPq (National Council for Scientific and Technological Development), FAPERJ (Carlos Chagas Filho Foundation for Research Support in the rio de Janeiro) and ANP-PRH8 (Brazilian National Agency for Oil, Gas and BioFuels –

Human Research Program number 8). The work was also partially funded under the National Recovery and Resilience Plan (NRRP) Mission 4 Component 2 Investment 1.5 - Call for tender No. 3277 of 30/12/2021 of Italian Ministry of University and Research funded by the European Union – NextGenerationEU (Award Number: Project code ECS00000033, Concession Decree No. 1052 of 23/06/2022 adopted by the Italian Ministry of, CUP D93C22000460001, "Ecosystem for Sustainable Transition in Emilia-Romagna" (Ecosister)).

References

[1] M. Marengo and V. S. Nikolayev. Pulsating Heat Pipes: Experimental Analysis, Design and Applications, World Scientific, Chapter 1, pp. 1–62, 2018.

[2] X. Han, X. Wang, H. Zheng, X. Xu and G. Chen. Review of the development of pulsating heat pipe for heat dissipation, Renewable and Sustainable energy Reviews, Vol. 59, pp. 692–709, 2016.

[3] Z. Su, Y. Hu, S. Zheng, T. Wu, K. Liu, M. Zhu and J. Huang. Recent advances in visualization of pulsating heat pipes: A review. Applied Thermal Engineering, 221, 119867, 2022.

[4] L. Cattani, D. Mangini, F. Bozzoli, L. Pietransanta, N. Miche, M. Mameli, S. Filippeschi, S. Rainieri and M. Marengo. An Original Look into Pulsating Heat Pipes: Inverse Heat Conduction Approach for Assessing the Thermal Behaviour. Thermal Science and Engineering Progress, vol. 10, pp. 317-326, 2019.

[5] L. Pagliarini, L. Cattani, F. Bozzoli, M. Mameli, S. Filippeschi, S. Rainieri and M. Marengo. Thermal Characterization of a Multi-Turn Pulsating Heat Pipe in Microgravity Conditions: Statistical Approach to the Local Wall-to-Fluid Heat Flux, International Journal of Heat and Mass Transfer, vol. 169, pp. 120930, 2021.

[6] M. J. Colaço, F. Bozzoli, L. Cattani and L. Pagliarini. Local heat flux estimation inside tubes through conjugate gradient method with adjoint operator: application to the pulsating heat pipes case. International Journal of Numerical Methods for Heat & Fluid Flow, Doi: 10.1108/HFF-09-2022-0547, 2022.

[7] L. Pagliarini, L. Cattani, M. Mameli, S. Filippeschi, F. Bozzoli and S. Rainieri. Global and local heat transfer behaviour of a three-dimensional Pulsating Heat Pipe: combined effect of the heat load, orientation and condenser temperature. Applied Thermal Engineering, 195, 117144, 2021.

[8] N. Iwata, F. Bozzoli, L. Pagliarini, L. Cattani, P. Vocale, M. Malavasi and S. Rainieri. Characterization of thermal behavior of a micro pulsating heat pipe by local heat transfer investigation, International Journal of Heat and Mass Transfer, vol. 196, pp. 123203, 2022.

[9] B. Ristic, S. Arulampalam and N. Gordon. Beyond the Kalman Filter: Particle Filters for Tracking Applications, Artech House, 2004.

[10] J. Kaipio and E. Somersalo. Statistical and Computational Inverse Problems, Applied Mathematical Sciences, Springer, vol. 160, 2004.

[11] C. Ji, P. Tuan and H. Jang. A recursive least-squares algorithm for on-line 1-D inverse heat conduction estimation. International Journal of Heat and Mass Transfer, vol. 40, 9, 2081-2096, 1996.

[12] N. Daouas and M. Radhouani. Version étendue du filtre de Kalman discret appliquée à un problème inverse de conduction de chaleur non linéaire. International Journal of Thermal Sciences, 39, (2), pp. 191–212, 2000.

[13] M. N. Ozisik, H. R. B. Orlande, M. J. Colaço and R. M. Cotta. Finite Difference Methods in Heat Transfer. CRC Press, Taylor & Francis Group, 2017.

[14] J. Kaipio and E Somersalo. Statistical inverse problems: Discretization, model reduction and inverse crimes. Journal of Computational and Applied Mathematics, 198, pp. 493-504, 2007.

[15] E. Anderson, Z. Bai, C. Bischof, S. Blackford, J. Demmel, J. Dongarra, J. Du Croz, A. Greenbaum, S. Hammarling, A. McKenney and D. Soresen. LAPACK Users' Guide (Third). Philadelphia, PA: Society for Industrial and Applied Mathematics, 1999.

[16] O. M. Alifanov. Inverse Heat Transfer Problems. 1st Edition. International series in heat and mass transfer. Springer Berlin, Heidelberg, 1994.

Parallel connected natural circulation loops using different working fluids: experimental results

Johan Augusto Bocanegra[1*], Annalisa Marchitto[1] and Mario Misale[1]

[1] DIME – Department of Mechanics, Energetics, Management and Transportation, University of Genoa, Genoa, 16145, Italy

* Corresponding author: augusto.bocanegra@edu.unige.it

Abstract. Natural Circulation Loops (NCLs) are closed-loop systems that transport heat from a source to a heat sink without a pump, relying on free convection of the working fluid. Previous research has focused on the stability and influence of different operative parameters on NCLs. This experimental study investigates the thermo-hydraulic performance of three parallel-connected NCLs with small inner diameters using three different working fluids: deionized water, glycol aqueous solution (50+50% wt), and FC-43 dielectric fluid. This study examines the steady-state behaviour of the NCLs at various heat sink temperatures and heat powers. The results indicate that the common one-dimensional model for a single loop's steady-state can be applied to this configuration and for all the tested fluids, and a proposed figure of merit can describe the working fluids and predict their steady-state behaviour.

1. Introduction

Natural circulation loops (NCLs) have been extensively studied due to their energy-efficient and cost-effective heat transfer capabilities. NCLs are closed-loop systems that use a working fluid that flows by free (natural) convection without the use of any external pumping system. The NCLSs have been used in various applications such as electronic cooling, solar panels, geothermic, and nuclear power plants.

The stability and influence of various operating parameters on NCLs have been extensively investigated in the past [1–4]. Analytical models have been proposed, including Vijayan's correlation, a simple formulation based on dimensionless numbers applicable to a single NCL [5]. Also numerical models have been used to study NCL behavior under various conditions, such as loop geometry (pipe diameter, length) [6], fluid properties [7], pressure losses[8], power steps or harmonic power excitation [9], Recently, complex loop configurations, such as series-connected [10,11] and parallel-connected loops [12], have also been studied numerically and experimentally. It has been found that even small inner-diameter loops are stable, even in parallel-connected configurations.

This work presents an experimental investigation of the thermo-hydraulic performance of parallel-connected NCLs with different working fluids. The study was conducted on a complex circuit consisting of three parallel-connected single-phase NCLs of small inner diameter, and the thermo-hydraulic performance at the steady-state was studied at different heat sink temperatures and heat powers. Three different fluids were tested: DI water, Glycol aqueous solution, and the dielectric fluid FC-43. The experimental setup and procedures are detailed in the following sections, and the results are analyzed and presented in the third section. This study aims to contribute to understanding the thermo-hydraulic behavior of parallel-connected NCLs with different working fluids and provide insights into their potential applications.

2. Methods

This section presents the experimental setup and procedures used to investigate the thermal-hydraulic behavior of different working fluids in three parallel connected natural circulation loops (NCLs). The study aimed to compare the performance of three fluids: DI water, Glycol aqueous solution, and FC-43, under different heat inputs and cooling conditions. The NCLs were constructed in a horizontal heater-horizontal cooler (HHHC) configuration, and the heat sink temperature was imposed using a cryostat. The experimental procedures involved measuring the temperature distribution along each loop, varying the heat power, and repeating the process at different heat sink temperatures. Additionally, one variable loop was used to study the mutual influence of the connected loops. The results obtained from this study are expected to provide valuable insights into the thermal-hydraulic behavior of natural circulation loops and aid in selecting the most appropriate working fluid for different applications.

2.1. Experimental setup

An experimental setup, represented in figure 1, consisting of three parallel connected NCLs, was used to measure the thermal-hydraulic behavior of various working fluids: DW Distilled Water ($Pr_{20°C}$ = 7.0), aqueous Glycol solution ($Pr_{20°C}$ = 50.9), and FC-43 ($Pr_{20°C}$ = 105.7). The circuit was constructed in a Horizontal Heater-Horizontal Cooler (HHHC) configuration. Each loop has an independent heater localized in the lower part of the circuit, and its power can be adjusted (the maximum power value was selected to avoid possible phase changes in the working fluid). The heat sink temperature is imposed using a cryostat. The loop was thermally isolated using several layers of a composite isolation material to reduce the heat exchange with the ambient. The temperature was measured at different points along each loop using T-type thermocouples (0.5 mm OD), and the data was stored by a data acquisition system (estimated uncertainty of ± 1.0 K).

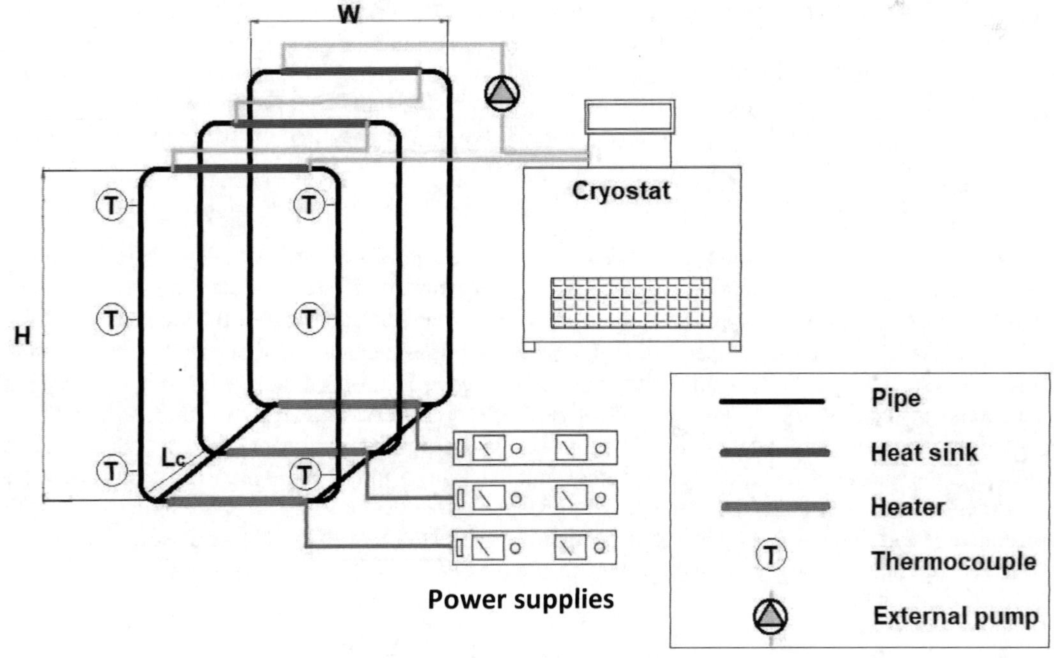

Figure 1. Sketch of the experimental setup.

2.2. Tested fluids

This section introduces the three fluids tested in the experiment: Distilled water (DW), Glycol aqueous solution (Glycol-water, 50%+50% wt), and FC-43 dielectric fluid with their properties and characteristics. DW water is a highly purified water commonly used in industrial applications due to its high purity and low conductivity. Glycol aqueous solutions are mixtures of water and glycol, typically used as coolants in HVAC systems and as antifreeze in automobile engines. FC-43 dielectric fluid is a fluorocarbon-based fluid with a low boiling point, low viscosity, and good chemical stability, making it suitable for high-performance cooling systems. The thermophysical properties of the fluids during the experiment were calculated using temperature correlations [13–15]. Table 1 presents the corresponding range of values (calculated during the experiments) for each physical property, including density (ρ), thermal conductivity (k), specific heat (c_p), thermal expansion coefficient (β), cinematic viscosity (γ), and Prandtl number (Pr).

Table 1. Fluid properties.

	Glycol-water	**FC-43**	**DW water**
ρ, kg m^{-3}	1102.0 - 1125.2	1778.5 – 1860.2	989.2 – 999.4
k, Wm^{-1}K^{-1}	0.413 - 0.434	0.063 – 0.065	0.585 – 0.632
c_p JK^{-1} kg^{-1}	3446 - 3640	1052 - 1110	4181 – 4185
β, K^{-1}	0.000490 - 0.000499	0.001172 – 0.001226	0.000128 – 0.000438
γ, m^2/s	1.63 10^{-6} – 5.51 10^{-6}	7.46 10^{-7} – 3.03 10^{-6}	5.80 10^{-7} – 1.20 10^{-6}
Pr, -	15.0 – 51.8	23.5 – 90.8	3.8 – 8.6

2.3. Experimental procedures

Two experimental procedures were employed during the experiments: Procedure A (single loop characterization) measures the temperature distribution along each loop, varying the provided heat power q from 20 W to 120 W (with 20 W steps). This procedure was implemented with every single loop working independently (disconnected from neighbor loops); Procedure B, one variable loop operates at different powers, and the neighbor loops remain at a fixed heat power (40 W). Procedure B was used to study the mutual influence of the connected loops. The experimental procedure was repeated at different heat sink temperatures.

3. Results

The temperature of each adiabatic vertical leg was obtained as an average of the three thermocouples positioned in each vertical section. The thermophysical behavior at steady-state was studied for the different fluids using ΔT_{avg}, the temperature difference between the adiabatic vertical legs of each NCL. It is possible to show that this temperature difference determines the mass flow rate of the working fluid. Moreover, as shown in Eq. (1), this temperature difference is proportional to the main operational parameters (in the first square root) and the figure of merit (FOM, that resumes the thermophysical properties of the working fluid). In Eq. (1), the gravity is g, the power q, the total loop length L, the loop height H, the diameter D, modified Grashof number Gr_m, and the geometrical factor N_G. The experimental results presented in figure 2 corroborate the dependency expressed in Eq. (1) for procedures A and B. Note that by increasing the power the first factor in Eq. (1) increases, however, the average temperature increases, which determines a smaller FOM (figure of merit) and therefore a negative effect on ΔT_{avg}.

$$\Delta T_{avg} = \sqrt{\frac{128\, q\, L}{\pi\, H\, D^4\, g}} \sqrt{\frac{k\, Pr}{c_p{}^2 \rho^2 \beta}} = \sqrt{\frac{Gr_m}{N_G}}\, FOM \tag{1}$$

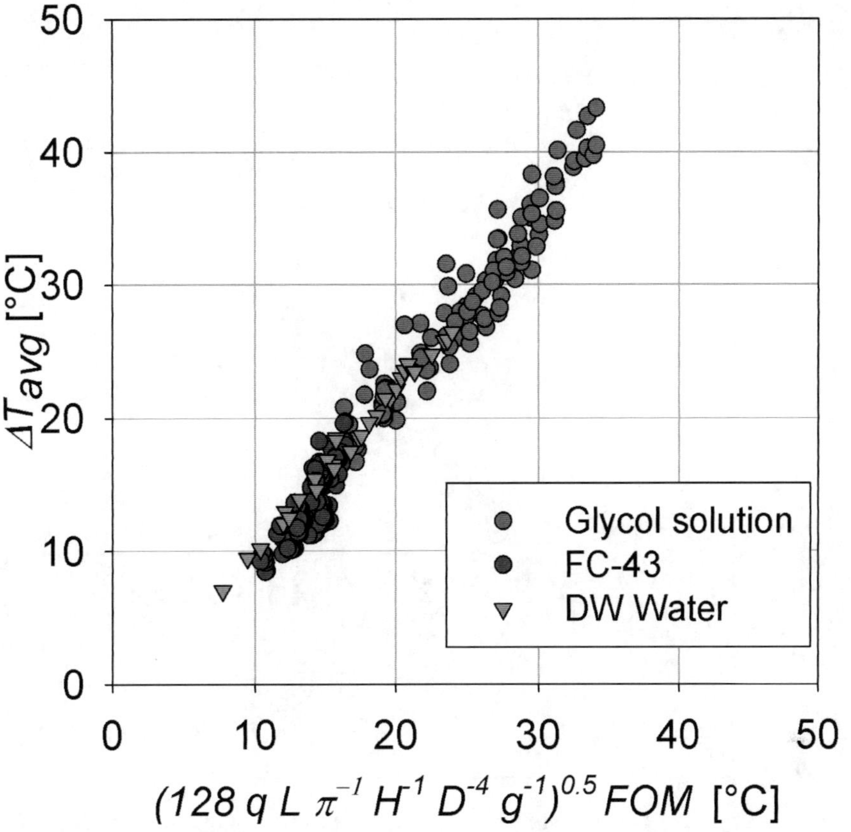

Figure 2. Experimental results: (a) relationship between the figure of merit FOM and ΔT_{avg}, Eq. (1).

Additionally, we corroborate Vijayan's correlation, Eq. (2), commonly used to describe the thermohydraulic behavior of a single NCL at the steady-state [5], see figure 2. This correlation links the Reynolds number at steady-state Re_{ss} with the fraction Gr_m/N_G (Gr_m represents the modified Grashof number and N_G a dimensionless geometrical parameter related with the friction factor).

$$Re_{ss} = 0.1768\sqrt{\frac{Gr_m}{N_G}} \tag{2}$$

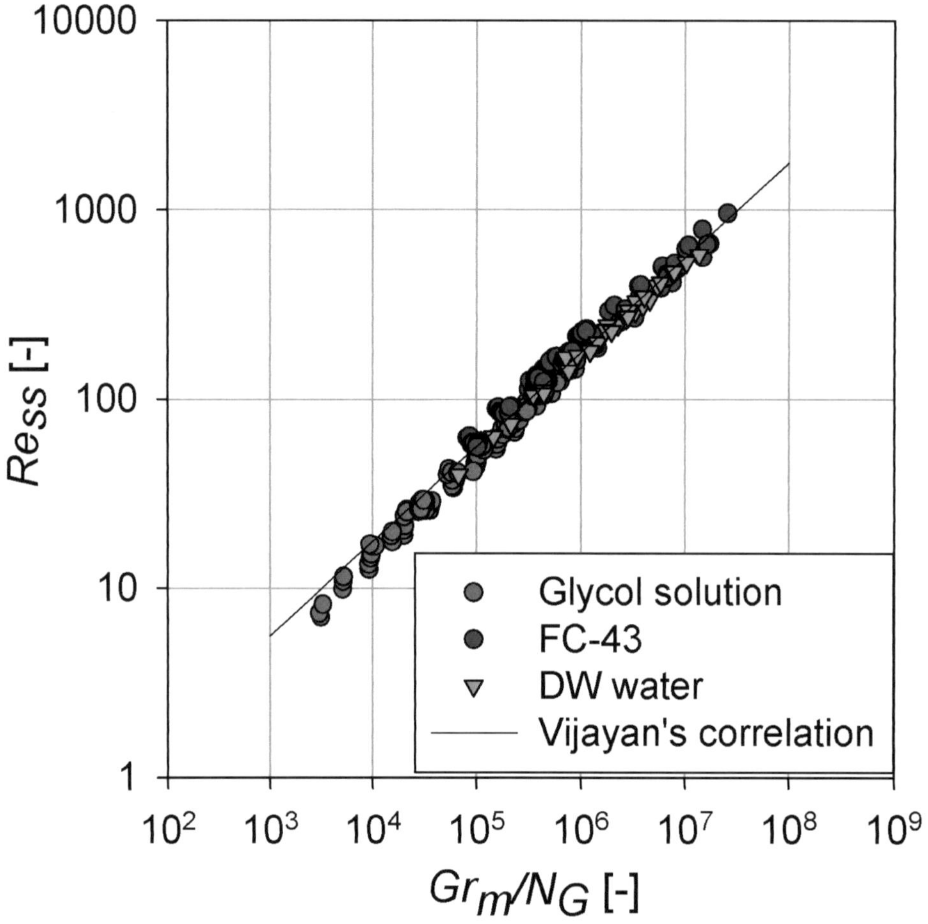

Figure 3. Relationship between dimensionless groups representing the thermo-hydraulic behavior of the parallel-connected loops at steady-state, comparison with Vijayan's correlation Eq. (2).

The results of Procedure B show that the neighboring loops' operating conditions slightly influence each neighbor loop's performance. For example, increasing the heat power in one loop can cause a decrease in the steady-state Reynolds number and thermal performance of the adjacent loops. However, the mutual influence is small and both equations, Eq. (1) and (2), correctly describes the behaviour of each NCL even when the power in one loop is changed by successive steps.

The results of this study demonstrate that the thermal-hydraulic behaviour of different working fluids can vary significantly depending on the operating conditions and the characteristics of the fluid. However, the same correlations apply to the circuits under the two experimental procedures, A and B. The data presented in this section can inform about the selection and design of heat transfer systems in various industrial applications. In future work, it may be useful to explore the behaviour of these fluids during transients, varying other parameters (such as NCL inclination's angle), and in more complex geometrical configurations. This will help further to refine our understanding of their thermal and hydraulic properties.

4. Conclusions

This study investigated the thermal-hydraulic behavior of three different working fluids: DW water, aqueous glycol solution, and FC-43, in parallel-coupled natural circulation loops (horizontal heater-horizontal cooler configuration). The results showed that the thermal performance of the NCLs is greatly affected by the properties of the working fluid and that the NCLs can provide effective cooling for a wide range of heat loads.

Overall, the obtained results indicate that the one-dimensional analytical model can describe the thermo-hydraulic behavior of the parallel-connected natural circulation loops for all the tested fluids. The analytical model was originally constructed for a single loop; however, the mutual influence is almost negligible for this complex NCL circuit (small inner diameter, parallel connected, different working fluids), and the analytical model is also applicable.

Our research has underlined the importance of selecting a proper working fluid to obtain the desired power range of operation. The figure of merit FOM represents a convenient way to synthesize the thermophysical properties of the working fluid; for example, the higher FOM values of the Glycol solution compared to DW water imply an operation at higher temperature differences and lower Reynolds numbers.

The findings of this study suggest that to fully understand the thermal performance of NCLs, future research should consider not only the steady-state behavior, but also the effects of the working fluid on transient behavior. A more comprehensive understanding of these dynamics will be crucial for developing more efficient and reliable NCLs in various industrial applications.

Acknowledgments

Funding: This research was funded by Ministero dell'Istruzione, dell'Universita e della Ricerca (MIUR, Italy), grant number PRIN-2017F7KZWS.

References

[1] Basu D N, Bhattacharyya S and Das P K 2013 Influence of geometry and operating parameters on the stability response of single-phase natural circulation loop *International Journal of Heat and Mass Transfer* **58** 322–34

[2] Satou A, Okamoto, Koji and Madarame H 1998 Instability of single phase natural circulation under double loop system *Chaos, Solitons and Fractals* **9** 1575–85

[3] Satou A, Madarame H and Okamoto K 2001 Unstable behavior of single-phase natural circulation under closed loop with connecting tube *Experimental Thermal and Fluid Science* **25** 429–35

[4] Vijayan P K, Sharma M and Saha D 2007 Steady state and stability characteristics of single-phase natural circulation in a rectangular loop with different heater and cooler orientations *Experimental Thermal and Fluid Science* **31** 925–45

[5] Vijayan P K 2002 Experimental observations on the general trends of the steady state and stability behaviour of single-phase natural circulation loops *Nuclear Engineering and Design* **215** 139–52

[6] du Toit C G 2021 Fundamental evaluation of the effect of pipe diameter, loop length and local losses on steady-state single-phase natural circulation in square loops using the 1D network code Flownex *Thermal Science and Engineering Progress* **22** 100840

[7] Bocanegra J A and Misale M 2023 Lattice Boltzmann model of a square natural circulation loop with small inner diameter: working fluid effects *J. Phys.: Conf. Ser.* **2509** 012007

[8] Misale M and Frogheri M 1999 Influence of pressure drops on the behavior of a single-phase natural circulation loop: Preliminary results *International Communications in Heat and Mass Transfer* **26** 597–606

[9] Ghosh Mazumder A, Saha R, Mondal S, Ghosh K, Mukhopadhyay A and Sen S 2022 Dynamics of a single-phase natural circulation system under harmonic excitation *Chaos* **32** 123125

[10] Elton D N, Arunachala U C and Vijayan P K 2022 Stability performance of series coupled natural circulation system with different operating procedures *International Journal of Thermal Sciences* **179** 107693

[11] Adinarayana K N V, Mangarjuna Rao P and Ali S M 2022 Numerical investigations on dynamic characteristics of series coupled single-phase natural circulation loops *International Journal of Thermal Sciences* **181** 107717
[12] Misale M, Bocanegra J A, Borelli D and Marchitto A 2020 Experimental analysis of four parallel single-phase natural circulation loops with small inner diameter *Applied Thermal Engineering* **180** 115739
[13] 3M Company 2019 *3M™ Fluorinert™ Electronic Liquid FC-43, Technical Data*
[14] Furukawa M 1995 Reduced Expressions for Thermophysical Properties of Fluid Loop Used Non-CFC Coolants *International Conference on Environmental Systems* p 951679
[15] Glycerine Producers' Association 1963 *Physical Properties of Glycerine and Its Solutions*

Instability of adiabatic shear flows in a channel

A Barletta[1,*], M Celli[1], S Lazzari[2], P V Brandão[1]

[1] Department of Industrial Engineering, Alma Mater Studiorum Università di Bologna, Bologna, Italy

[2] Department of Architecture and Design, University of Genoa, Stradone S. Agostino 37, 16123 Genoa, Italy

[*] email: antonio.barletta@unibo.it

Abstract. The combined forced and free convection flow of a Newtonian fluid in a horizontal plane-parallel channel is examined. The boundary walls are considered as adiabatic, so that the only thermal effect acting in the fluid is the viscous dissipation due to the nonzero shear flow. As the shear flow may be caused by either an imposed horizontal pressure gradient or an imposed velocity difference between the bounding walls, one may envisage two scenarios where the stationary basic flow is Poiseuille-like or Couette-like, respectively. Both cases are surveyed with a special focus on practically significant cases where the Gebhart number is considered as very small, though nonzero. Furthermore, the Prandtl number is assumed as extremely large, thus pinpointing a scenario of creeping buoyant flow with a fluid having a very large viscosity. Within such a framework, the instability of the basic flow is analysed versus small-amplitude perturbations.

Keywords: *Channel flow, Buoyancy, Viscous dissipation, Shear flows, Linear stability*

1. Introduction

The viscous dissipation contribution to the local energy balance for a fluid can influence the heat transfer and flow features for internal flows. It is well-known in the literature that significant effects of viscous dissipation may be present in either forced or mixed convection. On the other hand, such effects are generally less intense in a regime of natural convection [1].

Starting from the paper by Joseph [2], the question on whether viscous dissipation might be the source of instability for internal fluid flows was posed. Joseph [2] built up a model where the instability mechanism arises from a combined action of the viscous dissipation term in the local energy balance and the viscous force in the local momentum balance activated through the temperature-dependence of the fluid viscosity. A different approach has been proposed more recently by Barletta [3]. The idea is that the buoyancy force in a mixed convection regime can be the term providing the thermal interaction within the local momentum balance. Roughly speaking, such a different approach replaces the variable-viscosity coupling with a variable-density coupling between the energy and momentum balances. In particular, the variable-density coupling is modelled via the classical Boussinesq approximation.

When the focus is on the onset of convection patterns superposed to a parallel flow in a channel, the cause of instability is generally an external forcing induced by the temperature boundary conditions at the walls. When this is the case, then the thermal instability is of the Rayleigh-Bénard type and the governing parameter driving the transition is the Rayleigh number, provided that other parameters, such as the Prandtl number and the Reynolds number, are prescribed. Another scenario is one where no external forcing due to the temperature boundary conditions exists, yet a temperature gradient is present due to the viscous dissipation and consequent heating of the flowing fluid. This cause of thermal instability is endogenous as it is determined by the impressed flow rate associated with the stationary parallel velocity field in a channel or duct.

An interesting aspect is the possible comparison between the thermal instability induced by viscous dissipation and the well-known hydrodynamic instability of internal flows, classically studied via the analysis and solution of the Orr-Sommerfeld eigenvalue problem. The hydrodynamic instability is defined by the threshold, or critical, value of the Reynolds number predicted by the linear stability analysis. A finite critical Reynolds number exists for the Poiseuille flow in a plane channel. For the Hagen-Poiseuille flow in a circular duct or the Couette flow in a plane channel, the nonlinear analysis predicts a threshold to instability in terms of the Reynolds number, while the linear theory predicts no instability [4, 5].

The viscous dissipation is the source of an instability completely different from the hydrodynamic instability. In fact, its effects can well emerge at Reynolds numbers lower than those needed for the hydrodynamic instability [6]. Despite the thermal origin of the dissipation instability, it can display its onset under conditions of adiabatic boundary walls and with the flow itself causing the instability as mediated by the thermal coupling in the local momentum balance equation [6].

This paper will survey the study presented in Barletta et al. [6], where the buoyancy-induced steady-state modification of the Poiseuille profile in a horizontal plane channel is considered. That study [6] assumes adiabatic boundary walls and a non-negligible effect of viscous dissipation as the sole cause of the buoyancy force. The resulting velocity profiles for the basic flow depend on the Gebhart number, which is usually less or much less than unity except when large scale flows are considered, as it happens in geophysics [6]. The Gebhart number is also well-known in the geophysics literature as the dissipation parameter. It comes up whenever the interplay between viscous dissipation and thermal buoyancy proves to be important. If the focus of Barletta et al. [6] is on the dissipation instability of the adiabatic Poiseuille-like flows in a horizontal plane channel, this paper will also survey the onset of the dissipation instability for the Couette-like flows in a horizontal plane channel [7]. In the latter case, the zero-velocity boundary conditions for the velocity field are turned into no-slip conditions with a relative velocity imposed across the boundaries. The linear stability analysis will be parametrised by the Gebhart number and will turn out to be subject to an overall flow scale given by the Péclet number A numerical solution of the stability eigenvalue problem will be carried out leading to the determination of the neutral stability curves and to the critical values of the Péclet number for the transition to instability.

2. Governing Equations

The flow in a horizontal plane channel with boundary walls separated by a distance H is described by the mass, momentum and energy local balance equations. They are written according to the Boussinesq approximation in the dimensionless form

$$\boldsymbol{\nabla} \cdot \mathbf{u} = 0, \tag{1}$$

$$\frac{1}{\mathrm{Pr}} \left[\frac{\partial \mathbf{u}}{\partial t} + (\mathbf{u} \cdot \boldsymbol{\nabla})\, \mathbf{u} \right] = -\boldsymbol{\nabla} p + T\, \hat{\mathbf{e}}_z + \nabla^2 \mathbf{u}, \tag{2}$$

$$\frac{\partial T}{\partial t} + \mathbf{u} \cdot \boldsymbol{\nabla} T = \nabla^2 T + \mathrm{Ge}\, \Phi, \tag{3}$$

where the z axis, with unit vector $\hat{\mathbf{e}}_z$, is taken as vertical and perpendicular to the boundary walls. The Cartesian coordinates, (x, y, z), time, t, velocity, $\mathbf{u} = (u, v, w)$, temperature, T, and pressure, p, have been made dimensionless by using, respectively, the scales:

$$H, \quad \frac{H^2}{\alpha}, \quad \frac{\alpha}{H}, \quad \frac{\alpha \nu}{g \beta H^3}, \quad \frac{\rho \alpha \nu}{H^2}, \tag{4}$$

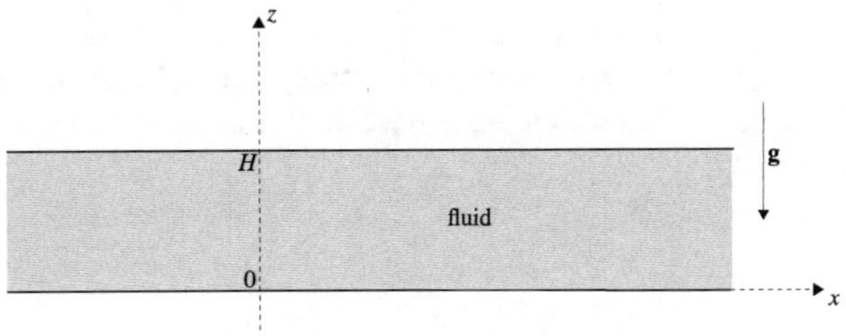

Figure 1. A sketch of the channel.

where the fluid properties α (thermal diffusivity), ν (kinematic viscosity), β (thermal expansion coefficient), ρ (density) are employed together with the modulus, g, of the gravitational acceleration, $\mathbf{g}$. The dimensionless parameters Pr and Ge are the Prandtl number and the Gebhart number. They are defined as

$$\mathrm{Pr} = \frac{\nu}{\alpha}, \quad \mathrm{Ge} = \frac{g\beta H}{c}, \tag{5}$$

with c the fluid specific heat. In Eq. (3), Φ denotes the dimensionless dissipation function,

$$\Phi = \frac{1}{2} \sum_{\mathrm{i,j}} \left(\frac{\partial u_{\mathrm{i}}}{\partial x_{\mathrm{j}}} + \frac{\partial u_{\mathrm{j}}}{\partial x_{\mathrm{i}}} \right)^2, \tag{6}$$

where u_{i} is the i-th component of $\mathbf{u}$ and x_{i} the i-th Cartesian coordinate.

The origin of the z axis, $z = 0$, is set at the lower boundary wall so that the upper boundary wall is at $z = 1$. A sketch of the channel with the dimensional quantities is displayed in Fig. 1.

3. Stationary Basic Shear Flows

A class of stationary two-dimensional solutions of Eqs. (1)-(3) exist in the xz plane. They can be expressed as

$$\mathbf{u}_b = \Big(\mathrm{Pe}\, F(z),\, 0,\, 0\Big), \quad T_b = \mathrm{Pe}\, A x + \mathrm{Pe}^2 G(z), \quad \boldsymbol{\nabla} p_b = \Big(\mathrm{Pe}\, F''(z),\, 0,\, T_b\Big), \tag{7}$$

where the primes denote derivatives with respect to z and subscript b stands for "basic" solution. The functions $F(z)$, $G(z)$ and the constants A, Pe are determined from the boundary conditions imposed at $z = 0, 1$. We always constrain the boundaries to be thermally insulated, so that

$$\frac{\partial T_b}{\partial z} = 0 \quad \text{for} \quad z = 0, 1. \tag{8}$$

On the other hand, we devise two cases where the momentum boundary conditions may yield either zero velocity at both boundaries,

$$u_b = 0 \quad \text{for} \quad z = 0, 1, \tag{9}$$

or a boundary velocity difference,

$$u_b = 0 \quad \text{for} \quad z = 0, \quad u_b = \mathrm{Pe} \quad \text{for} \quad z = 1. \tag{10}$$

Eq. (9) leads to a Poiseuille-like class of solutions, where the Pèclet number Pe defines the dimensionless flow rate across the channel,

$$\mathrm{Pe} = \int_0^1 u_b \, \mathrm{d}z. \tag{11}$$

Eq. (10) leads to a Couette-like class of solutions, where the Péclet number yields the dimensionless velocity difference between the boundaries. The flow rate is set to zero in the inertial reference frame where the relative velocity difference between the boundaries is zero. In other words, instead of Eq. (11), we prescribe in this case

$$\frac{\mathrm{Pe}}{2} = \int_0^1 u_b \, \mathrm{d}z. \tag{12}$$

In either cases, $F(z)$ and $G(z)$ depend both on Ge and on the constant A, while they are independent of Pe. Both $F(z)$ and $G(z)$ are polynomials in z. Due to the Neumann boundary conditions for the temperature (8), $G(z)$ is determined only up to an arbitrary additive constant. Such an arbitrariness can be broken by setting $G(0) = 0$ and, hence, attributing to $G(z)$ the meaning of dimensionless temperature difference between a given z and the lower boundary, $z = 0$, for a given $x =$ constant channel cross-section.

4. Linear Instability

One can inspect the onset of the instability by perturbing the basic flow (7), namely by expressing the flow fields $(\mathbf{u}, T, p)$ as the sum of the basic state contribution $(\mathbf{u}_b, T_b, p_b)$ plus a perturbation $(\mathbf{U}, \Theta, P)$, with $\mathbf{U} = (U, V, W)$. By substituting this decomposition into the governing equations (1)-(3) and by linearising with respect to the perturbations, one obtains

$$\boldsymbol{\nabla} \cdot \mathbf{U} = 0, \tag{13}$$

$$\frac{1}{\mathrm{Pr}} \left[\frac{\partial \mathbf{U}}{\partial t} + (\mathbf{u}_b \cdot \boldsymbol{\nabla}) \mathbf{U} + (\mathbf{U} \cdot \boldsymbol{\nabla}) \mathbf{u}_b \right] = -\boldsymbol{\nabla} P + \Theta \hat{\mathbf{e}}_z + \nabla^2 \mathbf{U}, \tag{14}$$

$$\frac{\partial \Theta}{\partial t} + (\mathbf{u}_b \cdot \boldsymbol{\nabla}) \Theta + (\mathbf{U} \cdot \boldsymbol{\nabla}) T_b = \nabla^2 \Theta + 2\,\mathrm{Ge} \left(\frac{\partial U}{\partial z} + \frac{\partial W}{\partial x} \right) u_b', \tag{15}$$

$$\mathbf{U} = 0, \quad \frac{\partial \Theta}{\partial z} = 0 \quad \text{for} \quad z = 0, 1, \tag{16}$$

where the boundary conditions (16) express the no-slip, impermeability and thermal insulation characteristics of the boundary walls.

A realistic regime for the onset of dissipation instability is one involving a very viscous fluid ($Pr \gg 1$). Then, the governing equations for the linear stability analysis lead to a partial differential boundary value problem in the unknowns (W, Θ),

$$\nabla^4 W + \frac{\partial^2 \Theta}{\partial y^2} = 0, \tag{17}$$

$$\frac{\partial \Theta}{\partial t} + \mathrm{Pe}^2 \, W \, G'(z) = \frac{\partial^2 \Theta}{\partial y^2} + \frac{\partial^2 \Theta}{\partial z^2}, \tag{18}$$

$$W = 0, \quad \frac{\partial W}{\partial z} = 0, \quad \frac{\partial \Theta}{\partial z} = 0 \quad \text{for} \quad z = 0, 1. \tag{19}$$

4.1. Normal Modes

According to the classical procedure, one now expresses the (W, Θ) perturbations as Fourier modes,

$$\big(W, \Theta\big) = \big(f(z), h(z)\big) \, e^{\lambda t} \, e^{iky}, \tag{20}$$

where the complex parameter $\lambda = \eta - i\,\omega$ defines, through its real and imaginary parts, the growth rate η and the angular frequency ω of the Fourier mode, while the real parameter k is the wavenumber. The threshold to linear instability, or neutral stability condition, is given by $\eta = 0$, while $\eta > 0$ means instability and $\eta < 0$ means linear stability. The forthcoming analysis will show that, both in the case of Poiseuille-like flows obtained by imposing Eqs. (9) and (11) and in the case of Couette-like flows obtained by imposing Eqs. (10) and (12), $G'(z)$ turns out to be proportional to the Gebhart number when this parameter is very small. Thus, we can define a function $q(z)$ as

$$\lim_{\mathrm{Ge}\to 0} \frac{G'(z)}{\mathrm{Ge}} = q(z). \tag{21}$$

One is lead to the conclusion that the realistic small-Ge regime is to be accomplished with a large Péclet number in order to describe the onset of the instability. More precisely, the double limit $\mathrm{Ge} \to 0$ and $\mathrm{Pe} \to \infty$ must be taken with the constraint that $\mathrm{Ge}\,\mathrm{Pe}^2$ is kept finite or, equivalently,

$$\mathscr{R} = \mathrm{Pe}\sqrt{\mathrm{Ge}} \sim O(1). \tag{22}$$

Bearing this result in mind, the $\mathrm{Ge} \to 0$ and $\mathrm{Pe} \to \infty$ the linear stability eigenvalue problem is given by

$$f'''' - 2k^2 f'' + k^4 f - k^2 h = 0, \tag{23}$$

$$h'' - \left(k^2 + \lambda\right) h - \mathscr{R}^2 \, q(z) \, f = 0, \tag{24}$$

$$f = 0, \quad f' = 0, \quad h' = 0 \quad \text{for} \quad z = 0, 1. \tag{25}$$

Eqs. (23)-(25) define the stability eigenvalue problem for the limiting case of small Gebhart number. Thus, the neutral stability condition may be expressed through a k-dependent value of $\mathscr{R}$, where the minimum of $\mathscr{R}$ over all the allowed wavenumbers k yields the critical condition: $k = k_c$ with $\mathscr{R} = \mathscr{R}_c$.

4.2. Numerical Solver

One can determine numerically the solution of Eqs. (23)-(25) by employing the shooting method (see, for instance, chapter 10 of Barletta [8]). This technique requires the original eigenvalue problem to be reshaped from its original homogeneous statement, Eqs. (23)-(25), into a non-homogeneous form that filters out the trivial solutions, $f(z) = 0$ and $h(z) = 0$, by breaking the scale-invariance of the eigenfunctions, *i.e.*, the functions $f(z)$ and $h(z)$. In practice, this step requires the introduction of a new boundary condition, say, $h(0) = 1$ and the exclusion of one of the conditions (25), say, $h'(1) = 0$. While the resulting inhomogeneous boundary value problem yields a non-trivial determination of the eigenfunctions $f(z)$ and $h(z)$ for every assignment of the parameters $(k, \lambda, \mathscr{R})$, one can employ the excluded condition $h'(1) = 0$ to determine the complex eigenvalue λ. The numerical determination of λ means the evaluation of the growth rate, η, and of the angular frequency, ω. If the former parameter allows one to place the locus of neutral stability ($\eta = 0$), the latter parameter results to be zero in every case of neutral stability examined, thus meaning that the longitudinal rolls do not travel along the y direction.

5. Poiseuille-like Flows

We refer the reader to Barletta et al. [6] for an explicit expression of $F(z)$ and $G(z)$ in the case of the Poiseuille-like flows when Eqs. (9) and (11) are employed. We just mention that we have a two-fold determination of the constant A for every fixed value of Ge, independently of the Péclet number,

$$A = A_{\pm} = 24\,\frac{15 \pm \sqrt{15\left(15 - \mathrm{Ge}^2\right)}}{\mathrm{Ge}}. \tag{26}$$

This result implies a two-fold determination of the velocity, temperature and pressure profiles (7) where, however, the branch $A = A_+$ turns out to be unphysical for the sufficiently small values of Ge characteristic of most engineering applications [6]. Thus, by taking $A = A_-$ and by assuming $\mathrm{Ge} \ll 1$, one gets the approximate expressions

$$F(z) \approx 6z(1-z) + \mathrm{Ge}\, z(1-z)(1-2z) + O\!\left(\mathrm{Ge}^2\right), \quad G(z) \approx -18\,\mathrm{Ge}\, z^2(1-z)^2 + O\!\left(\mathrm{Ge}^2\right). \tag{27}$$

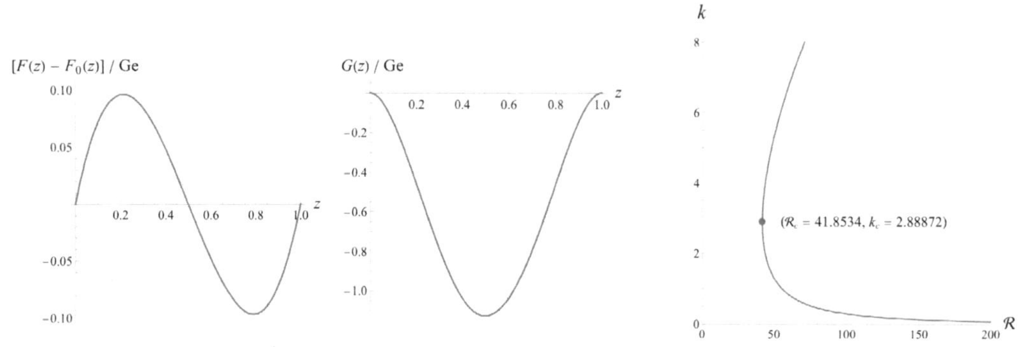

Figure 2. Poiseuille-like flows: incremental velocity profile, temperature profile for Ge $\ll$ 1 and neutral stability curve in the $(\mathscr{R}, k)$ plane.

It is evident from Eq. (27) that, when Ge $\to 0$, Eq. (7) yields the classical Poiseuille velocity profile for the isothermal fluid flow in a channel, $F(z) = F_0(z) = 6z(1-z)$ with $G(z) = 0$ and $A = 0$. Discrepancies with respect to such isothermal flow conditions are illustrated in Fig. 2 for a regime where Ge $\ll 1$. In particular, one may figure out that the effect of viscous dissipation produces an asymmetry of the velocity profile with the maximum slightly shifted from the midplane $z = 1/2$. Furthermore, Fig. 2 reveals a potentially unstable thermal stratification in the lower half-channel, with the midplane temperature smaller than the boundary temperature at a fixed x.

On account of Eqs. (21) and (27), one obtains, in this case, the expression of function $q(z)$,

$$q(z) = -36z(1-z)(1-2z). \tag{28}$$

One may substitute Eq. (28) into the eigenvalue problem (23)-(25) and use the numerical solution technique described in Section 4.2. Thus, as already anticipated in Section 4.2, one obtains that the angular frequency ω of the neutrally stable modes is zero. Furthermore, the neutral stability curve can be plotted in the $(\mathscr{R}, k)$ plane as shown in Fig. 2. As displayed in Fig. 2, the minimum of $\mathscr{R}$ along the neutral stability curve is achieved with $\mathscr{R} = \mathscr{R}_c$ and $k = k_c$, where

$$\mathscr{R}_c = 41.8534, \quad k_c = 2.88872. \tag{29}$$

Eq. (29) identifies the critical point for the onset of the linear instability [6]. An important feature of the neutral stability curve displayed in Fig. 2 is that there is no finite limit for $\mathscr{R}$ at neutral stability when $k \to 0$.

6. Couette-like Flows

In the case where Eqs. (10) and (12) are imposed, we are led to Couette-like flows [7]. In particular, we have

$$A = A_\pm = 12\,\frac{15 \pm \sqrt{5\left(45 - \mathrm{Ge}^2\right)}}{\mathrm{Ge}}. \tag{30}$$

As for the Poiseuille-like flows discussed in Section 5, also in this case we have just one physically acceptable solution, namely $A = A_-$, for the regime of small Gebhart numbers. With this choice of the constant A, the approximate solution for Ge $\ll 1$ is given by

$$F(z) \approx z + \frac{1}{6}\mathrm{Ge}\,z(1-z)(1-2z) + O\!\left(\mathrm{Ge}^2\right), \quad G(z) \approx -\frac{1}{6}\mathrm{Ge}\,z^2(3-2z) + O\!\left(\mathrm{Ge}^2\right). \tag{31}$$

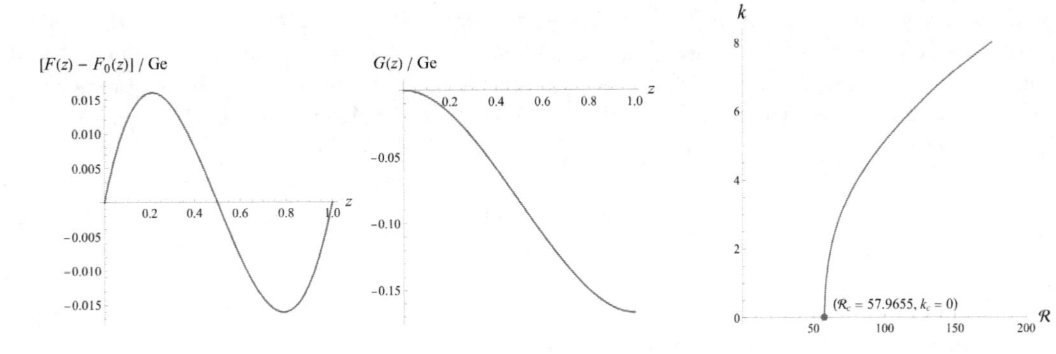

Figure 3. Couette-like flows: incremental velocity profile, temperature profile for $\mathrm{Ge} \ll 1$ and neutral stability curve in the $(\mathscr{R}, k)$ plane.

The obvious consequence of Eqs. (7) and (31) is that the limit $\mathrm{Ge} \to 0$ of the basic velocity profile is the classical isothermal Couette flow in a plane channel, defined by $F(z) = F_0(z) = z$ with $G(z) = 0$ and $A = 0$. The linear velocity profile of the Couette flow is slightly altered by the buoyancy effect of viscous dissipation with larger local values of velocity in the lower half-channel $(z < 1/2)$ and slightly smaller velocities in the upper half-channel $(z > 1/2)$. This behaviour is displayed in Fig. 3, where also the temperature profile in a $x = constant$ cross-section is displayed. Differently from the case of the Poiseuille-like flows illustrated in Section 5, Fig. 3 shows that the z component of the temperature gradient is always negative. The consequence is a potentially unstable thermal stratification throughout the whole channel.

By employing Eqs. (21) and (31), one may determine function $q(z)$ needed for a complete formulation of the stability eigenvalue problem (23)-(25), namely

$$q(z) = -z(1-z). \tag{32}$$

Again, one may employ the numerical solver presented in Section 4.2. As pointed out in Section 4.2, one evaluates again a zero angular frequency ω for all the neutrally stable modes. Moreover, one can display the neutral stability curve in the $(\mathscr{R}, k)$ plane as plotted in Fig. 3. Differently from the case of the Poiseuille-like flows, the minimum of $\mathscr{R}$ along the neutral stability curve is achieved when $k \to 0$, so that

$$\mathscr{R}_c = 57.9655, \quad k_c = 0. \tag{33}$$

Interestingly enough, the infinite wavelength modes, ineffective for the onset of the instability in the case of the Poiseuille-like flows, are the most unstable when one considers the Couette-like flows.

7. Critical Reynolds Number

The stability analysis carried out so far identifies $\mathscr{R}$, defined by Eq. (22), as the parameter governing the transition from the basic parallel flow to the cellular convection flow in the channel. A more inspiring statement of the transition to instability can be made in terms of the Reynolds number, $\mathrm{Re} = \mathrm{Pe}/\mathrm{Pr}$, as this parameter allows an easy comparison with the hydrodynamic instability driven by the Tollmien-Schlichting modes [9]. Eqs. (29) and (33) yield a critical condition for the Reynolds number expressed as

$$\mathrm{Re}_c = \begin{cases} 41.8534 \ \mathrm{Pr}^{-1} \mathrm{Ge}^{-1/2}, & \text{Poiseuille-like flows,} \\ 57.9655 \ \mathrm{Pr}^{-1} \mathrm{Ge}^{-1/2}, & \text{Couette-like flows.} \end{cases} \tag{34}$$

Obviously, any comparison with the hydrodynamic instability is possible only when Pr and Ge are prescribed. Due to their definitions, Eq. (5), this is possible only for a given fluid and a given value of the vertical width H. Then, we can imagine an experimental setup where $H = 10^{-2}$ m and the working fluid is glycerine at 300 K (data provided in Ref. [10]). By using Eq. (5), we obtain $\mathrm{Ge} = 1.94 \times 10^{-8}$, so that Eq. (34) yields

$$\mathrm{Re}_c = \begin{cases} 44.3, & \text{Poiseuille-like flow,} \\ 61.4, & \text{Couette-like flow.} \end{cases} \tag{35}$$

The linear transition to hydrodynamic instability happens with $\mathrm{Re}_c = 5772.22$ for the plane Poiseuille flow, while Re_c is infinite for the plane Couette flow [9]. Thus, Eq. (35) reveals that the dissipation instability induced by the buoyancy force acts at significantly lower Reynolds number with respect to the hydrodynamic instability.

8. Conclusions

A family of stationary and parallel shear flows in a horizontal plane channel have been studied. Such flows are solution of the governing balance equations for fluid flow under the Boussinesq approximation by taking into account the viscous dissipation effect. The channel boundaries are considered as adiabatic and subject to either Poiseuille-like or Couette-like velocity conditions. In the first case, the flow is sustained by an imposed horizontal pressure gradient. In the second case, the cause of the flow is the velocity difference between the boundaries. In all cases, the only thermal effect generating a nonzero temperature gradient is the viscous dissipation due to the fluid flow. The onset of the linear instability of the parallel shear flows has been investigated under the assumptions of a very small Gebhart number, Ge, and a very large Prandtl number, Pr. The results surveyed in this paper are important for several engineering applications where the dissipation of energy as heat can have significant implications for the design and performance of devices. In fluid transport systems, like pipelines, viscous dissipation can cause a temperature rise in the fluid. This temperature increase can affect the thermophysical properties of the fluid, such as the viscosity and density, which can, in turn, impact the performance and require additional thermal design solutions. In minifluidic or microfluidic devices, where the characteristic length scales are small, viscous dissipation can become a significant factor. It can lead to non-uniform temperature distributions within the fluid, affecting the chemical reactions, the flow rates, and the mixing processes.

Acknowledgement

The authors acknowledge financial support from Italian Ministry of Education, University and Research (MIUR) grant number PRIN 2017F7KZWS.

References

[1] Gebhart B 1962 *Journal of Fluid Mechanics* **14** 225–232
[2] Joseph D D 1964 *The Physics of Fluids* **7** 1761–1771
[3] Barletta A 2015 *International Journal of Thermal Sciences* **88** 238–247
[4] Joseph D D and Carmi S 1969 *Quarterly of Applied Mathematics* **26** 575–599
[5] Romanov V A 1973 *Functional Analysis and its Applications* **7** 137–146
[6] Barletta A, Celli M and Rees D A S 2023 *Physics of Fluids* **35** 033111
[7] Barletta A, Celli M, Lazzari S and Brandão P V 2023 *International Journal of Thermal Sciences* **194** 108571
[8] Barletta A 2019 *Routes to Absolute Instability in Porous Media* (Springer)
[9] Drazin P G and Reid W H 2004 *Hydrodynamic Stability* 2nd ed (Cambridge University Press)
[10] Bergman T L, Lavine A S, Incropera F P and DeWitt D P 2011 *Fundamentals of Heat and Mass Transfer* 7th ed (John Wiley & Sons)

Numerical study of the Thermo-hydrodynamic behavior of a non-Newtonian nanofluid in a backward facing step

A. Mokhefi[1], E. Rossi di Schio[2,*], P. Valdiserri[2], C. Biserni[2] and D. Derbal[1]

[1]Mechanics, Modeling and Experimentation Laboratory L2ME, Faculty of Sciences and Technology, Bechar University, B.P.417, 08000, Bechar, ALGERIA.
[2]Alma Mater Studiorum – University of Bologna, Department of Industrial Engineering DIN. Viale Risorgimento 2, I-40136 Bologna, ITALY.

* Corresponding author: eugenia.rossidischio@unibo.it

Abstract. In the present work, a numerical simulation of a laminar non-isothermal flow of a non-Newtonian nanofluid in a backward facing step (BFS) is presented. It deals with Cu-water nanofluid, where the mixture shows a shear thinning behavior flowing from the restricted part of the duct with a fully developed velocity and a cold temperature. The lower part of the extended area of the backward facing step is maintained at a hot temperature, while all the other boundaries are considered thermally insulated. Moreover, a uniform magnetic field according to different angle is applicated on the nanofluid flow. The numerical simulation is based on the resolution of the mass, momentum and energy balance equations using Comsol Multiphysics. The aim of the sensitivity study is to highlight the impact of the Reynolds number, the nanoparticles concentration, the Hartmann number and the angle of the magnetic field on the flow and the thermal behaviours, as well as on the Nusselt number. Surprisingly, the results show that an increase in the Hartmann number, corresponding to a more intense magnetic field, resulted in a significant reduction in flow intensity.

1. Introduction

The Backward Facing Step (BFS), also known as a backward step, is a geometric configuration commonly studied in fluid mechanics due to its many engineering applications. In fact, studies on the BFS help to better understand flow characteristics in real-life situations such as flow around vehicles, cooling systems, pipelines and heat exchangers. This configuration provides valuable insights into phenomena like boundary layer separation, vortices, unsteady flows, turbulence, and heat transfer. The sudden expansion in BFS causes the separation and reattachment of flows, and this expansion is a determining factor in the hydrodynamic structure of the flow, which has a significant impact on heat transfer. Certainly, the separation and reattachment of the flow at a backward-facing step have garnered significant attention, leading to numerous studies dedicated to this phenomenon [1-4]. To achieve a more precise forecast of flow behavior, it is essential to thoroughly describe fluid parameters and accurately identify the rheological consequences involved. Mahmoud et al. [5] employed the finite element method to examine the flow behavior of Newtonian fluids as well as power law fluids that exhibit shear thinning or thickening characteristics. Many researchers are focused on improving heat transfer to meet the demands of various processes. To enhance convective heat transfer, a novel approach has emerged, which involves the introduction of nanoparticles into the base fluid. In recent studies, researchers have conducted several investigations to demonstrate the efficacy of utilizing nanoparticles in base fluids [6-7]. Selimefendigil et al. [8] examined the impact of nanoparticles' volume fraction, inlet oscillation frequency, and Reynolds number on fluid flow and thermal transfer characteristics. Their findings revealed that heat transfer was enhanced with higher volume fractions

of nanoparticles, increased Reynolds numbers, and frequencies of oscillation. Through experimental analysis, they observed that the degree of heat transfer in a backward-facing step duct increased as the concentration of nanoparticles in the fluid rose. In their numerical study, Selimefendigil and Öztop [9] investigated the forced convection heat transfer in a backward-facing step duct with a baffle positioned on the top wall under pulsating laminar flow conditions. The researchers examined the influence of various crucial parameters and compared the results to a steady flow without a baffle. Their findings indicated that, in the scenario where the lower wall was located downstream of the enlargement, the presence of a deflector did not yield significant improvements in heat transfer. Selimefendigil et al. [10] examined the laminar forced convection of a pulsed nanofluid flow over a backward-facing step (BES). that the rate of heat transfer increased with the addition of nanoparticles in the nanofluid.
The novelty of the present analysis is the investigation of the effect of a magnetic field, with different possible angles, applied to the nanofluid flow facing, in fully developed forced convection regime, the BFS.

2. Problem description

The study domain is a Backward Facing Step (BFS), reduced in two-dimensional geometry as shown in figure 1. The water-CuO nanofluid displaying a shear-thinning behavior flows from the narrowed part of the Backward Facing Step with a fully developed velocity profile of maximum U_0 and a fully cold temperature (T_c). In addition, the lower wall of the widened portion is maintained at a hot temperature (T_h). On the other hand, the other walls of the BFS are considered thermally insulated. The flow of the nanofluid inside the BFS is subject to a magnetic field of uniform flux density B, described by an angle γ between the magnetic field B and the horizontal axis.
The thermophysical properties of the studied nanoparticles and base fluid are shown in Table 1.

According to the values of the volume fraction of the nanoparticles, the values of the non-Newtonian water-CuO nanofluid consistency as well as the index of the pseudoplastic behavior n are experimentally determined and shown in Table 2. Moreover, the two-parameter Ostwald–de Waele model has been used to govern the flow of this non-Newtonian shear thinning nanofluid.

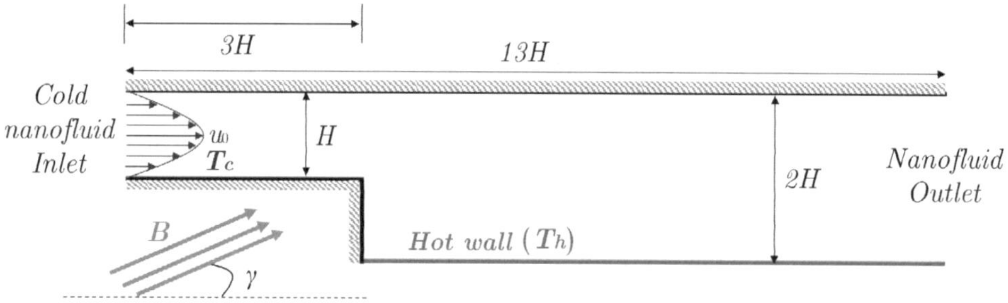

Figure 1. Backward Facing Step geometry

Table 1. Thermophysical proprieties of the nanoparticles and the base fluid.

Properties	Density [kg/m^3]	Thermal capacity [J/(kg.K)]	Thermal conductivity [W/(m.K)]	Dynamic viscosity [kg/(m.s)]	Thermal expansion [10^{-5}.1/K]	Electrical conductivity [S/K]
Eau (f)	997.1	4179	0.613	0.001003	21	0.05
CuO(p)	6500	540	18	-	0.85	2.7×10^{-8}

2.1. Mathematical model

The Navier-Stokes and heat equations are used to describe the flow of the non-Newtonian nanofluid flowing inside the BFS in a laminar regime. Let x, y be the Cartesian coordinates, u, v the velocities along the x and y axis respectively, p the pressure of the nanofluid, T the temperature and τ the shear stress rate. The governing equations are:

$$\frac{\partial u}{\partial x}+\frac{\partial v}{\partial y}=0 \tag{1}$$

$$\rho_{nf}\left(u\frac{\partial u}{\partial x}+v\frac{\partial u}{\partial y}\right)=-\frac{\partial p}{\partial x}+\left(\frac{\partial \tau_{xx}}{\partial x}+\frac{\partial \tau_{xy}}{\partial y}\right)+\sigma_{nf}B^2(v\cos\gamma\sin\gamma-u\sin\gamma) \tag{2}$$

$$\rho_{nf}\left(u\frac{\partial v}{\partial x}+v\frac{\partial v}{\partial y}\right)=-\frac{\partial p}{\partial y}+\left(\frac{\partial \tau_{yx}}{\partial x}+\frac{\partial \tau_{yy}}{\partial y}\right)+\sigma_{nf}B_0^2(u\cos\gamma\sin\gamma-v\cos\gamma) \tag{3}$$

$$\rho_{nf}Cp_{nf}\left(u\frac{\partial T}{\partial x}+v\frac{\partial T}{\partial y}\right)=k_{nf}\left(\frac{\partial^2 T}{\partial x^2}+\frac{\partial^2 T}{\partial y^2}\right) \tag{4}$$

where ρ, μ, k and Cp represents respectively the density, the dynamic viscosity, the heat conductivity and the specific heat, and the subscript "*nf*" denotes the nanofluid. The dynamic viscosity of the nanofluid is given as a function of the shear stress rate as follows:

$$\mu_{nf}=m_{nf}\left[2\left(\frac{\partial u}{\partial x}\right)^2+2\left(\frac{\partial v}{\partial y}\right)^2+\left(\frac{\partial u}{\partial y}+\frac{\partial v}{\partial x}\right)^2\right]^{n_{nf}-1} \tag{5}$$

The nanofluid shear stress is given by:

$$\tau_{ij}=\mu_{nf}\left(\frac{\partial u_i}{\partial x_j}+\frac{\partial u_j}{\partial x_i}\right) \tag{6}$$

Let us introduce the Reynolds and Hartmann number, namely:

$$\mathrm{Re}=\frac{\rho_f U_0^{2-n_{nf}}H^{n_{nf}}}{m_f} \text{ and } Ha=BH\sqrt{\frac{\sigma_f}{\mu_f}} \tag{7}$$

Table 2. Value of the nanofluid consistency and behavior index (m, n) Santra et al. (2009).

Volume fraction (%)	Consistency (N.s^n.m^{-2})	Behavior index (-)
0.5	0.00187	0.880
1.0	0.00230	0.830
1.5	0.00283	0.780
2.0	0.00347	0.730
2.5	0.00426	0.680
3.0	0.00535	0.625
3.5	0.00641	0.580
4.0	0.00750	0.540
4.5	0.00876	0.500
5.0	0.01020	0.460

In Eq.(7), m_f is taken as the dynamic viscosity of the base fluid (0.001 Pa s). The other thermophysical proprieties of the nanofluid, i.e. density, thermal conductivity, specific heat and electrical conductivity, are calculated as function of the properties of nanoparticles (p) and base fluid (f) as:

$$\rho_{\mathrm{nf}}=\varphi\rho_{\mathrm{p}}+(1-\varphi)\rho_{\mathrm{f}} \tag{8}$$

$$\frac{k_{\mathrm{nf}}}{k_{\mathrm{f}}} = \frac{k_{\mathrm{p}} + 2k_{\mathrm{f}} - 2\varphi(k_{\mathrm{f}} - k_{\mathrm{p}})}{k_{\mathrm{p}} + 2k_{\mathrm{f}} + \varphi(k_{\mathrm{f}} - k_{\mathrm{p}})} \tag{9}$$

$$(\rho Cp)_{\mathrm{nf}} = \varphi(\rho Cp)_{\mathrm{p}} + (1-\varphi)(\rho Cp)_{\mathrm{f}} \tag{10}$$

$$\sigma_{\mathrm{nf}} = \varphi\sigma_{p} + (1-\varphi)\sigma_{\mathrm{f}} \tag{11}$$

The boundary conditions used in this study are given as follows:

$$\text{Inlet: } u = U_0\left[-4\left(\frac{y}{H}\right)^2 + 4\left(\frac{y}{H}\right)\right],\ v = 0 \text{ and } T = T_c \tag{12}$$

$$\text{Outlet: } p = 0 \text{ and } \frac{\partial T}{\partial x} = 0 \tag{13}$$

$$\text{Hot wall: } u = v = 0 \text{ and } T = T_h \tag{14}$$

$$\text{Other walls: } u = v = 0 \text{ and } \frac{\partial T}{\partial N} = 0 \tag{15}$$

The Nusselt number is calculated along the hot wall as follows:

$$\mathrm{Nu}_l = -\frac{k_{nf}}{k_f}\frac{H}{T_h - T_c}\left(\frac{\partial T}{\partial y}\right)_{\text{hot wall}} \tag{16}$$

The average Nusselt number at this wall is given by:

$$\mathrm{Nu} = \frac{1}{10H}\int_{hot\ wall} Nu\ \mathrm{d}x \tag{17}$$

3. Numerical solution

The numerical resolution of the physical problem has been performed by using a finite element method based on the Galerkin discretization under Comsol Multiphysics 6.0 Software. The computational domain has been discretized with an unstructured triangular mesh. Moreover, close to the BFS walls, a rectangular boundary layer mesh has been introduced. After several trials of the mesh test for all the cases studied, we have opted for a grid with 134956 elements beyond which the values of the average Nusselt number remain unchanged.

4. Results and discussion

In this section, we examine the influence of Hartmann number, magnetic field angle, and nanoparticle volume concentration on heat transfer and vortex development within the backward facing step (BFS). The results are presented in the form of temperature contours, stream function contours, and local Nusselt number graphs along the hot wall.

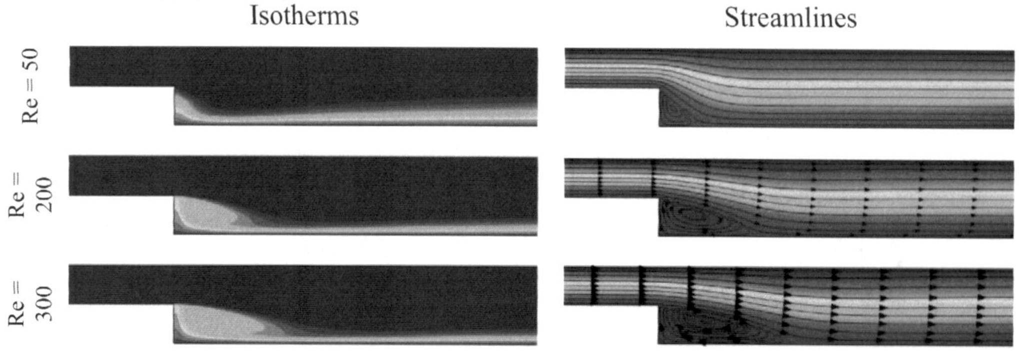

Figure 2. Isotherms and streamlines for different Reynolds numbers.

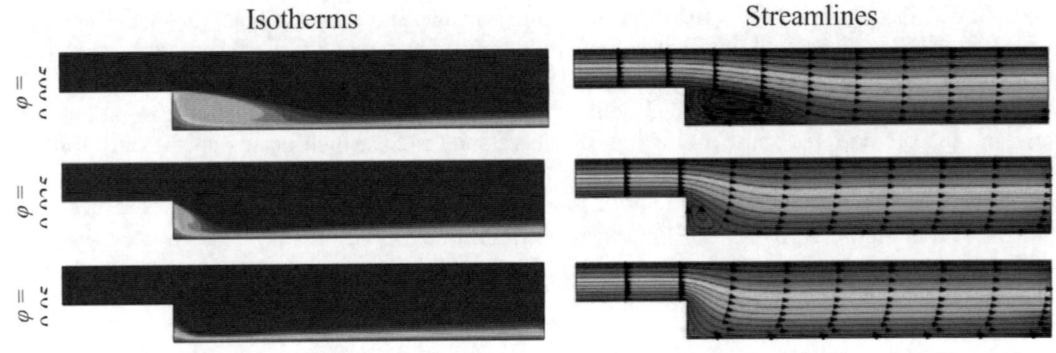

Figure 3. Isotherms and streamlines for different nanoparticle volume fraction.

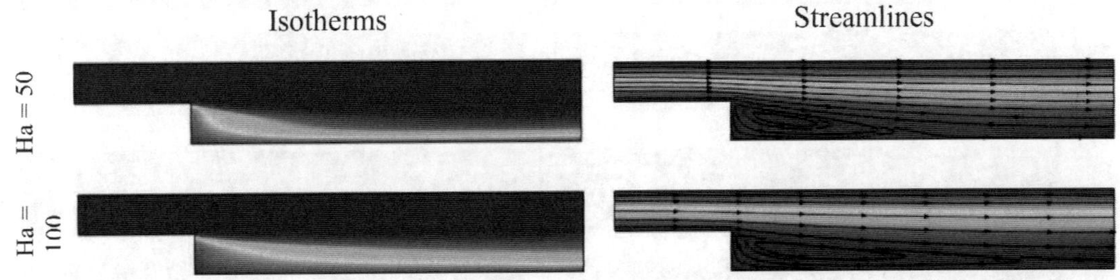

Figure 4. Isotherms and streamlines for Ha=50 and Ha=100.

In Figure 2, the temperature distribution and the streamlines of the non-Newtonian nanofluid inside the downward step are reported for different Reynolds numbers. With increasing values of Re, the vortex zone expands while the thermal boundary layer decreases in size. Indeed, the thickness of the boundary layer decreases along the hot wall with increasing Re, except in the vortex region where an increase in the thickness of the boundary layer is observed. As an explication, as Re increases, inertial forces become dominant compared to viscous forces. This leads to an intensification of the laminar flow and an immediate flow reattachment downstream of the expansion zone. In the vortex region, the vortices generated by the flow result in an increase in the thickness of the thermal boundary layer. However, outside of this region, turbulent flow promotes heat dispersion and leads to a decrease in the thickness of the thermal boundary layer along the hot wall.

The impact of volume fraction on the flow of non-Newtonian nanofluid inside the BFS is shown in Figure 3. The influence of volume fraction of nanoparticles, which leads to the shear thinning behavior of the nanofluid, coincides with the influence of the structural index and consistency of the nanofluid, especially with the increase in its apparent dynamic viscosity. However, unlike the case of using the Newtonian single-phase model for nanofluids, the influence of volume fraction on the flow in the case of non-Newtonian nanofluid is clear and significant. Indeed, a considerable reduction in the flow was observed with increasing volume fraction of nanoparticles. Furthermore, the size of the vortex zone decreases significantly with increasing volume fraction, also leading to a decrease in the thermal boundary layer in the vortex region. These observations suggest that the addition of nanoparticles in the non-Newtonian nanofluid can disrupt the formation and stability of vortices, thereby reducing the size of the vortex zone.

The effect of a horizontal magnetic field on the flow and heat transfer of the shear thinning nanofluid is shown in Figure 4. Contrary to expectations, increasing the Hartmann number results in a significant reduction in the flow intensity of the nanofluid. However, this reduction in flow does not lead to the disappearance of the vortex zone, but rather to its substantial longitudinal expansion and thus to a significant increase in the thickness of the thermal boundary layer along the hot wall.

Journal of Physics: Conference Series **2685** (2024) 012074 doi:10.1088/1742-6596/2685/1/012074

The effect of the magnetic field angle on the hydrodynamic and thermal behavior of the nanofluid inside the BFS is illustrated in Figure 5, with a reference angle of 0°. The figure shows that introducing a magnetic field with a different angle than 0° leads to an intensification of the flow. Moreover, the vortex zone is completely eliminated for angles of −90° and 90°, due to the interplay between the Lorentz force and the direction of the velocity field, which occurs specifically for these two angles. A decrease in the thickness of the thermal boundary layer is observed when the magnetic field inclination angle deviates from the horizontal position, particularly as it approaches angles of 90° and −90°. Furthermore, for $\gamma = 45°$, a slight increase in the thickness of this layer was observed near the outlet while for $\gamma = -45°$ the thickness of the boundary layer is larger near the expansion zone.

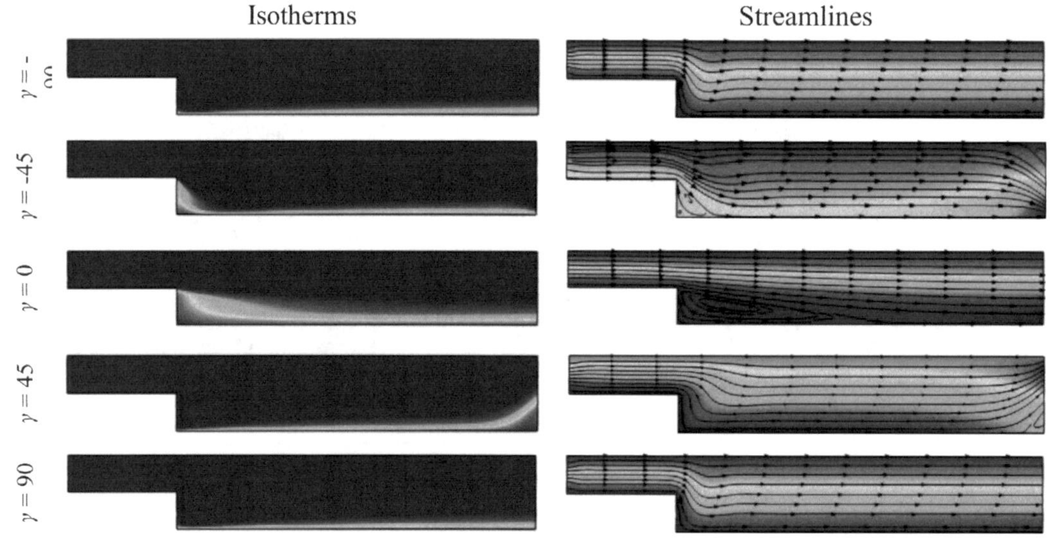

Figure 5. Isotherms and streamlines for different magnetic field inclination angle.

In Figure 6, the local Nusselt number along the hot wall is reported for different for different values of the Reynolds numbers (a), nanoparticles volume fraction (b), Hartmann number (c), magnetic field inclination angle (d). Regarding the effect of Re, the figure shows that the Nusselt number increases along the hot wall until it reaches a maximum corresponding to the limit of the vortex zone. Beyond this zone, the Nusselt number starts to decrease. The increase in the Nusselt number along the hot wall is due to improved convective heat transfer. With increasing Re the flow becomes more intense, promoting better thermal dispersion and more efficient heat exchange with the hot wall. Vortices tend to disturb the thermal boundary layer, thereby reducing convective heat transfer near the hot wall. Therefore, despite the increase in Reynolds number, the presence of the vortex limits the continuous rise of the Nusselt number in this zone. Despite the negative impact of increasing Reynolds number on heat transfer in the vortex region, the overall convective heat transfer is significantly improved. Examining the average Nusselt number data presented in Table 3 (a), we observe a significant increase in the heat transfer rate with increasing Reynolds number. In particular, an enhancement percentage of 73% was observed at a Reynolds number of 300 compared to 50.
Concerning Figure 6(b), i.e. the variation of the local Nusselt number as a function of the volume fraction of CuO nanoparticles, increasing the volume fraction leads to a significant increase in the Nusselt number in the vortex zone, due to the dominant inertial and viscous effects in this region. However, outside the vortex zone a relatively small decrease in Nu is noted. The average Nusselt number presented in Table 3(b) increases up to 17% for a volume fraction of 0.05 of CuO nanoparticles, compared to a volume fraction of 0.005. Indeed, the addition of CuO nanoparticles leads to a significant improvement in the average heat transfer rate.

Figure 6(c) highlights the local variations of Nu along the hot wall, as a function of Ha, showing that in the considered measurement direction increasing Ha leads to a substantial reduction in the heat transfer rate. Furthermore, a decrease of approximately 68% in the average Nusselt number for Ha=100 is noted in Table Table 3 (c), compared to the case where no magnetic field is present (Ha = 0). This decrease is attributed to the magnetic forces acting on the movement of the nanofluid, thus limiting convection and heat transfer.

Figure 6(d) depicts the variation of the local Nusselt number along the hot wall as a function of different magnetic field inclination angles. According to this figure, angles different from 0° resulted in significantly higher heat transfer rates: angles of 45°, 90° and -90° exhibited the best convective heat transfer rates. Although the magnetic field hinders thermal activity, its angle has a positive impact: the heat transfer associated with angles of 90° and -90° (which are equivalent) showed the highest rate. An increase of approximately 203% was observed compared to the case of a horizontal magnetic field. Non-zero inclination angles significantly improve the convective heat transfer rate along the hot wall, see Table 3(d).

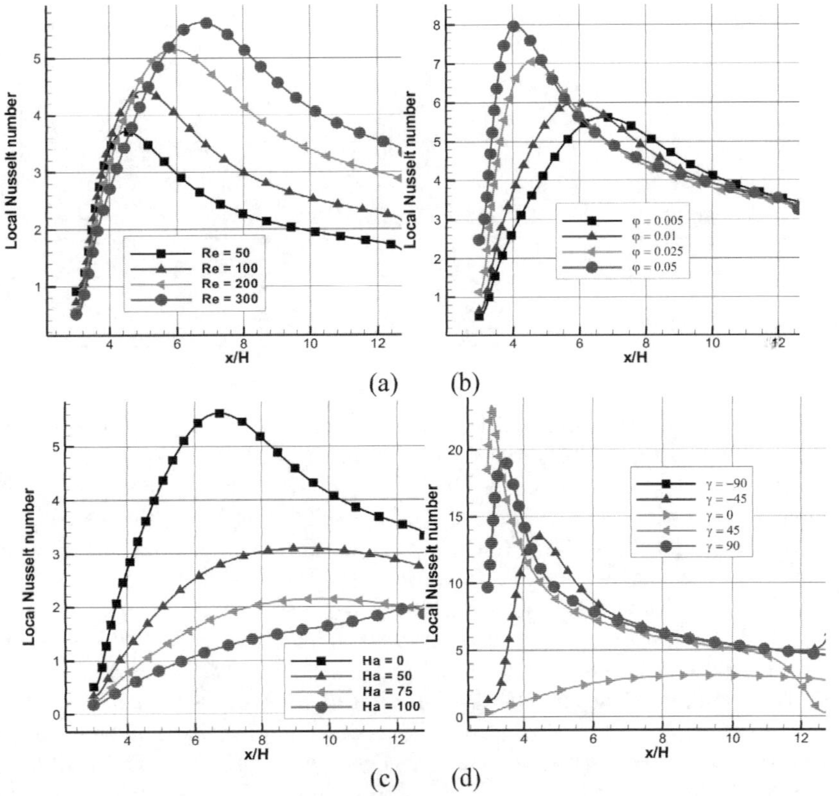

Figure 6. Local Nusselt number for different values of Reynolds numbers (a), nano-particle volume fraction (b), Hartmann number (c), magnetic field inclination angle (d).

5. Conclusion

This paper investigated the hydrodynamic and thermal behavior of non-Newtonian nanofluid flow inside a backward-facing step (BFS) under various influencing factors. The findings shed light on the intricate interplay between flow characteristics, heat transfer, and the presence of magnetic fields in the BFS domain. The main findings can be summarized as follows.

- As the Reynolds number increases, the vortex zone expands while the thermal boundary layer thickness along the hot wall decreases, except within the vortex region where it increases.

- An increase in nanoparticle volume fraction leads to a considerable reduction in flow intensity and vortex size, as well as to a decrease in the thermal boundary layer in the vortex region: the addition of nanoparticles disrupts the formation and stability of vortices.
- Surprisingly, an increase in the Hartmann number resulted in a significant reduction in flow intensity. However, the vortex zone was not eliminated but rather longitudinally expanded, leading to an increase in the thermal boundary layer thickness along the hot wall.
- The complete elimination of the vortex zone was observed at angles of −90° and 90°, probably due to the interaction between the Lorentz force and the velocity field direction. Moreover, changes in the thermal boundary layer thickness were observed as the inclination angle deviated from the horizontal position: angles of 45°, 90° and −90° exhibited the highest convective heat transfer rates.

Table 3. Average Nusselt number for different values of the Reynolds number (a), nanoparticle volume fraction (b), Hartmann number (c), magnetic field inclination angle (d).

	Re	50	100	200	300
(a)	Nu_{ave}	2.377	2.990	3.700	4.124
	% Heat rate	-	73.45	55.64	73.45
	φ	0.005	0.010	0.025	0.050
(b)	Nu_{ave}	4.124	4.301	4.567	4.846
	% heat rate	-	4.29	10.73	17.51
	Ha	0	50	75	100
(c)	Nu_{ave}	4.124	2.523	1.708	1.307
	% heat rate	-	-38.81	-58.57	-68.29
	γ	0	-90	90	-45	45
(d)	Nu_{ave}	2.523	7.668	7.668	6.868	7.050
	% heat rate	-	203.8	203.8	172.1	179.3

References

[1] Eturk E 2008 Numerical solutions of 2-D steady incompressible flow over a backward-facing step, Part I: High Reynolds number solutions *Comput Fluids* **37** 633-655
[2] Lan H, Armaly B F, Drallmeier J A 2009 Three-dimensional simulation of turbulent forced convection in a duct with backward-facing step *Int J Heat Mass Transf* **52** 1690-1700
[3] Iwai H, Nakabe K, Suzuki K 2000 Flow and heat transfer characteristics of backward-facing step laminar flow in a rectangular duct *Int J Heat Mass Transf* **43** 457-471
[4] Star S, Stabile G, Rozza G, Degroote J 2021 A POD-Galerkin reduced order model of a turbulent convective buoyant flow of sodium over a backward-facing step *Appl Math Model*, **89** 486-503
[5] Mahmood R, Bilal S, Majeed A H, Khan I., Sherif E S M 2020 A comparative analysis of flow features of Newtonian and power law material: a new configuration *J Mater Res Technol* **9** 1978
[6] Barletta A, Rossi di Schio E, Celli M 2015 Convection and instability phenomena in nano-fluid-saturated porous media *Heat Transfer Enhancement with Nanofluids* CRC Press FL 341-364
[7] Rossi di Schio E, Impiombato A N, Mokhefi A, Biserni C 2022 Theoretical and Numerical Study on Buongiorno's Model with a Couette Flow of a Nanofluid in a Channel with an Embedded Cavity *Appl Sci* **12** 7751.
[8] Selimefendigil F and Öztop H F 2013 Identification of forced convection in pulsating flow at a backward facing step with a stationary cylinder subjected to nanofluid *Int Commun Heat Mass Transf* **45** 111-121
[9] Selimefendigil F and Öztop H F 2013b Numerical analysis of laminar pulsating flow at a backward facing step with an upper wall mounted adiabatic thin fin *Comput Fluids* **88** 93-107
[10] Selimefendigil F and Öztop H F 2017 Forced convection and thermal predictions of pulsating nanofluid flow over a backward facing step with a corrugated bottom wall *Int J Heat Mass Transf* **110** 231-247

AUTHOR INDEX